Applied Numerical Methods for Digital Computation

Applied Numerical Methods for Digital Computation with FORTRAN and CSMP

Second Edition

M. L. James
G. M. Smith
J. C. Wolford

University of Nebraska

THOMAS Y. CROWELL
HARPER & ROW, PUBLISHERS
New York Hagerstown San Francisco London

APPLIED NUMERICAL METHODS FOR DIGITAL COMPUTATION
with FORTRAN and CSMP, Second Edition

Published simultaneously in Canada by Fitzhenry & Whiteside, Ltd., Toronto.

Library of Congress Cataloging in Publication Data

James, Merlin L
Applied numerical methods for digital computation with FORTRAN and CSMP.

Published in 1967 under title: Applied numerical methods for digital computation with FORTRAN.
Includes bibliographical references and index.
1. Numerical analysis—Data processing. 2. FORTRAN (Computer program language) 3. CSMP (Computer program) I. Smith, Gerald M., joint author. II. Wolford, J. C., joint author. III. Title.
QA297.J3 1977 519.4 76-58402
ISBN 0-7002-2499-8

Contents

2 Roots of Algebraic and Transcendental Equations

3 Solution of Simultaneous Algebraic Equations

4 Curve Fitting

5 Numerical Integration and Differentiation

6 Numerical Integration of Ordinary Differential Equations: Initial-Value Problems

7 Ordinary Differential Equations: Boundary-Value Problems

8 Partial Differential Equations

9 Introduction to Digital Computer Simulation Using CSMP (Continuous System Modeling Program)

Appendixes

A International System of Units (SI Units)

B Matrix Algebra

C Polynomial Interpolation and Application of Polynomial Approximation to Numerical Integration and Differentiation

Preface

Many of the revisions incorporated in the Second Edition are a result of carefully studying and evaluating suggestions from students and faculty who have used the First Edition during the past ten years. The major revisions include the addition of a chapter on curve fitting, a chapter on the Continuous System Modeling Program (CSMP), and an extensive appendix on interpolating polynomials with applications. Others include Bairstow's method for obtaining the roots of polynomials and Cholesky's (Crout's) method for solving simultaneous linear algebraic equations. Throughout the text other topics and areas have been modified or reinforced—for example, determining the roots of algebraic and transcendental equations and the solutions of eigenvalue problems. Since most students now taking a course in numerical methods have some knowledge of FORTRAN, Chapters 1 and 2 of the First Edition have been shortened and combined into one chapter.

As in the First Edition, the arrangement and scope of the material permit the use of the text at different course levels. The only mathematics required for the first five chapters is calculus and elementary matrix algebra, and the basic concepts and operations of matrix algebra are reviewed in Appendix B. For study beyond Chapter 5, a background in differential equations is essential.

The new edition continues to reflect the philosophy that the most effective approach to learning numerical methods is to integrate theoretical study with solving physical problems on the computer. Such problems have physical characteristics which enable the student to recognize whether he is obtaining results consistent with a correct solution, or obtaining unreasonable results, indicating an error in his mathematical model or in his program for a given problem. The student also obtains practice in interpreting and analyzing numerical results, and gains a "feel" for the effect of various parameters on the behavior of particular physical systems.

Since many engineers, educators, scientists, and technicians will be using the International Systems of Units (SI units) in the near future, some problems and examples using SI units have been included. A brief discussion of the SI system and factors for converting from the customary U.S. units to SI units are included in Appendix A.

Throughout the text, theoretical discussions of numerical methods are complemented by the formulation of FORTRAN programs for the solution of typical problems from the various engineering disciplines.

Chapter 1 presents the elements of FORTRAN IV. This chapter may serve as a reference, a review, or an introduction for those not already familiar with FORTRAN IV. Chapter 2 presents the most commonly used methods for obtaining the roots of algebraic and transcendental equations. In addition, Graeffe's method and Bairstow's method for obtaining roots to polynomials, along with synthetic division, are discussed in considerable detail. Chapter 3 presents the most commonly used methods for the solution of simultaneous linear algebraic equations, and also includes discussions of the solution of nonlinear simultaneous equations and the eigenvalue problem. Chapter 4 presents a general matrix formulation of the basic equations for curve fitting of experimental data, and Chapters 5 through 7 present many of the most commonly used methods for numerical integration and differentiation, and integration of ordinary differential equations involving initial-value problems and boundary-value problems. Chapter 8 introduces the numerical solution of partial differential equations, and the concluding Chapter 9 presents the Continuous System Modeling Program (CSMP) with numerous examples to illustrate the power and simplicity of CSMP.

The authors wish to acknowledge the many individuals who have contributed valuable suggestions for this revised edition. We are particularly grateful to Professors Corley, Landis, Soehngen, and Umholtz for their suggestions. We are also grateful to our wives, Jane, Eunice, and Joan, for their patience and understanding during the preparation of this text.

Applied Numerical Methods for Digital Computation

1 Digital Computer Principles and FORTRAN IV

1-1 / INTRODUCTION

Primitive man did his counting and simple arithmetic with the aid of his fingers, pebbles, and sticks, by relating these objects to other objects such as sheep, goats, cattle, and so on. His pile of sticks or stones constituted the earliest primitive digital computer. The abacus, developed some 3000 years ago, represents the first great advance in digital computation.

In 1642 Blaise Pascal, a French religious philosopher, scientist, and mathematician, developed the first mechanical adding machine which was similar in principle to present-day machines. This machine is now on display in a French museum. In 1671 the German philosopher and mathematician Gottfried Wilhelm von Leibniz independently conceived a more advanced mechanical calculator, which he completed in 1694.

Automatic mechanical calculators were suggested as early as the eighteenth century, but the technical knowledge required to build them was lacking. Joseph Jacquard, a French loom designer, perfected an automatic pattern loom in 1804 which contributed to the development of computers. The sequence of operations for the loom was controlled by punched cards, and the loom made intricate patterns as easily as other looms made plain cloth.

About 1833 the British mathematician Charles Babbage conceived and designed on paper the first automatic digital computer, which he called an "analytical engine." It had many features of modern computers. It was proposed to use a variation of the control mechanism of the Jacquard loom to control the sequence of arithmetic operations. These operations were to be specified, in advance, by means of punched cards. All calculations were to be done mechanically, and numbers were to be stored by the position of counter wheels. Unfortunately, this automatic computer was never built, owing to technical and financial difficulties, and an embittered Babbage died in 1871 without completing a working model, and with few people having even taken his work seriously.

It was not until the 1940s that computers, similar in principle to the one conceived by Babbage, were actually built. The first of these, known as the Automatic Sequence-Controlled Calculator, or Mark I, was designed by Howard Aiken of Harvard University, and was completed by the International Business Machine (IBM) Company in 1944. The Mark I was capable of performing a fixed sequence of operations controlled by punched tape. It consisted mainly of mechanical and electromechanical components and had a memory capable of storing 72 numbers of 24 digits each. World War II gave a definite impetus to the development of computers, and the Electronic Numerical Integrator and Calculator (ENIAC) was built under United States Army contract. It was completed in 1946 by engineers at the University of Pennsylvania, and was similar to the Mark I, except that it was predominantly electronic in operation and therefore performed its computations much more quickly. Its input and output were in the form of IBM punched cards.

A significant advance in computers was made in 1945 when John von Neumann, of the Princeton Institute for Advanced Study, and H. H. Goldstine, of the Army Ordnance Department, proposed storing in the computer memory the sequence of operations to be performed, along with the numbers which were to be operated on. This proposal was made in an Army Ordnance Department report, after an exchange of ideas with J. P. Eckert and John Mauchly of the Moore School of Electrical Engineering at the University of Pennsylvania. This feature enabled the computer to "branch"; that is, to follow either of two alternate sequences of steps, depending upon some condition existing at the time of execution of the branching instruction. It also made "looping" much easier; the repetitive execution of portions of the internally stored program of instructions could be easily accomplished. Thus, this important idea greatly increased the computing facility of the digital computer.

Digital-computer installations were constructed at only a moderate rate in the early 1950s, since it was felt that all the computing requirements of this country could be handled at a few large installations. However, the demand for computing facilities grew much faster than had been anticipated, and, at the present time, literally thousands of digital computers are being installed annually. It appears that this growth will surely continue, so it is imperative that most graduating engineers have some degree of familiarity with the programming and operation of digital computers.

Many equations in engineering problems, even though they can be solved analytically in closed form, require a great deal of tiresome and time-consuming work which can be eliminated by programming the equations to some type of computer. Other equations cannot be solved analytically, and, although their approximate solutions may be obtained by various numerical methods, these often involve large numbers of calculations which are time-consuming when performed manually. Furthermore, in some instances an engineer may wish to know the effect of changing certain design parameters on the behavior of a system, which necessitates the solving of the problem many times with different sets of data. A digital computer can be employed to perform the large number of calculations required, and, since they are executed at tremendous speeds, solutions are obtained quickly as well as accurately.

The digital computer performs arithmetic operations upon discrete numbers in a defined sequence of steps. Owing to modern developments in programming techniques, it is not necessary to be familiar with the details of the internal computing processes of a digital computer to be able to use one. It is helpful, however, to have a general concept of the operational characteristics of the computer, in order to understand its capabilities and limitations. Therefore, a brief review of the history of computers and of their components, functions, and operations is included here, along with a review of most of the more commonly used features of FORTRAN IV.

1-2 / DIGITAL-COMPUTER COMPONENTS

The modern electronic digital computer performs its functions by utilizing sequences of numerical operations performed upon discrete numbers. Five basic components are employed: (1) the *input* unit, which is used to provide data and instructions to the computer; (2) the *memory* or *storage* unit, in which data and instructions are stored; (3) the *arithmetic-logic* unit, which performs the arithmetic operations and provides the "decision-making" ability, or logic, of the computer; (4) the *control* unit, which controls the computer operations; and (5) the *output* unit, from which the computer results are obtained. The control unit and arithmetic-logic unit are often considered as a single unit, called the *central-processing* unit (CPU). Sometimes main memory is also considered as part of the CPU since the three components are often physically housed together as a unit. The relationships of these components are shown symbolically in Fig. 1-1, where the lines indicate the flow of instructions and data or control information.

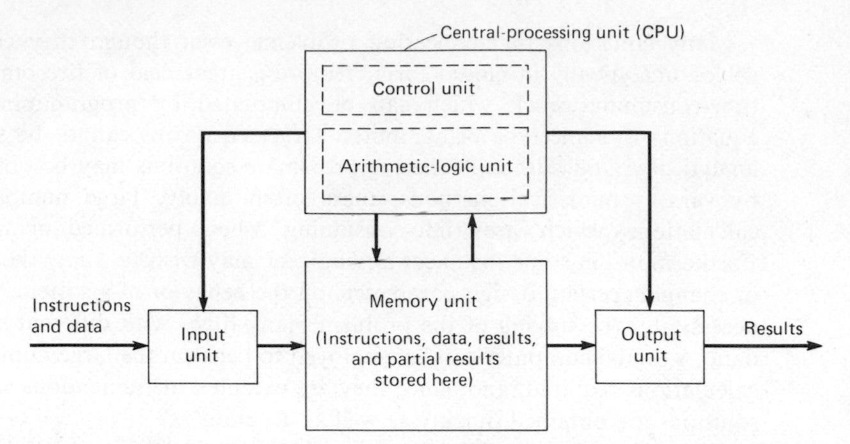

Fig. 1-1 / *Basic components of a digital computer.*

Input Unit

The input unit generally consists of one of the following devices: a card reader, a paper-tape reader, a magnetic-tape unit, a console keyboard or special typewriter (printer-keyboard), a remote-terminal keyboard, or in some instances specialized input devices such as magnetic-ink-character readers or optical-character readers. Magnetic disk and drum drives and data-cell drives also serve as input (and output) units, but they are generally thought of as being auxiliary storage devices. The input device converts the information being read to a form used internally in the computer and sends it to the main storage unit.

Memory or Main Storage Unit

The internal storage unit must be capable of storing the numbers and other characters which represent instructions and data. Although many devices have been employed for storage, the use of magnetic cores for main memory is predominant because of the rapidity with which the stored information can be located and transferred to another component. The numbers stored are in some form of the binary number system. Numbers can be represented in this system by just two digits (0 and 1), corresponding very conveniently to the two polarity states available with the magnetic cores.

The memory unit receives instructions and data and stores them in separately assigned locations. Each of these storage locations is designated by a numerical address which identifies the location to the computer. Once the data and instructions have been stored, they are subsequently always referred to by their numerical address. In addition to storing data and instructions, the memory unit is also utilized to store intermediate and final results for subsequent use or readout.

Arithmetic-Logic Unit

The arithmetic-logic unit consists of all the electronic circuitry necessary for performing the various arithmetic operations and for supplying computer logic. The latter is usually accomplished by performing tests on certain conditions existing in the computer, to decide which of alternate sequences of steps will be followed.

Control Unit

The control unit consists of the circuitry required to take the instructions from memory, interpret them, and cause them to be executed. Normally, the instructions in memory are followed in sequential order, unless the control unit is instructed by the program to interrupt the sequence.

Output Unit

The results obtained as a consequence of the arithmetic operations performed by the computer must be communicated to the user. This is accomplished by some type of output unit. Output units usually consist of one of the following devices: a high-speed printer, a magnetic-tape unit, a card punch, a paper-tape punch, a visual display unit, a special typewriter, or a graphical plotter. Time is saved if the output data are punched on tape or cards or put on magnetic tape or in a disk file, and the results plotted or printed "off line," since the computer can often perform its operations faster than the results of the operations can be tabulated in final numerical form.

1-3 / PREPARING A DIGITAL-COMPUTER PROGRAM

The first phase in programming a problem to the digital computer is to decide on the approach or method to be used in obtaining the solution. Since the digital computer is capable of performing only arithmetic operations, problems which cannot be solved by arithmetic procedures, when in their usual form, must be put in a form consistent with such procedures. Obviously, a problem in addition can be solved per se, since its solution is inherently an arithmetic operation. However, the closed-form solution of a differential equation, for example, is not an arithmetic operation. An arithmetic procedure for solving such an equation must be developed. Such procedures are discussed in detail in subsequent chapters.

Once a method of solution has been selected, the problem must be programmed to the computer as a definite step-by-step procedure. This is called an *algorithm*. The second phase of programming is to establish a general outline of the steps involved in the algorithm. One method is to write the steps in a numbered sequence. More frequently, however, the algorithm is stated graphically by means of a *flow chart*, in which the steps are displayed in the form of a block diagram. A flow chart may show the required steps in considerable detail, or it may merely outline the general procedure necessary to obtain a computer solution to the problem. In general, excessively detailed procedures should be avoided in the flow chart, since its primary purpose is merely to outline the overall process, which will be implemented by the more specific steps appearing in the finished computer program. The use of a flow chart is illustrated in Ex. 1-1 (Fig. 1-3).

The computer must be instructed very specifically for it to perform its operations. In fact, each step must be initiated by an instruction. The instructions which the computer interprets and executes must be transmitted to it in a *coded* form which it can interpret. Thus the third phase of programming is to write the *specific* program steps in coded form, following the outline provided by the flow chart. This procedure is called *coding*, and the resulting program is called a *machine-language* program.

Coding is a tedious task if the program is a long one, but, fortunately, it is no longer necessary to code the steps manually. The computer itself can be programmed to write the machine-language program. The desired program is first written in a less-detailed and much simpler language very similar to ordinary English and algebra. The simpler program is then translated into machine language by the computer. To accomplish this translation, the computer must first have a specially written machine-language program read into its memory. Such a special program is called a *translator*, a *processor*, or a *compiler*. The FORTRAN processor, or compiler, is an example of such a special program.

If a compiler is not used and the machine-language program is written manually, it must next be transcribed to a suitable medium, such as tape or cards, so that it may be fed to the computer through the appropriate type of input unit.

If the machine-language program is compiled by the computer, it is usually placed in a direct-access auxiliary storage space such as disk storage for immediate execution, or punched on cards or put on magnetic tape for later execution. This completes the programming and translation phases of the process. The execution phase, involving the actual computation, follows.

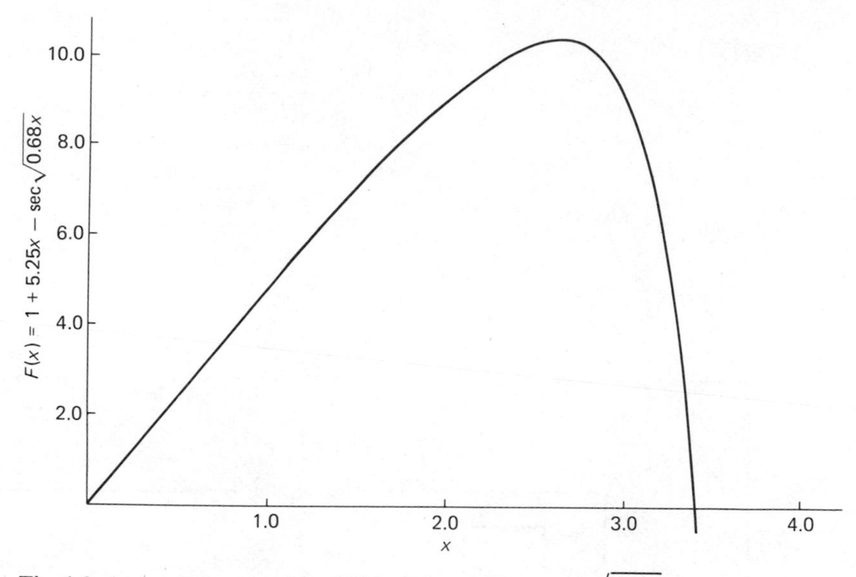

Fig. 1-2 / *Approximate graph of* $F(x) = 1 + 5.25x - \sec\sqrt{0.68x}$.

EXAMPLE 1-1

As an example of flow charting, let us consider the following problem. It is desired to find the first positive nonzero solution of the equation

$$1 + 5.25x - \sec\sqrt{0.68x} = 0$$

An approximate graph of the function $F(x) = 1 + 5.25x - \sec\sqrt{0.68x}$ is shown in Fig. 1-2. We must determine the first point at which the graph of the function crosses the x axis, as shown. For small values of x, the function will be positive. The technique will consist of calculating $F(x)$ for successively larger values of x, starting at $x = \Delta x$, and incrementing x by successive values of $\Delta x = 0.1$, until the function becomes negative. We will then know that the x value for which the function will be zero has been exceeded. We then revert to the immediately preceding x value and again increase x, this time incrementing by a smaller amount, say, one-tenth of the previously used Δx. Incrementing continues until the function again becomes negative. Again, we revert to the immediately preceding value of x and begin to increase x with still smaller increments. This procedure continues until the value of x which represents the first positive nonzero solution is obtained with the desired accuracy. The result is then printed.

As crude as this method may appear to be, a computer solution can rapidly be obtained by its use, since the computer performs the indicated

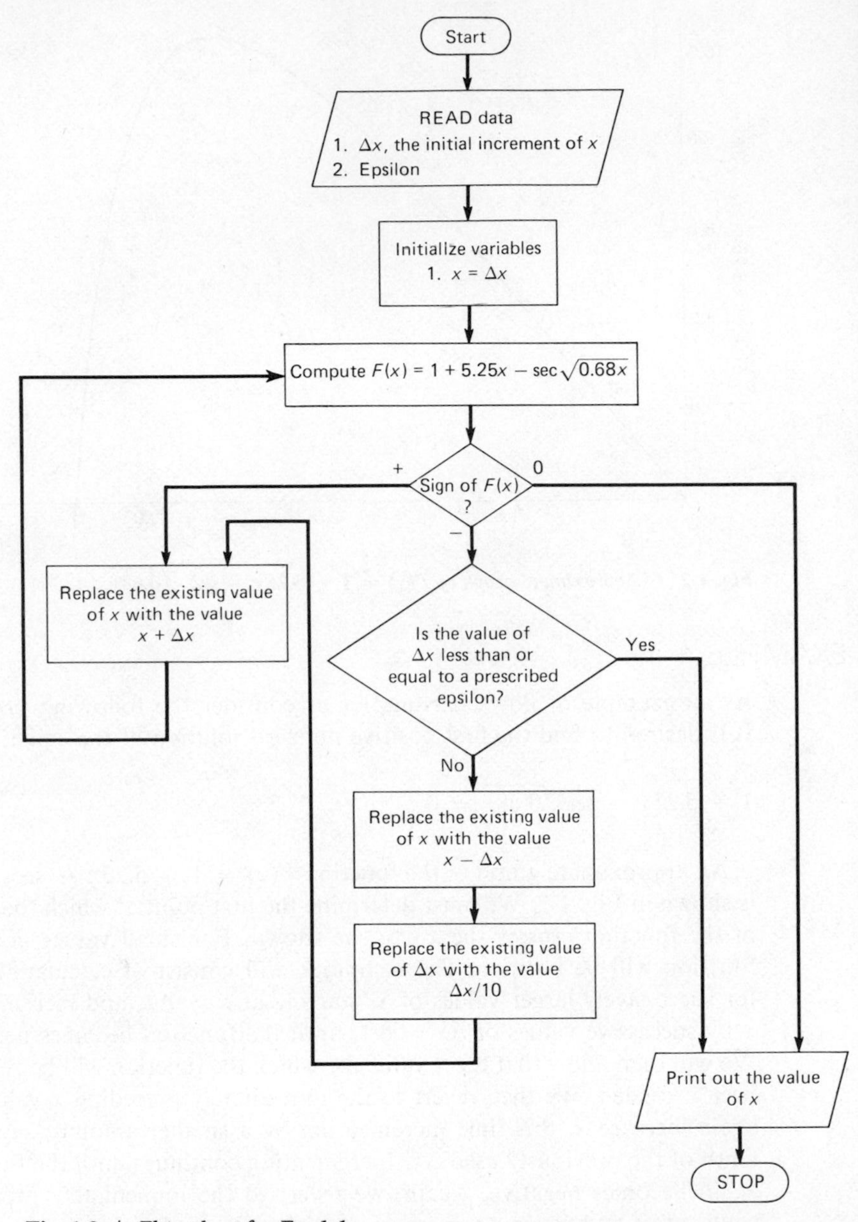

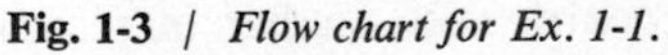
Fig. 1-3 / *Flow chart for Ex. 1-1.*

operations so swiftly. Convergence to a solution can be obtained even more rapidly if we refine the preceding technique by writing a slightly more complicated program, using, for example, the Newton-Raphson method (illustrated in Sec. 2-6).

A flow chart for the foregoing solution is shown in Fig. 1-3. At this

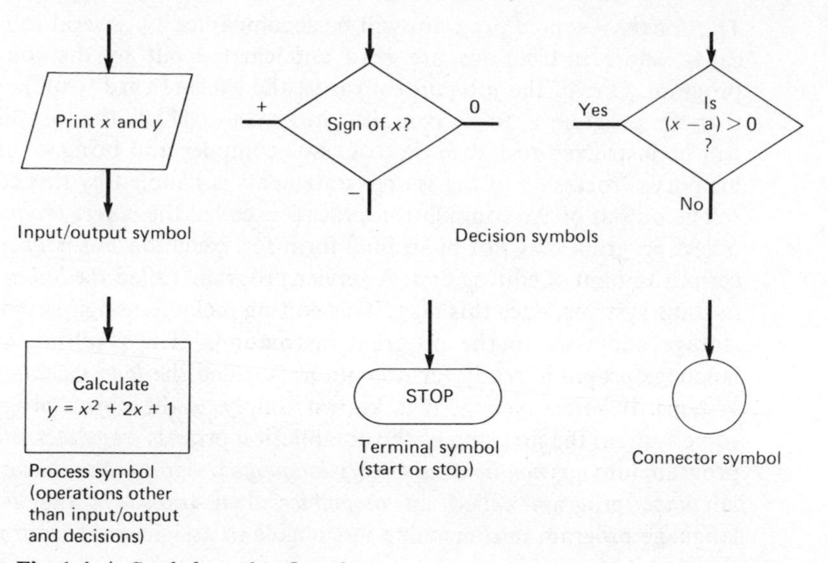

Fig. 1-4 / *Symbols used in flow charts.*

point, the reader should study the flow chart and correlate the notes in the blocks with the preceding problem discussion. Note that the flow chart merely outlines the general procedure to be followed. The actual program would be written by following this outline and detailing to the computer the specific instructions necessary to implement the procedures outlined in the blocks. The block symbols used for this flow chart, which will also be used in subsequent discussions in the text, are described in Fig. 1-4.

1-4 / LARGE COMPUTER OPERATING SYSTEMS

Most large computers operate under the control of an *operating system.* The user communicates with the operating system by means of a special language called, in some systems, the job-control language (JCL). The operating system consists of a number of machine-language programs stored in an auxiliary storage device, from which they can be brought into main memory. Typical system programs include program loaders, a job-control program for reading job-control language, a supervisor program for general supervisory tasks, the compilers for the various languages, and various service programs.

As an abbreviated example of how the operating system works, let us assume a user has written a program in FORTRAN, the high-level language used by most engineers. This FORTRAN program is called the *source program.* We assume that the program has been punched on cards, and that any data which the program requires have been punched on *data cards.*

The FORTRAN source program will be accompanied by several job-control cards whose instructions are read and carried out by the job-control program. One of the job-control cards, the execute card (EXEC), specifies that the program is to be compiled in FORTRAN. Thus the system loader will be instructed to locate the FORTRAN compiler and bring it into main memory. Processing of the source statements is handled by this compiler.

The output of the compilation process is called the *object program*. The object program may not be in final form for execution but may require a certain amount of editing first. A service program, called the *linkage-editor* in some systems, does this task. This editing includes assigning final main storage addresses to the program instructions. The resulting machine-language program, ready for execution, is called the *load module* in some systems. In other systems it is known simply as the object program. In some systems the first step of the compilation process translates the source program into a *symbolic* or *assembly-language* program. Another machine-language program called an *assembler* then translates the symbolic-language program into machine language and assigns final main memory locations.

The final load module, or object program, may be punched on cards or put on magnetic tape for later execution or, if it is a "compile-and-go" operation, the load module is brought from auxiliary storage into main storage, control is transferred to it, and the execution phase proceeds.

At many computer installations jobs are not run in the order in which they are received and read in by the system input device. Instead they go to a queue on an auxiliary storage device from which they are processed according to some arbitrary priority system.

1-5 / FORTRAN

As we have seen, the modern method of solving engineering problems on the digital computer involves the use of compilers to simplify the programming procedures. One of the most frequently used compilers is FORTRAN, first developed for the IBM 704 computer in 1956. The name FORTRAN is derived from FORmula TRANslator. It was designed primarily for use by engineers and scientists in handling problems involving considerable amounts of numerical computation. The use of the FORTRAN compiler permits the engineer to write a program in a simple language closely resembling that of ordinary English and algebra, and involving a minimum of programming steps. The computer itself thus performs the tedious task of coding the program into the machine language understood by the computer. FORTRAN has evolved considerably since 1956, with earlier versions being supplanted by newer and more sophisticated versions.

Currently the most widely used version is FORTRAN IV, and it is this version which is reviewed in the sections which follow.

The IBM System/360 computer is considered typical of modern large-scale computing equipment. The FORTRAN IV version written for this system contains some features not found, for example, in USA Standard FORTRAN. Many specific references to System/360 FORTRAN are made in the discussions which follow.

1-6 / THE ELEMENTS OF FORTRAN IV

The sentences of FORTRAN are called *statements*. Most versions list from five to nine different types of statements. In IBM 360 FORTRAN IV, statements are classified into nine different types. These statements will be discussed later, but first we will review the elements used to form such statements. These include constants, variables, subscripted variables, operation symbols, expressions, and mathematical functions.

1-7 / CONSTANTS

A constant is a fixed, unchanging quantity. There are three principal types of constants: (1) *numerical*, (2) *logical*, and (3) *literal*. Numerical constants may be further subdivided into *integer* constants, *real* constants, and *complex* constants. Both real and complex constants may be *single-precision* or *double-precision*. We will now discuss each of the above in the order mentioned.

Integer Constants

An integer constant is a number written without a decimal point and can thus be only a whole number. The use of a sign is optional except when the number is negative. Embedded commas are not permitted, and there is a maximum magnitude which cannot be exceeded, the magnitude depending upon the computer being used. The following are examples of valid integer constants:

```
0      216      1      -42963
```

Real Constants

Real constants are numbers written with a decimal point. As with integer constants, embedded commas are not permitted. There are two basic forms, *ordinary decimal* form such as −983.12, and *exponent* form, which consists of the use of a real constant followed by an E and either a positive or negative power of 10. The following are examples of valid exponent forms:

```
1.573E+13        -2.746E+11        -1.86E-12        20E+4
```

The first one, for example, specifies that the real constant 1.573 is multiplied by 10 to the 13th power. The last one illustrates that if the real constant is a whole number, it is permissible to omit the decimal point in using the exponent form. The maximum number of significant digits varies with the FORTRAN version used, but is usually from seven to nine. The minimum and maximum magnitudes of real constants also vary with the FORTRAN version used. In IBM 360 FORTRAN IV, real constants may vary from approximately 10^{-78} to 10^{75}. The real constants discussed in this paragraph are *single-precision.*

Double-Precision Real Constants

Double-precision real constants, called simply double-precision constants in some FORTRAN versions, use the letter D in the exponent instead of E. The maximum number of significant digits varies from approximately 14 to 29, but is typically about 16 or 17. For example,

```
987654321.234567D-26
```

is a valid double-precision real constant. In some versions constants may be double-precision by virtue of the fact that they contain more digits than are valid for real constants, and the D exponent is not required to make it double-precision. However, it is recommended that double-precision constants always be written with a D exponent, since double-precision is sometimes lost when it is not used, even though the system reference manual may imply that it will not be. The minimum and maximum magnitudes of double-precision real constants are the same as for single-precision as mentioned in the preceding paragraph.

Complex Constants

Complex constants are represented by an ordered pair of real constants. The two real constants are separated by a comma and enclosed in parentheses. The first real constant is the *real* part of the complex constant, and the second is the *imaginary* part. For example,

`(5.71,-2.95)` represents 5.71 − 2.95i

Both parts should be of the same precision. Double-precision complex constants are not available in many versions, but are in IBM 360 FORTRAN.

Logical Constants

Logical constants specify the logical values "true" and "false." When appearing in a source program, they are written as

```
.TRUE.          and          .FALSE.
```

The periods shown are required.

Literal Constants

A literal constant is a string of alphabetic, numeric, or special characters, which are either enclosed in apostrophes or preceded by *w*H, where *w* is the number of characters in the string. When the characters are preceded by *w*H, they are often referred to as a *Hollerith constant*. For example,

```
38HDYNAMIC CHARACTERISTICS OF A MECHANISM
```

illustrates the Hollerith form, and

```
'DISPLACEMENT VELOCITY ACCELERATION'
```

shows the use of apostrophes.

1-8 / VARIABLES

A FORTRAN variable is a symbolic name for a quantity that is stored in computer memory. It consists of from one to six alphabetic and numeric (alphanumeric) characters, the first of which must be alphabetic. FORTRAN IV provides for integer, real, complex, and logical variables. Real and complex variables may be either single-precision or double-precision.

Integer Variables

The first character of an integer variable name is I, J, K, L, M, or N. A different character than those listed may be used if the name appears in an INTEGER explicit type specification statement, or if the variable is specified as integer by an IMPLICIT type specification statement (these statements are discussed later under Specification Statements). If the variable is specified as integer by virtue of its first character being I, J, K, L, M, or N, its type is said to be specified by the *predefined specification*. JOBNO is a valid integer variable name, for example.

Real Variables

If a real variable has its type specified by the predefined specification, its name will begin with any alphabetic character other than I, J, K, L, M, or N. An example would be VELOC.

Double-Precision Real, Complex, Double-Precision Complex, and Logical Variables

These variables must have their type specified by either the explicit or the IMPLICIT type specification, as there is no predefined specification for these variables. As mentioned earlier, these statements are discussed later under Specification Statements.

1-9 / ARRAYS

A FORTRAN *array* is a set of variables identified by a single name. The different elements of the array are differentiated from one another by using subscripts with the array name. Thus the array elements are *subscripted variables*. The rules for choosing the array name are the same as those that govern nonsubscripted variables. They may also be the same modes (integer, real, and so forth) as nonsubscripted variables. The number of elements in an array must be specified by giving the number of subscripts and the maximum value of each subscript in a DIMENSION, COMMON, or explicit type specification statement. For example, the statement

```
DIMENSION AMATR(50,50)
```

indicates that there are two subscripts, each having a maximum value of 50. The array name is AMATR, and AMATR(10,25) would be a particular element of the array. AMATR(I,J) would represent any particular element corresponding to the values of I and J. Arrays are stored in computer memory in ascending storage locations, with the value of the first subscript increasing most rapidly and the value of the last subscript increasing least rapidly. Most FORTRAN versions permit up to three subscripts (three-dimensional arrays), but in some versions, including IBM 360 FORTRAN IV, up to seven-dimensional arrays are possible.

1-10 / SUBSCRIPTS

Subscripts are most often used in one of the following forms:

Symbolic Form	FORTRAN Form
v	JMAX
c	3
$v + c$	I+1
$v - c$	J-1

where v represents an unsigned, nonsubscripted integer variable, and c represents an unsigned integer constant. The numerical value of the subscript must be positive.

Most versions of FORTRAN also permit the following forms:

Symbolic Form	FORTRAN Form
$c*v$	3*JOB
$c*v + c'$	6*I+7
$c*v - c'$	4*N-2

In IBM 360 FORTRAN IV, considerably more freedom is allowed in the choice of subscript forms. Function names, subscripted variables, and all the arithmetic operators may appear in expressions used as subscripts. It is also permissible in this version to mix real and integer modes in subscript expressions. The values of all expressions used in subscripts are converted to integer mode after evaluation if they are not already integer, and the resulting subscript values must always be greater than or equal to 1.

1-11 / ARITHMETIC EXPRESSIONS

The most simple arithmetic expression is a single constant, variable, or function. More complicated expressions are formed by combining constants, variables, and functions with arithmetic operators. These operators are

Operator	Operation	Hierarchy Level
**	Exponentiation	1st
* and /	Multiplication and division	2nd
+ and –	Addition and subtraction	3rd

The hierarchy determines which of two consecutive operations is performed first. A group of operations of the same hierarchy are performed

Operations +, −, *, /	Integer (2)	Integer (4)	Real (4)	Real (8)	Complex (8)	Complex (16)
Integer (2)	Integer (2)	Integer (4)	Real (4)	Real (8)	Complex (8)	Complex (16)
Integer (4)	Integer (4)	Integer (4)	Real (4)	Real (8)	Complex (8)	Complex (16)
Real (4)	Real (4)	Real (4)	Real (4)	Real (8)	Complex (8)	Complex (16)
Real (8)	Real (8)	Real (8)	Real (8)	Real (8)	Complex (16)	Complex (16)
Complex (8)	Complex (8)	Complex (8)	Complex (8)	Complex (16)	Complex (8)	Complex (16)
Complex (16)	Complex (16)	Complex (16)	Complex (16)	Complex (16)	Complex (16)	Complex (16)

Fig. 1-5 / *Modes of expressions containing mixed types of operands in* IBM 360 FORTRAN IV (*operations* +, −, *, /).

from left to right, except that in IBM 360 FORTRAN IV the evaluation of consecutive exponentiations is from right to left. Parentheses may also be used to form groupings and thus control the order of operations just as they are used in ordinary algebra.

In many FORTRAN versions, modes cannot be mixed within an expression except for permitting real quantities to be raised to integer powers. Some versions allow mixed-mode expressions. In IBM 360 FORTRAN IV, almost complete mixing of modes is permissible. Figure 1-5 shows the resulting types of expressions for combinations of various types of operands in System/360 FORTRAN. The numbers 2, 4, 8, and 16 refer to the number of bytes of storage used in storing the various values. One byte is a group of 8 magnetic cores capable of storing 8 binary digits (bits). Thus Real(8) means double-precision real, and Real(4) means single-precision real. Complex(16) refers to double-precision complex, while Complex(8) means single-precision complex. Figure 1-6 shows the various valid combinations of operands for the arithmetic operator **.

Care should be exercised in using any pure integer quotient in an expression. For example, the quotient J/2 gives a value of 3 for J=7 since integer quotients are truncated. Using J/2. (assuming mixed-mode expressions are permitted) would result in a value of 3.5 for the quotient.

Base	Exponent
Integer (2), Integer (4) Real (4), or Real (8)	Integer (2), Integer (4) Real (4), or Real (8)
Complex (8) or Complex (16)	Integer (2) or Integer (4)

Fig. 1-6 / *Valid combinations of* IBM 360 FORTRAN IV *expressions with respect to the operator* **.

Negative quantities cannot be raised to a real power even though the power is a whole number.

A positive or negative sign (to be distinguished from a plus or minus indicating addition or subtraction) is treated in the hierarchy the same as the plus or minus in addition and subtraction. Thus C = −A*B is treated as C = −(A*B).

No two arithmetic operators may appear sequentially in an expression. Thus A*−B must be written as A*(−B). A typical valid FORTRAN expression is

```
((A+B)*C-D)/F
```

which represents the algebraic expression

$$\frac{(a + b)c - d}{f}$$

1-12 / LOGICAL EXPRESSIONS

A logical expression always has the value "true" or "false." One of the more common uses of logical expressions is in the Logical IF statement which is discussed later. The simplest form of a logical expression is a single logical constant, a single logical variable, or a single logical function. More complicated logical expressions state conditions which are either true or false. One such logical expression is called a *relational* expression, which simply makes a declaration regarding the relative magnitudes of two arithmetic expressions. For example, it might state that A+B is greater than C*D. In FORTRAN IV this is stated by writing the two arithmetic expressions separated by the relational operator .GT. which gives

```
A+B .GT. C*D
```

The relational operators available in FORTRAN IV are as follows:

Relational Operator	Definition
.GT.	Greater than
.GE.	Greater than or equal to
.EQ.	Equal to
.NE.	Not equal to
.LT.	Less than
.LE.	Less than or equal to

Only single- and double-precision real and integer arithmetic expressions may be combined by relational operators. In some FORTRAN versions arithmetic expressions of different modes cannot be combined by relational operators, although most permit the mixing of single- and double-precision real modes across the relational operator. In IBM 360 FORTRAN mixing of all three modes mentioned above is permissible across the relational operator. Complex and logical quantities cannot be used in relational expressions.

More complicated logical expressions can be formed by combining relational expressions, logical constants, logical functions, and logical variables with the logical operators .AND., .OR., and .NOT.. Suppose X and Y represent any of the logical quantities mentioned above (relational expressions, logical constants, and so forth). Then

```
.NOT. X
```

is true if X is false, or false if X is true. The logical expression

```
X .AND. Y
```

is true if and only if both X and Y are true. The logical expression

```
X .OR. Y
```

is true if either X or Y or both are true, and false if *both* X and Y are false.

Still more complicated logical expressions involving several logical operators are possible. Parentheses may be used to control the order in which operations are performed in logical expressions. When parentheses are not used, a hierarchy determines the order of performing operations which is as follows:

Operation	Hierarchy
Evaluation of functions	1st
**	2nd
* and /	3rd
+ and –	4th
Application of relational operators .GT., .LT., etc.	5th
Application of .NOT.	6th
Application of .AND.	7th
Application of .OR.	8th

1-13 / FORTRAN-SUPPLIED MATHEMATICAL FUNCTION SUBPROGRAMS

Certain mathematical functions have had their procedures for evaluation preprogrammed, and they are automatically available to the FORTRAN processor. There are two kinds of these supplied function subprograms.

The first are called *intrinsic*, *built-in*, or *in-line* functions. These require only a few machine-language instructions, so these instructions are placed in the object program each time the function name is used in the source program. The second kind are called *external*, *library*, or *out-of-line* functions. The preprogrammed procedures for the functions are in a library (such as on a magnetic disk), and only one set of machine-language instructions would be placed in the object program, even though the function is used several times.

It may be recalled that in order to use these functions, it is necessary only to write the function name in the expression in which it is required, and to supply the argument or arguments. Arguments are expressions in general, but often are as simple as a single variable. In the statement

```
Y=SQRT(X)+B
```

SQRT is the name of a FORTRAN-supplied function, and X is the argument which has been supplied by the programmer. When there is more than one argument, they are separated by commas as illustrated in the statement

```
M=MOD(I,J)+1
```

Some commonly used IBM 360 FORTRAN out-of-line mathematical function subprograms are given in the following table.

Function Name	**Number of Arguments**	**Function Type**	**Argument Type**	**Definition (a is argument)**
EXP	1	Real(4)	Real(4)	
DEXP	1	Real(8)	Real(8)	e^a
CEXP	1	Complex(8)	Complex(8)	
CDEXP	1	Complex(16)	Complex(16)	
ALOG	1	Real(4)	Real(4)	
DLOG	1	Real(8)	Real(8)	$\log_e (a)$
CLOG	1	Complex(8)	Complex(8)	
CDLOG	1	Complex(16)	Complex(16)	
ALOG10	1	Real(4)	Real(4)	$\log_{10} (a)$
DLOG10	1	Real(8)	Real(8)	
ARSIN	1	Real(4)	Real(4)	arc sin (a)
DARSIN	1	Real(8)	Real(8)	(in radians)
ARCOS	1	Real(4)	Real(4)	arc cos (a)
DARCOS	1	Real(8)	Real(8)	(in radians)

continued overleaf

Function Name	Number of Arguments	Function Type	Argument Type	Definition (a is argument)
ATAN	1	Real(4)	Real(4)	arc tan (a)
ATAN2	2	Real(4)	Real(4)	arc tan (a_1/a_2)
DATAN	1	Real(8)	Real(8)	arc tan (a)
DATAN2	2	Real(8)	Real(8)	arc tan (a_1/a_2)
SIN	1	Real(4)	Real(4)	
DSIN	1	Real(8)	Real(8)	sin (a)
CSIN	1	Complex(8)	Complex(8)	(a is in radians)
CDSIN	1	Complex(16)	Complex(16)	
COS	1	Real(4)	Real(4)	
DCOS	1	Real(8)	Real(8)	cos (a)
CCOS	1	Complex(8)	Complex(8)	(a is in radians)
CDCOS	1	Complex(16)	Complex(16)	
TAN	1	Real(4)	Real(4)	tan (a)
DTAN	1	Real(8)	Real(8)	(a is in radians)
COTAN	1	Real(4)	Real(4)	cotan (a)
DCOTAN	1	Real(8)	Real(8)	(a is in radians)
SQRT	1	Real(4)	Real(4)	
DSQRT	1	Real(8)	Real(8)	$\sqrt{a}$
CSQRT	1	Complex(8)	Complex(8)	
CDSQRT	1	Complex(16)	Complex(16)	
TANH	1	Real(4)	Real(4)	tanh (a)
DTANH	1	Real(8)	Real(8)	
SINH	1	Real(4)	Real(4)	sinh (a)
DSINH	1	Real(8)	Real(8)	
COSH	1	Real(4)	Real(4)	cosh (a)
DCOSH	1	Real(8)	Real(8)	

Some commonly used IBM 360 FORTRAN in-line mathematical function subprograms are

Function Name	Number of Arguments	Function Type	Argument Type	Definition (a is argument)
ABS	1	Real(4)	Real(4)	
IABS	1	Integer(4)	Integer(4)	$\lvert a \rvert$
DABS	1	Real(8)	Real(8)	

INT AINT IDINT	1 1 1	Integer(4) Real(4) Integer(4)	Real(4) Real(4) Real(8)	Truncates argument to largest integer having an absolute value less than or equal to argument. Sign same as sign of argument
AMOD MOD DMOD	2 2 2	Real(4) Integer(4) Real(8)	Real(4) Integer(4) Real(8)	Defined as $a_1 - [a_1/a_2]a_2$ where $[a_1/a_2]$ is truncated value of the quotient
REAL	1	Real(4)	Complex(8)	Obtains the real part of a complex argument
AIMAG	1	Real(4)	Complex(8)	Obtains the imaginary part of complex argument
CMPLX DCMPLX	2 2	Complex(8) Complex(16)	Real(4) Real(8)	Expresses two real arguments in complex form
CONJG DCONJG	1 1	Complex(8) Complex(16)	Complex(8) Complex(16)	Obtains conjugate of a complex argument

1-14 / FORTRAN STATEMENTS

A FORTRAN source program consists of a sequence of statements written in FORTRAN language. USA Standard FORTRAN specifies eight types of statements. Of these, three types are *executable*. They are

1 Arithmetic- and logical-assignment statements
2 Control statements
3 Input/output statements

There are five types of nonexecutable statements:

1 FORMAT statements
2 DATA initialization statements
3 Specification statements
4 Statement function definition statements
5 Subprogram statements

IBM 360 FORTRAN provides a sixth type of nonexecutable statement, the NAMELIST statement. In the next four sections we will discuss *executable statements.*

1-15 / ARITHMETIC-ASSIGNMENT STATEMENTS

The arithmetic-assignment statement consists of a single subscripted or nonsubscripted variable to the left of an equal sign and an arithmetic expression on the right of the equal sign. In executing the statement, the arithmetic expression is evaluated, and the resulting value is assigned to the variable to the left of the equal sign. The following is a typical arithmetic-assignment statement:

```
X=(-B+SQRT(B*B-4*A*C))/(2.*A)
```

If X has a value prior to the execution of the statement, this value will be *replaced* by the value calculated during the execution. In FORTRAN the equality sign does not have the customary mathematical meaning of "is equal to." Instead it means "is to be replaced by." For example, the FORTRAN statement

```
J=J+2
```

is a perfectly valid arithmetic-assignment statement, with the FORTRAN interpretation of the equality sign given above. This statement instructs the computer to add 2 to the existing value of J, and to then replace the existing value of J by the sum obtained.

It is sometimes necessary or convenient to use different modes on both sides of an equality sign, and such a procedure is valid. That is, the expression on the right side of the equality sign need not be in the same mode as the variable on the left side. For example, if we have the statement

```
A=I+1
```

the value of the expression I+1 will be calculated in integer mode, and this value will then be stored in the location referred to by the variable name A in real mode. In the statement

```
J=X**2-B
```

the expression on the right is evaluated in real mode, and the result is then truncated to the next lowest integer value and stored in integer mode in the location referred to by the variable name J.

1-16 / LOGICAL-ASSIGNMENT STATEMENTS

Statements consisting of a logical variable on the left of an equal sign and a logical expression on the right side are called logical-assignment statements. The expression on the right is evaluated as either "true" or "false," and this logical value is then assigned to the variable on the left. The following are examples of valid logical-assignment statements, where the variables on the left have been specified as logical variables:

```
LV1=.TRUE.
G=X.GT.Y
F=A.GT.B .AND. C.EQ.D
```

When logical quantities are involved, no mixing of modes across the equal sign is permitted.

1-17 / CONTROL STATEMENTS

Control statements perform a variety of functions. They provide a means for branching, looping, temporarily suspending execution of a program until operator intervention instructs it to resume execution, and terminating execution of the object program. It should be recalled that in all transfer of control statements, the statement to which control is transferred must be executable. Various control statements are discussed in the following paragraphs.

Unconditional GO TO Statement

This statement, referred to as an unconditional branch or transfer statement, is used to interrupt the usual sequential execution of statements by an unconditional transfer of control to a specified *numbered* statement. For example, the statement

```
GO TO 16
```

instructs the computer to execute immediately the statement numbered 16, regardless of the position of the statement in the program sequence. Any executable statement immediately following a GO TO statement must be numbered, or it can never be executed.

Computed GO TO Statement

This statement is a conditional branch or transfer type of control statement, since the transfer of control depends on the value of a nonsubscripted integer variable which is a part of the computed GO TO statement. For example, the statement

```
GO TO(3,8,11,24),J
```

will transfer control to statement 3 if the value of J is 1, to statement 8 if J is 2, and so on. Any number of statements to which control can be transferred is permissible. If the integer variable in the computed GO TO statement (J in the example above) has a value greater than the number of statements referred to in the parentheses, the next statement is executed.

The Assigned GO TO and ASSIGN Statements

The general form of the assigned GO TO statement is

GO TO i,(n_1, n_2, n_3, . . . , n_m)

where i represents an integer variable (nonsubscripted) and the n's are *statement numbers*. The integer variable must have appeared previously in an ASSIGN statement having the general form

ASSIGN n to i

where n represents a statement number and i represents the integer variable which appears subsequently in the assigned GO TO statement. Thus the appearance in a program of the statements

```
  .          .
  .          .
  .          .
ASSIGN 10 TO INT
  .          .
  .          .
GO TO INT,(5,10,36,42)
  .          .
  .          .
```

would cause transfer of control to statement 10 listed in the assigned GO TO statement. The number assigned to the integer variable in the ASSIGN statement must always be one of the statement numbers listed in the assigned GO TO statement, or the combination of the two statements will be invalid.

Arithmetic IF Statement

The arithmetic IF statement is a conditional branching type of control statement. It provides for control to be transferred to any one of three numbered statements, depending on the value of the expression enclosed in parentheses which is part of the IF statement. For example, the statement

```
IF(X-A)14,13,7
```

transfers control to statement 14 if the quantity (X−A) is *negative*, to statement 13 if (X−A) is equal to *zero*, and to statement 7 if (X−A) is *positive*. The parenthetical portion of the IF statement may contain any legitimate arithmetic expression other than a complex expression, including a single variable name.

Logical IF Statement

The logical IF statement has the general form

```
IF(e)s
```

where e represents a *logical expression* and s represents any executable FORTRAN statement except a DO or another logical IF. For example, if A and B are real variables and LV1, LV2, and LV3 are logical variables, the following are valid logical IF statements:

```
IF(A .GT. B)GO TO 16
IF(LV1)Y=A/B
IF(LV2 .AND. A .LE. B)LV3=.TRUE.
```

If the logical expression in parentheses (symbolized by e in the general expression) is true, the statement following (symbolized by s) is executed. The statement following the logical IF statement is then executed unless statement s is a transfer-of-control statement. In the latter instance, control is transferred as specified by statement s. If the logical expression e is false, control passes to the next statement in the program following the logical IF statement. In the first of the examples of valid logical IF statements shown above, control would be transferred to statement 16 if A were greater than B. If A were not greater than B, control would transfer to the statement in the program following the logical IF statement.

DO Statement

The DO statement causes the computer to repeat a specific sequence of statements, within a given program, a prescribed number of times. The statements in the sequence to be repeated are known as the *range* of the DO statement. This type of statement is convenient to use when a known number of repetitions, or loops, is required for a given sequence of statements. It may also be used when a maximum number of repetitions is specified, with a transfer from the DO loop *possibly* occurring before the maximum number of repetitions specified has been completed. The IF statement, discussed previously, may also be used to accomplish the same type of repetitive operation, but it generally is not so convenient as the DO statement.

The general form of the DO statement is

DO n i = m_1, m_2, m_3

in which n represents the number of the last statement in the range of the DO, i represents a nonsubscripted integer variable known as the *index* or DO *variable*, and m_1, m_2, and m_3 represent unsigned integer constants or nonsubscripted integer variables called *indexing parameters*. Individually, they are called, respectively, the *initial value*, the *test value*, and the *increment*.

The DO statement causes the statements within the range to be executed repeatedly. For the first execution of the range, the index has the value of m_1, the initial value. After each execution of the statements in the range, the index is incremented by the value of m_3, the increment. The index is then compared with the value of m_2, the test value. If it is less than or equal to the test value, execution of the statements in the range is carried out again. If the index value exceeds the test value, the execution of the DO loop ceases, and control passes to the first statement following the end of the range of the DO. The DO statement is then said to be satisfied, and the exit from the DO loop is known as a *normal exit*.

Let us consider the DO statement

```
DO 6I=1,100,2
```

as an example. The number 6 identifies statement number 6 as the last statement of the group of statements which are to be executed repeatedly. The integer variable I is the index in this example. The index may be used in statements within the range for any purpose for which any other integer variable would be permissible. It is most commonly used as a subscript for a subscripted variable. However, it is not necessary that the index be used in any of the statements in the range. If it is not, it merely serves as a counter and is incremented by the value of m_3 after each execution of the range. The index in this example would have a value of 1 during the first execution. Following each execution of the loop, the index would be incremented by the value of 2. When the value of the index reached 99, an additional execution of the loop would be made. The index would then assume a value of 101. This value would then be compared with the test value of 100, and, since it exceeds the test value, execution of the DO loop would cease.

After a normal exit, the integer variable assigned as the index may or may not be defined and available for use in a subsequent portion of the program, depending upon the version of FORTRAN being used. Usually, it is not. If control is transferred from a statement in the DO loop to a statement outside the DO loop before the DO statement is satisfied (a transfer of this type is permitted at any time), the index remains defined and may be used, as so defined, in a subsequent portion of the program.

It is permissible to have one or more other DO statements within the range of a DO loop. In this instance, the set of DO loops is referred to as nested DOs. If the range of a DO statement includes another DO statement, the former is called the *outer* or *major* DO, and the latter is called the *inner* or *minor* DO. All the statements in the range of an inner DO must also be in the range of the outer DO. This rule allows the inner and outer DO loops to end with a statement common to each loop, if this is desired. A set of nested DOs, as a portion of an overall program, might appear as follows:

```
      .....
      READ(5,2)M
    2 FORMAT (I4)
      N=M+1
      DO 9I=2,M
      DO 9J=2,N
      L=J-1
      SUM=0.
      DO 8K=1,L
    8 SUM=SUM+B(I,K)*T(K,J)
    9 B(I,J)=A(I,J)-SUM
      .....
      .....
```

In the normal sequence of operations of a set of nested DOs, control is first passed to the outer DO from the preceding portion of the program. The statements in the outer DO range preceding the inner DO statement are then executed in some type of sequence until that sequence arrives at the inner DO statement, to which control then passes. After the inner DO is satisfied, or after exit from it occurs by a transfer, any remaining statements in the outer DO are executed, completing the first cycle of the outer DO. Unless exit occurs from an inner DO by a transfer, it cycles the specified number of times and is satisfied for each separate execution of the range of the outer DO.

It is not permissible to transfer from outside the range of a DO to a statement inside the range of the DO, except that a transfer may be made into the *innermost* DO loop of a nest of DOs provided there has previously been a transfer *out* of this innermost DO loop, and the index or indexing parameters have not been changed by statements executed between the transfer out and the transfer back in. However, it is permissible to transfer out of any DO loop to a *subprogram*, and to then return from it. That is, a subprogram may be used within any DO loop in a nest of DOS.

If the constant or variable in the DO statement, defining the increment, is omitted (along with the comma which would precede it), the index will be incremented by 1 after each execution.

There are certain restrictions placed on the statements in the range of the DO which must be observed. They are

1 No statement within the range of the DO should redefine the index or any of the indexing parameters.
2 The last statement in the range of a DO loop must be an executable statement, but it cannot be a GO TO statement (unconditional, computed or assigned), an arithmetic IF statement, another DO statement, a PAUSE, STOP, or RETURN statement, or a logical IF containing any of these statements.

CONTINUE Statement

The CONTINUE statement is a dummy statement that may be placed anywhere in the source program without having any effect on the execution of the program. However, it is usually used as the last statement in the range of a DO loop to avoid the error of ending the DO loop with one of the prohibited types of statements outlined in the previous paragraph.

The use of the CONTINUE statement is illustrated in the following partial program for finding the largest element of an array A consisting of 100 elements.

```
  ..................
  N=1
  BIG=ABS(A(1))
  DO 4I=2,100
  IF(BIG-ABS(A(I)))3,4,4
3 BIG=ABS(A(I))
  N=I
4 CONTINUE
  ..................
```

If the absolute value of A(I) is not bigger than the value of BIG in the IF statement, statement 3 and the statement following it must be by-passed. To increment the index, control must pass to the last statement in the range of the DO. We might be tempted to use a GO TO statement as this last statement, to transfer control back to the first statement in the range of the DO. However, this is not permissible, so a CONTINUE statement is used as the last statement in the range. When control is transferred to the

CONTINUE statement, the index I is incremented by 1, and either the execution of the range is repeated, or a normal exit is made from the DO loop, depending upon whether or not the DO statement has been satisfied.

PAUSE Statement

The PAUSE statement has one of the forms

```
PAUSE   or   PAUSE n
```

where n represents a string of one through five decimal digits. The statement causes a temporary termination of execution, and operator intervention is required to resume execution. Either PAUSE n or PAUSE 00000 is displayed to the operator, depending upon whether the parameter n was used or not. In IBM 360 FORTRAN the form PAUSE 'message' may be used, where 'message' is a literal constant. If this form is used, PAUSE 'message' is displayed to the console operator. It should be noted that the PAUSE statement cannot be used at some computer installations.

STOP Statement

This statement has the form

```
STOP   or   STOP n
```

where n is a string of one through five decimal digits. The statement terminates execution of the object program. If the form STOP n is used, n is displayed to the console operator. At some installations CALL EXIT is used instead of the STOP statement.

END Statement

The END statement, although classified as a control statement, is not an executable statement. It produces no machine-language instructions, but is used to indicate the end of a source program or subprogram to the FORTRAN compiler. Physically it is always the last statement of a source program or subprogram.

1-18 / INPUT AND OUTPUT STATEMENTS

The input/output statements control the transmission of data between the computer and the input or output devices used. The following discussion will be concerned with one basic input statement, the READ statement, and one basic output statement, the WRITE statement. In some versions of FORTRAN IV, there are different forms of these two statements available, but we will confine our discussion to forms for *formatted* READ and WRITE (data transmitted under control of a FORMAT statement) which are available in all versions of FORTRAN IV.

The basic form of the formatted READ statement is

```
READ (r,n)list
```

where n is the statement number of an accompanying FORMAT statement (to be discussed later), and r represents an unsigned integer constant or a nonsubscripted integer variable specifying the unit to be used for input. It is sometimes called a *symbolic unit number* or *device code*. In IBM 360 FORTRAN IV, it is known as a *data set reference number*. In this version, the data set reference number will correspond to a number appearing on a JCL data definition card (DD card). This DD card "points to" a particular device. Data sets to be read in by that device will have a number for r in the READ statement that is the number appearing on the JCL data definition card. In the I/0 examples which follow, it will be assumed that data set reference number 5 is used for data which are to be read by the card reader, and number 6 is used for data which are to be printed on the line printer. A typical READ statement might appear as

```
READ(5,13)A, DISPL, X
```

This statement would cause three data values to be read and assigned to the three variable names in the list. A FORMAT statement having statement number 13 would have to be used in conjunction with this READ statement to specify the form of the data on the card(s).

The basic form of the formatted WRITE statement is

```
WRITE (r,n)list
```

where the r and n have the same meaning as they do in the READ statement. The *list* is a list of variables whose values are to be transmitted out of the computer (usually printed on the line printer). A typical WRITE statement is

```
WRITE(6,14)J, ARRAY(3), YDIST
```

The values of the three variables in the list would be printed by the line printer. The form of the data would be specified by the accompanying FORMAT statement having statement number 14.

Input and Output of Entire Arrays

In illustrating the input and output of arrays, FORMAT statements will be used even though they have not been discussed. For the reader using this material for review or reference, this will not present a problem. A reader learning FORTRAN for the first time may wish to review this material on transmission of arrays after studying FORMAT statements.

One possible way of transmitting (reading into memory or printing out from memory) an entire one-dimensional array or a part of it is to use a single DO loop. Similarly, a two-dimensional array can be transmitted by using a double DO loop. For example, the appearance in a program of the statements

```
47 FORMAT(1Hᵇ,F10.2)
   DO 6J=1,N
   DO 6I=J,N
 6 WRITE(6,47) A(I,J)
```

would cause the printout of the elements

$$\begin{matrix} a_{11} & & & & \\ a_{21} & a_{22} & & & \\ a_{31} & a_{32} & a_{33} & & \\ \vdots & & \vdots & & \\ a_{n1} & \cdots & a_{n4} & \cdots & a_{nn} \end{matrix}$$

of an array. These elements would not appear in the form shown above but would be printed out, one element to a line, in one long column, their order being $a_{11}, a_{21}, a_{31}, \ldots, a_{n1}, a_{22}, a_{32}, \ldots, a_{n2}, a_{33}, a_{43}, \ldots, a_{n3}$, and so forth. If the entire array were to be printed out, instead of just the elements shown, the statements referred to could be used by simply changing the initial value (first indexing parameter), in the second DO statement, from a J to a 1.

Another method of printing an entire array is by using a set of statements such as

```
   DIMENSION B(10)
   ..........
49 FORMAT(1Hᵇ,F10.2)
   WRITE(6,49)B
```

in which the array name B appears in the list in nonsubscripted form. In printing the one-dimensional array B, all ten elements would be printed consecutively in a column, with one element to a line. By changing statement 49 to FORMAT(1H^b,10F10.2) all ten elements of the array would be printed out consecutively on a single line. The input of an entire array may be handled in a similar manner.

If the array to be handled is two-dimensional or greater, a set of statements, such as those in the preceding paragraph, will read in or print out the elements in an order in which the *first* subscript progresses most rapidly, the second subscript progresses the next most rapidly, and so forth. For example, the elements of a 3-by-3-by-3, three-dimensional array would print out as a_{111}, a_{211}, a_{311}, a_{121}, a_{221}, a_{321}, a_{131}, a_{231}, a_{331}, a_{112}, a_{212}, a_{312}, a_{122}, a_{222}, a_{322}, and so forth.

Indexing Within an Input or Output List

Arrays in an input or output list may have their subscripts incremented without the use of DO statements. For example, the statements

```
27 FORMAT(10F7.1)
   READ(5,27)(A(I), I=1,10)
```

would cause the reading of ten quantities from a data card, and these values would be assigned to A(1), A(2), . . . , A(10), respectively. This indexing of subscripted variables within an input or output list is referred to as an *implied* DO. It has the advantage over the explicit use of the DO statement such as

```
27 FORMAT(F7.1)
   DO 7I=1,10
 7 READ(5,27)A(I)
```

in that, in the explicit use of the DO, each input value must be in a separate input record (be punched on a separate card). Changing the FORMAT specification to 10F7.1, for example, would not alter this requirement. Only the first F7.1 specification would be used each time the READ statement was executed. On the other hand, in the implied DO example shown, all ten values would be in the same input record (be punched on a single card).

Implied DOs may also be nested. For example, the statements

```
49 FORMAT(1Hᵇ,10F7.1)
   WRITE(6,49)((A(I,J),J=1,10),I=1,10)
```

would cause the printout of the a_{ij} array in the form

$$\begin{matrix} a_{11} & a_{12} & a_{13} & \cdots & a_{1,10} \\ a_{21} & a_{22} & a_{23} & & \vdots \\ a_{31} & & & & \vdots \\ \vdots & & & & \vdots \\ a_{10,1} & a_{10,2} & \cdots & \cdots & a_{10,10} \end{matrix}$$

This form is obtained since the subscript J progresses the fastest because it is in the inner implied DO in the WRITE statement shown. Note that this is not the order in which the elements would print out if the array were transmitted by using the array name in the list in nonsubscripted form, as discussed under input and output of entire arrays.

In the examples thus far, the increment of the index has been assumed to be 1, but it can be given any other value by using a third indexing parameter, just as in the DO statement. The statements

```
29 FORMAT(1Hᵇ,5F9.2)
   WRITE(6,29)(A(I),I=2,10,2)
```

would cause the values of A(2), A(4), A(6), A(8), and A(10) to be printed out on one line.

Manipulative I/O Statements

Three manipulative statements are used with auxiliary memory devices such as tape units or disks. The statement

```
BACKSPACE 2
```

for example, would cause device number 2 (or the data set associated with the number 2) to backspace one record. The statement

```
REWIND 3
```

would direct device number 3 to the start of the file (data set) so that a subsequent READ or WRITE statement referring to device number 3 would read data from or write data into the first record of the file. The statement

```
ENDFILE 4
```

would define the end of the file (data set) associated with the number 4. If the number 4 refers to a data set being transmitted by a magnetic-tape unit, for example, an end-of-file mark would be written on the tape.

1-19 / NONEXECUTABLE FORTRAN STATEMENTS

In the next sections, we will discuss nonexecutable FORTRAN statements. As mentioned previously, there are five types of nonexecutable statements:

1 FORMAT statements
2 DATA initialization statements
3 Specification statements
4 Statement function definition statements
5 Subprogram statements

IBM 360 FORTRAN provides a sixth type of nonexecutable statement, the NAMELIST statement. A transfer of control cannot be made to any of these nonexecutable statements. None of them may appear as the last statement in the range of a DO loop, nor may any of them appear as the last statement which is part of the logical IF statement. They will now be discussed in the order given.

1-20 / FORMAT STATEMENTS

FORMAT statements are used with both READ and WRITE statements to supply information to the processor regarding the structure of the data *fields* and *records* which are to be read in or printed out. Fields and records will be defined later. A complete listing of the order in which the programmer should arrange the various kinds of statements in his source program will be given after all of the kinds of statements have been discussed, but it should be noted here that FORMAT statements may be placed anywhere in the source program after the IMPLICIT statement, if one is used, and before the END statement, except where the use of a nonexecutable statement is prohibited.

The general form of a FORMAT statement is

n FORMAT $(s_1,s_2,s_3,\ldots,s_m)$

where n is the statement number consisting of from one to five decimal digits, and s_1, s_2, s_3, . . . , s_m are *format specifications* (also called *format codes* or *field specifications*). We will now discuss the eleven different types of format specifications used to describe data fields. They are:

Type	General Form	Example	Remarks
I	aIw	3I6	Describes data fields for transmission of integer quantities
F	aFw.d	2F10.3	Describes data fields for transmission of real quantities
E	aEw.d	2E14.7	Describes data fields for transmission of real quantities
D	aDw.d	4D20.13	Describes data fields for transmission of double-precision real quantities
G	aGw.s	3G14.7	A generalized specification for transmission of integer, real, complex, or logical quantities
X	wX	5X	Indicates a field of w card columns is to be skipped on input, or a field of w printer columns is to be filled with blanks on output
H	*w*Htext	7HDENSITY	Used for transmission of w characters of literal data indicated by text in general form
literal	'text'	'DENSITY'	Used for transmission of literal data indicated by text in general form (not available in all versions of FORTRAN IV)
A	aAw	5A4	Used for transmission of alphameric characters
L	aLw	5L4	Used for transmission of logical quantities
T	Tr	T35	Used to specify the record position where transmission of data begins (not available in all versions of FORTRAN IV)

In the general form, a is a repeat constant. That is, the specification 3I6 is equivalent to I6,I6,I6. The field width is symbolized by the letter w, and the number of places to the right of the decimal point by the letter d. In the G specification, the s is used only for output and symbolizes the number of significant digits to be printed. In the T specification, the r symbolizes an integer constant designating the position in a FORTRAN record at which the transmission of data is to begin. In the following paragraphs, we will discuss the various types of format specifications in more detail.

I-Type Specification (Output)

The statements

```
   WRITE(6,17)NUM
17 FORMAT(1Hᵇ,I5)
```

would cause the value of NUM to be printed by the line printer. Five spaces (printer columns) would be allocated for the printout of the digits and the sign (if negative) because of the field width of 5 specified in the I5 specification. These five spaces constitute a *field* in which the value of NUM is printed. If less than five spaces are required, the spaces to the left of the field are filled with blanks. If more than five are required, there would either be a deletion of the sign and/or of some digits. In some FORTRAN versions asterisks would be printed instead of a value. The latter method, or some printout equivalent to it, in which the printout of an incorrect value is avoided, is characteristic of the newer processors. When the repeat constant is 1, as it is in the above example, it may be omitted.

In the example above, there is an H specification (also called a Hollerith specification) preceding the I specification. This type has not yet been discussed, but its purpose in the FORMAT statement shown is for carriage control. It makes the first character of the output record (record will be defined subsequently) a blank. This first character is not printed, but is used for carriage control. A blank causes the printer to advance one line before printing, giving, in general, normal single spacing. Carriage control will be discussed in more detail after the H and literal specifications are explained.

F-Type Specification (Output)

This specification is used in a FORMAT statement accompanying a WRITE statement when a real value is to be printed out in ordinary decimal form. The statements

```
   WRITE(6,10)XDIST
10 FORMAT(1Hᵇ,F10.3)
```

would cause the value of XDIST to be printed out. There would be three digits after the decimal point, and a total field of ten spaces (printer columns) would be used. The ten spaces must be enough to accommodate the digits of the number and the decimal point, as well as the sign if the output value is negative. If the field width is insufficient, there may be a deletion of the sign/or some of the digits. With the more modern processors, an incorrect value will not be printed when the error is due to too small a field width. Typically the field would be filled with asterisks.

E-Type Specification (Output)

This specification is used for printing out real quantities in exponent form. Exponent form is usually required if the value is very large or very small, or if the programmer has no way of estimating the order of magnitude. The statements

```
  WRITE(6,7)X2
7 FORMAT(1Hᵇ,E14.7)
```

might cause the value of X2 to be printed out as

```
-0.1234567Eᵇ03
```

if the value of X2 were

```
-123.4567
```

A sign is usually printed only if the quantity is negative, and a leading zero (a zero to the left of the decimal point) is generally printed only if sufficient field width has been provided for it. It is specified in the example format specification that seven digits will be printed out after the decimal point. The most significant digit is immediately to the right of the decimal point in most FORTRAN versions, providing that a scale factor has not been used (scale factors will be discussed later). In general, four places are required for the exponent, one for the decimal point, one for a sign if required, and one for the zero. Thus a field width $w \geq d+7$ should be used in most FORTRAN versions, where d is the number of digits to be printed after the decimal point. If the field width provided is insufficient, there will be some characters deleted, or perhaps, as in IBM 360 FORTRAN, asterisks will be printed in the field.

I-Type Specification (Input)

The statements

```
  READ(5,2)LP
2 FORMAT(I4)
```

cause an integer value to be read from a card (assuming that the input data is on cards), and the value is assigned to the variable LP. The value to be read would be punched in the first four columns of the card. Since leading, trailing, or embedded blanks are generally interpreted as zeros, the number should be punched so that it is as far as possible to the right in the field on the card.

F-Type Specification (Input)

If the statements

```
  READ(5,3)XDOT
3 FORMAT(F10.3)
```

appear in a program, the value of XDOT must be punched in the first ten card columns of a data card. It can be punched with three places after the decimal point as specified by the 3 in the F specification. However, the actual punched location of the decimal point overrides the specification. Therefore, many programmers would use the specification (F10.0) in this example and punch a decimal point in the value on the data card wherever needed. If no decimal point is punched in the value on the data card, it is located by the d value (3 in the example) in the F specification. That is, in the example given, the decimal point would precede the last three digits. If the number is not as far as possible to the right in the data card field, the tailing blanks are generally interpreted as zeros, but this causes no change in the value of the number if a decimal point has been punched.

E-Type Specification (Input)

As with the F-type specification, the data being read in with this specification can be punched on the data card just as it would print out if it were being printed out using the E-type specification. For example, if 0.1257691×10^4 were to be read in and assigned to a variable BMAX by the statements

```
  READ(5,4)BMAX
4 FORMAT(E14.7)
```

the value could be punched in the first 14 columns of a data card as follows:

```
ᵇ0.1257691Eᵇ04
```

which is the form in which the number would be printed out in many FORTRAN versions. However, it could also be punched in the data card in a number of other forms such as

```
 ᵇᵇ.1257691Eᵇ04            ᵇᵇ.1257691E+04
 ᵇᵇᵇ.1257691E+4            ᵇᵇᵇᵇ1257691E+4
ᵇᵇᵇᵇ.1257691+4            ᵇᵇᵇᵇᵇ1257691E4
```

The important facts are to punch the number in the field of 14 card columns, and as far to the right in the field as possible so that one or more zeros are not placed after the desired exponent. In the above examples, where no decimal point is punched, it is located by the value of d in the specification.

D-Type Specification (Output and Input)

The D specification is similar to the E specification except that it is used to transmit double-precision quantities. The variable associated with a D specification should be defined as a double-precision variable in a type statement. The number of significant digits which can be read in or printed out for the value of a double-precision variable varies with different computers, from about 14 to 29 digits. In IBM 360 FORTRAN, up to 17 significant digits can be used for double-precision values.

D, E, and F Format Specifications in IBM 360 FORTRAN (Input)

In this version the D, E, and F specifications are interchangeable for input. A card field might contain the number

```
29.132E2
```

with the specification describing the field as E8.3. However, the programmer could also use, for example, F8.3 or D8.3.

G-Type Specification (Input and Output)

This is a generalized specification that may be used for input or output of integer, real (single- or double-precision), complex, or logical data. If a number being read in is real or complex (transmission of complex quantities is discussed later), the G may be thought of as replacing the E or D, and the rules are the same as for those specifications. If the variable in the input list is integer or logical, the s part of the specification (see previously given general form) is ignored if it is given, and therefore may be omitted.

For output of integer or logical quantities, the s part of the specification may be omitted and the G thought of as replacing the I or the L (the L specification is discussed later) with output printed according to the rules for those specifications. For output of real or complex quantities, the s determines the number of digits to be printed and whether the number will be printed in ordinary decimal form or exponent form. For example, in IBM 360 FORTRAN, if the value being printed is greater than or equal to 0.1, but is less than 10^s, it will be printed in ordinary decimal form. If the value is outside this range, it will appear in exponent form (E or D, depending on whether the corresponding variable is single- or double-precision, respectively). If the number is printed in E or D form, the field width specification must include enough room for the exponent (four print positions), the decimal point, and the sign if it is negative.

Transmission of Several Variable Values—Definition of a Record

When the values of several variables are to be transmitted, there must be a format specification for each variable. If different format specifications are used, they are separated by commas. If the same format specification is used for each variable, a repeat constant can be used with a single specification. For example, the FORMAT statement

```
12 FORMAT(1Hᵇ,3F10.3)
```

is equivalent to

```
12 FORMAT(1Hᵇ,F10.3,F10.3,F10.3)
```

Let us consider the FORMAT statement below which is used to describe the output of four variable values.

```
   WRITE(6,13)Y1,Y2,Y3,N
13 FORMAT(1Hᵇ,E15.7,2F10.3,I6)
```

Between the parentheses of the FORMAT statement, one *record* of output data is specified. A record of output data is printed on one line. In the above example there are 41 characters in the output record plus one blank for carriage control. A maximum of 132 output characters plus one for carriage control may usually be specified, although this maximum will vary with different computers using different line printers. The maximum number of characters should not exceed the line length of the printer.

If more variables were added to the above example without adding corresponding format specifications, control would return to the first specification, and the added variable values would be in a new output record and printed out on a new line.

In a FORMAT statement used with a READ statement, a maximum of 80 input characters can be specified corresponding to the 80 columns on a card.

The following is a typical READ statement, with its associated FORMAT statement, containing all three types of field specifications:

```
  READ(5,3)X,Y,XDOT,YDOT,N
3 FORMAT(F3.0,F4.1,E8.3,E10.4,I3)
```

Figure 1-7 shows a data card as it might be punched for read-in by the above READ statement. The vertical lines would not actually appear on the card; they have been added to identify clearly the various fields.

The FORMAT statement

```
16 FORMAT(1Hᵇ,E14.7,3(F10.3,F5.2))
```

contains a *group* format specification. The group is within the inner parentheses and is preceded by the repeat constant 3. The specifications

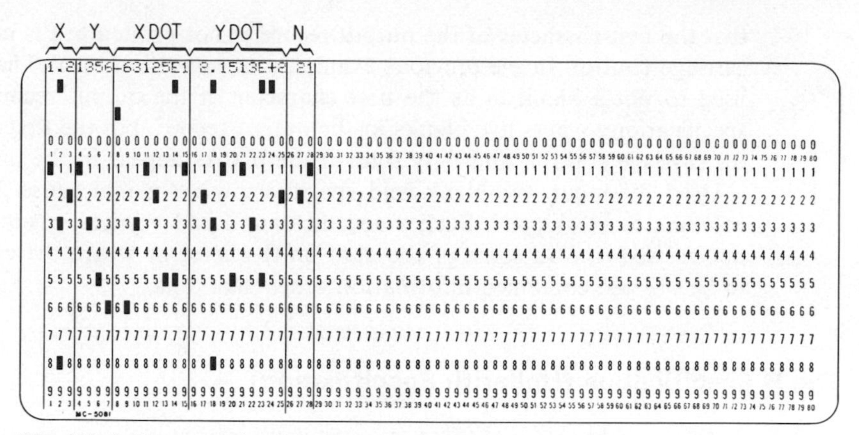

Fig. 1-7 / *Data card for given input statements.*

F10.3, F5.2 are repeated three times. If there are more variables in the list than there are specifications in such a FORMAT statement, control returns to the last group specification, and a new record is begun.

The X Specification (Blank Field Specification)

When the programmer wishes to leave blank spaces (blank printer columns, not blank lines) in the printout, or when card columns are to be ignored on input, the X specification is used in the FORMAT statement. Suppose, for example, that the values of time, displacement, velocity, and acceleration of a rocket are to be printed out for specified time increments over a specified time interval, and that all four values at each time increment are to be printed on a single line. A more readable printout would be obtained if a number of blank spaces were used to separate the various columns of figures and their headings.

The X specification has the general form wX, where w is an integer constant specifying the number of blank spaces to be provided in the output record. As an example, the FORMAT statement

```
3 FORMAT(5X,F6.2,6X,F5.1)
```

associated with a WRITE statement listing two variables might cause printout as

```
bbbb -34.61 bbbbbb b89.1
```

where the fields are indicated by the brackets shown. It should be noted that while the first blank field specification has a w value of 5, only four blank spaces appear in the first field. It has been mentioned previously that the first character of the output record is not printed but is used for carriage control. In the previous examples, the specification 1H^b has been used to put a blank in as the first character of the output record. The specification 5X puts five blanks in the output record, but the first is used for carriage control, and only four of the blanks appear in the printout.

Used for input, the blank-field specification 20X would cause 20 card columns to be skipped. Such a specification could be used where it was not desired to read some of the data on the card, or where the columns to be skipped contained information other than data.

The H Specification (Hollerith Specification)

The H specification is used for printing out alphameric (alphabetic, numeric, and special characters) text. Titles, names, dates, and so forth, are printed out with this specification. It may also be used for input of alphameric characters.

The H format specification has the general form

*w*Htext

in which *w* is an integer constant indicating the number of alphameric characters in the text to be printed out (the field width). The text, a literal constant, also called Hollerith text or literal data, follows the letter H. Any desired blank spaces in the text are included in the field width count. The statements

```
  WRITE(6,2)
2 FORMAT(1Hb,34HSOLUTION OF SIMULTANEOUS EQUATIONS)
```

appearing in a source program would result in a printout of the text shown following the letter H.

Hollerith text can also be printed along with the numerical values of variables in the printed output. This is best illustrated by an example such as

```
  WRITE(6,3)DISPL,VEL,ACCEL
3 FORMAT(1Hb,F7.1,3HbFT,F9.2,7HbFT/SEC,F7.2,11HbFT/SEC/SEC)
```

where the b's indicate blanks. Hollerith text is printed following each variable value so that the printed output might appear as

```
b1326.5 bFT bbb165.21 bFT/SEC bb28.61 bFT/SEC/SEC
```

where the brackets (which do not appear in the printout) are added to indicate the conformance of the numerical printout to the field specifications shown in the FORMAT statement.

The H specification for input is used to replace Hollerith text in a FORMAT statement with data read from a card. This is illustrated by the following example. Let us assume the two statements

```
   READ(5,33)
33 FORMAT(16HᵇUSERSᵇNAMEᵇHERE)
```

appear in a source program. These statements would cause 16 characters to be read from a card. They would replace, in storage, the 16 characters of the FORMAT statement following the H. Suppose the 16 characters read from the card were

```
ᵇWILLIAMᵇBROWNᵇᵇ
```

A subsequent statement

```
WRITE(6,33)
```

in the source program would cause printout of the name WILLIAM BROWN. The first blank in the field read in would not be printed out, but would be used for carriage control.

The Literal Specfication

The literal specification may be used in place of the H specification. It has the advantage that the number of characters in the field need not be counted in making the specification. The literal data in this specification are simply enclosed in apostrophes. The statements

```
  WRITE(6,4)
4 FORMAT(1Hᵇ,'DYNAMIC CHARACTERISTICS OF A ROCKET')
```

would result in the printout of the literal data.

The literal specification may be used in a FORMAT statement associated with a READ statement in a manner similar to the use of the H specification for input.

Carriage Control

As has been noted previously, the first character of the output record is not actually printed by the printer but is used for carriage control. The following tabulation shows some normally used control characters, the corresponding H and literal specifications (either of which could be used to insert the control character into the output record), and the control function performed by each control character.

Control Character	H Specification	Literal Specification	Carriage Advance Before Printing
blank	1Hᵇ	'b'	One line
0 (zero)	1H0	'0'	Two lines
+	1H+	'+'	No advance
1	1H1	'1'	To first line of next page

The X specification, 1X, may also be used to put a blank into the output record for carriage control.

The T Specification

IBM 360 FORTRAN includes the T specification which may be used in a FORMAT statement for output to indicate the position in a record at which the output of data is to begin. It has the general form

Tr

where r specifies the record position. If the output is being *printed*, the print position will be r−1, since the first character of the output record is used for carriage control of the printer. For example, used in conjunction with a literal specification, the statements

```
17 FORMAT(T1,'bDISPLACEMENT',T38,'ACCELERATION',T21,
                                                'VELOCITY')
   WRITE(6,17)
```

would cause the following printout:

```
DISPLACEMENTbbbbbbbVELOCITYbbbbbbbbbACCELERATION
|                  |                 |
└─print position 1 └─print           └─print position 37
                     position 20
```

In this example the literal specification causing the printout of VELOCITY follows the literal specification for ACCELERATION in the FORMAT statement. However, printout begins in the position specified, less 1. DISPLACEMENT begins in position 1, because the first character of the literal specification is a blank for carriage control of the printer.

With input, the T specification can be used to indicate at which card column the reading will begin for the following field. Thus the statements

```
   READ(5,29)A,B
29 FORMAT(T15,E17.8,T1,E12.0)
```

would cause the reading of a 17-character field containing the value of A, beginning in card column 15. Beginning in column 1, a 12-character field containing the value of B would be read.

The A Specification

The A format specification may be used for the input and output of alphameric characters. It differs from the H and literal specifications in that groups of characters transmitted by the A format are assigned to variable names. One application of the A specification which is discussed later is its use in storing FORMAT information which is read in at the time the program is executed.

The general form of the A specification is

Aw

where w is an integer constant specifying the number of characters to be read into or written from the memory location, or locations, associated with a particular variable name. Corresponding to each variable name, some maximum number of characters m can be stored. The value of m ranges from about 4 to 16 for various computers. In IBM 360 FORTRAN, m depends upon the number of bytes of storage associated with the corresponding variable in the I/0 list. Thus m would be 4 for single-precision real variables and integer variables of standard length (those integer variables which have not been specified in a type statement as having their values stored in 2 bytes). In the case of double-precision real variables, m would be 8. Normally w will not exceed m for input, but, if it does, only the m characters the farthest to the right in the input field of width w will be read in. If w is less than m, the storage location will be filled with blanks *following* the w characters. For output, if w exceeds m, only m characters will be written. These characters will be preceded by w − m blanks. If w is less than m, the first w characters in the storage location (the w characters farthest to the left) will be written.

If the alphameric data to be stored contain more characters than can be assigned to one variable name (more than m characters), several successive A fields can be specified in the usual manner of specifying successive fields having the same format. That is, the specification 10A4 would indicate ten successive A fields, each with a field width of four spaces, and ten variable names would be required to refer to the storage locations in which the alphameric characters were stored. A subscripted-variable name could be conveniently used for this purpose. Thus, the ten groups of characters might be stored in locations referred to by the names C(1), C(2), . . . , C(10). The statements

```
20 FORMAT(10A4)
   READ(5,20)(C(I),I=1,10)
```

could then be used to read the 40 alphameric characters to be stored as the values of the ten variables C(I).

Suppose that one wished to print out the contents of just the storage locations referred to by the variable names C(5) through C(10). This could be accomplished with the use of the statements

```
16 FORMAT(1Hᵇ,6A4)
   WRITE(6,16)(C(I)I=5,10)
```

The L Specification

This specification is used for input and output of logical quantities. The general form is

Lw

where w represents an integer constant which specifies the data field width. On input, the first T or F encountered in the w characters of the input field would cause a value of "true" or "false," respectively, to be assigned to the logical variable in the I/0 list which is associated with the L specification. If the input field is entirely filled with blanks, a value of "false" is assigned.

For output, a specification of L4 would allot four spaces (printer columns) for the value of the logical variable. If the value of the logical variable is "true," a T is printed in the extreme right printer position of the field. If the value of the logical variable is "false," an F is printed in that position. A specification of L1 would obviously suffice, but larger data fields might be desired for spacing and for the appearance of the data.

Format Specifications for Input and Output of Complex Quantities

As discussed previously, a complex constant is represented by an ordered pair of real constants. The first real constant is the real part of the complex number and the second is the imaginary part. The input and output of complex quantities requires a field specification for each of the real numbers representing the real and imaginary parts of the complex number. For example, if COMP is a complex variable, the statements

```
   WRITE(6,12)COMP
12 FORMAT(1Hᵇ,F8.3,F9.2)
```

might be used for output. The specification F8.3 would be used for the real part of COMP and the specification F9.2 for the imaginary part. As another example, if COMP1, COMP2, and COMP3 are complex variables, and A is a real variable, the statements

```
   WRITE(6,17)COMP1,COMP2,COMP3,A
17 FORMAT(1Hᵇ,E10.4,E12.4,2(E13.5,F9.2),F7.1)
```

might be used for output. The specifications E10.4 and E12.4 would be associated with the real and imaginary parts, respectively, of COMP1. The specifications E13.5 and F9.2 would be associated with the real and imaginary parts, respectively, of COMP2 and COMP3, and the specification F7.1 would be associated with the real variable A.

The Use of Slashes and the Multiple Use of Field Specifications

In a FORMAT statement, a slash (/) can be used to indicate the end of a record of input or output data. A slash is used to separate field specifications in a FORMAT statement when the FORMAT statement is associated with a READ or WRITE statement which is to read in or print out more than one record of data. Thus, if the FORMAT statement

```
3 FORMAT(1Hᵇ,F9.2/1Hᵇ,F10.4)
```

were associated with a WRITE statement having two variables in its list, the values of these variables would be printed out on separate lines. If the FORMAT statement

```
4 FORMAT(F9.2/F10.4)
```

were associated with a READ statement listing two variables, the 2 values to be read in would be punched on separate cards.

Suppose that the FORMAT statement

```
9 FORMAT(2F6.2,3E14.8/I5)
```

were used with an input statement which was to be executed a number of times. This FORMAT statement would then be used over and over for the input of a number of records. It would specify a format of (2F6.2,3E14.8) for the first and succeeding odd-numbered records and a format of (I5) for the second and succeeding even-numbered records.

Consecutive slashes can be used to create blank output records (leave blank lines) or to skip input records. Using n consecutive slashes at the beginning or end of the format specifications in the FORMAT statement would create n blank lines upon output if the carriage control character specifies that the carriage advance one line before printing. On input from cards, n cards would be skipped. If n slashes appear between format specifications, n − 1 blank lines are created or that number of cards skipped on output and input, respectively. Thus the statements

```
  WRITE(6,2)A,B
2 FORMAT(1Hᵇ,F9.2///1Hᵇ,F10.4)
```

would cause two blank lines to be introduced between the values of A and B.

Scale Factors

In discussing the output of the value of a real variable using the E specification, it was stated that the value had the form of a number between 0.1 and 1.0 multiplied by some power of 10 in most FORTRAN versions. It if is desired that the value have the form of a number between 1.0 and 10 multiplied by a power of 10, a 1P is added in front of the E specification. The 1 preceding the P is called the *scale factor*. If it is desired that the value have the form of a number between 10 and 100 multiplied by a power of 10, a 2P precedes the E specification. For example, the statements

```
  WRITE (6,2)A,B
2 FORMAT(1Hᵇ,1P2E13.6)
```

might cause the following printout:

```
-3.123456E-01ᵇ7.543212E 06
```

A positive scale factor multiplies the number by $(10)^{\text{scale factor}}$ and decreases the exponent a corresponding amount. The scale factor applies to all E and F field specifications which follow it within the FORMAT statement. This applies to multiple-record formats as well as single-record formats. To change back to a zero scale factor for a subsequent and following specifications in the FORMAT statement, a 0P (zero followed by P) may be used preceding the specification.

Scale factors have no effect on I-type specifications. With output, a scale factor will cause data values with F-type specifications to be multiplied by $(10)^{\text{scale factor}}$. Such multiplication is generally not desirable, although it is conceivable that it might be desired to convert units such as amperes to milliamperes or meters to centimeters, and so forth, by this method. For example, a value, with the units of amperes, stored in memory could be printed out as the value in milliamperes by using a scale factor of 3.

With input, scale factors may be used only with the F-type specification. The scale factor is then defined as follows:

$$(\text{External quantity})(10)^{-\text{scale factor}} = (\text{internal quantity})$$

For example, if the FORMAT statement

```
2 FORMAT(2PF6.2)
```

were used in conjunction with a READ statement, a data value 312.24 would be read into the computer and stored as 3.1224.

Reading in FORMAT Statements at Object Time

Many programs are general in nature and are not written just for a specific set of input data. The magnitude of the variable values read in as data, and the number of significant digits required for these values, may vary considerably from one program application to another. The same may be true of variable values which are to be printed out. It may not be practical to write FORMAT statements in a program which are suitable for all possible applications to which the program may be put. FORTRAN IV handles this problem by making it possible to read FORMAT statements directly into memory at object time (at the time the program is executed) using the A field specification. The word FORMAT and the statement number are omitted in this process. Only the FORMAT specifications and the left and right parentheses enclosing them are read into memory. This alphameric information is stored in locations referred to by a subscripted variable. The READ or WRITE statement then uses the array name in place of a FORMAT-statement number. Thus instead of writing

```
READ(5,64)A,B,C
```

with the FORMAT information given in statement 64, the statements

```
  DIMENSION SPC(5)
2 FORMAT(5A4)
  READ(5,2)(SPC(I),I=1,5)
  READ(5,SPC)A,B,C
```

could be used. The first READ statement reads five fields of four alphameric characters each, and stores them in locations referred to by the names SPC(1), SPC(2), . . . , SPC(5). The assumption is made here that the computer being used can store four alphameric characters per storage location as is the case with the IBM 360 computer. The alphameric characters must be the desired FORMAT specifications for the second READ statement and must include the left and right parentheses. The card read by the first READ statement might contain, for example, (F10.3, E14.7, F12.4). Then the array name SPC is used in place of a FORMAT statement number in the second READ statement. The array name may be chosen arbitrarily, and it must be dimensioned in a DIMENSION, COMMON, or explicit type specification statement, even if there is only one element in the array.

1-21 / THE DATA INITIALIZATION STATEMENT

In many programs, variables are assigned values which do not vary throughout the execution of the program. Thus the value of the acceleration of gravity, 32.17 ft/sec^2, might be assigned to the name G. The variable G, rather than the longer constant 32.17, would then be used in any calculation involving the acceleration of gravity. Other variables in a program may have values which vary during the execution but which always have the same value at the start of execution each time the program is run. These variables, whose values are either invariant throughout the program or at the start of every run of the program, are usually not assigned values which are read in as data; instead, arithmetic-assignment statements of the form

```
G=32.17
TERM=160.
T=0.
X=0.
Y=0.
YD=125.
```

are used.

The DATA statement of FORTRAN IV provides an alternate method of assigning values to the sort of variables just discussed. It may also be used to assign alphameric characters to variable names without reading A fields from cards. The DATA statement is a nonexecutable statement having the general form

DATA $v_1,v_2,\ldots,v_n/d_1,d_2,\ldots,d_n/, v_a,v_b,\ldots v_l,/d_a,d_b,\ldots,d_l/,\ldots$

in which the v's represent variable names and the d's are their respective values. The d's are constants, but are not necessarily numerical constants.

They may also be logical or literal constants. Logical constants may be written as .TRUE. and .FALSE. or just as T and F. There is a one-to-one correspondence between the variables in a list and the constants which follow between a given pair of slashes.

As a simple example, let us enter the data (previously shown as being entered by arithmetic-assignment statements) by the use of a DATA statement. This statement might appear as

```
DATA G,TERM,T,X,Y,YD/32.17,160.0,0.0,0.0,0.0,125.0/
```

The following two statements are alternate DATA statements which could be used to enter the same information:

```
DATA G/32.17/,TERM/160./,T/0./,X/0./,Y/0./,YD/125./
```

or

```
DATA G,TERM/32.17,160./, T,X/0.,0./, Y,YD/0.,125./
```

In using DATA statements, the assigned values are stored in memory upon loading the object program. The variables listed in a DATA statement may subsequently take on other values as the execution proceeds, but control cannot revert to the DATA statement for reassigning the original values, since the DATA statement is not an executable statement.

Any of the variables in the list of a DATA statement may be subscripted with integer constants, or an array name may be used without subscripts where the values of all the elements of the array must be listed. For example, in the statements

```
DIMENSION B(4),C(9)
    ....
DATA C(4),B/8.4,3.2,6.7,5.3,9.7/
```

only the value of the particular element C(4) of the C array would be listed (8.4), but all four values of the B array would have to be listed as shown. Some versions of FORTRAN IV permit the use of the implied DO in the list to specify particular elements of an array.

Any of the constants may be preceded by k*, giving it the form

```
k*d
```

in which k is an unsigned integer constant and d is the numerical, logical, or literal constant to be assigned. This form means that k consecutive list variables are to be assigned the particular value d.

For example, the statement

```
DATA X,Y,YD/3*0.0/
```

would cause values of zero to be assigned to the variables X, Y, and YD.

The following example illustrates the use of the implicit DO in the list of a DATA statement, and the use of literal constants (this form of literal constant is also called a Hollerith constant):

```
DATA(CODE(I), I=1,4)/1H*,1H.,1H-,1H+/
```

Using this statement, an asterisk would be stored in a location referred to by the name CODE(1). It would be stored in a memory location such that, on printout of the contents of the memory location, the asterisk would be on the left, followed by blanks. A period would be stored in the location referred to by the name CODE(2), and so forth.

Each literal constant in the above example contains a single character. Several characters (the exact number depends upon the computer used) may be stored as the "value" of a variable. Suppose that for a particular computer, four alphameric characters may be assigned to a variable name (this would be the case for IBM 360 FORTRAN if the variable is single-precision real). If a particular literal constant contained 16 characters, four variable names would have to be used to store that constant.

The statements

```
DIMENSION TITLE(6)
DATA TITLE/21HACCELERATION ANALYSIS/
```

would cause the literal constant to be stored in locations referred to by the names TITLE(1), TITLE(2), . . . , TITLE(6). TITLE(6) would be assigned the last alphabetic character, S, and three blanks. The statements

```
DIMENSION TITLE(6)
DATA TITLE/'ACCELERATION ANALYSIS'/
```

could be used for the same purpose.

A DATA initialization statement cannot precede any specification statements (an explicit type specification statement, for example) that refer to any of the variables listed in the DATA statement. It also cannot precede an IMPLICIT statement. Otherwise, for most computers, it can appear anywhere in the program except where a nonexecutable statement is prohibited. A few compilers require that DATA statements precede all executable statements.

1-22 / SPECIFICATION STATEMENTS

Specification statements are nonexecutable statements which provide certain types of information for the compiler. The specification statements of FORTRAN IV include DIMENSION, IMPLICIT type specification, explicit type specification, COMMON, EQUIVALENCE, and EXTERNAL statements. COMMON and EXTERNAL statements will be discussed later with the presentation of subprograms, since they are used with subprograms. Specification statements are often required to be placed before the executable statements and statement function definitions, although there are variations from this requirement.

DIMENSION Statements

These statements are used to specify the maximum size of arrays. This information is required by the compiler for the allocation of storage. The array size must be represented by integer constants if the DIMENSION statement is in a main program. *Adjustable dimensions,* in which the array size is represented by integer variables, may be used in subprograms and will be discussed along with the later presentation of subprograms.

As an example of a DIMENSION statement, suppose that a two-dimensional array T(I,J) is to be used in a program in which I and J are to have maximum values of 10 and 20, respectively. The statement

```
DIMENSION T(10,20)
```

in the source program would allocate storage space for 200 array element values.

A single DIMENSION statement may be used for as many array names as desired. When more than one array name appears in the DIMENSION statement, they are separated by commas. For example, the statement

```
DIMENSION A(15,10),B(15,15),T(10,10),X(15)
```

reserves space for four arrays.

Type Statements

The type specification of variables by means of the predefined specification has been presented previously. It should be recalled that with this specification the variable is integer if the first character of the name is I, J, K, L, M, or N, and is real if the name begins with any other alphabetic character. This specification applies unless it is overridden by one of the two kinds of type statements: the IMPLICIT statement or the explicit specification statement. IMPLICIT or explicit type specification statements *must* be used to specify double-precision real, single or double-precision complex, or logical variables, since these types are not included in the predefined specification.

The IMPLICIT Statement

This statement is not available in standard FORTRAN IV, but is available in the widely used IBM 360 FORTRAN IV. It is used to specify types of variables which begin with certain designated letters. Thus type specification by use of the IMPLICIT statement is similar to type specification by the predefined specification except that the programmer can select which initial letters are to specify a certain type of variable. The IMPLICIT statement can be used to specify initial letters for all types of variables—integer, real, complex, and logical—whereas the predefined specification applies only to integer and single-precision real variables. In addition, the IMPLICIT statement can specify the number of bytes of storage for variables of all types. For each type of variable, the programmer can specify a standard or optional length (number of bytes) as follows:

Variable Type	Standard Length	Optional Length
Integer	4	2
Real	4	8
Complex	8	16
Logical	4	1

If 8 bytes are specified for real variables, they are double-precision real, and if 16 bytes are specified for complex variables, they are double-precision complex.

The general form of the IMPLICIT statement is

```
IMPLICIT type*s1(a1,a2...),...,type*s2(b1,b2,...)
```

The word "type" in the general form symbolizes one of the words INTEGER, REAL, COMPLEX, or LOGICAL. The letters $s_1, \ldots, s_2$ represent standard or optional lengths, and the a's and b's represent single alphabetic characters separated by commas, or a range of alphabetic characters. A range of alphabetic characters—for example, B, C, D, E, F—is denoted by B−F. The first and last characters of the range are given, and are separated by a "minus" symbol. For example, the statement

```
IMPLICIT REAL*8(A-K, O-Z)
```

specifies that all variables in the program beginning with the letters A through K and O through Z are real, double-precision variables. The statement

```
IMPLICIT INTEGER*2(I,J,K)
```

specifies that all variables beginning with the letters I, J, and K are integer variables, occupying only two memory locations each. The last statement above might also have been written as

```
IMPLICIT INTEGER*2(I-K)
```

Several different type specifications can be included in a single IMPLICIT statement, and if a length specification is not included with the type of variable, the standard length is specified. For example, the statement

```
IMPLICIT INTEGER(A,B,D-F),REAL*8(G-K),COMPLEX(C),LOGICAL(L-N)
```

specifies that all variables beginning with the letters A, B, and D through F are integer variables with a standard number of memory locations for each; that all variables beginning with the letters G through K are real, double-precision variables with eight memory locations reserved for each; that all variables beginning with the letter C are complex with a standard number of eight memory locations for each; and that all variables beginning with the letters L through N are logical variables with a standard number of memory locations reserved. All initial letters not included in the IMPLICIT statement follow the predefined convention.

An IMPLICIT statement, if one is used, must be the first in a main program and the second in a subprogram (in which it follows immediately after the function or subroutine statement, as discussed later). There can be only one IMPLICIT statement in a main program or in a subprogram. IMPLICIT specifications override predefined specifications.

Explicit Specification Statements

Types of variables may be specified explicitly in standard FORTRAN IV as shown by the following examples:

```
REAL I,MASS,FORCE
INTEGER X,Y,Z,A(15,15)
DOUBLE PRECISION A,B
COMPLEX AMP,VOLT
LOGICAL LOG1, LOG2
```

As many variables as necessary may be included in a single type statement. In the first statement above, the variables I, MASS, and FORCE are specified as real. Note that the variable FORCE would not need to be included in the explicit specification statement since it is already real by the predefined specification. However, it is permissible to include it as shown.

The array A is dimensioned in the second type statement shown. Arrays can be dimensioned in any of the kinds of explicit type statements shown, but if a variable is dimensioned in a type statement, it may not be dimensioned elsewhere (such as in a DIMENSION statement). IBM 360 FORTRAN explicit specification statements may be more general than those illustrated, although the ones shown are all acceptable in that version. The more general explicit specification statement has the form

Type*s a*$s_1(k_1)/x_1/$, b*$s_2(k_2)/x_2/$, c*$s_3(k_3)/x_3/$, . . .

in which the word "type" symbolizes one of the words INTEGER, REAL, LOGICAL, or COMPLEX; the s's symbolize standard lengths and are optionally included; a, b, c, . . . symbolize variable, array, or function names; the k's symbolize dimensions and are optional, but may be used if the associated variable name is an array name; and the x's represent initial values of the variables or array elements. Let us use some examples to illustrate this general form of the explicit specification statement.

The type statement

```
REAL*8 DBL1, DBL2, DBL3
```

specifies the listed variables as real, double-precision variables since the *8 following REAL specifies that eight memory locations are to be reserved for each variable, rather than the standard four locations. If the *8 were deleted from the above statement, the absence of this length specification would provide the standard number of memory locations for each variable.

The type statement

```
REAL INV, JOB
```

would specify the listed variables as real variables, overriding the predefined convention, and four memory locations would be provided for each variable.

If some real variables in a list are to be double-precision, while others are not, a type statement such as

```
REAL*8A,B,NUM*4,MATR(10,10)
```

might be used. The variables A, B, and MATR are specified as real, double-precision variables since eight memory locations are specified. The variable NUM is specified a real, single-precision variable since the standard four memory locations are specified for it.

When a length specification is associated with the type specification, it applies to each variable in the list, unless a length specification is associated with a particular variable, in which case the former length specification is overridden. For example, in the last type statement shown above, the length specification *4 associated with the variable NUM overrides the length specification *8, and specifies NUM as a single-precision variable. If *4 were deleted from the above statement, NUM would be specified as double-precision.

If it is known that the magnitudes of certain integer variables will never exceed 32,767($2^{15}-1$), memory space can be served by specifying that these variables are to occupy only two locations in memory. The statement

```
INTEGER*2I,J,LL,JOB*4(10,10)
```

will reserve two storage locations for the values of the variables I, J, and LL. The elements of the array JOB, however, will each have four storage locations reserved. Note that the array JOB is dimensioned as part of the type statement. Since 100 elements are indicated in this array, 400 memory locations will be reserved for the array values.

The following statements illustrate the assignation of initial values to variables in IBM 360 FORTRAN IV type statements:

```
COMPLEX*16D,TAU/(3.92,1.54)/,CC*8
REAL A(10,10)/90*0.0,10*1.0/,B*8/1.1E6/,C(5)/5*0.0/,G/32.17/
```

In the first statement, D and TAU are specified as complex, double-precision variables, with 16 memory locations specified for each. The variable TAU is also assigned the initial value (3.92, 1.54) as shown. The variable CC is specified as single-precision complex, with 8 memory locations reserved, 4 for the real part and 4 for the imaginary part.

In the second statement, the first 90 values of the 100-element array A are assigned initial values of 0.0, while the last 10 elements are initialized as 1.0. (The order of the array elements in memory are A(1,1), A(2,1), A(3,1), . . . , A(1,2), A(2,2), A(3,2), . . . , A(1,3), A(2,3), A(3,3), . . . , and so forth, up to A(10,10), with the first subscript varying the fastest.) The variable B is specified as double-precision, and has the initial value 1.1E6. Array C will have space reserved for 5 elements, each occupying the standard 4 locations, and the initial value of all 5 elements will be 0.0. The variable G has the standard memory locations reserved, and an initial value of 32.17.

The EQUIVALENCE Statement

In most programs, the values of variables are normally assigned to unique locations in memory. In a large program, where there is often a shortage of memory space, it frequently is desirable, where feasible, to let different variables share locations in memory. With the use of the EQUIVALENCE statement, such an arrangement is possible. Since the storage of large arrays is a common source of memory shortage, subscripted variables are more frequently equivalenced (made to share memory locations) than are nonsubscripted variables.

To take advantage of such a technique, the program must obviously be such that it is never necessary for the values of different variables, which share the same locations, to be stored in memory at the same time. For example, if an array is used only in the early portion of a program, and another array is used only in a later part of the same program, the two arrays could be stored in the same locations in memory but at different times. However, if values from both arrays must be available in memory throughout the program, they obviously cannot share the same memory space, unless they happen to have identical values.

The general form of the EQUIVALENCE statement is

```
EQUIVALENCE (a, b, . . .),(e, f, . . .),(x, y, . . .)
```

In this general form the characters in the parentheses represent a series of variables which are equivalenced; that is, the variable values within a set of parentheses will be stored in the same location in memory but at different times. The order of the variables is not important. Any required number of equivalences (sets of parentheses) may be used. For example, the statement

```
EQUIVALENCE(A,B,C),(X,Y),(I,N)
```

would cause the values of A, B, and C to be stored in a single location in memory at different times, the values of X and Y would share another memory location in the same manner, and the values of I and N would share yet another location.

Equivalence variables may be nonsubscripted, as in the above example, but, as mentioned earlier, subscripted variables are most often necessarily equivalenced. When the latter are equivalenced, they must be single-subscripted variables or else the number of subscripts must equal the number of dimensions of the array in which the variables appear. Some FORTRAN versions *require* that a single subscript be used. In this case the double-subscripted array

```
B(1,1), B(2,1), B(3,1), B(1,2), B(2,2), B(3,2), B(1,3),
                                                B(2,3), B(3,3)
```

which is shown in what is termed its *usual order* (with the first subscript progressing the fastest), would be single-subscripted by taking advantage of the order in which the elements appear in the *usual order*. For example, the element B(3,2), shown above, would be listed in the equivalence statement as B(6), since it is the 6th element in the order shown above. Such a procedure allows the identification of the elements of a higher-subscripted array by the use of single subscripts. Using this technique, the statements

```
DIMENSION A(5,5),B(10)
        ....
EQUIVALENCE(A(21),B(1))
```

would cause the values of A(1,5) and B(1) to share the same memory location. (The former is the 21st element of the 5-by-5 array *in its usual order*, and the latter is the 1st element of the one-dimensional array B.) It would also cause the values of A(2,5) and B(2), A(3,5) and B(3), and so forth (through the 25th element of the A array and the 5th element of the B array), to share memory locations. Thus, the two arrays would be stored in memory in the following staggered form

```
A(1,1)
A(2,1)
A(3,1)
  .
  .
  .
A(3,4)
A(4,4)
A(5,4)
A(1,5)    B(1)
A(2,5)    B(2)
A(3,5)    B(3)
A(4,5)    B(4)
A(5,5)    B(5)
          B(6)
            .
            .
            .
          B(10)
```

where only the five matching elements shown would share memory space.

Some versions of FORTRAN IV permit the EQUIVALENCE statement to appear anywhere in the source program, but it is often required that they precede all executable statements and any statement function definitions.

1-23 / SUBPROGRAMS

Sometimes a particular routine is used so frequently in a program, or perhaps in many programs, that it becomes convenient to write the routine as a subprogram and simply refer to this subprogram whenever the routine is required. Some subprograms are supplied as part of the FORTRAN library. These include the mathematical function subprograms discussed in Sec. 1-13. Other library subprograms may be written by programmers at a particular computer installation. In addition, there are two classes of subprograms which the individual programmer may write and use with other programs which refer to them. These are called *function* subprograms and *subroutine* subprograms. Statement functions are similar in their use to function subprograms, so they will be discussed along with subprograms.

Statement Functions

Although not generally classified as subprograms, statement functions are used like function subprograms. They must be capable of being defined by a single statement. This statement function definition is part of a program unit rather than being a separate program unit with its own END statement. It is in this respect that it differs from a true subprogram. A statement function can be referenced (a function is referenced when its name, followed by its arguments in parentheses, appears in a FORTRAN expression) only in the program unit of which it is a part. A subprogram can be referenced from any program unit.

A statement function definition has the general form

name(a, b, . . .) = expression

where "name" represents the function name, the letters a, b, . . . represent the dummy arguments of the function, and "expression" represents a FORTRAN expression which is a function of the arguments.

The same rules apply in choosing a function name as in choosing a variable name. Type declaration of statement function names is also the same as the type declaration of variable names. If the predefined convention is not used, the type statement declaring the type of a statement function name must appear prior to the function definition. Statement function definitions immediately precede the executable statements of the program.

The arguments of the function in the definition are "dummy" arguments which are replaced by the values of the actual arguments which appear in the function reference. For example, the statement function definition

```
DRAG(S,V)=COEFF*RHO*S*V**2/2.
```

might appear in a source program to define a function with the name DRAG. The variables in the expression on the right, which are in addition to the arguments S and V, are known as *parameters* of the function. During compilation, the FORTRAN compiler sets up the necessary machine-language instructions to calculate the value of the function DRAG. These instructions are then used whenever the function is called for later in the program in a FORTRAN expression, using the values of whatever quantities are supplied for the dummy arguments S and V. For example, the arithmetic-assignment statement

```
F=FP-D1-DRAG(AREA,VEL)
```

might appear later in the source program. The value of DRAG would be calculated by using the current values of the arguments AREA and VEL for the dummy arguments S and V, respectively, in the function definition. The variables COEFF and RHO also appear in the definition, and their current values are used in the calculation. This value of DRAG would then be included in the calculation of F. The advantage of using arguments in the statement function definition rather than parameters is that the former are more flexible. For example, the current value of X+DELTX/2. might be the actual argument which replaces a dummy argument X. If X were a parameter rather than a dummy argument, only the current value of X could be used in the calculation of the statement function.

The dummy arguments S and V in the statement function definition could be used in other statement function definitions, if desired, with no relationship existing between these same variables in the different function definitions. They could also be used elsewhere in the program unit of which the statement function definition is a part, as names for other variables if desired. No relationship would exist between the latter and the dummy arguments. That is, the dummy arguments are said to be *local* to the definition statements in which they appear.

The dummy arguments of a statement function cannot be subscripted, since the expression on the right of the equal sign in the function definition cannot contain subscripted variables. However, when the statement function is referenced in a FORTRAN expression, the actual arguments may be expressions in general, and may contain subscripted variables. For example, the arithmetic-assignment statement

```
D2=DRAG(A(2), VEL+100.)
```

would constitute a valid use of the previously defined function DRAG, where A(2) and VEL+100. are the actual arguments corresponding to the dummy arguments S and V, respectively. *The actual arguments in a function reference must agree in number, order, and type with the dummy arguments.*

It is possible for the actual arguments used in the function reference to be the same as the dummy arguments used in the definition. For example, the statement function FUNC might be defined as

```
FUNC(X,Y)=0.5*(1+X)*Y**3
```

and might be referenced in the program unit in the statement

```
AK1=DELTX*FUNC(X,Y)
```

using actual arguments X and Y. The values of X and Y in the program replace the dummy arguments X and Y in the definition, and a value of FUNC is obtained which is then used in the calculation of AK1. To again point out the desirability of using X and Y as dummy arguments rather than parameters in the statement function definition (aside from the fact that a statement function must have at least one argument), note that the function FUNC might be referenced again as

```
AK2=DELTX*FUNC(X+DELTX/2.,Y+AK1/2.)
```

If X and Y were parameters rather than dummy arguments, only the current values of X and Y could be sent to the statement function definition for calculations.

Statement functions are defined only in the program unit in which the definition appears. A subprogram, for example, cannot reference a statement function defined in a main program.

Function Subprograms

A function subprogram is similar in many ways to the statement function just discussed, except that it is written as a separate program unit with its own END statement and may be referenced by any other program unit (main program or subprogram). In addition, the function subprogram definition is not limited to one statement. For example, a statement function definition

```
AVG(A,B,C,D)=(A+B+C+D)/4.
```

might appear in a program unit to be referenced whenever it were necessary to find the average of four quantities. Alternatively, a function subprogram could be used which might be written as

```
FUNCTION AVG(A,B,C,D)
SUM=A+B+C+D
AVG=SUM/4.
RETURN
END
```

This subprogram could be referenced from any program unit. For both the statement function and the function subprogram the values of the actual arguments replace the dummy arguments A, B, C, and D when the function is referenced, and a value of AVG is determined. Both are referenced in the same manner, by using the name followed by the arguments in parentheses in the same manner as for the supplied mathematical functions SIN, COS, ATAN2, and so forth. Variables such as SUM in the above example will not have their values available to the referencing program. The same name could be used in the referencing program, for example, or in any other program unit, with a completely different value assigned to it. Variables (and also statement numbers) of a subprogram are *local* to that subprogram. The general form of a function subprogram is

```
Type FUNCTION name (a1, a2, . . . , am)
  .
  .
  .
name = expression
RETURN
END
```

Type in the general statement above would actually be INTEGER, REAL, COMPLEX, or LOGICAL. The function name has a type like a variable name, and its type must be specified in both the referencing program and in the subprogram. If the predefined specification is used, then the type specification may be omitted in the FUNCTION statement. The specification may also be made by an IMPLICIT or explicit specification within the subprogram in most FORTRAN versions, in which case the type in the FUNCTION statement is generally omitted. However, if the function type is specified as shown in the general form above, it overrides a conflicting specification within the subprogram.

The word "name" in the general form of the FUNCTION statement is the name assigned to the function. It must receive a value in the subprogram. The function name could get its value by appearing in a READ statement or as an argument in a CALL statement (discussed later), but it usually gets its value by appearing to the left of an equal sign as shown in the general form of the function subprogram given above. In IBM 360 FORTRAN the name may optionally have a length specification with it. For example, a FUNCTION statement

```
REAL FUNCTION INT*8(XMIN, XMAX, N, DUMMYF)
```

would specify the name INT to be double-precision real.

The dummy arguments are represented in the general form by $a_1, a_2, \ldots, a_m$. They must be single nonsubscripted variable or array names, or dummy names of subprograms. The use of subprogram names as arguments of other subprograms will be discussed later.

A function subprogram must have at least one argument. The actual arguments in the function reference must correspond in number, order, and type to the corresponding dummy arguments in the FUNCTION statement, just as when referencing a statement function. If one of the dummy arguments is an array name, a suitable DIMENSION statement must be used in the subprogram. The corresponding actual argument in the function reference must also be an array name, and must be dimensioned in the referencing program. Both the dummy and the actual arrays are generally given the same size in their respective DIMENSION statements, except when adjustable dimensions (discussed later) are used. Some FORTRAN versions require that the dummy and actual arrays have the same dimensions specified, but others require only that the size of the dummy array not exceed the size of the actual array.

When the RETURN statement is reached in the subprogram, control reverts to the referencing program at the point where the function subprogram was referenced, with the function name having the value assigned to it in the subprogram. More than one RETURN statement may appear if branching in the subprogram results in the function values being determined at one of several possible places in the subprogram.

The END statement is always the last statement in the subprogram.

Let us look at an example of the use of a function subprogram called by a main program. Only the pertinent statements of the main program are shown.

Calling Program	**Function Subprogram**
`...`	`  FUNCTION PRESS(T,D)`
`...`	`  RATIO=T/D`
`STR=PRESS(THICK,DI)*D/(4.*THICK)`	`  IF(RATIO .LE. .023)GO TO 3`
`...`	`  PRESS=86670.*RATIO-1386.`
`...`	`  RETURN`
`P=PRESS(.025,1.25)`	`3 RAD=1.-1600.*RATIO**2`
`...`	`  PRESS=1000.*(1.-SQRT(RAD))`
`...`	`  RETURN`
	`  END`

The function PRESS is called two times, in the calling program, by writing the function name and listing the actual arguments to be used in evaluating the function. When the value of the function PRESS is calculated the first time, the value of the variable THICK would be used where T appears in the defining statements, and the value of the variable DI would be used where D appears. For the second calculation of the value of PRESS, .025 would be used for T and 1.25 for D. Note that in the first reference the actual arguments are variables, but in the second they are constants. Any suitable real expressions would be valid arguments in this example. Their mode would have to be real, since T and D are real. Note that the function subprogram PRESS calls on the library function SQRT.

Function subprograms are generally used when only a single value is to be returned to the referencing program, that being the value assigned to the function name. However, it is possible to return several values. This may be done by including one or more undefined variables in the actual argument list. Then the corresponding dummy arguments have values assigned to them in the subprogram, and the actual arguments take on these values in the referencing program and are available for use in the referencing program after the function has been referenced. For example, if we have a function subprogram defined as follows:

```
FUNCTION AVG(A,B,C,D,SUM)
SUM=A+B+C+D
AVG=SUM/4.
RETURN
END
```

and this subprogram is referenced with the statement

```
GRADE=AVG(HW,T1,T2,FIN,TOTAL)
```

the variable TOTAL in the referencing program takes on the value calculated as SUM in the subprogram. Generally, when values are to be returned to a calling program through the arguments as illustrated here, subroutine subprograms are used. Such subprograms are discussed next.

Subroutine Subprograms

These programs are similar to function subprograms in many respects. The rules for naming the subprograms are the same, with the exception that the name of a subroutine subprogram does not have a type. They both may have dummy arguments. They both consist of a set of commonly used statements which are used many times in general, and they both require an END statement as the last statement. They both return control to the calling program by use of a RETURN statement. The dummy arguments of a subroutine subprogram, like those of a function subprogram, may be nonsubscripted variables or array names, or dummy names of subroutine or function subprograms. As with function subprograms, the actual arguments of a subroutine subprogram must agree in number, order, and type with the dummy arguments. If array names are used as arguments, they must be dimensioned in both the calling program and the subprogram, and in some FORTRAN versions with the same dimensions. In other FORTRAN versions, it is only necessary that the dummy array dimensions do not exceed the size of the actual array dimensions. Subroutine subprograms do, however, differ from function subprograms in three important respects which follow.

1 Function subprograms usually return just one value to the referencing program, this being the value of the function name. Subroutine subprograms, on the other hand, may return many values, or no values at all. When the subroutine subprogram is called, usually some of the real arguments have values and some do not. The values replace corresponding dummy arguments in the subroutine subprogram, and thus become inputs to the subprogram. The other dummy arguments become outputs of the subprogram and return values to the calling program. Their values are usually determined in the subprogram by their appearance to the left of an equal sign in an assignment statement. However, their values may also be determined by being in an input list somewhere in the subprogram, or as an argument of a CALL statement or function reference used in the subprogram. The corresponding actual arguments in the calling program, which prior to the subroutine subprogram being called had no value, take on the values, calculated or otherwise specified, in the subprogram. *The name of the subroutine subprogram does not receive a value*, whereas a function subprogram name always does.

2 A subroutine subprogram is called differently than is a function subprogram. A subroutine subprogram cannot be called by simply listing its name in a FORTRAN expression as a function subprogram can, since its name does not take on a value. A special CALL statement (discussed later) is used.

3 Since a subroutine subprogram name has no value associated with it, it cannot be of a particular type, such as real or integer. Thus the first letter of a subroutine subprogram name has no special significance.

The subroutine subprogram begins with a SUBROUTINE statement and ends with an END statement. The general form is

SUBROUTINE n ($a_1, a_2, a_3, \ldots, a_m$)

```
 . . .
(FORTRAN statements)
 . . .
RETURN
END
```

where n represents the subprogram name and the a's represent dummy arguments which correspond to the actual arguments listed in the CALL statement.

The CALL statement, used to call a subroutine subprogram, has the general form

CALL n ($x_1, x_2, x_3, \ldots, x_m$)

where n represents the subprogram name; the x's represent the actual arguments. Those arguments which have values may be expressions, while those arguments which are to receive values from the subprogram are single variables, subscripted variables, or array names.

To illustrate the operation of a subroutine subprogram, let us consider the following simple example. Suppose that the addition of two n-by-m matrices is a very common operation in some particular program, and it is desired to use a subprogram to perform this operation each time it is required. The following might be used:

Calling Program

```
. . . . . . . . . . .
DIMENSION X(15,15), Y(15,15), Z(15,15)
. . . . . . . . . . .
CALL ADD(X,Y,Z,K,L)
 . . . . . . . . . . .
```

Subroutine Subprogram

```
  SUBROUTINE ADD(A,B,C,N,M)
  DIMENSION A(15,15), B(15,15), C(15,15)
  DO 7I=1,N
  DO 7J=1,M
7 C(I,J)=A(I,J)+B(I,J)
  RETURN
  END
```

Arrays X and Y are available in the calling program. They have K rows and L columns. The values of the X and Y array elements replace the elements of A and B in the subprogram, and the values of K and L replace N and M, respectively. The values of array C are calculated in the subprogram and are stored in memory, where they are available to the calling program by use of the variable name Z. These values of Z are available after the CALL statement has brought the subprogram into operation. Control returns from the subprogram to the statement following the CALL statement in the calling program.

A subroutine subprogram need not return any values to the calling program, as mentioned previously. If there also were no input to the subprogram, no arguments would be listed in the CALL or SUBROUTINE statements. Such a subprogram might be used, for example, to print out column headings at the top of each page for a program having many pages of output data.

When the program execution reaches the RETURN statement, control is transferred back to the calling program, which is the statement immediately after the CALL statement. This is the normal sequence of operation. Some FORTRAN versions, however, provide for returning to any numbered statement in the calling program by the use of a RETURN statement having the general form

```
RETURN i
```

where i represents an integer constant or variable whose value denotes the position of a statement number in the actual argument list. For example, if the subprogram

```
      SUBROUTINE QDRTIC(A,B,C,R1,R2,*,*)
      IF(B**2-4.*A*C)30,20,10
   10 R1=...
      R2=...
      RETURN
   20 RETURN 1
   30 RETURN 2
      END
```

is called by the CALL statement shown below

```
C     MAIN PROGRAM
             .
             .
             .
      CALL QDRTIC(A1,B1,X+Y,ROOT1,ROOT2,&25,&50)
             .
             .
             .
      END
```

and if B**2−4.*A*C is negative, the return will be to statement number 50 of the calling program, since RETURN 2 is executed following the IF statement. The 2 indicates the second statement number in the actual argument list, which is the number 50. The statement numbers in the actual argument list are preceded by the ampersand symbol, and the corresponding positions in the dummy argument list contain asterisks.

1-24 / THE COMMON STATEMENT

The COMMON statement is most often used in linking a main program with its subprograms, without using arguments in the CALL and SUBROUTINE statements. Before discussing the applications of the COMMON statement, let us first consider the statement itself. It has the general form

COMMON a, b, c, . . .

where a, b, c, d, . . . are variables, including array names that may be dimensioned. As an example, consider the statement

```
COMMON DISPL,VEL,ACCEL,N
```

appearing in a source program. The values of the variables listed would be stored in memory in the sequence in which they appear in the list in the statement. The block of memory in which these variables would be stored is referred to as a COMMON *block*. COMMON blocks can be labeled with names assigned by the programmer, but, in the example above, no name is assigned, and the block is known as a *blank* COMMON. The blank COMMON block is used more frequently, so it will be discussed first.

It is impossible for two variables listed in a COMMON statement to be equivalenced, since the variables listed could not have their values stored sequentially in separate locations and, at the same time, have two of them share the same location. This would be somewhat like attempting to equivalence two elements of an array, which is also invalid. It is also invalid to equivalence two variables in different COMMON blocks.

The size of a COMMON block, indicated by the number of variables listed in the COMMON statement, can be increased when an EQUIVALENCE statement is used in the following manner:

```
DIMENSION A(4)
 . . . . .
COMMON X,Y,Z,
EQUIVALENCE(Z,A(1))
```

The variable values would occupy storage locations as shown below, with X in the lowest numbered location and Z and A(1) sharing the same location in memory:

```
X, Y, Z
      A(1), A(2), A(3), A(4)
```

In this case the COMMON block has been extended to higher locations and includes six memory locations instead of three. An extension cannot be made to lower locations. For example, the statements

```
DIMENSION A(4)
COMMON X,Y,Z
EQUIVALENCE (X,A(4))
```

would be invalid.

If a blank COMMON statement appears in both a main program and a subprogram or subprograms, the variables in these COMMON statements share the same COMMON block in memory. For example, if the statement

```
COMMON E,F,G,N
```

appears in a main program, and the statement

```
COMMON P,Q,R,I
```

appears in a subprogram of the main program, the values of E and P, F and Q, and so forth, share locations in memory. (The variable names E and

P, F and Q, and so forth, refer to the same memory locations.) If the statement

```
CALL SUB(E,F,G,N)
```

appears in a main program, and the statement

```
SUBROUTINE SUB(P,Q,R,I)
```

appears in a subprogram of the main program, the variables, in some FORTRAN versions, again share locations in memory, as just described. (In IBM 360 FORTRAN IV, the *values* of the actual arguments rather than their addresses are brought from the calling program to the subprogram, unless the arguments are each enclosed in slashes.) Thus, in both instances, the effect of equivalencing is obtained, but without the use of the EQUIVALENCE statement. The same effect could not be obtained by equivalencing E and P, F and Q, and so forth, with the use of the EQUIVALENCE statement, since only variables appearing in the *same* program or subprogram can be equivalenced.

Using the statements shown in either of the last two examples, if E is a variable which has been defined in the main program, the appearance of P in the subprogram will cause the value stored as E to be used in the subprogram calculation, since the variable names E and P refer to the same memory location. Similarly, if the value of R is defined in the subprogram, this value will be stored in memory and will be available for use in the main program by using the name G. Since both sets of statements in the last two examples accomplish the same result, the use of two COMMON statements in a main program and a subprogram, respectively, allows the omission of the lists of arguments in the CALL and SUBROUTINE statements. Such a use of the COMMON statements results in a more efficient object program.

Another advantage of using COMMON statements is in dimensioning array names while linking elements between programs. For example, if the statement

```
COMMON A,B,C(50)
```

appears in a main program, and the statement

```
COMMON X,Y,Z(50)
```

appears in a subprogram of the main program, the values of A and X and of B and Y share memory locations, as described before, and the values of C(1) through C(50) share memory locations with Z(1) through Z(50), respectively. When the COMMON statement is used for dimensioning, the arrays dimensioned in it must *not* be listed in a DIMENSION statement or a type statement containing dimension information.

Common statements are often used when a long program is written in segments consisting of a main program and a number of subprograms, so that the individual segments can be checked independently. In this instance a variable defined in one segment may be required for use in one or more other segments. The variable would not be automatically available in different segments, since the variables used in a main program and in individual subprograms are independent of each other, even if the same variable names are used in the different programs. The variables can be made available where required by using a COMMON statement which lists all of the variables of the various program units which are used by other program units. Identical COMMON cards can be used in each segment of the overall program, thus avoiding the chance of errors which might arise in using arguments with the various CALL and SUBROUTINE statements (the arguments of the CALL and SUBROUTINE statements are omitted if COMMON statements are used in this way).

To avoid listing all the variables of a program in the COMMON statement of each subprogram, when any particular subprogram does not require the use of all the variables, a number of COMMON blocks may be set up in memory, each with its own name and each containing certain variables appropriate to the requirements for using the variables in the various subprograms. Such blocks are known as *labeled* COMMON blocks. Before illustrating the preceding discussion, let us consider the statements used to set up labeled COMMON blocks. The statement has the general form

COMMON / n / $a_1, a_2, \ldots, a_m$/ n_1/ $b_1, b_2, \ldots, b_m \ldots$

where the n's represent names assigned to the block of variables following them. These block names should consist of one to six alphameric characters, the first of which is alphabetic. The a's and b's are the variables listed for each block, and they appear in the block in the order given in the list.

To illustrate the labeled COMMON block and its use in avoiding the listing of all program variables in each subprogram, let us consider the following:

Main Program
Variables A, C, and E defined;
variables B, D, and F required

```
...
COMMON/NAMA/A,B/NAMB/C,D/NAMC/E,F
...
A=...
CALL SUB1
...
C=...
CALL SUB2
...
E=...
CALL SUB3
...
TILT=B*D/F
...
```

First Subprogram
Variable A required;
variable B defined
	`SUBROUTINE SUB1`
	`COMMON/NAMA/A,B`
	`...`
	`B=A+...`
	`RETURN`
	`...`
	`B=2.*A+...`
	`RETURN`
	`END`

Second Subprogram
Variable C required;
variable D defined
	`SUBROUTINE SUB2`
	`COMMON/NAMB/C,D`
	`...`
	`D=C**2+...`
	`RETURN`
	`...`
	`D=C**3+...`
	`RETURN`
	`END`

Third Subprogram
Variable E required;
variable F defined
	`SUBROUTINE SUB3`
	`COMMON/NAMC/E,F`
	`...`
	`F=5.*E+...`
	`RETURN`
	`...`
	`F=10.*E+...`
	`RETURN`
	`END`

The COMMON statement in the main program would cause three blocks to be set up in memory, having the names NAMA, NAMB, and NAMC. The labeled COMMON statements in each subprogram contain only the variables whose values are required or calculated in that subprogram. For example, the first subprogram uses the value of the variable A, as defined in the main program, and calculates a value of the variable B which is used, in turn, to calculate the value of TILT in the main program.

The COMMON statement in the first subprogram lists only variables A and B. Thus by using labeled COMMON blocks, certain variables of a main program might share a certain block of memory with the same variables in one or more subprograms, variables common to several subprograms might share a block, and so forth. As should be evident, there is more chance of introducing an error in using the labeled COMMON in this manner than in using identical blank COMMON statements in each segment. Usually, only professional programmers make extensive use of the labeled COMMON.

Since COMMON statements are specification statements, they must precede any statement function definitions, and the executable statements.

1-25 / ADJUSTABLE (OBJECT-TIME) DIMENSIONS

It has been pointed out previously that the arguments of both function subprograms and subroutine subprograms may contain array names. These arrays must be dimensioned in both the calling program and the subprogram. It should be recalled that some FORTRAN versions require that the dimensions be the same in the calling program and subprogram. Since a subprogram may be used with various calling programs, it would be impossible to specify explicitly the size of the arrays in the subprogram for use with various calling programs. To avoid having to specify explicitly the size of an array in a subprogram, FORTRAN IV provides for *adjustable dimensions* in the subprogram. The use of adjustable dimensions would not be as necessary in those FORTRAN versions in which the DIMENSION statement in the subprogram is not used for allocation of storage, but only to identify names as being array names. Such versions require only that the dimensions of the arrays in the subprogram be less than or equal to the dimensions of the corresponding arrays declared in the calling program.

To give adjustable dimensions to an array appearing in a subprogram, integer variables are used in place of integer constants in the DIMENSION statement of the subprogram. When the subprogram is called, the integer variables which specify the array dimensions of the subprogram array receive values from the calling program. This is accomplished by placing the integer variables in the argument list of the SUBROUTINE statement in the subprogram, and specifying values for these variables in the CALL statement of the calling program (or in the FUNCTION statement of the subprogram and function reference, respectively).

As an example, consider the following:

Calling Program

```
...
DIMENSION X(5,5)
CALL MATI(...,X,5,5,...)
...
```

Subprogram

```
SUBROUTINE MATI(...,XS,I,J,...)
DIMENSION XS(I,J)
...
...
RETURN
END
```

With the dimensions of the array which is to be used in the subprogram known, these dimensions are specified in the CALL statement of the calling program, as shown above. The array XS of the subprogram corresponds to the array X of the calling program. Each of the variables I and J in the subprogram receives the value of 5 from the calling program, specifying the size of the array which is to be used in the subprogram.

Adjustable dimensions cannot be altered within a subprogram, and arrays listed in a COMMON statement may not have adjustable dimensions. If the adjustable dimension feature is used in a FORTRAN version which does not require that the dimensions of arrays in a subprogram be the same as those for corresponding arrays in the calling program, it is only necessary that the dimensions assigned to the subprogram arrays at object time be less than or equal to the dimensions specified in the DIMENSION statement of the calling program.

1-26 / SUBPROGRAM NAMES AS ARGUMENTS OF OTHER SUBPROGRAMS—THE EXTERNAL STATEMENT

It has been mentioned before that, in FORTRAN IV, a function subprogram or subroutine subprogram name can be used as an argument of another subprogram. To distinguish a subprogram name in an argument list from an ordinary variable name, it must appear in an EXTERNAL statement. This statement has the general form

```
EXTERNAL a, b, c, . . .
```

where a, b, c, . . . represent the names of subprograms used as arguments of other subprograms. In the statements below, for example, the function name COS would be listed in an EXTERNAL statement to distinguish it as a subprogram name in the CALL statement shown.

```
EXTERNAL COS
...
CALL SUBX(E,COS,G)
...
```

To illustrate the use of a function name as an argument, consider the following:

Calling Program

```
...
EXTERNAL FUNC
...
CALL CALC(X,FUNC,Y,Z)
...
```

First Subprogram

```
SUBROUTINE CALC(A,F,C,D)
G=F(C,D)
...
A=...
...
RETURN
END
```

Second Subprogram

```
  FUNCTION FUNC(G,H)
  IF(G-H)1,1,2
1 FUNC=2.*G*H
  RETURN
2 FUNC=5.*G/H
  RETURN
  END
```

The function name FUNC replaces the dummy variable F in the first subprogram. The values of the variables Y and Z are used as the arguments of the function FUNC in the calculation of its value in the second subprogram. The function FUNC is shown in this example as a function subprogram with two arguments, written by the programmer. The supplied function subprograms (mathematical functions) EXP, ALOG, ALOG10, SIN, COS, TANH, SQRT, ATAN, CABS, and so forth, may also be used as arguments of other subprograms if they are listed in an EXTERNAL statement.

The EXTERNAL statement is a specification statement, and thus must precede statement function definitions and the executable statements.

1-27 / THE NAMELIST STATEMENT

This statement is not available in standard FORTRAN IV and will be discussed only briefly here. It provides a means of obtaining data input and output without using a FORMAT statement. It is most commonly used as a simple method of getting output which is only used for debugging purposes.

The NAMELIST statement has the general form

NAMELIST/name/list of variables and array names/name/list

in which "name" represents a FORTRAN name assigned to the list of variables and array names which follow. The variables of the list are separated by commas. The FORTRAN name of the list can then be used in a WRITE statement in place of a FORMAT statement number to print out the values of all the variables in the list. For example, suppose that the following NAMELIST statement appears in a source program as

```
NAMELIST/DBUG1/X,Y,XDOT
```

A subsequent WRITE statement

```
WRITE(6,DBUG1)
```

might cause printout as follows:

```
&DBUG1
X=  2.0000000      ,Y=  13.216999      ,XDOT=  135.92000
&END
```

As an input example, consider the following partial program:

```
DIMENSION A(2,2)
NAMELIST/ILIST/A,B,I
.
.
.
READ(5,ILIST)
```

The data card containing the four values of the array A and the values of B and I must be prepared in a special manner. The first column on each card must be blank. The second card column of the first card (there may be more than one card) must contain an ampersand (&) which is followed by the NAMELIST name. A required blank follows the name, and then the data assignments separated by commas. At the end of the data assignments on the last card, &END must appear. For example, the data card for the above example might appear as

```
ᵇ&ILISTᵇA=1.,0.,0.,1.,B=-.32E-3,I=5,&END
```

Array values are assigned in the order in which the elements occur in storage. The comma preceding &END is optional. Embedded blanks are not permitted in names or constants. Trailing blanks after integer constants and exponents should be avoided, since they are treated as zeros. If more than one card is used, column 1 of each card must be blank, and all cards after the first must begin with a complete variable or array name, or a constant.

1-28 / THE FORTRAN SOURCE PROGRAM

Having reviewed the various types of FORTRAN statements, let us next review how these statements are written and punched. Preferably, a FORTRAN source program is written on a *coding form* such as the one shown in Fig. 1-8. Coding-form columns 1 through 80 correspond to the 80 columns on a standard punch card. Each line of the coding form corresponds to one card.

As previously mentioned, some statements must be numbered. The statement numbers are written in columns 1 through 5 of the coding form and, in most FORTRAN versions, consist of from one to five digits written without a decimal point. The numbers assigned to statements need not have any particular sequence as long as a statement number is not repeated. The number 0 is not permitted as a statement number. The statements proper are written, one to a line, in columns 7 through 72. Blanks may be used, as desired, to increase readability, but they are ignored by the compiler. If a statement is too long to write on a single line, it may be continued on successive lines by placing any character

PROBLEM *Dynamic Characteristics of a Rattrap* DATE *Feb. 25, 1967* PAGE *1* OF *1* PROGRAMMER *Ralph Jones*

FORTRAN STATEMENT

```
      WRITE(6,1)
    1 FORMAT(1H1,23H TIME     THETA     ANG V//)
      READ(5,2)SK,EYE,TORCL,THCL,DELT,DTPR
    2 FORMAT(F4.3,F5.4,F4.3,F5.3,2F5.4)
      T=0.
      TH=0.
      THD=0.
      TPR=DTPR
    3 THDDT=(TORCL+SK*THCL)/EYE-SK*TH/EYE
      THD1=THD+THDDT*DELT
      TH=TH+(THD+THD1)/2.*DELT
      THD=THD1
      T=T+DELT
      IF(T-TPR)6,4,4
    4 WRITE(6,5)T,TH,THD
    5 FORMAT(1H ,F7.4,F8.4,F8.2)
      TPR=TPR+DTPR
    6 IF(TH-THCL)3,7,7
    7 WRITE(6,5)T,TH,THD
      STOP
      END
```

Fig. 1-8 / *A* FORTRAN *source program written on a* FORTRAN *coding form.*

(other than a blank or a zero) in column 6 of each continuation line. Columns 73 through 80 are ignored by the compiler and may be used for numbering the cards, identifying the program, miscellaneous information, and so forth.

In punching the source deck, the statements are punched in the cards exactly as they are written on the coding form. Each column of the coding form corresponds to a card column. A card, with a statement from the source program of Fig. 1-8 punched in it, is shown in Fig. 1-9.

Comments may be included in the source program merely to provide information to the reader about the program. Comments are not translated

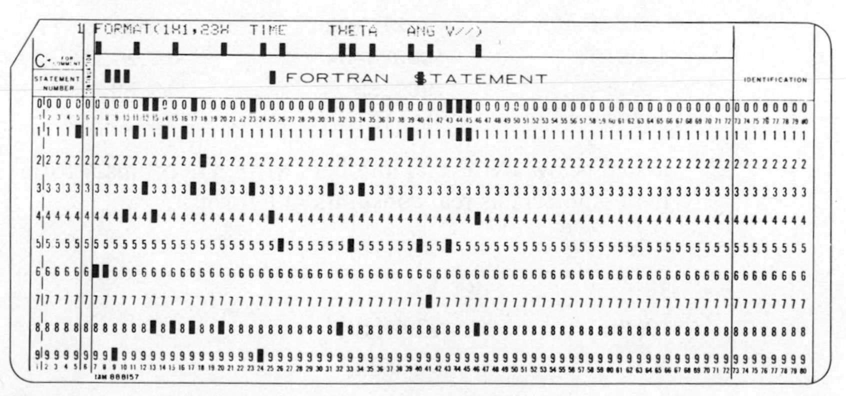

Fig. 1-9 / *A punched card for a* FORTRAN *source deck.*

by the compiler, but are printed out in the source program listing obtained during compilation. Comments are identified by placing a C in column one. In some versions of FORTRAN IV such as IBM 360 FORTRAN, comments may occupy card columns 2 through 80, whereas in others the comments cannot extend past column 72.

The Order of Statements in a Program or Subprogram

The following ordering of various types of statements in a source program is required in IBM 360 FORTRAN but is also applicable to many other versions of FORTRAN IV.

1 FUNCTION or SUBROUTINE statement if the program unit is a subprogram.
2 IMPLICIT statement, if used.
3 Other specification statements such as DIMENSION, COMMON, explicit type specification statements, EQUIVALENCE, and EXTERNAL statements.
4 Statement function definitions, if used.
5 The executable statements.
6 The END statement.

FORMAT, DATA definition, and NAMELIST statements, in general, can appear anywhere in a program between the IMPLICIT and END statements where nonexecutable statements are not specifically prohibited. However, DATA statements must follow any specification statements that contain variable or array names which are used in the DATA statements. A NAMELIST statement must precede the input or output statement with which it is used.

Problems

1-1 Listed below are some FORTRAN real constants. Rewrite these constants in ordinary form.

```
.07E-3        -3.21E-02     .00E5
+.9E+04       .0002E-5      -10.E-10
-10.E-1       .9999E9       +10.E0
.86E5         .1111E1       1234.E-03
```

1-2 Listed below are several numbers written in ordinary form. Rewrite these numbers as real constants in E form.

1,000,000	$3.67(10)^{-2}$
10^{15}	8.04
$10(10)^{-3}$	$32.61(10)^2$
1	9.54602
.000001	10

1-3 Listed below are some FORTRAN double-precision constants. Rewrite these in ordinary engineering form.

```
-.09D+1                     3408655.1392
+10.1D0                     .00363D2
 .2D-03                     1.D+25
 .192837465438 9D+09        5.D02
```

1-4 Listed below are several numbers written in ordinary engineering form. Rewrite these as FORTRAN double-precision constants.

$-.003$	$.1$
$5{,}921{,}364.559$	$+10$
10^{10}	$.0000001$
$-10(10)^{-3}$	2.5900283171

1-5 Write the quantities below as FORTRAN complex constants.

$\sqrt{-169} + 12$	$-95.2 - \sqrt{-49}$
$-\sqrt{169}$	$9 + 6i$
$10^{13} + (7.21)^9\sqrt{-1}$	35.127
$135.2i$	$3 + 4i$

1-6 For each of the names listed below, indicate whether the name is an acceptable real-variable name, an acceptable integer-variable name, or an invalid name. Base your determination upon the FORTRAN version which you are using. For the names which are invalid, state why.

```
Y          MIN$S       JOHN       AB
(Y)        22B4        ABLE       B(B)
I2         XCUBED      BAKER      VALUE
I.1        1MORE       CHARLIE    ACCELER
KKG        VELOC       -TEN       JERK
B707       STRESS      A*B        E*EPSI
```

1-7 Write the following mathematical expressions as FORTRAN expressions:

$\dfrac{(A + B)C}{D}$	$\dfrac{AB}{CD}$	$(A + B)^2$
$A - B + C$	$\dfrac{X + Y}{W + Z}$	$\dfrac{X}{Y + \dfrac{A}{B}}$
$A^{(B^C)}$		
$A + B^2$	$A + \dfrac{BC}{D}$	$(A + B)/(C/D)$
	$-3A + (B - C)$	$\dfrac{A}{B}(10)^{-2}$

1-8 Each FORTRAN expression below is written to represent the equivalent mathematical expression shown on the left. Point out the errors, if any, in each case and write the correct FORTRAN expression.

Mathematical Expression	FORTRAN Expression
$A + \frac{BC}{D} + F^E - G$	`(A+B)*C/D+F**E-G`
$\frac{A}{BC}$	`A/B*C`
$(A + B)C$	`A+B*C`
$(A + I)^2$	`(A+I)**2`
$X^2 + 3X + 4$	`X*X+3.(X)+4.`
$\frac{EHC}{AB}$	`E*H/A/B/C`
πR^2	`(3.1416*(R**2)`
I^A	`I**A`

1-9 The type statements

```
DOUBLE PRECISION DBL1, DBL2,N
COMPLEX COMP1, COMP2,X
```

appear in a program in which the expressions containing double-precision and complex quantities (shown below) are used. Indicate the validity or invalidity of each expression. If the expression is invalid, state why. If the expression is valid, state its mode.

```
N+1
(A+.14159D0)*DIAM
N/DBL1+3.
COMP1*2.2
DBL2+COMP1-6.75D+08
.314159D+01*LRAD**2
COMP2/(.153, .43E-2)
DBL2-.74328D-4
X*(3,4)
X**2
(DBL1-N)**X
I**J
24.4151**2.D1
COMP1**X
```

1-10 Determine the value of each of the following logical expressions, assuming that A > B, LG1 = .TRUE., LG2 = .TRUE., LG3 = .FALSE., and LG4 = .FALSE..

(a) `A .EQ. B .AND. LG2`
(b) `LG1 .AND. .NOT. A .LT. B`
(c) `A .EQ. B .OR. LG3`
(d) `A .LT. B .OR. LG4`
(e) `.NOT. A .LT. B .AND. LG3`
(f) `.NOT.(A .LT. B .AND. LG3)`
(g) `.NOT. (A .LT. B .OR. LG1)`
(h) `LG3 .AND. LG2 .OR. LG1`
(i) `.NOT. B .LE. A .OR. LG1 .AND. LG4`

1-11 Some of the arithmetic statements listed below contain errors. Rewrite the invalid statements in correct form.

```
A-B=C+D          A*B=C/D
N=(X+Y)          X(2+I,3)=-2.*(A-B)
Z=I+J-2.         6.=R
-X=-(A+B)        J(A)=A-10.
X(I,3)=I3        A(J)=10.-A
```

1-12 Indicate the validity or invalidity of each of the statements listed below. If invalid, state why. Assume that COMP1, COMP2, COMP3, and COMP4 have been declared to be complex variables in a type statement. Also assume that DBL1, DBL2, and DBL3 have been declared to be double-precision variables in a type statement, and that LG1, LG2, LG3, and LG4 have been declared logical variables in a type statement.

```
COMP1=A+COMP2
B=COMP1+(6.5, .0)
COMP3=2.51D+5+A+B
DBL1=A/B+C*D
I=C+C+COMP1
J=DBL1+F/H
X=DBL1*3.7154981668
DBL2=J
DBL3=COMP1/3.1416+2.99
COMP4=J
COMP2=A*B/C+F/G
LG1=DBL2 .GT. DBL3
LG2=COMP1 .NE. COMP2
A=I .LT. X
LG3=LG1 .EQ. .TRUE.
X=C+(3.,4.)
COMP4=X/Y
LG1=X*C/D .LE. I+J
```

1-13 One hundred numbers, X(1) through X(100), are available for use from a previous portion of a program. Continue the program by writing the statements required to find the *mean* of the value of these numbers (the sum of the numbers divided by their number). Store the result in AVG.

1-14 Two hundred numbers, X(1) through X(100) and F(1) through F(100), are stored in memory. Write the statements necessary to find

$$\frac{1}{100}\sum_{I=1}^{100}[X(I)][F(I)]$$

and store the result in COFG.

1-15 The values of X(1) through X(50) are available from a previous portion of a program. Write the statements necessary to calculate

$$\sum_{I=1}^{50}\left(X(I)-\frac{1}{50}\sum_{J=1}^{50}X(J)\right)^2$$

and store the result in a location SMIN.

1-16 The elements of a square matrix A(I,J) of order n have been read in as data. The order n has also been read in as data. Write the statements necessary to calculate the trace of the matrix ($A_{11} + A_{22} + A_{33} + \cdots + A_{nn}$) and store the result in a location defined by the name TRACE.

1-17 The elements of a square matrix B(I,J) of order n have been read in as data. The value of n has also been entered as data. There are available, from the previous portion of the program, n values of X(J), from X(1) through X(N). Write the statements necessary to compute the n elements of a new array, named A(I), whose elements are given by

$$A(I) = \sum_{j=1}^{n}(B(I, J))(X(J))$$

In Probs. 1-18 through 1-21, a partial flow chart is to be made for each problem, and the FORTRAN statements which accomplish the given requirements are to be written as partial source programs.

1-18 The algebraically smallest of the three variables A, B, and C is to be given the name SMALL. The algebraically largest of the same variables is to be given the name BIG.

1-19 If $X > Y$ and $W > Z$, N is to be given the value 1. If $X \leq Y$ and $W \leq Z$, N is to be given the value of Z; otherwise, set $N = 3$.

1-20 If $E \leq 100$ and either $F \leq 10$ or $G \leq 1$ or both, set $N = 1$; otherwise, set $N = 2$.

1-21 Give the smallest of the three variables A, B, and C the name SMALL, using only two arithmetic IF statements.

1-22 Making use of the logical IF statement, write a statement or statements to carry out the requirements of parts (a) through (h), below. Assume that each statement you write is part of a complete program in which all variables have been assigned values.

(a) If THETA−THETA1<360., transfer to statement 5. Otherwise, continue in sequence.
(b) If X>Y, set A=X/2. Otherwise, set A=X.
(c) If GAMMA+PHI>180., transfer to statement 28. Otherwise, transfer to statement 53.
(d) If Y is an even number, go to statement 10. Otherwise, continue in sequence.
(e) If X−XMAX is <0, go to statement 55. Otherwise, stop.
(f) If |PTH1−CTH1|<EPSI, go to statement 69. Otherwise, continue in sequence.
(g) If A>0., and if B>10. or C<20., let X=(C+B)/A.
(h) If A is the largest of the variables A, B, and C, set BIGEST equal to A and then go to statement 15. Otherwise, go directly to statement 15.

1-23 A one-dimensional array $A(N)$ is to be read into computer memory. The array has a maximum of 50 elements. Write a portion of a FORTRAN program which will read in the array. Assume that the value of each element will be on a separate card in the first ten card columns. Assume that the decimal points are punched and that no exponents are used. The cards will be in order, with the value of A(1) on the card which is read in first.

1-24 A one-dimensional array $A(N)$ is to be read into computer memory. The array has a maximum of 50 elements. Write a portion of a FORTRAN program which will read in the array. Assume that the cards will not be in any particular order, so that the first two card columns will be used to identify the element, with the next ten card columns used for the value of the element. Also assume that the decimal points are punched and that no exponents are used. Devise a method of indicating to the program that the last card of the array has been read.

1-25 Write a portion of a FORTRAN program using an implied DO for reading the elements of a triangular matrix into computer memory. The matrix has the form

$$\begin{matrix} a_{11} & a_{12} & \cdots & a_{1n} \\ & a_{22} & a_{23} \cdots & a_{2n} \\ & & a_{33} \cdots & a_{3n} \\ & & & \vdots \\ & & & a_{nn} \end{matrix}$$

where the value of n is to be read in as data.

1-26 Write a portion of a FORTRAN program for printing out the temperature values at the mesh points on a square plate. The temperature values are stored as the values of the elements of an 11-by-11 array $T(I, J)$, where I is the value of the row number of the mesh points, and J is the value of the column number. The temperature values are to be printed on the page in the same positions as those occupied by the mesh points on the plate; that is, there will be 11 values in one row on the paper, and there will be 11 rows. All temperatures will lie between 100°C and 0°C. One digit after the decimal point will provide a sufficiently accurate printout. The line printer is under program control.

1-27 Write a source program for calculating the amplitude of a forced vibration. The amplitude is given by

$$A = \frac{Q/K}{\sqrt{\left(1 - \frac{\omega^2}{\omega_n^2}\right)^2 + \left(\frac{C^2\omega^2}{K^2}\right)}}$$

where

A = amplitude
Q = maximum value of periodic disturbing force
K = spring constant
ω = 2π times the forcing frequency
ω_n = 2π times the natural frequency
C = damping coefficient

The values of Q, K, ω, ω_n, and C are to be read from data cards. The value of the the amplitude is to be printed as output.

1-28 Write a program for determining the acceleration of the slider of a slider-crank mechanism. The formula for the acceleration is

$$a = r\omega^2\left[\cos\theta + \frac{c\cos 2\theta + c^3\sin^4\theta}{(1 - c^2\sin^2\theta)^{3/2}}\right]$$

where

a = acceleration of slider
c = ratio of crank length to connecting-rod length
ω = angular velocity of crank
r = crank length
θ = angle of crank from a line through the fixed pivot of the crank and the joint connecting the slider and connecting rod

The values for c, ω, r, and θ are to be read from data cards. The value of the acceleration and angle θ are to be printed as output.

1-29 Write a FORTRAN source program for calculating the radius of gyration of a rectangular hollow beam about a horizontal axis. The equation for the radius of gyration is

$$k = \sqrt{\frac{b_1 d_1^3 - b_2 d_2^3}{12(b_1 d_1 - b_2 d_2)}}$$

The dimensions b_1 and d_1 are the outside width and depth of the section, respectively, and b_2 and d_2 are the corresponding inside dimensions, all in inches. The numerical values of these dimensions are to be read in from cards, and all dimensions and the corresponding radius of gyration are to be printed out.

1-30 Draw a flow chart and write a complete FORTRAN source program for calculating the theoretical adiabatic horsepower of a quantity of gas (the horsepower necessary to compress Q cubic feet of gas per second from p_1 to p_2). The formula is

$$\text{Hp} = \frac{144k}{500(k - 1)} p_1 Q \left[\left(\frac{p_2}{p_1} \right)^{(k-1)/k} - 1 \right]$$

where

Q = quantity of gas, cu ft/sec
p_1 = initial pressure of gas, lb/in.^{2}
p_2 = final pressure of gas, lb/in.^{2}
k = ratio of specific heat at constant pressure to specific heat at constant volume

The values of k, p_1, p_2, and Q are to be entered as data from cards, and the ratio of p_2/p_1 is to vary from 1.5 to 10, in increments of 0.5.

1-31 Write a complete FORTRAN source program for determining the acceleration of the slider of a slider-crank mechanism (see Prob.

1-28) for a number of different proportions of the linkage and over 180 deg of crank rotation. Let the ratio of crank length to connecting-rod length vary from 0.1 to 0.9, in increments of 0.1. Both the angular velocity of the crank and the crank length are to be entered as data, using cards. Print out the acceleration for every 5 deg of crank rotation for each value of crank-to-connecting-rod ratio.

1-32 Write statements defining the following arithmetic statement functions:

(a) $f(x, y) = \sqrt{x^2 + y^2}$
(b) $f(x) = a + bx + cx^2 + dx^3 + ex^4$
(c) $f(x, a, b, c, d) = a + bx + cx^2 + dx^3$
(d) $f(x, a, b, c, d) = a + b \cos x + c \cos^2 x + d \cos^3 x$

1-33 Write a statement defining the statement function FORCE as given by

$$\text{FORCE} = H\left(\frac{7.64}{\tau_2} - \frac{7.95}{\tau_1}\right)$$

Let τ_1 and τ_2 be arguments of the function, and let H be a parameter.

1-34 Write a statement defining the statement function RADGYN given by

$$\text{RADGYN} = R\sqrt{\frac{\alpha + \sin\alpha\cos\alpha - 2\sin^2\alpha/\alpha}{2\alpha}}$$

Let α and R be arguments of the function.

1-35 Write a statement defining the statement function RADIUS as given by

$$\text{RADIUS} = 0.721\sqrt[3]{(P)(D)\left[\frac{1 - v_1^2}{E_1} + \frac{1 - v_2^2}{E_2}\right]}$$

Let P and D be the arguments of the function, and let v_1, v_2, and E_1 and E_2 be parameters.

1-36 Write a statement defining the statement function WEIGHT as given by

$$\text{WEIGHT} = 0.4722A\sqrt{\frac{P_1}{v_s}\left(\frac{1 - (1/x^2)}{N + \log_e x}\right)}$$

Let P_1, x, and v_s be the arguments of the function, and let A and N be parameters.

1-37 Write a statement defining the statement function THICK as given by

$$\text{THICK} = \tfrac{1}{2}D[\sqrt{(S + 0.4p)/(S - 1.3p)} - 1]$$

Let D and p be arguments of the function, and let S be a parameter.

1-38 Write a statement defining the statement function VOLUM as given by

$$\text{VOLUM} = \frac{1}{4}\pi d^2 \sqrt{\left(\frac{gd}{4fl}\right)\frac{1}{wp_1}(p_1^2 - p_2^2)}$$

where $g = 32.2$, and $f = 0.0028(1.0 + 3.6/d)$. Let p_1, p_2, and w be arguments of the function, and let d and l be parameters.

1-39 Write a FUNCTION subprogram defining the function CRTICL where

$$\text{CRTICL} = \frac{0.669b^3d\sqrt{(1 - 0.63b/d)EG}}{l^2}\left[1 - \frac{a}{2l}\sqrt{\frac{E}{G(1 - 0.63b/d)}}\right]$$

Let a, b, d, l, E, and G be the function arguments.

1-40 Write a FUNCTION subprogram defining the function FOFY as given by

$$\text{FOFY} = R_1 \cos x - R_2 \cos y + R_3 - \cos(x - y)$$

where

$$R_1 = D/C$$

$$R_2 = D/A$$

$$R_3 = \frac{D^2 + A^2 - B^2 + C^2}{2CA}$$

The function arguments are A, B, C, D, x, and y.

1-41 Write a FUNCTION subprogram defining the complex function ROOT. If the roots are real, the subprogram is to determine the algebraically largest root of the quadratic equation

$$ax^2 + bx + c = 0$$

and the function is to take on the value of this root. If the roots are complex, the value of the function is to take on the complex value

$$\frac{-b + i\sqrt{-(b^2 - 4ac)}}{2a}$$

1-42 Write a FUNCTION subprogram defining a function PROD which is assigned the value of the product of the elements on the main diagonal of an n-by-n matrix.

1-43 Write a FUNCTION subprogram defining a function ONES which is assigned a value by the subprogram equal to the number of 1's in an m-by-n matrix whose elements all have the value 0 or 1.

1-44 Write a FUNCTION subprogram defining a function MAXA which is assigned the value of the largest element in an m-by-n array

$$\begin{array}{lllll} a_{11} & a_{12} & a_{13} & \cdots & a_{1n} \\ a_{21} & a_{22} & & \cdots\cdots & \vdots \\ a_{31} & a_{32} & & \cdots\cdots & \vdots \\ \vdots & \vdots & & & \vdots \\ a_{m1} & a_{m2} & & \cdots\cdots & a_{mn} \end{array}$$

1-45 Write a SUBROUTINE subprogram for finding the largest element in an m-by-n array such as that shown in Prob. 1-44. The largest element value is to be returned to the calling program as the value of an argument.

1-46 Write a SUBROUTINE subprogram which determines the number of 0's, 1's, and 2's in any m-by-n matrix, all of whose elements are known to be 0's, 1's, or 2's.

1-47 Write a SUBROUTINE subprogram which will interchange row k and row l of an m-by-n matrix a_{ij}.

1-48 Write a SUBROUTINE subprogram for finding the product of two matrices a_{ij} and b_{jk}, where the former is an m-by-n matrix, and the latter is a matrix having n rows and j columns.

1-49 Write a SUBROUTINE subprogram for determining the largest element of a one-dimensional array, $a_1, a_2, a_3, \ldots, a_n$.

1-50 Write a SUBROUTINE subprogram which will sort the elements of an array $a_1, a_2, \ldots, a_n$ into ascending order. That is, a_1 will be the name associated with the smallest element, a_2 will be the name associated with the next-to-the-smallest element, and so forth. Do the sorting by interchanging array elements a_i and a_{i+1} if a_i is greater than a_{i+1}, for $i = 1, 2, \ldots, k$, where k is initially equal to $n - 1$. This procedure will assign the largest value of the array to a_n. Then reduce k by 1 and repeat. This will assign the next-to-the-largest value in the array to a_{n-1}. Continue interchanging in this manner until and including $k = 2$.

2 Roots of Algebraic and Transcendental Equations

2-1 / INTRODUCTION

A problem commonly encountered in engineering is that of determining the roots of an equation of the form

$$f(x) = 0 \tag{2-1}$$

In this chapter we will consider six methods of finding the *real* roots of such algebraic and transcendental equations: (1) the incremental-search method, (2) the bisection method, (3) the method of false position, (4) the secant method, (5) the Newton-Raphson method, and (6) Newton's second-order method.

If $f(x)$ is a polynomial in x with real coefficients, it may have both real and complex roots. Of the various methods available for finding both real and complex roots, Graeffe's root-squaring method and Bairstow's iterative method will be presented.

2-2 / THE INCREMENTAL-SEARCH METHOD

In this approach we determine values of $f(x)$ for successive values of x in some interval to be searched until a sign change occurs for $f(x)$. A sign change occurs between x_i and $x_i + \Delta x$ if $f(x_i)f(x_{i+1}) < 0$. The sign change generally indicates that a root has been passed (it could also indicate a discontinuity in the function as shown in Fig. 2-1). A closer approximation to the value of the root may then be obtained by reverting

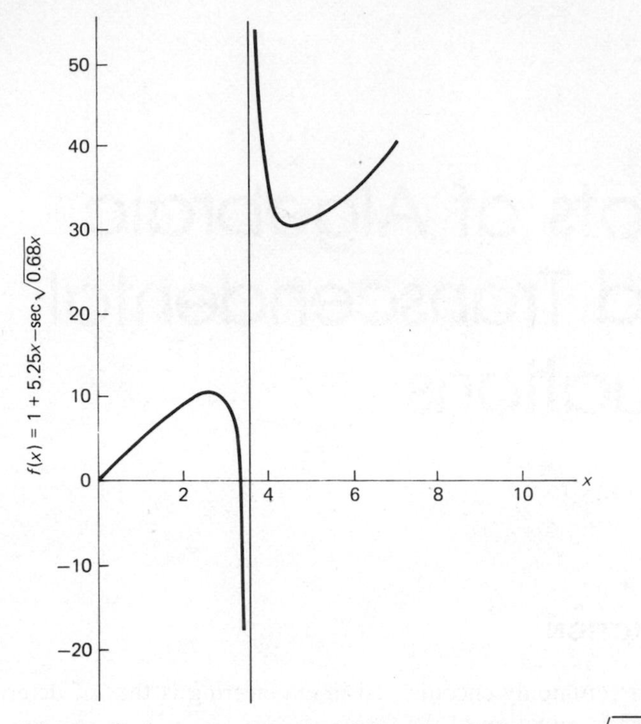

Fig. 2-1 / *Approximate graph of* $f(x) = 1 + 5.25x - \sec\sqrt{0.68x}$.

to the last x value preceding the sign change and, beginning with this x value, again determining values of $f(x)$ for successive values of x, using a smaller increment than was used initially, until the sign of $f(x)$ changes again. This procedure is repeated with progressively smaller increments of x until a sufficiently accurate value of the root is obtained. If additional roots are desired, the incrementation of x can be continued until the next root is approximately located by another sign change of $f(x)$, and so on.

Care must be exercised in selecting the initial value by which x is to be incremented, so that roots are not by-passed in an instance when two roots are close together in value. This is usually not a problem, if fairly small increments are used in the initial sequence. The following example will illustrate the use of the incremental-search method just described.

EXAMPLE 2-1

It is desired to determine the first positive nonzero root of the equation

$$1 + 5.25x - \sec\sqrt{0.68x} = 0 \tag{2-2}$$

by the searching method. A plot of the function is shown in Fig. 2-1. The first step is to sketch a flow chart outlining the general procedure

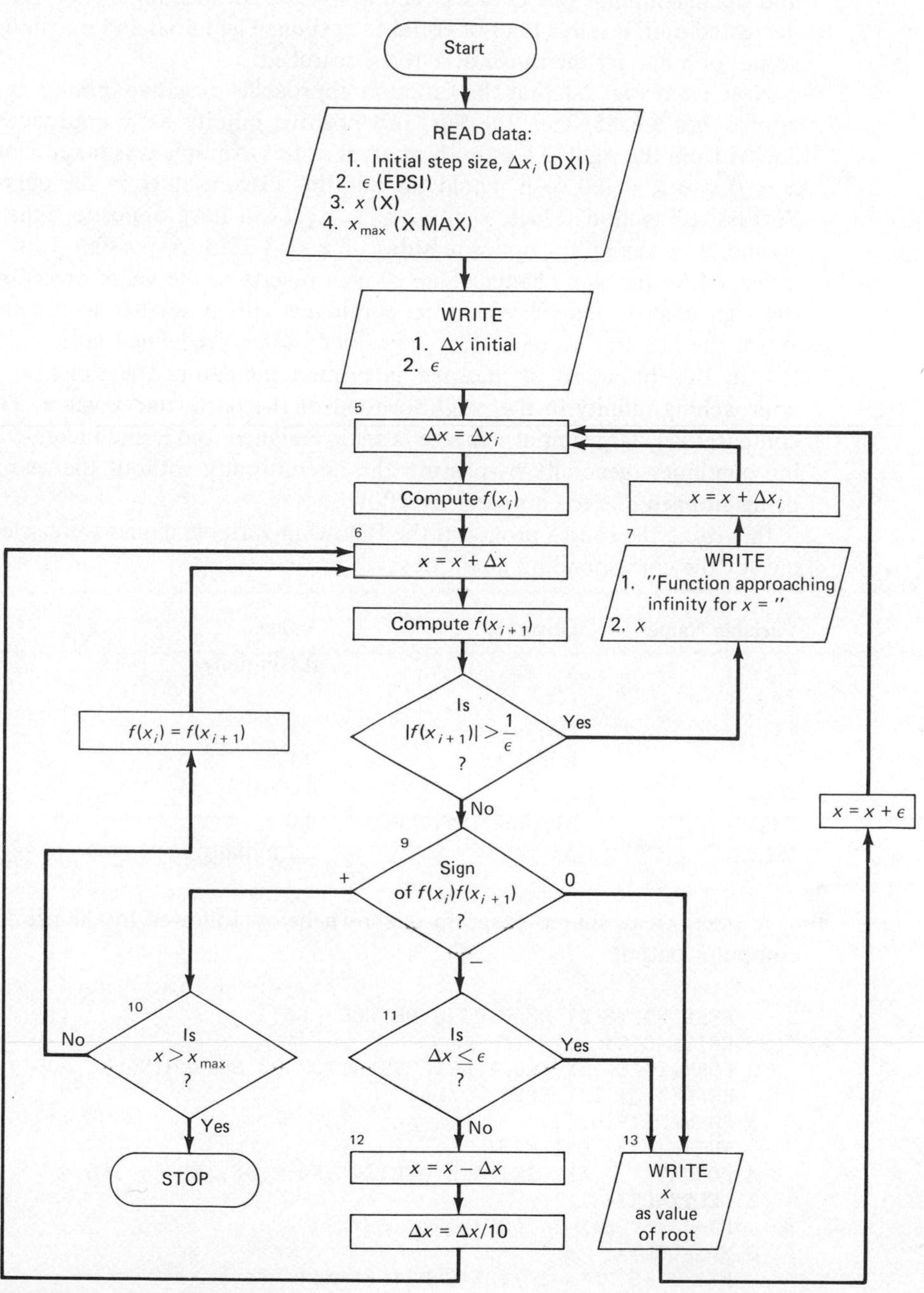

Fig. 2-2 / *Flow chart of Ex. 2-1.*

(Fig. 2-2). The initial value of Δx, an epsilon value, the initial value of x, and the maximum value of x are read in as data. In locating a root, Δx is decreased until it is less than or equal to epsilon. The initial and maximum values of x specify the range of x to be searched.

Note from Fig. 2-1 that the function approaches negative infinity as x approaches 3.6285 from the left and positive infinity as x approaches 3.6285 from the right. The search interval in this example was taken from $x = 0.1$ to $x = 4.0$ so it would include this discontinuity in the curve. Successive function values, $f(x_i)$ and $f(x_{i+1})$, will have opposite signs if x_i and $x_i + \Delta x$ lie on opposite sides of $x = 3.6285$. A possible root is indicated by the sign change. The x value reverts to the value preceding the sign change and incrementing continues with a smaller increment. When the absolute value of $f(x_{i+1})$ exceeds some predefined large value ($1/\varepsilon$ in this program), a message is printed indicating the function is approaching infinity in the neighborhood of the particular x value. The comparatively large initial step size is again assigned to Δx, and incrementing continues, generally by-passing the discontinuity without the search being stopped due to computer overflow.

In writing the source program, the following variable names were selected for the corresponding quantities:

Variable Name	Quantity	Value
X	x	0.1 initially
FX	$f(x_i)$	
FX1	$f(x_{i+1})$	
DXI	Initial Δx	0.1
EPSI	ε	0.0001
XMAX	Maximum x value	4.0
DELTX	Δx	0.1 initially

A FORTRAN IV source program is shown below followed by the printed computer output.

```
C     REAL ROOTS BY SEARCHING METHOD
      WRITE(6,2)
    2 FORMAT('1',5X,'ROOT DETERMINATION BY SEARCHING METHOD'//)
      READ(5,3) DXI,EPSI,X,XMAX
    3 FORMAT(4F10.0)
      WRITE(6,4) DXI,EPSI
    4 FORMAT(' ',5X,'INITIAL DELTX=',F4.2,9X,'EPSI=',F6.4//)
    5 DELTX=DXI
      FX=1.+5.25*X-1./COS(SQRT(.68*X))
    6 X=X+DELTX
      FX1=1.+5.25*X-1./COS(SQRT(.68*X))
      IF(ABS(FX1).GT.1./EPSI) GO TO 7
      GO TO 9
    7 WRITE(6,8) X
```

```
    8 FORMAT(' ','FUNCTION APPROACHING INFINITY FOR X=',F7.4)
      X=X+DXI
      GO TO 5
    9 IF(FX*FX1) 11,13,10
   10 IF(X.GT.XMAX) STOP
      FX=FX1
      GO TO 6
   11 IF(DELTX-EPSI) 13,13,12
   12 X=X-DELTX
      DELTX=DELTX/10.
      GO TO 6
   13 WRITE(6,14) X
   14 FORMAT(' ','X=',F8.4, ' IS A REAL ROOT')
      X=X+EPSI
      GO TO 5
      END
```

```
     ROOT DETERMINATION BY SEARCHING METHOD

     INITIAL DELTX=0.10                EPSI=0.0001

 X=  3.3867 IS A REAL ROOT
 FUNCTION APPROACHING INFINITY FOR X= 3.6288
```

2-3 / THE BISECTION METHOD

After the searching method described in Sec. 2-2 has revealed a change in sign of the function and thus a possible root, there are several ways of converging more rapidly to the root than by continuing the search as described in that section. One of the more reliable of these is the *bisection method*, also known as the half-interval method and the Bolzano method.

We begin with the assumption that an interval between x_i and x_{i+1} has been found such that $f(x_i) \cdot f(x_{i+1}) < 0$. This interval is next cut into two subintervals, the first extending from x_i to $x_{i+1/2}$ and the second from $x_{i+1/2}$ to x_{i+1} where

$$x_{i+1/2} = 0.5(x_i + x_{i+1}) \tag{2-3}$$

These two subintervals are shown in Fig. 2-3. The subinterval containing the root (the left subinterval in Fig. 2-3) must be selected for further bisection. This is easily done by checking the sign of $f(x_i) \cdot f(x_{i+1/2})$. If it is negative, the root lies in the first subinterval. If it is positive, the root lies in the second subinterval. If the product is exactly equal to zero, $x_{i+1/2}$ is an exact root. In general the root will be in one of the subintervals and this subinterval is then bisected, and the half containing the root is again selected. Bisection continues until the interval becomes small enough

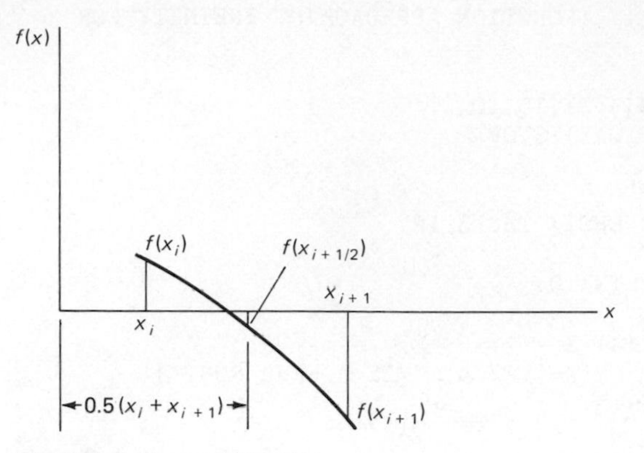

Fig. 2-3 / *Interval bisection.*

that the final x value resulting from bisection of the interval is sufficiently close to a root.

It should be noted that after n bisections, the original interval has been reduced by a factor of 2^n. Thus a large number of bisections are not required to converge to an accurate value of the root. For example, 20 bisections will reduce an interval by a factor of more than one million ($2^{20} = 1{,}048{,}576$). Accuracy greater than that given by 20 bisections would seldom be required

EXAMPLE 2-2

To illustrate how we might search over a range of x by the incremental-search method of Sec. 2-2 until an interval is found containing a possible root, and then switch to the bisection method for fairly rapid convergence to the root, we again consider the function given by Eq. 2-2 and shown in Fig. 2-1. Our region of search will again be from 0.1 to 4.0, and possible roots will be indicated by sign changes of the function in the neighborhood of 3.38 and 3.63. However, only the first interval (around $x = 3.38$) contains a root as we see from Fig. 2-1. The bisection method applied to an interval containing $x = 3.6285$ will result in overflow if provision is not made to avoid it. We will do this by checking the magnitude of $f[0.5(x_i + x_{i+1})]$ each time it is calculated. If it exceeds some predetermined large value, we assume we are not converging to a root and print a message that the function is approaching infinity for the x value equal to $0.5(x_i + x_{i+1})$.

The flow chart for the combined incremental-search method of Sec. 2-2 and the bisection method of this section is shown in Fig. 2-4. A thorough study of the flow chart will help in understanding how the successive bisections are carried out, with the root always being in the subinterval to be bisected next.

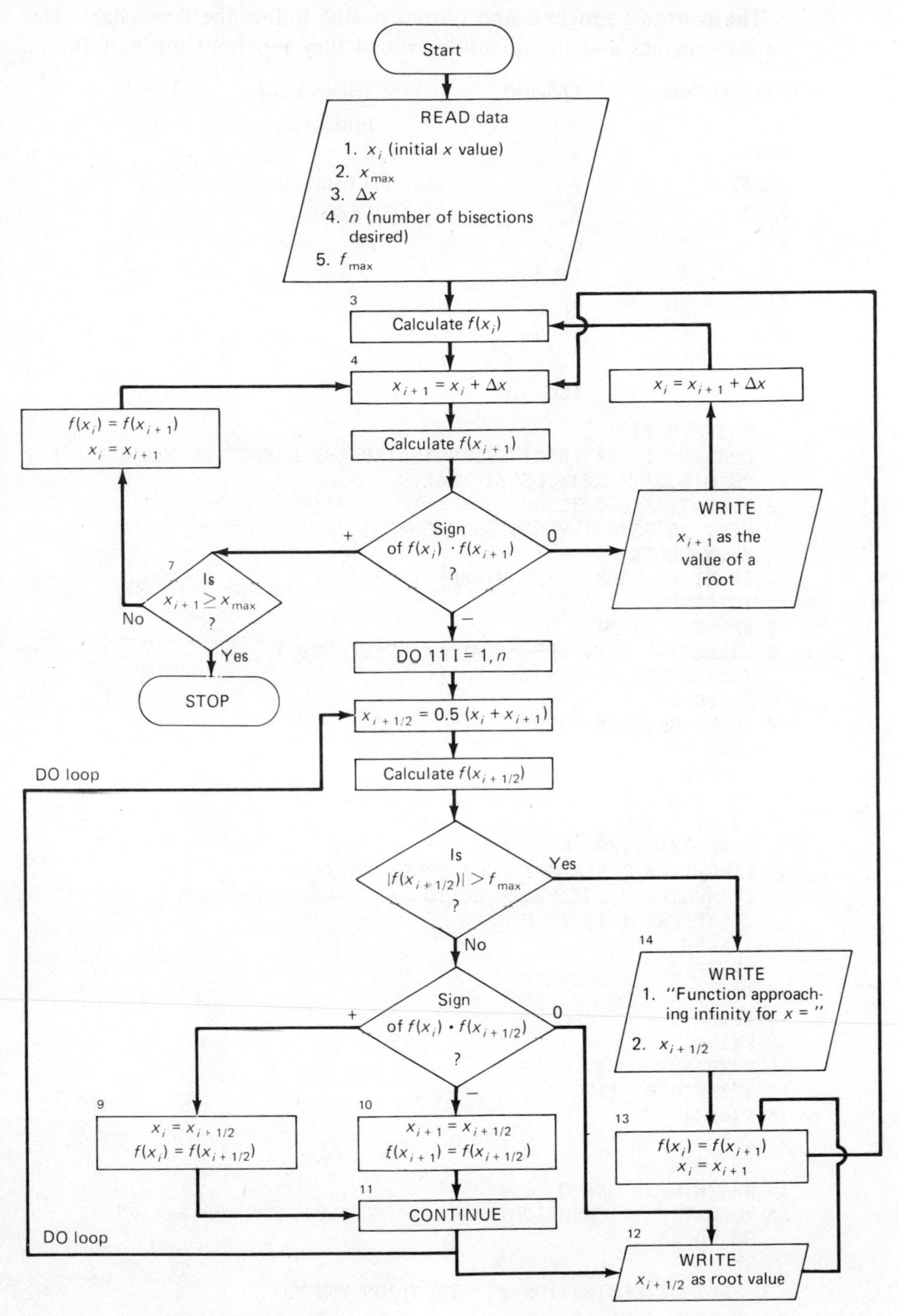

Fig. 2-4 / *Flow chart for Ex. 2-2.*

The FORTRAN program and output results follow the flow chart. The variable names and the quantities which they represent are as follows:

Variable Name	Quantity	Values Used
X	x_i	Initial value, 0.1
XMAX	x_{max}	4.0
DELTX	Δx	0.1 initially
FMAX	f_{max}	10,000
N	n	20
FX	$f(x_i)$	
X1	x_{i+1}	
FX1	$f(x_{i+1})$	
XAVG	$0.5(x_i + x_{i+1})$	
FAVG	$f(x_{i+1/2})$	

```
   WRITE(6,1)
 1 FORMAT('1',5X,'ROOT DETERMINATION BY BISECTION METHOD'//)
   READ(5,2) X,XMAX,DELTX,FMAX,N
 2 FORMAT(4F10.0,I2)
 3 FX=1.+5.25*X-1./COS(SQRT(.68*X))
 4 X1=X+DELTX
   FX1=1.+5.25*X1-1./COS(SQRT(.68*X1))
   IF(FX*FX1) 8,5,7
 5 WRITE(6,6) X1
 6 FORMAT(' ','X=',F8.4,' IS A REAL ROOT')
   X=X1+DELTX
   GO TO 3
 7 IF(X1.GE.XMAX) STOP
   X=X1
   FX=FX1
   GO TO 4
 8 DO 11 I=1,N
   XAVG=(X+X1)/2.
   FAVG=1.+5.25*XAVG-1./COS(SQRT(.68*XAVG))
   IF(ABS(FAVG).GT.FMAX) GO TO 14
   IF(FX*FAVG) 10,12,9
 9 X=XAVG
   FX=FAVG
   GO TO 11
10 X1=XAVG
   FX1=FAVG
11 CONTINUE
12 WRITE(6,6) XAVG
13 FX=FX1
   X=X1
   GO TO 4
14 WRITE(6,15) XAVG
15 FORMAT(' ','FUNCTION APPROACHING INFINITY FOR X=',F7.4)
   GO TO 13
   END
```

```
     ROOT DETERMINATION BY BISECTION METHOD
X=  3.3866 IS A REAL ROOT
FUNCTION APPROACHING INFINITY FOR X= 3.6288
```

2-4 / THE METHOD OF FALSE POSITION (LINEAR INTERPOLATION)

After the incremental-search method of Sec. 2-2 has found an interval which contains a real root of the function $f(x) = 0$, the method of false position, or linear interpolation, may be used to converge to the root. Convergence may be more rapid than by the bisection method discussed in Sec. 2-3, but there is no assurance that this will always be the case. The number of iterations required for satisfactory convergence will depend on the shape of the graph of the function in the interval which has been found to contain a root.

Let us assume that x_i and x_{i+1} are two values of x which bound an interval containing a real root of the function, as shown in Fig. 2-5. For convenience in this discussion, these two x values will be referred to as x_1 and x_2, respectively. A straight line is passed through points p_1 and p_2, and the x value (x_3) where this straight line intersects the x axis is a closer approximation to the root than either x_1 or x_2. The value of x_3 may be obtained from the following relationship based on similar triangles:

$$\frac{f(x_2) - f(x_1)}{x_2 - x_1} = \frac{-f(x_1)}{x_3 - x_1}$$

or

$$x_3 = \frac{x_1 f(x_2) - x_2 f(x_1)}{f(x_2) - f(x_1)} \tag{2-4}$$

Now either x_1 and x_3 or x_3 and x_2 will bound an interval containing the root. We see from Fig. 2-5 that the subinterval containing the root is bounded by x_3 and x_2. In a computer program, the interval containing the root may be determined by noting the sign of $f(x_1) \cdot f(x_3)$. If $f(x_1) \cdot f(x_3) < 0$, x_1 and x_3 bound the interval containing the root,

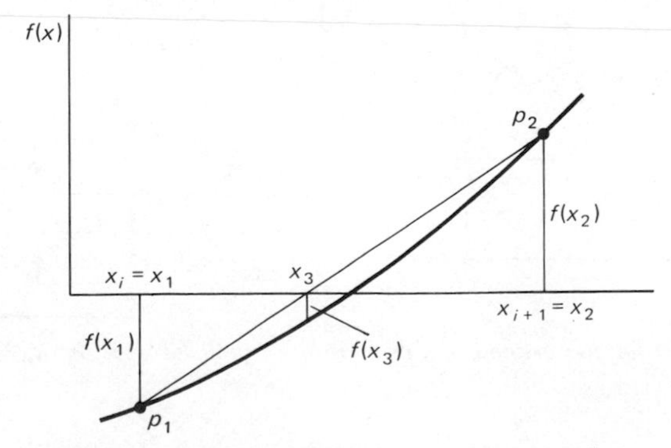

Fig. 2-5 / *Root approximation by linear interpolation.*

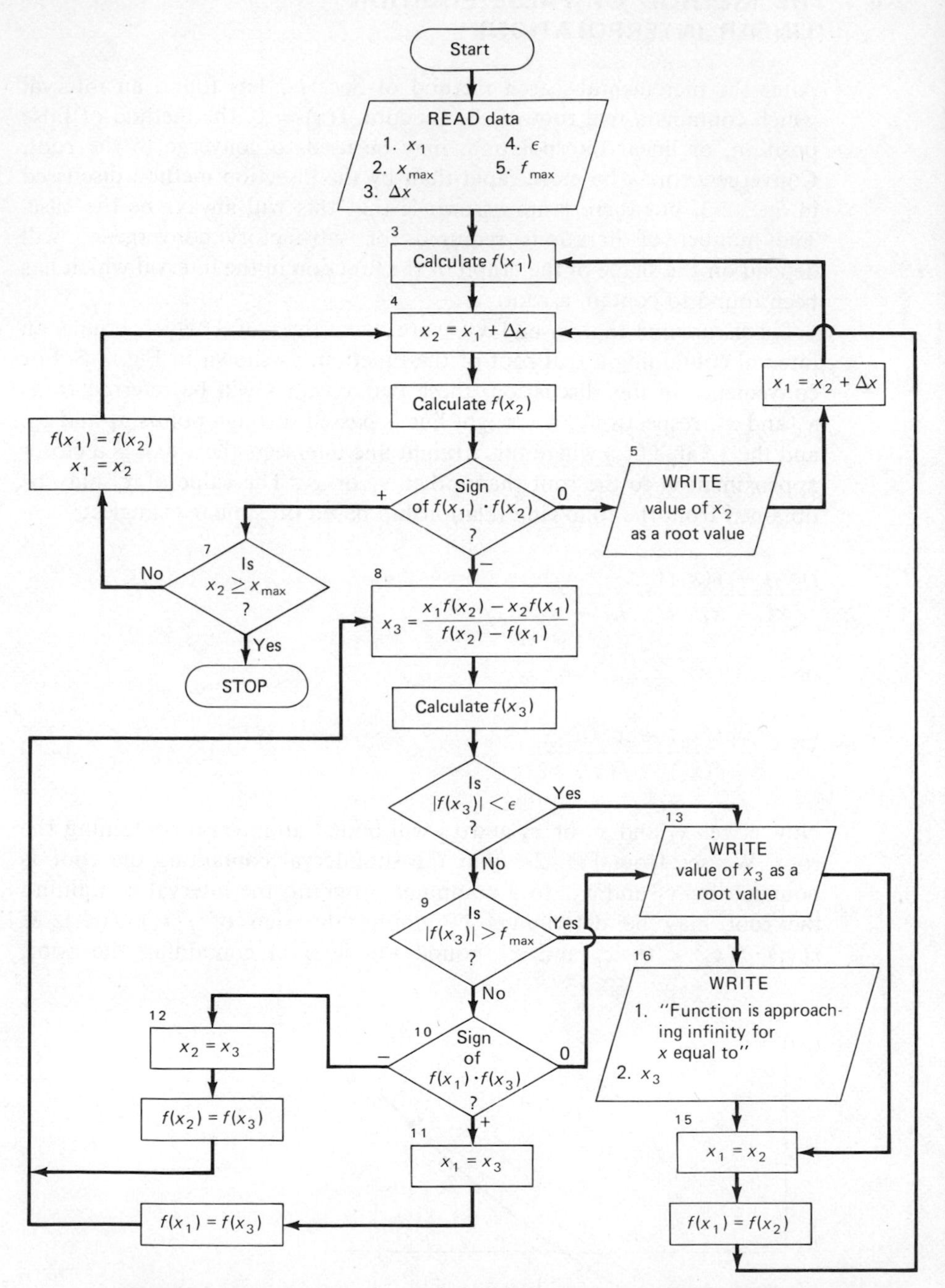

Fig. 2-6 / *Flow chart for the method of false position combined with the incremental-search method.*

whereas if $f(x_1) \cdot f(x_3) > 0$, x_3 and x_2 bound the interval containing the root. In the latter case, which is illustrated in Fig. 2-5, interpolation continues by assigning the value of x_3 to the name x_1 and the value of $f(x_3)$ to the name $f(x_1)$ and reapplying Eq. 2-4 to obtain a still better approximation to the root. If $f(x_1) \cdot f(x_3) < 0$, the value of x_3 is assigned to the name x_2 and the value of $f(x_3)$ is assigned to the name $f(x_2)$, and Eq. 2-4 is reapplied for an improved approximation to the root. Iteration continues in this manner until $|f(x_3)| < \varepsilon$, where ε is some small number such as perhaps 0.0001.

A flow chart for applying the incremental-search method to find an interval containing a possible root and then applying the method of false position for convergence to the root is shown in Fig. 2-6. Study of the flow chart will reveal that discontinuities in the function such as the one shown in Fig. 2-1 will probably be by-passed without causing overflow, since when $|f(x_3)|$ is larger than some preassigned large number f_{max}, interpolation ceases and the searching method is resumed to search for another interval in which the function changes sign. The range of x searched for roots is from the initial value read in for x_1 to x_{max}.

It is left as an exercise for the reader to write a computer program based on the flow chart of Fig. 2-6 and a particular function (Prob. 2-1c).

2-5 / THE SECANT METHOD

The secant method is similar to the method of false position except that the two *most recent* x values and their corresponding function values are used in obtaining a new approximation to the root instead of always using two x values which bound a subinterval containing the root. The method is illustrated in Fig. 2-7 where we assume we have found an interval between x_1 and x_2 which contains a root. A new approximation to the root is then obtained by the use of Eq. 2-4. In the renaming process for iteration,

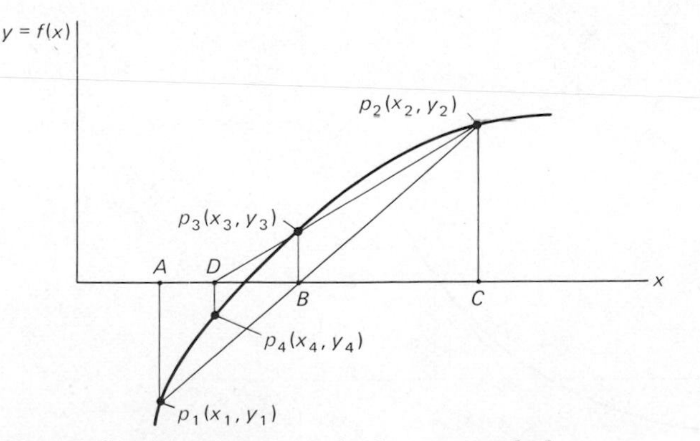

Fig. 2-7 / *Root determination by the secant method.*

the value of x_2 is always assigned to the name x_1 and the value of x_3 is assigned to the name x_2. Equation 2-4 is then reapplied. Iteration continues until $|f(x)| < \varepsilon$, as in the method of false position.

We see from Fig. 2-7 that part of the time we are using linear interpolation and part of the time linear *extrapolation* (when both x values lie on the same side of the root as is the case when we use the straight line, or secant line, joining points p_2 and p_3 to obtain x_4).

It is left as a problem for the reader to revise the flow chart of Fig. 2-6 so that the secant method is used, and to write a computer program based on this flow chart (Prob. 2-6).

2-6 / NEWTON-RAPHSON METHOD (NEWTON'S METHOD OF TANGENTS)

This method is very useful for improving a first approximation to a root of an equation of the form $f(x) = 0$, which might have been obtained by the searching method of Sec. 2-2, by an approximate graph of the function, or by some other means.

Consider the graph of $f(x)$ versus x, shown in Fig. 2-8, and assume that x_n is a first approximation of a root. If we draw a tangent line to the curve at $x = x_n$, the tangent line will intersect the x axis at a value x_{n+1}, which is an improved approximation to the root. It can be seen (Fig. 2-8) that the slope of the tangent line is

$$f'(x_n) = \frac{f(x_n)}{x_n - x_{n+1}} \tag{2-5}$$

from which

$$x_{n+1} = x_n - \frac{f(x_n)}{f'(x_n)} \tag{2-6}$$

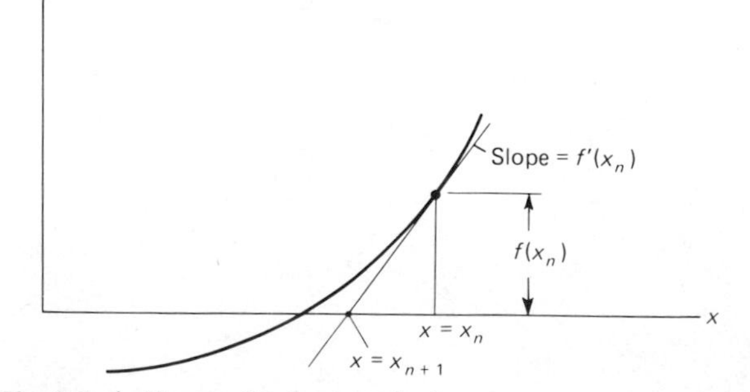

Fig. 2-8 / *Newton-Raphson method.*

The value of the function and the value of the derivative of the function are determined at $x = x_n$, and the new approximation to the root, x_{n+1}, is obtained by using Eq. 2-6. The same procedure is repeated, with the new approximation, to get a still better approximation to the root. This continues until successive values of the approximate root differ by less than a prescribed small epsilon which controls the allowable error in the root, or until the value of the function becomes less than some prescribed small value.

The Newton-Raphson method is widely used in practice because of its generally rapid convergence. However, there are cases in which convergence does not occur. One such example is shown in Fig. 2-9(a) where $f''(x)$ changes sign near the root. A second case in which convergence may not occur is illustrated in Fig. 2-9(b). In this example, the initial approximation to the root was not sufficiently close to the true value, and the

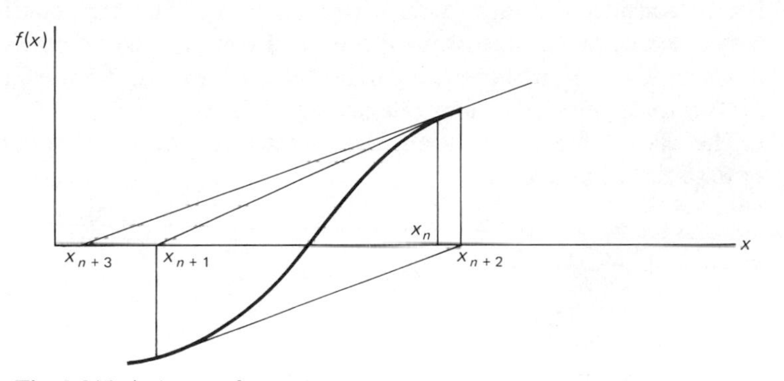

Fig. 2-9(a) / *A case of no convergence.*

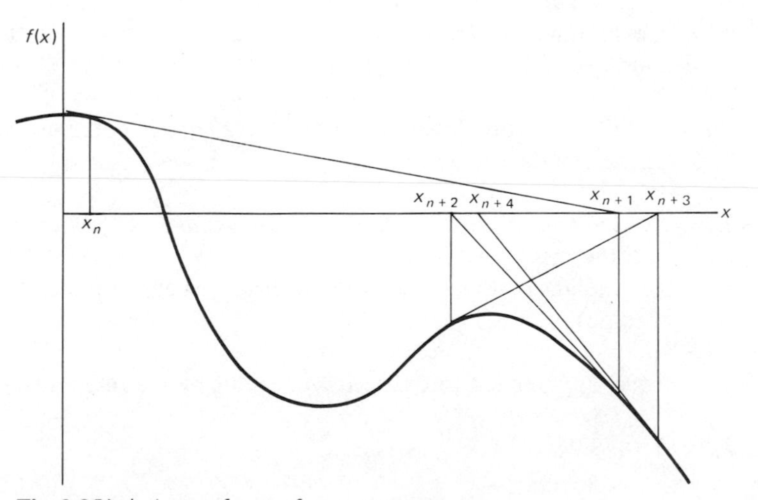

Fig. 2-9(b) / *A second case of no convergence.*

tangent to the curve for x_n has a very small slope, resulting in x_{n+1} being far to the right where a local maximum in the curve causes the difficulty (oscillation about the local maximum). Other functions could be illustrated in which there is a jump to a root other than the one nearest to the first approximation. These difficulties can be avoided by having the initial approximation sufficiently close to the root value, but sometimes this is not possible.

As we have seen, the Newton-Raphson method requires the differentiation of the function. If the derivative of the function is very complicated, it may be advantageous to substitute a finite-difference approximation of the derivative for the actual derivative (see Sec. 5-6 for such approximations).

EXAMPLE 2-3

Let us consider the rattrap shown in Fig. 2-10. The trap consists of a movable jaw, trip arm, torsion spring, and trip pan. Upon release of the trip arm owing to a slight disturbance of the trip pan, the torque of the torsion spring closes the movable jaw and kills the rat.

The relationship giving the angular displacement of the jaw as a function of time can be determined as[1]

$$\theta = \frac{T_0}{k}\left(1 - \cos\sqrt{\frac{k}{I_O}}\,t\right) \tag{2-7}$$

where

θ = angular displacement of jaw from "set" position, radians
T_0 = torque exerted by spring on jaw at $\theta = 0$, lb-ft
k = torsional spring constant, lb-ft/radian
I_O = mass moment of inertia of jaw about axis of rotation, lb-ft-sec^2
t = time, sec

With the following trap dimensions (A = 1.125 in., B = 0.5 in., and R = 3.75 in.), we can determine that

θ_k = 2.97 radians or 170.4 deg (angular displacement of jaw upon contact with the rat)
θ_c = 3.27 radians or 187.7 deg (angular displacement of jaw in the closed position)

The torque T_0 is related to the spring constant by the relationship

$$T_0 = T_c + 3.27k \tag{2-8}$$

[1] The derivation is given in Sec. 6-3.

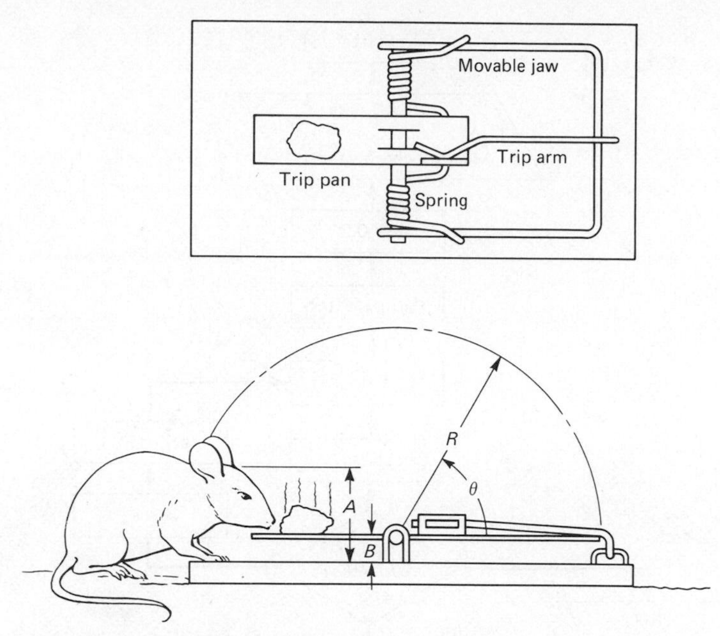

Fig. 2-10 / *Rattrap shown in open position.*

as indicated in Fig. 2-11 where T_c is the torque when the jaw is in the unset or closed position ($\theta = 3.27$ radians).

The manufacturer of the trap has received complaints that the trap he is marketing closes too slowly (0.0382 sec to strike the rat) and allows too many rats to escape after tripping the trap. Therefore, it is desired to install a new torsion spring which will close the trap on the rat twice as quickly as before, but which will still have the same torque in the closed position ($T_c = 0.625$ lb-ft), since the existing trap is easily set with this initial torque. It is also desired to maintain the same mass moment of inertia (0.0006 lb-ft-sec^2) for the jaw, so that increased energy will also be

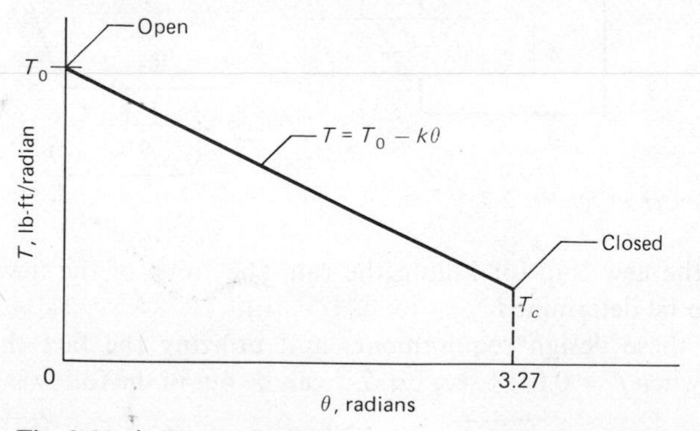

Fig. 2-11 / *Torque-displacement characteristic of torsion spring.*

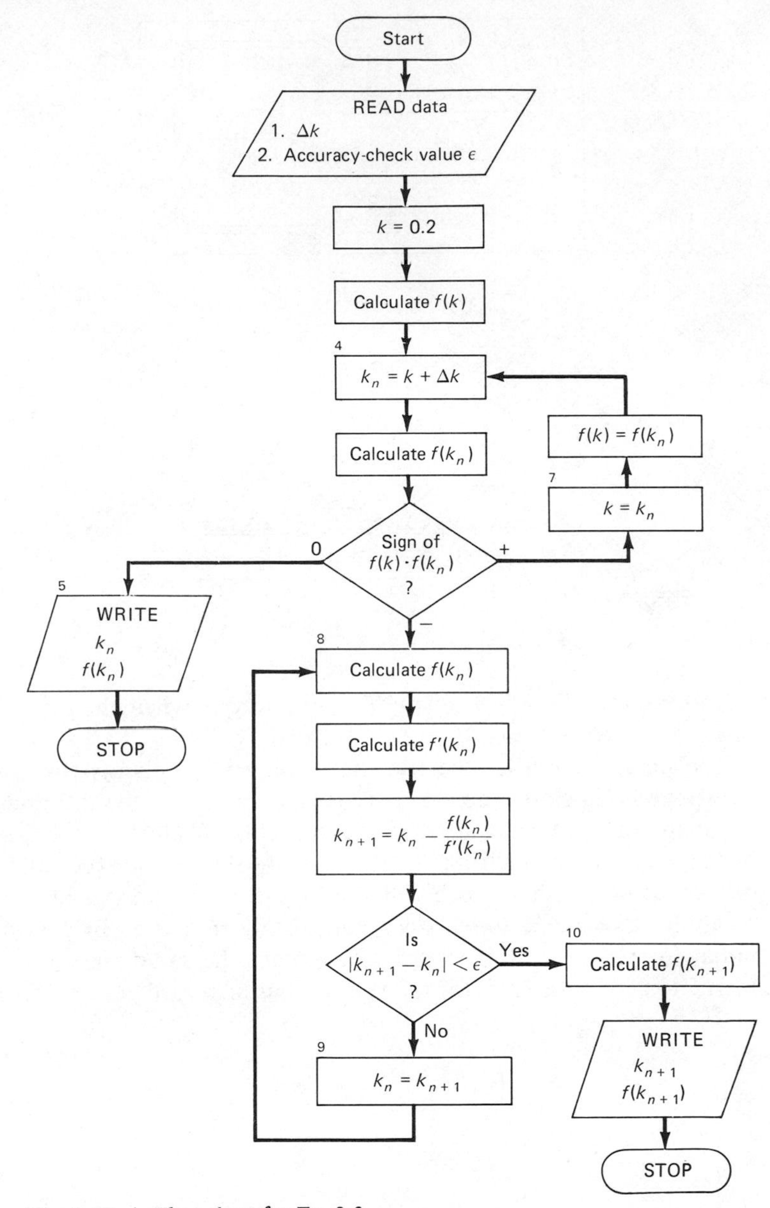

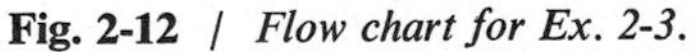

Fig. 2-12 / *Flow chart for Ex. 2-3.*

available in the new trap for killing the rat. The value of the new spring constant is to be determined.

Observing these design requirements and utilizing the fact that $\theta = 2.97$ radians when $t = 0.0191$ sec, Eq. 2-7 can be put in the following form:

$$\frac{0.625 + 0.30k}{0.625 + 3.27k} - \cos\left(\sqrt{\frac{k}{0.0006}}\,(0.0191)\right) = 0 \tag{2-9}$$

A value of $k = 0$ is obviously a root of this equation, but it yields a trivial solution with no physical significance. The solution we desire is that given by the first positive nonzero root.

The incremental-search method of Sec. 2-2 is used to find an approximation to the smallest positive root, and then the Newton-Raphson method is used to converge rapidly to a more accurate approximation. The flow chart for this approach is shown in Fig. 2-12. The FORTRAN variable names represent the quantities shown below.

Variable Name	Quantity	Data Values
K, KN, KN1	Spring constant, k	
DELK	Δk	0.2 lb-ft/radian
EPSI	Accuracy-check value, ε	0.0001 lb-ft/radian
FOFK	$f(k)$	
FOFKN	$f(k_n)$	
FOFKN1	$f(k_{n+1})$	
A	Part of $f'(k_n)$	
B	Part of $f'(k_n)$	
DFOFKN	$f'(k_n)$	

The FORTRAN program is as follows:

```
C REAL ROOT BY SEARCHING WITH CONVERGENCE BY NEWTON-RAPHSON
      REAL K,KN,KN1
      WRITE(6,2)
    2 FORMAT('1','ROOT DETERMINATION BY SEARCHING WITH
                                                   CONVERGENCE BY')
      WRITE(6,20)
   20 FORMAT(' ',15X,'NEWTON-RAPHSON METHOD'//)
      READ(5,3) DELK,EPSI
    3 FORMAT(2F10.0)
      K=0.2
      FOFK=(.625+.30*K)/(.625+3.27*K)-COS(SQRT(K/.0006)*.0191)
    4 KN=K+DELK
      FOFKN=(.625+.30*KN)/(.625+3.27*KN)
                                       -COS(SQRT(KN/.0006)*.0191)
      IF(FOFK*FOFKN) 8,5,7
    5 WRITE(6,6) KN,FOFKN
    6 FORMAT(' ','NEW SPRING CONSTANT=',F5.3,
                                   ' LB-FT/RAD',5X,'FOFKN=',
     *F6.4)
      STOP
    7 K=KN
      FOFK=FOFKN
      GO TO 4
    8 FOFKN=(.625+.30*KN)/(.625+3.27*KN)
                                       -COS(SQRT(KN/.0006)*.0191)
      A=((.625+3.27*KN)*.30-(.625+.30*KN)*3.27)/
                                             (.625+3.27*KN)**2
      B=.0191/(2.*.0006)*SQRT(.0006/KN)
                                       *SIN(SQRT(KN/.0006)*.0191)
      DFOFKN=A+B
      KN1=KN-FOFKN/DFOFKN
      IF(ABS(KN1-KN)-EPSI) 10,10,9
```

```
    9 KN=KN1
      GO TO 8
   10 FOFKN1=(.625+.30*KN1)/(.625+3.27*KN1)
                                     -COS(SQRT(KN1/.0006)*.0191)
      WRITE(6,11) KN1,FOFKN1
   11 FORMAT(' ','NEW SPRING CONSTANT=',F5.3,
                                    ' LB-FT/RAD',5X,'FOFKN1=',
     *F6.4)
      STOP
      END
```

```
ROOT DETERMINATION BY SEARCHING WITH CONVERGENCE BY
                                          NEWTON-RAPHSON METHOD

NEW SPRING CONSTANT=3.362 LB-FT/RAD         FOFKN1=-.0000
```

EXAMPLE 2-4

As another example of the Newton-Raphson method, let us consider the problem of relating the input and output crank angles of a four-bar mechanism. These angles, θ and ϕ, respectively, are measured from the line of the fixed pivots, as shown in Fig. 2-13. The moving links are a, b, and c, and the fixed link is d. Considering the links as vectors, as indicated by the arrows shown in the figure, it follows that their vector sum must always be equal to zero, since they constitute a closed polygon. Setting the sum of the x components and the y components equal to zero, respectively, yields the following equations:

$$b \cos \beta - c \cos \phi + d + a \cos \theta = 0 \tag{2-10}$$

$$b \sin \beta - c \sin \phi + a \sin \theta = 0 \tag{2-11}$$

Solving Eqs. 2-10 and 2-11 for $b \cos \beta$ and $b \sin \beta$, respectively, and squaring both sides of the resulting equations yields

$$b^2 \cos^2 \beta = (c \cos \phi - d - a \cos \theta)^2 \tag{2-12}$$

$$b^2 \sin^2 \beta = (c \sin \phi - a \sin \theta)^2 \tag{2-13}$$

Adding Eqs. 2-12 and 2-13 gives

$$b^2 = c^2 + d^2 + a^2 - 2dc \cos \phi - 2ca \cos \phi \cos \theta - 2ca \sin \phi \sin \theta + 2da \cos \theta \tag{2-14}$$

Dividing both sides of Eq. 2-14 by $2ca$ and letting

$$R_1 = d/c$$

$$R_2 = d/a$$

$$R_3 = (d^2 + a^2 - b^2 + c^2)/2ca$$

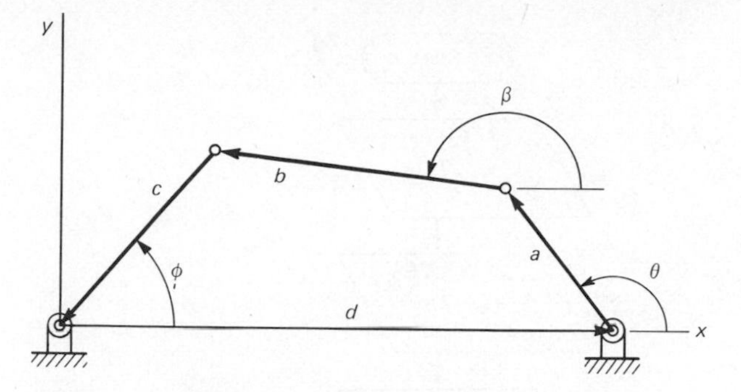

Fig. 2-13 / *Reference diagram for development of the relation between input and output crank angles of a four-bar mechanism.*

we can write Eq. 2-14 in simpler form as

$$R_1 \cos\theta - R_2 \cos\phi + R_3 - \cos(\theta - \phi) = 0 \tag{2-15}$$

This, known as *Freudenstein's equation*, is probably the most useful form of the relation between input and output crank angles of a four-bar mechanism. It should be noted that this equation cannot be readily solved directly for ϕ as a single-valued function of θ.

Let us select a particular four-bar mechanism so that R_1, R_2, and R_3 are defined, and program a computer solution of Eq. 2-15 which will yield the output angle ϕ for each corresponding input angle θ over the full range of motion of the mechanism, using successive values of θ varying by increments of 5 deg. If we select the following linkage dimensions,

a = length of input crank = 1 in.

b = length of coupler link = 2 in.

c = length of output crank = 2 in.

d = length of fixed link = 2 in.

the proportions of these dimensions define a *crank-and-lever* mechanism in which the input crank has 360 deg of motion while the output lever oscillates.

Our approach to the problem will consist of substituting successive values of θ into Eq. 2-15 and determining the root (the value of ϕ which satisfies the equation) of each resulting equation by the Newton-Raphson method. The initial approximation of ϕ required in this method may be obtained graphically, for the first value assigned to θ, by sketching the mechanism. The value of ϕ, for this first value of θ, is then calculated to the desired degree of accuracy by the steps outlined in the discussion of the Newton-Raphson method. This calculated value of ϕ_1 is then used as the

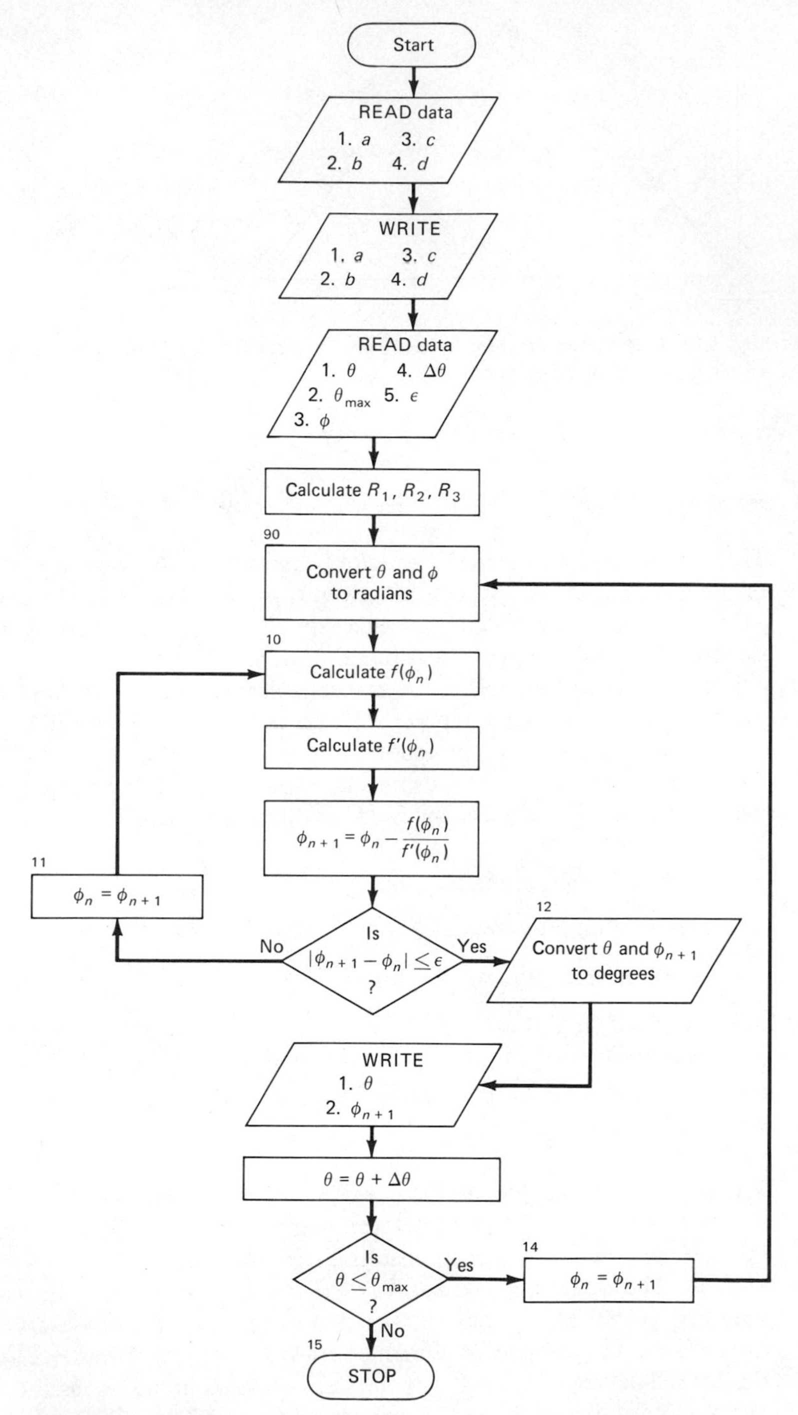

Fig. 2-14 / *Flow chart for Ex. 2-4.*

initial approximation of ϕ_2 in the calculations involved in determining the value of ϕ for the second assigned value of θ. This procedure continues until θ has passed through a 360-deg range of values. For each position of the mechanism (each value of θ), the iterative process involved in converging on the root value continues until two successive values of ϕ differ by less than the value of some prescribed small epsilon. The flow chart outlining the procedure is shown in Fig. 2-14.

The FORTRAN names assigned to the various quantities are shown below:

FORTRAN Name	Quantity	Given Value
A	Length of input crank a, in.	1
B	Length of coupler link b, in.	2
C	Length of output lever c, in.	2
D	Length of fixed link d, in.	2
DELTH	Increment of input angle $\Delta\theta$, deg	5
TH	Value of input angle θ, deg	Initial value = 0
THMX	Maximum value of input angle θ, deg	360
R1,R2,R3	Constants calculated from link lengths	
PHI	Value of output angle ϕ, deg and radians	Initial guess 41 deg
PHI1	Improved value of output angle ϕ, deg and radians	
FOFPH	$f(\phi) = R_1 \cos\theta - R_2 \cos\phi + R_3 - \cos(\theta - \phi)$	
DFOFPH	$f'(\phi) = R_2 \sin\phi - \sin(\theta - \phi)$	
EPSI	Accuracy-check value, radians	0.00001

The program for this example is as follows:

```
C FOUR BAR LINKAGE ANALYSIS PROGRAM
      WRITE(6,2)
    2 FORMAT('1',12X,'4-BAR LINKAGE INPUT-OUTPUT'/)
      WRITE(6,3)
    3 FORMAT(' ',9X,'NEWTON-RAPHSON METHOD APPLIED TO')
      WRITE(6,4)
    4 FORMAT(' ',14X,'FREUDENSTEIN EQUATION'//)
      READ(5,5) A,B,C,D
    5 FORMAT(4F10.0)
      WRITE(6,6)
    6 FORMAT(' ',18X,'LINK LENGTHS')
      WRITE(6,7) A,B,C,D
    7 FORMAT(' ','A=',F7.3,7X,'B =',F7.3,7X,'C =',
                                        F7.3,7X,'D =',F7.3//)
      WRITE(6,8)
    8 FORMAT(' ',20X,'THETA',11X,'PHI'/)
      READ(5,9) TH,THMAX,PHI,DELTH,EPSI
```

```
  9 FORMAT(5F10.0)
    R1=D/C
    R2=D/A
    R3=(D*D+A*A-B*B+C*C)/(2.*C*A)
 90 TH=TH*.01745329
    PHI=PHI*.01745329
 10 FOFPH=R1*COS(TH)-R2*COS(PHI)+R3-COS(TH-PHI)
    DFOFPH=R2*SIN(PHI)-SIN(TH-PHI)
    PHI1=PHI-FOFPH/DFOFPH
    IF(ABS(PHI1-PHI)-EPSI) 12,12,11
 11 PHI=PHI1
    GO TO 10
 12 TH=TH/.01745329
    PHI1=PHI1/.01745329
    WRITE(6,13) TH,PHI1
 13 FORMAT(' ',21X,F4.0,10X,F5.1)
    TH=TH+DELTH
    IF(TH-THMAX) 14,14,15
 14 PHI=PHI1
    GO TO 90
 15 STOP
    END
```

The output results are as follows (for simplicity of printing, the format of the output shown does not completely conform to the FORMAT statements of the program):

```
4-BAR LINKAGE INPUT-OUTPUT

NEWTON-RAPHSON METHOD APPLIED TO
FREUDENSTEIN EQUATION

LINK LENGTHS
A= 1.000          B= 2.000          C= 2.000          D= 2.000

THETA          PHI          THETA          PHI

  0.           41.4          185.           70.4
  5.           43.1          190.           65.4
 10.           45.0          195.           60.5
 15.           46.9          200.           56.0
 20.           48.9          205.           51.8
 25.           51.0          210.           48.2
 30.           53.2          215.           44.9
 35.           55.5          220.           42.1
 40.           57.9          225.           39.7
 45.           60.3          230.           37.6
 50.           62.7          235.           35.9
 55.           65.2          240.           34.3
 60.           67.7          245.           33.1
 65.           70.2          250.           32.0
 70.           72.7          255.           31.1
```

```
 75.        75.3            260.       30.4
 80.        77.7            265.       29.9
 85.        80.2            270.       29.4
 90.        82.6            275.       29.2
 95.        84.9            280.       29.0
100.        87.1            285.       29.0
105.        89.2            290.       29.0
110.        91.1            295.       29.2
115.        92.8            300.       29.5
120.        94.3            305.       29.9
125.        95.6            310.       30.4
130.        96.5            315.       31.0
135.        97.1            320.       31.7
140.        97.2            325.       32.5
145.        96.7            330.       33.4
150.        95.7            335.       34.5
155.        94.1            340.       35.6
160.        91.7            345.       36.9
165.        88.6            350.       38.3
170.        84.8            355.       39.8
175.        80.4            360.       41.4
180.        75.5
```

2-7 / NEWTON'S SECOND-ORDER METHOD

When it is necessary to determine very accurately the value of a root of an equation, Newton's second-order method has the advantage of converging rapidly to a solution, and an extremely close approximation of the value of the root may be obtained with a minimum of calculations. However, this method is limited, in a practical sense, to use on equations which have fairly simple higher-order derivatives (second order, at least), since the time consumed in obtaining and programming involved derivatives outweighs the advantage of rapid convergence. This will be understood more clearly after the following discussion.

Consider, once again, an equation of the form

$$f(x) = 0$$

A graph of the function plotted against x is shown in Fig. 2-15. Suppose that an approximate value of the root, $x = x_n$, has been determined by some method such as a graphical approximation. Expanding $f(x)$ in a Taylor series about $x = x_n$ gives

$$f(x_{n+1}) = f(x_n) + f'(x_n)(\Delta x) + \frac{f''(x_n)(\Delta x)^2}{2!} + \frac{f'''(x_n)(\Delta x)^3}{3!} + \cdots \quad (2\text{-}16)$$

If Δx were the particular increment of x which, added to x_n, would result in a zero value for the series, then the quantity $(x_n + \Delta x)$ would be the

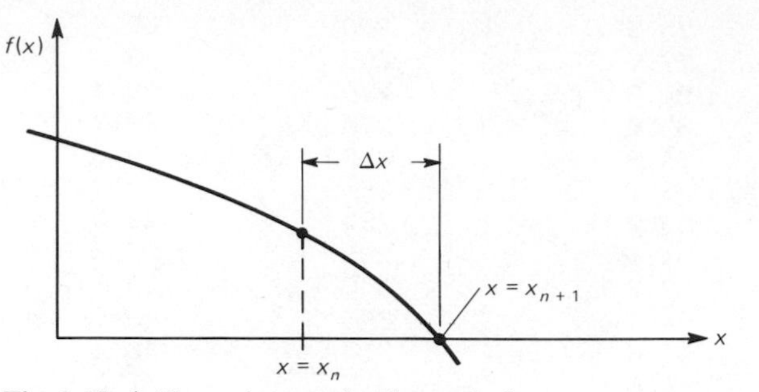

Fig. 2-15 / *Newton's second-order method.*

exact root. Since we are interested in a practical means of determining a value of Δx which will make the series sum to zero, let us set the right side of Eq. 2-16 equal to zero, using just three terms of the series. We obtain the approximate equality

$$f(x_n) + \Delta x\left[f'(x_n) + \frac{f''(x_n)(\Delta x)}{2}\right] = 0 \tag{2-17}$$

A value of Δx determined from Eq. 2-17, when added to x_n, will not yield the exact value of the root, since only three terms of an infinite series were utilized in evaluating Δx. However, a much closer approximation to the value of the root will be obtained. Inspection reveals that Eq. 2-17 is a quadratic in Δx and can be solved as such. However, such a solution involves the problem of determining which root is the correct one. This can be avoided, at the cost of slightly slower convergence, by substituting the following expression (from Eq. 2-5) for the $\Delta x(\Delta x = x_{n+1} - x_n)$ within the brackets of Eq. 2-17,

$$\Delta x = -\frac{f(x_n)}{f'(x_n)}$$

and writing Eq. 2-17 in the new form

$$f(x_n) + \Delta x\left[f'(x_n) - \frac{f''(x_n)f(x_n)}{2f'(x_n)}\right] = 0 \tag{2-18}$$

Solving Eq. 2-18 for Δx, we obtain

$$\Delta x = -\left[\frac{f(x_n)}{f'(x_n) - \left(\dfrac{f''(x_n)f(x_n)}{2f'(x_n)}\right)}\right] \tag{2-19}$$

Since $\Delta x = x_{n+1} - x_n$, we may rewrite Eq. 2-19 as

$$x_{n+1} = x_n - \left[\frac{f(x_n)}{f'(x_n) - \left(\dfrac{f''(x_n)f(x_n)}{2f'(x_n)}\right)}\right] \tag{2-20}$$

which may be used to obtain successively closer approximations to a root by successive applications.

EXAMPLE 2-5

Let us determine the positive real root of the equation

$$2\cos x - e^x = 0 \tag{2-21}$$

Inspection of Eq. 2-21 reveals that it has an infinite number of negative real roots but only one positive real root. This can best be seen by sketching separately the graphs of $y_1 = 2\cos x$ and $y_2 = e^x$, as shown in Fig. 2-16, and noting the intersection of the curves. The positive real root lies between $x = 0.5$ and $x = 0.6$. To demonstrate the rapidity of convergence of this method, even when the initial approximation is not too close, we will take $x = 0.4$ as a first approximation and apply Eq. 2-20 successively until we obtain a solution of the desired accuracy. The results, calculated to three decimal places, are shown herewith.

x	$f(x) =$ $2\cos x - e^x$	$f'(x) =$ $-2\sin x - e^x$	$f''(x) =$ $-2\cos x - e^x$	FR*
$x_n = 0.4$	0.350	−2.270	−3.334	−0.139
$x_{n+1} = 0.539$	0.003	−2.741	−3.433	−0.001
$x_{n+2} = 0.540$	0.000			

* FR is the bracketed expression in Eq. 2-20.

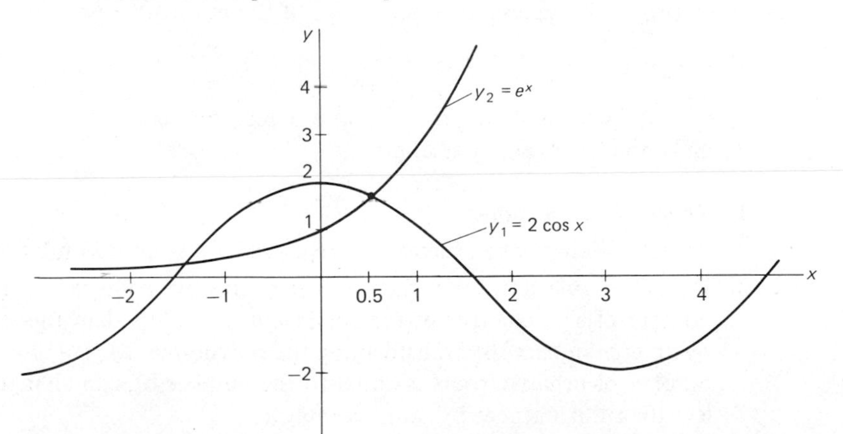

Fig. 2-16 / *Approximate location of the positive real root of the transcendental equation* $2\cos x - e^x = 0$.

2-8 / ROOTS OF POLYNOMIALS

The analysis of certain types of physical systems involve the solution of polynomial equations of the form

$$x^n + a_1x^{n-1} + a_2x^{n-2} + a_3x^{n-3} + \cdots + a_n = 0 \tag{2-22}$$

where $a_1, a_2, \ldots, a_n$ are real coefficients. For example, the frequency equation of a lumped-mass system with n degrees of freedom is an nth-degree polynomial. The roots of second-degree polynomials are readily determined by the use of the quadratic formula. Various formulas exist for determining the roots of third- and fourth-degree polynomials, but, in general, they are difficult to apply.

There are a number of methods available for finding the roots of polynomials. Unfortunately, each method has certain limitations or disadvantages. For example, the Newton-Raphson method, previously discussed in Sec. 2-6, may be used to find real and distinct roots but requires the use of complex arithmetic for finding complex roots. The quotient-difference method may be used to find complex roots, but it is slow to converge and as a result the method is not very efficient.

In Secs. 2-9 and 2-10, we will discuss Graeffe's root-squaring method and Bairstow's iterative method, which are two of the more commonly used methods today. Both these are capable of finding both real and complex roots of polynomials. The procedure of Graeffe's root-squaring method is quite simple, and the roots of a fifth- or sixth-degree polynomial can be determined in a relatively short time with today's electronic pocket calculators. Bairstow's method is better suited for use on the digital computer, but convergence does not always occur and more than one trial may be necessary to obtain the roots.

To establish a background for the discussions in Secs. 2-9 and 2-10, let us review the general characteristics of the nth-degree polynomial

$$x^n + a_1x^{n-1} + a_2x^{n-2} + a_3x^{n-3} + \cdots + a_n = 0$$

where the coefficients $a_1, a_2, \ldots, a_n$ are real numbers. For such polynomials, the following facts apply:

1. There will be n roots.
2. There will always be at least one real root if n is an odd integer.
3. Descartes' rule states that the number of positive roots is equal to the number of sign changes of the coefficients *or* is less than this number by an *even* integer. By transforming the polynomial to $f(-x) = 0$, the number of negative roots is equal to the number of sign changes *or* is less than this number by an *even* integer.
4. It is possible that equal roots exist.
5. When complex roots exist, they occur in conjugate pairs.

Synthetic Division

It is recalled from algebra that the division of a polynomial, $f(x)$, by $(x - r)$ can be carried out rapidly by *synthetic division*. If r is not a root, then the remainder resulting from the synthetic division is the value of the polynomial for $x = r$. If r is a root of the polynomial, then the remainder is, of course, zero since $f(r) = 0$.

The value of the derivative, $f'(x)$, of a polynomial for $x = r$ is equal to the remainder obtained by a second synthetic division of the result obtained in the first synthetic division. Therefore, the use of synthetic division facilitates the evaluation of $f(x_n)$ and $f'(x_n)$ when using the Newton-Raphson method (Sec. 2-6) to find the roots of a polynomial.

Since synthetic division is discussed in most college algebra books, we will merely review the process by way of the following example.

EXAMPLE 2-6

Divide the polynomial below by the expression $(x + 4)$ and determine the values of $f(x)$ and $f'(x)$ for $x = -4$ using synthetic division.

$$f(x) = x^5 - 3x^4 - 10x^3 + 10x^2 + 44x + 48$$

In dividing the polynomial by $(x + 4)$, the remainder is the value of the function at $x = -4$. The synthetic division process for dividing the polynomial by $(x + 4)$ is carried out as indicated in the following table.

−4	1	−3	−10	10	44	48
		−4	28	−72	248	−1168
	1	−7	18	−62	292	(−1120)

The first row consists of the coefficients of the polynomial. The second row is generated from left to right by multiplying −4 times the elements generated in the third row, the latter being simply the sum of the first- and second-row elements. The value of the polynomial for $x = -4$ is the remainder (−1120). Thus

$$f(-4) = -1120$$

It is easy to verify by long division that the numbers in the third row are actually the coefficients of the fourth-degree polynomial, with a remainder of −1120, resulting from division of the fifth-degree polynomial by $(x + 4)$. That is,

Quotient: $x^4 - 7x^3 + 18x^2 - 62x + 292$; with remainder of −1120

To obtain the value of the derivative, $f'(-4)$, of the fifth-degree polynomial for $x = -4$, a second synthetic division is applied to the third row (quotient) obtained in the first synthetic division process to evaluate $f(-4)$. Hence

-4	1	-7	18	-62	292
		-4	44	-248	1240
	1	-11	62	-310	(1532)

The remainder of 1532 is the value of $f'(-4)$ in which

$$f'(x) = 5x^4 - 12x^3 - 30x^2 + 20x + 44$$

2-9 / GRAEFFE'S ROOT-SQUARING METHOD

Graeffe's method for determining the roots of a polynomial is based upon the formation of a new polynomial, of the same degree as the original polynomial but having roots which are some large *even* power m of the roots of the original polynomial. The new polynomial will have relationships between its coefficients and its roots which may be solved to obtain the values of the roots. The roots of the original polynomial are then obtained by utilizing the exponential relationships existing between the roots of the *original* polynomial and the *derived* polynomial.

In developing Graeffe's method, let us first discuss a root-squaring procedure for obtaining a derived polynomial whose roots are the *negative* of some *even* power of the roots of the original polynomial. For the sake of simplicity, let us at present consider the third-degree polynomial

$$f(x) = 0 = x^3 + a_1x^2 + a_2x + a_3 \tag{2-23}$$

having the roots x_1, x_2, and x_3. A polynomial having the roots $-x_1$, $-x_2$, and $-x_3$ would be

$$f(-x) = 0 = -x^3 + a_1x^2 - a_2x + a_3 \tag{2-24}$$

Multiplying Eq. 2-23 by Eq. 2-24 yields

$$f(x)f(-x) = 0 = -x^6 + (a_1^2 - 2a_2)x^4 + (-a_2^2 + 2a_1a_3)x^2 + a_3^2 \tag{2-25}$$

Letting $y = -x^2$, Eq. 2-25 may be written as

$$y^3 + (a_1^2 - 2a_2)y^2 + (a_2^2 - 2a_1a_3)y + a_3^2 = 0 \tag{2-26}$$

Equation 2-26 is a *derived* polynomial of the same degree as the *original* polynomial of Eq. 2-23, and it has roots which are the *negative* of the squares of the roots of the original polynomial ($-x_1^2$, $-x_2^2$, and $-x_3^2$). If this root-squaring procedure is again applied, beginning with the derived polynomial of Eq. 2-26, another third-degree polynomial will be obtained whose roots will be the negative of the squares of the roots of the derived polynomial of Eq. 2-26 and, at the same time, the negative of the *fourth powers* of the roots of the original polynomial. Successive applications of the root-squaring process will thus yield successive derived polynomials having roots which are the negative of successively higher even powers m of the roots of the original polynomial.

If the root-squaring process is applied to an nth-degree polynomial, the derived polynomials will have the general form of

$$\left.\begin{aligned} & y^n + (a_1^2 - 2a_2)y^{n-1} + (a_2^2 - 2a_1a_3 + 2a_4)y^{n-2} \\ & \quad + (a_3^2 - 2a_2a_4 + 2a_1a_5 - 2a_6)y^{n-3} + \cdots + a_n^2 = 0 \quad \text{(a)} \\ & \text{or} \\ & y^n + \begin{Bmatrix} a_1^2 \\ -2a_2 \end{Bmatrix} y^{n-1} + \begin{Bmatrix} a_2^2 \\ -2a_1a_3 \\ +2a_4 \end{Bmatrix} y^{n-2} + \begin{Bmatrix} a_3^2 \\ -2a_2a_4 \\ +2a_1a_5 \\ -2a_6 \end{Bmatrix} y^{n-3} \\ & \qquad + \cdots + a_n^2 = 0 \quad \text{(b)} \end{aligned}\right\} \quad (2\text{-}27)$$

Inspection of Eq. 2-27 reveals that each coefficient of the polynomial being formed consists of the sum of the square of the corresponding coefficient in the preceding polynomial and twice each product which can be formed by multiplying coefficients that are symmetrically located on each side of the corresponding coefficient in the preceding polynomial. The signs of the product terms are alternatively negative and positive.

Thus, when the coefficients of a given polynomial are substituted into Eq. 2-27, a new polynomial, having new coefficients and roots which are the negatives of the squares of the original polynomial, is obtained. If the coefficients of the derived polynomial are then, in turn, substituted into Eq. 2-27, still another derived polynomial, having roots which are the negatives of the squares of the roots of the polynomial from which it was derived, is obtained. These roots are, at the same time, the negatives of the fourth powers of the roots of the original polynomial, and so on. Each successive derived polynomial will have roots which are further and further apart in magnitude. This may be seen, for example, by considering an original polynomial with the roots 1, -2, and 4. After just three applications of Eq. 2-27, the last derived polynomial would have roots of -1, -256, and $-65{,}536$. The root-squaring process is terminated, in Graeffe's method, when the roots of the last derived polynomial are very widely separated. The separation necessary can be determined from the

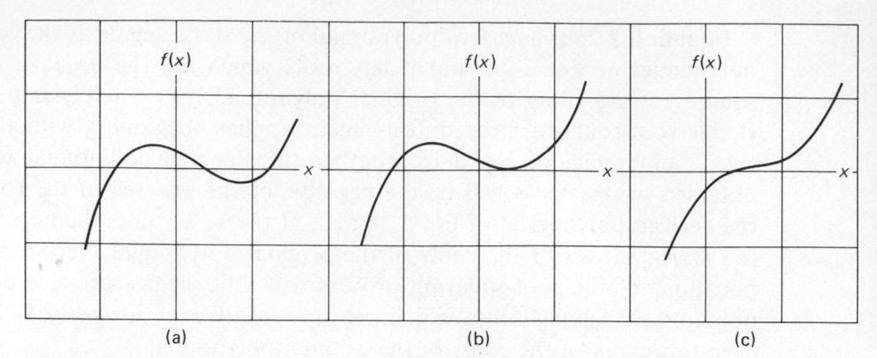

Fig. 2-17 / *Examples of third-degree polynomials. (a) Real and distinct roots. (b) Real and equal roots. (c) Real and complex roots.*

relationships observed between the values of corresponding coefficients in successively derived polynomials obtained in the root-squaring process. Such relationships, for various types of roots, will be studied later in this section.

Having now discussed the root-squaring process, which is an integral part of Graeffe's method, let us go on to the method itself. Three kinds of polynomials will be considered individually: (1) those having only real and distinct roots, (2) those whose real roots include equal roots, and (3) those whose roots include conjugate pairs of complex roots. Examples of third-degree polynomials having each type of root are shown in Fig. 2-17.

Real and Distinct Roots

Consider again the general third-degree polynomial

$$x^3 + a_1x^2 + a_2x + a_3 = 0 \tag{2-28}$$

which has the roots x_1, x_2, and x_3. Equation 2-28 may be expressed in factored form as

$$f(x) = 0 = (x - x_1)(x - x_2)(x - x_3)$$

Carrying out the multiplication indicated in the latter expression yields

$$x^3 - (x_1 + x_2 + x_3)x^2 + (x_1x_2 + x_1x_3 + x_2x_3)x - x_1x_2x_3 = 0 \tag{2-29}$$

Comparing Eqs. 2-28 and 2-29, we obtain the following relationships between the coefficients and the roots of the polynomial:

$$\begin{aligned} a_1 &= -(x_1 + x_2 + x_3) \\ a_2 &= x_1x_2 + x_1x_3 + x_2x_3 \\ a_3 &= -x_1x_2x_3 \end{aligned} \tag{2-30}$$

At this point, let us introduce the *Enke roots* of Eq. 2-28. The Enke roots of a polynomial are the *negatives* of the roots of the polynomial. Thus, letting r be the designation for an Enke root, we may write

$$x_1 = -r_1$$
$$x_2 = -r_2$$
$$\cdots$$
$$x_n = -r_n$$

If these relationships are substituted into Eq. 2-30, the following relationships are obtained between the coefficients of the polynomial and its Enke roots:

$$\begin{aligned} a_1 &= r_1 + r_2 + r_3 \\ a_2 &= r_1r_2 + r_1r_3 + r_2r_3 \\ a_3 &= r_1r_2r_3 \end{aligned} \tag{2-31}$$

Remembering that, in the root-squaring process, the derived polynomials have roots which are the *negatives* of even powers of the roots of the original polynomial $(-x_1^m, -x_2^m, \ldots, -x_n^m)$, we see that the Enke roots of the derived polynomials will be $r_1^m, r_2^m, \ldots, r_n^m$. Thus the coefficients b_i, determined for successive derived polynomials in the root-squaring process, will always be positive (with one significant exception which will be discussed when complex roots are considered).

Let us return now to a consideration of the relationships given by Eq. 2-31. This set of simultaneous nonlinear algebraic equations would be difficult to solve. To avoid working with this set of equations, let us now consider the application of the root-squaring process in Graeffe's method. Applying the process to the original polynomial of Eq. 2-28 by repeated applications of Eq. 2-27, we determine a final derived polynomial

$$y^3 + b_1y^2 + b_2y + b_3 = 0 \tag{2-32}$$

having the Enke roots r_1^m, r_2^m, and r_3^m. As pointed out, the derived polynomials have the same relationships between their roots and coefficients as those given for the original polynomial in Eq. 2-31, so that

$$\begin{aligned} b_1 &= r_1^m + r_2^m + r_3^m \\ b_2 &= r_1^mr_2^m + r_1^mr_3^m + r_2^mr_3^m \\ b_3 &= r_1^mr_2^mr_3^m \end{aligned} \tag{2-33}$$

For purposes of derivation, the *absolute* values of the roots of the original polynomial are considered to be related so that the Enke roots will have the relationship $|r_1| > |r_2| > |r_3|$. After sufficient applications of the root-squaring process, the Enke roots of the last derived polynomial will then be related by $r_1^m \gg r_2^m \gg r_3^m$. With the roots thus widely separated in magnitude, only the dominant terms of Eq. 2-33 need be retained, and we may write

$$\begin{aligned} b_1 &\cong r_1^m \\ b_2 &\cong r_1^m r_2^m \\ b_3 &= r_1^m r_2^m r_3^m \end{aligned} \tag{2-34}$$

where the approximations will be very close with m sufficiently large.

The expressions of Eq. 2-34 are easily solved for the Enke roots r_i^m. The roots of the original polynomial may then be obtained as

$$x_i = \pm(r_i^m)^{1/m}$$

The correct signs of the roots must be determined by substituting each possibility into the original polynomial and selecting the respective signs which allow the equation to be satisfied.

To determine the point at which the root-squaring process may be terminated, again consider Eq. 2-34. As stated earlier, these relationships between the coefficients and the roots of a derived polynomial exist only when the root-squaring process has been applied a sufficient number of times (when m has become large enough to develop dominant terms). The coefficients b_i shown are then the coefficients of the last derived polynomial which it is necessary to obtain. To determine when this point has been reached, consider *one more* application of the root-squaring process. Using the coefficients of Eq. 2-34 and applying Eq. 2-27, the coefficients b_i' of the new polynomial will be

$$\begin{aligned} b_1' &= r_1^{2m} - 2r_1^m r_2^m \\ b_2' &= r_1^{2m} r_2^{2m} - 2r_1^{2m} r_2^m r_3^m \\ b_3' &= r_1^{2m} r_2^{2m} r_3^{2m} \end{aligned} \tag{2-35}$$

Considering the dominant terms of each of the new coefficients,

$$\begin{aligned} b_1' &\cong r_1^{2m} \\ b_2' &\cong r_1^{2m} r_2^{2m} \\ b_3' &= r_1^{2m} r_2^{2m} r_3^{2m} \end{aligned} \tag{2-36}$$

Comparing Eqs. 2-34 and 2-36, it can be seen that each new coefficient is essentially the square of the corresponding preceding coefficient. Thus when m is sufficiently large to develop dominant terms, consecutive applications of the root-squaring process will yield coefficients with values such that *each* new coefficient obtained is essentially the square of the corresponding preceding coefficient, and the process may then be terminated. The relationship between consecutive values of a coefficient, just described, is known as a *regular* relationship and will be referred to as such subsequently.

The coefficient relationships just discussed provide not only the criteria for stopping the root-squaring process but also the information necessary to determine the type of roots to expect when solving a given polynomial. Sometimes the physical problem, from which the polynomial stems, will indicate the kind of roots contained in the root system of the polynomial. However, when little is known about what sorts of roots to expect from the polynomial being solved, the numerical relationships existing between the consecutive coefficients obtained in the root-squaring process will indicate what type, or types, of roots comprise the root system of the polynomial. In fact, as will be seen later, the relationships between consecutive coefficient values must be analyzed to determine each kind of root present before the root values can be determined. For example, if *all* the coefficients assume a regular relationship at the end of the root-squaring process, the polynomial being solved will be found to have all real and distinct roots, as was indicated by the relationships between Eqs. 2-34 and 2-36. However, when the root system of the polynomial being solved contains equal roots, complex roots, or combinations of all three types of roots, not all the coefficient relationships will be regular. The relationships which do exist will be discussed when the solutions of polynomials having such roots are taken up later.

In a manner similar to that detailed earlier for the specific case of a third-degree polynomial, it can be demonstrated that, for an nth-degree polynomial, the coefficients and Enke roots of the derived polynomials are related by[2]

$$
\begin{aligned}
b_1 &= r_1^m + r_2^m + r_3^m + \cdots + r_n^m && \text{(a)} \\
b_2 &= (r_1r_2)^m + (r_1r_3)^m + \cdots + (r_1r_n)^m + (r_2r_3)^m \\
&\quad + (r_2r_4)^m + \cdots + (r_2r_n)^m + \cdots + (r_{n-1}r_n)^m && \text{(b)} \\
b_3 &= (r_1r_2r_3)^m + (r_1r_2r_4)^m + \cdots + (r_1r_2r_n)^m \\
&\quad + (r_2r_3r_4)^m + (r_2r_3r_5)^m + \cdots + (r_2r_3r_n)^m && \text{(2-37)} \\
&\quad + \cdots + (r_{n-2}r_{n-1}r_n)^m && \text{(c)} \\
&\vdots \\
b_n &= (r_1r_2r_3 \cdots r_n)^m && \text{(d)}
\end{aligned}
$$

[2] (a) Sum of all Enke roots; (b) all possible combinations of products of two Enke roots; (c) all possible combinations of products of three Enke roots; . . . (d) product of all Enke roots.

After repeated applications of the root-squaring process, so that m is large enough to provide dominant terms, we may write (from Eq. 2-37), for polynomials having only real and distinct roots,

$$\begin{aligned} b_1 &\cong r_1^m \\ b_2 &\cong (r_1 r_2)^m \\ b_3 &\cong (r_1 r_2 r_3)^m \\ b_4 &\cong (r_1 r_2 r_3 r_4)^m \\ &\cdots \\ b_n &= (r_1 r_2 r_3 r_4 \cdots r_n)^m \end{aligned} \tag{2-38}$$

Inspection of Eqs. 2-37 and 2-38 indicates that the coefficients of the derived polynomials will always be positive, since m is always an even power. (A significant exception occurs when complex roots are present, as will be seen later.)

From Eq. 2-38 we may write

$$\begin{aligned} r_1^m &\cong b_1 & & & r_1 &\cong (b_1)^{1/m} \\ r_2^m &\cong \frac{b_2}{b_1} & & & r_2 &\cong \left(\frac{b_2}{b_1}\right)^{1/m} \\ r_3^m &\cong \frac{b_3}{b_2} & &\text{or} & r_3 &\cong \left(\frac{b_3}{b_2}\right)^{1/m} \\ &\cdots & & & &\cdots \\ r_n^m &\cong \frac{b_n}{b_{n-1}} & & & r_n &\cong \left(\frac{b_n}{b_{n-1}}\right)^{1/m} \end{aligned} \tag{2-39}$$

The form of Eq. 2-39 is well suited for applying Graeffe's method on a digital computer, and the calculations are easily performed on an electronic pocket calculator as well.

EXAMPLE 2-7

In determining the roots of a polynomial having only real and distinct roots, let us consider the fourth-degree polynomial

$$x^4 - 6x^3 + 8x^2 + 2x - 1 = 0 \tag{2-40}$$

To obtain a feeling for the root-squaring process, before reducing it to a computer routine, let us apply the procedure manually, utilizing an electronic pocket calculator for the computations. Equation 2-27 is applied to determine the coefficients of the successively derived polynomials.

	$a_0^2 = 1$	a_1^2 $-2a_2$	a_2^2 $-2a_1a_3$ $+2a_4$	a_3^2 $-2a_2a_4$	a_4^2
m	x^4	a_1x^3	a_2x^2	a_3x	a_4x^0
1	**1**	**−6**	**8**	**2**	**−1**
	1	36 −16	64 24 −2	4 16	1
2	**1**	**20**	**86**	**20**	**1**
	1	400 −172	7396 −800 2	400 −172	1
4	**1**	**228**	**6598**	**228**	**1**
	1	$5.20(10)^4$ $-1.32(10)^4$	$4353(10)^4$ $-10(10)^4$ 2	$5.20(10)^4$ $-1.32(10)^4$	1
8	**1**	**$3.88(10)^4$**	**$4343(10)^4$**	**$3.88(10)^4$**	**1**
	1	$15.05(10)^8$ $0.87(10)^8$	$18.86(10)^{14}$ ---- ----	$15.05(10)^8$ $0.87(10)^8$	1
16	**1**	**$14.18(10)^8$**	**$18.86(10)^{14}$**	**$14.18(10)^8$**	**1**
	1	$2.01(10)^{18}$ ---- ----	$3.56(10)^{30}$ ---- ----	$2.01(10)^{18}$	1
32	**1**	**$2.01(10)^{18}$**	**$3.56(10)^{30}$**	**$2.01(10)^{18}$**	**1**

Fig. 2-18 / *Root-squaring table for Ex. 2-7.*

To establish an orderly procedure, a table such as the one in Fig. 2-18 is prepared. At the top of the table the pertinent terms of Eq. 2-27, for the given polynomial, are listed for easy reference. The coefficients of the original polynomial are shown in the first row of the table under the appropriate powers of x. These coefficients are then used in Eq. 2-27 to determine the coefficients of a new polynomial having roots which are the negatives of the squares of the roots of the original polynomial ($m = 2$). These new coefficients are then used, in turn, to determine the coefficients of yet another derived polynomial ($m = 4$), and so on. The numerical values of the squared and product terms used to determine the new coefficients are displayed in that order in the respective columns immediately above the values of the new coefficients.

As shown in Fig. 2-18, the root-squaring process was terminated when $m = 32$, since, at this point, each new coefficient obtained was essentially the square of the corresponding preceding coefficient. Since *each* coefficient has a regular relationship, the root-squaring process indicates that all the roots are real and distinct. The coefficients of the last derived polynomial are thus

$$b_1 = 2.01(10)^{18}$$
$$b_2 = 3.56(10)^{30}$$
$$b_3 = 2.01(10)^{18}$$
$$b_4 = 1$$

From Eq. 2-39 we obtain

$$r_1 = (b_1)^{1/m} = [2.01(10)^{18}]^{1/32} = \pm 3.732$$

$$r_2 = \left(\frac{b_2}{b_1}\right)^{1/m} = \left[\frac{3.56(10)^{30}}{2.01(10)^{18}}\right]^{1/32} = \pm 2.414$$

$$r_3 = \left(\frac{b_3}{b_2}\right)^{1/m} = \left[\frac{2.01(10)^{18}}{3.56(10)^{30}}\right]^{1/32} = \pm 0.414$$

$$r_4 = \left(\frac{b_4}{b_3}\right)^{1/m} = \left[\frac{1}{2.01(10)^{18}}\right]^{1/32} = \pm 0.268$$

Using the plus and minus values of each root for synthetic divisions of Eq. 2-40, the roots are found to be

$$x_1 = 3.732$$
$$x_2 = 2.414$$
$$x_3 = -0.414$$
$$x_4 = 0.268$$

Descartes' rule, when applied to Eq. 2-40, predicts one or three positive roots, owing to the three sign changes in the equation. Note that three positive roots were obtained.

Real and Equal Roots

In the preceding discussion it was found that, when all the roots of a polynomial are real and distinct, *all* the coefficients determined in the root-squaring process assume regular relationships. However, when two or more real and equal roots are part of the root system of a polynomial, the coefficient relationships obtained in the root-squaring process are not all regular. (Equal roots, in Graeffe's method, are considered to be roots which are equal in *absolute* value.)

To find the kind of relationships that do exist, let us again consider a third-degree polynomial—this one having all real roots, two of which are equal. The general relationships between the coefficients and the Enke roots of third-degree derived polynomials are given by Eq. 2-33 and are repeated here for convenience.

$$\begin{aligned} b_1 &= r_1^m + r_2^m + r_3^m \\ b_2 &= r_1^m r_2^m + r_1^m r_3^m + r_2^m r_3^m \\ b_3 &= r_1^m r_2^m r_3^m \end{aligned} \tag{2-41}$$

If $|r_2| = |r_3|$ and the Enke roots are defined as being related by $|r_1| > |r_2| = |r_3|$, Eq. 2-41 becomes

$$\begin{aligned} b_1 &= r_1^m + 2r_2^m \\ b_2 &= 2r_1^m r_2^m + r_2^{2m} \\ b_3 &= r_1^m r_2^{2m} \end{aligned} \tag{2-42}$$

When the root-squaring process has been repeated enough times so that dominant terms exist ($r_1^m \gg r_2^m = r_3^m$), a consideration of the dominant terms of Eq. 2-42 yields the following relationships:

$$\begin{aligned} b_1 &\cong r_1^m \\ b_2 &\cong 2r_1^m r_2^m \\ b_3 &= r_1^m r_2^{2m} \end{aligned} \tag{2-43}$$

where the coefficients b_i are the coefficients of the last derived polynomial which it is necessary to obtain.

If we now consider *one more* application of the root-squaring process, using the coefficients of Eq. 2-43 and applying Eq. 2-27, we obtain the new coefficients

$$\begin{aligned} b_1' &= r_1^{2m} - 4r_1^m r_2^m \\ b_2' &= 4r_1^{2m} r_2^{2m} - 2r_1^{2m} r_2^{2m} \\ b_3' &= r_1^{2m} r_2^{4m} \end{aligned} \tag{2-44}$$

Considering only the dominant terms, we obtain

$$
\begin{aligned}
b'_1 &= r_1^{2m} \\
b'_2 &= 2r_1^{2m}r_2^{2m} \\
b'_3 &= r_1^{2m}r_2^{4m}
\end{aligned}
\tag{2-45}
$$

Comparing Eqs. 2-43 and 2-45, it can be seen that, when the root-squaring process terminates, two of the new coefficients (b'_1 and b'_3) are essentially the square of the corresponding preceding coefficients b_1 and b_3, whereas the new coefficient b'_2 is essentially *one-half* the square of b_2.

Thus when the root-squaring process terminates, the presence of a coefficient whose final value is essentially one-half the square of its preceding value indicates that the polynomial has two equal roots. Note that, with $|r_2| = |r_3|$ as initially assumed, the coefficient b_2 assumes the *half-squared* relationship just discussed.

If it had been assumed initially that $|r_1| = |r_2|$ in the preceding development, with $|r_1| = |r_2| > |r_3|$, the following relationships would have been obtained instead of those shown in Eq. 2-43:

$$
\begin{aligned}
b_1 &= 2r_1^{m} \\
b_2 &\cong r_1^{2m} \\
b_3 &= r_1^{2m}r_3^{m}
\end{aligned}
\tag{2-46}
$$

In this case, one more application of the root-squaring process would show that the coefficient b_1 is the one which has a final value essentially equal to one-half the square of its preceding value. Thus, in general, if the coefficient b_i has the half-squared relationship, the roots r_i and r_{i+1} will be the equal roots.

If a polynomial has three real and equal roots, the procedure used for the case of two equal roots will show that two of the coefficients will approach final values such that the last derived coefficients are *one-third* of the square of the preceding ones, and that, if b_i and b_{i+1} are these coefficients, the roots r_i, r_{i+1}, and r_{i+2} will be the equal roots.

EXAMPLE 2-8

To solve a polynomial which has a pair of equal roots, let us consider the third-degree polynomial

$$
x^3 + 3x^2 - 4 = 0 \tag{2-47}
$$

Utilizing Eq. 2-27 in the root-squaring process, the coefficient table of Fig. 2-19 is obtained. For convenience, the pertinent terms of Eq. 2-27 are listed at the top of the table. The first row shows the coefficients of the original polynomial, and the last row gives the coefficient values of the last derived polynomial with $m = 32$, at which point the root-squaring process was terminated.

Inspection of the table reveals that the coefficient b_1 terminates in a half-squared relationship, the last value of $85.85(10)^8$ being essentially one-half of the square of the preceding value of $13.1(10)^4$. This indicates that the roots r_1 and r_2 are the equal roots. The other two coefficients terminate in a regular relationship, indicating that the third root of the polynomial is real and distinct. Knowing that $|r_1| = |r_2|$ and that r_3 is real, we refer to the general coefficient-root relationships, given by Eq. 2-37, and substitute the fact that $r_1 = r_2$ into the given expressions. Considering only the dominant terms of the resulting expressions (by convention, we assume that $r_1^m = r_2^m \gg r_3^m$), we find that

$$\begin{aligned} b_1 &\cong 2r_1^m \\ b_2 &\cong r_1^{2m} \\ b_3 &\cong r_1^{2m} r_3^m \end{aligned} \qquad (2\text{-}48)$$

Utilizing Eq. 2-48 and the coefficient values shown in Fig. 2-19, we find that

$$\begin{aligned} b_1 &= 2(r_1)^{32} &&= 85.61(10)^8 \\ b_2 &= (r_1)^{64} &&= 1849(10)^{16} \\ b_3 &= (r_1)^{64}(r_3)^{32} &&= 1840(10)^{16} \end{aligned}$$

The value of r_1 may be determined from the above relationship involving *either* b_1 or b_2, after which the value of r_3 is determined. These values are found to be

$$\begin{aligned} r_1 &= r_2 = \pm 2 \\ r_3 &= \pm 1 \end{aligned}$$

Using the plus and minus values of each root for synthetic division of Eq. 2-47, the roots are found to be

$$\begin{aligned} x_1 &= x_2 = -2 \\ x_3 &= 1 \end{aligned}$$

	$a_0{}^2 = 1$	a_1^2 $-2a_2$	a_2^2 $-2a_1a_3$	a_3^2
m	x^3	a_1x^2	a_2x	a_3x^0
1	1	3	0	−4
2	1	9	24	16
4	1	33	288	256
8	1	513	$6.61(10)^4$	$6.55(10)^4$
16	1	$13.1(10)^4$	$43.0(10)^8$	$42.9(10)^8$
32	1	$85.61(10)^8$	$1849(10)^{16}$	$1840(10)^{16}$

Fig. 2-19 / *Root-squaring table for Ex. 2-8.*

Complex Roots

When the root system of a polynomial contains complex roots, they occur in conjugate pairs. The presence of complex roots is indicated, in the root-squaring process, when the signs of one or more of the coefficients fluctuate during the calculations (one coefficient will fluctuate in sign for each distinct conjugate pair present).

To investigate the procedure for determining the complex roots of a polynomial, let us consider a fourth-degree polynomial having two real and distinct roots x_1 and x_2, and a conjugate pair x_3 and x_4. Assuming that the roots of the complex pair have a magnitude of R, they may be expressed as

$$\begin{aligned} x_3 &= Re^{i\theta} = R(\cos\theta + i\sin\theta) = u + iv \\ x_4 &= Re^{-i\theta} = R(\cos\theta - i\sin\theta) = u - iv \end{aligned} \tag{2-49}$$

where $i = \sqrt{-1}$ and $R = \sqrt{u^2 + v^2}$.

Referring to Eq. 2-37, we may write the following relationships between the coefficients and the Enke roots of the derived polynomials:

$$\begin{aligned} b_1 &= r_1^m + r_2^m + R^m(e^{i\theta m} + e^{-i\theta m}) \\ b_2 &= (r_1r_2)^m + (r_1Re^{i\theta})^m + (r_1Re^{-i\theta})^m + (r_2Re^{i\theta})^m \\ &\quad + (r_2Re^{-i\theta})^m + R^{2m} \\ b_3 &= (r_1r_2Re^{i\theta})^m + (r_1r_2Re^{-i\theta})^m + (r_1R^2)^m + (r_2R^2)^m \\ b_4 &= (r_1r_2R^2)^m \end{aligned} \tag{2-50}$$

Utilizing the trigonometric relationships of Eq. 2-49 and simplifying, we may express Eq. 2-50 as

$$
\begin{aligned}
b_1 &= r_1^m + r_2^m + 2R^m \cos m\theta \\
b_2 &= (r_1 r_2)^m + 2R^m(r_1^m + r_2^m) \cos m\theta + R^{2m} \\
b_3 &= 2(r_1 r_2 R)^m \cos m\theta + R^{2m}(r_1^m + r_2^m) \\
b_4 &= (r_1 r_2 R^2)^m
\end{aligned}
\tag{2-51}
$$

Again assuming that $|r_1| > |r_2| > |R|$, the root-squaring process will result in $r_1^m \gg r_2^m \gg R^m$, and only the dominant terms of Eq. 2-51 need be considered. This establishes the following relationships:

$$
\begin{aligned}
b_1 &\cong r_1^m \\
b_2 &\cong (r_1 r_2)^m \\
b_3 &\cong 2(r_1 r_2 R)^m \cos m\theta \\
b_4 &= (r_1 r_2 R^2)^m
\end{aligned}
\tag{2-52}
$$

It is apparent, from the relationships of Eq. 2-52, that the sign of the coefficient b_3 will fluctuate owing to the trigonometric function $\cos m\theta$, since m is doubled each time a new coefficient value is determined in the root-squaring process. The sign fluctuations will not occur in a regular pattern, since successive values of $m\theta$ are not necessarily separated by π. However, sign fluctuations of successive values of a coefficient b_i occur *only* when complex roots exist, so the fluctuations, per se, are significant.

If b_3 is a coefficient with a fluctuating sign, the Enke roots r_3 and r_4 constitute a conjugate pair of complex roots. If the sign of b_2 fluctuates, the roots r_2 and r_3 comprise a complex pair. In general, then, if the sign of the coefficient b_i fluctuates, the roots r_i and r_{i+1} constitute a conjugate pair of complex roots.

When there are complex double roots, their presence will be indicated by the appearance of three adjacent irregular coefficients in the table, the outside two of which fluctuate in sign. The center coefficient will not follow any apparent law of variation.

After it has been determined which roots are complex, equations similar to Eqs. 2-51 and 2-52 must be written. In general, the modulus of a complex root can be found from the quotient of the coefficients on either side of the group of three irregular coefficients. After the real roots are determined, the real part of the complex roots are obtained from the relations between the roots and coefficients of the original equation. The imaginary parts of the complex roots are then found from the relationship

$$R^2 = u^2 + v^2$$

EXAMPLE 2-9

To solve a polynomial having a root system containing complex roots, let us consider the fourth-degree polynomial

$$x^4 + x^3 - 6x^2 - 14x - 12 = 0 \tag{2-53}$$

Utilizing Eq. 2-27 in the root-squaring process, the table of Fig. 2-20 is obtained. For convenience, the pertinent terms of Eq. 2-27 are given at the top of the table. The first row shows the coefficients of the original polynomial, and the last row gives the coefficient values of the last derived polynomial with $m = 64$. The sign fluctuations evident in the column of b_3 values indicate that the roots r_3 and r_4 are complex. The regular relationships observed for the coefficients b_1 and b_2 indicate that the roots r_1 and r_2 are real and distinct.

To determine the magnitudes of the roots, we first consider the real and distinct roots r_1 and r_2. From Eq. 2-52 and the coefficient values In Fig. 2-20, we find that

$$r_1^{64} = b_1 = 34.30(10)^{29}$$

$$r_2^{64} = \frac{b_2}{b_1} = \frac{63.33(10)^{48}}{34.29(10)^{29}}$$

	$a_0^2 = 1$	a_1^2 $-2a_2$	a_2^2 $-2a_1a_3$ $+2a_4$	a_3^2 $-2a_2a_4$	a_4^2
m	x^4	a_1x^3	a_2x^2	a_3x	a_4x^0
1	1	1	−6	−14	−12
2	1	13	40	52	144
4	1	89	536	−8816	20736
8	1	6849	$189.8(10)^4$	$55.49(10)^6$	$4.300(10)^8$
16	1	$43.11(10)^6$	$2.843(10)^{12}$	$14.47(10)^{14}$	$18.49(10)^{16}$
32	1	$18.52(10)^{14}$	$7.958(10)^{24}$	$10.42(10)^{29}$	$3.419(10)^{34}$
64	1	$3.430(10)^{30}$	$63.33(10)^{48}$	$54.16(10)^{58}$	$11.69(10)^{68}$

Fig. 2-20 / *Root-squaring table for Ex. 2-9.*

The magnitude or modulus of the complex roots is next determined as

$$R^{128} = \frac{b_4}{b_2} = \frac{11.69(10)^{68}}{63.33(10)^{48}}$$

Solving the preceding expressions, we find that

$$r_1 = \pm 3$$

$$r_2 = \pm 2$$

$$R = \pm 1.414 = \pm\sqrt{2}$$

Using the plus and minus values of each root for synthetic division of Eq. 2-53, the roots are found to be

$$x_1 = 3 \quad \text{and} \quad x_2 = -2$$

We next determine the real part of each of the complex roots. Referring to Eq. 2-30, which illustrates the relationships between the coefficients and the roots of the original polynomial, we extend the first of these relationships to a fourth-degree polynomial and write

$$a_1 = -(x_1 + x_2 + x_3 + x_4) \tag{2-54}$$

With the roots x_1 and x_2 determined, the value of a_1 given in Eq. 2-53, and Eq. 2-49 showing that the sum of the roots x_3 and x_4 is $2u$, Eq. 2-54 may be written as

$$1 = -(3 - 2 + 2u)$$

from which

$$u = -1$$

The imaginary parts of the complex roots x_3 and x_4 may now be determined from the relationship

$$R^2 = u^2 + v^2$$

so that

$$v = \sqrt{2 - 1} = 1$$

Thus the roots of Eq. 2-53 are

$$x_1 = 3$$

$$x_2 = -2$$

$$x_3 = -1 + i$$

$$x_4 = -1 - i$$

We have now found that the relationships between successive derived coefficient values, in the root-squaring process, indicate the type of roots composing the root system of the polynomial being solved and that, furthermore, the coefficient b_i having a particular relationship indicates which of the roots $r_i, r_{i+1}, \ldots, r_n$ are of the kind associated with that relationship.

Recalling how the coefficient relationships were developed for various types of roots, and realizing that many different combinations of root types are possible in higher degree polynomials, it should be evident that one general rule for interpreting the results of the root-squaring process for any nth-degree polynomial is a practical impossibility. It is possible to formulate rules for particular combinations of roots—two real and equal roots, three real and equal roots, equal pairs of complex roots, and so on—in the manner by which the coefficient relationships were developed for various types of roots in the preceding sections. This involved the assumption of a particular combination of root types, the substitution of the assumption into Eq. 2-37, and the consideration of the dominant terms resulting from the root-squaring process. The use of such rules should be predicated upon a thorough understanding of the process from which they are determined, since once the procedure is understood, any combination of root types may be analyzed.

Programming Graeffe's Method

It should now be evident that a *general* computer program for solving any nth-degree polynomial would involve very complex logic, since the program would have to be capable of *interpreting* the many combinations of coefficient relationships which might be encountered in the root-squaring process, before a solution could be obtained. Programming is further complicated by the possibility of encountering derived coefficient values having magnitudes which exceed the limits of the computer being used. If the roots of the polynomial are fairly large, the coefficient values of the derived polynomials will obviously be large also (see Eq. 2-33). If several roots are very close in magnitude, the derived coefficient values also become very large, since more applications of the root-squaring process than usual must be made to establish dominant terms in the coefficient-root relationships. (If r_i and r_{i+1} are close in magnitude, m must be very large to make $r_i \gg r_{i+1}$).

The problem of excessively large derived coefficient values can be alleviated by *scaling* the polynomial which is to be solved. However, the program logic required in writing a general program for obtaining complete solutions presents a challenge. Fortunately, a practical consideration of the solution of polynomials in certain engineering applications can often reduce this difficulty, to some extent. The polynomials which stem from various types of physical systems often have certain kinds of root systems which are characteristic of the system. For example, the polynomials defining the modes of free undamped vibration, in systems with various degrees of freedom, characteristically have real and distinct roots, since the roots of the polynomial yield the natural circular frequencies of vibration of the system. Thus it can be seen that, if one were working in a particular area such as the one just described, a general computer program for determining the roots of polynomials known to have *only* real and distinct roots would be considerably less difficult to formulate.

Another practical approach may be made by combining human logic with a rather general program. The program is written so that the interpretation of the results of the root-squaring process can be supplemented by human logic. The computer process is interrupted at key points, and the programmer performs some degree of interpretation which is provided as input to the program at that point, after which the computer process is resumed.

In still another method, the computer performs the tedious task of determining the coefficients in the root-squaring process, after which the programmer interprets the tabulated results and performs the relatively easy job of determining the root values. An application of the last approach will be illustrated after we have discussed the scaling of a polynomial.

Scaling of Polynomials

As mentioned earlier, if the roots of a polynomial are large or if several of the roots are close together, very large coefficient values may be encountered in the root-squaring process. The scaling of such a polynomial will reduce the magnitude of the derived coefficient values so that they can be handled by many of the widely used computers.

Consider the third-degree polynomial

$$x^3 + a_1x^2 + a_2x + a_3 = 0 \tag{2-55}$$

having the roots x_1, x_2, and x_3 and the Enke roots r_1, r_2, and r_3.[3] The polynomial is scaled by scaling its roots, so, letting α be the *scale factor*, we write

$$\begin{aligned} x_1 &= \alpha\bar{x}_1 & \qquad r_1 &= \alpha\bar{r}_1 \\ x_2 &= \alpha\bar{x}_2 \quad \text{and} & \qquad r_2 &= \alpha\bar{r}_2 \\ x_3 &= \alpha\bar{x}_3 & \qquad r_3 &= \alpha\bar{r}_3 \end{aligned} \tag{2-56}$$

[3] The coefficient a_0 of x^3 is unity in Eq. 2-55. If $a_0 \neq 1$ for the given polynomial, the first step in the scaling process is to reduce a_0 to unity by dividing each coefficient of the polynomial by a_0.

where the quantities with bars over them are the roots and the Enke roots, respectively, of the *scaled polynomial*. The relationships between the coefficients and the Enke roots of the scaled polynomial are

$$\begin{aligned} \bar{a}_1 &= \bar{r}_1 + \bar{r}_2 + \bar{r}_3 \\ \bar{a}_2 &= \bar{r}_1\bar{r}_2 + \bar{r}_1\bar{r}_3 + \bar{r}_2\bar{r}_3 \\ \bar{a}_3 &= \bar{r}_1\bar{r}_2\bar{r}_3 \end{aligned} \tag{2-57}$$

Substituting the relationships of Eq. 2-56 into Eq. 2-57 yields

$$\begin{aligned} \alpha\bar{a}_1 &= r_1 + r_2 + r_3 \\ \alpha^2\bar{a}_2 &= r_1r_2 + r_1r_3 + r_2r_3 \\ \alpha^3\bar{a}_3 &= r_1r_2r_3 \end{aligned} \tag{2-58}$$

Comparing Eqs. 2-58 and 2-31, we see that the coefficients of the scaled polynomial $\bar{a}_i$ are related to the coefficients of the given polynomial by

$$\bar{a}_1 = \frac{a_1}{\alpha}, \qquad \bar{a}_2 = \frac{a_2}{\alpha^2}, \qquad \bar{a}_3 = \frac{a_3}{\alpha^3} \tag{2-59}$$

To determine a magnitude for the scale factor α, consider the relationships of Eq. 2-57. The final derived coefficient values, obtained by applying the root-squaring process to the scaled polynomial, may be kept to a minimum by letting the absolute value of the product of the scaled roots equal unity.[4] That is,

$$|\bar{r}_1\bar{r}_2\bar{r}_3| = 1$$

so that the absolute value of the scaled coefficient $|\bar{a}_3| = 1$. Since α is always positive, it can be seen, from the last relationship of Eq. 2-59, that the magnitude of the scale factor is determined from

$$\alpha = |a_3|^{1/3}$$

Having found the value of α from the above relationship, the other scaled-coefficient values may be determined from Eq. 2-59. This process yields a scaled polynomial

$$\bar{x}^3 + \frac{a_1}{\alpha}\bar{x}^2 + \frac{a_2}{\alpha^2}\bar{x} + 1 = 0 \tag{2-60}$$

[4] If $|\bar{r}_1\bar{r}_2\bar{r}_3| < 1$ were selected, $(\bar{r}_1\bar{r}_2\bar{r}_3)^m$ could become *small* enough to exceed the limit of the computer.

In general, for scaling an nth-degree polynomial,

$$\alpha = |a_n|^{1/n} \tag{2-61}$$

and the scaled coefficients are

$$\bar{a}_i = a_i/\alpha^i \qquad (i = 1, 2, \ldots, n) \tag{2-62}$$

After solving the scaled polynomial for the scaled Enke roots, by Graeffe's method, the Enke roots of the original polynomial are determined from Eq. 2-56. The roots of the given polynomial are then found by substituting the plus and minus values of the Enke roots into the original polynomial, as explained before.

EXAMPLE 2-10

A general computer program for performing the root-squaring process for any nth-degree polynomial is to be written so that the original polynomials are scaled, as described above, before the root-squaring process is applied. (It might be well, at this point, for the reader to review Eqs. 2-27, 2-61, and 2-62, since these are the pertinent equations for scaling a polynomial and for applying the root-squaring process.)

The flow chart for the program is shown in Fig. 2-21. The FORTRAN variable names used and the quantities that they represent are as follows:

FORTRAN Name	Quantity
N	Integer name for degree of polynomial
DEG	Real name for degree of polynomial
ITMAX	Number of iterations which the computer is to perform (number of applications of root-squaring process)
A(I)	Polynomial coefficients, scaled-polynomial coefficients, and derived coefficients before determining new derived coefficients
SCF	Scale factor used in scaling original polynomial
IT	Index of a DO loop controlling the number of iterations
M	Power of roots of original polynomial resulting from root-squaring process
S	Quantity having value of ± 1.0 used to control sign of product terms
K	Subscript of variable A
J	Subscript of variable A
SUM	Sum of products of coefficients
B(I)	Derived coefficients

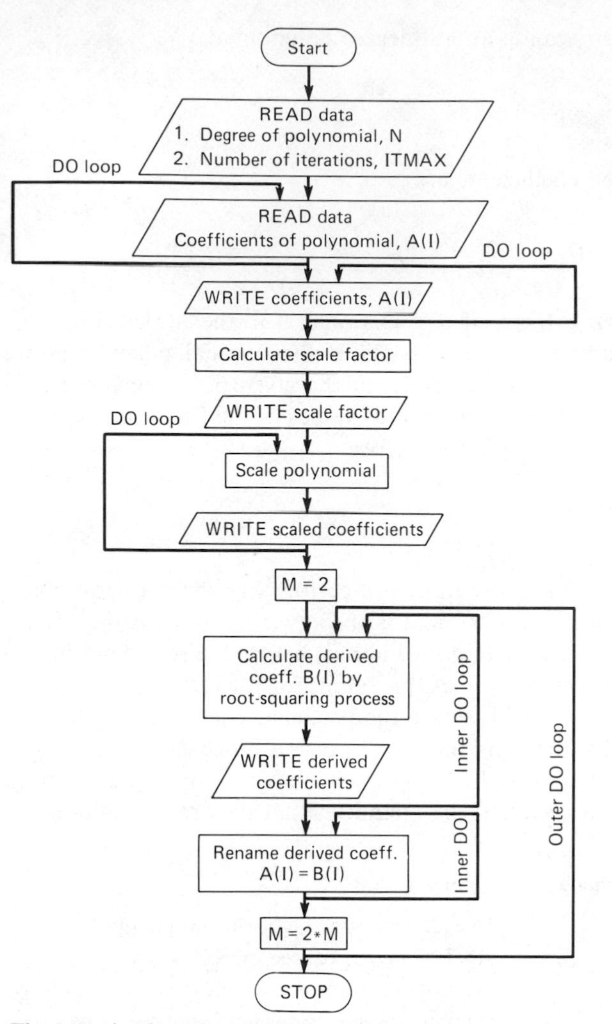

Fig. 2-21 / *Flow chart for root-squaring process.*

Utilizing the flow chart of Fig. 2-21 as a guide, the following FORTRAN program is obtained, showing the specific steps necessary to get the derived coefficients of the scaled polynomial by the root-squaring process.

```
C GRAEFFE TABLE FOR ROOTS OF A POLYNOMIAL
      DIMENSION A(10),B(10)
      READ(5,1) N,ITMAX
    1 FORMAT(2I3)
    9 READ(5,2) (A(I),I=1,N)
    2 FORMAT(8F10.0)
      WRITE(6,3)
    3 FORMAT('1','COEFFICIENTS OF ORIGINAL POLYNOMIAL'/)
      DO 10 I=1,N
```

```
10 WRITE(6,8) A(I)
 8 FORMAT(' ',4X,E14.7)
   DEG=N
   SCF=ABS(A(N))**(1./DEG)
   WRITE(6,4) SCF
 4 FORMAT('0','SCALE FACTOR=',E14.7///)
   WRITE(6,5)
 5 FORMAT(' ',3X,'DERIVED COEFFICIENTS')
   WRITE(6,6)
 6 FORMAT('0','SCALED COEFFICIENTS M= 1'/)
   DO 11 I=1,N
   A(I)=A(I)/SCF**I
11 WRITE(6,8) A(I)
   M=2
   DO 20 IT=1,ITMAX
   WRITE(6,7) IT,M
 7 FORMAT('0',4X,'ITER. NO.',I3,3X,'M=',I4/)
   DO 18 I=1,N
   S=1.
   K=I
   J=I
   SUM=0.
12 K=K-1
   J=J+1
   IF(K) 15,15,13
13 IF(J-N) 14,14,17
14 S=-S
   SUM=SUM+2.*S*A(K)*A(J)
   GO TO 12
15 IF(J-N) 16,16,17
16 B(I)=A(I)**2+SUM-2.*S*A(J)
   GO TO 18
17 B(I)=A(I)**2+SUM
18 WRITE(6,8) B(I)
   DO 19 I=1,N
19 A(I)=B(I)
20 M=2*M
   STOP
   END
```

This program is capable of handling polynomials of any reasonable degree. The degree of the particular polynomial under consideration is specified in the program by an input data value assigned to N. The dimension statement also reflects the degree of the polynomial being considered, since the higher the degree of the polynomial, the more memory space must be reserved for storing coefficient values. (The dimension statement in the program shown has reserved memory space for handling up to and including a tenth-degree polynomial.)

The value assigned to ITMAX specifies the number of times the root-squaring process is to be applied. (Note its use as the *test value* of a DO statement.) For most polynomials, six iterations ($m = 64$) are usually sufficient to establish the regular, half-squared, or fluctuating sign relationships of the derived coefficients. However, since the computing time of each iteration is small, additional iterations could be made to obviate any chance of terminating the root-squaring process too soon. Overflow will occur, however, if m is too large.

Note that the program specifies that the printout of the coefficient values is to be in column form rather than in rows, as shown in preceding discussions. This is done to avoid exceeding the limit of characters allowable in a given row for the printout device, in case a polynomial of high degree is handled.

EXAMPLE 2-11

The program of the preceding example will be used to provide a root-squaring table for the sixth-degree polynomial

$$x^6 - 1389x^4 + 5(10)^5x^2 - 37.1(10)^6 = 0 \tag{2-63}$$

The computer printout of the table is displayed in the columns shown:

```
COEFFICIENTS OF ORIGINAL POLYNOMIAL
    0.0
   -0.1389000E 04
    0.0
    0.5000000E 06
    0.0
   -0.3710000E 08
SCALE FACTOR= 0.1826256E 02

  DERIVED COEFFICIENTS

SCALED COEFFICIENTS  M= 1

    0.0
   -0.4164655E 01
    0.0
    0.4494937E 01
    0.0
   -0.1000010E 01

   ITER. NO.  1   M=  2

    0.8329309E 01
    0.2633421E 02
    0.3943973E 02
    0.2853383E 02
    0.8989959E 01
    0.1000019E 01

ITER. NO.  2   M=  4

 0.1670897E 02
 0.9354715E 02
 0.2004204E 03
 0.1577256E 03
 0.2375061E 02
 0.1000038E 01

ITER. NO.  3   M=  8

 0.9209515E 02
 0.2368885E 04
 0.1145046E 05
 0.1554425E 05
 0.2486282E 03
 0.1000076E 01

ITER. NO.  4   M= 16

 0.3743746E 04
 0.3533639E 07
 0.5751384E 08
 0.2359347E 09
 0.3072509E 05
 0.1000153E 01
```

```
ITER. NO.  5   M= 32

  0.6948356E 07
  0.1205644E 14
  0.1640425E 16
  0.5566166E 17
  0.4720896E 09
  0.1000305E 01

ITER. NO.  6   M= 64

  0.2416677E 14
  0.1453350E 27
  0.1348834E 31
  0.3098221E 34
  0.1115113E 18
  0.1000610E 01

ITER. NO.  7   M= 128

  0.2933628E 27
  0.2112226E 53
  0.9187915E 60
  0.9598970E 67
  0.6234548E 34
  0.1001221E 01
```

Inspection of the printout reveals that no coefficients have fluctuating signs, so all the roots of the polynomial are real. Further examination of the coefficient values in the last two iterations reveals the following relationships between corresponding coefficients:

$$\begin{aligned} \bar{b}_1' &\cong \frac{(\bar{b}_1)^2}{2} \\ \bar{b}_2' &\cong (\bar{b}_2)^2 \\ \bar{b}_3' &\cong \frac{(\bar{b}_3)^2}{2} \\ \bar{b}_4' &\cong (\bar{b}_4)^2 \\ \bar{b}_5' &\cong \frac{(\bar{b}_5)^2}{2} \\ \bar{b}_6' &= (\bar{b}_6)^2 \end{aligned} \tag{2-64}$$

Equation 2-64 indicates the three pairs of equal roots

$$\bar{r}_1 = \bar{r}_2 \qquad \bar{r}_3 = \bar{r}_4 \qquad \bar{r}_5 = \bar{r}_6$$

Utilizing these relationships in Eq. 2-37, we find that

$$\begin{aligned} \bar{b}_1 &= 2\bar{r}_1^m + 2\bar{r}_3^m + 2\bar{r}_5^m \\ \bar{b}_2 &= \bar{r}_1^{2m} + 2(\bar{r}_1\bar{r}_3)^m + 2(\bar{r}_1\bar{r}_6)^m + 2(\bar{r}_2\bar{r}_3)^m + \cdots + r_5^{2m} \\ \bar{b}_3 &= 2\bar{r}_1^{2m}\bar{r}_3^m + 2\bar{r}_1^{2m}\bar{r}_5^m + \cdots + \bar{r}_4^m\bar{r}_5^{2m} \\ \bar{b}_4 &= \bar{r}_1^{2m}\bar{r}_3^{2m} + 2\bar{r}_1^{2m}\bar{r}_3^m\bar{r}_5^m + 2\bar{r}_2^m\bar{r}_3^{2m}\bar{r}_5^m + r_3^{2m}r_5^{2m} \\ \bar{b}_5 &= 2\bar{r}_1^{2m}\bar{r}_3^{2m}\bar{r}_5^m + \bar{r}_2^m\bar{r}_3^{2m}\bar{r}_5^{2m} \\ \bar{b}_6 &= \bar{r}_1^{2m}\bar{r}_3^{2m}\bar{r}_5^{2m} \end{aligned} \tag{2-65}$$

Considering the dominant terms ($r_1^m = r_2^m \gg r_3^m = r_4^m \gg r_5^m = r_6^m$), we obtain

$$\begin{aligned}
\bar{b}_1 &\cong 2\bar{r}_1^m \\
\bar{b}_2 &\cong \bar{r}_1^{2m} \\
\bar{b}_3 &\cong 2\bar{r}_1^{2m}\bar{r}_3^m \\
\bar{b}_4 &\cong \bar{r}_1^{2m}\bar{r}_3^m\bar{r}_4^m \\
\bar{b}_5 &\cong 2\bar{r}_1^{2m}\bar{r}_3^{2m}\bar{r}_5^m \\
\bar{b}_6 &= \bar{r}_1^{2m}\bar{r}_3^{2m}\bar{r}_5^{2m}
\end{aligned} \tag{2-66}$$

Using the coefficient values of the last iteration in the table, we solve Eq. 2-66 for the Enke roots of the scaled polynomial and obtain

$$\begin{aligned}
\bar{r}_1 &= \pm 1.601 \\
\bar{r}_3 &= \pm 1.141 \\
\bar{r}_5 &= \pm 0.547
\end{aligned}$$

Noting (in the printout) that $\alpha = 18.2625$, Eq. 2-56 is next utilized to determine the value of the Enke roots of the original polynomial. Thus

$$\begin{aligned}
r_1 &= 18.2625(\pm 1.6) = \pm 29.2 \\
r_3 &= 18.2625(\pm 1.141) = \pm 20.8 \\
r_5 &= 18.2625(\pm 0.547) = \pm 10.0
\end{aligned}$$

Substituting the plus and minus values obtained into the original polynomial, we find that the polynomial is satisfied for both signs of each coefficient, so the roots of the polynomial are

$$\begin{aligned}
x_1 &= +29.2 \\
x_2 &= -29.2 \\
x_3 &= +20.8 \\
x_4 &= -20.8 \\
x_5 &= +10.0 \\
x_6 &= -10.0
\end{aligned}$$

2-10 / BAIRSTOW'S METHOD

Bairstow's method is an iterative method which involves finding quadratic factors, $f(x) = x^2 + ux + v$, of the polynomial[5]. From appropriate starting values for u and v, the iteration process converges to the correct values of u and v so that two roots (r_1, r_2) may be determined from the quadratic factor. That is,

$$r_{1,2} = \frac{-u \pm \sqrt{u^2 - 4v}}{2} \tag{2-67}$$

The first quadratic factor obtained from the iteration process gives two roots of the polynomial. Two more roots are then obtained by repeating the iteration process to obtain a quadratic factor of the $n - 2$ degree polynomial resulting from the extraction of the quadratic factor for the first two roots. Thus the formulation of a quadratic factor of the polynomial resulting from the extraction of a prior quadratic factor will yield additional roots. This procedure may be continued to find all the roots of the original polynomial.

If the quantity under the radical of Eq. 2-67 is negative, the roots, of course, form a pair of complex conjugates.

In a computer program it must be determined whether the value of $u^2 - 4v$ is positive, zero, or negative. In the case of a negative quantity, the sign is changed and the square root of the positive quantity divided by 2 is assigned the value of the complex part of the pair of complex conjugates. If $u^2 - 4v = 0$, then the roots are real and equal. Two real and distinct roots occur when $u^2 - 4v > 0$.

Unfortunately, the iteration process may not converge for some starting (initial) values selected for u and v. It is common practice to select starting values of zero for both u and v, and if convergence does not occur after a number of iterations, new starting values are selected for the iteration procedure. Frequently, suitable starting values of u and v may be determined from $u = a_{n-1}/a_{n-2}$ and $v = a_n/a_{n-2}$, where the a's are the coefficients of the nth-degree polynomial.

We will now develop the equations for the iteration procedure. The extraction of a quadratic factor $x^2 + ux + v$ from an nth-degree polynomial may be written in the form

$$(x^2 + ux + v)(x^{n-2} + b_1x^{n-3} + b_2x^{n-4} + \cdots + b_{n-3}x + b_{n-2}) + \text{remainder} \tag{2-68}$$

[5]The polynomial is of the form given by Eq. 2-22. If the coefficient of x^n is not 1, the equation should be divided through by the coefficient of x^n to make it so.

Dividing an nth-degree polynomial by the quadratic factor gives the following expressions for the b's and *remainder:*

$$\left.\begin{aligned} b_0 &= 1 \\ b_1 &= a_1 - u \\ b_2 &= a_2 - b_1 u - v \\ b_3 &= a_3 - b_2 u - b_1 v \\ b_4 &= a_4 - b_3 u - b_2 v \\ &\vdots \\ b_k &= a_k - b_{k-1} u - b_{k-2} v \\ &\vdots \\ b_{n-1} &= a_{n-1} - b_{n-2} u - b_{n-3} v \\ b_n &= a_n - b_{n-1} u - b_{n-2} v \end{aligned}\right\} \tag{2-69}$$

and

$$\text{remainder} = (x + u)b_{n-1} + b_n$$

If u and v have values such that the quadratic factor $x^2 + ux + v$ contains two roots to the nth-degree polynomial, then the remainder is zero. Therefore, b_n and b_{n-1} must both be zero for the remainder to be zero.

Before proceeding with the development of Bairstow's method, let us show that the b's given by Eq. 2-69 can also be obtained by a synthetic division procedure. For illustration purposes, we will use a fifth-degree polynomial ($n = 5$), Eq. 2-70, with $u = v = 1$.

$$x^5 - 3x^4 - 10x^3 + 10x^2 + 44x + 48 = 0 \tag{2-70}$$

$(n = 5)$

	a_0	a_1	a_2	a_3	a_4	a_5
$-u = -1$	1	-3	-10	10	44	48
		-1	4	7	-21	-30
$-v = -1$		—	-1	4	7	-21
	1	-4	-7	21	30	-3
	b_{n-5}	$b_1 = b_{n-4}$	$b_2 = b_{n-3}$	$b_3 = b_{n-2}$	$b_4 = b_{n-1}$	$b_5 = b_n$

It is noted that $b_0 = 1$ and that $b_1 = a_1 - u$. The procedure for the synthetic division is evident if we compare the columns of the table above with the expressions for the b's in Eq. 2-69. From the results of the table we can write the fifth-degree polynomials as

$$(x^2 + x + 1)(x^3 - 4x^2 - 7x + 21) + (x + 1)30 - 3$$

where the remainder is $(x + 1)30 - 3$ or $30x + 27$. As we will see later, the iteration process will be carried out until b_{n-1} and b_n become essentially zero.

The basic concept of Bairstow's method is to reduce the remainder terms to as near zero as required for a satisfactory approximation of the roots. This means that b_{n-1} and b_n must be reduced to near zero. However, we see from Eq. 2-69 that the b's are functions of both u and v. If we consider Δu and Δv as increments to be added to u and v, respectively, to make the remainder zero, we can express b_n and b_{n-1} in the form of a Taylor-series expansion for a function of two variables. Assuming that Δu and Δv are small so that the higher order terms are negligible, the Taylor-series expansions give

$$\begin{aligned} b_n(u + \Delta u, v + \Delta v) &= 0 \cong b_n + \frac{\partial b_n}{\partial u}\Delta u + \frac{\partial b_n}{\partial v}\Delta v \\ b_{n-1}(u + \Delta u, v + \Delta v) &= 0 \cong b_{n-1} + \frac{\partial b_{n-1}}{\partial u}\Delta u + \frac{\partial b_{n-1}}{\partial v}\Delta v \end{aligned} \tag{2-71}$$

where the terms on the right are evaluated for values of u and v and not for values of $u + \Delta u$ and $v + \Delta v$.

The required partial derivatives in Eq. 2-71 may be obtained from differentiating the expressions of Eq. 2-69. Furthermore, the partial derivatives can be expressed in terms of a set of c's which can be determined from the b's by the synthetic division procedure previously demonstrated for obtaining the b's from the a's, or from a set of equations similar to Eq. 2-69 which may be used for obtaining the b's from the a's. Taking partial derivatives of the expressions of Eq. 2-69 with respect to u and defining these in terms of c's, we obtain the following:

$$\begin{aligned} \frac{\partial b_1}{\partial u} &= -1 = -c_0 \\ \frac{\partial b_2}{\partial u} &= u - b_1 = -c_1 \\ \frac{\partial b_3}{\partial u} &= -b_2 + c_1 u + v = -c_2 \\ \frac{\partial b_4}{\partial u} &= -b_3 + c_2 u + c_1 v = -c_3 \\ &\vdots \qquad\qquad \vdots \qquad\qquad \vdots \\ \frac{\partial b_{n-1}}{\partial u} &= -b_{n-2} + c_{n-3} u + c_{n-4} v = -c_{n-2} \\ \frac{\partial b_n}{\partial u} &= -b_{n-1} + c_{n-2} u + c_{n-3} v = -c_{n-1} \end{aligned} \tag{2-72}$$

In a similar manner, taking the partial derivatives with respect to v results in the following:

$$
\begin{aligned}
\frac{\partial b_1}{\partial v} &= 0 \\
\frac{\partial b_2}{\partial v} &= -1 = -c_0 \\
\frac{\partial b_3}{\partial v} &= u - b_1 = -c_1 \\
\frac{\partial b_4}{\partial v} &= -b_2 + c_1 u + v = -c_2 \\
\vdots \qquad & \qquad \vdots \qquad \vdots \\
\frac{\partial b_{n-1}}{\partial v} &= -b_{n-3} + c_{n-4} u + c_{n-5} v = -c_{n-3} \\
\frac{\partial b_n}{\partial v} &= -b_{n-2} + c_{n-3} u + c_{n-4} v = -c_{n-2}
\end{aligned}
\tag{2-73}
$$

From either Eq. 2-72 or 2-73, we see that

$$
c_k = b_k - c_{k-1} u - c_{k-2} v \qquad (k = 2, 3, \ldots, n-1) \tag{2-74}
$$

where $c_0 = 1$ and $c_1 = b_1 - u$. Equation 2-74 is analogous to the expression for b_k in terms of a_k, u, and v of Eq. 2-69. Thus the partial derivatives, or c's, can be obtained from the b's by the synthetic division procedure previously shown as well as from Eq. 2-74.

From Eqs. 2-72 and 2-73, it is apparent that Eq. 2-71 may be written in the form

$$
\left.
\begin{aligned}
b_n &= c_{n-1}\,\Delta u + c_{n-2}\,\Delta v \\
b_{n-1} &= c_{n-2}\,\Delta u + c_{n-3}\,\Delta v
\end{aligned}
\right\}
\tag{2-75}
$$

Solving these two simultaneous equations gives

$$
\Delta u = \frac{\begin{vmatrix} b_n & c_{n-2} \\ b_{n-1} & c_{n-3} \end{vmatrix}}{\begin{vmatrix} c_{n-1} & c_{n-2} \\ c_{n-2} & c_{n-3} \end{vmatrix}} \tag{2-76}
$$

and

$$\Delta v = \frac{\begin{vmatrix} c_{n-1} & b_n \\ c_{n-2} & b_{n-1} \end{vmatrix}}{\begin{vmatrix} c_{n-1} & c_{n-2} \\ c_{n-2} & c_{n-3} \end{vmatrix}} \tag{2-77}$$

In Eqs. 2-76 and 2-77, n denotes the degree of the polynomial from which the quadratic factor is being extracted.

Bairstow's method may now be summarized by the following step-by-step procedure:

1 Select initial values for u and v. Most frequently these are taken as $u_1 = v_1 = 0$.

2 Calculate $b_1, b_2, \ldots, b_n$ from

$$b_k = a_k - b_{k-1}u - b_{k-2}v \qquad (k = 2, 3, \ldots, n) \tag{2-78}$$

where $b_0 = 1$ and $b_1 = a_1 - u$ as shown in Eq. 2-69 and the synthetic division procedure.

3 Calculate $c_1, c_2, c_3, \ldots, c_{n-1}$ from

$$c_k = b_k - c_{k-1}u - c_{k-2}v \qquad (k = 2, 3, \ldots, n - 1) \tag{2-79}$$

where $c_0 = 1$ and $c_1 = b_1 - u$ as shown in Eq. 2-73.

4 Calculate Δu and Δv from Eqs. 2-76 and 2-77.

5 Increment u and v by Δu and Δv:

$$\left.\begin{aligned} u_{i+1} &= u_i + \Delta u_i \\ v_{i+1} &= v_i + \Delta v_i \end{aligned}\right\} \tag{2-80}$$

where i denotes the iteration number.

6 Return to step 2 and repeat the procedure until Δu and Δv approach zero to within some preassigned value ε such that

$$|\Delta u_{i+1}| + |\Delta v_{i+1}| < \varepsilon \tag{2-81}$$

For machine computation, an upper limit of iterations should be specified to protect against nonconvergence. If convergence to the desired result does not occur after the specified number of iterations, then new starting values should be selected. Frequently, suitable starting values of u and v may be determined from $u_1 = a_{n-1}/a_{n-2}$ and $v_1 = a_n/a_{n-2}$.

7 Calculate the two roots from the quadratic formula, Eq. 2-67.

8 Obtain additional roots by starting with step 1 and the reduced polynomial which is formulated from the final values of the b's obtained at the end of step 6.

EXAMPLE 2-12

To illustrate Bairstow's method, the roots will be determined for the polynomial

$$x^5 - 3x^4 - 10x^3 + 10x^2 + 44x + 48 = 0$$

To facilitate the required computations in using an electronic pocket calculator, the synthetic division scheme previously explained will be used to determine the b's and c's.

In starting with the initial values of $u_1 = v_1 = 0$, the computations with an electronic pocket calculator did not appear to be converging (convergence did occur with these starting values in a subsequent computer program after 14 iterations). Therefore, we select starting values of $u_1 = v_1 = 1.5$ which result in convergence after 5 iterations on the pocket calculator. With these starting values, the synthetic division scheme results in the following for the first iteration:

$(n = 5)$

	a_0	a_1	a_2	a_3	a_4	a_5
$-u_1 = -1.5$	1	-3	-10	10	44	48
		-1.5	6.75	7.13	-35.82	-22.97
$-v_1 = -1.5$		—	-1.5	6.75	7.13	-35.82
	1	-4.5	-4.75	23.88	15.31 (b_{n-1})	-10.79 (b_n)
		-1.5	9.00	-4.13	-43.13	
		—	-1.50	9.00	-4.13	
	1 (c_0)	-6.0 (c_1)	2.75 (c_2)	28.75 (c_3)	-31.95 ($c_{n-1} = c_4$)	

Substituting the appropriate values into Eqs. 2-76 and 2-77 gives

$$\Delta u_1 = \frac{\begin{vmatrix} -10.79 & 28.75 \\ 15.31 & 2.75 \end{vmatrix}}{\begin{vmatrix} -31.95 & 28.75 \\ 28.75 & 2.75 \end{vmatrix}} = \frac{-469.84}{-914.4} = 0.51$$

$$\Delta v_1 = \frac{\begin{vmatrix} -31.95 & -10.79 \\ 28.75 & 15.31 \end{vmatrix}}{-914.4} = \frac{-178.94}{-914.4} = 0.20$$

$$u_2 = 1.5 + 0.51 = 2.01$$

$$v_2 = 1.5 + 0.20 = 1.70$$

For the second iteration, we obtain

	a_0	a_1	a_2	a_3	a_4	a_5
-2.01	1	-3	-10	10	44	48
		-2.01	10.07	3.28	-43.80	-5.97
-1.70		—	-1.70	8.51	2.77	-37.04
	1	-5.01	-1.63	21.79	2.97	4.99
		-2.01	14.11	-21.67	-24.22	
		—	-1.70	11.93	-18.33	
	1	-7.02	10.78	12.05	-39.58	

$$\Delta u_2 = \frac{\begin{vmatrix} 4.99 & 12.05 \\ 2.97 & 10.78 \end{vmatrix}}{\begin{vmatrix} -39.58 & 12.05 \\ 12.05 & 10.78 \end{vmatrix}} = \frac{18.00}{-571.87} = -0.03$$

$$\Delta v_2 = \frac{\begin{vmatrix} -39.57 & 4.99 \\ 12.05 & 2.97 \end{vmatrix}}{-571.87} = \frac{-177.65}{-571.87} = 0.31$$

$$u_3 = 2.01 - 0.03 = 1.98$$

$$v_3 = 1.70 + 0.31 = 2.01$$

The third and fourth iterations yield the following:

$$u_4 = 1.98 + 0.01 = 1.99$$

$$v_4 = 2.01 + 0.00 = 2.01$$

$$u_5 = 1.99 + 0.01 = 2.00$$

$$v_5 = 2.01 - 0.01 = 2.00$$

Since the final b values are required to continue the process to find additional roots, the fifth and last iteration is shown.

	a_0	a_1	a_2	a_3	a_4	a_5
-2.0	1	-3	-10	10	44	48
		-2	10	4	-48	0
-2.0		—	-2	10	4	-48
	1	-5	-2	24	0	0
		-2	14	-20	-36	
		—	-2	14	-20	
	1	-7	10	18	-56	

$$\Delta u_6 = 0 \text{ and } \Delta v_6 = 0$$

Hence the values of u and v are

$$u = 2 \quad \text{and} \quad v = 2$$

Substituting these values into Eq. 2-67 gives

$$r_1 = -1 + i \quad \text{and} \quad r_2 = -1 - i$$

It is noted that the remainder converged to zero since $b_4 = b_5 = 0$. Thus the original polynomial may be written as

$$(x^2 + 2x + 2)(x^3 - 5x^2 - 2x + 24) = 0$$

To obtain the next two roots, we determine the quadratic factor for the third-degree polynomial, $x^3 - 5x^2 - 2x + 24 = 0$. Selecting starting values of $u_1 = v_1 = 0$ and noting that $n = 3$, the first iteration results in the following:

$(n = 3)$

	a_0	a_1	a_2	a_3
$u_1 = 0$	1	-5	-2	24
		0	0	0
$v_1 = 0$		—	0	0
	1	-5	-2	24 ← $b_n = b_3$
		0	0	
		—	0	
	1	-5	-2 ← $c_{n-1} = c_2$	

$$\Delta u_1 = \frac{\begin{vmatrix} 24 & -5 \\ -2 & 1 \end{vmatrix}}{\begin{vmatrix} -2 & -5 \\ -5 & 1 \end{vmatrix}} = \frac{14}{-27} = -0.52$$

$$\Delta v_1 = \frac{\begin{vmatrix} -2 & 24 \\ -5 & -2 \end{vmatrix}}{-27} = \frac{124}{-27} = -4.59$$

$$u_2 = 0.0 - 0.52 = -0.52$$
$$v_2 = 0.0 - 4.59 = -4.59$$

After two more iterations, the fourth and last iteration results in the following:

$(n = 3)$

	a_0	a_1	a_2	a_3
$-u_4 = 1.0$	1	-5	-2	24
		1.0	-4	0.03
$-v_4 = 6.03$		—	6.03	-24.12
	1	-4	0.03	-0.09
		1.0	-3.00	
		—	6.03	
	1	-3.0	3.06	

$$u = u_5 = -1.0 + 0.0 = -1.0$$
$$v = v_5 = -6.03 + 0.03 = -6.0$$

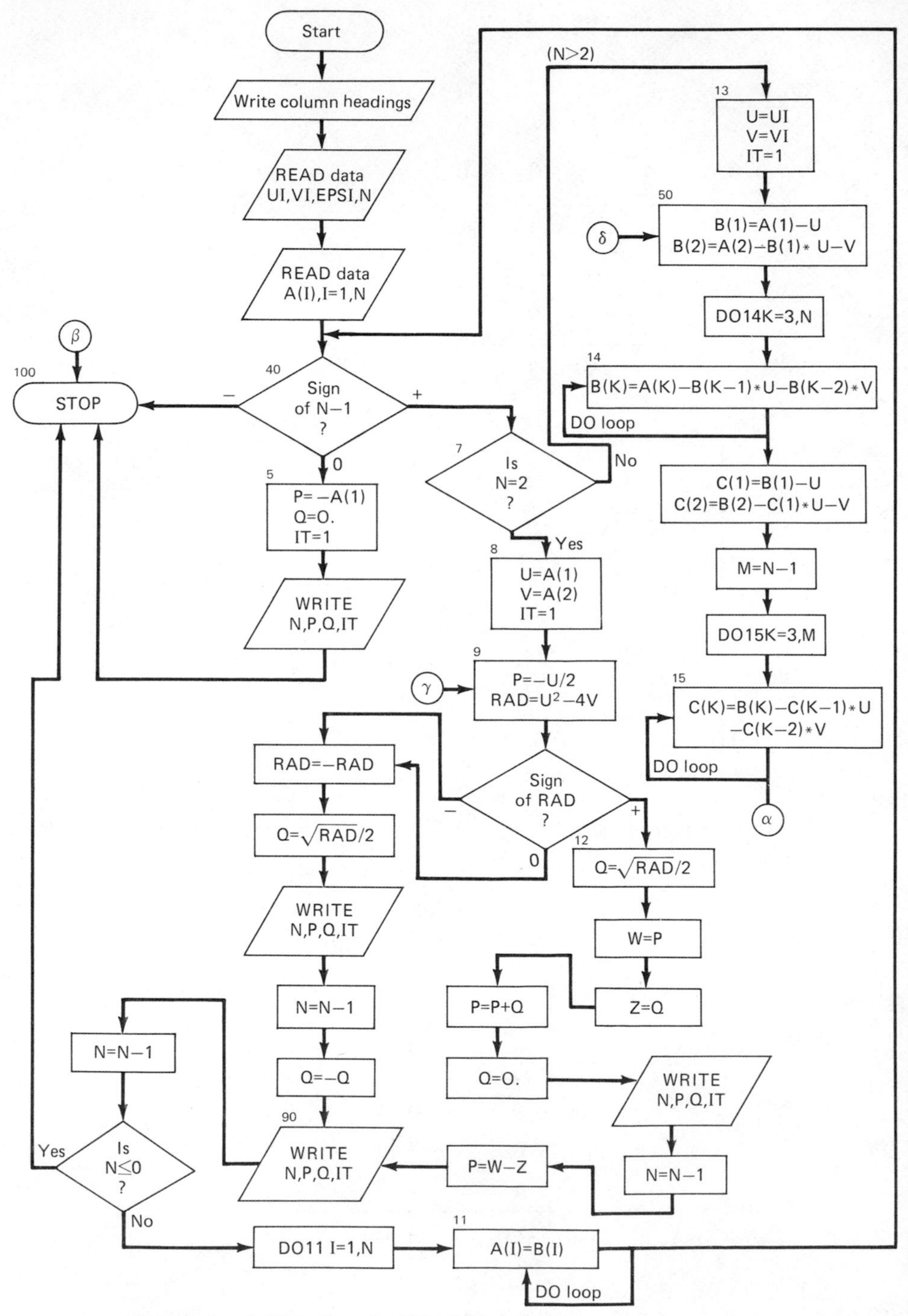

Fig. 2-22(a) / *Flow chart for Bairstow's method.*

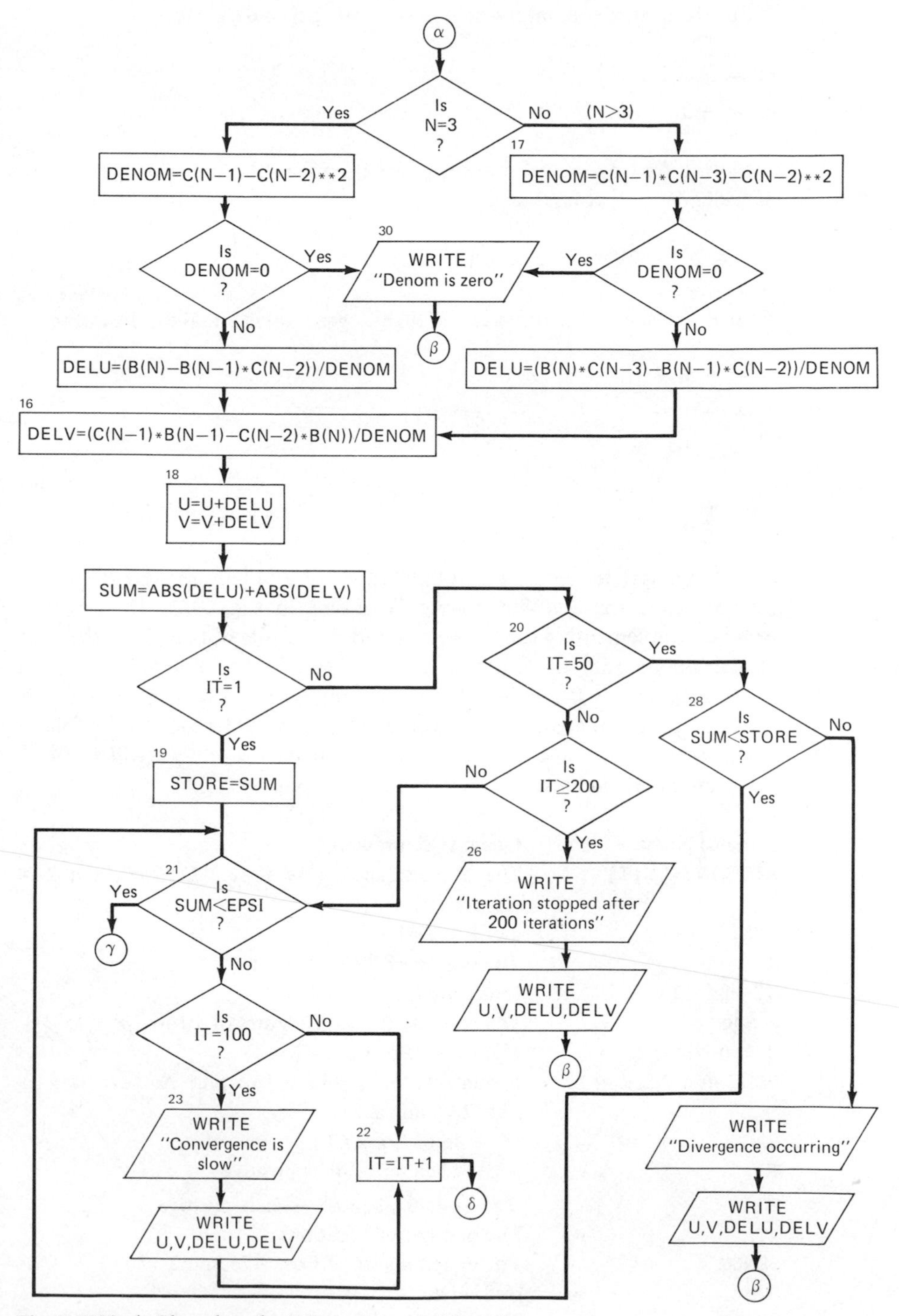

Fig. 2-22(b) / *Flow chart for Bairstow's method* (*cont.*).

Substituting these values of u and v into Eq. 2-67 gives

$$r_3 = 3$$
$$r_4 = -2$$

Noting that $b_1 = -4$ from the last iteration, the factored form of the fifth-degree polynomial is

$$(x^2 + 2x + 2)(x^2 - x - 6)(x - 4) = 0$$

Thus $r_5 = 4$, and the roots to the fifth-degree polynomial are found to be

$$r_1 = -1 + i$$
$$r_2 = -1 - i$$
$$r_3 = 3$$
$$r_4 = -2$$
$$r_5 = 4$$

A flow chart for finding all of the roots of a polynomial by Bairstow's method using the digital computer is shown in Fig. 2-22. The FORTRAN program implementing the flow chart and the results obtained for the roots of the polynomial of Example 2-12 follow. In the FORTRAN program $u_1 = 0$ and $v_1 = 0$. Many of the variable names which appear in the FORTRAN program are also used in the flow chart. Reference to the following variable names and the quantities which they represent will facilitate study of the flow chart.

Variable Name	Quantity Represented
A(I),B(I),C(I)	The a's, b's, and c's of Eqs. 2-22, 2-69, and 2-74, respectively
EPSI	ε of Eq. 2-81
N	The degree of the polynomial
UI and VI	Initial values of u and v
P and Q	The real and imaginary parts of a root, $p + iq$
U and V	The u and v of Eq. 2-68
DELU and DELV	Increments of u and v, Δu and Δv, as given by Eqs. 2-76 and 2-77
RAD	$u^2 - 4v$ of Eq. 2-67
W	$-u/2$ when $u^2 - 4v$ is positive
Z	$\sqrt{u^2 - 4v}/2$ when $u^2 - 4v$ is positive
IT	The number of iterations
DENOM	The denominator of Eqs. 2-76 and 2-77
SUM	$\lvert\Delta u\rvert + \lvert\Delta v\rvert$
STORE	The first value of $\lvert\Delta u\rvert + \lvert\Delta v\rvert$ calculated

It should be noted in studying the flow chart that the first value of $|\Delta u| + |\Delta v|$ calculated is stored and later compared with the fiftieth value, if that many iterations are used. If the fiftieth value exceeds the first value, a message is printed that divergence is occurring and execution stops after printing current values of u, v, Δu, and Δv. If divergence is not occurring but the condition $|\Delta u| + |\Delta v| < \varepsilon$ has not been satisfied after 100 iterations, a message is printed that convergence is slow, current values of u, v, Δu, and Δv are printed, and iterating continues. If the condition $|\Delta u| + |\Delta v| < \varepsilon$ has not been satisfied after 200 iterations, a message is printed that iterating is being stopped after 200 iterations, and current values of u, v, and Δu, and Δv are printed.

```
C ROOTS OF POLYNOMIAL BY BAIRSTOWS METHOD
      DIMENSION A(30),B(30),C(30)
      WRITE(6,2)
    2 FORMAT('1',9X,'REAL PART',3X,'IMAGINARY PART',3X,
                                                  'ITERATIONS'//)
      READ(5,3) UI,VI,EPSI,N
    3 FORMAT(3F10.0,I2)
      READ(5,4) (A(I),I=1,N)
    4 FORMAT(10F8.0)
C SEE IF N=0,N=1,OR N IS GREATER THAN 1
   40 IF(N-1) 100,5,7
    5 P=-A(1)
      Q=0.
      IT=1
      WRITE(6,6) N,P,Q,IT
    6 FORMAT(' ','X(',I2,') =',2X,F8.4,6X,F8.4,10X,I3)
      GO TO 100
C SEE IF N=2 OR IF N IS GREATER THAN 2
    7 IF(N.EQ.2) GO TO 8
      GO TO 13
    8 U=A(1)
      V=A(2)
      IT=1
    9 P=-U/2.
      RAD=U**2-4.*V
C CHECK THE SIGN OF U**2-4.*V
      IF(RAD.GT.0.) GO TO 12
      RAD=-RAD
      Q=SQRT(RAD)/2.
      WRITE(6,6) N,P,Q,IT
      N=N-1
      Q=-Q
   90 WRITE(6,6) N,P,Q,IT
   10 N=N-1
C CHECK TO SEE IF N IS GREATER THAN ZERO
      IF(N.LE.0) GO TO 100
      DO 11 I=1,N
   11 A(I)=B(I)
      GO TO 40
```

Continued overleaf

```
   12 Q=SQRT(RAD)/2.
      W=P
      Z=Q
      P=P+Q
      Q=0
      WRITE(6,6) N,P,Q,IT
      N=N-1
      P=W-Z
      GO TO 90
   13 U=UI
      V=VI
      IT=1
C CALCULATE THE B VALUES
   50 B(1)=A(1)-U
      B(2)=A(2)-B(1)*U-V
      DO 14 K=3,N
   14 B(K)=A(K)-B(K-1)*U-B(K-2)*V
C CALCULATE THE C VALUES
      C(1)=B(1)-U
      C(2)=B(2)-C(1)*U-V
      M=N-1
      DO 15 K=3,M
   15 C(K)=B(K)-C(K-1)*U-C(K-2)*V
C CALCULATE DELU AND DELV
      IF(N.GT.3) GO TO 17
      DENOM=C(N-1)-C(N-2)**2
      IF(DENOM.EQ.0.) GO TO 30
      DELU=(B(N)-B(N-1)*C(N-2))/DENOM
   16 DELV=(C(N-1)*B(N-1)-C(N-2)*B(N))/DENOM
      GO TO 18
   17 DENOM=C(N-1)*C(N-3)-C(N-2)**2
      IF(DENOM.EQ.0) GO TO 30
      DELU=(B(N)*C(N-3)-B(N-1)*C(N-2))/DENOM
      GO TO 16
C CALCULATE NEW U AND V VALUES
   18 U=U+DELU
      V=V+DELV
      SUM=ABS(DELU)+ABS(DELV)
C STORE THE FIRST SUM CALCULATED
      IF(IT.EQ.1) GO TO 19
      GO TO 20
   19 STORE=SUM
      GO TO 21
   20 IF(IT.EQ.50) GO TO 28
      IF(IT.GE.200) GO TO 26
   21 IF(SUM.LE.EPSI) GO TO 9
      IF(IT.EQ.100) GO TO 23
   22 IT=IT+1
      GO TO 50
   23 WRITE(6,24)
   24 FORMAT(' ',10X,'CONVERGENCE IS SLOW')
      WRITE(6,25) U,V,DELU,DELV
```

```
   25 FORMAT(' ','U=',E14.7,3X,'V=',E14.7,3X,'DELU=',
                                                  E14.7,3X,'DELV=',
     *E14.7)
      GO TO 22
   26 WRITE(6,27)
   27 FORMAT(' ',10X,'ITERATING STOPPED AFTER 200 ITERATIONS')
      WRITE(6,25) U,V,DELU,DELV
      GO TO 100
C SEE IF SUM AFTER 50 ITERATIONS EXCEEDS FIRST SUM STORED
   28 IF(SUM.LT.STORE) GO TO 21
      WRITE(6,29)
   29 FORMAT(' ',10X,'DIVERGENCE OCCURRING')
      WRITE(6,25) U,V,DELU,DELV
      GO TO 100
   30 WRITE(6,31)
   31 FORMAT(' ',10X,'DENOM IS ZERO')
      GO TO 100
  100 STOP
      END
```

The roots of the polynomial determined by the computer program are

	REAL PART	IMAGINARY PART	ITERATIONS
X(5) =	3.0000	0.0	14
X(4) =	−2.0000	0.0	14
X(3) =	−1.0000	1.0000	3
X(2) =	−1.0000	−1.0000	3
X(1) =	4.0000	0.0	1

Problems

2-1 The equation

$$x^2(1 - \cos x \cosh x) - \gamma \sin x \sinh x = 0$$

is associated with the flexural vibrations of a missile subjected to a thrust T. The thrust is directly related to the parameter γ appearing in the equation, and the positive roots of the equation determine the configurations and natural frequencies of oscillation of a missile in flight. This information is pertinent to the design of guidance and control systems for such missiles.

Determine the first three positive (nonzero) roots of the equation for a range of γ from 0.1 to 3.0 ($0.1 \leq \gamma \leq 3.0$). Use increments of 0.1 for γ until $\gamma = 1.0$, after which increments of 0.5 should be used until the final value of $\gamma = 3.0$ is reached.

Beginning with $\gamma = 0.1$, determine the first three positive roots of the equation. Then increment γ, and again determine the first three positive roots of the equation, and so forth. It is obvious from inspection of the equation that $x = 0$ is a root, so an initial value of $x = 0.1$ may be used to start the search for the roots required.

Use the incremental-search method of Sec. 2-2 to locate approximate values of the roots, and then converge to the root values using

a. A continuation of the incremental-search method.

b. The bisection method.

c. The false-position method.

d. The secant method.

Compare your results with the curves given in the figure accompanying this problem.

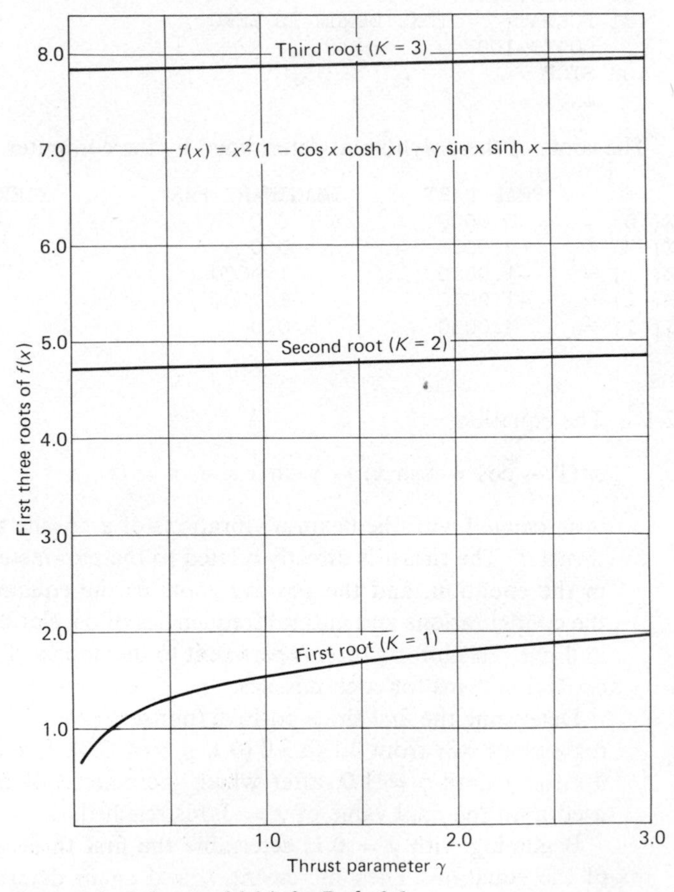

Prob. 2-1 / ***Roots of $f(x)$ for Prob. 2-1.***

2-2 Determine the first three positive nonzero roots of the equation

$$\cos x \cosh x = 1$$

using

a. The incremental-search method.

b. The incremental-search method combined with the bisection method.

c. The incremental-search method combined with the false position method.

d. The incremental-search method combined with the secant method.

2-3 Determine the first four positive roots of the equation

$$\cos x \cosh x = -1$$

using the incremental-search method to locate approximate values of the roots, and the Newton-Raphson method for convergence to the root values.

2-4 An automatic washer transmission uses the linkage shown in the accompanying figure to convert rotary motion from the drive motor to a large oscillating output of the agitator shaft G. It consists of two 4-bar linkages in series.

Determine the angle β for values of θ from 0 to 360 deg, using 5-deg increments. Begin with $\theta = 0$ deg, and from a rough sketch

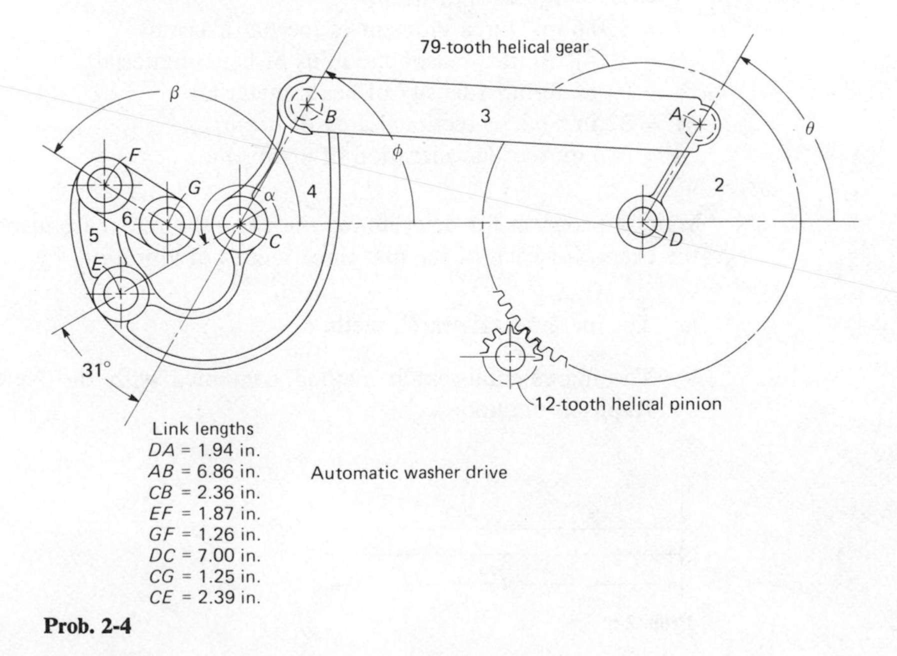

Prob. 2-4

of the linkage in this position, guess at the angles ϕ and β. Then using Freudenstein's equation and the Newton-Raphson method (see Ex. 2-4), converge to the correct value of ϕ. Add a constant to ϕ to determine α, and again using Freudenstein's equation and the Newton-Raphson method, converge to the correct value of β. For the $\theta = 5$-deg position of the linkage, the first approximations to the values of ϕ and β will be the final values obtained for the 0-deg position of the mechanism. This procedure is used for the full cycle of the mechanism.

2-5 The frequency equation of the vibrating beam shown in the accompanying figure is

$$\cos kl \cosh kl = -1$$

where

$$k^2 = \frac{p}{a}$$

$$a^2 = \frac{EIg}{A\gamma}$$

p = natural circular frequency of beam, radians/sec
l = 120 in. (length of beam)
I = 170.6 in.^{4} (area moment of inertia of beam)
E = $3(10)^6$ lb/in.^{2} (elastic modulus of beam material)
γ = 0.066 lb/in.^{3} (density of beam material)
A = 32 in.^{2} (cross-sectional area of beam)
g = 386 in./sec^2 (acceleration of gravity)

Write a program for determining the natural circular frequency of the beam, for each of the first three modes of vibration, by

a. The incremental-search method.

b. The incremental-search method combined with the Newton-Raphson method.

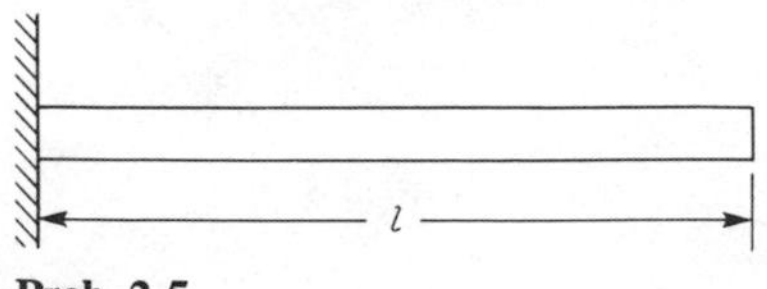

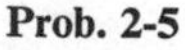

Prob. 2-5

2-6 Revise the flow chart of Fig. 2-6 so that the secant method is used, and write a computer program based on this flow chart for solving Ex. 2-1.

2-7 Make a rough plot of the function $2x^2 + 1$ versus x. On the same graph make a rough plot of the function e^x versus x. From the intersection of the two curves, make a rough guess of the value of the nonzero root of the function

$$2x^2 + 1 - e^x = 0$$

Using a pocket calculator and applying Newton's second-order method, find an accurate value of the root.

2-8 Make a rough plot of the function $x^2 + 1$ versus x, and the function $2e^{-x}$ versus x, on the same graph. From the intersection of the two curves make a rough estimate of the positive root of the function

$$x^2 + 1 - 2e^{-x} = 0$$

Using a pocket calculator and applying Newton's second-order method, determine an accurate value of the root.

2-9 Example 2-4 illustrated the application of the Newton-Raphson method in solving Freudenstein's equation. This method was used in the example to determine the angle ϕ which the output crank of a four-bar mechanism made with the line of fixed pivots for each corresponding angle θ which the input crank made with the same line.

Write a computer program, patterned after the one shown in Ex. 2-4 but using Newton's second-order method for solving Freudenstein's equation.

2-10 The frequency equations for elastic beams with various boundary conditions may be found in most texts in the field of vibrations. The frequency equation for a beam of length l, fixed at $x = 0$ and pinned at $x = l$, is

$$\tan kl = \tanh kl$$

By determining the real positive roots of this equation, the natural circular frequencies p of the various beam modes of a particular beam could then be determined from the relation

$$p = k^2 \sqrt{\frac{EI}{\gamma}}$$

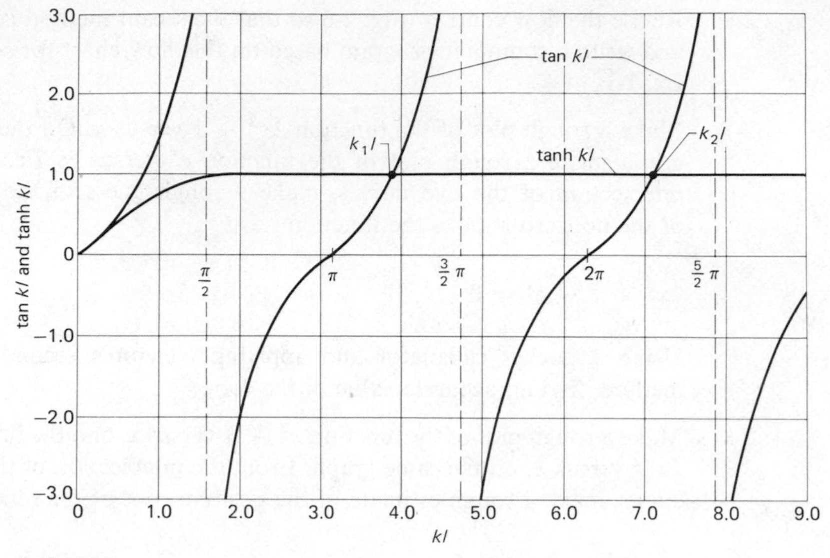

Prob. 2-10

where

E = modulus of elasticity of beam
I = moment of inertia of cross section of beam
γ = mass of beam per unit length

Write a FORTRAN program for determining the first three real positive roots of the given frequency equation. Approximate values of the first two real positive roots are shown in the accompanying figure. Combine the incremental-search method with a method of your choice (or a method assigned) for converging to the roots. Incorporate in your program a feature which will prevent overflow for kl values near $\pi/2$, $3\pi/2$, and so forth.

2-11 Determine the first nonzero positive root of the transcendental equation

$$\tan kl = kl$$

using a pocket calculator and the method of your choice. Begin by roughly plotting the functions $\tan kl$ and kl versus kl to obtain an approximate value of the root.

2-12 An interesting problem in space dynamics involves the interception and rendezvous of two vehicles orbiting the earth. One of the simplest cases is concerned with the rendezvousing of two vehicles which are initially orbiting in the same circular orbit.

Referring to the figure accompanying this problem, it can be seen that vehicles 1 and 2 are moving in a circular orbit of radius r_o, each with a velocity v_o. Vehicle 2 is leading vehicle 1 by the angle ϕ_{12}, as shown. If vehicle 1 is to intercept vehicle 2 at some position 3, vehicle 1 must be transferred into a new orbit (usually hyperbolic), as indicated by the dashed line in the figure. Vehicle 1 is placed in the transfer orbit by an impulsive thrust which causes the velocity increment Δv. The velocity v_1 of vehicle 1 on the transfer orbit is the vector sum of v_o and Δv, as shown. At position 3, vehicle 1 is given another impulsive thrust, causing a resulting Δv such that vehicle 1 again assumes a velocity v_o in the same direction as the velocity of vehicle 2, and the two vehicles move off together.

To determine Δv at either position 1 or position 3, for given values of r_o, ϕ_{12}, and ϕ_{23}, the eccentricity e of the transfer orbit must be known. This eccentricity is determined from the fact that the time t_{23} for vehicle 2 to move to position 3 must equal the time t_{13} for vehicle 1 to move to position 3. Using the equations of motion of a body in a central force field, and the geometry of conic

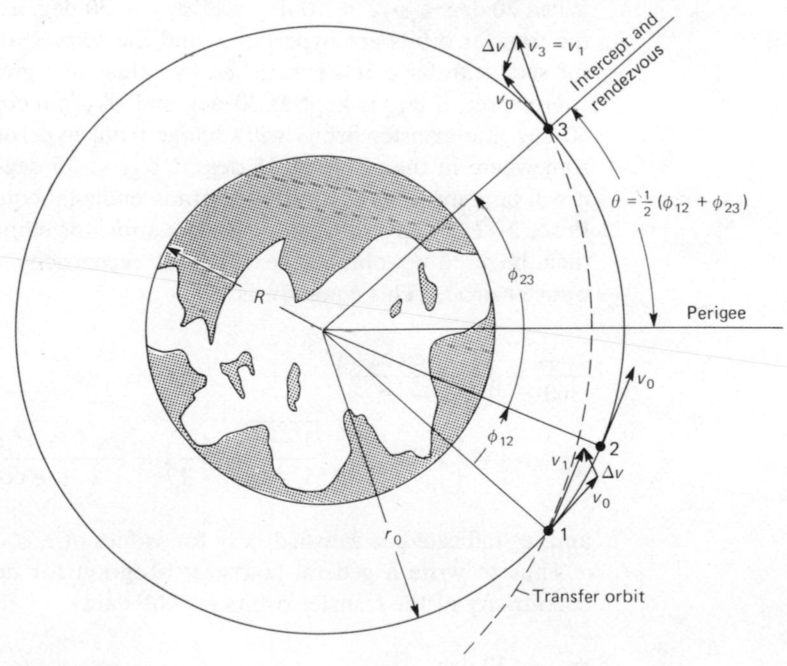

Prob. 2-12

sections, the transcendental equation resulting from making $t_{13} = t_{23}$ is

$$\frac{\pi\phi_{23}}{360} = \sqrt{\left(\frac{1 + e\cos\theta}{e^2 - 1}\right)^3} \times \left[\frac{e\sqrt{e^2 - 1}}{1 + e\cos\theta}\sin\theta - \ln\left(\frac{\sqrt{e+1} + \sqrt{e-1}\tan\theta/2}{\sqrt{e+1} - \sqrt{e-1}\tan\theta/2}\right)\right] \qquad (e > 1)$$

where $\theta = (\phi_{12} + \phi_{23})/2$, and a hyperbolic transfer orbit has been used.

Select a suitable method for solving the transcendental equation for e, and write a general FORTRAN program for obtaining values of e for the following data:

$\phi_{12} = 30$ deg
$\phi_{23} = 20$, 30, 40, and 50 deg

The eccentricity e will be greater than 1 for hyperbolic orbits.

2-13 (The background material for this problem is given in Prob. 2.12.) When 20 deg $\leq \phi_{23} \leq$ 50 deg and $\phi_{12} = 30$ deg, as in Prob. 2-12, the transfer orbits are hyperbolic, and the transcendental equation for such transfer orbits is satisfied by values of e greater than 1.

However, if ϕ_{12} is kept at 30 deg and ϕ_{23} is increased to above 50 deg, the transfer orbits will change from hyperbolic to elliptic, somewhere in the range of 55 deg $< \phi_{23} <$ 75 deg. At this point it will be impossible to satisfy the transcendental equation, given in Prob. 2-12, by any value of e. An equation for elliptic orbits must then be used to obtain the required eccentricity of the elliptic transfer orbits. This equation is

$$\frac{\pi\phi_{23}}{360} = \left(\frac{1 + e\cos\theta}{1 - e^2}\right)^{3/2} \times \left[2\tan^{-1}\left(\sqrt{\frac{1-e}{1+e}}\tan\frac{\theta}{2}\right) - \frac{e\sqrt{1 - e^2}\sin\theta}{1 + e\cos\theta}\right] \qquad (e < 1)$$

and, as indicated, is satisfied only for values of $e \leq 1$.

Thus to write a general FORTRAN program for determining the eccentricity of the transfer orbits for the data

$\phi_{12} = 30$ deg
20 deg $\leq \phi_{23} \leq$ 120 deg (use 10-deg increments)

it will be necessary to utilize both the equation given in Prob. 2-12 and the equation given above.

Write a general FORTRAN program for determining the eccentricity of the transfer orbits for the range of data given. (*Hint:* Since $e < 1$ for elliptic orbits and is thus bounded by 0 and 1, while e may approach infinity for hyperbolic orbits, begin with $\phi_{23} =$ 120 deg and then decrease by 10-deg increments.

2-14 In this problem both negative and positive real roots may be present. All root values of the polynomials are less than 10 in absolute magnitude. Determine the roots of the polynomials

a. $x^4 - 10x^3 + 35x^2 - 50x + 24 = 0$

b. $x^5 - 15x^4 + 85x^3 - 225x^2 + 274x - 120 = 0$

c. $x^5 - 9x^4 + 25x^3 - 5x^2 - 26x + 24 = 0$

d. $x^5 - 3x^4 - 23x^3 + 55x^2 + 74x - 120 = 0$

by the incremental-search method combined with either the bisection method, the false-position method, the secant method, or the Newton-Raphson method. Program your root-finder routine as a subroutine subprogram capable of finding the roots of any function defined by a function subprogram. The function name in the root finder will be a dummy name. It will appear as an argument in the root-finder argument list, and will be replaced by the real function name. Store the roots as they are determined in a one-dimensional array for return to the calling program. The calling program will print out the roots for the polynomials given above.

In Probs. 2-15 through 2-22, utilize the computer program of Ex. 2-10 to determine the Graeffe table for the given polynomial. After obtaining the table from the computer output, determine the roots of the polynomial using a pocket calculator.

2-15 $x^4 - 5x^3 + 13x^2 - 19x + 10 = 0$

2-16 $x^5 - 2x^4 - 2x^3 + 20x^2 - 47x + 30 = 0$

2-17 $x^6 + x^5 - 8x^4 + 14x^3 + 13x^2 - 111x + 90 = 0$

2-18 $x^4 - 6x^3 + 29x^2 - 52x + 80 = 0$

2-19 $x^4 - 2x^3 + 3x^2 + 4x + 4 = 0$

2-20 $x^8 + x^7 - 9x^6 + 13x^5 + 21x^4 - 125x^3 + 77x^2 + 111x - 90 = 0$

2-21 $x^6 - 8x^5 + 25x^4 - 32x^3 - x^2 + 40x - 25 = 0$

2-22 $x^8 - 7x^7 + 11x^6 + 41x^5 - 183x^4 + 231x^3 + 21x^2 - 265x + 150 = 0$

2-23 In the Bairstow derivation of this chapter, an nth-degree polynomial is divided by a quadratic factor $x^2 + ux + v$ to give an $n - 2$ degree polynomial plus a remainder. The remainder has the form

$$(x + u)b_{n-1} + b_n$$

as given in Eq. 2-69. If u and v have values such that the quadratic factor contains two roots of the nth-degree polynomial, the remainder must equal zero. Values of u and v are found which make b_n and b_{n-1} go to zero. An alternate approach is to express the remainder as

$$xb_{n-1} + (ub_{n-1} + b_n) = rx + s$$

and find values of u and v which make r and s go to zero. This may be done by expanding $r(u + \Delta u, v + \Delta v)$ and $s(u + \Delta u, v + \Delta v)$ in a Taylor series similar to the expansions of Eq. 2-71. Values of Δu and Δv are then determined which make r and s approximately zero. That is,

$$\begin{aligned} r(u + \Delta u, v + \Delta v) &= b_{n-1}(u + \Delta u, v + \Delta v) \\ &= 0 \cong b_{n-1} + \frac{\partial b_{n-1}}{\partial u}\Delta u + \frac{\partial b_{n-1}}{\partial v}\Delta v \end{aligned} \tag{a}$$

$$\begin{aligned} s(u + \Delta u, v + \Delta v) &= ub_{n-1}(u + \Delta u, v + \Delta v) + b_n(u + \Delta u, v + \Delta v) \\ &= 0 \cong ub_{n-1} + b_n + \frac{\partial[ub_{n-1} + b_n]}{\partial u}\Delta u \\ &\quad + \frac{\partial[ub_{n-1} + b_n]}{\partial v}\Delta v \end{aligned} \tag{b}$$

From Eqs. 2-72 and 2-73, Eq. a becomes

$$b_{n-1} = c_{n-2}\,\Delta u + c_{n-3}\,\Delta v \tag{c}$$

Equation b may be written as

$$\begin{aligned} 0 = ub_{n-1} + b_n &+ \left[u\frac{\partial b_{n-1}}{\partial u} + b_{n-1} + \frac{\partial b_n}{\partial u}\right]\Delta u \\ &+ \left[u\frac{\partial b_{n-1}}{\partial v} + \frac{\partial b_n}{\partial v}\right]\Delta v \end{aligned} \tag{d}$$

Replacing the partial derivatives with their corresponding c values from Eqs. 2-72 and 2-73, and replacing ub_{n-1} with

$$u[c_{n-2}\,\Delta u + c_{n-3}\,\Delta v]$$

from Eq. c, we obtain

$$b_n = [c_{n-1} - b_{n-1}]\,\Delta u + c_{n-2}\,\Delta v \tag{e}$$

Solving for Δu and Δv from Eqs. c and e, we have

$$\Delta u = \frac{\begin{vmatrix} b_{n-1} & c_{n-3} \\ b_n & c_{n-2} \end{vmatrix}}{\begin{vmatrix} c_{n-2} & c_{n-3} \\ c_{n-1} - b_{n-1} & c_{n-2} \end{vmatrix}} \tag{f}$$

and

$$\Delta v = \frac{\begin{vmatrix} c_{n-2} & b_{n-1} \\ c_{n-1} - b_{n-1} & b_n \end{vmatrix}}{\begin{vmatrix} c_{n-2} & c_{n-3} \\ c_{n-1} - b_{n-1} & c_{n-2} \end{vmatrix}} \tag{g}$$

Rewrite the Bairstow computer program given in Chap. 2 to incorporate Eqs. f and g for finding u and v instead of the expressions given by Eqs. 2-76 and 2-77. Find the roots of polynomials chosen from those given in Probs. 2-15 through 2-22 with the program you have written.

The following problems involve SI units.

2-24 Refer to Prob. 2-5 and use the following data:

$$l = 4 \text{ m}$$
$$I = 8 \times 10^{-5} \text{ m}^4$$
$$E = 200 \text{ GPa}$$
$$\gamma = 76 \times 10^3 \text{ N/m}^3$$
$$A = 0.02 \text{ m}^2$$
$$g = 9.81 \text{ m/s}^2$$

2-25 Refer to Ex. 2-3 and convert the problem parameter values to the SI system of units and obtain Eq. 2-9 so that k has the units of N · m/rad.

2-26 Referring to the equation derived in Prob. 2-25, modify the computer program of Ex. 2-3 and obtain the value of k.

3 Solution of Simultaneous Algebraic Equations

3-1 / INTRODUCTION

Engineers frequently encounter problems involving the solution of sets of simultaneous algebraic equations. Very frequently the equations are *linear*, so our discussion in this chapter will be primarily concerned with methods applicable for solving sets of linear equations. However, the last section of this chapter (Sec. 3-13) will be devoted to a discussion of the more difficult problem of solving *nonlinear* simultaneous equations.

Problems involving simultaneous linear equations arise in the areas of elasticity, electric-circuit analysis, heat transfer, vibrations, and so forth We will see later that the numerical integration of some types of ordinary and partial differential equations may be reduced to the solution of such a set of equations.

From a study of algebra, we are familiar with two common methods of solving simultaneous equations: the elimination of unknowns by combining equations, and the use of determinants (Cramer's rule). When three simultaneous equations are to be solved, Cramer's rule appears to have an advantage over the elimination method. However, the method of determinants with expansion by minors is completely impractical when large numbers of equations must be solved simultaneously. Forsythe[1] points out that the solution of n simultaneous equations by Cramer's rule, evaluating determinants in the usual manner of expansion by minors, requires $(n - 1)(n + 1)!$ multiplications. Thus the solution of ten simul-

[1] Edwin F. Beckenbach, ed., *Modern Mathematics for the Engineer*, University of California Engineering Extension Series (New York: McGraw-Hill Book Company, 1956), p. 436.

taneous equations by determinants would require 359,251,200 multiplications. With the multiplications performed on the computer at the rate of 2600/sec, approximately 38 hr would be required to obtain a solution. For $n = 26$, $3(10)^{18}$ *years* would be required to obtain a solution. Yet in some engineering problems it may be necessary to solve hundreds, or even thousands, of simultaneous equations. Obviously a solution involving the use of determinants with expansion by minors is not practical. Even if the determinants are evaluated in a more efficient manner such as by the use of an elimination method (see Sec. 3-2), the use of Cramer's rule is very inefficient.

In this chapter we will consider the solution of both *homogeneous* and *nonhomogeneous* sets of linear algebraic equations, since both types appear in engineering problems. The sets of equations with which we will be concerned will involve n equations in n unknowns having the general form

$$\begin{aligned}
a_{11}x_1 + a_{12}x_2 + \cdots + a_{1n}x_n &= C_1 \quad &\text{(a)}\\
a_{21}x_1 + a_{22}x_2 + \cdots + a_{2n}x_n &= C_2 \quad &\text{(b)}\\
a_{31}x_1 + a_{32}x_2 + \cdots + a_{3n}x_n &= C_3 \quad &\text{(c)}\\
\vdots \qquad\quad \vdots \qquad\qquad\qquad &\quad\ \vdots &\vdots\\
a_{n1}x_1 + a_{n2}x_2 + \cdots + a_{nn}x_n &= C_n
\end{aligned} \tag{3-1}$$

If the C's are not all zero, the set of equations is *nonhomogeneous*, and we will find that all the equations must be independent to obtain *unique* solutions. If the C's are all zero, the set of equations is *homogeneous*, and we will find that *nontrivial* solutions exist only if all the equations are *not* independent.

The *coefficient matrix* of the set of equations in Eq. 3-1 is

$$\mathbf{A} = \begin{bmatrix} a_{11} & a_{12} \cdots a_{1n} \\ a_{21} & \cdots a_{2n} \\ \vdots & \vdots \\ a_{n1} & \cdots a_{nn} \end{bmatrix} \tag{3-2}$$

A comparison of Eqs. 3-1 and 3-2 shows that a coefficient matrix is a matrix whose elements are the coefficients of the unknowns in the set of equations. If the constants of Eq. 3-1 are added to the coefficient matrix as a column of elements in the position shown in Eq. 3-3, the *augmented matrix* of Eq. 3-1 is formed as

$$\mathbf{A} = \begin{bmatrix} a_{11} & a_{12} \cdots a_{1n} & C_1 \\ a_{21} & a_{22} \cdots a_{2n} & C_2 \\ \vdots & & \vdots \\ a_{n1} & \cdots a_{nn} & C_n \end{bmatrix} \tag{3-3}$$

In many instances in computer programming, it may be found convenient to express the column of C's (the constants) as simply an additional column of a_{ij}'s. In such an instance the matrix of Eq. 3-3 might be expressed as

$$\mathbf{A} = \begin{bmatrix} a_{11} & a_{12} \cdots a_{1n} & a_{1,n+1} \\ a_{21} & a_{22} \cdots a_{2n} & a_{2,n+1} \\ \vdots & & \vdots \\ a_{n1} & \cdots a_{nn} & a_{n,n+1} \end{bmatrix} \tag{3-4}$$

This form will be used in the discussions in the sections to follow.

We first consider the solution of nonhomogeneous equations, and five methods commonly used on the computer will be discussed. There are two general types of methods available. Methods of the first type are called elimination methods, reduction methods, or direct methods. Of these we will consider Gauss's elimination method, the Gauss-Jordan elimination method, Cholesky's (Crout's) method, and the use of matrix inversion.

Methods of the second type are called indirect or iterative methods. Of these only the widely used Gauss-Seidel method will be presented in this chapter.

Since a nonhomogeneous set of equations, such as Eq. 3-1, describing a physical system, generally consists of linearly independent equations, it will be assumed that such is the case for the sets of equations considered, so that unique solutions exist. A unique solution exists for such a set of simultaneous equations if the coefficient matrix of the set (see Eq. 3-2) is *nonsingular*; that is, if it has linearly independent rows (and columns) or, stated in another way, if the *determinant* of the coefficient matrix is *nonzero*. We will now consider the first of the five methods.

3-2 / GAUSS'S ELIMINATION METHOD

The first method usually presented in algebra for the solution of simultaneous linear algebraic equations is one in which the unknowns are eliminated by combining equations. Such a method is known as an *elimination* method. It is called *Gaussian elimination* if a particular systematic scheme, attributed to Gauss, is used in the elimination process.

Using Gauss's method, a set of n equations in n unknowns is reduced to an *equivalent* triangular set (an equivalent set is a set having identical solution values), which is then easily solved by "back substitution," a simple procedure which will be illustrated in the following explanation.

Gauss's scheme begins by reducing a set of simultaneous equations, such as those given by Eq. 3-1, to an equivalent triangular set such as

$$\begin{aligned}
a_{11}x_1 + a_{12}x_2 + a_{13}x_3 + a_{14}x_4 + \cdots \cdots + a_{1n}x_n &= C_1 \\
a'_{22}x_2 + a'_{23}x_3 + a'_{24}x_4 + \cdots + a'_{2n}x_n &= C'_2 \\
a''_{33}x_3 + a''_{34}x_4 + \cdots + a''_{3n}x_n &= C''_3 \\
\cdots\cdots\cdots\cdots\cdots\cdots\cdots\cdots\cdots & \\
a^{n-2}_{n-1,n-1}x_{n-1} + a^{n-2}_{n-1,n}x_n &= C^{n-2}_{n-1} \\
a^{n-1}_{nn}x_n &= C^{n-1}_n
\end{aligned} \tag{3-5}$$

where the prime superscripts indicate the new coefficients which are formed in the reduction process. The actual reduction is accomplished in the following manner:

1 Equation 3-1a is divided by the coefficient of x_1 in that equation to obtain

$$x_1 + \frac{a_{12}}{a_{11}}x_2 + \frac{a_{13}}{a_{11}}x_3 + \cdots + \frac{a_{1n}}{a_{11}}x_n = \frac{C_1}{a_{11}} \tag{3-6}$$

Equation 3-6 is next multiplied by the coefficient of x_1 in Eq. 3-1b, and the resulting equation is subtracted from Eq. 3-1b, thus eliminating x_1 from Eq. 3-1b. Equation 3-6 is then multiplied by the coefficient of x_1 in Eq. 3-1c, and the resulting equation is subtracted from Eq. 3-1c to eliminate x_1 from Eq. 3-1c. In a similar manner, x_1 is eliminated from all equations of the set except the first, so that the set assumes the form

$$\begin{aligned}
a_{11}x_1 + a_{12}x_2 + a_{13}x_3 + \cdots + a_{1n}x_n &= C_1 \quad \text{(a)} \\
a'_{22}x_2 + a'_{23}x_3 + \cdots + a'_{2n}x_n &= C'_2 \quad \text{(b)} \\
a'_{32}x_2 + a'_{33}x_3 + \cdots + a'_{3n}x_n &= C'_3 \quad \text{(c)} \\
\vdots \qquad\quad \vdots \qquad\qquad \vdots \quad &\quad \vdots \qquad \vdots \\
a'_{n2}x_2 + a'_{n3}x_3 + \cdots + a'_{nn}x_n &= C'_n
\end{aligned} \tag{3-7}$$

The equation used to eliminate the unknowns in the equations which follow it is called the *pivot equation* (Eq. 3-1a in the preceding steps). In the pivot equation the coefficient of the unknown which is to be eliminated from subsequent equations is known as the *pivot coefficient* (a_{11} in the preceding steps).

2 Following the above steps, Eq. 3-7a becomes the pivot equation, and

the steps of part 1 are repeated to eliminate x_2 from all the equations following this pivot equation. This reduction yields

$$\begin{aligned} a_{11}x_1 + a_{12}x_2 + a_{13}x_3 + \cdots + a_{1n}x_n &= C_1 \quad \text{(a)} \\ a'_{22}x_2 + a'_{23}x_3 + \cdots + a'_{2n}x_n &= C'_2 \quad \text{(b)} \\ a''_{3n}x_3 + \cdots + a''_{3n}x_n &= C''_3 \quad \text{(c)} \\ a''_{43}x_3 + \cdots + a''_{4n}x_n &= C''_4 \quad \text{(d)} \\ \vdots \qquad\qquad \vdots \quad &\quad \vdots \qquad \vdots \\ a''_{n3}x_3 + \cdots + a''_{nn}x_n &= C''_n \end{aligned} \tag{3-8}$$

3 Equation 3-8c is next used as the pivot equation, and the procedure described is used to eliminate x_3 from all equations following Eq. 3-8c. This procedure, using successive pivot equations, is continued until the original set of equations has been reduced to a triangular set, such as that given by Eq. 3-5.

4 After the triangular set of equations has been obtained, the last equation in this equivalent set yields the value of x_n directly (see Eq. 3-5). This value is then substituted into the next-to-the-last equation of the triangular set to obtain a value of x_{n-1}, which is, in turn, used along with the value of x_n in the second-to-the-last equation to obtain a value of x_{n-2}, and so on. This is the *back-substitution* procedure referred to earlier.

To illustrate the method with a numerical example, let us apply these procedures to solving the following set of equations:

$$\begin{aligned} 2x_1 + 8x_2 + 2x_3 &= 14 \\ x_1 + 6x_2 - x_3 &= 13 \\ 2x_1 - x_2 + 2x_3 &= 5 \end{aligned} \tag{3-9}$$

Using the first equation as the pivot equation (the pivot coefficient is 2), we obtain

$$\begin{aligned} 2x_1 + 8x_2 + 2x_3 &= 14 \\ 2x_2 - 2x_3 &= 6 \\ -9x_2 + (0)x_3 &= -9 \end{aligned} \tag{3-10}$$

Next, using the second equation of Eq. 3-10 as the pivot equation, and repeating the procedure, the following triangular set of equations is obtained:

$$\begin{aligned} 2x_1 + 8x_2 + 2x_3 &= 14 \\ 2x_2 - 2x_3 &= 6 \\ -9x_3 &= 18 \end{aligned} \tag{3-11}$$

Finally, through back substitution, beginning with the last of Eqs. 3-11, the following values are obtained:

$$x_3 = -2$$

$$x_2 = 1$$

$$x_1 = 5$$

In the above example the solution yielded values which were exact, since only whole numbers were encountered in the elimination process. In most instances, however, fractions will be encountered in the reduction process. The computer handles fractions in decimal form to a certain limited number of decimal places, and, in handling fractions which transform to nonterminating decimals, an error is introduced in the computer solution. This is called *roundoff error*.

When only a small number of equations is to be solved, the roundoff error is small and usually does not substantially affect the accuracy of the results, but if many equations are to be solved simultaneously, the *cumulative* effect of roundoff error can introduce relatively large solution errors. For this reason, the number of simultaneous equations which can be satisfactorily solved by Gauss's elimination method, using seven to ten significant digits in the arithmetic operations, is generally limited to 15 to 20 when most or all of the unknowns are present in all of the equations (the coefficient matrix is dense). On the other hand, if only a few unknowns are present in each equation (the coefficient matrix is sparse), many more equations may be satisfactorily handled.

The number of equations which can be accurately solved also depends to a great extent on the *condition* of the system of equations. If a small relative change in one or more of the coefficients of a system of equations results in a small relative change in the solution, the system of equations is called a *well-conditioned* system. If, however, a small relative change in one or more of the coefficient values results in a large relative change in solution values, the system of equations is said to be *ill conditioned.* Ill-conditioned equations generally result from physical systems in which small changes in one or more system parameters result in large changes in system performance. Since small changes in the coefficients of an equation may result from roundoff error, the use of double-precision arithmetic and *partial pivoting* (described later in this section) or *complete pivoting* becomes very important in obtaining meaningful solutions of such sets of equations. If coefficient errors result from experimental error, then no amount of accuracy in the solution method can be expected to yield meaningful results.

As an example of an extremely ill-conditioned set, consider the following equations:

$$\begin{aligned} 1.000000x + 1.000000y &= 0 \\ 1.000000x + 0.999999y &= 1 \end{aligned} \qquad (3\text{-}12)$$

with solution

$$\begin{aligned} y &= -10^6 \\ x &= +10^6 \end{aligned}$$

Now, if we make a change of 0.000002 (only 0.0002 percent) in the coefficient of y in the second equation, making the set

$$\begin{aligned} 1.000000x + 1.000000y &= 0 \\ 1.000000x + 1.000001y &= 1 \end{aligned} \qquad (3\text{-}13)$$

the solution becomes

$$\begin{aligned} y &= +10^6 \\ x &= -10^6 \end{aligned}$$

One method that has been used to measure the condition of a set of equations is to divide each equation by the square root of the sum of the squares of its coefficients, and then evaluate the determinant of the coefficient matrix. The smaller the magnitude in comparison with ± 1, the more ill conditioned the set is. An extensive discussion of the solution of ill-conditioned linear equations is given by Wilkinson.[2]

Let us now turn our attention to the implementation of Gauss's elimination method on the computer. Since the computer can handle only numerical data, the use of matrices is required in programming. Referring again to the set of equations given in Eq. 3-9, we can write the augmented matrix of this set as

$$\mathbf{A} = \begin{bmatrix} 2 & 8 & 2 & 14 \\ 1 & 6 & -1 & 13 \\ 2 & -1 & 2 & 5 \end{bmatrix} \qquad (3\text{-}14)$$

[2] *Mathematical Methods for Digital Computers,* Vol. 2, ed. Anthony Ralston and Herbert S. Wilf (New York: John Wiley and Sons, Inc., 1967), pp. 65–93.

The computer solution is involved with the reduction of this augmented matrix to the augmented matrix of the equivalent triangular set of equations in Eq. 3-11. The necessary successive matrix reductions are accomplished by the same procedures given in steps 1 through 3, above, although in working with matrices we are now concerned with the *pivot row* and *pivot element* rather than with the pivot equation and pivot coefficient, as before.

The first matrix reduction should result in the augmented matrix of Eq. 3-10, which is

$$\mathbf{A}' = \begin{bmatrix} 2 & 8 & 2 & 14 \\ 0 & 2 & -2 & 6 \\ 0 & -9 & 0 & -9 \end{bmatrix} \tag{3-15}$$

Reviewing the procedure outlined in step 1, it can be seen that the elements of the reduced matrix $\mathbf{A}'$ can be written directly from the original matrix $\mathbf{A}$, using the following formula:

$$a'_{ij} = a_{ij} - \frac{a_{ik}}{a_{kk}}(a_{kj}) \qquad \begin{Bmatrix} k \leq j \leq m \\ k + 1 \leq i \leq n \end{Bmatrix} \tag{3-16}$$

where

a = an element of original matrix $\mathbf{A}$
a' = an element of reduced matrix $\mathbf{A}'$
i = row number of matrices
j = column number of matrices
k = number identifying pivot row
n = number of rows in matrices
m = number of columns in matrices

At this point the reader should utilize Eq. 3-16 with $k = 1$ to confirm several of the elements of the matrix $\mathbf{A}'$ given by Eq. 3-15. Then confirm several of the elements of the next reduced matrix

$$\mathbf{A}'' = \begin{bmatrix} 2 & 8 & 2 & 14 \\ 0 & 2 & -2 & 6 \\ 0 & 0 & -9 & 18 \end{bmatrix} \tag{3-17}$$

which is determined from the matrix $\mathbf{A}'$ utilizing Eq. 3-16 with $k = 2$. Note that the matrix of Eq. 3-17 is the augmented matrix of Eq. 3-11.

After obtaining the augmented matrix of the equivalent triangular set of equations, the x_i values are obtained by back substitution. (The procedure will be generalized in a following paragraph.)

Having studied the general procedure for reducing matrices, let us now concern ourselves with an efficient way of programming Gauss's elimination method for the computer. We can rewrite Eq. 3-16 as

$$a_{ij}^{k} = a_{ij}^{k-1} - \frac{a_{kj}^{k-1}}{a_{kk}^{k-1}}(a_{ik}^{k-1}) \qquad \begin{Bmatrix} k+1 \le j \le m \\ k+1 \le i \le n \end{Bmatrix} \tag{3-18}$$

where the i's, j's, k's, and so on, are as previously defined. The superscripts shown merely correspond to the primes used in identifying successive reduced matrices, in preceding discussions, and are not needed in a computer program. Note that in Eq. 3-18 the lower limit of j is $k + 1$ instead of k, as in Eq. 3-16. This limit is used because there is no need to calculate the initial zero values which occur in each row owing to the reduction process, since we know the sequence of these zero values. Furthermore, these zero values are not pertinent to the back-substitution process, so they need not appear as values in the computer.

The back-substitution procedure may be generalized in the form of the following set of equations:

$$\begin{aligned} x_n &= \frac{a_{nm}}{a_{nn}} \\ x_i &= \frac{a_{im} - \sum_{j=i+1}^{n} a_{ij}x_j}{a_{ii}} \qquad i = n-1, n-2, \ldots, 1 \end{aligned} \tag{3-19}$$

The reader should pause here to use Eq. 3-19, with the augmented matrix given by Eq. 3-17, to confirm the values of x_1, x_2, and x_3 given following Eq. 3-11.

There are two important points we have not yet considered. First, it has been tacitly assumed, thus far, that every pivot element encountered in the reduction process has been a nonzero element. If this is not the case, the procedure, as discussed, must be modified. If the pivot row has a zero pivot element, the row may be interchanged with any row *following* it which, upon becoming the pivot row, will not have a zero pivot element. For example, suppose that the pivot element of the matrix of Eq. 3-15, a_{22}, were zero instead of 2, as shown. This row could be interchanged with the last row, which would provide a new pivot row with a pivot element having a value of -9, and the procedure could then be continued.

If a pivot element should theoretically have a value of zero but actually retains a very small nonzero value, due to roundoff error, we will find that it will still be desirable to utilize row interchanges. This leads us to the second important point, namely, the effect of the magnitude of the pivot elements on the accuracy of the solution. It can be demonstrated that if the magnitude of the pivot element is appreciably smaller than the magnitude, in general, of the other elements in the matrix, the use of the small pivot element will cause a decrease in solution accuracy.[3] Therefore, for overall accuracy, each reduction should be made by using as a pivot row the row having the largest pivot element. For example, suppose that we were ready to reduce the augmented matrix of the set of equations in Eq. 3-7. The value of a'_{22} would be checked against the values of $a'_{32} \cdots a'_{n2}$, and a row interchange would be made if one of these latter values were larger than a'_{22}, so that the row containing the largest of these elements would become the pivot row for that reduction. This procedure is known as *partial pivoting*. Such a provision should always be incorporated in a computer program that is to solve fairly large numbers of simultaneous equations. *Complete pivoting* consists of interchanging columns as well as rows. This is illustrated in the program at the end of Sec. 3-10 in connection with the determination of eigenvectors.

The results shown below indicate the improvement in solution accuracy obtained in a particular case by using partial pivoting. The exact solutions are the whole numbers 1 through 10.

	Without Largest Pivot Element	With Largest Pivot Element
$x_1 =$	1.0000000	0.9999996
$x_2 =$	1.9999955	1.9999992
$x_3 =$	2.9999983	3.0000001
$x_4 =$	4.0000302	4.0000109
$x_5 =$	5.0000021	4.9999999
$x_6 =$	5.9999964	6.0000039
$x_7 =$	6.9999956	7.0000094
$x_8 =$	8.0000311	8.0000036
$x_9 =$	8.9999557	8.9999822
$x_{10} =$	9.9999531	9.9999913

Gaussian Elimination Used in Determinant Evaluation

Although the use of Cramer's rule is not recommended for the solution of linear simultaneous equations, it may at times be necessary to evaluate a determinant. For larger determinants, the use of Gaussian elimination is much more efficient than expansion by minors.

From the definition of a determinant and the rules of expansion by minors, it may be shown that

[3] The magnitudes referred to in this paragraph are the absolute values of the elements.

1 A determinant is changed only in sign by interchanging any two of its rows (or columns).
2 A determinant is not changed in value if we add (or subtract) a constant multiple of the elements of a row (or column) to the corresponding elements of another row (or column).

From the second statement above, we see that we can apply Eq. 3-18 with

$$k + 1 \le j \le n$$
$$k + 1 \le i \le n$$

to the elements of a determinant without changing its value. Thus by repeated application of Eq. 3-18 with $k = 1, 2, \ldots, n - 1$ we can reduce a determinant to upper triangular form without changing its value. The value of the determinant is then simply the product of the diagonal elements (this may easily be verified by expanding an upper triangular determinant by minors). Partial pivoting may be used to avoid having a zero pivot element and to increase the accuracy, but the resulting sign changes as indicated by the first statement above must be corrected.

The Gaussian reduction procedure is illustrated by evaluating the determinant of the coefficient matrix of Eq. 3-9. The determinant is

$$\begin{vmatrix} 2 & 8 & 2 \\ 1 & 6 & -1 \\ 2 & -1 & 2 \end{vmatrix} \tag{3-20}$$

Applying Eq. 3-18 with $k = 1$ and simply replacing the second and third elements of column 1 with zeros, we obtain

$$\begin{vmatrix} 2 & 8 & 2 \\ 1 & 6 & -1 \\ 2 & -1 & 2 \end{vmatrix} = \begin{vmatrix} 2 & 8 & 2 \\ 0 & 2 & -2 \\ 0 & -9 & 0 \end{vmatrix} \tag{3-21}$$

Now, applying Eq. 3-18 with $k = 2$ and replacing the third element of the second column with zero, we obtain

$$\begin{vmatrix} 2 & 8 & 2 \\ 0 & 2 & -2 \\ 0 & -9 & 0 \end{vmatrix} = \begin{vmatrix} 2 & 8 & 2 \\ 0 & 2 & -2 \\ 0 & 0 & -9 \end{vmatrix} \tag{3-22}$$

The value of the determinant given by the product of the main diagonal elements is -36.

EXAMPLE 3-1

Let us write a general FORTRAN program for solving up to 15 simultaneous linear algebraic equations (limited to 15 only by the DIMENSION statement).

The FORTRAN names used in the program, and the quantities they represent, are as follows:

FORTRAN Name	Quantity
N	Number of simultaneous equations (number of rows in the augmented matrix)
M	Number of columns in the augmented matrix
L	$N - 1$
A(I,J)	Elements of the augmented matrices
I	Matrix row number
J	Matrix column number
JJ	Takes on values of the row numbers which are possible pivot rows, eventually taking on the value identifying the row having the largest pivot element
BIG	Takes on values of the elements in the column containing possible pivot elements, eventually taking on the value of the pivot element used
TEMP	Temporary name used for the elements of the row selected to become the pivot row, before the interchange is made
K	Index of a DO loop taking on values from 1 to $n - 1$; it identifies the column containing possible pivot elements
KP1	$K + 1$
AB	Absolute value of a_{ik}
QUOT	Quotient a_{ik}/a_{kk}
X(I)	Unknowns of the set of equations being solved
SUM	$\sum_{j=i+1}^{n} a_{ij}x_j$
NN	Index of a DO loop taking on values from 1 to $n - 1$
IP1	$I + 1$

The flow chart for the program is shown in Fig. 3-1. The FORTRAN IV program is as follows:

```
C SOLUTION OF SIMULTANEOUS EQUATIONS BY GAUSSIAN ELIMINATION
      DIMENSION A(15,16),X(15)
      READ(5,3) N
    3 FORMAT(I3)
      M=N+1
      L=N-1
    5 READ(5,4) ((A(I,J),J=1,M),I=1,N)
```

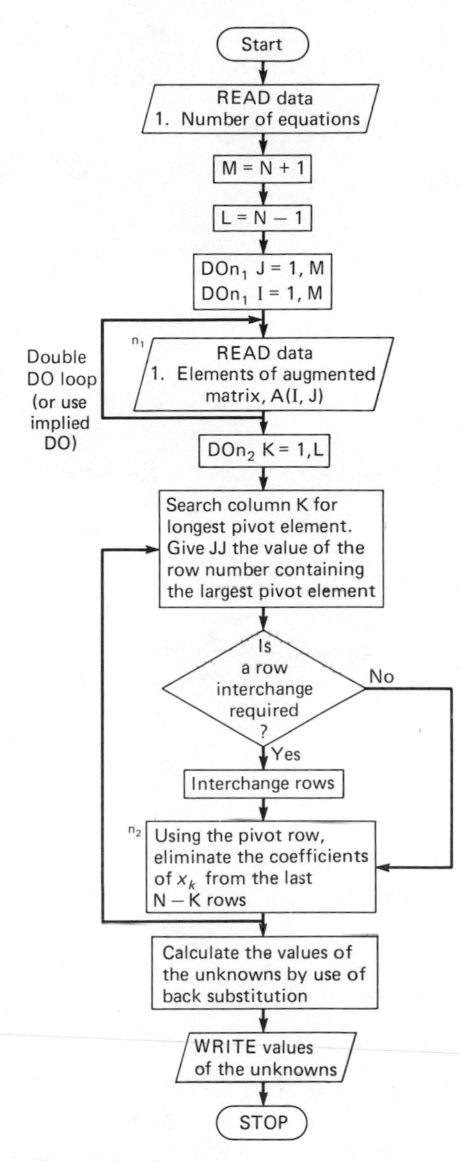

Fig. 3-1 / *Flow chart of Ex. 3-1.*

```
    4 FORMAT(8F10.0)
      DO 12 K=1,L
      JJ=K
      BIG=ABS(A(K,K))
      KP1=K+1
C
C SEARCH FOR LARGEST POSSIBLE PIVOT ELEMENT
      DO 7 I=KP1,N
      AB=ABS(A(I,K))
      IF(BIG-AB) 6,7,7
```

```
    6 BIG=AB
      JJ=I
    7 CONTINUE
C
C DECISION ON NECESSITY OF ROW INTERCHANGE
      IF(JJ-K) 8,10,8
C
C ROW INTERCHANGE
    8 DO 9 J=K,M
      TEMP=A(JJ,J)
      A(JJ,J)=A(K,J)
    9 A(K,J)=TEMP
C
C CALCULATION OF ELEMENTS OF NEW MATRIX
   10 DO 11 I=KP1,N
      QUOT=A(I,K)/A(K,K)
      DO 11 J=KP1,M
   11 A(I,J)=A(I,J)-QUOT*A(K,J)
      DO 12 I=KP1,N
   12 A(I,K)=0.
C
C FIRST STEP IN BACK SUBSTITUTION
      X(N)=A(N,M)/A(N,N)
C
C REMAINDER OF BACK-SUBSTITUTION PROCESS
      DO 14 NN=1,L
      SUM=0.
      I=N-NN
      IP1=I+1
      DO 13 J=IP1,N
   13 SUM=SUM+A(I,J)*X(J)
   14 X(I)=(A(I,M)-SUM)/A(I,I)
      WRITE(6,1)
    1 FORMAT('1','X(1) THROUGH X(N)'/)
      DO 15 I=1,N
   15 WRITE(6,2) X(I)
    2 FORMAT(' ',E14.8)
      STOP
      END
```

In the above program, various groups of statements have been identified with respect to their functions by means of comments. Note the correlation of this breakdown with the flow chart of Fig. 3-1. The reader may find it helpful to make such correlations in analyzing the various programs which follow. An application of this general program will be made in the following example.

EXAMPLE 3-2

The five equations in five unknowns appearing below result from the design of a mechanism which is a component of an automatic packaging machine. In this device it is required that a point on the mechanism link

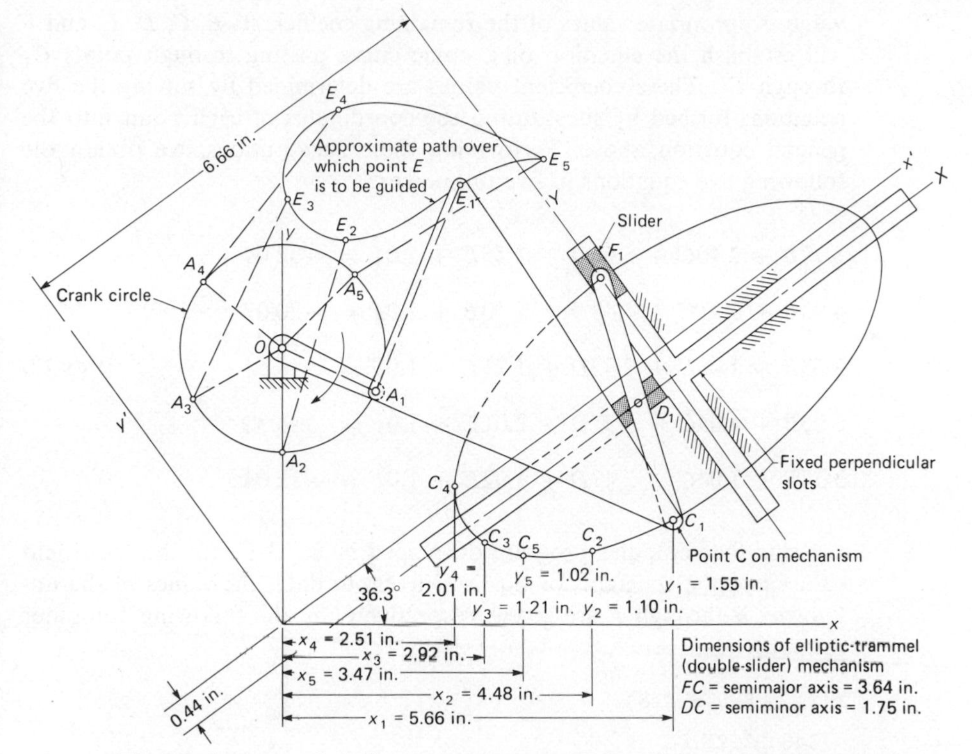

Fig. 3-2 / *Design of a linkage to guide a point E through five specified positions.*

be constrained to move over a path having the approximate shape shown in Fig. 3-2. The five points E_1 through E_5 are selected on this curve, with the requirement that the driven point E must pass through each of these five points. A crank length OA is selected, the crank circle is drawn, and for a suitable length EA (arbitrary within limits), the five positions of line EA are determined. A point C is then arbitrarily selected on the link containing line EA. The five positions of point C will in general lie on a conic section since five points determine a conic section. To complete the design of the mechanism, the equation of the conic section passing through the five positions of point C must be determined.[4]

From analytical geometry we recall that the general equation of a conic section is

$$Ax^2 + Bxy + Cy^2 + Dx + Ey + F = 0$$

The coefficient A may be considered to have a magnitude of unity, since the entire equation may be divided by the coefficient of x^2 to establish this

[4] For complete synthesis of the mechanism, see M. L. James, G. M. Smith, and J. C. Wolford, *Applied Numerical Methods for Digital Computation*, 1st ed. (New York: International Textbook Company, 1967), pp. 195–198.

value. Appropriate values of the remaining coefficients B, C, D, E, and F will establish the equation of a conic curve passing through points C_1 through C_5. These coefficient values are determined by solving the five equations formed by substituting the coordinates of each point into the general equation above. Performing these substitutions, we obtain the following five equations in five unknowns:

$$
\begin{aligned}
8.77B + 2.40C + 5.66D + 1.55E + 1.0F &= -32.04 \\
4.93B + 1.21C + 4.48D + 1.10E + 1.0F &= -20.07 \\
3.53B + 1.46C + 2.92D + 1.21E + 1.0F &= -8.53 \\
5.05B + 4.04C + 2.51D + 2.01E + 1.0F &= -6.30 \\
3.54B + 1.04C + 3.47D + 1.02E + 1.0F &= -12.04
\end{aligned}
\tag{3-23}
$$

Using the computer program developed in Ex. 3-1 with the coefficient values and the constants of Eq. 3-23 as input data, the values of the unknowns B through F are given, respectively, by the following computer printout:

```
X(1) THROUGH X(N)

-.14649792E 01
0.14580536E 01
-.60047874E 01
-.22087317E 01
0.14719152E 02
```

3-3 / GAUSS-JORDAN ELIMINATION METHOD

This method, which is a variation of the Gaussian elimination method, is suitable for solving as many as 15 to 20 simultaneous equations, with 7 to 10 significant digits used in the arithmetic operations of the computer. This procedure varies from the Gaussian method in that, when an unknown is eliminated, it is eliminated from *all* the other equations, that is, from those preceding the pivot equation as well as those following it. For example, in the preceding section describing Gauss's method, Eq. 3-7b was used as the pivot equation in eliminating x_2 from all the equations below it in obtaining Eq. 3-8. In the Gauss-Jordan method, the pivot equation would be used to eliminate x_2 from Eq. 3-7a as well. In the subsequent step, in which Eq. 3-8c was used as the pivot equation, x_3 would be eliminated from Eqs. 3-8a and 3-8b as well as from the equations below the pivot equation. This eliminates the necessity of using the back-substitution process employed in Gauss's method.

To illustrate the Gauss-Jordan elimination method, let us solve the following set of equations:

$$\begin{aligned} 2x_1 - 2x_2 + 5x_3 &= 13 && \text{(a)} \\ 2x_1 + 3x_2 + 4x_3 &= 20 && \text{(b)} \\ 3x_1 - x_2 + 3x_3 &= 10 && \text{(c)} \end{aligned} \qquad (3\text{-}24)$$

We begin by dividing the first equation of the set, Eq. 3-24a, by the coefficient of the *first* unknown in that equation, which gives us Eq. 3-25a below. We then multiply Eq. 3-25a, respectively, by the coefficient of the *first* unknown in each of the remaining equations (Eqs. 3-24b and c) to get

$$\begin{aligned} x_1 - x_2 + \tfrac{5}{2}x_3 &= \tfrac{13}{2} && \text{(a)} \\ 2x_1 - 2x_2 + 5x_3 &= 13 && \text{(b)} \\ 3x_1 - 3x_2 + \tfrac{15}{2}x_3 &= \tfrac{39}{2} && \text{(c)} \end{aligned} \qquad (3\text{-}25)$$

Now we subtract Eq. 3-25b from Eq. 3-24b and Eq. 3-25c from Eq. 3-24c, and we let Eq. 3-25a become Eq. 3-26c, obtaining

$$\begin{aligned} 5x_2 - x_3 &= 7 && \text{(a)} \\ 2x_2 - \tfrac{9}{2}x_3 &= -\tfrac{19}{2} && \text{(b)} \\ x_1 - x_2 + \tfrac{5}{2}x_3 &= \tfrac{13}{2} && \text{(c)} \end{aligned} \qquad (3\text{-}26)$$

Next, we divide Eq. 3-26a by the coefficient of the *first* unknown in that equation (the coefficient of x_2) to obtain Eq. 3-27a. We then multiply Eq. 3-27a, respectively, by the coefficient of the x_2 term in each of Eqs. 3-26b and c. This yields

$$\begin{aligned} x_2 - \tfrac{1}{5}x_3 &= \tfrac{7}{5} && \text{(a)} \\ 2x_2 - \tfrac{2}{5}x_3 &= \tfrac{14}{5} && \text{(b)} \\ -x_2 + \tfrac{1}{5}x_3 &= -\tfrac{7}{5} && \text{(c)} \end{aligned} \qquad (3\text{-}27)$$

We next subtract Eq. 3-27b from Eq. 3-26b and Eq. 3-27c from Eq. 3-26c, and we let Eq. 3-27a become Eq. 3-28c, obtaining

$$\begin{aligned} -\tfrac{41}{10}x_3 &= -\tfrac{123}{10} && \text{(a)} \\ x_1 + \tfrac{23}{10}x_3 &= \tfrac{79}{10} && \text{(b)} \\ x_2 - \tfrac{1}{5}x_3 &= \tfrac{7}{5} && \text{(c)} \end{aligned} \qquad (3\text{-}28)$$

Repeating the procedure yields

$$x_3 = 3 \qquad \text{(a)}$$
$$\tfrac{23}{10}x_3 = \tfrac{69}{10} \qquad \text{(b)} \qquad \text{(3-29)}$$
$$-\tfrac{1}{5}x_3 = -\tfrac{3}{5} \qquad \text{(c)}$$

and finally

$$x_1 = 1 \qquad \text{(a)}$$
$$x_2 = 2 \qquad \text{(b)} \qquad \text{(3-30)}$$
$$x_3 = 3 \qquad \text{(c)}$$

which constitutes the desired solution of the simultaneous equations.

Having solved Eq. 3-24 by the procedure shown, let us see how we can obtain the same result by working with just the coefficients and constants of these equations. Associated with Eq. 3-24 is a matrix **A**, which is the *augmented* matrix of the set of equations. This matrix is

$$\mathbf{A} = \begin{bmatrix} 2 & -2 & 5 & 13 \\ 2 & 3 & 4 & 20 \\ 3 & -1 & 3 & 10 \end{bmatrix}$$

We next associate an augmented matrix **B** with Eq. 3-26, in which the first column of coefficients (0, 0, and 1) is omitted, and we write

$$\mathbf{B} = \begin{bmatrix} 5 & -1 & 7 \\ 2 & -\frac{9}{2} & -\frac{19}{2} \\ -1 & \frac{5}{2} & \frac{13}{2} \end{bmatrix}$$

Having established matrix **B** from Eq. 3-26, we note, from the way in which Eq. 3-26 was obtained from Eq. 3-24 in the first procedure, that we can write the elements of **B** directly from the elements of **A** by using the following formulas:

$$b_{i-1,j-1} = a_{ij} - \frac{a_{1j}a_{i1}}{a_{11}} \begin{cases} 1 < i \le n \\ 1 < j \le m \\ a_{11} \ne 0 \end{cases} \qquad \text{(3-31)}$$

$$b_{n,j-1} = \frac{a_{1j}}{a_{11}} \begin{cases} 1 < j \le m \\ a_{11} \ne 0 \end{cases} \qquad \text{(3-32)}$$

Equation 3-31 is used to find all elements of the new matrix **B** except those making up the last row of that matrix. For determining the elements of the last row of the new matrix, Eq. 3-32 is used. In these equations,

i = row number of old matrix **A**
j = column number of old matrix **A**
n = maximum row number
m = maximum column number
a = an element of old matrix **A**
b = an element of new matrix **B**

Now, if we write the augmented matrix of Eq. 3-28, this time omitting the first two columns (elements 0, 1, 0 and 0, 0, 1), we get

$$\mathbf{C} = \begin{bmatrix} -\frac{41}{10} & -\frac{123}{10} \\ \frac{23}{10} & \frac{79}{10} \\ -\frac{1}{5} & \frac{7}{5} \end{bmatrix}$$

Again, we note that the elements of **C** can be obtained directly from the elements of **B** by utilizing Eqs. 3-31 and 3-32, where **C** is now the *new* matrix and **B** is the *old* matrix.

The augmented matrix of Eq. 3-30, with the first three columns (1, 0, 0; 0, 1, 0; and 0, 0, 1) omitted, is simply the matrix whose elements are the solution of the set of simultaneous equations with which we started,

$$\mathbf{D} = \begin{bmatrix} 1 \\ 2 \\ 3 \end{bmatrix}$$

Just as with the previous matrices, matrix **D** can be obtained from matrix **C** by application of Eqs. 3-31 and 3-32. The elements shown in **D** are associated, respectively, with x_1, x_2, and x_3 by the above procedure.

Thus it can be seen that the roots of a set of n simultaneous equations can be obtained by successive applications of Eqs. 3-31 and 3-32 to get n new matrices, the last of which will be a *column* matrix whose elements are the roots of the set of simultaneous equations.

In a computer program employing the method just described, only the FORTRAN names A and B would be used for pairs of successive matrices, rather than calling them A, B, C, D, and so forth, as was done in describing the method. After obtaining the **B** matrix from the **A** matrix, the elements of **B** would be assigned to the name A, and a new matrix with the name B obtained. This will be illustrated in the program shown in Ex. 3-3 which follows later in this section.

It is possible that the pivot element a_{11} of one or more of the matrices obtained during the elimination process could have a zero value, in which case divisions by a_{11} in Eqs. 3-31 and 3-32 would be invalid. If this occurs, the first row can be interchanged with one of the first $m - 1$ rows of the matrix which has a first element not equal to zero, where m is the number of columns in the current matrix. The use of a row beyond row $m - 1$ as a pivot row in the elimination process would reintroduce one of the previously eliminated unknowns into the equations. The elimination process then continues as before, until a solution is reached.

In instances in which a_{11} is not zero but is very small in comparison with the general magnitude of the other elements of the matrix, its use could cause a decrease in solution accuracy, as discussed in Gauss's elimination method. For improved accuracy the applicable row having the largest potential pivot element (the row among the first $m - 1$ rows having the largest pivot element) should be placed in the first or pivot position if it is not already there. This would be accomplished by checking the absolute value of a_{11} against the absolute values of $a_{21}, a_{31}, \ldots, a_{m-1,1}$ and making the appropriate row interchange if one of these latter values were larger than $|a_{11}|$. This procedure, called *partial pivoting*, should always be incorporated in a computer program for solving fairly large numbers of simultaneous equations.

It should be noted that, if it were desired to solve a second set of simultaneous equations which differed from a first set only in the constant terms which appeared, both sets could be solved at the same time by representing each set of constants as a separate column in augmenting the coefficient matrix. Actually, two or more sets of simultaneous equations, differing only in their constant terms, can be solved in a single elimination procedure by placing the constants of each set in a separate column to the right of the coefficient columns in the augmented matrix, and applying Eqs. 3-31 and 3-32 until a reduced matrix is obtained which has the same number of columns as the number of *sets* of simultaneous equations being solved. Such a solution will be illustrated in Sec. 3-6.

An Improved Gauss-Jordan Algorithm

In the computer program of Ex. 3-3 at the end of this section, storage space is reserved for both the **A** and **B** arrays. To provide core space for both matrices would generally not be a problem. However, if a fairly large number of equations were being solved on a smaller computer, or if the program were to be frequently used on a computer where charges were based in part on the amount of core used, it would be desirable to store the new matrix generated by Eqs. 3-31 and 3-32 in the same location as the original matrix. To see how these equations must be modified to accomplish this core-saving feature, let us consider the system of equations given earlier by Eqs. 3-9 and 3-14, which are repeated here for convenience.

$$\begin{aligned} 2x_1 + 8x_2 + 2x_3 &= 14 \\ x_1 + 6x_2 - x_3 &= 13 \\ 2x_1 - x_2 + 2x_3 &= 5 \end{aligned} \tag{3-33}$$

and augmented matrix

$$\mathbf{A} = \begin{bmatrix} 2 & 8 & 2 & 14 \\ 1 & 6 & -1 & 13 \\ 2 & -1 & 2 & 5 \end{bmatrix} \tag{3-34}$$

The first step in the Gauss-Jordan elimination method is to divide the pivot equation (the first of Eq. 3-33) through by the coefficient of x_1 in that equation. The pivot row in matrix **A** is identified by k, and $k = 1$ in the first reduction. The pivot equation is then multiplied by the coefficients of x_1 in the second and third equations, and the resulting pivot equations are subtracted, respectively, from the second and third equations to yield an equivalent set in which x_1 has been eliminated from the second and third equations. In the Gauss-Jordan procedure previously given, the pivot equation (or row) divided by the pivot element was brought to the bottom position so that the new pivot equation (or row) was always in the top position. However, if we wish to store the new reduced matrix in the same location as the old matrix, it is not convenient to do this. The new matrix of the first reduction, with the pivot equation of this reduction divided through by the pivot element a_{kk}, is

$$\mathbf{A}' = \begin{bmatrix} 1 & 4 & 1 & 7 \\ 0 & 2 & -2 & 6 \\ 0 & -9 & 0 & -9 \end{bmatrix} \tag{3-35}$$

We observe from the manner in which the new equivalent set is determined that the elements of **A**′ may be obtained from the elements of **A** by use of the equations

$$\left.\begin{aligned} a'_{kj} &= \frac{a_{kj}}{a_{kk}} \qquad \text{(a)} \\ a'_{ij} &= a_{ij} - a_{ik}a'_{kj} \qquad \text{(b)} \end{aligned}\right\} \tag{3-36}$$

where Eq. 3-36a is applied first, and

$1 \leq i \leq 3$ (except that $i \neq k$)

$1 \leq j \leq 4$

$k = 1$

If the elements of $\mathbf{A}'$ were to be stored in locations different from the elements of $\mathbf{A}$, the use of Eq. 3-36 would present no problem. However, with the elements of $\mathbf{A}'$ actually replacing the elements of $\mathbf{A}$ in core, the following difficulties are noted:

1 When Eq. 3-36a is applied with $j = k$ (1 in this case), the pivot element a_{kk} is replaced with a new value (unity). This new pivot value would be used in subsequent calculations of a'_{kj} (with $j = 2, 3,$ and 4 in this case), whereas the old one should be used.

2 In applying Eq. 3-36b, the element a_{ik} ($k = 1$, $i = 2$) is used in the calculation of all of the elements a'_{ij} of row 2, and a_{ik} ($k = 1$, $i = 3$) is used in calculating all a'_{ij} of row 3. These elements a_{ik} would be replaced by their new values (zero) whenever $j = k$ (1 in this case). The use of the zero values instead of the old values of 1 and 2 (a_{21} and a_{31}, respectively) would give incorrect values for subsequent a'_{ij} values calculated in this step.

These difficulties may be overcome by observing that the elements of column 1 of the $\mathbf{A}'$ matrix, Eq. 3-35, are never used in any later steps of the reduction process. Therefore, they can be left just as they were in the $\mathbf{A}$ matrix (which they need to be for the present step), without affecting calculations in subsequent steps. This may be accomplished by letting the range of j be given by

$$2 \leq j \leq 4$$

in Eq. 3-36, so that the $\mathbf{A}'$ matrix as actually stored is

$$\mathbf{A}' = \begin{bmatrix} (2) & 4 & 1 & 7 \\ (1) & 2 & -2 & 6 \\ (2) & -9 & 0 & -9 \end{bmatrix} \tag{3-37}$$

The elements of the first column are shown in parentheses since they would normally be 1, 0, 0 at the end of the first reduction step. It should be recalled that in the previous Gauss-Jordan procedure, this column was not carried in the $\mathbf{B}$ matrix.

The $\mathbf{A}'$ matrix of Eq. 3-36 will now be considered an $\mathbf{A}$ matrix, and a new $\mathbf{A}'$ matrix determined by using the second row as the pivot row ($k = 2$). Thus

pivot element

$$\mathbf{A} = \begin{bmatrix} (2) & 4 & 1 & 7 \\ (1) & 2 & -2 & 6 \\ (2) & -9 & 0 & -9 \end{bmatrix} \tag{3-38}$$

The pivot equation is now used to eliminate x_2 from the first and third equations. The pivot equation is first divided through by the pivot element 2. It is then multiplied by the coefficients of x_2 in the first and third equations, and the resulting equations subtracted, respectively, from the first and third equations of the set to give a new equivalent set in which x_2 has been eliminated from the first and third equations. With the pivot equation of the second reduction divided through by the pivot element 2, this equivalent set is

$$\begin{aligned} 1x_1 + 0x_2 + 5x_3 &= -5 \\ 0x_1 + 1x_2 - 1x_3 &= 3 \\ 0x_1 + 0x_2 - 9x_3 &= 18 \end{aligned} \tag{3-39}$$

with augmented matrix

$$\mathbf{A}' = \begin{bmatrix} (2) & 0 & 5 & -5 \\ (1) & 1 & -1 & 3 \\ (2) & 0 & -9 & 18 \end{bmatrix} \tag{3-40}$$

Again we note that the $\mathbf{A}'$ matrix of Eq. 3-40 could be obtained from the $\mathbf{A}$ matrix of Eq. 3-38 by the use of Eq. 3-36 with

$$\begin{aligned} &k = 2 \\ &1 \le i \le 3 \qquad \text{(except that } i \ne k) \\ &2 \le j \le 4 \end{aligned}$$

The same difficulties previously described again occur when the elements of the second column of **A**, Eq. 3-38, are replaced by the new elements shown in column 2 of the matrix $\mathbf{A}'$, Eq. 3-40, before the new elements of columns 3 and 4 are calculated. As before, we simply change the range of j so that it becomes

$$3 \le j \le 4$$

and the elements of column 2 are left unchanged. The $\mathbf{A}'$ matrix as it is actually stored, now called the **A** matrix in preparation for the final step, is

pivot element

$$\mathbf{A} = \begin{bmatrix} (2) & (4) & 5 & -5 \\ (1) & (2) & -1 & 3 \\ (2) & (-9) & -9 & 18 \end{bmatrix} \tag{3-41}$$

The final equivalent set is determined using the third equation as the pivot equation. It is divided through by the pivot element -9 in Eq. 3-41, and then multiplied by the coefficients of x_3 in the first and second equations. The resulting pivot equations are subtracted, respectively, from the first and second equations, and the pivot equation divided by the pivot element to give the final equivalent set

$$\begin{aligned} 1x_1 + 0x_2 + 0x_3 &= 5 \\ 0x_1 + 1x_2 + 0x_3 &= 1 \\ 0x_1 + 0x_2 + 1x_3 &= -2 \end{aligned} \tag{3-42}$$

with augmented matrix

$$\mathbf{A}' = \begin{bmatrix} (2) & (4) & 0 & 5 \\ (1) & (2) & 0 & 1 \\ (2) & (-9) & 1 & -2 \end{bmatrix} \tag{3-43}$$

This matrix $\mathbf{A}'$ can be obtained from the $\mathbf{A}$ matrix of Eq. 3-41 by the application of Eq. 3-36 with

$$k = 3$$

$$1 \leq i \leq j \qquad \text{(except that } i \neq k\text{)}$$

$$3 \leq j \leq 4$$

As before, however, we change the range of j to

$$4 \leq j \leq 4 \qquad (\text{or } j = 4)$$

and do not actually calculate the new elements for the third column of $\mathbf{A}'$, Eq. 3-43. The final $\mathbf{A}'$ matrix as actually stored is

$$\mathbf{A}' = \begin{bmatrix} (2) & (4) & (5) & 5 \\ (1) & (2) & (-1) & 1 \\ (2) & (-9) & (-9) & -2 \end{bmatrix} \tag{3-44}$$

The values of the unknowns, x_1, x_2, and x_3 are 5, 1, and -2, respectively (the elements of the fourth column of the final $\mathbf{A}'$ matrix).

Equation 3-36 can be made general for all steps of the reduction process as follows:

$$\left.\begin{aligned} a'_{kj} &= \frac{a_{kj}}{a_{kk}} & &\text{(a)} \\ a'_{ij} &= a_{ij} - a_{ik}a'_{kj} & &\text{(b)} \\ 1 &\leq i \leq n & &\text{(except that } i \neq k\text{)} \\ k + 1 &\leq j \leq n + 1 \end{aligned}\right\} \quad k = 1, 2, 3, \ldots, n \tag{3-45}$$

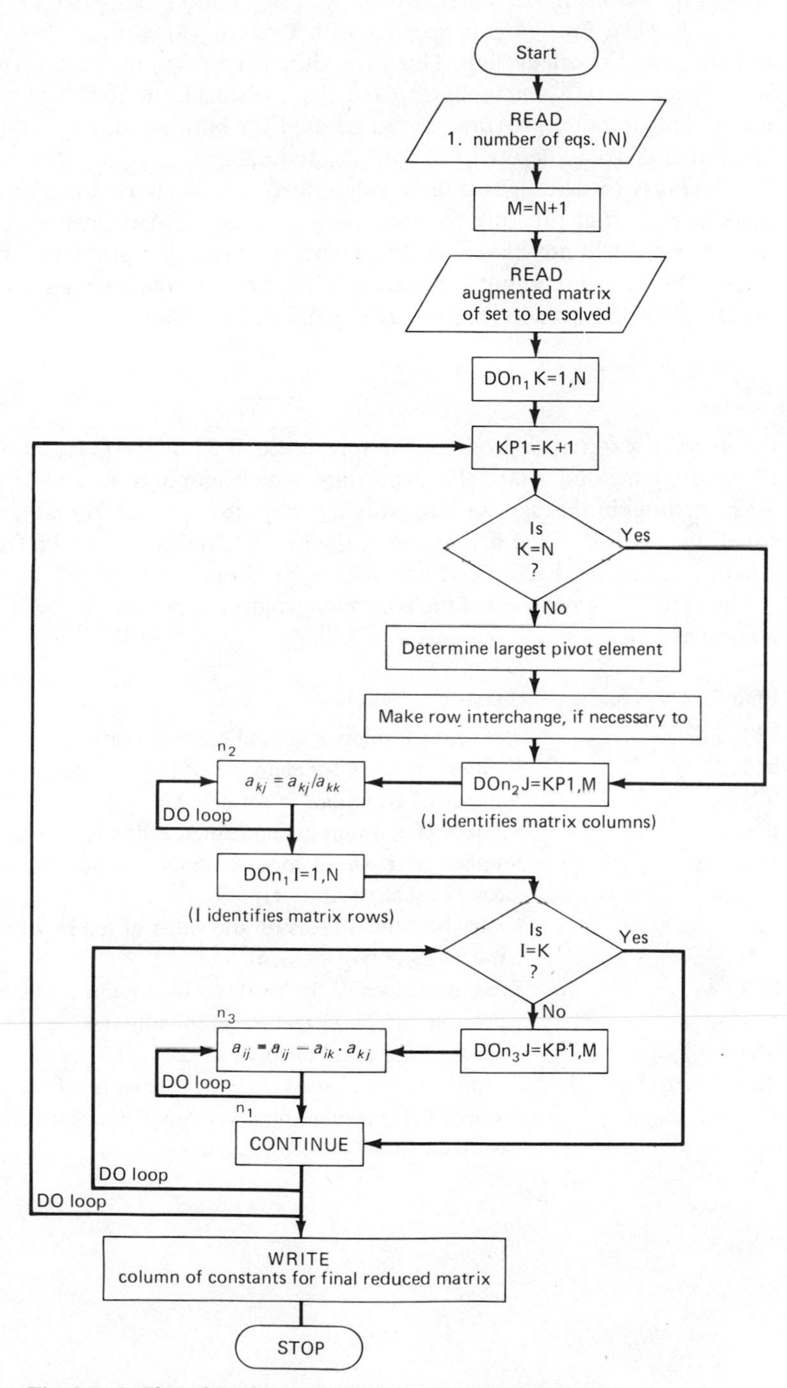

Fig. 3-3 / *Flow chart for improved Gauss-Jordan algorithm.*

where Eq. 3-45a is applied first with $k = 1$, say, and j cycling from $k + 1$ to $n + 1$. Then Eq. 3-45b is applied with k still equal to the same value and with i and j both cycling. This procedure is repeated for each k value. Since the a' elements actually replace the a elements in core, the name A(I, J) in a computer program would be used for both a_{ij} and a'_{ij}, and the name A(K, J) would be used for both a_{kj} and a'_{kj}.

The Gauss-Jordan algorithm just described is flow-charted in Fig. 3-3. It includes partial pivoting for improved accuracy and to insure that a pivot element will not have a value of zero. It is left as a problem for the student to write a computer program utilizing this space-saving Gauss-Jordan algorithm (see Prob. 3-10 at the end of this chapter).

EXAMPLE 3-3

Let us write a FORTRAN program for solving Eq. 3-23 by the Gauss-Jordan elimination method using the procedure which employs Eqs. 3-31 and 3-32. Although the set we are solving contains only 5 simultaneous equations, we will write a program capable of handling up to 15 simultaneous equations. Partial pivoting will be used in the program.

The variable names used in the FORTRAN program represent the following quantities:

FORTRAN Name	**Quantity**
A(I,J)	Elements of augmented and reduced matrices
B(I,J)	Temporary name for elements of reduced matrices
N	Number of equations in set being solved
M	Number of columns in augmented and reduced matrices
K	Number of rows in matrix which are possible pivot rows (equal to $M - 1$)
JJ	A variable which takes on the value of the row having the largest pivot element
BIG	Takes on values of the elements in column 1 which are possible pivot elements, eventually taking on the value of the pivot element used
TEMP	Temporary name used for the elements of the row selected to become the pivot row before interchange is made

The FORTRAN program is as follows:

```
C SOLUTION OF SIMULTANEOUS EQUATIONS BY GAUSS-JORDAN METHOD
C THIS PROGRAM USES PARTIAL PIVOTING
      DIMENSION A(15,16),B(15,15)
      WRITE(6,1)
    1 FORMAT('1','SOLUTION OF SIMULTANEOUS EQUATIONS')
      WRITE(6,2)
    2 FORMAT(' ','BY GAUSS-JORDAN ELIMINATION METHOD'/)
      WRITE(6,3)
    3 FORMAT(' ',7X,'X(1) THROUGH X(N)'/)
C READ NUMBER OF EQUATIONS TO BE SOLVED
      READ(5,4) N
    4 FORMAT(I2)
      M=N+1
C READ AUGMENTED MATRIX OF SET TO BE SOLVED
      READ(5,5) ((A(I,J),J=1,M),I=1,N)
    5 FORMAT(8F10.0)
C SEARCH FOR LARGEST PIVOT ELEMENT
    6 K=M-1
      IF(K.EQ.1) GO TO 11
      JJ=1
      BIG=ABS(A(1,1))
      DO 8 I=2,K
      AB=ABS(A(I,1))
      IF(BIG-AB) 7,8,8
    7 BIG=AB
      JJ=I
    8 CONTINUE
C MAKE DECISION ON ROW INTERCHANGE
      IF(JJ-1) 9,11,9
C MAKE ROW INTERCHANGE
    9 DO 10 J=1,M
      TEMP=A(JJ,J)
      A(JJ,J)=A(1,J)
   10 A(1,J)=TEMP
C CALCULATE THE ELEMENTS OF THE B MATRIX
   11 DO 12 J=2,M
      DO 12 I=2,N
   12 B(I-1,J-1)=A(I,J)-A(1,J)*A(I,1)/A(1,1)
      DO 13 J=2,M
   13 B(N,J-1)=A(1,J)/A(1,1)
C REDUCE COLUMN COUNTER BY ONE
      M=M-1
C ASSIGN THE B MATRIX TO THE NAME A
      DO 14 J=1,M
      DO 14 I=1,N
   14 A(I,J)=B(I,J)
C CHECK TO SEE IF THE A MATRIX CONSISTS OF JUST ONE COLUMN
      IF(M-1) 6,15,6
C PRINT THE SINGLE COLUMN OF THE A MATRIX
   15 WRITE(6,16) (A(I,1),I=1,N)
   16 FORMAT(' ',10X,E14.7)
      STOP
      END
```

The output results of the program are shown below:

```
SOLUTION OF SIMULTANEOUS EQUATIONS
BY GAUSS-JORDAN ELIMINATION METHOD
         X(1) THROUGH X(N)
            -0.1464931E 01
             0.1458158E 01
            -0.6004849E 01
            -0.2209171E 01
             0.1471957E 02
```

3-4 / CHOLESKY'S METHOD

Cholesky's method, also known as Crout's method, the method of matrix decomposition, and the method of matrix factorization, is more economical of computer time than other elimination methods. As a result it has been used extensively in some of the larger structural analysis programs.[5]

For illustrative purposes, let us discuss the solution of three simultaneous equations in three unknowns. The set is represented by the matrix equation[6]

$$\begin{bmatrix} a_{11} & a_{12} & a_{13} \\ a_{21} & a_{22} & a_{23} \\ a_{31} & a_{32} & a_{33} \end{bmatrix} \begin{Bmatrix} x_1 \\ x_2 \\ x_3 \end{Bmatrix}^* = \begin{Bmatrix} c_1 \\ c_2 \\ c_3 \end{Bmatrix}^* \tag{3-46}$$

(*The braces shown will be used henceforth in the text to signify a column matrix.)

If we let **A** represent the coefficient matrix, $\{X\}$ the column matrix of the unknowns, and $\{C\}$ the column matrix of the constants, we can replace Eq. 3-46 by

$$\mathbf{A}\{X\} - \{C\} = 0 \tag{3-47}$$

If we could reduce the system of equations to an equivalent system of the form

$$\begin{bmatrix} 1 & u_{12} & u_{13} \\ 0 & 1 & u_{23} \\ 0 & 0 & 1 \end{bmatrix} \begin{Bmatrix} x_1 \\ x_2 \\ x_3 \end{Bmatrix} = \begin{Bmatrix} d_1 \\ d_2 \\ d_3 \end{Bmatrix} \tag{3-48}$$

[5] William Weaver, Jr., *Computer Programs for Structural Analysis* (Princeton, N.J.: D Van Nostrand Co., Inc., 1967), pp. 56–64.

[6] For a review of matrix multiplication, see App. B.

they could readily be solved by back substitution as discussed under Gaussian elimination. The upper unit triangular matrix in Eq. 3-48 will be represented by **U**. Then Eq. 3-48 can be written as

$$\mathbf{U}\{X\} - \{D\} = 0 \tag{3-49}$$

Suppose there exists a lower triangular matrix

$$\begin{bmatrix} l_{11} & 0 & 0 \\ l_{21} & l_{22} & 0 \\ l_{31} & l_{32} & l_{33} \end{bmatrix}$$

which we represent by **L** and which has the following property: When we premultiply the left side of Eq. 3-49 by **L**, it will give us the left side of Eq. 3-47. The existence of such a matrix **L** has not been proved at this point, but in the discussion which follows we demonstrate its existence whenever matrix **A** is nonsingular. In accordance with the above property of **L**, we have

$$\mathbf{L}(\mathbf{U}\{X\} - \{D\}) = \mathbf{A}\{X\} - \{C\} \tag{3-50}$$

Therefore,

$$\mathbf{LU} = \mathbf{A} \tag{3-51}$$

and

$$\mathbf{L}\{D\} = \{C\} \tag{3-52}$$

Equations 3-51 and 3-52 may be combined into the single matrix equation

$$\begin{bmatrix} l_{11} & 0 & 0 \\ l_{21} & l_{22} & 0 \\ l_{31} & l_{32} & l_{33} \end{bmatrix} \left[\begin{array}{ccc|c} 1 & u_{12} & u_{13} & d_1 \\ 0 & 1 & u_{23} & d_2 \\ 0 & 0 & 1 & d_3 \end{array}\right] = \left[\begin{array}{ccc|c} a_{11} & a_{12} & a_{13} & c_1 \\ a_{21} & a_{22} & a_{23} & c_2 \\ a_{31} & a_{32} & a_{33} & c_3 \end{array}\right] \tag{3-53}$$

For the sake of convenience, let the c's be represented by a_{i4} and the d's by u_{i4} to obtain Eq. 3-53 as

$$\begin{bmatrix} l_{11} & 0 & 0 \\ l_{21} & l_{22} & 0 \\ l_{31} & l_{32} & l_{33} \end{bmatrix} \begin{bmatrix} 1 & u_{12} & u_{13} & u_{14} \\ 0 & 1 & u_{23} & u_{24} \\ 0 & 0 & 1 & u_{34} \end{bmatrix} = \begin{bmatrix} a_{11} & a_{12} & a_{13} & a_{14} \\ a_{21} & a_{22} & a_{23} & a_{24} \\ a_{31} & a_{32} & a_{33} & a_{34} \end{bmatrix} \tag{3-54}$$

The l_{ij} and u_{ij} elements may be determined from Eq. 3-54. The augmented **U** matrix of Eq. 3-54 is the augmented matrix of the equivalent triangular set given by Eq. 3-48. Therefore, with the u_{ij} determined, the unknown x_i values may be determined by back substitution.

From the rules of matrix multiplication, we note that the first column of matrix **L** is identical with the first column of the augmented **A** matrix. That is,

$$l_{i1} = a_{i1} \tag{3-55}$$

We also note that

$$l_{11}u_{12} = a_{12}$$
$$l_{11}u_{13} = a_{13}$$
$$l_{11}u_{14} = a_{14}$$

Therefore, we can obtain the first row of the **U** matrix as the first row of **A** divided by l_{11} (or a_{11}). That is,

$$u_{1j} = \frac{a_{1j}}{l_{11}} \tag{3-56}$$

We have determined the first column of **L** and the first row of **U**. We can now proceed to determine the second column of **L**, followed by the second row of **U**.

From matrix multiplication, we have

$$l_{21}u_{12} + l_{22} \cdot 1 = a_{22}$$
$$l_{31}u_{12} + l_{32} \cdot 1 = a_{32}$$

from which we obtain

$$\begin{aligned} l_{22} &= a_{22} - l_{21}u_{12} \\ l_{32} &= a_{32} - l_{31}u_{12} \end{aligned} \tag{3-57}$$

Having the second column of **L**, we get the second row of **U** from

$$l_{21}u_{13} + l_{22}u_{23} = a_{23}$$
$$l_{21}u_{14} + l_{22}u_{24} = a_{24}$$

or

$$u_{23} = \frac{a_{23} - l_{21}u_{13}}{l_{22}}$$
$$u_{24} = \frac{a_{24} - l_{21}u_{14}}{l_{22}} \tag{3-58}$$

Next we get the third column of **L** followed by the third row of **U**. The third column of **L** is obtained from

$$l_{31}u_{13} + l_{32}u_{23} + l_{33} = a_{33}$$

or

$$l_{33} = a_{33} - [l_{31}u_{13} + l_{32}u_{23}] \tag{3-59}$$

The element u_{34}, the only element in the third row of **U** except the 1, is obtained from

$$l_{31}u_{14} + l_{32}u_{24} + l_{33}u_{34} = a_{34}$$

or

$$u_{34} = \frac{a_{34} - [l_{31}u_{14} + l_{32}u_{24}]}{l_{33}} \tag{3-60}$$

The general equations for the elements of the **L** and augmented **U** matrices, for n equations in n unknowns, are

$$l_{i1} = a_{i1} \qquad \text{for} \begin{cases} i = 1, 2, \ldots, n \\ j = 1 \end{cases}$$

$$u_{1j} = \frac{a_{1j}}{a_{11}} \qquad \text{for} \begin{cases} j = 2, 3, \ldots, n+1 \\ i = 1 \end{cases}$$

$$l_{ij} = a_{ij} - \sum_{k=1}^{j-1} l_{ik}u_{kj} \qquad \text{for} \begin{cases} j = 2, 3, \ldots, n \\ i = j, j+1, \ldots, n \\ \text{(for each value of } j) \end{cases} \tag{3-61}$$

$$u_{ij} = \frac{a_{ij} - \sum_{k=1}^{i-1} l_{ik}u_{kj}}{l_{ii}} \qquad \text{for} \begin{cases} i = 2, 3, \ldots, n \\ j = i+1, i+2, \ldots, n+1 \\ \text{(for each value of } i) \end{cases}$$

After the first and second equations of the above set are applied over their ranges of i and j, respectively, the third equation is applied with $j = 2$ and i ranging from j to n. Then the fourth equation is applied with $i = 2$ and j ranging from $i + 1$ to $n + 1$. Then the third equation is applied with $j = 3$ and i ranging from j to n, the fourth with $i = 3$ and j ranging from $i + 1$ to $n + 1$, and so on.

The back-substitution formulas are

$$x_n = u_{n,n+1}$$

$$x_i = u_{i,n+1} - \sum_{j=i+1}^{n} u_{ij}x_j \qquad \text{for } i = n-1, n-2, \ldots, 1 \tag{3-62}$$

If we make sure that a_{11} in the original coefficient matrix is nonzero, then the divisions of Eq. 3-61 will always be defined since the l_{ii} values will be nonzero. This may be seen by noting that

$$\mathbf{LU} = \mathbf{A}$$

and therefore the determinant of **L** times the determinant of **U** equals the determinant of **A**. That is,

$$|\mathbf{L}|\,|\mathbf{U}| = |\mathbf{A}| \tag{3-63}$$ [7]

We are assuming independent equations, so the determinant of **A** is nonzero. Therefore, the determinant of **L** must be nonzero. Since the determinant of a triangular matrix is the product of the main diagonal elements, the l_{ii} elements are all nonzero.

For $n > 2$, Cholesky's method requires fewer arithmetic operations than either the Gaussian or Gauss-Jordan methods, making it the fastest of the basic elimination methods. It may also be made economical of core storage in the computer by overlaying the **U** and **L** matrices on the **A** matrix (in the same storage locations). This may be done since there is no need to store the zeros and ones of the **U** and **L** matrices. The original a_{ij} values of the augmented **A** matrix are first stored in the computer. Then, calling both the u_{ij} and l_{ij} elements simply a_{ij}, the old a_{ij} values are replaced with new a_{ij} values. Then the first of Eq. 3-61 is automatically satisfied with no computer execution. The other equations become

$$\begin{aligned}
a_{1j} &= \frac{a_{1j}}{a_{11}} && \text{for } j = 2, 3, \ldots, n+1 \\
a_{ij} &= a_{ij} - \sum_{k=1}^{j-1} a_{ik}a_{kj} && \text{for } \begin{cases} j = 2, 3, \ldots, n \\ i = j, j+1, \ldots, n \\ \quad \text{(for each } j \text{ value)} \end{cases} \\
a_{ij} &= \frac{a_{ij} - \sum_{k=1}^{i-1} a_{ik}a_{kj}}{a_{ii}} && \text{for } \begin{cases} i = 2, 3, \ldots, n \\ j = i+1, i+2, \ldots, n+1 \\ \quad \text{(for each } i \text{ value)} \end{cases}
\end{aligned} \tag{3-64}$$

[7] Frank Ayres, Jr., *Matrices* (New York: Schaum Publishing Co., 1962), p. 33.

Following the application of the first of Eq. 3-64, the second and third equations are applied in the manner previously described for the *third* and *fourth* of Eq. 3-61. A careful study of the computer program which follows will clarify how this may be accomplished. The Cholesky program is given as a subroutine subprogram since this subprogram will be used in Chap. 4 in connection with curve fitting.

If the **A** matrix is symmetrical as it often is in structural-analysis problems, the number of calculations of Cholesky's method may be further reduced.[8] The Cholesky subroutine subprogram is as follows:

```
      SUBROUTINE CHLSKY (A,N,M,X)
      DIMENSION A(N,M),X(N)
C CALCULATE FIRST ROW OF UPPER UNIT TRIANGULAR MATRIX
      DO 3 J=2,M
    3 A(1,J)=A(1,J)/A(1,1)
C CALCULATE OTHER ELEMENTS OF U AND L MATRICES
      DO 8 I=2,N
      J=I
      DO 5 II=J,N
      SUM=0.
      JM1=J-1
      DO 4 K=1,JM1
    4 SUM=SUM+A(II,K)*A(K,J)
    5 A(II,J)=A(II,J)-SUM
      IP1=I+1
      DO 7 JJ=IP1,M
      SUM=0.
      IM1=I-1
      DO 6 K=1,IM1
    6 SUM=SUM+A(I,K)*A(K,JJ)
    7 A(I,JJ)=(A(I,JJ)-SUM)/A(I,I)
    8 CONTINUE
C SOLVE FOR X(I) BY BACK SUBSTITUTION
      X(N)=A(N,N+1)
      L=N-1
      DO 10 NN=1,L
      SUM=0.
      I=N-NN
      IP1=I+1
      DO 9 J=IP1,N
    9 SUM=SUM+A(I,J)*X(J)
   10 X(I)=A(I,M)-SUM
      RETURN
      END
```

[8] See William Weaver, Jr., *Computer Programs for Structural Analysis* (Princeton, N.J.: D. Van Nostrand Co., Inc., 1967), pp. 56–58; or see M. G. Salvadori and M. L. Baron, *Numerical Methods in Engineering*, 2d ed. (Englewood Cliffs, N.J.: Prentice-Hall, Inc., 1961), p. 29.

3-5 / THE USE OF ERROR EQUATIONS

Error equations are used to increase the accuracy with which the roots of simultaneous equations may be determined. For example, when the elimination method, just discussed, is used to find the roots of simultaneous equations, there may be a considerable loss of accuracy in the results, owing to *roundoff error* which accumulates during the large number of arithmetic operations performed in obtaining a solution. (Roundoff error is defined as the error which results from replacing a number having more than n digits by a number having n digits.)

Suppose that the computer were used to solve, for example, 15 to 20 simultaneous equations by the elimination method, and that seven significant digits were retained in the result of each calculation. The final results might have only two- or three-digit accuracy, owing to the cumulative errors introduced by rounding off numbers in the hundreds of arithmetic operations performed.

Error equations can be used to reduce the error introduced by rounding off. Their use is explained in the following discussion.

Consider a set of equations of the form

$$\begin{aligned}
a_{11}x_1 + a_{12}x_2 + a_{13}x_3 + \cdots + a_{1n}x_n &= C_1 \\
a_{21}x_1 + a_{22}x_2 + a_{23}x_3 + \cdots + a_{2n}x_n &= C_2 \\
\vdots \qquad \vdots \qquad \vdots \qquad \qquad \vdots \quad &\quad \vdots \\
a_{n1}x_1 + a_{n2}x_2 + a_{n3}x_3 + \cdots + a_{nn}x_n &= C_n
\end{aligned} \tag{3-65}$$

Suppose that we have obtained the approximate roots $x_1', x_2', \ldots, x_n'$ by the elimination method. Upon substituting these values into Eq. 3-65, we find that the constant values $C_1', C_2', \ldots, C_n'$ obtained vary from the respective original values $C_1, C_2, \ldots, C_n$, since the roots substituted were not exact. Such a substitution may be expressed in equation form as

$$\begin{aligned}
a_{11}x_1' + a_{12}x_2' + \cdots + a_{1n}x_n' &= C_1' \\
a_{21}x_1' + a_{22}x_2' + \cdots + a_{2n}x_n' &= C_2' \\
\vdots \qquad \vdots \qquad \qquad \vdots \quad &\quad \vdots \\
a_{n1}x_1' + a_{n2}x_2' + \cdots + a_{nn}x_n' &= C_n'
\end{aligned} \tag{3-66}$$

If $\Delta x_1, \Delta x_2, \ldots, \Delta x_n$ are the corrections which must be added to the approximate root values to obtain the *exact* root values $\bar{x}_1, \bar{x}_2, \ldots, \bar{x}_n$, it follows that

$$\begin{aligned}
\bar{x}_1 &= x_1' + \Delta x_1 \\
\bar{x}_2 &= x_2' + \Delta x_2 \\
\vdots \quad & \quad \vdots \qquad \vdots \\
\bar{x}_n &= x_n' + \Delta x_n
\end{aligned} \tag{3-67}$$

If we substitute these expressions for the exact roots into Eq. 3-65, we obtain

$$\begin{aligned} a_{11}(x'_1 + \Delta x_1) + a_{12}(x'_2 + \Delta x_2) + \cdots + a_{1n}(x'_n + \Delta x_n) &= C_1 \\ a_{21}(x'_1 + \Delta x_1) + a_{22}(x'_2 + \Delta x_2) + \cdots + a_{2n}(x'_n + \Delta x_n) &= C_2 \\ \vdots \qquad\qquad \vdots \qquad\qquad\qquad \vdots \quad &\quad \vdots \\ a_{n1}(x'_1 + \Delta x_1) + a_{n2}(x'_2 + \Delta x_2) + \cdots + a_{nn}(x'_n + \Delta x_n) &= C_n \end{aligned} \tag{3-68}$$

If Eq. 3-66 is then subtracted from Eq. 3-68, we obtain the following set of simultaneous equations, involving the root corrections:

$$\begin{aligned} a_{11}\Delta x_1 + a_{12}\Delta x_2 + \cdots + a_{1n}\Delta x_n &= (C_1 - C'_1) = e_1 \\ a_{21}\Delta x_1 + a_{22}\Delta x_2 + \cdots + a_{2n}\Delta x_n &= (C_2 - C'_2) = e_2 \\ \vdots \qquad \vdots \qquad\qquad \vdots \quad &\quad \vdots \qquad\quad \vdots \\ a_{n1}\Delta x_1 + a_{n2}\Delta x_2 + \cdots + a_{nn}\Delta x_n &= (C_n - C'_n) = e_n \end{aligned} \tag{3-69}$$

Inspection of Eq. 3-69 reveals that the required root corrections are, themselves, the roots of a set of equations which differs from the original set (Eq. 3-65) only in the respective constant values. The constants e_i of Eq. 3-69 express the error resulting from inaccurate root values, and the equations are called *error equations.*

In a computer solution of a set of simultaneous equations by an elimination method, in which error equations are to be employed, the augmented matrix of the original set of equations should be left stored in memory during its use in obtaining the approximate roots. The corrections to the roots may then be obtained by simply replacing the values in the right-hand column of this matrix by the e_i values calculated, and repeating the reduction of the matrix by the elimination method used. The corrections obtained are then added, respectively, to the approximate root values found in the original elimination process, to obtain more accurate root values.

If still greater accuracy is desired, corrections may be made in the values obtained as the root corrections from solving Eq. 3-69. If we designate the corrections to the corrections by $\Delta^2 x_i$, then the more exact roots $\bar{x}_i$ are given by

$$\bar{x}_i = x'_i + \Delta x_i + \Delta^2 x_i \tag{3-70}$$

This process of using error equations to correct the roots of other error equations can be carried as far as necessary to obtain roots of the desired accuracy. The use of error equations is illustrated in Ex. 3-4.

EXAMPLE 3-4

The following set of simultaneous equations defines the motion of a dwell-return-dwell polynomial cam in which the C's are unknown polynomial coefficients.

$$\begin{aligned} C_4 + C_5 + C_6 + C_7 &= -1 \\ 4C_4 + 5C_5 + 6C_6 + 7C_7 &= 0 \\ 6C_4 + 10C_5 + 15C_6 + 21C_7 &= 0 \\ 12C_4 + 30C_5 + 60C_6 + 105C_7 &= 0 \end{aligned} \tag{3-71}$$

The roots of this set of equations have been determined by the Gauss-Jordan elimination method and have the following values:

$$\begin{aligned} C_4 &= -35 \\ C_5 &= 84 \\ C_6 &= -70 \\ C_7 &= 20 \end{aligned}$$

Since all the coefficients of Eq. 3-71 are whole numbers, and since the arithmetic operations used in solving the roots did not involve any numbers with sufficient significant digits to exceed the computer capacity, no roundoff error occurred in obtaining the roots, which are whole numbers themselves and are thus the exact roots of Eq. 3-71.

However, for the purpose of illustrating the use of error equations, let us suppose that roundoff has occurred and that the roots have been found to be

$$\begin{aligned} C_4' &= -35.1 \\ C_5' &= 83.9 \\ C_6' &= -70.3 \\ C_7' &= 20.2 \end{aligned}$$

If we substitute these roots into Eq. 3-71 to determine how accurate they are, we find that

$$\begin{aligned} C_4 + C_5 + C_6 + C_7 &= -1.3 \\ 4C_4 + 5C_5 + 6C_6 + 7C_7 &= -1.3 \\ 6C_4 + 10C_5 + 15C_6 + 21C_7 &= -1.9 \\ 12C_4 + 30C_5 + 60C_6 + 105C_7 &= -1.2 \end{aligned} \tag{3-72}$$

A comparison of the constant values of Eq. 3-72 with the respective constant values of Eq. 3-71 reveals that considerable error exists in the roots. Therefore, we introduce the use of error equations to determine more acurate values. Referring to Eqs. 3-69 and 3-72, we obtain the following set of error equations:

$$\begin{aligned}
\Delta C_4 + \Delta C_5 + \Delta C_6 + \Delta C_7 &= -1.0 - (-1.3) = 0.3 \\
4\Delta C_4 + 5\Delta C_5 + 6\Delta C_6 + 7\Delta C_7 &= 0.0 - (-1.3) = 1.3 \\
6\Delta C_4 + 10\Delta C_5 + 15\Delta C_6 + 21\Delta C_7 &= 0.0 - (-1.9) = 1.9 \\
12\Delta C_4 + 30\Delta C_5 + 60\Delta C_6 + 105\Delta C_7 &= 0.0 - (-1.2) = 1.2
\end{aligned} \tag{3-73}$$

Using the matrix-reduction method (the matrix form of the Gauss-Jordan elimination method), we find that the augmented and reduced matrices are

$$\mathbf{A} = \begin{bmatrix} 1 & 1 & 1 & 1 & 0.3 \\ 4 & 5 & 6 & 7 & 1.3 \\ 6 & 10 & 15 & 21 & 1.9 \\ 12 & 30 & 60 & 105 & 1.2 \end{bmatrix}$$

$$\mathbf{B} = \begin{bmatrix} 1 & 2 & 3 & 0.10 \\ 4 & 9 & 15 & 0.10 \\ 18 & 48 & 93 & -2.40 \\ 1 & 1 & 1 & 0.30 \end{bmatrix}$$

$$\mathbf{C} = \begin{bmatrix} 1 & 3 & -0.30 \\ 12 & 39 & -4.20 \\ -1 & -2 & 0.20 \\ 2 & 3 & 0.10 \end{bmatrix}$$

$$\mathbf{D} = \begin{bmatrix} 3 & -0.6 \\ 1 & -0.1 \\ -3 & 0.7 \\ 3 & -0.3 \end{bmatrix}$$

$$\mathbf{E} = \begin{Bmatrix} 0.1 \\ 0.1 \\ 0.3 \\ -0.2 \end{Bmatrix}$$

Using the corrections shown in the column matrix, the corrected values of the roots are

$$
\begin{aligned}
C_4 &= C_4' + \Delta C_4 = -35.1 + 0.1 = -35.0 \\
C_5 &= C_5' + \Delta C_5 = 83.9 + 0.1 = 84.0 \\
C_6 &= C_6' + \Delta C_6 = -70.3 + 0.3 = -70.0 \\
C_7 &= C_7' + \Delta C_7 = 20.2 + (-0.2) = 20.0
\end{aligned}
\tag{3-74}
$$

Since no roundoff error was introduced in solving the error equations, the corrections are exact corrections and, when added to the approximate values of the roots assumed, yield the exact roots determined in the actual solution.

As was evident at the beginning of this discussion, this example does not involve equations whose solution requires the use of error equations. Such an example was used, for the sake of simplicity, to illustrate the actual matrix reductions which resulted in the root corrections. It should be obvious to the reader, at this point, why error equations would be necessary in solving a set of, say, 30 simultaneous equations, involving coefficients as follows, on a computer having the capacity of handling perhaps only seven significant digits:

$$
\begin{aligned}
0.86746C_1 + 0.97121C_2 + 1.67543C_3 + \cdots + 0.23456C_{30} &= 1.76408 \\
0.34657C_1 + 0.78645C_2 + \cdots \qquad\qquad &= 2.67823 \\
\vdots \qquad\qquad \vdots \qquad\qquad\qquad\qquad &\quad \vdots \\
0.54236C_1 + \cdots \qquad\qquad &= 1.53266
\end{aligned}
\tag{3-75}
$$

The procedure discussed in this example would be used to obtain roots of the desired accuracy for Eq. 3-75, except that the whole procedure would be programmed to let the computer perform all the tedious reduction operations, which are complicated by the numerical values of the coefficients.

3-6 / MATRIX-INVERSION METHOD

When it is necessary to solve a large number of different *sets* of simultaneous equations which differ only by the constant values appearing in the respective equations, the matrix-inversion method may be used to advantage in reducing the number of operations required.

Before proceeding with this method, the reader may wish to review the rules governing the multiplication of matrices given in App. B.

Let us consider again the set of simultaneous equations given by Eq. 3-65. From the rules of matrix multiplication, it can be seen that Eq. 3-65 may be expressed by the single matrix equation

$$\begin{bmatrix} a_{11} & a_{12} \cdots a_{1n} \\ a_{21} & a_{22} \cdots a_{2n} \\ \vdots & \vdots \quad \vdots \\ & \quad a_{nn} \end{bmatrix} \begin{Bmatrix} x_1 \\ x_2 \\ \vdots \\ x_n \end{Bmatrix} = \begin{Bmatrix} c_1 \\ c_2 \\ \vdots \\ c_n \end{Bmatrix} \tag{3-76}$$

If we let **A** represent the coefficient matrix, $\{X\}$ the column matrix of the unknowns, and $\{C\}$ the column matrix of the constants, we can express Eq. 3-76 as

$$\mathbf{A}\{X\} = \{C\} \qquad \text{or} \qquad \mathbf{AX} = \mathbf{C} \tag{3-77}$$

If the given set of equations (Eq. 3-65) has a unique solution, the coefficient matrix **A** is nonsingular, and, as such, there exists for **A** an inverse matrix $\mathbf{A}^{-1}$ such that

$$\mathbf{A}^{-1}\mathbf{A} = \mathbf{I}$$

where **I** is the *identity* or *unit* matrix. The identity matrix, which is an *n*-by-*n* matrix with its main diagonal consisting of *ones* and zeros everywhere else (see Eq. 3-80), is to matrix algebra what the identity number (1) is to ordinary algebra; that is, $\mathbf{IA} = \mathbf{A}$ or $\mathbf{AI} = \mathbf{A}$.

Premultiplying both sides of Eq. 3-77 by $\mathbf{A}^{-1}$ gives

$$\begin{aligned} (\mathbf{A}^{-1}\mathbf{A})\{X\} &= \mathbf{A}^{-1}\{C\} \\ \mathbf{I}\{X\} &= \mathbf{A}^{-1}\{C\} \\ \{X\} &= \mathbf{A}^{-1}\{C\} \end{aligned} \tag{3-78}$$

Inspection of Eq. 3-78 reveals that if $\mathbf{A}^{-1}$ is known, the elements of $\{X\}$ can easily be determined for any number of different $\{C\}$ matrices by merely premultiplying the particular constant matrix by the inverse matrix.

The inverse of **A** can be determined from the equation

$$\mathbf{AA}^{-1} = \mathbf{I} \tag{3-79}$$

If we let a_{ij} be the general element of **A** and b_{ij} be the general element of $\mathbf{A}^{-1}$, Eq. 3-79 may be expressed as

$$\begin{bmatrix} a_{11} & a_{12} \cdots & a_{1n} \\ a_{21} & a_{22} \cdots & a_{2n} \\ \vdots & \vdots & \vdots \\ a_{n1} & a_{n2} \cdots & a_{nn} \end{bmatrix} \begin{bmatrix} b_{11} & b_{12} \cdots & b_{1n} \\ b_{21} & b_{22} \cdots & b_{2n} \\ \vdots & \vdots & \vdots \\ b_{n1} & b_{n2} \cdots & b_{nn} \end{bmatrix} = \begin{bmatrix} 1 & 0 \cdots & 0 \\ 0 & 1 & 0 \\ \vdots & & \vdots \\ 0 & 0 \cdots & 1 \end{bmatrix} \tag{3-80}$$

where the identity matrix on the right side of the equation has the same n-by-n order as the $\mathbf{A}$ and $\mathbf{A}^{-1}$ matrices.

From the rules of matrix multiplication, Eq. 3-80 is equivalent to the following n *sets* of simultaneous equations:

$$\left.\begin{array}{l} a_{11}b_{11} + a_{12}b_{21} + \cdots + a_{1n}b_{n1} = 1 \\ a_{21}b_{11} + a_{22}b_{21} + \cdots + a_{2n}b_{n1} = 0 \\ \vdots \qquad \vdots \qquad\qquad \vdots \qquad \vdots \\ a_{n1}b_{11} + a_{n2}b_{21} + \cdots + a_{nn}b_{n1} = 0 \end{array}\right\} (1)$$

$$\left.\begin{array}{l} a_{11}b_{12} + a_{12}b_{22} + \cdots + a_{1n}b_{n2} = 0 \\ a_{21}b_{12} + a_{22}b_{22} + \cdots + a_{2n}b_{n2} = 1 \\ \vdots \qquad \vdots \qquad\qquad \vdots \qquad \vdots \\ a_{n1}b_{12} + a_{n2}b_{22} + \cdots + a_{nn}b_{n2} = 0 \end{array}\right\} (2) \qquad (3\text{-}81)$$

$$\vdots \qquad\qquad \vdots$$

$$\left.\begin{array}{l} a_{11}b_{1n} + a_{12}b_{2n} + \cdots + a_{1n}b_{nn} = 0 \\ a_{21}b_{1n} + a_{22}b_{2n} + \cdots + a_{2n}b_{nn} = 0 \\ \vdots \qquad \vdots \qquad\qquad \vdots \qquad \vdots \\ a_{n1}b_{1n} + a_{n2}b_{2n} + \cdots + a_{nn}b_{nn} = 1 \end{array}\right\} (n)$$

Inspection of Eq. 3-81 reveals that all the n sets of equations have identical known coefficients a_{ij}, and that each set contains one column of elements of the inverse matrix $\mathbf{A}^{-1}$ as unknowns and the corresponding column of the identity matrix $\mathbf{I}$ (see Eq. 3-80) as constants. Thus we have n sets of simultaneous equations which differ only by the constants associated with each set.

It was stated in Sec. 3-3 that the Gauss-Jordan elimination method could be utilized to solve simultaneously n sets of simultaneous equations, differing only in their constants, by including the column of constants associated with each set of equations on the right side of the coefficient matrix in the augmented matrix. Utilizing the Gauss-Jordan elimination method, we first form the augmented matrix

$$\begin{bmatrix} a_{11} & a_{12} \cdots a_{1n} & 1 & 0 \cdots 0 \\ a_{21} & a_{22} \cdots a_{2n} & 0 & 1 \cdots 0 \\ \vdots & \vdots \qquad \vdots & \vdots & \vdots \qquad \vdots \\ a_{n1} & a_{n2} \cdots a_{nn} & 0 & 0 \cdots 1 \end{bmatrix}$$

The elements of the inverse matrix (the b_{ij}'s) may then be obtained by successive applications of Eqs. 3-31 and 3-32, continuing the reduction process until n columns remain, or by application of Eq. 3-45, modified for $2n$ columns instead of $n + 1$ columns. The following example utilizes the first reduction method for matrix inversion.

EXAMPLE 3-5

Let us solve Eq. 3-71 of Ex. 3-4 using the matrix-inversion method. The coefficient matrix is

$$\mathbf{A} = \begin{bmatrix} 1 & 1 & 1 & 1 \\ 4 & 5 & 6 & 7 \\ 6 & 10 & 15 & 21 \\ 12 & 30 & 60 & 105 \end{bmatrix}$$

We begin the matrix reduction with the augmented matrix

$$\begin{bmatrix} 1 & 1 & 1 & 1 & 1 & 0 & 0 & 0 \\ 4 & 5 & 6 & 7 & 0 & 1 & 0 & 0 \\ 6 & 10 & 15 & 21 & 0 & 0 & 1 & 0 \\ 12 & 30 & 60 & 105 & 0 & 0 & 0 & 1 \end{bmatrix}$$

Successive reductions, applying Eqs. 3-31 and 3-32, yield the inverted matrix

$$\mathbf{A}^{-1} = \begin{bmatrix} 35 & -15 & 5 & -\frac{1}{3} \\ -84 & 39 & -14 & 1 \\ 70 & -34 & 13 & -1 \\ -20 & 10 & -4 & \frac{1}{3} \end{bmatrix}$$

The correctness of the inverse matrix can easily be checked by multiplying it by the matrix **A**. If it is correct, the resulting product will be the identity matrix.

If $\{K\}$ represents the column matrix of the constants of Eq. 3-71, and $\{C\}$ represents the column matrix of the unknowns C_4, C_5, C_6, and C_7, then, from Eq. 3-78,

$$\{C\} = \mathbf{A}^{-1}\{K\}$$

or

$$\{C\} = \begin{bmatrix} 35 & -15 & 5 & -\frac{1}{3} \\ -84 & 39 & -14 & 1 \\ 70 & -34 & 13 & -1 \\ -20 & 10 & -4 & \frac{1}{3} \end{bmatrix} \begin{Bmatrix} -1 \\ 0 \\ 0 \\ 0 \end{Bmatrix} = \begin{Bmatrix} -35 \\ 84 \\ -70 \\ 20 \end{Bmatrix}$$

These are the same values which were obtained much more simply by the elimination method illustrated in Ex. 3-4. Comparing the two methods, it is evident that the matrix-inversion method is not practical for solving a single set (or even two or three sets) of simultaneous equations, because of the amount of calculation involved in determining the inverse matrix. If, however, 20 sets of ten simultaneous equations, differing only in their constants, were to be solved, an augmented matrix containing 20 columns of constants would be used in the elimination method, and the matrix-inversion method could be used to advantage.

Matrix Inversion in Place

In accomplishing the matrix reductions on the computer by either of the Gauss-Jordan algorithms given in Sec. 3-3, it would not be necessary to read in the elements of the appended identity matrix as data values. If they are to be stored in memory at all (and they need not be, at the cost of a slightly more complicated inversion program), they can be readily generated by statements in the program. Such statements might be as follows:

```
    ...
    DO 8J=NP1,M
    DO 8I=1,N
    IF(J-N-I)7,6,7
  6 A(I,J)=1.
    GO TO 8
  7 A(I,J)=0.
  8 CONTINUE
    ...
```

where NP1 = N + 1 and M is the number of columns in the augmented matrix.

If fairly large matrices are being inverted, and it is important to save as much core storage as possible, it is preferable to use the second Gauss-Jordan algorithm given in Sec. 3-3, which overlays the reduced matrix over the original augmented matrix in core. A further saving of memory space can be accomplished by not storing the identity matrix. The matrix to be inverted is stored in memory and is replaced by its inverse. This is called "inversion in place."

A problem arises in the use of partial pivoting with inversion in place. When rows of a matrix are interchanged, the rows containing zeros and ones in the appended matrix must also be interchanged. If the identity matrix is not present in core, this required interchange must be taken into account. A discussion of this problem will be postponed until we have first studied *inversion in place* without row interchange.

Let us consider inversion of the matrix

$$\begin{bmatrix} 3 & -2 & -1 \\ -4 & 1 & -1 \\ 2 & 0 & 1 \end{bmatrix}$$

With the appended identity matrix, it becomes

$$\begin{bmatrix} 3 & -2 & -1 & 1 & 0 & 0 \\ -4 & 1 & -1 & 0 & 1 & 0 \\ 2 & 0 & 1 & 0 & 0 & 1 \end{bmatrix}$$

Applying Eq. 3-45 with $2n$ columns instead of $n + 1$ columns, the first step in the reduction process results in

$$\begin{bmatrix} 3 & -2 & -1 & 1 & 0 & 0 \\ -4 & 1 & -1 & 0 & 1 & 0 \\ 2 & 0 & 1 & 0 & 0 & 1 \end{bmatrix} \to \begin{bmatrix} 1 & -\frac{2}{3} & -\frac{1}{3} & \frac{1}{3} & 0 & 0 \\ 0 & -\frac{5}{3} & -\frac{7}{3} & \frac{4}{3} & 1 & 0 \\ 0 & \frac{4}{3} & \frac{5}{3} & -\frac{2}{3} & 0 & 1 \end{bmatrix}$$

The first column of the reduced matrix consists of zeros and a one, and they are not used in further steps of the reduction process. Therefore, the fourth column could be stored in its place. Only three columns are required to store the vital elements resulting from this reduction step. With matrix inversion in place, the matrix in memory at this stage would be

$$\begin{bmatrix} \frac{1}{3} & -\frac{2}{3} & -\frac{1}{3} \\ \frac{4}{3} & -\frac{5}{3} & -\frac{7}{3} \\ -\frac{2}{3} & \frac{4}{3} & \frac{5}{3} \end{bmatrix}$$

The second step in the reduction process results in

$$\begin{bmatrix} 1 & -\frac{2}{3} & -\frac{1}{3} & \frac{1}{3} & 0 & 0 \\ 0 & -\frac{5}{3} & -\frac{7}{3} & \frac{4}{3} & 1 & 0 \\ 0 & \frac{4}{3} & \frac{5}{3} & -\frac{2}{3} & 0 & 1 \end{bmatrix} \to \begin{bmatrix} 1 & 0 & \frac{3}{5} & -\frac{1}{5} & -\frac{2}{5} & 0 \\ 0 & 1 & \frac{7}{5} & -\frac{4}{5} & -\frac{3}{5} & 0 \\ 0 & 0 & -\frac{1}{5} & \frac{2}{5} & \frac{4}{5} & 1 \end{bmatrix}$$

Again, the fourth column could be stored where the first column is, and the fifth column could replace the second column. With matrix inversion in place, we would have in memory

$$\begin{bmatrix} -\frac{1}{5} & -\frac{2}{5} & \frac{3}{5} \\ -\frac{4}{5} & -\frac{3}{5} & \frac{7}{5} \\ \frac{2}{5} & \frac{4}{5} & -\frac{1}{5} \end{bmatrix}$$

After one more reduction step, the final matrix would be

$$\begin{bmatrix} 1 & 0 & 0 & 1 & 2 & 3 \\ 0 & 1 & 0 & 2 & 5 & 7 \\ 0 & 0 & 1 & -2 & -4 & -5 \end{bmatrix}$$

or for inversion in place, simply

$$\begin{bmatrix} 1 & 2 & 3 \\ 2 & 5 & 7 \\ -2 & -4 & -5 \end{bmatrix}$$

We note that the matrix reduction in place can be accomplished with the following equations applied in the order listed; first with $k = 1$ for all equations and j and i cycling as indicated for each equation; then applied again in the same manner with $k = 2$, and so forth, up to $k = n$, where n is the order of the matrix.

$$\begin{aligned} &1)\quad a'_{kj} = \frac{a_{kj}}{a_{kk}} && \text{for } j = 1, 2, \ldots, n \text{ (except } j \neq k) \\ &2)\quad a'_{kk} = \frac{1}{a_{kk}} && \\ &3)\quad a'_{ij} = a_{ij} - a'_{kj}a_{ik} && \text{for } \begin{cases} i = 1, 2, \ldots, n \text{ (except } i \neq k) \\ j = 1, 2, \ldots, n \text{ for each } i \text{ (except } j \neq k) \end{cases} \\ &4)\quad a'_{ik} = 0 - a_{ik}a'_{kk} && \text{for } i = 1, 2, \ldots, n \text{ (except } i \neq k) \end{aligned} \tag{3-82}$$

To aid in the study of matrix inversion in place, a computer program applying Eq. 3-82 is presented next. The student should correlate this program with the preceding discussion. Partial pivoting is not used, so the computer would stop execution because of overflow upon attempted division by zero in the event that a pivot element were very small or exactly zero. Partial pivoting, which is used to avoid this problem, is discussed in the next subsection.

```
C MATRIX INVERSION USING GAUSS-JORDAN REDUCTION
C INVERTED MATRIX OVERLAYS ORIGINAL MATRIX IN CORE
C PARTIAL PIVOTING IS NOT USED
C STATEMENT 8 SHOULD BE CHANGED AS REQUIRED BY ORDER OF MATRIX
      DIMENSION A(15,15)
C READ IN ORDER OF MATRIX
      READ(5,2) N
    2 FORMAT(I2)
C READ IN MATRIX TO BE INVERTED
      READ(5,3) ((A(I,J),J=1,N),I=1,N)
    3 FORMAT(8F10.0)
C CALCULATE ELEMENTS OF REDUCED MATRIX
      DO 6 K=1,N
C CALCULATE NEW ELEMENTS OF PIVOT ROW
      DO 4 J=1,N
      IF(J.EQ.K) GO TO 4
      A(K,J)=A(K,J)/A(K,K)
    4 CONTINUE
C CALCULATE ELEMENT REPLACING PIVOT ELEMENT
      A(K,K)=1./A(K,K)
C CALCULATE NEW ELEMENTS NOT IN PIVOT ROW OR PIVOT COLUMN
      DO 5 I=1,N
      IF(I.EQ.K) GO TO 5
      DO 5 J=1,N
      IF(J.EQ.K) GO TO 5
      A(I,J)=A(I,J)-A(K,J)*A(I,K)
    5 CONTINUE
C CALCULATE REPLACEMENT ELEMENTS FOR PIVOT COLUMN-EXCEPT
                                                      PIVOT ELEMENT
      DO 6 I=1,N
      IF(I.EQ.K) GO TO 6
      A(I,K)=-A(I,K)*A(K,K)
    6 CONTINUE
C WRITE OUT INVERTED MATRIX
      WRITE(6,7)
    7 FORMAT('1','INVERTED MATRIX'/)
      WRITE(6,8) ((A(I,J),J=1,N),I=1,N)
    8 FORMAT(' ',3F16.4)
      STOP
      END
```

Matrix Inversion In Place with Partial Pivoting

Let us again suppose that the matrix

$$\begin{bmatrix} -3 & -2 & -1 \\ -4 & 1 & -1 \\ 2 & 0 & 1 \end{bmatrix}$$

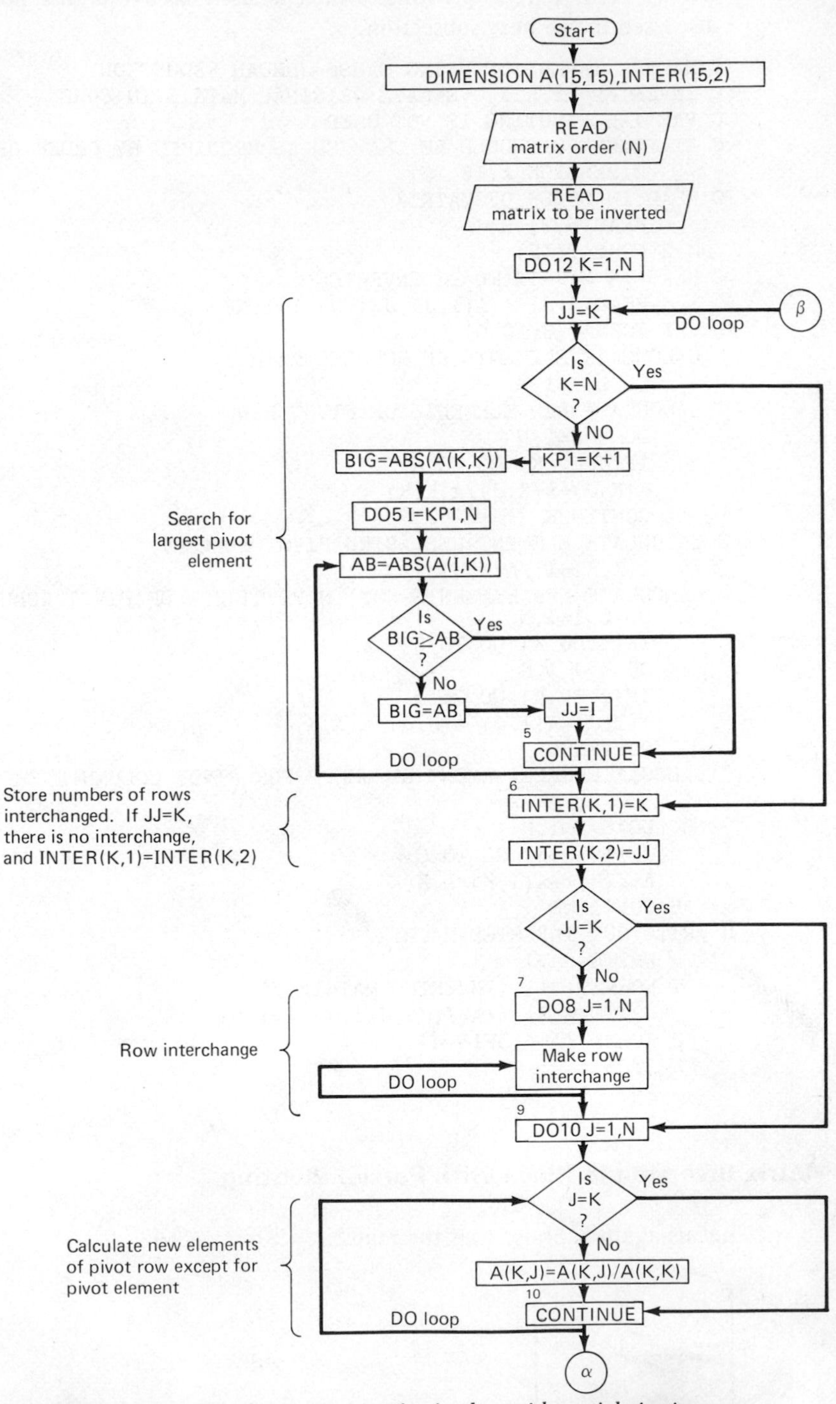

Fig. 3-4(a) / *Flow chart for matrix inversion in place with partial pivoting.*

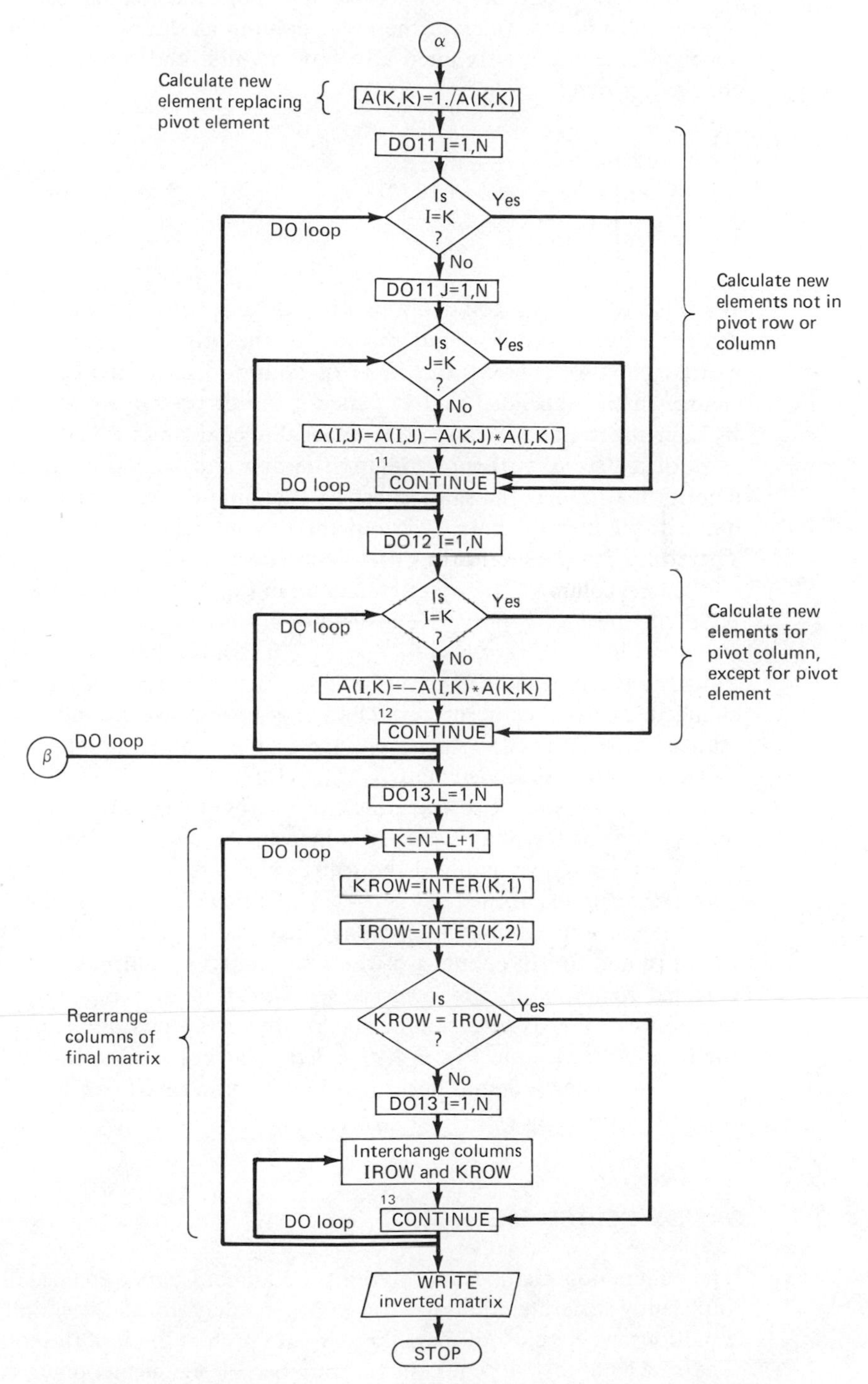

Fig. 3-4(b) / *Flow chart for matrix inversion in place with partial pivoting* (*cont.*).

is to be inverted, but that we wish to interchange the first and second rows to utilize the largest value in the pivot column as the pivot element. The appended identity matrix must also have its first and second rows interchanged, giving

$$\begin{bmatrix} 0 & 1 & 0 \\ 1 & 0 & 0 \\ 0 & 0 & 1 \end{bmatrix}$$

We observe that we could obtain the same appended matrix by interchanging the first and second *columns* of the original appended identity matrix. Next we consider the effect of failing to make the column interchange in the appended matrix (since it is not present in core storage). By referring to Eqs. 3-80 and 3-81, we see that changing the first and second rows of matrix **A**, without changing the first and second columns of the identity matrix, has the same effect as changing the first and second columns of the identity matrix without the row interchange in the **A** matrix. Therefore, the question arises as to the effect of changing only the first and second column of the identity matrix. In Eq. 3-81(1), we would actually be solving for $b_{12}, b_{22}, b_{32}, \ldots$ instead of $b_{11}, b_{21}, b_{31}, \ldots$. In Eq. 3-81(2) we would be solving for $b_{11}, b_{21}, b_{31}, \ldots$ instead of for $b_{12}, b_{22}, b_{32}, \ldots$. Thus the inverted matrix would have its first and second columns interchanged. This is also the effect of changing the first and second rows of the matrix to be inverted, with no change in the rows containing the zeros and one in the appended matrix. Therefore, if partial pivoting is used in the program, we must keep track of all row interchanges and reorder the columns of the final matrix obtained in the reduction process. If the last row interchange involved the fourth and sixth rows, then the fourth and sixth columns of the final matrix would have to be interchanged first. If the next-to-the-last row interchange involved the third and fourth rows, the third and fourth columns of the final matrix would have to be interchanged *following the interchange of columns* 4 and 6. The complete procedure of matrix inversion in place with partial pivoting is outlined in the flow chart given in Fig. 3-4. It is left as a problem for the student to write a computer program implementing this flow chart (see Prob. 3-11 at the end of the chapter).

3-7 / GAUSS-SEIDEL METHOD

The elimination methods of solving simultaneous equations usually yield sufficiently accurate solutions for approximately 15 to 20 simultaneous equations, where most of the unknowns are present in all of the equations. The exact number depends on the number of significant digits retained in the results of the arithmetic operations, and the actual equations being solved. When the coefficient matrix is sparse, a considerably larger number of equations can be handled by the elimination methods. Error equations can be used to increase the number which can be accurately solved, but elimination methods are generally impractical when many hundreds or thousands of equations must be solved simultaneously.

There are, however, several techniques which can be used to solve large numbers of simultaneous equations. One of the most useful is the Gauss-Seidel method. None of the several approaches is completely satisfactory, and the Gauss-Seidel method has the disadvantages of not always converging to a solution and of sometimes converging very slowly when it does converge. However, this method will always converge to a solution when the magnitude of a coefficient of a different unknown, in each equation of the set, is sufficiently dominant with respect to the magnitudes of the other coefficients in that equation. It is difficult to define the exact minimum margin by which such a coefficient must dominate the other coefficients to ensure convergence, and it is even more difficult to predict the rate of convergence for some combination of coefficient values when convergence exists. However, when the absolute value of the dominant coefficient for a different unknown in each equation is larger than the sum of the absolute values of the other coefficients in that equation, convergence is assured. Such a set of linear simultaneous equations is known as a *diagonal* system. A diagonal system is *sufficient* to ensure convergence but is not *necessary*. Fortunately, the linear simultaneous equations which derive from many engineering problems are of the type in which dominant coefficients are present.

As a simple example of a set of simultaneous equations which are solvable by the Gauss-Seidel method, let us consider

$$\begin{aligned} 10x_1 + x_2 + 2x_3 &= 44 \\ 2x_1 + 10x_2 + x_3 &= 51 \\ x_1 + 2x_2 + 10x_3 &= 61 \end{aligned} \tag{3-83}$$

It can be seen that the coefficient of x_1 is dominant in the first equation of the set, and that the coefficients of x_2 and x_3 are dominant in the second and third equations, respectively. Since the dominant coefficient in each equation is larger than the sum of the other coefficients in that equation, Eq. 3-83 represents a diagonal system, and convergence is assured.

The sequence of steps constituting the Gauss-Seidel method is as follows:

1 Assign an initial value for each unknown appearing in the set. If it is possible to make a reasonable assumption of these values, do so. If not, any arbitrarily selected values may be assigned. The initial values used will not affect the convergence, as such, but will affect the number of iterations required for convergence.

2 Starting with the first equation, solve that equation for a new value of the unknown which has the largest coefficient in that equation, using the assumed values for the other unknowns.

3 Go to the second equation and solve it for the unknown having the largest coefficient in that equation, using the value calculated for the unknown in step 2 and the assumed values for the remaining unknowns.

4 Proceed with the remaining equations, always solving for the unknown having the largest coefficient in the particular equation, and always using the *last calculated* values for the other unknowns in the equation. (During the first iteration, assumed values must be used for the unknowns until a calculated value has been obtained.) When the final equation has been solved, yielding a value for the last unknown, one iteration is said to have been completed.

5 Continue iterating until the value of each unknown determined in a particular iteration differs from its respective value obtained in the preceding iteration by an amount less than some arbitrarily selected epsilon. The procedure is then complete.

Referring to step 5, the smaller the magnitude of the epsilon selected, the greater will be the accuracy of the solution. However, the magnitude of epsilon does not specify the error which may exist in the values obtained for the unknowns, as this is a function of the rate of convergence. The faster the rate of convergence, the greater will be the accuracy obtained in the values of the unknowns for a given epsilon.

As a simple illustration of the steps just discussed, let us solve Eq. 3-83, using a value of $\varepsilon = 0.02$. Assuming initial values of zero for all three unknowns, the steps appear as follows:

$$\begin{bmatrix} 10x_1 + 0 + 0 = 44 \\ x_1 = 4.40 \end{bmatrix} \quad (1)$$

$$\begin{bmatrix} 2(4.40) + 10x_2 + 0 = 51 \\ x_2 = 4.22 \end{bmatrix} \quad (2)$$

$$\begin{bmatrix} 4.40 + 2(4.22) + 10x_3 = 61 \\ x_3 = 4.81 \end{bmatrix} \quad (3)$$

The first iteration has now been completed. A second iteration yields

$$\begin{bmatrix} 10x_1 + 4.22 + 2(4.81) = 44 \\ x_1 = 3.01 \end{bmatrix} \quad (1)$$

$$\begin{bmatrix} 2(3.01) + 10x_2 + 4.81 = 51 \\ x_2 = 4.01 \end{bmatrix} \quad (2)$$

$$\begin{bmatrix} 3.01 + 2(4.01) + 10x_3 = 61 \\ x_3 = 4.99 \end{bmatrix} \quad (3)$$

Comparing the last values with the respective values obtained in the previous iteration, it can be seen that Δx_1, Δx_2, and Δx_3 are all greater than the ε chosen. Therefore, a third iteration is indicated. This iteration yields

$$x_1 = 3.00$$
$$x_2 = 4.00$$
$$x_3 = 5.00$$

Comparing these values with the respective values obtained in the previous iteration, it can be seen that Δx_1, Δx_2, and Δx_3 are all less than ε, indicating that only three iterations are required. In this example, a fourth iteration would be found to give no change in the values of the unknowns, indicating that an exact solution has been reached.

In obtaining the above solution, only three digits were retained in the result of each calculation. This was done by arbitrarily lopping off all the excess digits which appeared. In a computer solution the computer might retain approximately seven to twelve digits in the result of each calculation, depending on the computer being used. Although this example is a simple one, it serves very well to illustrate the Gauss-Seidel method, since the solution of a much larger set of simultaneous equations would proceed in exactly the same manner, differing only in that many more calculations would be required.

The Gauss-Siedel method is used frequently in the solution of Laplace's and Poisson's partial differential equations, and an example of an engineering problem in which this method is employed will be discussed in Sec. 8-2.

3-8 / HOMOGENEOUS ALGEBRAIC EQUATIONS—EIGENVALUE PROBLEMS

In the preceding sections several methods were presented for solving n nonhomogeneous simultaneous linear algebraic equations in n unknowns. It was assumed that the equations were such that their solutions yielded unique values for the unknowns, since that is generally the case for nonhomogeneous equations describing the characteristics of physical systems. It was noted that the determinants of the coefficient matrices of the sets of equations had to be nonzero (all the equations of the set had to be linearly independent) before unique solutions could be obtained.

Let us now consider the solution of *homogeneous* simultaneous linear algebraic equations which have the general form

$$\begin{aligned}
a_{11}x_1 + a_{12}x_2 + a_{13}x_3 + \cdots + a_{1n}x_n &= 0 \\
a_{21}x_1 + a_{22}x_2 + \cdots\cdots\cdots + a_{2n}x_n &= 0 \\
\vdots \qquad\qquad\qquad\qquad \vdots \;\; & \;\; \vdots \\
a_{n1}x_1 + \cdots\cdots\cdots\cdots + a_{nn}x_n &= 0
\end{aligned} \tag{3-84}$$

or, in matrix notation,

$$\begin{bmatrix} a_{11} & a_{12} \cdots a_{1n} \\ a_{21} & a_{22} \cdots a_{2n} \\ \vdots & \vdots \\ a_{n1} & a_{n2} \cdots a_{nn} \end{bmatrix} \begin{Bmatrix} x_1 \\ x_2 \\ \vdots \\ x_n \end{Bmatrix} = \mathbf{0} \tag{3-85}$$

In linear algebra it is proved that any system of m linear algebraic equations in n unknowns has a solution if, and only if, the coefficient matrix and the augmented matrix of the set have the same *rank*.[9] Therefore, a set of homogeneous equations such as Eq. 3-85 always has a solution (that is, they are consistent), since the augmented matrix and the coefficient matrix of the set are always necessarily of the same rank.

Any system of homogeneous linear algebraic equations always has a trivial solution ($x_1 = x_2 = \cdots = x_n = 0$). If the rank r of the coefficient matrix of the set of equations is equal to the order n, only a trivial solution will exist. *Nontrivial solutions* exist for a set of homogeneous equations if, and only if, the rank r of the coefficient matrix of the set is *less* than the order n. For such a set of equations, the determinant of the coefficient matrix is *zero*, and the set will consist of r linearly independent equations and $n - r$ dependent equations which are linear combinations of the independent equations. In obtaining nontrivial solutions of a set of homogeneous equations, unique values are not obtained for the unknowns. Rather, *relationships* are established between the unknowns (the x's of Eq. 3-84). Any combination of x_i values which satisfies these relationships constitutes a solution. Such solutions will be discussed further later in this section.

To understand the application of homogeneous equations in the analysis of engineering problems, let us consider the solution of *eigenvalue problems* (*characteristic-value* problems) which occur in the areas of vibration analysis, electric-circuit analysis, theory of elasticity, and so on. In developing the mathematical models of systems in these areas, the equations generally have the form

$$\begin{aligned} (a_{11} - \lambda)x_1 + \quad a_{12}x_2 + \quad a_{13}x_3 + \cdots + a_{1n}x_n &= 0 \\ a_{21}x_1 + (a_{22} - \lambda)x_2 + \quad a_{23}x_3 + \cdots + a_{2n}x_n &= 0 \\ a_{31}x_1 + \quad a_{32}x_2 + (a_{33} - \lambda)x_3 + \cdots + a_{3n}x_n &= 0 \\ \vdots \qquad\qquad\qquad\qquad\qquad & \;\vdots \\ a_{n1}x_1 + \cdots\cdots\cdots + (a_{nn} - \lambda)x_n &= 0 \end{aligned} \tag{3-86}$$

[9] The rank of a matrix is the order of the largest nonzero determinant which can be obtained considering all minors of the matrix. Or, stated in another way, it is the number of linearly independent rows (or columns) of the matrix. The procedure for finding the rank of a matrix is discussed in Sec. 3-10.

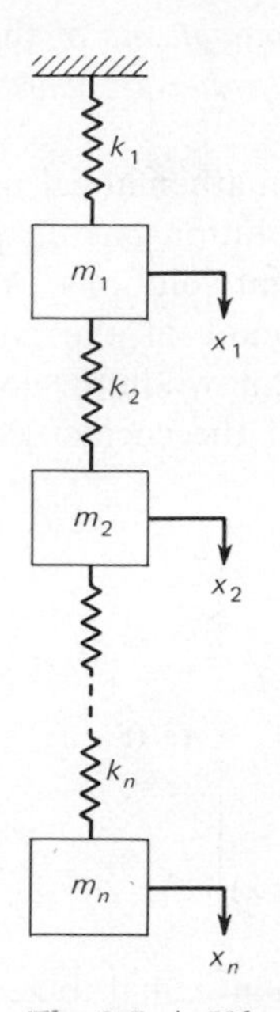

Fig. 3-5 / *Vibrating system with multiple degrees of freedom.*

where the coefficients a_{ij} are real, the x's are system variables, and λ is a particular parameter of the system having unknown values. For example, in a vibrating system having several degrees of freedom, such as the one shown in Fig. 3-5, the coefficient values would derive from the m_i and k_i values (masses and spring constants, respectively), and x_i would be the displacements of the respective masses, and the values of λ would be the squares of the natural circular frequencies of the system.

In matrix notation Eq. 3-86 is expressed as

$$\begin{bmatrix} (a_{11}-\lambda) & a_{12} & a_{13} \cdots\cdots & a_{1n} \\ a_{21} & (a_{22}-\lambda) & a_{23} \cdots\cdots & a_{2n} \\ a_{31} & a_{32} & (a_{33}-\lambda) \cdots & a_{3n} \\ \vdots & \vdots & \vdots & \vdots \\ a_{n1} & \cdots\cdots & \cdots\cdots & (a_{nn}-\lambda) \end{bmatrix} \begin{Bmatrix} x_1 \\ x_2 \\ x_3 \\ \vdots \\ x_n \end{Bmatrix} = \mathbf{0} \tag{3-87}$$

or more compactly as

$$(\mathbf{A} - \lambda\mathbf{I})\mathbf{X} = 0 \tag{3-88}$$

where the use of the *identity* matrix **I** allows us to use $(\mathbf{A} - \lambda\mathbf{I})$ as a coefficient matrix. The column matrix **X** is frequently referred to as an *eigen-*

vector, with $x_1, x_2, x_3, \ldots, x_n$ considered as the *components* of the eigenvector. The values obtained for λ are known as *eigenvalues* or *characteristic values* of the matrix **A**.

Thinking of Eq. 3-86 or Eq. 3-87 as being the mathematical model of a physical system, it is obvious that the trivial solution has no physical significance and that we are interested in nontrivial solutions. As stated earlier, nontrivial solutions exist if the determinant of the coefficient matrix is equal to zero. With λ appearing as an unknown in the coefficient matrix, we may arbitrarily set the determinant of the coefficient matrix equal to zero,

$$|D| = \begin{vmatrix} (a_{11} - \lambda) & a_{12} & a_{13} & a_{1n} \\ a_{21} & (a_{22} - \lambda) & a_{23} & a_{2n} \\ a_{31} & a_{32} & (a_{33} - \lambda) & a_{3n} \\ \vdots & \vdots & \vdots & \vdots \\ a_{n1} & \cdots & \cdots & (a_{nn} - \lambda) \end{vmatrix} = 0$$

and find values of λ which will make the determinant equal to zero.

The expansion of such a determinant results in an nth-degree polynomial

$$\lambda^n + b_1\lambda^{n-1} + b_2\lambda^{n-2} + \cdots + b_n = 0 \tag{3-89}$$

which is solved to obtain the λ_i values that will make $|D| = 0$. (There will be n values of λ appearing as the n roots of the polynomial.) This polynomial is referred to as the *characteristic equation* of the matrix **A**, and the roots of the polynomial are known as *eigenvalues* or *characteristic values*.

After determining the eigenvalues, these values may be substituted, one value at a time, back into the given set of equations to obtain a corresponding *set of relationships* between the unknowns x_i for each substitution. The relationships obtained will depend upon the rank r of the coefficient matrix $(\mathbf{A} - \lambda\mathbf{I})$. If $r = n - 1$, the relationships will be such that the assumption of a value for one unknown will yield a corresponding value for each of the remaining unknowns; if $r = n - 2$, the relationships will be such that values will have to be assumed for two unknowns in order to obtain a corresponding value for each of the remaining unknowns; and so forth.

In many engineering applications the rank of the coefficient matrix is one less than the order (we will see later that this corresponds to only one linearly *dependent* equation in the set), and the relationships between the unknowns may be obtained as the ratios $x_1/x_2 = \alpha_1, x_1/x_3 = \alpha_2, \ldots, x_1/x_n = \alpha_{n-1}$. The unknowns x_i are known as *eigenvector components*. They will be discussed in more detail in Sec. 3-10.

When n is small ($n = 2$ or 3), the expansion of the determinant by minors to obtain the polynomial in λ is not difficult, nor is the subsequent determination of the roots of the polynomial. However, when n becomes larger, the determination of the coefficients of the polynomial becomes more difficult, and a procedure more practical than an expansion by minors must be employed. (Such a procedure will be discussed in Sec. 3-10.) These higher degree polynomials may be solved by one of the methods discussed in Chap. 2 (preferably Bairstow's method).

EXAMPLE 3-6

To make the preceding discussion a little more specific, let us consider a simple set of just two equations,

$$\begin{aligned} (a_{11} - \lambda)x_1 + \qquad a_{12}x_2 &= 0 \\ a_{21}x_1 + (a_{22} - \lambda)x_2 &= 0 \end{aligned} \tag{3-90}$$

which may be written in matrix notation as

$$\begin{bmatrix} (a_{11} - \lambda) & a_{12} \\ a_{21} & (a_{22} - \lambda) \end{bmatrix} \begin{Bmatrix} x_1 \\ x_2 \end{Bmatrix} = \mathbf{0} \tag{3-91}$$

Using Cramer's rule to solve for x_1 and x_2, we obtain

$$x_1 = \frac{0}{\begin{vmatrix} (a_{11} - \lambda) & a_{12} \\ a_{21} & (a_{22} - \lambda) \end{vmatrix}} \quad \text{and} \quad x_2 = \frac{0}{\begin{vmatrix} (a_{11} - \lambda) & a_{12} \\ a_{21} & (a_{22} - \lambda) \end{vmatrix}} \tag{3-92}$$

Looking at Eq. 3-92 it is obvious that if the determinant of the coefficient matrix appearing in the denominator of each equation is nonzero, only a trivial solution will be obtained. However, if $|D| = 0$,

$$x_1 = \frac{0}{0} \quad \text{and} \quad x_2 = \frac{0}{0} \tag{3-93}$$

which shows that nontrivial solutions *may* exist. We know from the previous discussion that when $|D| = 0$, nontrivial solutions *do* exist.

Expanding the determinant yields the characteristic equation

$$\lambda^2 - (a_{11} + a_{22})\lambda + (a_{11}a_{22} - a_{21}a_{12}) = 0 \tag{3-94}$$

which is a second-degree polynomial in this simple example. The roots λ_1 and λ_2 (the eigenvalues) of Eq. 3-94 are the two values of λ which will make the determinant of the coefficient matrix of Eq. 3-91 equal to zero, thus yielding a nontrivial solution.

After obtaining λ_1 and λ_2, their substitution into either Eq. 3-90 or Eq. 3-91 yields two relationships between the unknowns x_1 and x_2. For example, using λ_1 in Eq. 3-90, we obtain

$$x_1 = \frac{-a_{12}x_2}{(a_{11} - \lambda_1)} = \frac{-(a_{22} - \lambda_1)x_2}{a_{21}} \tag{3-95}$$

Similarly, using λ_2 in Eq. 3-90 yields another relationship

$$x_1 = \frac{-a_{12}x_2}{(a_{11} - \lambda_2)} = \frac{-(a_{22} - \lambda_2)x_2}{a_{21}} \tag{3-96}$$

It is apparent, from an inspection of Eqs. 3-95 and 3-96, that unique values of x_1 and x_2 cannot be determined. However, any *combination* of values of x_1 and x_2 constitutes a solution of Eq. 3-90, as long as the combination satisfies Eq. 3-95 when λ_1 is used, or Eq. 3-96 when λ_2 is used.

EXAMPLE 3-7

To illustrate the solution of a physically significant characteristic-value problem, let us determine the principal stresses at a point in a body in a state of plane stress.

Figure 3-6(a) shows an infinitesimal element of material subjected to the normal stresses σ_x and σ_y and the shear stress τ_{xy}. Since all stresses in the z direction are zero, the element is in a state of plane stress. In deriving the equilibrium equations, we consider a wedge-shaped portion of the element

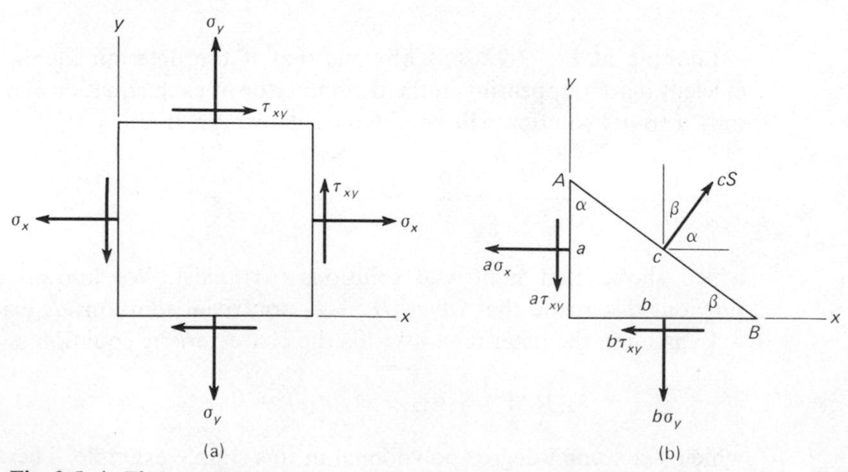

Fig. 3-6 / *Element of material in a state of plane stress.*

formed by cutting the element with an inclined plane at some angle α to the yz plane. This plane AB is arbitrarily designated as a principal plane upon which one of the principal normal stresses acts. We know, from mechanics-of-materials theory, that the shear stress is zero on a principal plane. Therefore, the only stress shown acting on plane AB is the principal normal stress, which will be designated by S. Considering the element to have a depth of unity, with sides of lengths a, b, and c, as shown, the corresponding areas of the sides upon which the stresses act are of magnitudes a, b, and c. The forces shown in Fig. 3-6(b) which maintain the wedge-shaped element in equilibrium are obtained by multiplying the stresses acting on these areas by the respective areas of the element.

Since the element is in equilibrium,

$$\sum F_x = 0 = cS\cos\alpha - a\sigma_x - b\tau_{xy}$$

from which

$$\frac{a}{c}\sigma_x - S\cos\alpha + \frac{b}{c}\tau_{xy} = 0 \qquad (3\text{-}97)$$

Also,

$$\sum F_y = 0 = cS\cos\beta - a\tau_{xy} - b\sigma_y$$

from which

$$\frac{a}{c}\tau_{xy} - S\cos\beta + \frac{b}{c}\sigma_y = 0 \qquad (3\text{-}98)$$

Denoting the direction cosines l and m of the normal to the principal plane as

$$l = \cos\alpha = \frac{a}{c}$$

$$m = \cos\beta = \frac{b}{c}$$

Eqs. 3-97 and 3-98 may be written as

$$\begin{aligned} (\sigma_x - S)l + \tau_{xy}m &= 0 \\ \tau_{xy}l + (\sigma_y - S)m &= 0 \end{aligned} \qquad (3\text{-}99)$$

or, in matrix notation, as

$$\begin{bmatrix} (\sigma_x - S) & \tau_{xy} \\ \tau_{xy} & (\sigma_y - S) \end{bmatrix} \begin{Bmatrix} l \\ m \end{Bmatrix} = 0 \qquad (3\text{-}100)$$

Comparing Eq. 3-100 with Eq. 3-87, we see that the normal and shear stresses correspond to the coefficients a_{ij}, the direction cosines correspond to the unknowns x_i, and the principal stresses S are λ's or the eigenvalues necessary to satisfy the equilibrium equations of Eq. 3-99.

Remembering that the determinant of the coefficient matrix of Eq. 3-100 must equal zero to obtain a nontrivial solution,

$$|D| = \begin{vmatrix} (\sigma_x - S) & \tau_{xy} \\ \tau_{xy} & (\sigma_y - S) \end{vmatrix} = 0 \tag{3-101}$$

Expanding the determinant yields

$$S^2 - (\sigma_x + \sigma_y)S + \sigma_x\sigma_y - \tau_{xy}^2 = 0 \tag{3-102}$$

Using the binomial theorem, the roots or eigenvalues of Eq. 3-102 are found to be

$$S_{1,2} = \frac{\sigma_x + \sigma_y}{2} \pm \sqrt{\left(\frac{\sigma_x - \sigma_y}{2}\right)^2 + \tau_{xy}^2} \tag{3-103}$$

Equation 3-103 is the familiar equation, found in mechanics of materials, relating the principal stresses to the stresses acting on any two orthogonal planes.

Suppose that the orthogonal stresses at the point are given by

$$\sigma_x = 1000 \text{ psi}$$

$$\sigma_y = 500 \text{ psi}$$

$$\tau_{xy} = 500 \text{ psi}$$

From Eq. 3-103 we can compute the principal stresses as

$$S_1 = 1310 \text{ psi}$$

$$S_2 = 190 \text{ psi}$$

Then the relationships between the direction cosines l and m may be determined by substituting the principal-stress values into Eq. 3-100. Using the value of S_2

$$\begin{bmatrix} (1000 - 190) & 500 \\ 500 & (500 - 190) \end{bmatrix} \begin{Bmatrix} l \\ m \end{Bmatrix} = \mathbf{0} \tag{3-104}$$

which yields the relationship

$$\frac{m}{l} = \frac{\cos\beta}{\cos\alpha} = -\frac{810}{500} \tag{3-105}$$

The direction-cosine relationships for the principal stress S_1 are determined similarly, where

$$\begin{bmatrix} (1000 - 1310) & 500 \\ 500 & (500 - 1310) \end{bmatrix} \begin{Bmatrix} l \\ m \end{Bmatrix} = \mathbf{0} \tag{3-106}$$

which yields

$$\frac{m}{l} = \frac{\cos\beta}{\cos\alpha} = \frac{310}{500} \tag{3-107}$$

In this particular example unique values may be determined for l and m, since an additional relationship exists in that the angles α and β are complementary angles. Thus since $\sin\alpha = \cos\beta$,

$$\tan\alpha = -\frac{810}{500} \qquad (\text{for } S_2)$$

$$\alpha = -58.3 \text{ deg}$$

$$\tan\alpha = \frac{310}{500} \qquad (\text{for } S_1)$$

$$\alpha = 31.7 \text{ deg}$$

These angles locate the principal planes (the planes upon which the respective principal stresses act) with α measured positive counterclockwise from the y axis and negative clockwise (see Fig. 3-6(b)).

It should be emphasized that the unique values which could be obtained for l and m resulted from the availability of the *additional* relationship between the angles α and β. In general, in characteristic-value problems, such additional relationships will not be available, and only ratios of the components of the eigenvector will be obtainable, as discussed before.

EXAMPLE 3-8[10]

As another example of a characteristic-value problem, let us analyze the free undamped vibrational characteristics of the three-degrees-of-freedom system shown schematically in Fig. 3-7. The system consists of the three

[10] SI system of units used; see App. A.

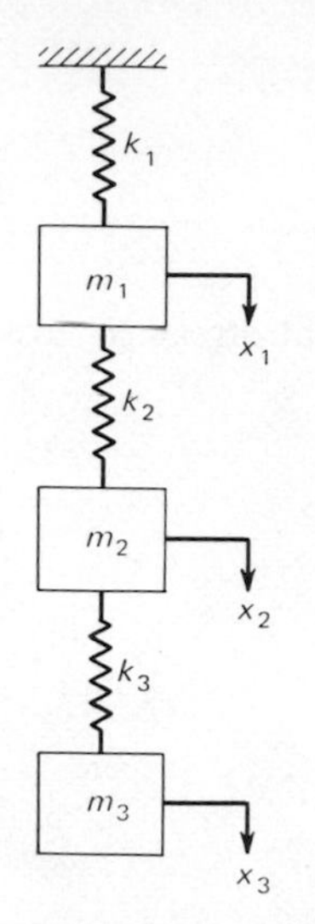

$m_1 = m_2 = m_3 = 14.59$ kg (kilograms)
$k_1 = k_2 = k_3 = 145.9$ N/m (newtons/meter)

Fig. 3-7 / *Three-degrees-of-freedom system.*

masses m_1, m_2, and m_3, connected by the three springs shown, with spring constants k_1, k_2, and k_3. The displacements of the masses are defined by the generalized coordinates x_1, x_2, and x_3, respectively, each displacement being measured from the static-equilibrium position of the respective mass.

Utilizing either Lagrange's equation or Newton's second law, the differential equations of motion of the system are found to be

$$\begin{aligned} m_1\ddot{x}_1 + (k_1 + k_2)x_1 - k_2x_2 &= 0 \\ m_2\ddot{x}_2 - k_2x_1 + (k_2 + k_3)x_2 - k_3x_3 &= 0 \\ m_3\ddot{x}_3 - k_3x_2 + k_3x_3 &= 0 \end{aligned} \qquad (3\text{-}108)$$

It is well known, from vibration theory, that the solution of Eq. 3-108 may be taken in the form

$$\begin{aligned} x_1 &= X_1 \sin pt \\ x_2 &= X_2 \sin pt \\ x_3 &= X_3 \sin pt \end{aligned} \qquad (3\text{-}109)$$

where X_1, X_2, and X_3 are the *amplitudes* of motion of the respective masses, and p denotes the natural circular frequencies corresponding to the principal modes of vibration of the system.

Substituting Eq. 3-109 and appropriate derivatives of these expressions into Eq. 3-108, and using the spring-constant and mass values shown in Fig. 3-7, we obtain the following set of homogeneous algebraic equations:

$$\begin{aligned} (20 - p^2)X_1 \quad -10X_2 \qquad\qquad &= 0 \\ -10X_1 + (20 - p^2)X_2 \quad -10X_3 &= 0 \\ -10X_2 + (10 - p^2)X_3 &= 0 \end{aligned} \tag{3-110}$$

To obtain a nontrivial solution of Eq. 3-110, we know that the determinant of the coefficients of X_i must be equal to zero, so that

$$|D| = \begin{vmatrix} (20 - p^2) & -10 & 0 \\ -10 & (20 - p^2) & -10 \\ 0 & -10 & (10 - p^2) \end{vmatrix} = 0 \tag{3-111}$$

Expansion of this determinant results in the characteristic polynomial

$$p^6 - 50p^4 + 600p^2 - 1000 = 0 \tag{3-112}$$

which may be written as a cubic equation in p^2 as

$$(p^2)^3 - 50(p^2)^2 + 600(p^2) - 1000 = 0 \tag{3-113}$$

The roots of Eq. 3-113 are found to be

$$p_1^2 = 1.98 \text{ s}^{-2}$$
$$p_2^2 = 15.5 \text{ s}^{-2}$$
$$p_3^2 = 32.5 \text{ s}^{-2}$$

These eigenvalues are the squares of the natural circular frequencies of the first, second, and third modes of vibration of the system, respectively.

Since Eq. 3-110 constitutes a homogeneous set of simultaneous equations, a unique set of values cannot be obtained for X_1, X_2, and X_3. However, various ratios of the amplitudes may be determined which will yield the configuration of the system for the various modes when the amplitude of any one of the masses is defined. For example, substituting $p^2 = 1.98$ into Eq. 3-110 yields the following configuration for the *first* mode:

$$\left.\begin{aligned} X_2 &= 1.80X_1 \\ X_3 &= 2.24X_1 \end{aligned}\right\} \quad \text{first mode} \tag{3-114}$$

Similarly, the second- and third-mode configurations, using $p_2^2 = 15.5$ and $p_3^2 = 32.5$, respectively, are

$$\left.\begin{aligned} X_2 &= 0.45X_1 \\ X_3 &= -0.82X_1 \end{aligned}\right\} \quad \text{second mode} \tag{3-115}$$

$$\left.\begin{aligned} X_2 &= -1.25X_1 \\ X_3 &= 0.555X_1 \end{aligned}\right\} \quad \text{third mode} \tag{3-116}$$

From the last three equations it can be seen that if the amplitude of any one of the three masses is known or assumed for a particular mode of vibration, the configuration of the system may be determined for that mode. Since Eqs. 3-114 through 3-116 consist of ratios of amplitudes X_i, the substitution of Eq. 3-109 into these equations reveals that the ratios shown are also the ratios of the *displacements*. For example, when m_1 has a displacement of 0.01 m and the system is vibrating in the second mode, the corresponding displacements of m_2 and m_3 will be 0.0045 m and -0.0082 m, respectively, and the motion of m_3 will be 180 deg out of phase with that of m_1. It might be added here that the configuration of a system, as given by the above ratios, also defines the *initial* displacements which would have to be given to the masses to have the system vibrate in the mode associated with that configuration, with no other harmonics present, when the system is released from rest.

3-9 / METHODS FOR SOLUTION OF EIGENVALUE PROBLEMS—GENERAL

The characteristic polynomial method was used in the previous section to develop an "insight" into the nature of the eigenvalue problem. Through the years, various methods and techniques for the solution of eigenvalue problems have evolved because of their importance in engineering, science, and mathematics. Some of the more widely used methods for the solution of eigenvalue problems include (1) Jacobi method, (2) Householder's method, (3) the polynomial method, and (4) the iteration, or power, method.

The Jacobi method is perhaps one of the most reliable methods but is not very efficient in comparison to some of the other methods. One advantage of the Jacobi method is that it obtains the eigenvectors along with the eigenvalues. Jacobi's method is limited to symmetric matrices.

Householder's method is based upon an orthogonal transformation which produces a large number of zeros in a given row so as to yield a tridiagonalized matrix. Various techniques are then used to determine the eigenvalues from the tridiagonalized matrix. Today it is generally recognized that the most efficient and accurate method for obtaining all the eigenvalues of very large matrices (n equal to several hundred) is best accomplished by the use of Householder's method and an algorithm called *QL*. This method is also limited to symmetric matrices.

Detailed discussions of the Jacobi method and Householder's method are beyond the scope of this text; the reader is referred to a discussion of the Jacobi method by Greenstadt[11] and a discussion of the Householder method by Ortega.[12]

The polynomial method previously mentioned will be discussed in some detail relative to computer computation in the next section since it is one of the most widely used methods for solving moderately sized matrices ($n \leq 30$). We will also discuss in some detail, in Sec. 3-11, an iteration, or power, method which is frequently used when only the *smallest* and/or *largest* eigenvalue(s) of a matrix are desired. An advantage of the iteration method over most other methods is that the eigenvalues and corresponding eigenvectors are obtained simultaneously in the iteration process. In Sec. 3-12, we will also present a procedure for "sweeping" out previously determined eigenvalues and corresponding eigenvectors so that intermediate eigenvalues and eigenvectors can also be obtained from the iteration method.

3-10 / POLYNOMIAL METHOD—EIGENVALUE PROBLEMS

In the examples presented in Sec. 3-8, the determination of the eigenvalues involved two major steps: (1) obtaining the characteristic polynomial by expanding the determinant of the coefficient matrix $(\mathbf{A} - \lambda\mathbf{I})$ and (2) solving for the roots of the polynomial. Both steps employed familiar elementary methods, since we were concerned with matrices of low order.

Computer Determination of Characteristic Polynomials

When the order of the matrices becomes larger, computer methods must be adopted to efficiently accomplish each of these steps. Since, as mentioned previously, the roots of the polynomial can be found by one of the methods already covered in Chap. 2, we will be concerned here with a computer method for determining the characteristic equation.[13]

[11] John Greenstadt, *Mathematical Methods for Digital Computers*, Vol. I, ed. A. Ralston and H. S. Wilf (New York: John Wiley & Sons, 1960), pp. 84–91.

[12] James Ortega, *Mathematical Methods for Digital Computers*, Vol. II, ed. A. Ralston and H. S. Wilf (New York: John Wiley & Sons, 1967), pp. 94–115.

[13] In Ex. 7-3 (Chap. 7) a characteristic-value problem arising from the finite-difference form of a differential equation is solved, in which Bairstow's method is used to find the roots of the characteristic polynomial.

Many methods have been proposed for generating the coefficients p_k of the characteristic equation

$$\lambda^n + p_1\lambda^{n-1} + p_2\lambda^{n-2} + p_3\lambda^{n-3} + \cdots + p_n = 0 \qquad (k = 1, 2, \ldots, n) \tag{3-117}$$

when the order n of the set of algebraic equations is large. Of these methods, we will discuss the *Faddeev-Leverrier method*, an efficient technique which is equally suitable for generating the polynomial coefficients whether the coefficient matrix of the equations is symmetrical or unsymmetrical. This method has an additional feature in that the *inverse* of the matrix for which the eigenvalues are desired is, essentially, obtained in the process of generating the coefficients of the characteristic polynomial. This is advantageous in instances where one may wish to check the accuracy of the eigenvalues obtained by the polynomial method, by using an iterative method, since an inversion of the matrix may be desired in an iterative procedure (see Sec. 3-11). Proof of the validity of the Faddeev-Leverrier method is beyond the intended scope of this text, but it may be found elsewhere.[14] (This reference also includes numerous other methods and their derivations.)

Before outlining the method we have chosen, let us define the *trace* of a matrix, since it is an integral part of the method. Given the matrix

$$\begin{bmatrix} a_{11} & a_{12} & a_{13} \cdots & a_{1n} \\ a_{21} & a_{22} & a_{23} \cdots & a_{2n} \\ a_{31} & a_{32} & a_{33} \cdots & a_{3n} \\ \vdots & \vdots & & \vdots \\ a_{n1} & a_{n2} & \cdots\cdots & a_{nn} \end{bmatrix} \tag{3-118}$$

the trace of the matrix, written as "tr **A**," is

$$\text{tr}\,\mathbf{A} = a_{11} + a_{22} + a_{33} + \cdots + a_{nn} \tag{3-119}$$

The Faddeev-Leverrier method generates the polynomial coefficients $p_k(k = 1, 2, 3, \ldots, n)$ from the matrix **A** of the given set of equations written as

$$(\mathbf{A} - \lambda\mathbf{I})\mathbf{X} = 0 \tag{3-120}$$

[14] See D. K. Faddeev and U. N. Faddeeva, *Computational Methods of Linear Algebra*, trans. Robert C. Williams (San Francisco: W. H. Freeman & Co., 1963). For other methods see Leon Lapidus, *Digital Computation for Chemical Engineers* (New York: McGraw-Hill Book Company, 1962).

by forming a *sequence* of matrices $\mathbf{B}_1, \mathbf{B}_2, \ldots, \mathbf{B}_n$ from which the p_k values are determined. These p_k values are then substituted in the following basic form of the characteristic polynomial:

$$(-1)^n(\lambda^n - p_1\lambda^{n-1} - p_2\lambda^{n-2} - p_3\lambda^{n-3} - \cdots - p_n) = 0 \tag{3-121}$$

where the $(-1)^n$ is used merely to give the terms of the polynomial the same signs that they would have if the polynomial were generated by expanding a determinant. The p_k values are determined as follows:

$$\begin{aligned}
\mathbf{B}_1 &= \mathbf{A} && \text{and} \quad p_1 = \operatorname{tr} \mathbf{B}_1 \\
\mathbf{B}_2 &= \mathbf{A}(\mathbf{B}_1 - p_1\mathbf{I}) && \text{and} \quad p_2 = \frac{1}{2} \operatorname{tr} \mathbf{B}_2 \\
\mathbf{B}_3 &= \mathbf{A}(\mathbf{B}_2 - p_2\mathbf{I}) && \text{and} \quad p_3 = \frac{1}{3} \operatorname{tr} \mathbf{B}_3 \\
&\ \vdots && \qquad\qquad \vdots \\
\mathbf{B}_k &= \mathbf{A}(\mathbf{B}_{k-1} - p_{k-1}\mathbf{I}) && \text{and} \quad p_k = \frac{1}{k} \operatorname{tr} \mathbf{B}_k \\
&\ \vdots && \qquad\qquad \vdots \\
\mathbf{B}_n &= \mathbf{A}(\mathbf{B}_{n-1} - p_{n-1}\mathbf{I}) && \text{and} \quad p_n = \frac{1}{n} \operatorname{tr} \mathbf{B}_n
\end{aligned} \tag{3-122}$$

Referring to the generation of the inverse of the matrix for which the eigenvalues are desired, mentioned earlier, Faddeev has shown that the inverse of $\mathbf{A}$ can be determined from

$$\mathbf{A}^{-1} = \frac{1}{p_n}(\mathbf{B}_{n-1} - p_{n-1}\mathbf{I}) \tag{3-123}$$

EXAMPLE 3-9

Let us suppose that the homogeneous set of algebraic equations

$$\begin{aligned}
(3 - \lambda)x_1 + 2x_2 + 4x_3 &= 0 \\
2x_1 + (0 - \lambda)x_2 + 2x_3 &= 0 \\
4x_1 + 2x_2 + (3 - \lambda)x_3 &= 0
\end{aligned} \tag{3-124}$$

has been obtained as the mathematical model of some physical system which we are analyzing. In matrix notation, Eq. 3-124 appears as

$$\begin{bmatrix} (3-\lambda) & 2 & 4 \\ 2 & (0-\lambda) & 2 \\ 4 & 2 & (3-\lambda) \end{bmatrix} \begin{Bmatrix} x_1 \\ x_2 \\ x_3 \end{Bmatrix} = \mathbf{0} \tag{3-125}$$

Our problem will be to find the characteristic polynomial, using the Faddeev-Leverrier method, and then to solve the resulting polynomial for its roots which are the eigenvalues of the matrix

$$\mathbf{A} = \begin{bmatrix} 3 & 2 & 4 \\ 2 & 0 & 2 \\ 4 & 2 & 3 \end{bmatrix}$$

of the equation

$$(\mathbf{A} - \lambda\mathbf{I})\mathbf{X} = 0 \tag{3-126}$$

Using the procedure indicated in Eq. 3-122

$$\mathbf{B}_1 = \mathbf{A} = \begin{bmatrix} 3 & 2 & 4 \\ 2 & 0 & 2 \\ 4 & 2 & 3 \end{bmatrix}$$

and

$$p_1 = \text{tr}\,\mathbf{B}_1 = 3 + 0 + 3 = 6$$

$$\mathbf{B}_2 = \mathbf{A}(\mathbf{B}_1 - p_1\mathbf{I}) = \begin{bmatrix} 3 & 2 & 4 \\ 2 & 0 & 2 \\ 4 & 2 & 3 \end{bmatrix} \left(\begin{bmatrix} 3 & 2 & 4 \\ 2 & 0 & 2 \\ 4 & 2 & 3 \end{bmatrix} - \begin{bmatrix} 6 & 0 & 0 \\ 0 & 6 & 0 \\ 0 & 0 & 6 \end{bmatrix} \right)$$

$$= \begin{bmatrix} 3 & 2 & 4 \\ 2 & 0 & 2 \\ 4 & 2 & 3 \end{bmatrix} \begin{bmatrix} -3 & 2 & 4 \\ 2 & -6 & 2 \\ 4 & 2 & -3 \end{bmatrix}$$

$$= \begin{bmatrix} 11 & 2 & 4 \\ 2 & 8 & 2 \\ 4 & 2 & 11 \end{bmatrix}$$

and

$$p_2 = \tfrac{1}{2} \text{ tr } \mathbf{B}_2 = \tfrac{1}{2}(11 + 8 + 11) = 15$$

$$\mathbf{B}_3 = \mathbf{A}(\mathbf{B}_2 - p_2\mathbf{I}) = \begin{bmatrix} 3 & 2 & 4 \\ 2 & 0 & 2 \\ 4 & 2 & 3 \end{bmatrix} \left(\begin{bmatrix} 11 & 2 & 4 \\ 2 & 8 & 2 \\ 4 & 2 & 11 \end{bmatrix} - \begin{bmatrix} 15 & 0 & 0 \\ 0 & 15 & 0 \\ 0 & 0 & 15 \end{bmatrix} \right)$$

$$\begin{bmatrix} 3 & 2 & 4 \\ 2 & 0 & 2 \\ 4 & 2 & 3 \end{bmatrix} \begin{bmatrix} -4 & 2 & 4 \\ 2 & -7 & 2 \\ 4 & 2 & -4 \end{bmatrix} = \begin{bmatrix} 8 & 0 & 0 \\ 0 & 8 & 0 \\ 0 & 0 & 8 \end{bmatrix}$$

and

$$p_3 = \tfrac{1}{3} \text{ tr } \mathbf{B}_3 = \tfrac{1}{3}(8 + 8 + 8) = 8$$

Substituting the values of p_1, p_2, and p_3 into Eq. 3-121, we obtain

$$(-1)^3(\lambda^3 - 6\lambda^2 - 15\lambda - 8) = 0 \tag{3-127}$$

The roots of Eq. 3-127 are simply obtained, since the polynomial may be factored as

$$(\lambda - 8)(\lambda + 1)(\lambda + 1) = 0 \tag{3-128}$$

from which the eigenvalues are

$$\begin{aligned} \lambda_1 &= 8 \\ \lambda_2 &= -1 \\ \lambda_3 &= -1 \end{aligned} \tag{3-129}$$

Reviewing the determination of the coefficient p_3 in the procedure just finished, we observe that

$$\mathbf{B}_3 = p_3\mathbf{I} \tag{3-130}$$

Faddeev has proved that for an nth-order matrix

$$\mathbf{B}_n = p_n\mathbf{I} \tag{3-131}$$

from which we see that we can always obtain p_n simply as

$$p_n = b_{11} = b_{22} = b_{33} = \cdots = b_{nn} \tag{3-132}$$

where the b_{ii} are *identical* elements composing the trace of $\mathbf{B}_n$.

At this point the interested reader might expand the determinant of the coefficient matrix of Eq. 3-125

$$|D| = \begin{vmatrix} (3-\lambda) & 2 & 4 \\ 2 & (0-\lambda) & 2 \\ 4 & 2 & (3-\lambda) \end{vmatrix} = 0$$

by minors and check the polynomial obtained in Eq. 3-127.

The generation of the coefficients of the characteristic polynomial, by the Faddeev-Leverrier method, on the computer is outlined in the flow chart of Fig. 3-8. The variable names and the quantities they represent in the FORTRAN source program are as follows:

FORTRAN Name	Quantity
N	Order of the $\mathbf{A}$ matrix for which the characteristic polynomial is to be determined
A(I,J)	Elements of the $\mathbf{A}$ matrix
B(I,J)	Elements of the $\mathbf{B}_k$ matrices formed by the Faddeev-Leverrier method; also used for elements of $(\mathbf{B}_k - p_k\mathbf{I})$
I	Row numbers of $\mathbf{A}$ and $\mathbf{B}_k$ matrices
J	Column numbers of $\mathbf{A}$ and $\mathbf{B}_k$ matrices
M	$N - 1$
TRACE	Traces of $\mathbf{B}_k$ matrices
P(K)	Coefficients of characteristic polynomial
K	Subscript of P(K) used as index of a DO loop
AK	Real name for the integer subscript K of P(K)
COLB(I)	Elements of the columns of the $(\mathbf{B}_{k-1} - p_{k-1}\mathbf{I})$ matrices

The source program, written to implement the flow chart of Fig. 3-8, is as follows:

```
      DIMENSION A(15,15),B(15,15),P(15),COLB(15)
      READ(5,1) N
    1 FORMAT(I3)
      READ(5,2) ((A(I,J),J=1,N),I=1,N)
    2 FORMAT(8F10.0)
      DO 3 I=1,N
      DO 3 J=1,N
    3 B(I,J)=A(I,J)
      M=N-1
      DO 10 K=1,M
      TRACE=0.
      DO 7 I=1,N
```

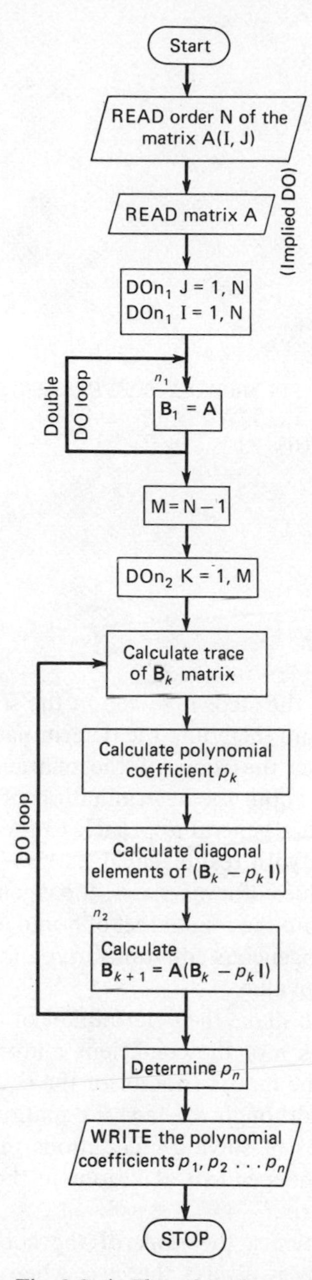

Fig. 3-8 / *Flow chart for generating coefficients of a characteristic polynomial.*

```
 7 TRACE=TRACE+B(I,I)
   AK=K
   P(K)=TRACE/AK
   DO 8 I=1,N
 8 B(I,I)=B(I,I)-P(K)
   DO 10 J=1,N
   DO 9 I=1,N
 9 COLB(I)=B(I,J)
   DO 10 I=1,N
   B(I,J)=0.
   DO 10 L=1,N
10 B(I,J)=B(I,J)+A(I,L)*COLB(L)
   P(N)=B(1,1)
   WRITE(6,4)
 4 FORMAT('1','THE CHARACTERISTIC POLYNOMIAL COEFFICIENTS')
   WRITE(6,5)
 5 FORMAT(' ',10X,'P(1) THROUGH P(N)'/)
   WRITE(6,6) (P(K),K=1,N)
 6 FORMAT(' ',14X,E14.7)
   STOP
   END
```

Computer Determination of Eigenvectors

So far we have discussed in some detail the steps involved in the solution of characteristic-value problems up to and including the determination of the eigenvalues which were obtained as the roots of the characteristic equation. However, little has been said about the determination of eigenvectors. Now we will be concerned with a general approach to the determination of the eigenvectors associated with real eigenvalues.[15] We have seen that after determining the eigenvalues of a matrix **A**, the eigenvalues may be substituted, one at a time, back into the original set of homogeneous equations to obtain a new set of n homogeneous equations in n unknowns (eigenvector components) for each eigenvalue.

Since the eigenvalues are values which make the determinant of the coefficient matrix $(\mathbf{A} - \lambda\mathbf{I})$ zero, the rank r of the coefficient matrix must necessarily be less than the order n of the matrix (not all of the equations obtained are independent). Therefore, although we have n equations in n unknowns, the problem is actually that of solving r equations in n unknowns, since the number of linearly independent equations in the set is equal to the rank of the coefficient matrix.

In the majority of engineering problems, the rank of the coefficient matrix is one less than the order. (This is always the case when all the eigenvalues are discrete.) In such problems, then, the set of homogeneous equations to be solved contains $n - 1$ independent equations for determining the n eigenvector components. Therefore, by assuming a value for

[15] In some problems the determination of eigenvalues constitutes the desired solution; in others, the solution includes the determination of the associated eigenvectors.

one eigenvector component, values may be obtained for the remaining components. There will be one eigenvector associated with each eigenvalue.

In solving such a set of equations on the computer, the Gaussian or Gauss-Jordan elimination method can be used to reduce the set of n equations to an equivalent set of $n - 1$ independent equations. (This will be illustrated later, using the Gauss-Jordan method.)

Infrequently, in engineering problems, the eigenvalues obtained from the characteristic equation may not all be discrete. For example, in solving for the principal stresses at a point in a body, suppose that two principal stresses were equal. Since the eigenvalues are the principal stresses in such a problem (see Ex. 3-7), the two eigenvalues obtained in the solution would be equal (have a multiplicity of 2). In general, when an eigenvalue obtained has a multiplicity of k, the rank of the coefficient matrix for that eigenvalue may be from 1 to k less than the order of the matrix, and there will be that same number of linearly independent eigenvectors associated with the eigenvalue having the multiplicity.

If the rank of the coefficient matrix is two less than the order, it is necessary to assume values for two eigenvector components to determine values for the remaining components. (There will be only $n - 2$ linearly independent equations in the set of n equations.) Two linearly independent eigenvectors can be obtained, and any linear combination of these two eigenvectors constitutes a possible solution vector. Thus the solution space is described as being two-dimensional. If the rank of the coefficient matrix is three less than the order, three eigenvector-component values must be assumed, three linearly independent eigenvectors are obtained, and the solution space is three-dimensional, and so forth.

Although the rank of the coefficient matrix is one less than the order in most engineering problems, a computer program for determining eigenvectors should be versatile enough to handle the occasional case in which the eigenvalues have some multiplicity, as discussed above. Before developing such a computer program, let us study several simple examples of the determination of eigenvectors. (It might be well, at this point, for the reader to review briefly the Gauss-Jordan elimination method, Sec. 3-3).

To begin with, let us determine the eigenvalues and associated eigenvectors for the set of homogeneous equations

$$\left.\begin{aligned}(4 - \lambda)x_1 + 2x_2 - 2x_3 &= 0 && \text{(a)}\\ -5x_1 + (3 - \lambda)x_2 + 2x_3 &= 0 && \text{(b)}\\ -2x_1 + 4x_2 + (1 - \lambda)x_3 &= 0 && \text{(c)}\end{aligned}\right\} \quad (3\text{-}133)$$

The characteristic equation of this set may be written as

$$\begin{vmatrix}(4 - \lambda) & 2 & -2\\ -5 & (3 - \lambda) & 2\\ -2 & 4 & (1 - \lambda)\end{vmatrix} = 0$$

or, in factored form, as

$$(\lambda - 1)(\lambda - 5)(\lambda - 2) = 0 \tag{3-134}$$

where the eigenvalues are $\lambda = 1$, $\lambda = 2$, and $\lambda = 5$.

Knowing the eigenvalues, we are next interested in obtaining the eigenvector associated with each discrete eigenvalue shown. To determine the eigenvector associated with $\lambda = 1$, we substitute this value into Eq. 3-133 to obtain the homogeneous set of equations

$$\left.\begin{aligned} 3x_1 + 2x_2 - \ 2x_3 &= 0 \qquad \text{(a)} \\ -5x_1 + 2x_2 + \ 2x_3 &= 0 \qquad \text{(b)} \\ -2x_1 + 4x_2 + (0)x_3 &= 0 \qquad \text{(c)} \end{aligned}\right\} \tag{3-135}$$

We next apply the Gauss-Jordan elimination process to reduce Eq. 3-135 to an *equivalent* set consisting only of *independent* equations. The number of independent equations obtained by this process will be equal to the rank of the coefficient matrix of Eq. 3-135. In the Gaussian elimination method, we performed an interchange of equations to get the largest coefficient in the pivot column into the pivot position. This resulted in improved accuracy and prevented the appearance of a zero pivot element. We could have obtained even greater accuracy, in cases in which the largest coefficient of the set of equations appeared in a column other than the pivot column, by interchanging the sequence of the unknowns in the equations (a column interchange) to get this coefficient into the pivot column. If necessary, an equation interchange would then place it in the pivot position. However, such column interchanges were not used in implementing either the Gaussian or the Gauss-Jordan method, in previous discussions, since this would have unnecessarily complicated the "book-keeping" in the computer program.

However, in using the Gauss-Jordan method to reduce a set of equations which are not all independent to an equivalent set which are all independent, we will find that, in many instances, we *must* interchange the sequence of the unknowns in the equations to avoid a zero pivot element. Therefore, to obtain maximum accuracy and to avoid the possibility of a zero pivot element, we will always place the largest applicable coefficient (the applicable coefficient having the largest *absolute value*) in the pivot position. This may require an interchange of equations, an interchange in the sequence of the unknowns, or both.

In Eq. 3-135 no change is required in the sequence of the unknowns, but parts a and b should be interchanged to make -5 the pivot element for eliminating x_1. Making this interchange, we obtain

$$\left.\begin{aligned} -5x_1 + 2x_2 + \ 2x_3 &= 0 \qquad \text{(a)} \\ 3x_1 + 2x_2 - \ 2x_3 &= 0 \qquad \text{(b)} \\ -2x_1 + 4x_2 + (0)x_3 &= 0 \qquad \text{(c)} \end{aligned}\right\} \tag{3-136}$$

Applying the Gauss-Jordan process in which we first divide Eq. 3-136a by -5, use the resulting equation to eliminate x_1 from parts b and c, and finally move Eq. 3-136a from the top to the bottom position, we obtain

$$\left.\begin{aligned}(0)x_1 + \tfrac{16}{5}x_2 - \tfrac{4}{5}x_3 &= 0 \qquad &\text{(a)}\\ (0)x_1 + \tfrac{16}{5}x_2 - \tfrac{4}{5}x_3 &= 0 \qquad &\text{(b)}\\ x_1 - \tfrac{2}{5}x_2 - \tfrac{2}{5}x_3 &= 0 \qquad &\text{(c)}\end{aligned}\right\} \quad (3\text{-}137)$$

In considering the elimination of x_2 or x_3 in the next step, we look for the coefficient, in Eq. 3-137a or b, having the largest absolute value for use as a pivot element. Since the coefficient of x_2 in Eq. 3-137a is the largest coefficient and is already in the pivot position, no equation interchange or change in the sequence of the unknowns is necessary, and we are ready to eliminate x_2. Note that the coefficients in Eq. 3-137c were not considered in the search for a pivot element. These coefficients cannot be considered, regardless of their magnitude, since the use of this equation to eliminate x_2 or x_3 would reintroduce the previously eliminated x_1 into the equations. In general, only the coefficients in the first $n - q$ equations are candidates for the pivot element, where q is the number of eliminations *already completed.*

Again applying the Gauss-Jordan elimination procedure, we obtain

$$\left.\begin{aligned}(0)x_1 + (0)x_2 + (0)x_3 &= 0 \qquad &\text{(a)}\\ x_1 + (0)x_2 - \tfrac{1}{2}x_3 &= 0 \qquad &\text{(b)}\\ (0)x_1 + x_2 - \tfrac{1}{4}x_3 &= 0 \qquad &\text{(c)}\end{aligned}\right\} \quad (3\text{-}138)$$

This final equivalent set of equations consists only of the independent equations b and c, and contains the three unknown eigenvector components x_1, x_2, and x_3. By assuming a value for any one of these three unknowns, values can be determined for the other two.

With two independent equations in the equivalent set, we know that the rank of the coefficient matrix of Eq. 3-133 is 2. The eigenvectors for the other two eigenvalues are determined by the identical process just described.

Now let us turn our attention to implementing this method on the computer. With the coefficient matrix of Eq. 3-135 in memory, the computer would perform the required row interchange and would then utilize the general equations

$$\begin{aligned} b_{i-1,j-1} &= a_{ij} - \frac{a_{1j}a_{i1}}{a_{11}} \qquad &\begin{Bmatrix} 1 < i \le n \\ 1 < j \le m \\ a_{11} \neq 0 \end{Bmatrix} \\ b_{n,j-1} &= \frac{a_{1j}}{a_{11}} \qquad &\begin{Bmatrix} 1 < j \le m \\ a_{11} \neq 0 \end{Bmatrix} \end{aligned} \qquad (3\text{-}139)$$

where

i = row number of old matrix **A**
j = column number of old matrix **A**
n = maximum row number
m = maximum column number
a = an element of old matrix **A**
b = an element of new matrix **B**

to obtain the new matrix

$$\mathbf{B} = \begin{bmatrix} \frac{16}{5} & -\frac{4}{5} \\ \frac{16}{5} & -\frac{4}{5} \\ -\frac{2}{5} & -\frac{2}{5} \end{bmatrix} \tag{3-140}$$

This matrix is the coefficient matrix of Eq. 3-137 with the column consisting only of *zeros* and a *one* omitted.

Before moving to the next elimination process, the computer would search the first two rows of the matrix of Eq. 3-140 for the coefficient having the largest absolute value, and would make any necessary interchanges before performing the next elimination. In general, with all columns containing only zeros and ones dropped from the matrices formed in each elimination, the computer would search only the first k rows of the matrix being considered, where k is the number of columns in that matrix. The use of any rows beyond k in eliminating the next unknown would reintroduce a previously eliminated unknown into the equations, as explained above.

Repeating the Gauss-Jordan procedure, the computer would obtain the column matrix

$$\mathbf{B}' = \begin{Bmatrix} 0 \\ -\frac{1}{2} \\ -\frac{1}{4} \end{Bmatrix} \tag{3-141}$$

which corresponds to Eq. 3-138 with the two columns containing only zeros and ones omitted. No further application of Eq. 3-139 is possible, and the number of columns in the final matrix, which is in computer memory, is equal to the difference between the rank and order of the coefficient matrix $(\mathbf{A} - \lambda\mathbf{I})$ of Eq. 3-133.

Thus it is apparent from Eq. 3-138 (a set of equations equivalent to the original set) that there are only two independent equations in the set. Therefore, the rank of the coefficient matrix is 2, while the order is 3. Since there are three unknowns in the two independent equations of the set of Eq. 3-138, the computer must be given an arbitrary value for one unknown so that it may solve the resulting two *nonhomogeneous* equations for the two remaining unknowns.

In determining the eigenvector on the computer, note the convenience of arbitrarily choosing a value of -1 for x_3. By choosing this value for x_3, a consideration of Eq. 3-138 reveals that the components x_1 and x_2 will have the respective values shown for the nonzero coefficients of x_3 in that equation which appears as elements of the matrix of Eq. 3-141, which is in computer memory. The computer need only place these matrix values and the assumed value for x_3 in the usual sequence

$$\begin{Bmatrix} x_1 \\ x_2 \\ x_3 \end{Bmatrix}$$

to obtain the desired eigenvector. In this manner the eigenvector is easily found from the final matrix in computer memory by instructing the computer, in this case, to replace the zero with -1 and to move it to the bottom position so that

$$\mathbf{X} = \begin{Bmatrix} -\frac{1}{2} \\ -\frac{1}{4} \\ -1 \end{Bmatrix}$$

or, more generally,

$$\mathbf{X} = C\begin{Bmatrix} -\frac{1}{2} \\ -\frac{1}{4} \\ -1 \end{Bmatrix} \tag{3-142}$$

where C is any arbitrary constant.

In the example just completed, the coefficients of x_3 appeared in the final column matrix of Eq. 3-141, and a procedure was described for utilizing these values in determining the eigenvector in its correct x_1, x_2, $x_3, \ldots,$ sequence. With the various column interchanges which could be required in arriving at this final matrix for equations with other coefficients, the final matrix could contain coefficients of any one of the x's. In each case, -1 would be substituted for the zero coefficient in the column matrix and would then be moved to the bottom position. After this basic step, different manipulations would be required, in each case, to arrange the eigenvector components in their proper sequence in representing the eigenvector. Therefore, a general computer program must include some provision for keeping track of the column interchanges performed during the elimination processes, and for utilizing this information later to place the eigenvector components in their proper sequence. (The relationship between the column interchanges and the resulting sequence of the eigenvector components in computer memory will be explained later in this section.)

As a further illustration of the method we will use on the computer to determine eigenvectors, let us consider an example in which there will be two linearly independent eigenvectors associated with an eigenvalue. Let us find the eigenvalues of the matrix

$$\mathbf{A} = \begin{bmatrix} 3 & 2 & 4 \\ 2 & 0 & 2 \\ 4 & 2 & 3 \end{bmatrix} \tag{3-143}$$

The characteristic equation is

$$\begin{vmatrix} 3-\lambda & 2 & 4 \\ 2 & 0-\lambda & 2 \\ 4 & 2 & 3-\lambda \end{vmatrix} = \mathbf{0} \tag{3-144}$$

or

$$(\lambda + 1)(\lambda + 1)(\lambda - 8) = 0$$

To determine the eigenvector(s) associated with $\lambda = -1$, we substitute this value in the equation $(\mathbf{A} - \lambda\mathbf{I})\mathbf{X} = \mathbf{0}$ and obtain

$$\begin{bmatrix} 4 & 2 & 4 \\ 2 & 1 & 2 \\ 4 & 2 & 4 \end{bmatrix} \begin{Bmatrix} x_1 \\ x_2 \\ x_3 \end{Bmatrix} = \mathbf{0} \tag{3-145}$$

We then apply the Gauss-Jordan elimination procedure. One application yields

$$\begin{bmatrix} 0 & 0 & 0 \\ 0 & 0 & 0 \\ 1 & \frac{1}{2} & 1 \end{bmatrix} \begin{Bmatrix} x_1 \\ x_2 \\ x_3 \end{Bmatrix} = \mathbf{0} \tag{3-146}$$

Using the computer, we would begin by storing the matrix

$$\mathbf{A} = \begin{bmatrix} 4 & 2 & 4 \\ 2 & 1 & 2 \\ 4 & 2 & 4 \end{bmatrix} \tag{3-147}$$

in memory. Then, with a single application of Eq. 3-139, the computer would obtain

$$\mathbf{B} = \begin{bmatrix} 0 & 0 \\ 0 & 0 \\ \frac{1}{2} & 1 \end{bmatrix} \tag{3-148}$$

The only applicable pivot elements for further elimination are

$$\begin{matrix} 0 & 0 \\ 0 & 0 \end{matrix}$$

so no further application of Eq. 3-139 is possible. There are two columns remaining in the matrix of Eq. 3-148, and it is apparent, from inspection of Eq. 3-146, that the rank of the coefficient matrix is 1, which is two less than the order. Thus in general, the *rank* of the *coefficient matrix* is equal to the order of the coefficient matrix minus the number of columns remaining in the final **B** matrix obtained from the applications of Eq. 3-139. Therefore, there will be two linearly independent eigenvectors associated with $\lambda = -1$. We must choose arbitrary values for two of the unknowns in determining each of these linearly independent eigenvectors. It is convenient to choose the values -1 and 0 for x_2 and x_3, respectively, in obtaining one eigenvector, and 0 and -1 for the same unknowns in obtaining the second eigenvector. This leads to the two linearly independent eigenvectors

$$\begin{Bmatrix} \frac{1}{2} \\ -1 \\ 0 \end{Bmatrix} \quad \text{and} \quad \begin{Bmatrix} 1 \\ 0 \\ -1 \end{Bmatrix}$$

It should be noted that these eigenvectors may be conveniently obtained from the matrix of Eq. 3-148, which is in computer memory, by replacing the first zero in the first column with -1 and the second zero in the second column with -1, followed by bringing the first and second rows, in turn, to the bottom position.

Any solution vector in the two-dimensional solution space can be formed by the combination of the linearly independent eigenvectors such as

$$C_1 \begin{Bmatrix} \frac{1}{2} \\ -1 \\ 0 \end{Bmatrix} + C_2 \begin{Bmatrix} 1 \\ 0 \\ -1 \end{Bmatrix}$$

where C_1 and C_2 are arbitrary constants.

To illustrate why it may be necessary to make column interchanges to prevent having a zero pivot element, let us find the eigenvalues of the matrix

$$\mathbf{A} = \begin{bmatrix} -2 & -8 & -12 \\ 1 & 4 & 4 \\ 0 & 0 & 1 \end{bmatrix} \tag{3-149}$$

without performing any column interchanges. The characteristic equation is

$$(\lambda - 2)(\lambda)(\lambda - 1) = 0 \tag{3-150}$$

To find the eigenvector associated with $\lambda = 2$, we write

$$\begin{bmatrix} -4 & -8 & -12 \\ 1 & 2 & 4 \\ 0 & 0 & -1 \end{bmatrix} \begin{Bmatrix} x_1 \\ x_2 \\ x_3 \end{Bmatrix} = \mathbf{0} \tag{3-151}$$

One application of the Gauss-Jordan procedure gives

$$\begin{bmatrix} 0 & 0 & 1 \\ 0 & 0 & -1 \\ 1 & 2 & 3 \end{bmatrix} \begin{Bmatrix} x_1 \\ x_2 \\ x_3 \end{Bmatrix} = \mathbf{0} \tag{3-152}$$

Thus the possible pivot elements, without considering column interchange, are

0

0

and no further elimination is possible because of division by zero. Therefore, a column interchange (a change in the sequence of the unknowns) would have to be made to continue the elimination process.

Proceeding again with the same problem but putting the largest applicable coefficient in the pivot position by means of a column interchange, Eq. 3-151 becomes

$$\begin{bmatrix} -12 & -8 & -4 \\ 4 & 2 & 1 \\ -1 & 0 & 0 \end{bmatrix} \begin{Bmatrix} x_3 \\ x_2 \\ x_1 \end{Bmatrix} = \mathbf{0} \tag{3-153}$$

Note that an interchange of the first and third columns of the coefficient matrix requires an interchange of the position of the elements x_3 and x_1 of the **X** matrix if the equations are to remain the same.

Applying the Gauss-Jordan elimination procedure to Eq. 3-153, we obtain

$$\begin{bmatrix} 0 & -\frac{2}{3} & -\frac{1}{3} \\ 0 & \frac{2}{3} & \frac{1}{3} \\ 1 & \frac{2}{3} & \frac{1}{3} \end{bmatrix} \begin{Bmatrix} x_3 \\ x_2 \\ x_1 \end{Bmatrix} = \mathbf{0} \tag{3-154}$$

One additional application yields

$$\begin{bmatrix} 0 & 0 & 0 \\ 1 & 0 & 0 \\ 0 & 1 & \frac{1}{2} \end{bmatrix} \begin{Bmatrix} x_3 \\ x_2 \\ x_1 \end{Bmatrix} = \mathbf{0} \tag{3-155}$$

In using the computer, the single-column matrix

$$\begin{Bmatrix} 0 \\ 0 \\ \frac{1}{2} \end{Bmatrix}$$

would be in memory. Referring to Eq. 3-155, we see that if we arbitrarily assume a value of -1 for the unknown in the third position in the column matrix of Eq. 3-155 (x_1), the values $x_2 = \frac{1}{2}$ and $x_3 = 0$ are obtained. If the computer replaces the first element of the column matrix by -1 and then moves it to the bottom position in the column, the following matrix is formed:

$$\begin{Bmatrix} 0 \\ \frac{1}{2} \\ -1 \end{Bmatrix}$$

These are the eigenvector components and they are now in the order shown in the column matrix of Eq. 3-155. The computer would next utilize the information stored referring to the column interchanges made during the elimination processes to arrange the eigenvector components in the usual x_1, x_2, x_3 sequence to obtain the general eigenvector

$$\mathbf{X} = C\begin{Bmatrix} -1 \\ \frac{1}{2} \\ 0 \end{Bmatrix} \tag{3-156}$$

It should be noted, at this point, that each time it is necessary to interchange the columns of a coefficient matrix in the elimination processes, it is also necessary to interchange the eigenvector components associated with these columns to maintain the correct matrix equation. The sequence of the eigenvector components appearing in the final matrix (after always replacing the zero in the top position with -1 and shifting it to the bottom position) will thus depend upon the column interchanges made during the successive elimination processes. The computer program must keep track, in memory, of these column interchanges, so that this information can be used to rearrange the eigenvector components in the usual $x_1, x_2, \ldots, x_n$ sequence in determining the eigenvector in the final step (Eq. 3-156, for example).

With the preceding discussion and examples as background, we are now in a position to formulate a computer program which will determine the eigenvector (or linearly independent eigenvectors, as the case may be) associated with any eigenvalue. The complete computer procedure is outlined in the flow chart of Fig. 3-9. The FORTRAN names and associated quantities used in the source program are as follows:

FORTRAN Name	**Quantity**
N	Order of **A** matrix
EIGEN	Eigenvalue for which the eigenvector or -vectors are to be determined
EPSI	A small quantity used in determining when a matrix element should be zero but is not, due to roundoff error or a small error in eigenvalue
A(I,J)	Elements of **A** matrix; also used for elements of coefficient matrix $\mathbf{A} - \lambda\mathbf{I}$
IX(I)	A variable used to keep track of the sequence of eigenvector components in the equation $(\mathbf{A} - \lambda\mathbf{I})\mathbf{X} = \mathbf{0}$. For example, if IX(3) has the value 3, then X(3) is in the third position. If IX(3) has the value 1, then X(3) is in the first position

K — The number of columns of the reduced matrix in memory (equivalent to original **A** matrix), *not counting columns of zeros and ones*

M — The column number of the pivot *column, counting columns of ones and zeros*

JJ — The column number in which the largest possible pivot element is found

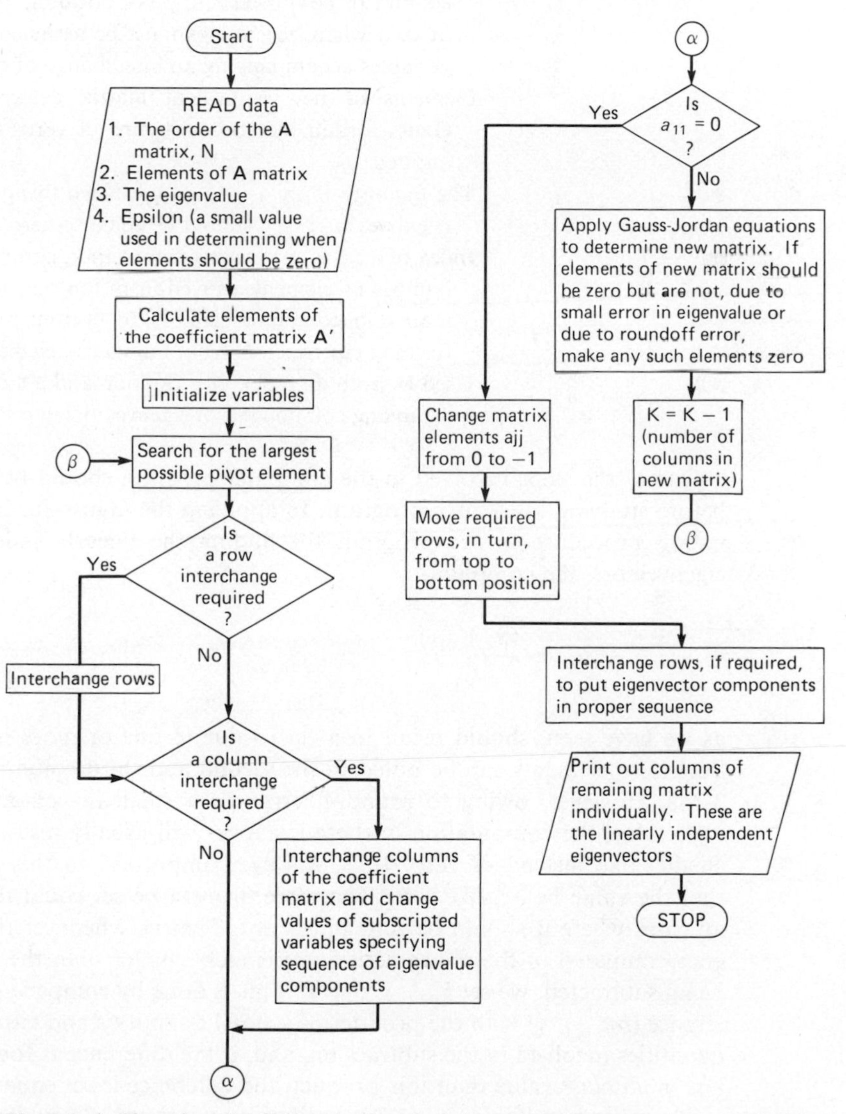

Fig. 3-9 / *Flow chart for finding the linearly independent eigenvectors associated with a particular eigenvalue of a matrix* **A**.

L	The column number of the column containing the largest possible pivot element, *counting columns of zeros and ones*
II	The row number in which the largest possible pivot element is found
TEMP	A name used, at different times in the program, to designate *temporarily* elements of the matrix row to be moved to pivot position, elements of the matrix column to be moved into pivot position, and values of IX(I) when new values must be assigned to these variables accompanying an interchange of columns
B(I,J)	Elements of new equivalent matrix determined by Gauss-Jordan method—columns of zeros and ones omitted
Y	The quantity $\|b_{i-1,\,j-1}\| - \varepsilon\|a_{ij}\|$; when this quantity is negative, $b_{i-1,\,j-1}$ should be equal to zero
MM	Index of a DO loop, varying from 1 to K, controlling the number of elements moved from top position of the matrix in computer memory to bottom position in forming eigenvectors from the matrix elements
NUM	Used as both an index of a DO loop and a subscript in arranging components of eigenvectors in proper order

One of the steps involved in the computer program should be clarified before studying the source program. In applying the Gauss-Jordan elimination procedure in the program for finding the linearly independent eigenvectors, the computation

$$b_{i-1,j-1} = a_{ij} - \frac{(a_{1j})(a_{i1})}{a_{11}}$$

as we have seen, should result in a zero value in one or more instances. Such zero elements can be noted in the second and third columns of Eq. 3-146. However, owing to roundoff error or a small inaccuracy in the eigenvalue, the computation in these instances will usually result in some small value instead of zero. It is, however, important in this program that the value be *exactly* zero. Therefore, it must be set equal to zero in all cases where it should be zero but is not. That is, whenever the difference computed in the above equation is much smaller than the numbers being subtracted, we set $b_{i-1,j-1} = 0$. This is done by comparing the difference $(b_{i-1,j-1})$ with the product of a small quantity ε and either of the quantities involved in the subtraction, and, if the difference is found to be less in *absolute* value than this product, the difference is set equal to zero.

Choosing a suitable value for epsilon may present a problem. If we have an eigenvalue that is obtained with slide-rule accuracy, we may need an epsilon as large as 0.01 or more to make the elements that should be zero actually zero. If we make epsilon too large, however, we may make some matrix elements zero which should not be zero, and a result will not be obtained. If the eigenvalue is exact, an epsilon value of 0.0001 or less may be suitable. Obviously, if we make epsilon either too small or too large we cannot obtain a result from the program. Instead, it prints out

```
RANK NOT LESS THAN ORDER. USE MORE ACCURATE
EIGENVALUE OR ADJUST EPSI
```

Several trial values of epsilon may have to be used to place it within permissible limits. If the eigenvalue is too inaccurate, there is *no* suitable range of epsilon, and a more accurate eigenvalue *must* be used.

The FORTRAN source program for finding the linearly independent eigenvectors associated with an eigenvalue is as follows:

```
C PROGRAM FOR FINDING THE LINEARLY INDEPENDENT
C EIGENVECTORS ASSOCIATED WITH AN EIGENVALUE
      DIMENSION A(11,10),B(10,9),X(10,3),IX(10)
C READ IN THE ORDER OF THE A MATRIX
      READ(5,1) N
    1 FORMAT(I2)
C READ IN THE A MATRIX BY COLUMNS
      READ(5,2) ((A(I,J),I=1,N),J=1,N)
    2 FORMAT(8F10.0)
C READ IN THE EIGENVALUE AND A SMALL VALUE EPSI
      READ(5,3) EIGEN,EPSI
    3 FORMAT(2F10.0)
C CALCULATE THE DIAGONAL ELEMENTS OF THE COEFFICIENT MATRIX
      DO 7 J=1,N
    7 A(J,J)=A(J,J)-EIGEN
C INITIALIZE THE SUBSCRIPTED VARIABLES SPECIFYING THE SEQUENCE
C OF THE EIGENVECTOR COMPONENTS
      DO 8 I=1,N
    8 IX(I)=I
C INITIALIZE THE VARIABLES K AND M
      K=N
    9 M=N-K+1
C SEARCH FOR THE LARGEST POSSIBLE PIVOT ELEMENT
      JJ=1
      II=1
      BIG=ABS(A(1,1))
      DO 11 J=1,K
      DO 11 I=1,K
      DIFF=BIG-ABS(A(I,J))
      IF(DIFF)10,11,11
   10 JJ=J
      II=I
      BIG=ABS(A(I,J))
   11 CONTINUE
C CHECK TO SEE IF ROW INTERCHANGES ARE REQUIRED
      IF(II-1)14,14,12
```

```
C MAKE ROW INTERCHANGES
   12 DO 13 J=1,K
      TEMP=A(II,J)
      A(II,J)=A(1,J)
   13 A(1,J)=TEMP
C CHECK TO SEE IF COLUMN INTERCHANGES ARE REQUIRED
   14 IF(JJ-1)17,17,15
C INTERCHANGE VALUES OF IX(M) AND IX(L) ACCOMPANYING A
                                        COLUMN INTERCHANGE
   15 TEMP=IX(M)
      L=JJ+N-K
      IX(M)=IX(L)
      IX(L)=TEMP
C MAKE COLUMN INTERCHANGE
      DO 16 I=1,N
      TEMP=A(I,JJ)
      A(I,JJ)=A(I,1)
   16 A(I,1)=TEMP
   17 CONTINUE
C CHECK TO SEE IF ELIMINATION PROCESS IS COMPLETE
      IF(A(1,1)) 18,23,18
C APPLY GAUSS-JORDAN ELIMINATION PROCEDURE
   18 DO 20 J=2,K
      DO 20 I=2,N
      B(I-1,J-1)=A(I,J)-A(1,J)*A(I,1)/A(1,1)
      Y=ABS(B(I-1,J-1))-EPSI*ABS(A(I,J))
      IF(Y)19,20,20
   19 B(I-1,J-1)=0.
   20 CONTINUE
      DO 21 J=2,K
   21 B(N,J-1)=A(1,J)/A(1,1)
      K=K-1
      IF(K)31,31,30
   30 DO 22 J=1,K
      DO 22 I=1,N
   22 A(I,J)=B(I,J)
      GO TO 9
C FORM LINEARLY INDEPENDENT EIGENVECTORS FROM REMAINING
                                        COLUMNS IN MEMORY
   23 DO 24 J=1,K
   24 A(J,J)=-1.
      DO 25 MM=1,K
      DO 25 J=1,K
      A(N+1,J)=A(1,J)
      DO 25 I=1,N
   25 A(I,J)=A(I+1,J)
      DO 27 J=1,K
      DO 27 I=1,N
      DO 27 NUM=1,N
      IF(IX(NUM)-I)27,26,27
   26 X(I,J)=A(NUM,J)
   27 CONTINUE
C PRINT OUT COMPONENTS OF LINEARLY INDEPENDENT EIGENVECTORS
      WRITE(6,4)
```

```
    4 FORMAT('1','COMPONENTS OF LINEARLY INDEPENDENT
                                              EIGENVECTORS'/)
      DO 29 J=1,K
      DO 28 I=1,N
   28 WRITE(6,5) X(I,J)
   29 WRITE(6,6)
    5 FORMAT(' ',16X,E14.7)
    6 FORMAT('0',' ')
      GO TO 34
   31 WRITE(6,32)
   32 FORMAT('1','RANK NOT LESS THAN ORDER. USE MORE ACCURATE'/,
     *' ','EIGENVALUE OR ADJUST EPSI')
   34 STOP
      END
```

3-11 / ITERATION METHOD—EIGENVALUE PROBLEMS

In addition to the polynomial method discussed in the preceding section, there are various iterative methods to determine eigenvalues. Of these, the one most commonly used in engineering problems is the *power method.*

An iterative method is used most frequently when only the *smallest* and/or *largest* eigenvalue(s) of a matrix are desired, but may also be used when intermediate eigenvalues and eigenvectors are to be found. For finding the intermediate eigenvalues and eigenvectors, a sweeping technique discussed in Sec. 3-12 may be used. An advantage of the iterative method is that the eigenvectors are obtained simultaneously with the associated eigenvalues rather than requiring separate operations, as in the polynomial method.

To illustrate the power method, let us first use it to determine the largest eigenvalue of a matrix. We will assume that both the elements of the matrix and its eigenvalues are real, although the use of iterative methods is not thus restricted.

Suppose that, in the analysis of a physical system, we have obtained as a mathematical model of the system the following set of simultaneous homogeneous linear algebraic equations:

$$\begin{aligned}
(a_{11} - \lambda)x_1 + \quad a_{12}x_2 + a_{13}x_3 + \cdots + \quad a_{1n}x_n &= 0 \\
a_{21}x_1 + (a_{22} - \lambda)x_2 + a_{23}x_3 + \cdots + \quad a_{2n}x_n &= 0 \\
\vdots \qquad\qquad \vdots \qquad\qquad & \vdots \\
a_{n1}x_1 + \quad a_{n2}x_2 + a_{n3}x_3 + \cdots + (a_{nn} - \lambda)x_n &= 0
\end{aligned} \tag{3-157}$$

which may be expressed more compactly, in matrix notation, as

$$(\mathbf{A} - \lambda\mathbf{I})\mathbf{X} = \mathbf{0} \tag{3-158}$$

To find the largest eigenvalue λ_1, we rewrite Eq. 3-158 in the form

$$\mathbf{AX} = \lambda\mathbf{X} \tag{3-159}$$

and carry out the following steps:

1 Arbitrarily assume values for the components of the eigenvector $\mathbf{X} = (x_1, x_2, \ldots, x_n)$. This assumed vector will subsequently be referred to as $\mathbf{X}_0$. Choosing all components of $\mathbf{X}_0$ equal to unity is generally satisfactory. Upon substituting the components of $\mathbf{X}_0$ into the left side of Eq. 3-159 as the elements of the matrix $\mathbf{X}$ and carrying out the matrix multiplication, a first approximation to the right side of Eq. 3-159 is obtained, where

$$\lambda\mathbf{X} = \begin{Bmatrix} \lambda x_1 \\ \lambda x_2 \\ \lambda x_3 \\ \vdots \\ \lambda x_n \end{Bmatrix} \tag{3-160}$$

2 Normalize the vector $\lambda\mathbf{X}$ obtained in step 1. This is done by dividing the vector by the magnitude of its first component, by its largest component, or by normalizing the vector $\mathbf{X}$ to a unit length. In this case, let us normalize by dividing the vector by the magnitude of the first component, reducing that component to unity.

3 Use the components of the normalized vector as improved values of x_i to be substituted into the left side of Eq. 3-159. Carrying out the matrix multiplication, a still better approximation of the right-hand side of Eq. 3-159 is obtained.

4 Repeat steps 2 and 3 until Eq. 3-159 is essentially satisfied; that is, until the eigenvector components in step 2 vary from the previously obtained values by less than some small preassigned ε value, in two successive iterations. The normalizing factor will be the *largest eigenvalue* λ_1, and the elements of $\mathbf{X}$ will be the components of the eigenvector associated with it.[16]

In carrying out the iteration procedure we are, in effect, forming a sequence of vectors $\mathbf{AX}_0, \mathbf{A}^2\mathbf{X}_0, \mathbf{A}^3\mathbf{X}_0, \ldots, \mathbf{A}^k\mathbf{X}_0$, where $\mathbf{X}_0$ is the arbitrary vector initially assumed. (The name "power method" comes from the fact that the sequence is made up of powers of the matrix **A**.)

[16] For a proof that convergence is to the largest eigenvalue, see Moshe F. Rubinstein. *Structural Systems Statics, Dynamics and Stability* (Englewood Cliffs, N.J.: Prentice Hall, Inc., 1970), pp. 216–217.

To illustrate the steps just outlined, let us consider the equations

$$\begin{aligned}(2 - \lambda)x_1 + 4x_2 &= 0\\ 3x_1 + (13 - \lambda)x_2 &= 0\end{aligned} \tag{3-161}$$

and determine the value of λ_1, the *largest eigenvalue*. Putting Eq. 3-161 in the form of Eq. 3-159, we have

$$\begin{bmatrix} 2 & 4 \\ 3 & 13 \end{bmatrix} \begin{Bmatrix} x_1 \\ x_2 \end{Bmatrix} = \lambda \begin{Bmatrix} x_1 \\ x_2 \end{Bmatrix}$$

Carrying out steps 1 and 2, we obtain

$$\begin{bmatrix} 2 & 4 \\ 3 & 13 \end{bmatrix} \begin{Bmatrix} 1 \\ 1 \end{Bmatrix} = \begin{Bmatrix} 6 \\ 16 \end{Bmatrix} = 6 \begin{Bmatrix} 1 \\ \frac{16}{6} \end{Bmatrix}$$

Applying step 3 gives

$$\begin{bmatrix} 2 & 4 \\ 3 & 13 \end{bmatrix} \begin{Bmatrix} 1 \\ \frac{16}{6} \end{Bmatrix} = \begin{Bmatrix} \frac{38}{3} \\ \frac{113}{3} \end{Bmatrix} = \frac{38}{3} \begin{Bmatrix} 1 \\ \frac{113}{38} \end{Bmatrix}$$

The next iteration yields

$$\begin{bmatrix} 2 & 4 \\ 3 & 13 \end{bmatrix} \begin{Bmatrix} 1 \\ \frac{113}{38} \end{Bmatrix} = \begin{Bmatrix} \frac{264}{19} \\ \frac{1538}{38} \end{Bmatrix} = 13.9 \begin{Bmatrix} 1 \\ 2.998 \end{Bmatrix}$$

Comparing the normalizing factors of the *second* and *third* iterations ($\frac{38}{3}$ and 13.9), it is apparent that the third iteration results in only a minor change of the normalizing factor obtained in the second iteration. Thus a good approximation to λ_1 is $\lambda_1 = 13.9$. The eigenvector associated with λ_1 is $\mathbf{X} = (1, 2.998)$. A fourth iteration would, of course, yield a still better approximation to λ_1 and the corresponding components of the eigenvector. Application of the polynomial method readily reveals that the exact values are $\lambda_1 = 14$ and $\mathbf{X} = (1, 3)$.

In many physical problems the *smallest* eigenvalue turns out to be the one of primary importance. To arrange the matrix equation in a form in which the iteration converges to the smallest eigenvalue, we first premultiply Eq. 3-159 by the inverse matrix $\mathbf{A}^{-1}$, obtaining

$$\mathbf{A}^{-1}\mathbf{A}\mathbf{X} = \mathbf{A}^{-1}\lambda\mathbf{X} = \lambda\mathbf{A}^{-1}\mathbf{X}$$

or

$$\mathbf{X} = \lambda\mathbf{A}^{-1}\mathbf{X} \tag{3-162}$$

Dividing Eq. 3-162 by λ gives

$$\mathbf{A}^{-1}\mathbf{X} = \frac{1}{\lambda}\mathbf{X} \tag{3-163}$$

Equation 3-163 is in a form which will result in convergence to the *smallest value* of λ.

Let us now find the smallest eigenvalue of the same matrix

$$\mathbf{A} = \begin{bmatrix} 2 & 4 \\ 3 & 13 \end{bmatrix}$$

for which we previously found the largest eigenvalue. Using the procedure discussed in Sec. 3-6, the inverse matrix $\mathbf{A}^{-1}$ is readily found to be

$$\mathbf{A}^{-1} = \begin{bmatrix} \frac{13}{14} & -\frac{2}{7} \\ -\frac{3}{14} & \frac{1}{7} \end{bmatrix}$$

Proceeding as outlined in steps 1 through 4, we obtain

$$\begin{bmatrix} \frac{13}{14} & -\frac{2}{7} \\ -\frac{3}{14} & \frac{1}{7} \end{bmatrix} \begin{Bmatrix} 1 \\ 1 \end{Bmatrix} = \begin{Bmatrix} \frac{9}{14} \\ -\frac{1}{14} \end{Bmatrix} = \frac{9}{14} \begin{Bmatrix} 1 \\ -\frac{1}{9} \end{Bmatrix}$$

$$\begin{bmatrix} \frac{13}{14} & -\frac{2}{7} \\ -\frac{3}{14} & \frac{1}{7} \end{bmatrix} \begin{Bmatrix} 1 \\ -\frac{1}{9} \end{Bmatrix} = \begin{Bmatrix} \frac{121}{126} \\ -\frac{29}{126} \end{Bmatrix} = \frac{121}{126} \begin{Bmatrix} 1 \\ -\frac{29}{121} \end{Bmatrix}$$

$$\begin{bmatrix} \frac{13}{14} & -\frac{2}{7} \\ -\frac{3}{14} & \frac{1}{7} \end{bmatrix} \begin{Bmatrix} 1 \\ -\frac{29}{121} \end{Bmatrix} = \begin{Bmatrix} \frac{1689}{1694} \\ -\frac{421}{1694} \end{Bmatrix} = \frac{1689}{1694} \begin{Bmatrix} 1 \\ -\frac{421}{1689} \end{Bmatrix}$$

Therefore, we find that the *smallest eigenvalue* $\lambda_2 \cong 1.003$, and the associated vector $\mathbf{X} \cong (1, -0.2493)$. The polynomial solution shows that the exact solution is $\lambda_2 = 1$ and $\mathbf{X} = (1, -0.25)$.

It should be apparent, from the discussion thus far, that the iteration method is ideally suited for computer calculation. In the solution of engineering problems there is rarely any difficulty in its use. If the largest and next-to-the-largest or the smallest and next-to-the-smallest eigenvalues have nearly the same value, convergence to the largest or smallest value, respectively, may be slow. If the largest eigenvalue has a multiplicity of 2 ($\lambda_1 = \lambda_2$), convergence is to the largest eigenvalue, but the components of the eigenvector converge to any of the eigenvectors (which usually form a two-dimensional space) associated with λ_1. The eigenvector obtained depends on the vector assumed initially.

In the event that the vector initially assumed is *orthogonal* to the eigenvector associated with the largest eigenvalue of the *transposed* matrix (their dot product is equal to zero), convergence will be to the *second-largest* eigenvalue instead of to the *largest*. To illustrate this point, consider the matrix

$$\mathbf{A} = \begin{bmatrix} 1 & 0 & -1 \\ 1 & 2 & 1 \\ 2 & 2 & 3 \end{bmatrix}$$

in which the second-largest and the largest eigenvalues are $\lambda_2 = 2$ and $\lambda_1 = 3$, respectively. The transpose of matrix **A**, formed by interchanging rows and columns, is

$$\mathbf{A}^{\mathrm{T}} = \begin{bmatrix} 1 & 1 & 2 \\ 0 & 2 & 2 \\ -1 & 1 & 3 \end{bmatrix}$$

The largest eigenvalue of the transposed matrix is $\lambda = 3$ with the associated eigenvector $(1, 1, \frac{1}{2})$. An arbitrary vector orthogonal to this eigenvector is $(4, -3, -2)$. That is, the dot product of the two orthogonal vectors is

$$1(4) + 1(-3) + \tfrac{1}{2}(-2) = 0$$

If we use the vector $(4, -3, -2)$ as the initially assumed vector in determining the largest eigenvalue of the matrix **A**, we write, for the first iteration

$$\begin{bmatrix} 1 & 0 & -1 \\ 1 & 2 & 1 \\ 2 & 2 & 3 \end{bmatrix} \begin{Bmatrix} 4 \\ -3 \\ -2 \end{Bmatrix} = 6 \begin{Bmatrix} 1 \\ -\frac{2}{3} \\ -\frac{2}{3} \end{Bmatrix}$$

After three more iterations we finally obtain

$$\begin{bmatrix} 1 & 0 & -1 \\ 1 & 2 & 1 \\ 2 & 2 & 3 \end{bmatrix} \begin{Bmatrix} 1 \\ -\frac{5}{9} \\ -\frac{8}{9} \end{Bmatrix} = \frac{17}{9} \begin{Bmatrix} 1 \\ -\frac{9}{17} \\ -\frac{16}{17} \end{Bmatrix}$$

which is converging to $\lambda_2 = 2$ with the associated eigenvector $(1, -\frac{1}{2}, -1)$ rather than to $\lambda_1 = 3$ with the associated eigenvector $(1, -1, -2)$.

Other special cases, which would be encountered only rarely in engineering analysis, are discussed by Faddeev and Faddeeva.[17]

[17] Faddeev and Faddeeva, *op. cit.*

3-12 / ITERATION FOR INTERMEDIATE EIGENVALUES AND EIGENVECTORS

The iteration procedure discussed in Sec. 3-11 can be extended to determine intermediate eigenvalues and eigenvectors between the smallest and the largest. This is accomplished by sweeping out the last eigenvector determined before proceeding with the iteration to obtain the next smallest or largest eigenvalue, whichever the case may be.

Since any vector $\mathbf{V}$ can be expressed as a linear combination of the orthogonal eigenvectors $\mathbf{X}_i$, then

$$\mathbf{V} = \begin{Bmatrix} v_1 \\ v_2 \\ \vdots \\ v_n \end{Bmatrix} = C_1 \begin{Bmatrix} x_1 \\ x_2 \\ \vdots \\ x_n \end{Bmatrix}_1 + C_2 \begin{Bmatrix} x_1 \\ x_2 \\ \vdots \\ x_n \end{Bmatrix}_2 + \cdots + C_n \begin{Bmatrix} x_1 \\ x_2 \\ \vdots \\ x_n \end{Bmatrix}_n$$

or

$$\mathbf{V} = \sum_{i=1}^{n} C_i \mathbf{X}_i \tag{3-164}$$

If $\mathbf{X}_1$ is the smallest eigenvector determined by the iteration procedure discussed in Sec. 3-11, then the smallest eigenvalue λ_1 and eigenvector $\mathbf{X}_1$ can be removed from the system of equations by making $C_1 = 0$. By sweeping out $\mathbf{X}_1$, the iteration process applied to Eq. 3-163 will converge to the next smallest eigenvalue λ_2 and corresponding eigenvector $\mathbf{X}_2$. After obtaining $\mathbf{X}_2$, both $\mathbf{X}_1$ and $\mathbf{X}_2$ can then be swept out of the equations by making $C_1 = C_2 = 0$, so that the iteration process converges to λ_3 and $\mathbf{X}_3$. This procedure of obtaining the eigenvalues and eigenvectors in ascending order can be continued until all of the eigenvalues and eigenvectors are obtained. However, the iteration method, with the sweeping technique just discussed, is subject to the accumulation of roundoff errors, and it may not be possible to obtain accurate values for all of the eigenvalues and eigenvectors of large matrices. For many engineering problems this is not a serious problem. For example, in vibration studies it is not a serious problem since the first few lowest eigenvalues (frequencies) and eigenvectors (mode shapes) are usually of primary importance in the dynamic analysis of vibrating systems.

The largest eigenvalues and eigenvectors can be obtained in descending order by sweeping out the higher modes and using Eq. 3-159.

Having discussed in general terms the determination of intermediate eigenvalues and eigenvectors, let us now discuss in more detail the procedure for obtaining them. We recall from Sec. 3-11 that the iteration process applied to

$$\mathbf{AX} = \lambda\mathbf{X} \tag{3-165}$$

converged to the largest eigenvalue and eigenvector, and the same process applied to

$$\mathbf{A}^{-1}\mathbf{X} = \frac{1}{\lambda}\mathbf{X} \tag{3-166}$$

converged to the smallest eigenvalue and eigenvector. To obtain intermediate eigenvalues and eigenvectors, the column matrix **X** of Eq. 3-165 or Eq. 3-166 is modified at each iteration step by new trial values for C_1, $C_2, \ldots, C_r$ associated with the eigenvectors which are to be swept out. Thus each iteration step yields new values for $C_1, C_2, \ldots, C_r$, which values should eventually converge to zero after a sufficient number of iterations.

The modified column matrix **X** for use in the left-hand side of Eq. 3-165 or Eq. 3-166 is obtained from the following equations:

$$\mathbf{X}_i^{\mathrm{T}}\mathbf{X}_j = \begin{Bmatrix} 0 & i \neq j \\ G_{ii} & i = j \end{Bmatrix} \quad \text{(A must be symmetric)} \tag{3-167}$$

$$C_i = \frac{1}{G_{ii}}\mathbf{X}_i^{\mathrm{T}}\mathbf{X}_k \tag{3-168}$$

$$\mathbf{X} = \mathbf{X}_k - \sum_{i=1}^{r} C_i\mathbf{X}_i \tag{3-169}$$

Equation 3-167 follows from the fact that the eigenvectors of Eq. 3-164 are orthogonal (see App. B). Since $\mathbf{X}^{\mathrm{T}}$ is a row matrix, $\mathbf{X}_i^{\mathrm{T}}\mathbf{X}_i = G_{ii}$ is simply the sum of the squares of the components of the ith eigenvector. In using Eq. 3-168 and Eq. 3-169 it is important to note that $r = 1, 2, \ldots, S$ (where S is the maximum number of eigenvectors to be swept out), and that for *each* r, $i = 1, 2, \ldots, r$.

In Eq. 3-168, $\mathbf{X}_k$ is any trial column computed from Eq. 3-165 or Eq. 3-166. The components of $\mathbf{X}_k$ may be taken as unity for the first iteration cycle. It should be noted that the initially assumed $\mathbf{X}_k$ vector corresponds to vector **V** of Eq. 3-164. That is, in general it is a linear combination of all the eigenvectors. The coefficient C_i for the ith eigenvector being swept out is a new trial value for calculating a new modified column matrix **X** of Eq. 3-169 for the next iteration cycle. The subtraction of $\sum_{i=1}^{r} C_i\mathbf{X}_i$ from $\mathbf{X}_k$ in Eq. 3-169, in effect, forces the vector **X** to exclude the first r eigenvectors which in turn forces **X**, or $\mathbf{X}_k$, to converge to the next lowest eigenvector $\mathbf{X}_{r+1}$ in the iteration process. As $\mathbf{X}$, or $\mathbf{X}_k$, approaches the $r + 1$ eigenvector, the C_i values of Eq. 3-168 approach zero, since $\mathbf{X}_i$ and $\mathbf{X}_{r+1}$ are orthogonal. It should be emphasized that Eq. 3-168 must be utilized in each iteration cycle. In sweeping out $\mathbf{X}_1$, $r = 1$, so that $C_1 \rightarrow 0$ in the iteration process to obtain $\mathbf{X}_2$. However, in the iteration process to obtain $\mathbf{X}_3$, C_1 is not initially zero, but must be calculated from Eq. 3-168 (as is C_2) for each iteration cycle. Thus the summation term in Eq. 3-169 consists of two terms with $r = 2$, so that both C_1 and C_2 approach zero as the iterations converge to $\mathbf{X}_3$. To obtain $\mathbf{X}_4$ after obtaining $\mathbf{X}_3$, the summation term in Eq. 3-169 includes C_1, C_2, and C_3, since $r = 3$. This procedure continues until the last desired eigenvector is swept out with $r = S$.

The steps for determining intermediate eigenvalues and eigenvectors between the smallest and the largest are as follows:

1 Calculate G_{ii} from Eq. 3-167 for each of the previously determined eigenvalues and eigenvectors which are to be swept out of Eq. 3-165 or Eq. 3-166, whichever the case may be.

2 Calculate the coefficients C_i from Eq. 3-168 by using $\mathbf{X}_k$ from the previous iteration cycle (components of $\mathbf{X}_k$ may be taken as unity for the first iteration cycle).

3 Calculate a new trial column matrix $\mathbf{X}$ from Eq. 3-169.

4 Multiply out the left-hand side of Eq. 3-165 or Eq. 3-166 using the new trial column $\mathbf{X}$ obtained in step 3. This gives a better approximation of the right-hand side of either Eq. 3-165 or 3-166.

5 Normalize the vector obtained in step 4 to obtain $\mathbf{X}_k$ for the next iteration cycle.

6 Repeat steps 2 through 5 until Eq. 3-165 or 3-166 is essentially satisfied. This may be accomplished by comparing successive values of each eigenvector component until they differ by less than some predetermined value ε. It may also be accomplished by iterating until each eigenvector component differs from its previous value by less than some predetermined value ε times the last eigenvector component value (see the "accuracy check" in the computer program given following Ex. 3-10).

EXAMPLE 3-10

Before writing a general computer program for determining eigenvalues and eigenvectors, let us determine by iteration, just using an electronic calculator, the two smallest eigenvalues and corresponding eigenvectors of the matrix

$$\mathbf{A}^{-1} = \begin{bmatrix} 0.1 & 0.1 & 0.1 \\ 0.1 & 0.2 & 0.2 \\ 0.1 & 0.2 & 0.3 \end{bmatrix}$$

Utilizing the iteration procedure of Sec. 3-11, the smallest eigenvalue λ_1 and corresponding eigenvector $\mathbf{X}_1$ are found to be

$$\lambda_1 = 1.98$$

$$\mathbf{X}_1 = \begin{Bmatrix} 1.00 \\ 1.80 \\ 2.24 \end{Bmatrix}$$

Having determined the smallest values, we can continue the iteration process utilizing the five steps given earlier in this section for determining the next smallest eigenvalue and corresponding eigenvector. For we are now ready to sweep out the smallest eigenvector $\mathbf{X}_1$ and proceed through the given steps as enumerated below, for the first iteration.

1) $(r = 1 \quad \text{and} \quad i = 1)$

$$G_{11} = 1^2 + (1.8)^2 + (2.24)^2 = 9.26$$

2) $$C_1 = \frac{1}{9.26}[1 \quad 1.80 \quad 2.24]\begin{Bmatrix} 1 \\ 1 \\ 1 \end{Bmatrix} = 0.544$$

3) $$\mathbf{X} = \begin{Bmatrix} 1 \\ 1 \\ 1 \end{Bmatrix} - 0.544\begin{Bmatrix} 1.00 \\ 1.80 \\ 2.24 \end{Bmatrix} = \begin{Bmatrix} 0.456 \\ 0.021 \\ -0.218 \end{Bmatrix}$$

4) $$\begin{bmatrix} 0.1 & 0.1 & 0.1 \\ 0.1 & 0.2 & 0.2 \\ 0.1 & 0.2 & 0.3 \end{bmatrix}\begin{Bmatrix} 0.456 \\ 0.021 \\ -0.218 \end{Bmatrix} = \begin{Bmatrix} 0.026 \\ 0.006 \\ -0.016 \end{Bmatrix}$$

5) $$\frac{1}{\lambda_2}\mathbf{X}_2 \cong 0.026\begin{Bmatrix} 1.00 \\ 0.23 \\ -0.62 \end{Bmatrix} \qquad \therefore \qquad \mathbf{X}_k = \begin{Bmatrix} 1.00 \\ 0.23 \\ -0.62 \end{Bmatrix}$$

For the second iteration, we start with step 2 and obtain the following:

2) $$C_1 = \frac{1}{9.26}[1 \quad 1.80 \quad 2.24]\begin{Bmatrix} 1.00 \\ 0.23 \\ -0.62 \end{Bmatrix} = 0.0027$$

$$3)\ \mathbf{X} = \begin{Bmatrix} 1.00 \\ 0.23 \\ -0.62 \end{Bmatrix} - 0.0027 \begin{Bmatrix} 1.00 \\ 1.80 \\ 2.24 \end{Bmatrix} = \begin{Bmatrix} 0.997 \\ 0.23 \\ -0.63 \end{Bmatrix}$$

$$4)\ \begin{bmatrix} 0.1 & 0.1 & 0.1 \\ 0.1 & 0.2 & 0.2 \\ 0.1 & 0.2 & 0.3 \end{bmatrix} \begin{Bmatrix} 0.997 \\ 0.23 \\ -0.63 \end{Bmatrix} = \begin{Bmatrix} 0.0597 \\ 0.0197 \\ -0.0433 \end{Bmatrix}$$

$$5)\ \frac{1}{\lambda_2}\mathbf{X}_2 \cong 0.0597 \begin{Bmatrix} 1.00 \\ 0.329 \\ -0.725 \end{Bmatrix} \qquad \therefore \qquad \mathbf{X}_k = \begin{Bmatrix} 1.00 \\ 0.33 \\ -0.73 \end{Bmatrix}$$

After five more such iterations, we finally obtain for the next smallest eigenvalue and corresponding eigenvector

$$\lambda_2 = 15.64 \qquad \mathbf{X}_2 = \begin{Bmatrix} 1.00 \\ 0.44 \\ -0.82 \end{Bmatrix}$$

If we were to continue on to obtain $\mathbf{X}_3$, both $\mathbf{X}_1$ and $\mathbf{X}_2$ would be swept out with $r = 2$ and $i = (1 \text{ and } 2)$ for Eqs. 3-168 and 3-169.

The computer program which follows for obtaining the smallest eigenvalues and eigenvectors in ascending order, utilizing Eqs. 3-166, 3-167, 3-168, and 3-169, will handle up to and including a 20th-order matrix ($n = 20$), and can sweep out up to 10 eigenvectors ($S = 10$). It should be noted in the FORTRAN program that the DO variable I in the DO loop ending with statement 100 plays the role of r in Eqs. 3-168 and 3-169. Also the DO variable II in statement 15 plays the role of i in the same equations. Other important variable names used in the program, and the quantities which they represent, are as follows:

Variable Name	**Quantity**
`AI`	The inverse of matrix **A**
`N`	The order of matrix **A** or $\mathbf{A}^{-1}$
`LAMBDA(I)`	The eigenvalues
`X(I)`	Components of smallest eigenvector
`D(I)`	Unnormalized components of smallest eigenvector
`Z(I,L)`	The L components of the Ith eigenvector
`XK1(IP1,L)` `XK(IP1,L)`	The L components of the (I+1)th eigenvector, which approach correct values with iteration

XX(IJ)	Unnormalized trial **X** values
C(IJ)	C_i values as defined by Eq. 3-168
G(I)	G_i values as defined by Eq. 3-167
EPSI	A small predetermined value (0.00001). Accuracy requirement is that all eigenvector components differ from previous value in the iteration by less than EPSI times the latest eigenvector component values.

The computer printout following the program shown below gives the values of all three eigenvalues and associated eigenvectors of the matrix of Ex. 3-10, in which the first two were solved for by use of an electronic calculator. The FORTRAN program is as follows:

```
C THIS PROGRAM FINDS THE SMALLEST EIGENVALUE OF A MATRIX A
C AND THE ASSOCIATED EIGENVECTOR COMPONENTS
C INTERMEDIATE EIGENVALUES ARE FOUND BY SWEEPING OUT
                                                    PREVIOUSLY
C DETERMINED EIGENVECTORS
C ASSOCIATED EIGENVECTOR COMPONENTS ARE FOUND FOR ALL
                                                   EIGENVALUES
      REAL LAMBDA
      INTEGER S,SP1
      DIMENSION AI(20,20),X(20),D(20),Z(10,20),XK(11,20),
                                               C(10),G(10),
     *SUMM(20),Y(20),X1(11,20),XX(20),XK1(11,20),LAMBDA(11)
C READ ORDER OF MATRIX A,NUMBER OF EIGENVECTORS TO BE SWEPT,
C AND A SMALL VALUE EPSI
      READ(5,2) N,S,EPSI
    2 FORMAT(2I2,F16.0)
C READ IN THE INVERSE OF MATRIX A
      READ(5,3) ((AI(I,J),J=1,N),I=1,N)
    3 FORMAT(8F10.0)
C BY THE ITERATION METHOD DETERMINE THE SMALLEST EIGENVALUE
C AND ASSOCIATED EIGENVECTOR
C ASSIGN VALUES TO COMPONENTS OF INITIALLY ASSUMED VECTOR X
      DO 4 I=1,N
    4 X(I)=1.
C CALCULATE COMPONENTS OF THE VECTOR 1/LAMBDA*X
      IT=0
    5 DO 6 I=1,N
      D(I)=0.
      DO 6 J=1,N
    6 D(I)=D(I)+AI(I,J)*X(J)
      IT=IT+1
C NORMALIZE THE VECTOR D(I)
      DO 7 I=1,N
    7 Z(1,I)=D(I)/D(1)
C CHECK TO SEE IF REQUIRED ACCURACY HAS BEEN OBTAINED
      DO 8 I=1,N
      DIFF=X(I)-Z(1,I)
      IF(ABS(DIFF)-EPSI*Z(1,I)) 8,9,9
```

```
      8 CONTINUE
        GO TO 11
      9 DO 10 I=1,N
     10 X(I)=Z(1,I)
        IF(IT.GE.50) GO TO 11
        GO TO 5
     11 LAMBDA(1)=1./D(1)
C THE SMALLEST EIGENVALUE AND ASSOCIATED VECTOR ARE
                                                  NOW AVAILABLE
C SWEEP OUT S EIGENVECTORS
        DO 100 I=1,S
        IP1=I+1
C ASSIGN THE VALUE 1 TO XK COMPONENTS
        DO 12 L=1,N
     12 XK(IP1,L)=1.
C CALCULATE G(I) VALUES
        G(I)=0.
        DO 14 L=1,N
     14 G(I)=G(I)+Z(I,L)**2
C CALCULATE VALUES OF C
        IT=0
     15 DO 99 II=1,I
        SUM=0.
        DO 16 L=1,N
     16 SUM=SUM+Z(II,L)*XK(IP1,L)
     99 C(II)=SUM/G(II)
        IT=IT+1
C CALCULATE SUM OF CX'S
        DO 17 L=1,N
     17 SUMM(L)=0.
        DO 19 K=1,I
        DO 18 L=1,N
        Y(L)=C(K)*Z(K,L)
     18 SUMM(L)=SUMM(L)+Y(L)
     19 CONTINUE
C CALCULATE NEW TRIAL X
        DO 20 L=1,N
     20 X1(IP1,L)=XK(IP1,L)-SUMM(L)
C MULTIPLY A INVERSE TIMES X
        DO 21 II=1,N
        XX(II)=0
        DO 21 KK=1,N
     21 XX(II)=XX(II)+AI(II,KK)*X1(IP1,KK)
C NORMALIZE TO GET XK FOR NEXT ITERATION,OR FINAL XK
        DO 22 L=1,N
     22 XK1(IP1,L)=XX(L)/XX(1)
C CHECK TO SEE IF REQUIRED ACCURACY HAS BEEN ATTAINED
        DO 23 L=1,N
        DIFF=XK(IP1,L)-XK1(IP1,L)
        IF(ABS(DIFF)-EPSI*XK1(IP1,L)) 23,24,24
     23 CONTINUE
        GO TO 26
```

```
C ASSIGN XK1 VALUES TO NAME XK
   24 DO 25 L=1,N
   25 XK(IP1,L)=XK1(IP1,L)
      IF(IT.GE.50) GO TO 26
      GO TO 15
C CALCULATE EIGENVALUE
   26 LAMBDA(IP1)=1./XX(1)
C ASSIGN EIGENVECTORS TO NAME Z
      DO 27 L=1,N
   27 Z(IP1,L)=XK1(IP1,L)
  100 CONTINUE
C WRITE OUT THE EIGENVALUES AND EIGENVECTORS
      SP1=S+1
      DO 104 I=1,SP1
      WRITE(6,101) I,LAMBDA(I)
  101 FORMAT('1','LAMBDA(',I2,')=',E14.7//)
      WRITE(6,102)
  102 FORMAT(' ','THE ASSOCIATED EIGENVECTOR COMPONENTS ARE'/)
      WRITE(6,103) (Z(I,L),L=1,N)
  103 FORMAT(' ',13X,E14.7)
  104 CONTINUE
      STOP
      END
```

```
LAMBDA( 1)= 0.1980628E 01


THE ASSOCIATED EIGENVECTOR COMPONENTS ARE

              0.1000000E 01
              0.1801937E 01
              0.2246979E 01


LAMBDA( 2)= 0.1554961E 02


THE ASSOCIATED EIGENVECTOR COMPONENTS ARE

              0.1000000E 01
              0.4450409E 00
             -0.8019413E 00


LAMBDA( 3)= 0.3246979E 02


THE ASSOCIATED EIGENVECTOR COMPONENTS ARE

              0.1000000E 01
             -0.1246983E 01
              0.5549539E 00
```

EXAMPLE 3-11

Frequently the matrix equations for real physical problems do not occur in the standard form of Eq. 3-165 or 3-166. So, let us consider a problem in structural dynamics in which the matrix equations are not of the standard form, and write a general FORTRAN program for obtaining the smallest eigenvalues (square of natural circular frequencies) and eigenvectors (mode configurations) in ascending order. The program will be written to handle up to a 20th-order matrix ($n = 20$) and for sweeping out up to 10 eigenvectors ($S = 10$). It will be left as an exercise for the reader to modify the program to determine the largest eigenvalues and eigenvectors in descending order (see Prob. 3-30 at the end of this chapter).

To illustrate the use of the program, the data for the four-story building shown in Fig. 3-10 will be used as input data. The eigenvalues λ_i and associated eigenvectors $\mathbf{X}_i$ will correspond to the square of the natural circular frequencies ω_i^2 and the mode configurations, respectively.

The four-story building is shown schematically in Fig. 3-10(a). It is assumed that the weight distribution of the building may be represented in the form of concentrated loads at each floor level, as shown. It is further assumed that the girders of the structure are infinitely rigid, in comparison with the supporting columns, so that a general configuration between floors will be obtained, as shown in Fig. 3-10(b). The spring constants in

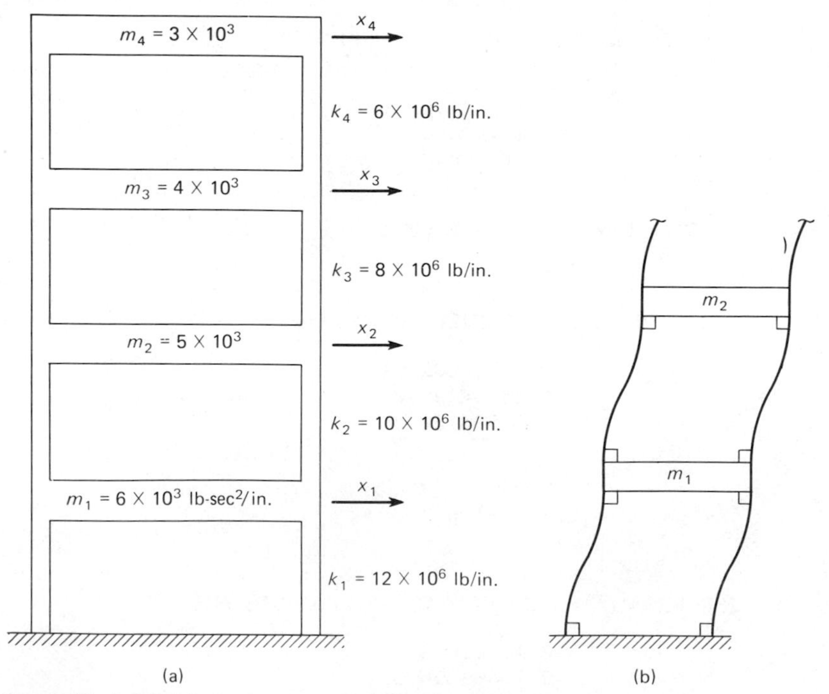

Fig. 3-10 / *Model of a four-story building.*

the figure are *equivalent* constants representing the aggregate stiffness of the number of columns supporting a given floor, and they were obtained by considering the columns as springs in parallel.

In structural dynamics, the matrix equations may be formulated in terms of a stiffness matrix **K** (symmetric) or in terms of a flexibility matrix **B** (symmetric) which is the inverse of **K**. The matrix equation in terms of **K** is

$$\mathbf{KX} = \lambda\mathbf{MX} \tag{3-170}$$

where λ is the natural circular frequency squared, ω^2, and **M** is the mass matrix containing only elements on the diagonal. That is,

$$\mathbf{M} = \begin{bmatrix} m_1 & 0 & 0 \cdots 0 \\ 0 & m_2 & 0 \cdots 0 \\ 0 & 0 & m_3 \cdots 0 \\ \cdots & \cdots & \cdots \\ 0 & 0 & 0 \cdots m_n \end{bmatrix}$$

The element k_{ij} of the stiffness matrix is the force required at station i to give a unit displacement at station j with all other stations held fixed. It should be apparent from the preceding discussion in this section that the iteration procedure applied to Eq. 3-170 will converge to the largest eigenvalues and associated eigenvectors in descending order.

The matrix equation in terms of the flexibility matrix is

$$\mathbf{BMX} = \frac{1}{\lambda}\mathbf{X} \tag{3-171}$$

in which the element b_{ij} of the flexibility matrix is the displacement at station i due to a unit force at station j. From the preceding discussion in this section, it should be apparent that the iteration procedure applied to Eq. 3-171 will converge to the smallest eigenvalues and associated eigenvectors in ascending order. Let us formulate the **B** matrix for input data to the FORTRAN program. From the definition of b_{ij} and the spring constants shown in Fig. 3-10, we obtain the following:

$$b_{11} = b_{12} = b_{13} = b_{14} = \frac{1}{k_1} = 8.33(10)^{-8}$$

$$b_{21} = b_{12} = 8.33(10)^{-8}$$

$$b_{22} = b_{23} = b_{24} = \frac{1}{k_1} + \frac{1}{k_2} = 18.33(10)^{-8}$$

$$b_{31} = b_{13} = 8.33(10)^{-8}$$

$$b_{32} = b_{23} = 18.33(10)^{-8}$$

$$b_{33} = b_{34} = \frac{1}{k_1} + \frac{1}{k_2} + \frac{1}{k_3} = 30.83(10)^{-8}$$

$$b_{41} = b_{14} = 8.33(10)^{-8}$$

$$b_{42} = b_{24} = 18.33(10)^{-8}$$

$$b_{43} = b_{34} = 30.83(10)^{-8}$$

$$b_{44} = \frac{1}{k_1} + \frac{1}{k_2} + \frac{1}{k_3} + \frac{1}{k_4} = 47.50(10)^{-8}$$

From Fig. 3-10, we see that the mass matrix **M** is

$$\mathbf{M} = (10)^3 \begin{bmatrix} 6 & 0 & 0 & 0 \\ 0 & 5 & 0 & 0 \\ 0 & 0 & 4 & 0 \\ 0 & 0 & 0 & 3 \end{bmatrix}$$

Since Eq. 3-171 contains the diagonal mass matrix **M**, the orthogonality principle for the normal modes (eigenvectors) takes the form (see App. B)

$$\mathbf{X}_i^T\mathbf{M}\mathbf{X}_j = \begin{Bmatrix} 0 & i \neq j \\ G_{ii} & i = j \end{Bmatrix} \tag{3-172}$$

instead of the form given by Eq. 3-167. This, in turn, leads to a modification of Eq. 3-168 so that

$$C_i = \frac{1}{G_{ii}} \mathbf{X}_i^T\mathbf{M}\mathbf{X}_k \tag{3-173}$$

There is no change in Eq. 3-169 which is repeated here for convenience,

$$\mathbf{X} = \mathbf{X}_k - \sum_{i=1}^{s} C_i\mathbf{X}_i \tag{3-174}$$

so Eqs. 3-171 through 3-174 constitute the system of equations for iteration to the smallest eigenvalues (natural circular frequencies squared) and associated eigenvectors (mode shapes) in ascending order for a structural dynamics system.

For our particular problem, the values of the elements b_{ij} of the **B** matrix and elements of the diagonal mass matrix will be used as input data to the FORTRAN program. The variable names used in the program, and the quantities they represent are as follows:

Variable Name	Quantity
B	Flexibility matrix
M	Mass matrix
A	Product of matrices **B** and **M**
N	Order of matrix **B**
LAMBDA(I)	The eigenvalues
X(I)	Components of smallest eigenvector
D(I)	Components of smallest eigenvector
Z(I,L)	The L components of the Ith eigenvector
XK1(IP1,L) XK(IP1,L)	The L components of the (I+1)th eigenvector which approach correct values with iteration
XX(II)	Unnormalized trial **X** values
C(II)	C_i values as defined by Eq. 3-173
G(I)	G_i values as defined by Eq. 3-172
EPSI	A small value (0.00001 in this program), where it is required that all eigenvector components differ from their previous values in the iteration by less than EPSI times the latest eigenvector component values

The program is as follows:

```
C THIS PROGRAM FINDS THE SMALLEST EIGENVALUE AND S
                                                 SUCCESSIVELY
C LARGER EIGENVALUES OF THE MATRIX EQUATION BMX=1/L*X
C ASSOCIATED EIGENVECTORS ARE FOUND FOR ALL EIGENVALUES
      REAL LAMBDA,M
      INTEGER S,SP1
      DIMENSION A(20,20),B(20,20),M(20),X(20),D(20),Z(10,20),
                                                    XK(11,20),
     *C(10),G(10),SUMM(20),Y(20),X1(11,20),
                                 XX(20),XK1(11,20),LAMBDA(11)
C READ ORDER OF MATRIX B,NUMBER OF EIGENVECTORS TO BE SWEPT,
C AND A SMALL VALUE EPSI
      READ(5,2) N,S,EPSI
    2 FORMAT(2I2,F16.0)
C READ IN THE MATRIX B
      READ(5,3) ((B(I,J),J=1,N),I=1,N)
    3 FORMAT(5E14.7)
C READ IN THE DIAGONAL MASS MATRIX INTO A ONE-DIMENSIONAL ARRAY
      READ(5,50) (M(I),I=1,N)
   50 FORMAT(8F10.0)
C MULTIPLY MATRIX B AND MATRIX M AND CALL PRODUCT MATRIX A
      DO 60 I=1,N
      DO 60 J=1,N
   60 A(I,J)=M(J)*B(I,J)
```

```
C BY THE ITERATION METHOD DETERMINE THE SMALLEST EIGENVALUE
C AND ASSOCIATED EIGENVECTOR
C ASSIGN VALUES TO COMPONENTS OF INITIALLY ASSUMED VECTOR X
      DO 4 I=1,N
    4 X(I)=1.
C CALCULATE COMPONENTS OF THE VECTOR 1/LAMBDA*X
      IT=0
    5 DO 6 I=1,N
      D(I)=0.
      DO 6 J=1,N
    6 D(I)=D(I)+A(I,J)*X(J)
      IT=IT+1
C NORMALIZE THE VECTOR D(I)
      DO 7 I=1,N
    7 Z(1,I)=D(I)/D(1)
C CHECK TO SEE IF REQUIRED ACCURACY HAS BEEN OBTAINED
      DO 8 I=1,N
      DIFF=X(I)-Z(1,I)
      IF(ABS(DIFF)-EPSI*Z(1,I)) 8,9,9
    8 CONTINUE
      GO TO 11
    9 DO 10 I=1,N
   10 X(I)=Z(1,I)
      IF(IT.GE.50) GO TO 11
      GO TO 5
   11 LAMBDA(1)=1./D(1)
C THE SMALLEST EIGENVALUE AND ASSOCIATED VECTOR ARE NOW
                                                         AVAILABLE
C SWEEP OUT S EIGENVECTORS
      DO 100 I=1,S
      IP1=I+1
C ASSIGN THE VALUE 1 TO XK COMPONENTS
      DO 12 L=1,N
   12 XK(IP1,L)=1.
C CALCULATE G(I) VALUES
      G(I)=0.
      DO 14 L=1,N
   14 G(I)=G(I)+M(L)*Z(I,L)**2
C CALCULATE VALUES OF C
      IT=0
   15 DO 99 II=1,I
      SUM=0.
      DO 16 L=1,N
   16 SUM=SUM+Z(II,L)*M(L)*XK(IP1,L)
   99 C(II)=SUM/G(II)
      IT=IT+1
C CALCULATE SUM OF CX'S
      DO 17 L=1,N
   17 SUMM(L)=0.
      DO 19 K=1,I
      DO 18 L=1,N
      Y(L)=C(K)*Z(K,L)
   18 SUMM(L)=SUMM(L)+Y(L)
   19 CONTINUE
```

```
C CALCULATE NEW TRIAL X
      DO 20 L=1,N
   20 X1(IP1,L)=XK(IP1,L)-SUMM(L)
C MULTIPLY MATRIX A TIMES X
      DO 21 II=1,N
      XX(II)=0
      DO 21 KK=1,N
   21 XX(II)=XX(II)+A(II,KK)*X1(IP1,KK)
C NORMALIZE TO GET XK FOR NEXT ITERATION,OR FINAL XK
      DO 22 L=1,N
   22 XK1(IP1,L)=XX(L)/XX(1)
C CHECK TO SEE IF REQUIRED ACCURACY HAS BEEN ATTAINED
      DO 23 L=1,N
      DIFF=XK(IP1,L)-XK1(IP1,L)
      IF(ABS(DIFF)-EPSI*XK1(IP1,L)) 23,24,24
   23 CONTINUE
      GO TO 26
C ASSIGN XK1 VALUES TO NAME XK
   24 DO 25 L=1,N
   25 XK(IP1,L)=XK1(IP1,L)
      IF(IT.GE.50) GO TO 26
      GO TO 15
C CALCULATE EIGENVALUE
   26 LAMBDA(IP1)=1./XX(1)
C ASSIGN EIGENVECTORS TO NAME Z
      DO 27 L=1,N
   27 Z(IP1,L)=XK1(IP1,L)
  100 CONTINUE
C WRITE OUT THE EIGENVALUES AND EIGENVECTORS
      SP1=S+1
      DO 104 I=1,SP1
      WRITE(6,101) I,LAMBDA(I)
  101 FORMAT('1','LAMBDA(',I2,')=',E14.7//)
      WRITE(6,102)
  102 FORMAT(' ','THE ASSOCIATED EIGENVECTOR COMPONENTS ARE'/)
      WRITE(6,103) (Z(I,L),L=1,N)
  103 FORMAT(' ',13X,E14.7)
  104 CONTINUE
      STOP
      END
```

The computer printout is as follows:

```
LAMBDA( 1)= 0.3176807E 03


THE ASSOCIATED EIGENVECTOR COMPONENTS ARE

              0.1000000E 01
              0.2009870E 01
              0.2873154E 01
              0.3415819E 01

LAMBDA( 2)= 0.1877433E 04
```

```
THE ASSOCIATED EIGENVECTOR COMPONENTS ARE

                  0.1000000E 01
                  0.1074018E 01
                 -0.9371233E-01
                 -0.1533693E 01


LAMBDA( 3)= 0.4233406E 04


THE ASSOCIATED EIGENVECTOR COMPONENTS ARE

                  0.1000000E 01
                 -0.3395634E 00
                 -0.1115582E 01
                  0.9986151E 00


LAMBDA( 4)= 0.6338262E 04


THE ASSOCIATED EIGENVECTOR COMPONENTS ARE

                  0.1000000E 01
                 -0.1602491E 01
                  0.1492490E 01
                 -0.6878729E 00
```

The computer printout shows that

$$\lambda_1 = \omega_1^2 = 317.6$$

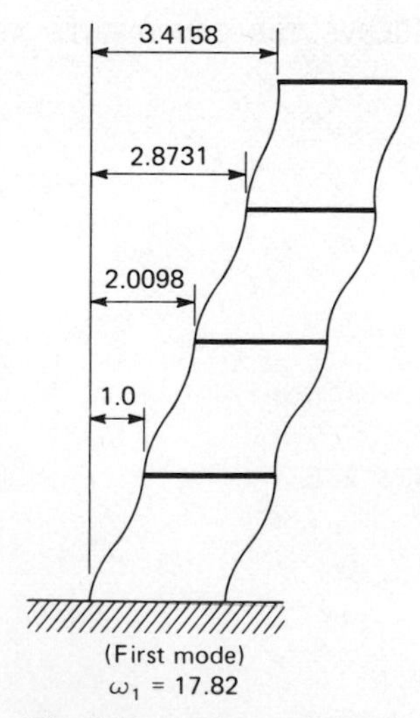

Fig. 3-11 / *First-mode configuration of the four-story building.*

Thus the natural circular frequency of the first (the fundamental) mode of vibration is

$$\omega_1 = \sqrt{317.6} = 17.82 \text{ rad/s}$$

The frequency in cycles per second is then

$$f_1 = \frac{\omega_1}{2\pi} = \frac{17.82}{2\pi} = 2.84$$

From the components of the eigenvector corresponding to λ_1, we obtain the configuration for the first mode of vibration shown in Fig. 3-11.

3-13 / NONLINEAR SIMULTANEOUS EQUATIONS

There are no direct methods available for solving nonlinear simultaneous equations such as are available for linear sets. In general, iterative approaches must be used, which require initial guesses or approximations of the solution values to get started. The iterative approach which we will consider makes use of Taylor's series for functions of several variables[18] to find real roots of nonlinear sets.

Let us consider the two nonlinear equations

$$\begin{aligned} f(x, y) &= 0 \\ g(x, y) &= 0 \end{aligned} \tag{3-175}$$

with $x = x_r$ and $y = y_r$ an exact solution to the set. If $x = x_i$, $y = y_i$ are approximations to the solution, a Taylor series expansion about x_i, y_i yields

$$\begin{aligned} f(x_r, y_r) &= f(x_i, y_i) + h \left.\frac{\partial f}{\partial x}\right)_i + j \left.\frac{\partial f}{\partial y}\right)_i + \cdots = 0 \\ g(x_r, y_r) &= g(x_i, y_i) + h \left.\frac{\partial g}{\partial x}\right)_i + j \left.\frac{\partial g}{\partial y}\right)_i + \cdots = 0 \end{aligned} \tag{3-176}$$

where

$$x_r = x_i + h \qquad y_r = y_i + j$$

[18] For a discussion of Taylor's series for functions of several variables, see any advanced calculus book. Also see Sec. 6-5.

and

$$\left.\frac{\partial f}{\partial x}\right)_i = \left.\frac{\partial f}{\partial x}\right]_{\substack{x=x_i\\y=y_i}}, \quad \left.\frac{\partial g}{\partial x}\right)_i = \left.\frac{\partial g}{\partial x}\right]_{\substack{x=x_i\\y=y_i}}, \quad \text{and so forth.}$$

If we truncate the Taylor series of Eq. 3-176 to only linear terms, we then have two linear equations in the two unknowns h and j. However, since the Taylor series has been truncated, h and j are only approximations to $x_r - x_i$ and $y_r - y_i$, respectively. Therefore, solving the two linear equations for h and j, we can only obtain improved values for x and y (assuming that x_i and y_i are close enough to x_r and y_r for convergence) rather than exact solution values. That is,

$$\begin{aligned} x_{i+1} &= x_i + h \\ y_{i+1} &= y_i + j \end{aligned} \tag{3-177}$$

where x_{i+1} and y_{i+1} are improved approximations to x_r and y_r. They are used as x_i and y_i, respectively, in obtaining still better solution values. Iteration continues until the values of x and y stabilize.

For three equations in three unknowns (say, x, y, and z), the three functions $f(x, y, z)$, $g(x, y, z)$, and $F(x, y, z)$ would be expanded in a Taylor series for functions of three variables. Proceeding as for two equations, three linear equations in the unknowns h, j, and k would be obtained. The remaining procedure would parallel the case discussed above for two equations. Extension may be made to more than three equations.

EXAMPLE 3-12

Let us find the roots of the set

$$\begin{aligned} f(x, y) &= x^2 + 2y^2 - 22.0 = 0 \\ g(x, y) &= -2x^2 + xy - 3y + 11.0 = 0 \end{aligned} \tag{3-178}$$

The partial derivatives needed in Eq. 3-176 are

$$\frac{\partial f}{\partial x} = 2x \qquad \frac{\partial g}{\partial x} = -4x + y$$

$$\frac{\partial f}{\partial y} = 4y \qquad \frac{\partial g}{\partial y} = x - 3$$

Suppose that our first approximation of a solution (perhaps arrived at by trial and error) is $x = 1.5$, $y = 2.0$. Evaluating $f(x, y)$, $g(x, y)$, and the above partial derivatives at $x_i = 1.5$, $y_i = 2.0$, we obtain

$$f(x_i, y_i) = (1.5)^2 + 2(2.0)^2 - 22.0 = -11.75$$

$$g(x_i, y_i) = -2(1.5)^2 + (1.5)(2.0) - 3(2.0) + 11.0 = 3.5$$

$$\left.\frac{\partial f}{\partial x}\right)_i = 2(1.5) = 3.0 \qquad \left.\frac{\partial f}{\partial y}\right)_i = 4(2.0) = 8.0$$

$$\left.\frac{\partial g}{\partial x}\right)_i = -4(1.5) + 2.0 = -4.0 \qquad \left.\frac{\partial g}{\partial y}\right)_i = 1.5 - 3 = -1.5$$

Then from Eq. 3-176

$$3h + 8j = 11.75$$

$$-4h - 1.5j = -3.5$$

from which

$$h = 0.377$$

$$j = 1.327$$

and

$$x_{i+1} = 1.5 + 0.377 = 1.877$$

$$y_{i+1} = 2.0 + 1.327 = 3.327$$

Using these values for x_i and y_i, one more iteration yields

$$h = 0.136$$

$$j = -0.313$$

and

$$x_{i+1} = 1.877 + 0.136 = 2.013$$

$$y_{i+1} = 3.327 - 0.313 = 3.014$$

The exact solution values are $x = 2.000$, $y = 3.000$.

Problems

3-1 A parabola having an equation of the form $y = Ax^2 + Bx + C$ goes through the points (1, 8), (2, 13), and (3, 20). *Manually* determine the equation of this parabola.

3-2 Determine the equation of the third-degree polynomial of the form $y = Ax^3 + Bx^2 + Cx + D$ which passes through the points (1, 10), (2, 26), (−1, 2), and (0, 4).

3-3 Determine the coefficients of the polynomial $y = Ax^3 + Bx^2 + Cx + D$ which passes through the points (1, 10), (2, 26), (−1, 2) and has a slope of 10 at the point (1, 10).

3-4 Determine the coefficients of the polynomial $y = Ax^4 + Bx^3 + Cx^2 + Dx + E$ which passes through the points (1, 8), (2, 44), (−1, 2), (−3, 44), and (−2, 8).

3-5 Determine the coefficients of the polynomial $y = Ax^5 + Bx^4 + Cx^3 + Dx^2 + Ex + F$ which passes through the points (1, −7), (2, 2), (−1, −7), (−2, −34), (3, 121), and (0, −8).

3-6 Determine the coefficients of the polynomial $y = Ax^5 + Bx^4 + Cx^3 + Dx^2 + Ex + F$ which passes through the points (−2, −34), (1, −7), (2, 2), (−1, −7), and has a slope of 0 at the points (1, −7), and (−1, −7).

3-7 The general equation of a conic section can be written as $x^2 + Bxy + Cy^2 + Dx + Ey + F = 0$. Determine the equation of the conic passing through the points $(0, \frac{1}{2})$, $(\frac{1}{2}, \frac{4}{9})$, $(-2, \frac{57}{2})$, $(1, \frac{3}{22})$, and $(-\frac{1}{2}, 0)$.

3-8 A rotating plate cam is to drive a reciprocating radial roller follower such as shown in part a of the accompanying figure. During the first 90 deg of cam rotation, the follower is to rise 1 in. and then be followed by a dwell as indicated by the displacement diagram in part b of the accompanying figure. When cams have moderate speeds, various standard motions, such as constant acceleration and deceleration or simple harmonic motion, are usually used for such displacements. For high-speed cams, certain restrictions may have to be placed on the acceleration or second-acceleration (jerk or pulse). Polynomial equations are often used to describe the displacement in such cases. The restrictions imposed in this problem are that the follower must have zero velocity, zero acceleration, and zero second-acceleration at $\theta = 0$ and at $\theta = \pi/2$. Further-

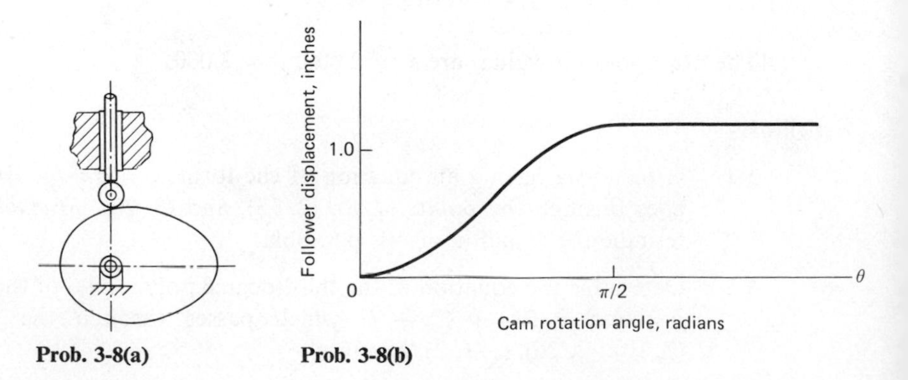

Prob. 3-8(a) Prob. 3-8(b)

more, when $\theta = \pi/4$, it is desired that the acceleration be zero. Since the angular velocity of the cam $d\theta/dt$ is to be constant, zero velocity, zero acceleration, and zero jerk of the follower imply that $dy/d\theta = 0$, $d^2y/d\theta^2 = 0$, and $d^3y/d\theta^3 = 0$. One further restriction to be imposed is that the value of $dy/d\theta$ should be $4/\pi$ in./radian at $\theta = \pi/4$. The restrictions are summarized as follows: when

a. $\theta = 0$: $y = 0$, $y' = 0$, $y'' = 0$, $y''' = 0$

b. $\theta = \pi/2$: $y = 1$ in., $y' = 0$, $y'' = 0$, $y''' = 0$

c. $\theta = \pi/4$: $y' = 4/\pi$ in./radian, $y'' = 0$

With ten conditions prescribed, it will be necessary to use ten constants in the polynomial describing the follower motion. That is,

$$y = C_0 + C_1\theta + C_2\theta^2 + C_3\theta^3 + C_4\theta^4 + C_5\theta^5 + C_6\theta^6 + C_7\theta^7 + C_8\theta^8 + C_9\theta^9$$

Differentiate the polynomial three times to obtain expressions for y', y'', and y'''. Applying conditions a, it will be obvious that $C_0 = C_1 = C_2 = C_3 = 0$. Substituting conditions b and c into the equations with the above zero constants, six equations in the remaining six C's will result. Solve the simultaneous equations by the method assigned, and plot curves of y, y', y'', and y''' versus θ.

3-9 Referring to the FORTRAN program written for Ex. 3-3, modify the program to bring the largest *applicable* element in the *matrix* into the pivot position by the use of row and/or column interchange. Use the program given at the end of Sec. 3-10 as a guide.

3-10 Write a computer program for solving simultaneous linear algebraic equations by the Gauss-Jordan method flow charted in Fig. 3-3.

3-11 Write a computer program for inverting a matrix in place using partial pivoting, following the flow chart given in Fig. 3-4.

3-12 Modify the program written in Prob. 3-11 so that a set of simultaneous equations is solved, and at the same time the coefficient matrix is inverted. Provision will have to be made for a column of constants to be appended to the coefficient matrix.

In Probs. 3-13 through 3-18, determine the coefficients of the characteristic polynomial of the matrix **A** by manual use of the Faddeev-Leverrier method.

3-13
$$\begin{bmatrix} (3-\lambda) & -3 & 1 \\ 4 & (3-\lambda) & -2 \\ 4 & 4 & (-2-\lambda) \end{bmatrix} \begin{Bmatrix} x_1 \\ x_2 \\ x_3 \end{Bmatrix} = \{0\}$$

3-14
$$\begin{bmatrix} (1-\lambda) & 2 & 3 \\ -10 & (0-\lambda) & 2 \\ -2 & 4 & (8-\lambda) \end{bmatrix} \begin{Bmatrix} x_1 \\ x_2 \\ x_3 \end{Bmatrix} = \{0\}$$

3-15
$$\begin{bmatrix} (0-\lambda) & 2 & 3 \\ -10 & (-1-\lambda) & 2 \\ -2 & 4 & (7-\lambda) \end{bmatrix} \begin{Bmatrix} x_1 \\ x_2 \\ x_3 \end{Bmatrix} = \{0\}$$

3-16
$$\begin{bmatrix} (5-\lambda) & 2 & -1 \\ 2 & (5-\lambda) & 1 \\ 1 & 4 & (5-\lambda) \end{bmatrix} \begin{Bmatrix} x_1 \\ x_2 \\ x_3 \end{Bmatrix} = \{0\}$$

3-17
$$\begin{aligned} 2x_1 + 2x_2 + 3x_3 &= \lambda x_1 \\ -10x_1 + x_2 + 2x_3 &= \lambda x_2 \\ -2x_1 + 4x_2 + 9x_3 &= \lambda x_3 \end{aligned}$$

3-18
$$\begin{aligned} 3x_1 - x_2 - x_3 &= \lambda x_1 \\ 4x_1 - x_2 - 2x_3 &= \lambda x_2 \\ 3x_1 - 2x_2 &= \lambda x_3 \end{aligned}$$

In Probs. 3-19 through 3-21, the eigenvalues are given below the equations. Using the method discussed in Sec. 3-10, *manually* determine the eigenvector associated with each eigenvalue given.

3-19
$$\begin{bmatrix} (1-\lambda) & 2 & 3 \\ -10 & (0-\lambda) & 2 \\ -2 & 4 & (8-\lambda) \end{bmatrix} \begin{Bmatrix} x_1 \\ x_2 \\ x_3 \end{Bmatrix} = \{0\}$$

$(\lambda_1 = 2, \lambda_2 = 3, \lambda_3 = 4)$

3-20
$$\begin{bmatrix} (0-\lambda) & 2 & 3 \\ -10 & (-1-\lambda) & 2 \\ -2 & 4 & (7-\lambda) \end{bmatrix} \begin{Bmatrix} x_1 \\ x_2 \\ x_3 \end{Bmatrix} = \{0\}$$

$(\lambda_1 = 1, \lambda_2 = 2, \lambda_3 = 3)$

3-21
$$\begin{bmatrix} (2-\lambda) & 2 & 3 \\ -10 & (1-\lambda) & 2 \\ -2 & 4 & (9-\lambda) \end{bmatrix} \begin{Bmatrix} x_1 \\ x_2 \\ x_3 \end{Bmatrix} = \{0\}$$

$(\lambda_1 = 3, \lambda_2 = 4, \lambda_3 = 5)$

3-22 If the voltage drop across an induction coil is $L\,di/dt$ and the voltage drop across a capacitor is $1/C \int i\,dt$, derive the differential equations, for the circuit shown, in terms of the charges q_1, q_2, and q_3 flowing in the respective loops. The currents and charges are related by $i_1 = dq_1/dt$, $i_2 = dq_2/dt$, and $i_3 = dq_3/dt$. (a) Using solutions of the form $q_i = Q_i \sin pt$, and letting $E = 0$, transform the set of simultaneous differential equations, obtained above, into a set of homogeneous linear algebraic equations. (b) Write the set of equations, obtained in part a, in matrix form, and determine, by the iteration method, the largest eigenvalue (p^2) and associated eigenvector, with $L_1 = L_2 = L_3 = 0.001$ henry and $C_1 = C_2 = C_3 = 0.001$ farad. It is suggested that the iterations be made by using an electronic calculator, in order to familiarize the reader with the process.

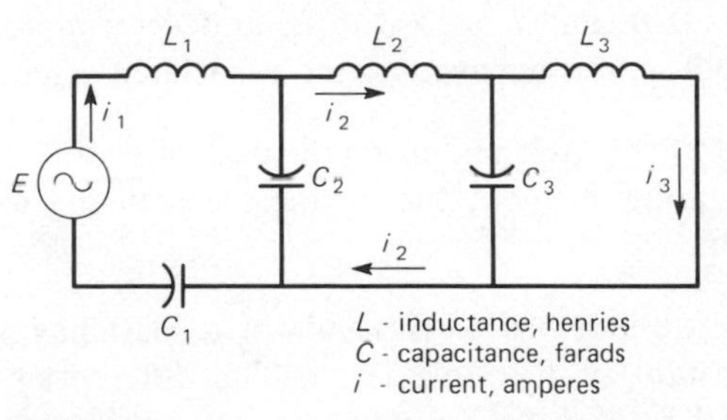

Prob. 3-22

3-23 Referring to Prob. 3-22 and using the data given there, determine, by the iteration method, the smallest eigenvalue (p^2) and associated eigenvector.

3-24 Referring to Prob. 3-22 and using the data given there, determine, by the polynomial method, the eigenvalues and corresponding eigenvectors for the electric circuit.

3-25 A cylinder, shown in the figure on page 278, rolls without slipping and has a mass m_1 and a moment of inertia $\bar{I}$. Determine (a) the differential equations of motion, in terms of the coordinates θ and x, and (b) the characteristic equation, by assuming a solution of the form $\theta = \Theta \sin pt$ and $x = X \sin pt$.

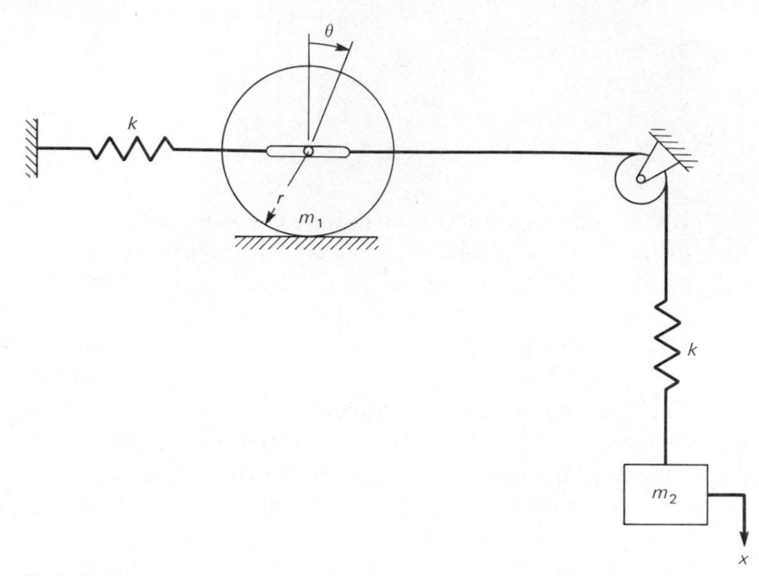

Prob. 3-25

3-26 Referring to Prob. 3-25, let $m_1 = m_2 = 2$ lb-sec^2/ft, $\bar{I} = 1.0$ lb-ft-sec^2, $r = 1.0$ ft, and $k = 200$ lb/ft, and determine, by the polynomial method, the eigenvalues and associated eigenvectors.

3-27 Referring to Prob. 3-25 and using the data of Prob. 3-26, determine, by the iteration method, the smallest eigenvalue and associated eigenvector.

3-28 Two disks are attached to the ends of a shaft having a torsional spring constant of k lb-in./radian. The differential equations of motion of the system are found to be

$$\bar{I}_1\ddot{\theta}_1 + k\theta_1 - k\theta_2 = 0$$

$$\bar{I}_2\ddot{\theta}_2 + k\theta_2 - k\theta_1 = 0$$

Assuming a solution of the form $\theta_i = \Theta_i \sin pt$ and noting that $\bar{I}_2 = 2\bar{I}_1$, we obtain the set of homogeneous equations

$$\left(\frac{k}{\bar{I}_1} - \lambda\right)\Theta_1 - \frac{k}{\bar{I}_1}\Theta_2 = 0$$

$$-\frac{k}{2\bar{I}_1}\Theta_1 + \left(\frac{k}{2\bar{I}_1} - \lambda\right)\Theta_2 = 0$$

where $\lambda_i = p_i^2$. Using the polynomial method, the roots of the characteristic equation are found to be

$$\lambda_1 = p_1^2 = \frac{3}{2}\frac{k}{\bar{I}_1}$$

$$\lambda_2 = p_2^2 = 0$$

The fact that $\lambda_2 = 0$ indicates that the system rotates as a rigid body $(\theta_1 = \theta_2)$.

Using the iteration method, and assuming values of $\Theta_1 = \Theta_2 = 1$ for the eigenvector components, show that convergence will be to the smallest eigenvalue instead of to the largest eigenvalue. Discuss the significance of this result, with respect to the assumed eigenvector components.

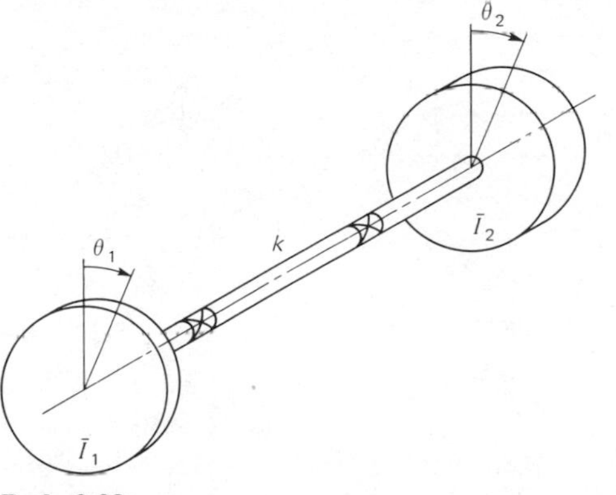

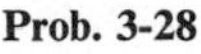
Prob. 3-28

3-29 Referring to Prob. 3-28, assume initial values of $\Theta_1 = -1$ and $\Theta_2 = 1$ for the eigenvector components in the iteration procedure, and determine the largest eigenvalue (p^2 when the disks are oscillating 180 deg out of phase with each other).

3-30 In Ex. 3-11, a general FORTRAN program was written for obtaining the smallest eigenvalues and associated eigenvectors in ascending order by the iteration method. (a) Modify the given program so that it may be used to obtain the largest eigenvalues and associated eigenvectors in descending order. (b) Using the modified program specified in part a of this problem, determine the eigenvalues (natural circular frequencies) and eigenvectors (mode shapes) of the four-story building discussed in Ex. 3-11.

3-31 Four disks are attached to a shaft which is supported by bearings, as shown in the accompanying figure. (a) Using Lagrange's equation

$$\frac{d}{dt}\left(\frac{\partial T}{\partial \dot{\theta}_i}\right) - \frac{\partial T}{\partial \theta_i} + \frac{\partial V}{\partial \theta_i} = 0$$

derive the differential equations of motion of the system, where T is the kinetic energy of the system, and V is the strain energy stored in the shaft. (b) Assuming a solution of the form $\theta_i = \Theta_i \sin pt$ for the differential equations derived in part a, transform them to a set of homogeneous algebraic equations. (c) Using the values of k_i and $\bar{I}_i$, shown in the accompanying figure, as input data to the computer, determine, by the polynomial method, the eigenvalues and associated eigenvectors for the four principal modes of vibration.

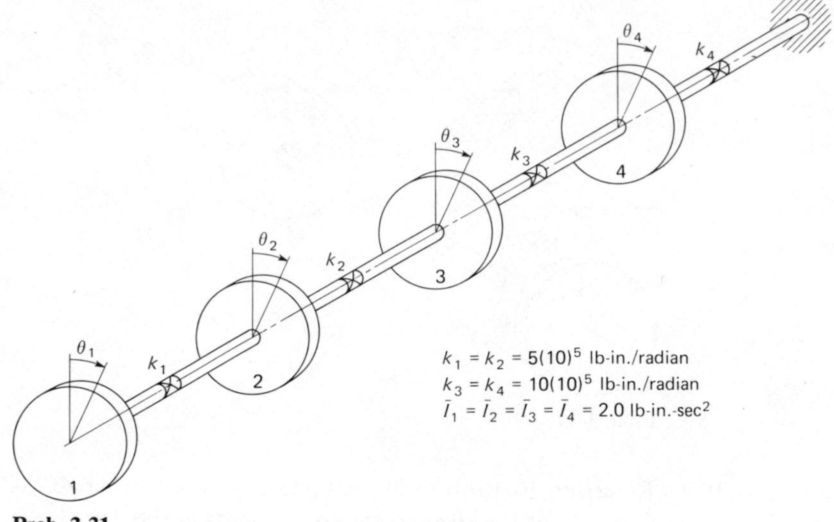

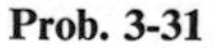

Prob. 3-31

The following problems involve SI units.

3-32 Refer to Prob. 3-26 and determine the eigenvalues and eigenvectors by the polynomial method by taking the system data as follows:

$$m_1 = m_2 = 50 \text{ kg}$$
$$\bar{I} = 2 \text{ kg·m}^2$$
$$r = 0.3 \text{ m}$$
$$k = 3500 \text{ N/m}$$

3-33 Refer to Prob. 3-31 and convert the system data given to the SI system of units and determine the eigenvalues and associated eigenvectors for the four principal modes of vibration by the polynomial method.

3-34 Refer to the four disks and shaft system of Prob. 3-31. Determine the eigenvalues and eigenvectors corresponding to the four principal modes of vibration by using the computer program shown in Ex. 3-11 for the iteration method used in the solution of the four-story building. Referring to Eq. 3-171, the element b_{ij} of the flexibility matrix **B** is the angular displacement at station i due to a unit moment at station j. The diagonal mass matrix **M** consists of the mass moment of inertias I_1, I_2, I_3, and I_4. For input data to the computer program of Ex. 3-11, formulate the flexibility matrix **B** and mass matrix **M** in terms of the SI system of units.

4 Curve Fitting

4-1 / INTRODUCTION

In fitting a curve to given data points, there are two possible approaches. One is to have the graph of the approximating function pass exactly through the given data points. The method of polynomial approximation presented in Appendix C utilizes this approach. If the data values are experimental, having the scatter frequently found in such data, this method may yield unsatisfactory results. The second approach, which is usually a more satisfactory one for experimental data, uses an approximating function which graphs as a smooth curve having the general shape suggested by the data values, but not in general passing exactly through all of the data points. This latter approach is presented in this chapter.

In engineering, curve fitting plays an important role in the analysis, interpretation, and correlation of experimental data with mathematical models formulated from fundamental engineering principles. In the most general sense, curve fitting involves the determination of a continuous function

$$y = f(x) \tag{4-1}$$

which results in the most "reasonable" or "best" fit of experimentally measured values of (x_1, y_1), (x_2, y_2), and so forth. The particular form of $f(x)$ will in many cases be known in advance from a consideration of the physical laws, or theory, associated with the x and y variables being measured. In other instances, the general appearance of the plotted data

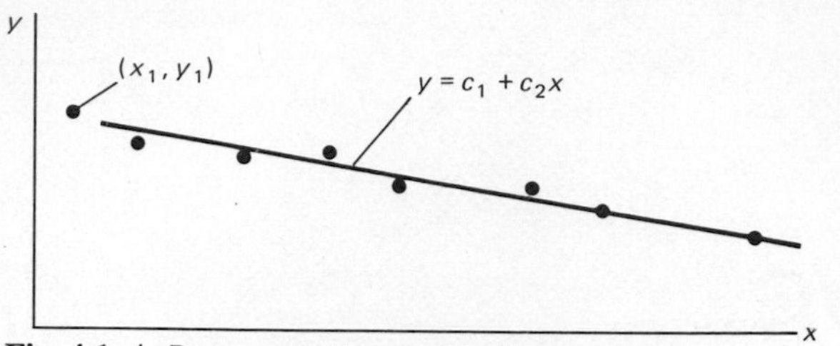

Fig. 4-1 / *Data suggesting straight line.*

may suggest the particular form of $f(x)$. For example, a plot of x and y values as shown in Fig. 4-1 would suggest that the measured data can best be represented by a straight line. That is,

$$y = f(x) = c_1 + c_2x \tag{4-2}$$

One can, of course, intuitively construct a straight line through the points as shown which will appear as a reasonable average of the plotted points. However, as we will see, the constants c_1 and c_2 can be determined from a "least-squares" criterion (discussed in Sec. 4-2) which will result in a "best-fit" straight-line representation of the experimental data. Other criteria besides least squares could be employed, such as minimizing the maximum error, but the least-squares criterion is the most widely used, and our discussion of curve fitting will be limited to obtaining a "best fit" in the least-squares sense.

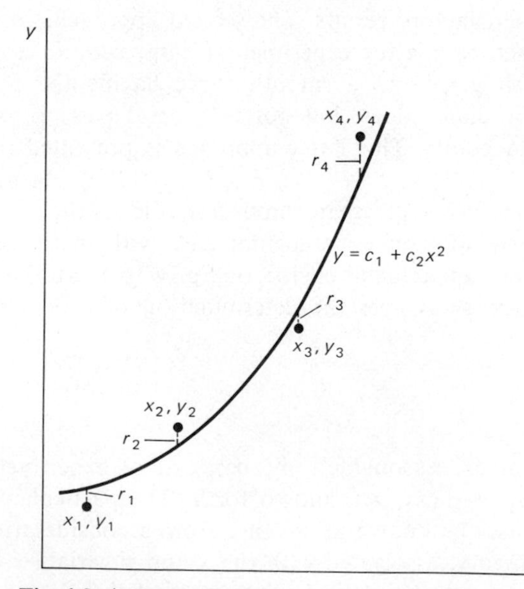

Fig. 4-2 / *Data suggesting a quadratic equation.*

As another example, a plot of experimentally determined x and y values as shown in Fig. 4-2 might suggest a curve fitted by the pure quadratic equation

$$y = c_1 + c_2x^2 \tag{4-3}$$

where c_1 and c_2 would be determined from the least-squares criterion for a best fit of the function to the measured data.

4-2 / METHOD OF LEAST-SQUARES FIT—LINEAR FORMS

Let us consider curve fitting of data with a curve having the form of

$$y = f(x, c_1, c_2, \ldots, c_m) = c_1f_1(x) + c_2f_2(x) + \cdots + c_mf_m(x)$$

in which the dependent variable y is linear with respect to constants $c_1, c_2, \ldots, c_m$. That is, the constants appear only as coefficients and never as part of the arguments (in the equations $y = c_1e^{c_2x}$ or $y = c_1 \sin c_2x + c_3 \sin c_4x$, for example, y *would not* be linear with respect to the constants). We will call these equations, such as Eqs. 4-2 and 4-3, *linear forms with undetermined constants.*

Although there are many ramifications and approaches to curve fitting, the application of the least-squares criterion can be applied to a wide variety of curve-fitting problems involving linear forms with undetermined constants. As we will soon see, the constants are determined to minimize the sum of the squares of the differences between the measured values $(y_1, y_2, \ldots, y_n)$ and the corresponding function $y = f(x)$ being used to fit a curve to these points.

Before formulating the general equations suitable for digital computation of the least-squares curve-fitting technique, let us first consider Fig. 4-2 and illustrate the general procedure for obtaining a best fit of the function $y = c_1 + c_2x^2$ to the measured data (x_1, y_1), (x_2, y_2), (x_3, y_3), and (x_4, y_4). From the four known values of x and y, we can write the equations

$$\begin{aligned} c_1 + c_2(x_1)^2 &= y_1 \\ c_1 + c_2(x_2)^2 &= y_2 \\ c_1 + c_2(x_3)^2 &= y_3 \\ c_1 + c_2(x_4)^2 &= y_4 \end{aligned} \tag{4-4}$$

Since c_1 and c_2 are the only unknowns, there are four equations with two unknowns. It should be apparent that, in general, the number of equations n is equal to the number of known data values (x_1, y_1), (x_2, y_2), . . . ,

(x_n, y_n), and that the number of unknowns m is equal to the number of unknown constants $(c_1, c_2, \ldots, c_m)$. In general, the number of data values known will be greater than the number of unknown constants, as is the case in Eq. 4-4, and no solution is possible.

However, the concept of least squares for a best fit circumvents the problem of having more equations than unknowns. To develop the least-squares concept, we first rewrite Eq. 4-4 in the form

$$\begin{aligned} c_1 + c_2(x_1)^2 - y_1 &= r_1 \\ c_1 + c_2(x_2)^2 - y_2 &= r_2 \\ c_1 + c_2(x_3)^2 - y_3 &= r_3 \\ c_1 + c_2(x_4)^2 - y_4 &= r_4 \end{aligned} \tag{4-5}$$

where r_1, r_2, and so on, are the "residuals" which would all be zero in the case of a perfect fit. However, the residuals will generally not be zero (see Fig. 4-2) and a best fit is obtained by making them as small as possible. This is accomplished by finding values for $c_1, c_2, \ldots, c_m$ such that the sum of the squares of the residuals is a minimum. That is,

$$\sum_{i=1}^{n} (r_i)^2 = \text{minimum} \tag{4-6}$$

Considering r_i as a function of the constants c_1, c_2, and so on, it follows from Eq. 4-6 that

$$\frac{\partial}{\partial c_k} \sum_{i=1}^{n} (r_i)^2 = 0$$

or

$$\sum_{i=1}^{n} r_i \frac{\partial r_i}{\partial c_k} = 0 \qquad k = 1, 2, \ldots, m \tag{4-7}$$

The application of Eq. 4-7 will yield m equations from which the m unknown constants $(c_1, c_2, \ldots, c_m)$ can be determined. For the particular case of Eq. 4-5, $m = 2$ and $n = 4$ and Eq. 4-7 gives

$$\begin{aligned} r_1 \frac{\partial r_1}{\partial c_1} + r_2 \frac{\partial r_2}{\partial c_1} + r_3 \frac{\partial r_3}{\partial c_1} + r_4 \frac{\partial r_4}{\partial c_1} &= 0 \\ r_1 \frac{\partial r_1}{\partial c_2} + r_2 \frac{\partial r_2}{\partial c_2} + r_3 \frac{\partial r_3}{\partial c_2} + r_4 \frac{\partial r_4}{\partial c_2} &= 0 \end{aligned} \tag{4-8}$$

Looking at Eq. 4-5 and taking each of the partial derivatives shown in Eq. 4-8, we obtain

$$\frac{\partial r_1}{\partial c_1} = \frac{\partial r_2}{\partial c_1} = \frac{\partial r_3}{\partial c_1} = \frac{\partial r_4}{\partial c_1} = 1$$

$$\frac{\partial r_1}{\partial c_2} = x_1^2, \quad \frac{\partial r_2}{\partial c_2} = x_2^2, \quad \frac{\partial r_3}{\partial c_2} = x_3^2, \quad \frac{\partial r_4}{\partial c_2} = x_4^2$$

Then, noting that Eq. 4-5 may be expressed as

$$r_i = c_1 + c_2 x_i^2 - y_i \qquad (i = 1, 2, 3, \text{and } 4)$$

we substitute this expression and the partial derivatives into Eq. 4-8 to obtain

$$\begin{aligned} 4c_1 + c_2 \sum_{i=1}^{4} x_i^2 - \sum_{i=1}^{4} y_i &= 0 \\ c_1 \sum_{i=1}^{4} x_i^2 + c_2 \sum_{i=1}^{4} x_i^4 - \sum_{i=1}^{4} y_i x_i^2 &= 0 \end{aligned} \tag{4-9}$$

which may be solved simultaneously for the unknown constants c_1 and c_2. The values of c_1 and c_2 resulting from Eq. 4-9 will give the best fit of $y = c_1 + c_2 x^2$ to the four known data values.

EXAMPLE 4-1

The experimental data points (1.0, 3.8), (2.5, 15.0), (3.5, 26.0), and (4.0, 33.0) shown in Fig. 4-3 indicate a curve having the form

$$y = c_1 + c_2 x^2$$

Determine the least-squares fit of this function to the data.

Solution. Substituting the data values into Eq. 4-9 gives

$$4c_1 + 35.5c_2 = 77.8$$

$$35.5c_1 + 446c_2 = 944$$

from which

$$c_1 = 2.27$$

$$c_2 = 1.93$$

Thus $y = 2.27 + 1.93x^2$ for a least-squares best fit to the given data points. The given data points and the resulting function are shown in Fig. 4-3.

It should be emphasized that while the application of the least-squares fit of a specific function, $f(x)$, gives the best fit of that function to the data points, it may be necessary to apply the least-squares criterion to several different functions to determine which function best represents the data. For example, a least-squares fit of a straight-line function, $y = c_1 + c_2x$, could be obtained for the data points shown in Fig. 4-3. However, it should be apparent that the straight-line function would not represent the data nearly as well as the quadratic equation determined in Ex. 4-1. In some cases, it may be necessary to obtain and compare the $\sum r_i^2$ for various assumed functions in order to determine which one best represents the given set of data points. The most desirable function would be, of course, the one which results in the smallest value of $\sum r_i^2$.

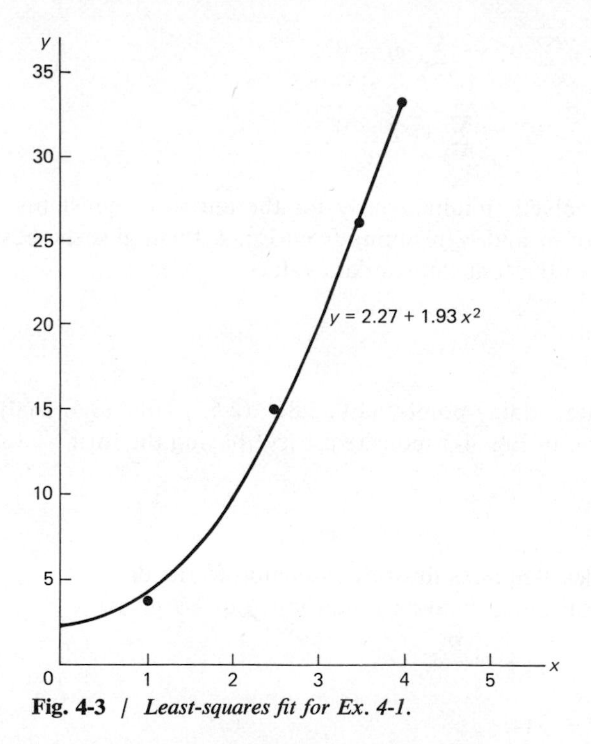

Fig. 4-3 / *Least-squares fit for Ex. 4-1.*

4-3 / MATRIX FORMULATION FOR LEAST-SQUARES PROCEDURE

From the discussion in Sec. 4-2, we found that a function $y = f(x)$ consisting of m functionally linear undetermined constants $c_1, c_2, \ldots, c_m$, for a least-squares fit to n known data values, involved the simultaneous solution of m linear algebraic equations. The least-squares procedure can be generalized in the form of a matrix equation to facilitate the generation of m linear algebraic equations which can be solved by one of the methods discussed in Chap. 3.

The function $y = f(x)$ can be written in the general form

$$y = f(x) = c_1 f_1(x) + c_2 f_2(x) + \cdots + c_m f_m(x) \tag{4-10}$$

in which known values for x and y correspond to experimental data. That is,

$$y_i = c_1 f_1(x_i) + c_2 f_2(x_i) + \cdots + c_m f_m(x_i)$$

or

$$y_i = \sum_{k=1}^{m} c_k f_k(x_i) \tag{4-11}$$

Considering the residuals $r_1, r_2, \ldots, r_n$, for a least-squares fit, it follows from Eq. 4-11 that

$$\left.\begin{array}{l} \sum_{k=1}^{m} c_k f_k(x_1) - y_1 = r_1 \\ \sum_{k=1}^{m} c_k f_k(x_2) - y_2 = r_2 \\ \cdots\cdots\cdots\cdots\cdots\cdots \\ \sum_{k=1}^{m} c_k f_k(x_n) - y_n = r_n \end{array}\right\} \tag{4-12}$$

We also have from Eq. 4-7 that

$$\sum_{i=1}^{n} r_i \frac{\partial r_i}{\partial c_k} = 0 \qquad k = 1, 2, \ldots, m$$

which may be written in the matrix form

$$[r_1 \quad r_2 \quad \cdots \quad r_n] \begin{bmatrix} \dfrac{\partial r_1}{\partial c_1} & \dfrac{\partial r_1}{\partial c_2} & \cdots & \dfrac{\partial r_1}{\partial c_m} \\ \dfrac{\partial r_2}{\partial c_1} & \dfrac{\partial r_2}{\partial c_2} & \cdots & \dfrac{\partial r_2}{\partial c_m} \\ \cdots & \cdots & \cdots & \cdots \\ \dfrac{\partial r_n}{\partial c_1} & \dfrac{\partial r_n}{\partial c_2} & \cdots & \dfrac{\partial r_n}{\partial c_m} \end{bmatrix} = [0 \quad 0 \quad \cdots \quad 0] \tag{4-13}$$

where $[r_1 \ r_2 \ \cdots \ r_n]$ and $[0 \ 0 \ \cdots \ 0]$ are row matrices.

From Eq. 4-12, we see that

$$\frac{\partial r_i}{\partial c_k} = f_k(x_i) \tag{4-14}$$

Thus Eq. 4-13 may be written in the form

$$[r_1 \quad r_2 \quad \cdots \quad r_n]\mathbf{F} = [0 \quad 0 \quad \cdots \quad 0] \tag{4-15}$$

where

$$\mathbf{F} = \begin{bmatrix} f_1(x_1) & f_2(x_1) & \cdots & f_m(x_1) \\ f_1(x_2) & f_2(x_2) & \cdots & f_m(x_2) \\ \cdots & \cdots & \cdots & \cdots \\ f_1(x_n) & f_2(x_n) & \cdots & f_m(x_n) \end{bmatrix} \tag{4-16}$$

We now denote $[r_1\ r_2 \cdots r_n] = \mathbf{R}$ and take the *transpose* (see Appendix B) of Eq. 4-15 to obtain

$$[\mathbf{RF}]^{\mathrm{T}} = [0 \quad 0 \quad \cdots \quad 0]^{\mathrm{T}} \tag{4-17}$$

Since $[\mathbf{RF}]^{\mathrm{T}} = \mathbf{F}^{\mathrm{T}}\mathbf{R}^{\mathrm{T}}$ and $[0\ 0 \cdots 0]^{\mathrm{T}}$ is a column matrix, Eq. 4-17 takes the form

$$\mathbf{F}^{\mathrm{T}} \begin{Bmatrix} r_1 \\ r_2 \\ \vdots \\ r_n \end{Bmatrix} = \begin{Bmatrix} 0 \\ 0 \\ \vdots \\ 0 \end{Bmatrix} \tag{4-18}$$

where

$$\mathbf{F}^{\mathrm{T}} = \begin{bmatrix} f_1(x_1) & f_1(x_2) & \cdots & f_1(x_n) \\ f_2(x_1) & f_2(x_2) & \cdots & f_2(x_n) \\ \cdots & \cdots & \cdots & \cdots \\ f_m(x_1) & f_m(x_2) & \cdots & f_m(x_n) \end{bmatrix} \tag{4-19}$$

and from Eq. 4-12

$$\begin{Bmatrix} r_1 \\ r_2 \\ \vdots \\ r_n \end{Bmatrix} = \begin{Bmatrix} \sum_{k=1}^{m} c_k f_k(x_1) \\ \sum_{k=1}^{m} c_k f_k(x_2) \\ \vdots \\ \sum_{k=1}^{m} c_k f_k(x_n) \end{Bmatrix} - \begin{Bmatrix} y_1 \\ y_2 \\ \vdots \\ y_n \end{Bmatrix} \tag{4-20}$$

From Eq. 4-20, it is apparent that Eq. 4-18 may be written in the form

$$\mathbf{F}^{\mathrm{T}}\begin{Bmatrix} \sum_{k=1}^{m} c_k f_k(x_1) \\ \sum_{k=1}^{m} c_k f_k(x_2) \\ \vdots \\ \sum_{k=1}^{m} c_k f_k(x_n) \end{Bmatrix} = \mathbf{F}^{\mathrm{T}}\begin{Bmatrix} y_1 \\ y_2 \\ \vdots \\ y_n \end{Bmatrix} \tag{4-21}$$

Finally, we note that

$$\begin{Bmatrix} \sum_{k=1}^{m} c_k f_k(x_1) \\ \sum_{k=1}^{m} c_k f_k(x_2) \\ \vdots \\ \sum_{k=1}^{m} c_k f_k(x_n) \end{Bmatrix} = \mathbf{F}\begin{Bmatrix} c_1 \\ c_2 \\ \vdots \\ c_m \end{Bmatrix} \tag{4-22}$$

Thus Eq. 4-21 may be written in the form

$$\mathbf{F}^{\mathrm{T}}\mathbf{F}\begin{Bmatrix} c_1 \\ c_2 \\ \vdots \\ c_m \end{Bmatrix} = \mathbf{F}^{\mathrm{T}}\begin{Bmatrix} y_1 \\ y_2 \\ \vdots \\ y_n \end{Bmatrix}$$

or more simply as

$$\mathbf{F}^{\mathrm{T}}\mathbf{F}\{c\} = \mathbf{F}^{\mathrm{T}}\{y\} \tag{4-23}$$

The use of Eq. 4-23 facilitates the formulation of a system of m linear algebraic equations for the least-squares curve-fitting criterion since the **F** matrix is readily generated from the selected $f(x)$ function and the known data values for x.

EXAMPLE 4-2

Use Eq. 4-23 to formulate the algebraic equations determined for the solution of Ex. 4-1.

Solution. From Ex. 4-1, we have $y = c_1 + c_2 x^2$ and the known data points (1.0, 3.8), (2.5, 15.0), (3.5, 26.0), and (4.0, 33.0) shown in Fig. 4-3. From the given $f(x)$ function, we note that

$$f_1(x) = 1 \quad \text{and} \quad f_2(x) = x^2$$

Thus Eq. 4-16 with the four known x values gives

$$\mathbf{F} = \begin{bmatrix} 1 & 1 \\ 1 & (2.5)^2 \\ 1 & (3.5)^2 \\ 1 & (4.0)^2 \end{bmatrix} = \begin{bmatrix} 1 & 1 \\ 1 & 6.25 \\ 1 & 12.25 \\ 1 & 16.0 \end{bmatrix}$$

Interchanging the rows and columns of the **F** matrix gives

$$\mathbf{F}^{\mathrm{T}} = \begin{bmatrix} 1 & 1 & 1 & 1 \\ 1 & 6.25 & 12.25 & 16.0 \end{bmatrix}$$

Hence Eq. 4-23 becomes

$$\begin{bmatrix} 1 & 1 & 1 & 1 \\ 1 & 6.25 & 12.25 & 16 \end{bmatrix} \begin{bmatrix} 1 & 1 \\ 1 & 6.25 \\ 1 & 12.25 \\ 1 & 16.0 \end{bmatrix} \begin{Bmatrix} c_1 \\ c_2 \end{Bmatrix} = \begin{bmatrix} 1 & 1 & 1 & 1 \\ 1 & 6.25 & 12.25 & 16 \end{bmatrix} \begin{Bmatrix} 3.8 \\ 15 \\ 26 \\ 33 \end{Bmatrix}$$

and the matrix multiplication yields

$$\begin{Bmatrix} 4c_1 + 35.5c_2 \\ 35.5c_1 + 446c_2 \end{Bmatrix} = \begin{Bmatrix} 77.8 \\ 944 \end{Bmatrix}$$

which agrees with the two simultaneous equations determined in Ex. 4-1.

4-4 / WEIGHTING FOR LEAST SQUARES

Sometimes certain data values obtained in an experiment may be more accurate than other data values because of the range of variables being measured relative to the full-scale accuracy of the equipment used. In such cases it may be desirable to give more "weight" to those data values which are considered to be the most accurately measured quantities. The weighting of certain data points may also be desirable to bring a selected function closer to certain data points. For example, the function $y = c_1 + c_2x^2$ can be brought into closer agreement with the data point (1.0, 3.8) than is shown in Fig. 4-3 by weighting the data point (1.0, 3.8). Usually the weighting value selected depends upon one's experience and judgment.

One of the simplest approaches to the weighting of a given data point is simply to consider the multiple use of the data point as if it were obtained from several different experimental runs. For example, a weighting factor of 2 can be given to the data point (1.0, 3.8) of Ex. 4-2 by considering that the known data points are (1.0, 3.8), (1.0, 3.8), (2.5, 15.0), (3.5, 26.0), and (4.0, 33.0).

The multiple use of data points for weighting purposes corresponds to minimizing the quantity

$$w_1r_1^2 + w_2r_2^2 + \cdots + w_nr_n^2 = \text{minimum} \tag{4-24}$$

where w_1, w_2, and so on, are weighting factors which are positive real numbers. Proceeding as in Sec. 4-3, we have

$$\frac{\partial}{\partial c_k}\sum_{i=1}^{n} w_ir_i^2 = \sum_{i=1}^{n} w_ir_i\frac{\partial r_i}{\partial c_k} = 0 \qquad k = 1, 2, \ldots, m \tag{4-25}$$

and

$$[w_1r_1 \quad w_2r_2 \quad \cdots \quad w_nr_n]\mathbf{F} = [0 \quad 0 \quad \cdots \quad 0] \tag{4-26}$$

Taking the transpose of Eq. 4-26 gives

$$\mathbf{F}^{\mathrm{T}}\begin{Bmatrix} w_1r_1 \\ w_2r_2 \\ \vdots \\ w_nr_n \end{Bmatrix} = \begin{Bmatrix} 0 \\ 0 \\ \vdots \\ 0 \end{Bmatrix}$$

which may be written in the form

$$\mathbf{F}^{\mathrm{T}}\left[\diagdown w_{i\diagdown}\right]\mathbf{F}\{c\} = \mathbf{F}^{\mathrm{T}}\left[\diagdown w_{i\diagdown}\right]\{y\} \tag{4-27}$$

where $\left[\diagdown w_{i\diagdown}\right]$ is a diagonal matrix in which the weighting factors w_1, w_2, and so on, appear on the diagonal with zeros elsewhere.

From the previous discussion, it should be apparent that weighting of data can be done by the use of either Eq. 4-23 or Eq. 4-27. In using Eq. 4-23, specific data points are repeated the desired number of times in

$\mathbf{F}$, $\mathbf{F}^T$, and $\{y\}$, to simulate additional known data. To illustrate the use of Eqs. 4-23 and 4-27, let us consider a weighting factor of 2 ($w_1 = 2$) for point (1.0, 3.8) of Ex. 4-2. Using the data point (1.0, 3.8) twice, we obtain for substitution into Eq. 4-23

$$\mathbf{F} = \begin{bmatrix} 1 & 1 \\ 1 & 1 \\ 1 & 6.25 \\ 1 & 12.25 \\ 1 & 16.0 \end{bmatrix}$$

$$\mathbf{F}^T = \begin{bmatrix} 1 & 1 & 1 & 1 & 1 \\ 1 & 1 & 6.25 & 12.25 & 16.0 \end{bmatrix}$$

and

$$\{y\} = \begin{Bmatrix} 3.8 \\ 3.8 \\ 15 \\ 26 \\ 33 \end{Bmatrix}$$

If Eq. 4-27 is used, then

$$\left[\diagdown w_i \diagdown\right] = \begin{bmatrix} 2 & 0 & 0 & 0 \\ 0 & 1 & 0 & 0 \\ 0 & 0 & 1 & 0 \\ 0 & 0 & 0 & 1 \end{bmatrix}, \qquad \{y\} = \begin{Bmatrix} 3.8 \\ 15 \\ 26 \\ 33 \end{Bmatrix}$$

and

$$\mathbf{F} = \begin{bmatrix} 1 & 1 \\ 1 & 6.25 \\ 1 & 12.25 \\ 1 & 16.0 \end{bmatrix}$$

which is the same as previously used in Ex. 4-2.

The repeated use of data points in Eq. 4-23 gives the same result as Eq. 4-27 in which weighting factors (whole numbers) are introduced by way of the $\left[\diagdown w_i \diagdown\right]$ matrix. It should be emphasized that the weighting factors w_i need not be whole integers when using Eq. 4-27, but may be any positive number such as 1.3, 2.5, and so on.

4-5 / EXPONENTIAL FUNCTIONS

Exponential functions quite frequently appear in the mathematical analysis of physical systems, so these functions may also be pertinent for fitting curves to experimental data. However, an undetermined constant will usually appear in the exponent, and as a result the function $f(x)$ for curve fitting will not be a linear form as discussed in Sec. 4-2. The nonlinear form of $f(x)$ can, however, be transformed to a linear form by the use of logarithms so that the least-squares method can be accomplished by the same procedure as described in Sec. 4-3.

Let us first consider the exponential function shown in Fig. 4-4

$$y = k_1 e^{-k_2 x} \tag{4-28}$$

in which k_1 and k_2 are *positive real numbers*. Taking the logarithm of both sides of Eq. 4-28 gives

$$\ln y = \ln k_1 - k_2 x \tag{4-29}$$

Letting $z = \ln y$, $c_1 = \ln k_1$, and $c_2 = -k_2$, Eq. 4-29 becomes

$$z = c_1 + c_2 x \tag{4-30}$$

which is the equation of a straight line in which the constants, c_1 and c_2, may be found by the procedure discussed in Sec. 4-3. Actually Eq. 4-30

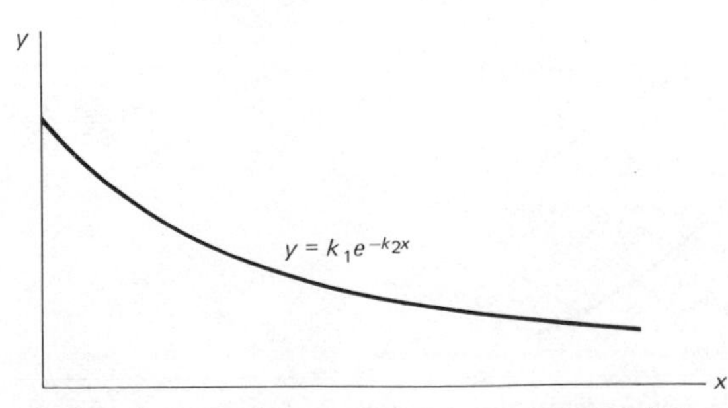

Fig. 4-4 / *General form of the function* $y = k_1 e^{-k_2 x}$.

tells us that Eq. 4-29 will plot as a straight line on semilog paper when the values of y are plotted as ordinates on the log scale. It should be emphasized that in the least-squares fit of Eq. 4-30, we are minimizing $\sum r_i^2$ in which $r_i = c_1 + c_2 x_i - \ln y_i$. This means that the error obtained between the data and the function $z = c_1 + c_2 x$ will be in the $\ln y_i$ and not in y_i.

Another exponential function which is frequently encountered in engineering is shown in Fig. 4-5. It is

$$y = k_1 x e^{-k_2 x} \tag{4-31}$$

in which k_1 and k_2 are positive real numbers. Taking the logarithm of both sides of Eq. 4-31 gives

$$\ln y = \ln (k_1 x) - k_2 x \tag{4-32}$$

which is a nonlinear form. To obtain a linear form, Eq. 4-32 may be written in the form

$$\ln \left(\frac{y}{x}\right) = \ln k_1 - k_2 x \tag{4-33}$$

Letting $z = \ln (y/x)$, $c_1 = \ln k_1$, and $c_2 = -k_2$, Eq. 4-33 becomes

$$z = c_1 + c_2 x \tag{4-34}$$

which is a straight-line plot on semilog paper when the values of y_i/x_i are plotted as ordinates on the log scale. In this case the error is in $\ln (y_i/x_i)$ since $\sum r_i^2$ is being minimized, where $r_i = c_1 + c_2 x_i - \ln (y_i/x_i)$.

We will now consider two examples to illustrate the use of the functions shown in Figs. 4-4 and 4-5.

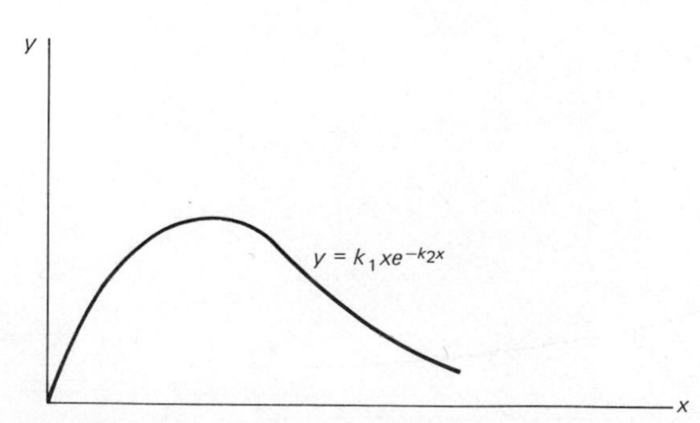

Fig. 4-5 / *General form of the function $y = k_1 x e^{-k_2 x}$.*

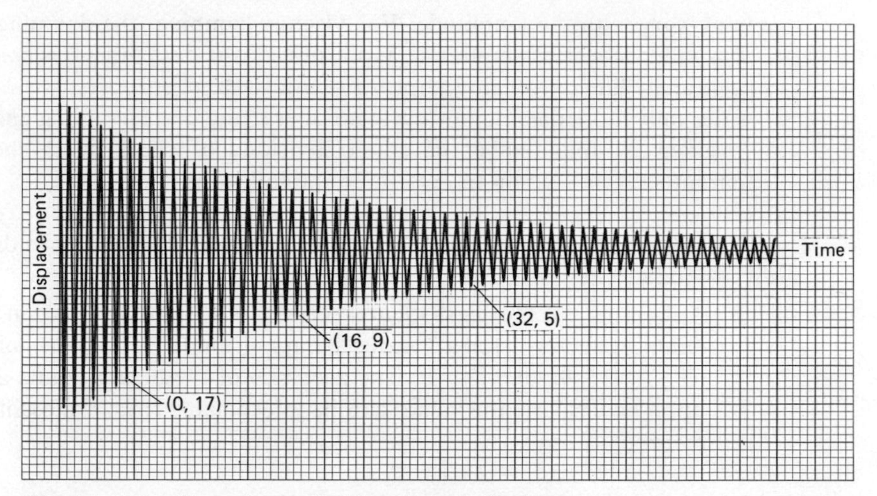

Fig. 4-6 / *Oscillograph record of free-damped vibrations.*

EXAMPLE 4-3

A photograph of an actual oscillograph record of the free-damped vibrations for a small structure is shown in Fig. 4-6. From vibration theory it is known that for viscous damping (damping proportional to velocity) the *envelope* of such a vibration (the curve through the peaks as shown in Fig. 4-7) is an exponential function of the form

$$y = k_1 e^{-2\pi\xi x} \tag{4-35}$$

where x is the cycle number, y is the corresponding amplitude, and ξ is a damping factor in which $\xi_c = 1$ for critical damping.

The free-damped vibrations of structural systems quite often appear to be that of a viscously damped system when actually the damping factor ξ

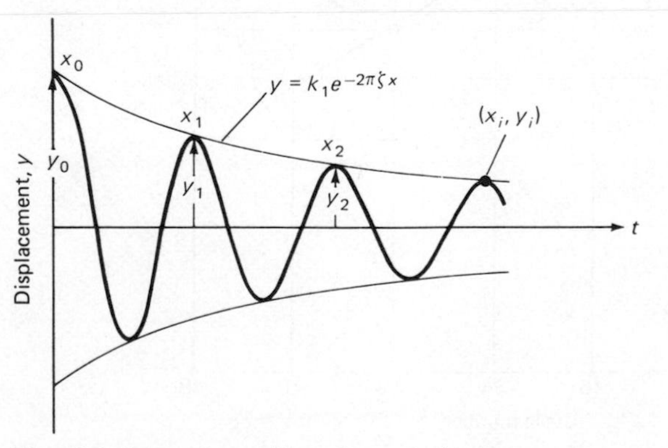

Fig.. 4-7 / *Envelope of free-damped vibrations.*

may vary with the amplitude. For viscous damping, the damping factor ξ is independent of amplitude, and a plot of the data (x_i, y_i) is essentially a straight line on semilog paper. For illustration purposes, we will consider only those data points (cycle number x and corresponding amplitude y given by the number of small divisions of chart lines) shown in parentheses in Fig. 4-6 to determine the following:

1. Whether the free-damped vibrations correspond to a viscous damping model.
2. A best fit by the least-squares criterion for the evaluation of the damping factor ξ for a viscous damping model. The data points are (0, 17), (16, 9), and (32, 5). The accuracy of these data points depends upon how closely the values can be interpolated from the oscillograph record.

Solution. To answer question 1, the three data values are plotted on semilog paper as shown in Fig. 4-8. Since the data points appear to be fairly close to a straight-line plot, the damping factor ξ is, for all practical purposes, independent of amplitude, which means that the system damping can be described by a viscous damping model. In this example we know the type of equation to use to fit the data from knowing Eq. 4-35. Our

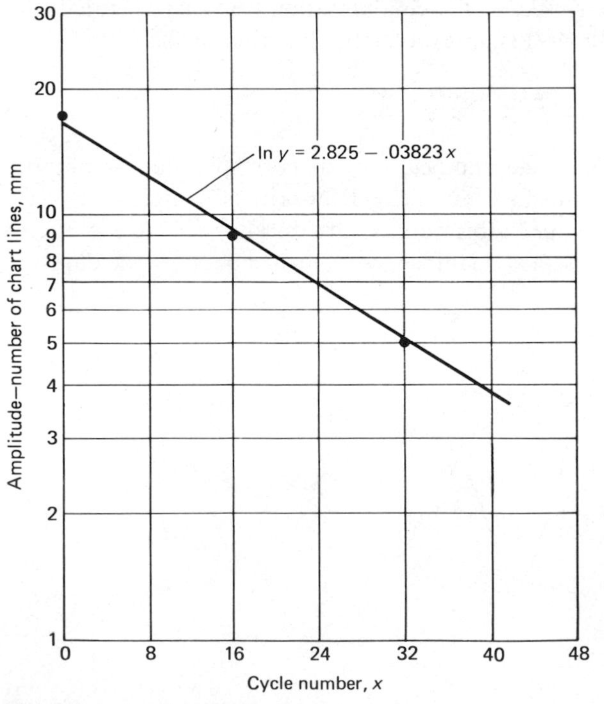

Fig. 4-8 / *Semilog plot for Ex. 4-3.*

purpose then is to determine values of k_1 and ξ to best fit the equation to the data.

For a least-squares fit, we first take the logarithm of Eq. 4-35. Thus

$$\ln y = \ln k_1 - 2\pi\xi x \tag{4-36}$$

Letting $z = \ln y$, $c_1 = \ln k_1$, and $c_2 = -2\pi\xi$, Eq. 4-36 becomes

$$z = c_1 + c_2 x \tag{4-37}$$

To formulate the two equations for the unknown constants, c_1 and c_2, we have from Eq. 4-23

$$\mathbf{F}^{\mathrm{T}}\mathbf{F}\{c\} = \mathbf{F}^{\mathrm{T}}\{z\} \tag{4-38}$$

Noting from Eq. 4-37 that $f_1(x) = 1$ and that $f_2(x) = x$, Eq. 4-38 with the given data values yields

$$\begin{bmatrix} 1 & 1 & 1 \\ 0 & 16 & 32 \end{bmatrix} \begin{bmatrix} 1 & 0 \\ 1 & 16 \\ 1 & 32 \end{bmatrix} \begin{Bmatrix} c_1 \\ c_2 \end{Bmatrix} = \begin{bmatrix} 1 & 1 & 1 \\ 0 & 16 & 32 \end{bmatrix} \begin{Bmatrix} \ln 17 \\ \ln 9 \\ \ln 5 \end{Bmatrix} \tag{4-39}$$

which yields

$$\left.\begin{aligned} 3c_1 + 48c_2 &= 6.6398 \\ 48c_1 + 1280c_2 &= 86.6576 \end{aligned}\right\} \tag{4-40}$$

The simultaneous solution of Eq. 4-40 gives

$$c_1 = \ln k_1 = 2.825$$

$$k_1 = e^{2.825} = 16.86$$

$$c_2 = -2\pi\xi = -0.03823$$

$$\xi = 0.00608$$

Hence the least-squares fit gives

$$y = 16.86e^{-2\pi(0.00608)x}$$

or

$$\ln y = 2.825 - 0.03823x$$

The straight line from the least-squares fit is shown in Fig. 4-8 in which the initial value of y is 16.86.

EXAMPLE 4-4

A plot of the stress-strain data obtained from the compression test of a 6×12-in. concrete cylinder is shown in Fig. 4-9. The general form of the data plot suggests that the exponential function shown in Fig. 4-5 may be a good function to describe the nonlinear stress-strain curve of concrete. Therefore, we will obtain a least-squares fit for the function

$$\sigma = k_1 \varepsilon e^{-k_2 \varepsilon} \tag{4-41}$$

where σ is the stress in psi and ε is the strain in in./in. To simplify the computation, we will use only the five data points shown as solid dark circles in Fig. 4-9. These data values are also shown in tabular form in Fig. 4-9.

Solution. Taking the logarithm of Eq. 4-41 in terms of $\ln(\sigma/\varepsilon)$ to obtain a linear form as previously discussed, we obtain

$$\ln \frac{\sigma}{\varepsilon} = \ln k_1 - k_2 \varepsilon \tag{4-42}$$

Letting $z = \ln(\sigma/\varepsilon)$, $c_1 = \ln k_1$, and $c_2 = -k_2$, Eq. 4-42 becomes

$$z = c_1 + c_2 \varepsilon \tag{4-43}$$

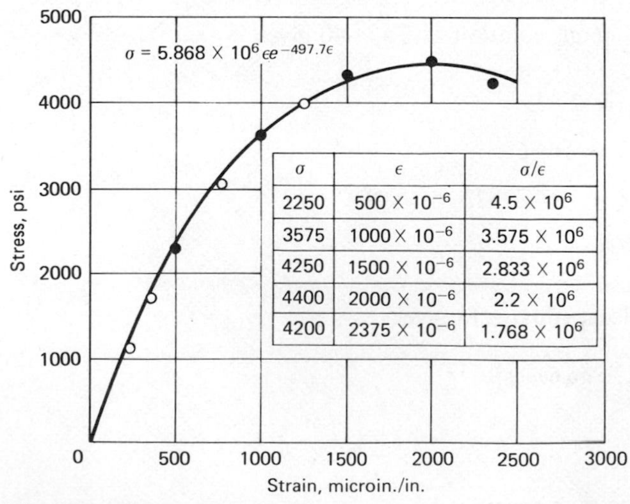

σ	ϵ	σ/ϵ
2250	500×10^{-6}	4.5×10^6
3575	1000×10^{-6}	3.575×10^6
4250	1500×10^{-6}	2.833×10^6
4400	2000×10^{-6}	2.2×10^6
4200	2375×10^{-6}	1.768×10^6

Fig. 4-9 / *Stress-strain plot for concrete.*

The two simultaneous equations in terms of the unknown constants, c_1 and c_2, are obtained from Eq. 4-23 which is repeated here and is

$$\mathbf{F}^T\mathbf{F}\{c\} = \mathbf{F}^T\{z\} \tag{4-44}$$

From the tabulated data of Fig. 4-9, and noting that $f_1(\varepsilon) = 1$ and that $f_2(\varepsilon) = \varepsilon$, we obtain the following for substitution into Eq. 4-44:

$$\mathbf{F} = \begin{bmatrix} 1 & 500 \times 10^{-6} \\ 1 & 1000 \times 10^{-6} \\ 1 & 1500 \times 10^{-6} \\ 1 & 2000 \times 10^{-6} \\ 1 & 2375 \times 10^{-6} \end{bmatrix}$$

$$\mathbf{F}^T = \begin{bmatrix} 1 & 1 & 1 & 1 & 1 \\ 500 \times 10^{-6} & 1000 \times 10^{-6} & 1500 \times 10^{-6} & 2000 \times 10^{-6} & 2375 \times 10^{-6} \end{bmatrix}$$

$$\{c\} = \begin{Bmatrix} c_1 \\ c_2 \end{Bmatrix}$$

and

$$\{z\} = \begin{Bmatrix} \ln 4.5 \times 10^6 \\ \ln 3.575 \times 10^6 \\ \ln 2.833 \times 10^6 \\ \ln 2.2 \times 10^6 \\ \ln 1.768 \times 10^6 \end{Bmatrix}$$

Carrying out the matrix multiplication of Eq. 4-44 gives

$$\begin{Bmatrix} 5c_1 + 73.75 \times 10^{-4}c_2 \\ 73.75 \times 10^{-4}c_1 + 1314.1 \times 10^{-8}c_2 \end{Bmatrix} = \begin{Bmatrix} 74.255 \\ 0.1084 \end{Bmatrix} \tag{4-45}$$

The solution of the simultaneous equations of Eq. 4-45 yields

$$c_1 = \ln k_1 = 15.585$$

$$k_1 = e^{15.585} = 5.868 \times 10^6$$

$$c_2 = -k_2 = -497.7$$

Using these values, the least-squares fit gives us the equation

$$\sigma = 5.868 \times 10^6 \varepsilon e^{-497.7\varepsilon}$$

A plot of this function in Fig. 4-9 shows that the function describes the stress-strain data quite well. For this reason, the exponential function of Eq. 4-41 is quite often used to analyze the flexure stresses in reinforced concrete beams.

4-6 / CURVE FITTING WITH FOURIER SERIES

The application of the least-squares criterion for the "best fit" of a Fourier series, or trigonometric series, to experimental data requires that one have a basic understanding of the representation of a function, $y = f(x)$, by a Fourier series. Therefore, let us first briefly review the basic concepts of Fourier series, and then illustrate the use of the matrix equation, Eq. 4-23, for obtaining the set of simultaneous linear algebraic equations involving the undetermined Fourier coefficients.

Any single-valued function $f(x)$ that has a finite number of discontinuities may be represented by a Fourier series in the interval of $(-\pi, \pi)$, or in the interval of $(-l, l)$. For a function $f(x)$ defined in the interval of $(-l, l)$, the trigonometric form of a Fourier series is defined as

$$f(x) = \frac{a_0}{2} + \sum_{n=1}^{\infty} a_n \cos \frac{n\pi x}{l} + \sum_{n=1}^{\infty} b_n \sin \frac{n\pi x}{l} \tag{4-46}$$

where a_n and b_n are the Fourier coefficients. For example, the function

$$f(x) = 0 \qquad 0 \leq x \leq l$$

$$f(x) = -x \qquad -l \leq x \leq l$$

as represented by the series of Eq. 4-46 is periodic and repeats itself as shown in Fig. 4-10. It is noted that the period is $2l$, which corresponds to 2π, and that $f(x + 2l) = f(x)$.

For a given function $f(x)$, the Fourier coefficients, a_n and b_n, may be

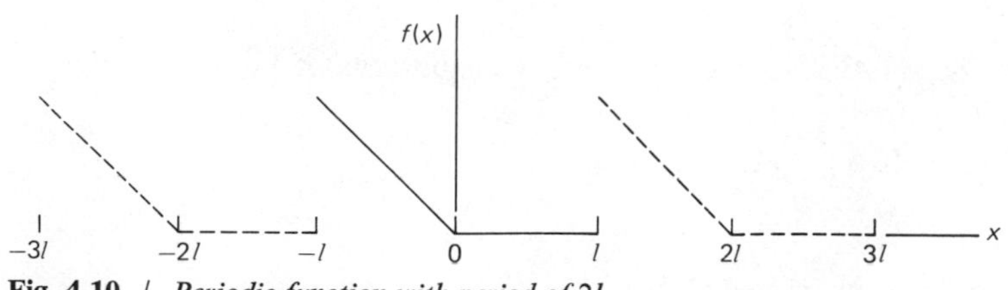

Fig. 4-10 / *Periodic function with period of 2l.*

determined from the following integral equations which are stated here without proof:

$$a_n = \frac{1}{l}\int_{-l}^{l} f(x)\cos\frac{n\pi x}{l}\,dx \qquad n = 0, 1, 2, 3, \ldots \tag{4-47}$$

$$b_n = \frac{1}{l}\int_{-l}^{l} f(x)\sin\frac{n\pi x}{l}\,dx \qquad n = 1, 2, 3, \ldots \tag{4-48}$$

In many instances it is desirable to use a *half-range* series over the interval $(0, l)$. If $f(-x) = -f(x)$, the function is an *odd function,* and the cosine terms of Eq. 4-46 vanish because each of the cosine terms is an *even function.* Conversely, if $f(x) = f(-x)$, the function is an *even function,* and the sine terms of Eq. 4-46 vanish because each of the sine terms is an *odd function.* Thus

$$f(x) = \frac{a_0}{2} + \sum_{n=1}^{\infty} a_n \cos\frac{n\pi x}{l} \qquad \text{(even)} \tag{4-49}$$

$$f(x) = \sum_{n=1}^{\infty} b_n \sin\frac{n\pi x}{l} \qquad \text{(odd)} \tag{4-50}$$

Consider the function $f(x) = -x + 1$ defined in the interval $(0, l)$ where $l = \frac{1}{2}$. The *Fourier cosine series* represents the *even* function as shown in Fig. 4-11(a), and the *Fourier sine series* represents the *odd* function as shown in Fig. 4-11(b). In either case, the series (cosine or sine) will converge to the same value in the interval $(0, l)$, except at $x = 0$ and $x = l$. For the even function, the cosine series converges to 1 at $x = 0$ and to $\frac{1}{2}$ at $x = \frac{1}{2}$. However, the odd function is discontinuous at $x = 0$ and $x = l$, and the sine series converges to the mean value of the function

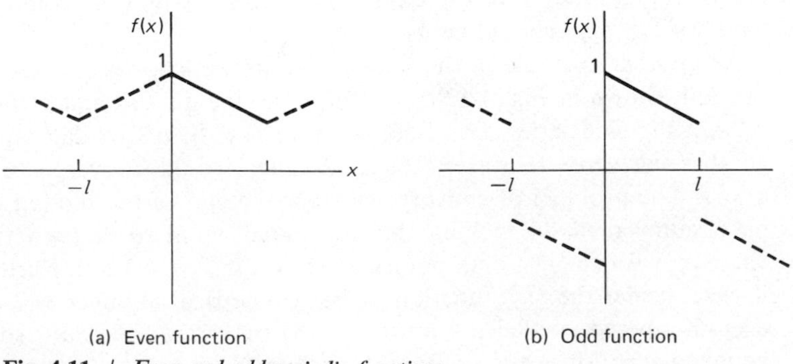

Fig. 4-11 / *Even and odd periodic functions.*

at the points of discontinuity. That is, the sine series will converge to zero at $x = 0$ and $x = l$. For half-range expansions, it is easily shown that

$$\left.\begin{aligned} a_n &= \frac{2}{l}\int_0^l f(x)\cos\frac{n\pi x}{l}\,dx \qquad n = 0, 1, 2, 3, \ldots \\ b_n &= 0, \end{aligned}\right\}\ \text{(even)} \tag{4-51}$$

$$\left.\begin{aligned} b_n &= \frac{2}{l}\int_0^l f(x)\sin\frac{n\pi x}{l}\,dx \qquad n = 1, 2, 3, \ldots \\ a_n &= 0 \end{aligned}\right\}\ \text{(odd)} \tag{4-52}$$

Least-Squares Procedure

In the least-squares procedure for the "best fit" of a trigonometric series to experimental data, the coefficients are not determined by the integration process indicated in Eqs. 4-47, 4-48, 4-51, and 4-52. Instead the coefficients for the first m terms of a series are determined by the procedure explained in Sec. 4-3. The number of terms m for a good representation of the data may not be initially known. However, a trial solution with the first three or four terms will usually indicate whether more terms are necessary. In general, the representation of data by a series improves with an increase in the number of terms of the series.

Whether we select the first m terms of Eq. 4-46, or the first m terms of Eq. 4-49, or the first m terms of Eq. 4-50, depends to some extent on the general form of the plotted data and the convergence required for a reasonable fit near the ends at $x = 0$ and $x = l$. For example, the plot shown in Fig. 4-12(a) clearly shows that the sine series, Eq. 4-50, is not suitable, since each of the sine terms vanishes at $x = 0$ and $x = l$. That is, the sine series for any number of terms converges to zero at $x = 0$ and $x = l$, which would result in a poor fit at the ends. The cosine series, Eq. 4-49, might be suitable, since the cosine terms do not vanish at $x = 0$ and $x = l$. If a trial solution of, say, three terms of a cosine series shows a poor representation of the data, then a combination of cosine and sine terms of Eq. 4-46 should be used.

As another example in the selection of an appropriate series, consider the data shown in Fig. 4-12(b). In this case, the data at first glance might suggest the sine series, Eq. 4-50, since an $f(x)$ fit of the data appears to go through zero. However, the sine series would converge to zero at $x = l$. The problem of convergence near the end of the plotted data can be circumvented by making the half-range equal to $2l$ (two times the value of x for the last data point), as shown in Fig. 4-12(c). Furthermore, if we consider the $f(x)$ function to be symmetrical about $x = l$, then the even-numbered coefficients must vanish; that is, b_2, b_4, and so on, are equal to zero, since the $\sin 2\pi x/2l$, $\sin 4\pi x/2l$, and so on, are asymmetric with respect to $x = l$.

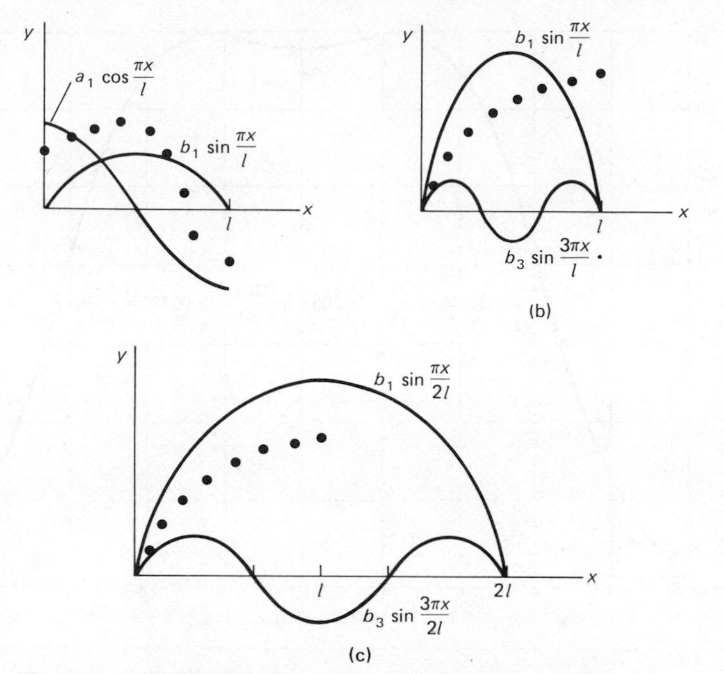

Fig. 4-12 / *Typical data plots with trigonometric function behavior.*

EXAMPLE 4-5

The following data represent the growth in height of a boy from 11 to 21 years of age.[1] Determine a "best fit" of a trigonometric series. To simplify the computations so that they may be done by an electronic pocket calculator, use the data denoted by the asterisks and a three-term trigonometric series.

Elapsed Time (yr)	Growth (cm)
0 (age 11)	0
0.8	0.74
*1.4	2.25
2.0	5.25
2.4	8.25
*3.2	15.00
4.0	21.38
*4.8	26.25
5.4	28.88
6.0	30.60
7.0	32.25
*8.0	33.00
*10.0 (age 21)	35.00

[1] Jean Demining, "Application of the Gompertz Curve to the Observed Pattern of Growth in Length of 48 Individual Boys and Girls During Adolescent Cycle of Growth," *Human Biology*, Vol. 29, No. 1, Feb., 1957.

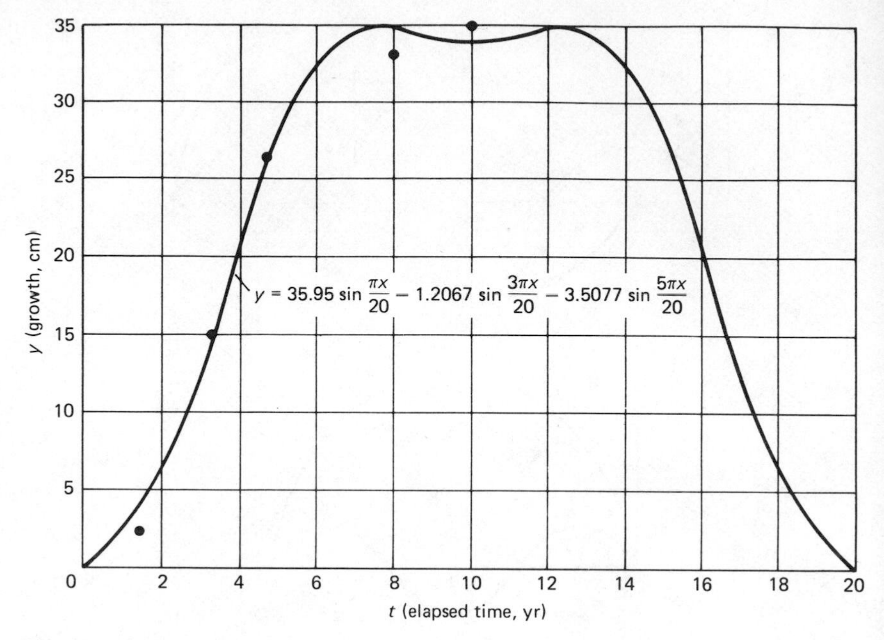

Fig. 4-13 / *Growth-time plot.*

Solution. The data (1.4, 2.25), (3.2, 15), (4.8, 26.25), (8.0, 33), and (10.0, 35) are first plotted as shown in Fig. 4-13. Based upon our previous discussion, we select the sine series with a half-range of $t = 20$ so that $f(t)$ is symmetrical about $t = 10$. Thus for the first three terms, we take

$$y = b_1 \sin \frac{\pi t}{20} + b_3 \sin \frac{3\pi t}{20} + b_5 \sin \frac{5\pi t}{20}$$

where y is the growth in centimeters and t is the elapsed time during growth in years.

For the generation of the $\mathbf{F}$ and $\mathbf{F}^T$ matrices of Eq. 4-23, we note that

$$f_1(t) = \sin \frac{\pi t}{20}$$

$$f_2(t) = \sin \frac{3\pi t}{20}$$

$$f_3(t) = \sin \frac{5\pi t}{20}$$

Calculating the values of these functions from the designated time data gives

$$\mathbf{F} = \begin{bmatrix} 0.218 & 0.613 & 0.891 \\ 0.482 & 0.998 & 0.588 \\ 0.685 & 0.771 & -0.588 \\ 0.951 & -0.588 & 0.00 \\ 1.00 & -1.00 & 1.00 \end{bmatrix} \tag{4-53}$$

and $\mathbf{F}^{\mathrm{T}}$ is

$$\mathbf{F}^{\mathrm{T}} = \begin{bmatrix} 0.218 & 0.482 & 0.685 & 0.951 & 1.00 \\ 0.613 & 0.998 & 0.771 & -0.588 & -1.00 \\ 0.891 & 0.588 & -0.588 & 0.00 & 1.00 \end{bmatrix} \tag{4-54}$$

We also note that

$$\{c\} = \begin{Bmatrix} b_1 \\ b_3 \\ b_5 \end{Bmatrix} \tag{4-55}$$

and that

$$\{y\} = \begin{Bmatrix} 2.25 \\ 15.00 \\ 26.25 \\ 33.00 \\ 35.00 \end{Bmatrix} \tag{4-56}$$

The matrix multiplication of Eqs. 4-53, 4-54, 4-55, and 4-56 in accordance with

$$\mathbf{F}^{\mathrm{T}}\mathbf{F} \begin{Bmatrix} b_1 \\ b_3 \\ b_5 \end{Bmatrix} = \mathbf{F}^{\mathrm{T}}\{y\}$$

yields

$$2.653b_1 - 0.416b_3 + 1.075b_5 = 92.08$$
$$-0.416b_1 + 3.312b_3 - 0.321b_5 = -17.816$$
$$1.075b_1 - 0.320b_3 + 2.484b_5 = 30.389$$

The simultaneous solution of these equations gives

$$b_1 = 35.94$$
$$b_3 = -1.2067$$
$$b_5 = -3.5077$$

Thus the best fit for a three-term sine series to the five data points used is

$$y = 35.94 \sin \frac{\pi t}{20} - 1.2067 \sin \frac{3\pi t}{20} - 3.5077 \sin \frac{5\pi t}{20}$$

A plot of this function in Fig. 4-13 shows that it is a reasonably good representation of the five data points used. In addition, it should be pointed out that by taking one or two more terms of the series and using all of the data points for computation of the unknown constants, a four- or five-term series would give a very good representation of the growth-time data.

4-7 / POLYNOMIALS

Polynomials are frequently used in curve fitting. In general, an $n - 1$ degree polynomial has the form

$$y = c_1 + c_2 x + c_3 x^2 + \cdots + c_n x^{n-1} \tag{4-57}$$

where

$$f_1(x) = 1$$
$$f_2(x) = x$$
$$\vdots \qquad \vdots$$
$$f_n(x) = x^{n-1}$$

The least-squares fit of an $n - 1$ degree polynomial to n data points will, of course, pass through all the given data points since all the residuals go to zero. However, the polynomial may fluctuate considerably between the data points, as shown in Fig. 4-14. To minimize such excessive deviations of a polynomial function from a smooth curve

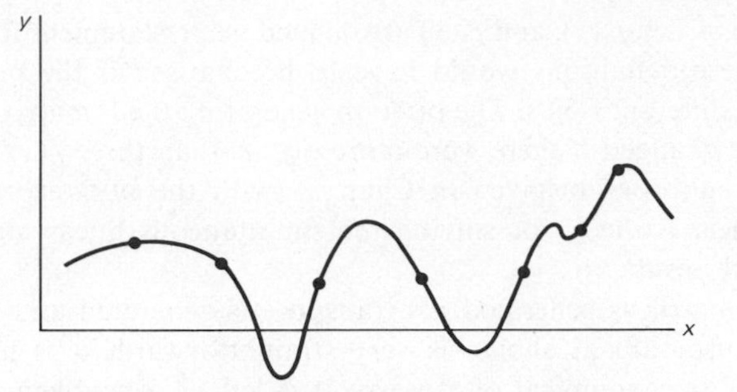

Fig. 4-14 / *Typical fluctuations of a polynomial between data points.*

between data points, the degree of the polynomial should be less than the number of data points used. Although no general rule exists for selecting the most suitable degree, a "rule of thumb" sometimes used is to select the degree of the polynomial somewhere in the neighborhood of $\frac{1}{2}$ to $\frac{3}{4}$ of the number of data points used. The smoothing effect between data points due to using a lower degree polynomial is, of course, generally accompanied by a curve having a poorer fit to the data points used in a least-squares fit.

4-8 / COMPUTER PROGRAM

A least-squares fit involving n data points and m unknown constants can be done fairly easily with today's electronic pocket calculators, if n is no greater than 6 or 7, and m is less than 4. However, the matrix multiplication of Eq. 4-23, or Eq. 4-27, involving a considerable number of n data values and/or the solution of more than three simultaneous equations, should be programmed for the digital computer.

EXAMPLE 4-6

The computer program in this example determines the coefficients c_1, c_2, and c_3 for the least-squares fit of the polynomial

$$y = c_1 + c_2x + c_3x^2$$

to the following data:

x	y
0.5	1.71
1.0	2.45
1.5	3.81
2.0	4.80
2.5	7.00
3.0	8.60

The functions $f_1(x)$, $f_2(x)$, and $f_3(x)$ are defined using statement function definitions. These definitions would have to be changed if the program were used with different $f(x)$'s. The DO loop generating the **F** matrix would also have to be changed if there were more or less than three $f(x)$'s used. The Cholesky subprogram given in Chap. 3 (with the DIMENSION statement changed) is "called" for solving the simultaneous linear algebraic equations which result.

After the **F** matrix is generated, its transpose is generated and stored. Although the program as shown is very straightforward, it is not one which is the most economical of storage. It is left as a problem for the student to revise the given program so that the matrix **F** is multiplied by its transpose without generating and storing the transpose (see Prob. 4-13 at the end of this chapter). The main program is as follows:

```
C LEAST SQUARES CURVE FITTING
      DIMENSION X(20),Y(20),F(20,6),FT(6,20),A(6,7),B(6),C(6)
C
C DEFINE THE FUNCTIONS
      F1(X)=1.
      F2(X)=X
      F3(X)=X*X
C
C READ IN THE NUMBER OF C'S AND NUMBER OF DATA POINTS
      READ(5,2)M,N
    2 FORMAT(2I2)
C
C READ X-Y VALUES OF DATA POINTS
      READ(5,3) (X(I),Y(I),I=1,N)
    3 FORMAT(8F10.0)
C
C GENERATE THE F MATRIX
      DO 4 I=1,N
      F(I,1)=F1(X(I))
      F(I,2)=F2(X(I))
    4 F(I,3)=F3(X(I))
C
C GENERATE THE TRANSPOSE OF THE F MATRIX
      DO 5 I=1,N
      DO 5 J=1,M
    5 FT(J,I)=F(I,J)
C
C DETERMINE COEFFICIENT MATRIX A OF SIMULTANEOUS EQUATION
                                                       SYSTEM
      CALL MATMPY(FT,F,A,M,N,M)
C
C DETERMINE THE COLUMN OF CONSTANTS FOR SIMULTANEOUS EQUATION
                                                       SYSTEM
      CALL MATMPY(FT,Y,B,M,N,1)
      DO 6 I=1,M
    6 A(I,M+1)=B(I)
C
```

```
C DETERMINE C VALUES BY SOLVING SIMULTANEOUS EQUATIONS USING
                                                  CHOLESKY METHOD
      MP1=M+1
      CALL CHLSKY(A,M,MP1,C)
C
C WRITE OUT THE C VALUES
      WRITE(6,7)
    7 FORMAT('1',4X,'C(1) THROUGH C(M)'/)
      WRITE(6,8) (I,C(I),I=1,M)
    8 FORMAT(' ',3X,'C(',I1,')=',E14.7)
      STOP
      END
```

The subroutine subprogram MATMPY for the required matrix multiplication is

```
      SUBROUTINE MATMPY(A,B,C,M,N,L)
C DETERMINES MATRIX C AS PRODUCT OF A AND B MATRICES
      DIMENSION A(6,20),B(20,6),C(6,7)
      DO 2 I=1,M
      DO 2 J=1,L
      C(I,J)=0.
      DO 2 K=1,N
    2 C(I,J)=C(I,J)+A(I,K)*B(K,J)
      RETURN
      END
```

The subroutine subprogram utilizing Cholesky's method for the solution of the simultaneous linear algebraic equations is

```
      SUBROUTINE CHLSKY (A,N,M,X)
      DIMENSION A(6,7),X(6)
C CALCULATE FIRST ROW OF UPPER UNIT TRIANGULAR MATRIX
      DO 3 J=2,M
    3 A(1,J)=A(1,J)/A(1,1)
C CALCULATE OTHER ELEMENTS OF U AND L MATRICES
      DO 8 I=2,N
      J=I
      DO 5 II=J,N
      SUM=0.
      JM1=J-1
      DO 4 K=1,JM1
    4 SUM=SUM+A(II,K)*A(K,J)
    5 A(II,J)=A(II,J)-SUM
      IP1=I+1
      DO 7 JJ=IP1,M
      SUM=0.
      IM1=I-1
      DO 6 K=1,IM1
    6 SUM=SUM+A(I,K)*A(K,JJ)
    7 A(I,JJ)=(A(I,JJ)-SUM)/A(I,I)
    8 CONTINUE
```

```
C SOLVE FOR X(I) BY BACK SUBSTITUTION
      X(N)=A(N,N+1)
      L=N-1
      DO 10 NN=1,L
      SUM=0.
      I=N-NN
      IP1=I+1
      DO 9 J=IP1,N
    9 SUM=SUM+A(I,J)*X(J)
   10 X(I)=A(I,M)-SUM
      RETURN
      END
```

The solution obtained is

```
 C(1) THROUGH C(M)
C(1)= 0.1096015E 01
C(2)= 0.8901110E 00
C(3)= 0.5471530E 00
```

Problems

4-1 In Ex. 4-2, the pure quadratic equation $y = c_1 + c_2x^2$ was used to obtain a least-squares fit to four data points. Obtain a least-squares fit to the four data points for

$$y = c_1 + c_2x + c_3x^2$$

and plot the result for comparison with the data points.

4-2 From the results of Ex. 4-2 and Prob. 4-1, determine which of the functions $y = c_1 + c_2x^2$ or $y = c_1 + c_2x + c_3x^2$ best represents the data.

4-3 Apply a weighting factor of 2 to the data point (1.0, 3.8) of Ex. 4-2 and obtain a least-squares fit using the function $y = c_1 + c_2x^2$. Compare the values of c_1 and c_2 obtained from the weighted data with the values of c_1 and c_2 obtained in Ex. 4-2.

4-4 Determine a least-squares fit of a trigonometric series for the stress-strain data of Ex. 4-4. From the plot of the data shown in Fig. 4-9, select the most suitable series from the following:

$$\sigma = \frac{a_0}{2} + a_1 \cos \frac{\pi\varepsilon}{2000} + a_2 \cos \frac{2\pi\varepsilon}{2000}$$

$$\sigma = b_1 \sin \frac{\pi\varepsilon}{4000} + b_2 \sin \frac{2\pi\varepsilon}{4000} + b_3 \sin \frac{3\pi\varepsilon}{4000}$$

$$\sigma = b_1 \sin \frac{\pi\varepsilon}{4000} + b_3 \sin \frac{3\pi\varepsilon}{4000} + b_5 \sin \frac{5\pi\varepsilon}{4000}$$

For the given stress-strain data, the ultimate stress and strain are 4400 psi and 2000×10^{-6} in./in., respectively. Plot the series function obtained for comparison with the stress-strain data.

4-5 Determine the coefficients b_1, b_3, and b_5 of the sine series used in Ex. 4-5 by using a weighting factor of 2 for the data points (8.0, 33.0) and (10.00, 35.0). Are the results obtained from the weighted data in any better agreement with the plotted data than the series obtained in Ex. 4-5?

4-6 The stress-strain data obtained from a compression test of a 6 × 12-in. concrete cylinder are as follows:

Stress (σ)	**Strain (ε)**
1025 psi	265×10^{-6} in./in.
1400	400
1710	500
2080	700
2425	950
2760	1360
3005	2080 (ultimate)
2850	2450
2675	2940

Plot the data and select a minimum of five data points, and obtain a least-squares fit for the function

$$\sigma = k_1 \varepsilon e^{-k_2 \varepsilon}$$

4-7 A functional relation between the mass density ρ of air and the altitude z above sea level is to be determined for the dynamic analysis of bodies moving within the earth's atmosphere. From the mass density-altitude data given, investigate the suitability of the following functions for representation of the data:

$$\rho = c_1 + \frac{c_2}{z}$$

$$\rho = c_1 e^{-c_2 z}$$

Altitude, z (ft)	**Mass Density, ρ (slugs/ft^3)**
0	0.002377
1,000	0.002308
2,000	0.002241
4,000	0.002117
6,000	0.001987
10,000	0.001755
15,000	0.001497
20,000	0.001267
30,000	0.000891
40,000	0.000587
50,000	0.000364
60,000	0.000224

4-8 The support of a slender hollow tube, which is immersed in water, is driven by an electromechanical shaker so that the motion of the support is $y = \lambda \sin \omega t$. SR-4 strain gages, mounted on the inside of the tube, monitor the strain at the base of the tube. A plot of actual strain, ε, and support amplitude, λ, data of one test, is shown with the tabulated data. Investigate the desirability of representing the data with the function

$$\varepsilon = k_1 \lambda^{c_2}$$

where k_1 and c_2 are constants to be determined from a least-squares fit.

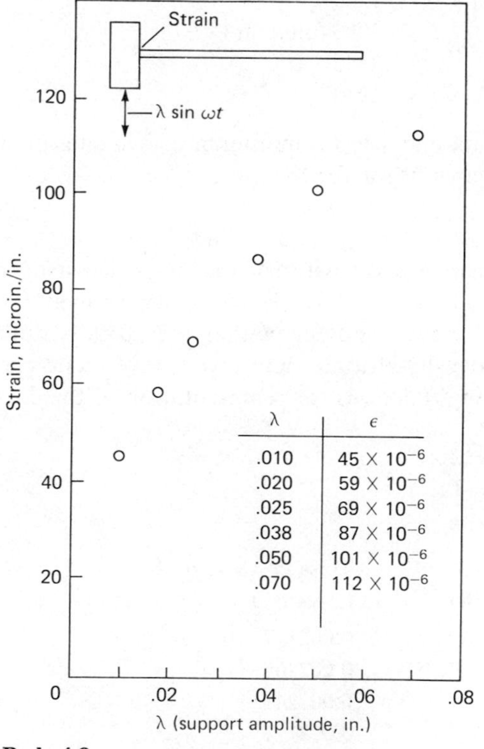

λ	ϵ
.010	45×10^{-6}
.020	59×10^{-6}
.025	69×10^{-6}
.038	87×10^{-6}
.050	101×10^{-6}
.070	112×10^{-6}

Prob. 4-8

4-9 The ultimate compressive strength σ of concrete is known to decrease with an increase in the water-cement ratio, w/c. The following compressive strengths were obtained by averaging the test results of three 6×12-in. concrete cylinders for each of the w/c ratios shown:

w/c (gal/sack)	Compressive Strength (psi)
$4\frac{1}{2}$	7000
5	6125
$5\frac{1}{2}$	5237
6	4665
$6\frac{1}{2}$	4123
7	3840
$7\frac{1}{2}$	3407
8	3070
$8\frac{1}{2}$	2580
9	2287

With $x = w/c$, compare a least-squares fit of the function

$$\sigma = c_1 e^{-c_2 x}$$

with the given data.

4-10 From the oscillograph record shown in the accompanying figure, determine the damping factor ξ as was done in Ex. 4-3.

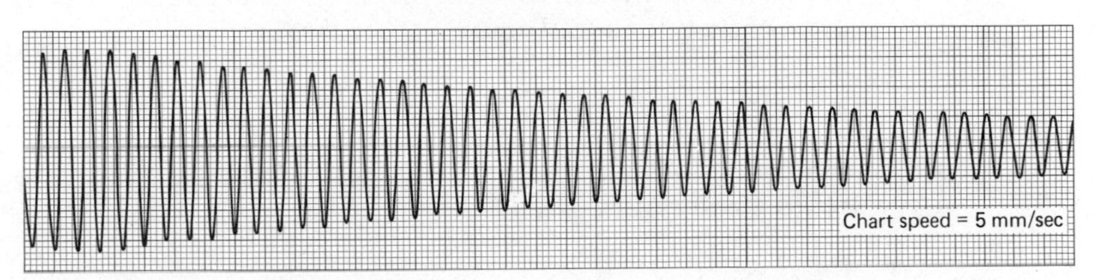

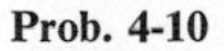
Prob. 4-10

4-11 Utilizing the computer program in Sec. 4-8 and the data points given in Ex. 4-5, determine the best fit for the growth-time curve using the polynomial

$$y = c_1 + c_2 t + c_3 t^2 + c_4 t^3$$

where $f_1(t) = 1, f_2(t) = t, f_3(t) = t^2, f_4(t) = t^3$. Compare the fit with the result obtained in Ex. 4-5.

4-12 Utilizing the computer program in Sec. 4-8 and the five data points given in Fig. 4-9, determine the best fit for the stress-strain curve using the polynomial

$$\sigma = c_1 + c_2 \varepsilon + c_3 \varepsilon^2 + c_4 \varepsilon^3$$

where $f_1(\varepsilon) = 1, f_2(\varepsilon) = \varepsilon, f_3(\varepsilon) = \varepsilon^2, f_4(\varepsilon) = \varepsilon^3$.

a. Compare the fit of the resulting polynomial to the given data points.
b. Determine several values between each of the data points used to check for excessive deviation in the polynomial between data points.

4-13 Revise the curve-fitting program given in Ex. 4-6 so that core space is saved by not storing the transpose of the **F** matrix. Generate the product of $\mathbf{F}^T\mathbf{F}$ without first generating $\mathbf{F}^T$ and storing it.

The following problems involve SI units.

4-14 Refer to Ex. 4-4 and convert the stress data shown in Fig. 4-9 to the SI system of units for a least-squares fit for the function

$$\sigma = k_1 \varepsilon e^{-k_2 \varepsilon}$$

Compare the values of k_1 and k_2 obtained from the least-squares fit using the SI system of units with the values found in Ex. 4-4.

4-15 Refer to Prob. 4-6 and convert the stress data given to the SI system of units for a least-squares fit for the function

$$\sigma = k_1 \varepsilon e^{-k_2 \varepsilon}$$

4-16 Work Prob. 4-7 converting the altitude, z, and mass density, ρ, data to the SI system of units.

5 Numerical Integration and Differentiation

5-1 / INTRODUCTION

Engineers are frequently confronted with the problem of differentiating or integrating functions which are defined in tabular or graphical form rather than as explicit functions. The interpretation of experimentally obtained data is a good example of this. A similar situation involves the integration of functions which have explicit forms that are difficult or impossible to integrate in terms of elementary functions. Graphical techniques, employing the construction of tangents to curves and the estimation of areas under curves, are commonly used in solving such problems, when great accuracy is not a prerequisite for the results.

However, there are occasions when a higher degree of accuracy is desired, and, for these, various numerical methods are available. It is with these techniques that this chapter will be concerned. Although both integration and differentiation formulas will be discussed, it should be pointed out that numerical differentiation is inherently much less accurate than numerical integration, and its application is generally avoided whenever possible. Nevertheless, it has been used successfully in certain applications; an example of such an application will be included later.

Numerical integration, or numerical *quadrature* as it is often called, consists essentially of finding a close approximation to the area under a curve of a function $f(x)$ which has been determined either from experimental data or from a mathematical expression. In this chapter the direct integration methods presented include only the trapezoid rule and Simpson's rule. Their derivations in this chapter will be by the most direct methods possible. For readers interested in a more general approach to the derivation of integration formulas using interpolating polynomials, and an alternate method of deriving the formulas for the derivatives, Appendix C may be studied preceding or along with the study of Chapter 5. In Appendix C it is shown that the trapezoid and Simpson formulas are special cases of what are known as Newton-Cotes quadrature formulas. The use of Newton-Cotes and other integration formulas of order higher than Simpson's rule is seldom necessary in most engineering applications. For those cases where extremely high accuracy is required, Romberg's extrapolation method presented in this chapter will usually be found satisfactory.

5-2 / INTEGRATION BY THE TRAPEZOID RULE

Consider the function $f(x)$, whose graph between $x = a$ and $x = b$ is shown in Fig. 5-1. A close approximation to the area under the curve is obtained by dividing it into n strips Δx in width and approximating the area of each strip by that of a trapezoid, as shown in the figure. Calling the ordinates y_i $(i = 1, 2, \ldots, n + 1)$, the areas of the trapezoids are

$$\begin{aligned} A_1 &= \Delta x\left(\frac{y_1 + y_2}{2}\right) \\ A_2 &= \Delta x\left(\frac{y_2 + y_3}{2}\right) \\ &\vdots \\ A_n &= \Delta x\left(\frac{y_n + y_{n+1}}{2}\right) \end{aligned} \tag{5-1}$$

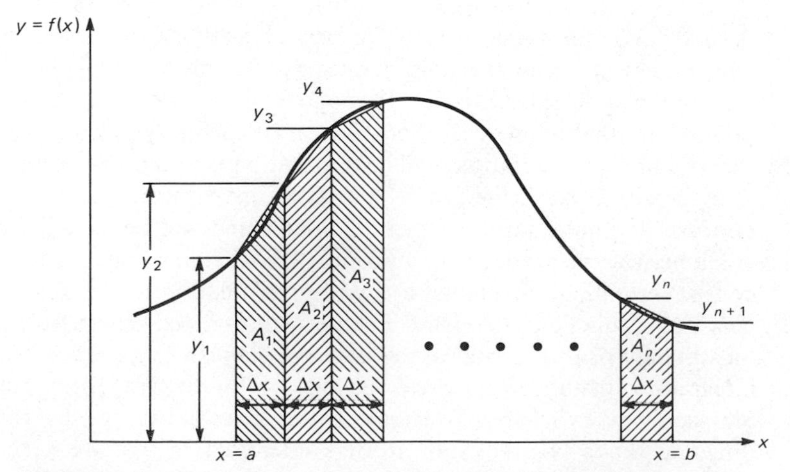

Fig. 5-1 / *Approximation, by the trapezoid rule, of the area under a curve.*

The total area lying between $x = a$ and $x = b$ is given by

$$A = \int_a^b f(x)\,dx \cong A_1 + A_2 + A_3 + \cdots + A_n$$

Substituting Eq. 5-1 into this expression yields

$$A = \int_a^b f(x)\,dx \cong \frac{\Delta x}{2}(y_1 + 2y_2 + 2y_3 + \cdots + 2y_n + y_{n+1}) \tag{5-2}$$

or, letting $\Delta x = h$,

$$A = \frac{h}{2}\left(y_1 + 2\sum_{i=2}^{n} y_i + y_{n+1}\right) \tag{5-3}$$

Equation 5-2 or 5-3 is referred to as the *trapezoid* rule.

A subroutine subprogram utilizing the trapezoid rule for integrating any FORTRAN-supplied function such as sin (x), or any function defined by a function subprogram, is given below. It integrates from a lower limit of XMIN to an upper limit of XMAX, using N strips. The subprogram uses the dummy name DUMMYF for the function being integrated. This is replaced by the actual function name appearing in the argument list of the CALL statement in the calling program. The function name, because it is used as the argument of a subprogram, must appear in an EXTERNAL statement in the calling program.

```
      SUBROUTINE TRAPZ(AREA,DUMMYF,XMIN,XMAX,N)
C THIS SUBPROGRAM CALCULATES AN INTEGRAL BY THE TRAPEZOID RULE
      H=(XMAX-XMIN)/N
      SUM=0.
      X=XMIN+H
      DO 4 I=2,N
      SUM=SUM+DUMMYF(X)
    4 X=X+H
      AREA=H/2*(DUMMYF(XMIN)+2.*SUM+DUMMYF(XMAX))
      RETURN
      END
```

Truncation Error Estimation in Integration by the Trapezoid Rule

If the function $f(x)$ can be expressed as a continuous mathematical function having continuous derivatives $f'(x)$ and $f''(x)$, the error resulting from approximating the true area in a strip under the curve of $f(x)$ between x_i and x_{i+1} by the area of a trapezoid can be shown to be

$$E_{T_i} = -\tfrac{1}{12}f''(\xi)(\Delta x)^3 \qquad x_i < \xi < x_{i+1}$$

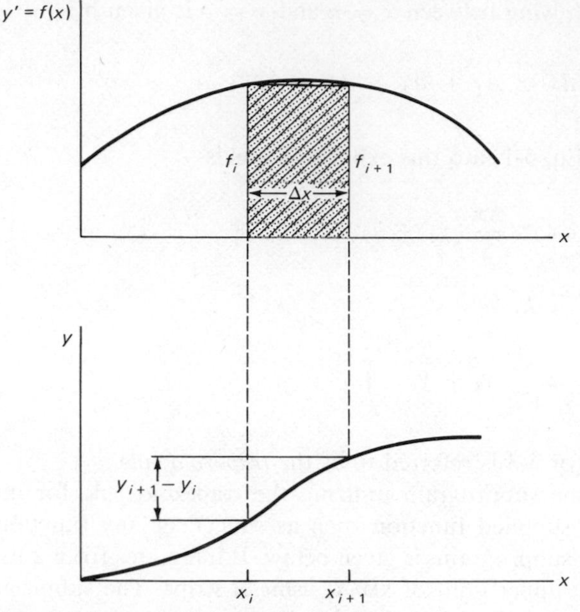

Fig. 5-2 / *Assumed curves for y′ and y.*

To show this, let us consider a graph of a function $y' = f(x)$ and one of the integral of y', both plotted against the independent variable x as in Fig. 5-2. The area in the crosshatched strip under the y' curve is given by $y_{i+1} - y_i$. We will express this area as that of the shaded trapezoidal area plus a remainder. The remainder represents the per-strip error in the trapezoid rule.

Let us first expand y_{i+1} in a Taylor series about x_i, letting $\Delta x = h$,

$$y_{i+1} = y_i + y_i'h + \frac{y_i''h^2}{2!} + \frac{y_i'''h^3}{3!} + \text{higher order terms} \tag{5-4}$$

Then the exact (crosshatched) area in the strip is

$$(y_{i+1} - y_i) = y_i'h + \frac{y_i''h^2}{2!} + \frac{y_i'''h^3}{3!} + \text{higher order terms} \tag{5-5}$$

or

$$(y_{i+1} - y_i) = f_ih + \frac{f_i'h^2}{2!} + \frac{f_i''h^3}{3!} + \text{higher order terms} \tag{5-6}$$

The Taylor series for f_{i+1} expanded about x_i is

$$f_{i+1} = f_i + f_i'h + \frac{f_i''h^2}{2} + \text{higher order terms}$$

from which we get

$$f_i'h = f_{i+1} - f_i - \frac{f_i''h^2}{2} - \text{higher order terms} \tag{5-7}$$

Substituting Eq. 5-7 into Eq. 5-6 yields

$$(y_{i+1} - y_i) = f_i h + \frac{[f_{i+1} - f_i - (f_i''h^2/2) - \text{higher order terms}]h}{2} + \frac{f_i''h^3}{3!} + \text{higher order terms} \tag{5-8}$$

or

$$\text{exact area} = \underbrace{\left(\frac{f_i + f_{i+1}}{2}\right)h}_{\text{trapezoidal area}} - \frac{1}{12}f_i''h^3 + \text{higher order terms} \tag{5-9}$$

If we define error as what must be added to the computed result to get the exact result, the per-strip truncation error of the trapezoid rule is expressed as

$$E_{T_i} = -\tfrac{1}{12} f_i''h^3 + \text{higher order terms} \tag{5-10}$$

This may be shown to be equivalent to

$$E_{T_i} = -\tfrac{1}{12} f''(\xi)h^3 \qquad x_i < \xi < x_{i+1} \tag{5-11}$$

Although this error cannot be evaluated exactly, a good estimate of its value for each strip can be obtained by assuming that f'' is fairly constant over the strip interval, and evaluating f'' at $\xi = x_i$ or at $\xi = x_{i+1/2}$.

Let us next determine an expression for the error in integrating a function by the trapezoid rule over an interval containing n strips. An error estimation can be made for each strip using Eq. 5-11, giving the total error as

$$E_T = -\tfrac{1}{12}[f''(\xi_1) + f''(\xi_2) + f''(\xi_3) + \cdots + f''(\xi_n)]h^3 \tag{5-12}$$

where ξ_1 is some x value in the first strip, ξ_2 is an x value in the second strip, and so on.

The total truncation error estimate for an integration can be made using Eq. 5-12. However, in many cases the second derivative of the function being integrated turns out to be a cumbersome expression, and this method is undesirable for such cases. A simple alternative approach based on our knowledge of the form of the error is as follows:

We define $f''(\xi)_{av}$ to be the arithmetic mean of the $f''(\xi)$ values. That is,

$$f''(\xi)_{av} = \frac{f''(\xi_1) + f''(\xi_2) + \cdots + f''(\xi_n)}{n} \tag{5-13}$$

or

$$[f''(\xi_1) + f''(\xi_2) + \cdots + f''(\xi_n)] = nf''(\xi)_{av} \tag{5-14}$$

Substituting Eq. 5-14 into Eq. 5-12, we get

$$E_T = -\tfrac{1}{12}(nf''(\xi)_{av})h^3 \tag{5-15}$$

But nh equals the total range of integration. If we are integrating from $x = a$ to $x = b$, then $nh = (b - a)$, and Eq. 5-15 becomes

$$E_T = -\tfrac{1}{12}(b - a)f''(\xi)_{av}h^2 \tag{5-16}$$

We see that $(b - a)$ is a constant and will not vary with the number of strips. The average second derivative, $f''(\xi)_{av}$, will be essentially constant so long as the number of strips does not become too small. Thus we may write that

$$E_T \cong Ch^2 \tag{5-17}$$

where C is some constant and h is any strip width. For $\Delta x = h_1$

$$E_T \cong Ch_1^2 \tag{5-18}$$

For a different strip width, $\Delta x = h_2$,

$$E_T \cong Ch_2^2 \tag{5-19}$$

Thus

$$I \cong I_{h_1} + Ch_1^2 \tag{5-20}$$

and

$$I \cong I_{h_2} + Ch_2^2 \tag{5-21}$$

where I is the exact integral value, I_{h_1} is the integral value obtained with $\Delta x = h_1$, and I_{h_2} is the integral value obtained with $\Delta x = h_2$. Subtracting Eq. 5-21 from Eq. 5-20, and solving for C, we get

$$C \cong \frac{I_{h_1} - I_{h_2}}{h_2^2 - h_1^2} \tag{5-22}$$

Then from Eq. 5-21,

$$I \cong I_{h_2} + \left[\frac{I_{h_1} - I_{h_2}}{h_2^2 - h_1^2}\right] h_2^2$$

or

$$I \cong I_{h_2} + \frac{I_{h_2} - I_{h_1}}{(h_1/h_2)^2 - 1} \tag{5-23}$$

The second term on the right side of Eq. 5-23 is the truncation error estimate for integration with $\Delta x = h_2$. Adding it to I_{h_2} will not give the exact integral value because the expression for C given by Eq. 5-22 is only approximate. However, it does give a much improved value of the integral.

Another error which is introduced in obtaining the approximate area of each strip is *roundoff error*. This arises from performing the required arithmetic operations with numerical values having a limited number of significant digits. Roundoff error can be minimized by the use of double-precision arithmetic.

From the preceding facts it can be seen that the total error over the desired interval of integration is the sum of the truncation and roundoff errors. If the total error were due only to truncation error, it could be made as small as desired by merely reducing the strip width sufficiently. For example, halving the strip width would double the number of per-strip truncation errors summed, but the expression given for the per-strip error reveals that each would be approximately one-eighth its previous size.

However, decreasing the strip width also affects the total error by increasing the total roundoff error due to the larger number of calculations in evaluating Eq. 5-3. Thus in decreasing the strip width to decrease the total error, there is an optimum point at which further decreases in strip width will cause the total error to increase rather than decrease as the roundoff error becomes dominant. The optimum strip width for a particular function is easily determined experimentally on the computer (assuming that the true area under the graph of the function can be evaluated) but is difficult to define analytically.

5-3 / ROMBERG INTEGRATION[1]

Romberg's method of integration is basically Richardson's extrapolation procedure.[2] Romberg's name is attached to the method because, according

[1] W. Romberg, "Vereinfachte numerische Integration," *Norske Vidensk, Selskab Forhandlinger*, Bd. 28, No. 7 (1955), pp. 30–36.
[2] L. F. Richardson and J. A. Gaunt, "The Deferred Approach to the Limit," *Trans. Royal Soc., London* Vol. 226A (1927), p. 300.

to Davis and Rabinowitz, he was the first to describe the algorithm in recursive form.[3]

We recall that the error estimate for trapezoidal integration can be expressed as

$$E_T = Ch^2 \tag{5-24}$$

in which h can take on variable values. It can be shown that, with the inclusion of higher order terms, the exact error can be expressed in the form

$$E_T = C_1h^2 + C_2h^4 + C_3h^6 + C_4h^8 + \cdots \tag{5-25}$$

For small h, the first term dominates. Equation 5-25 can also be written as

$$E_T = C_1h^2 + O(h^4) \tag{5-26}$$

where $O(h^4)$ is a quantity of "order h^4." That is, the h^4 term dominates the higher order terms as $h \to 0$. If we obtain an integral value with a particular strip width, and then halve the strip width to get a second integral value, and apply Eq. 5-23, the C_1h^2 term is removed from the error. This leaves an error $O(h^4)$, which might also be expressed as $C_2h^4 + O(h^6)$. If we now determine a third integral value by again halving the strip width, we can combine it with the second to obtain another estimate containing an error expressed by $C_2h^4 + O(h^6)$. From the two values containing errors of $C_2h^4 + O(h^6)$, we can compute (from an equation similar to Eq. 5-23) a new value in which the C_2h^4 term is removed. (A derivation parallel to that for Eq. 5-23, but using Ch^4 instead of Ch^2, gives an equation similar to Eq. 5-23 except that the exponent 2 becomes a 4.)[4] This process can continue, next removing the C_3h^6 term from the error, until the desired accuracy is obtained. An example will be given to clarify the procedure, but let us first convert Eq. 5-23 to a form more commonly used in Romberg integration.

$$I_{\text{improved}} = \frac{I_{h_2}\left[\left(\dfrac{h_1}{h_2}\right)^2 - 1\right] + I_{h_2} - I_{h_1}}{\left(\dfrac{h_1}{h_2}\right)^2 - 1} = \frac{I_{h_2}\left(\dfrac{h_1}{h_2}\right)^2 - I_{h_1}}{\left(\dfrac{h_1}{h_2}\right)^2 - 1} \tag{5-27}$$

If the second integration is obtained by halving the strip width, $h_1/h_2 = 2$, and Eq. 5-27 can be written as

$$I_{\text{improved}} = \frac{(2^2)I_{h_2} - I_{h_1}}{2^2 - 1} \tag{5-28}$$

[3] P. J. Davis and P. Rabinowitz, *Numerical Integration* (Waltham, Mass.: Blaisdell Publishing Company, 1967), p. 166.

[4] See the error analysis for Simpson's rule in Sec. 5-4.

or

$$I_{\text{improved}} = \frac{(2^2)(I_{\text{more accurate}}) - I_{\text{less accurate}}}{2^2 - 1} \tag{5-29}$$

Equation 5-29 is used for what is called first-order extrapolation. For second-order extrapolation, the equation becomes

$$I_{\text{improved}} = \frac{(2^4)(I_{\text{more accurate}}) - I_{\text{less accurate}}}{2^4 - 1} \tag{5-30}$$

A slight change in form is useful in writing a computer program for Romberg integration. It is

$$I_{\text{improved}} = \frac{(4^n)(I_{\text{more accurate}}) - I_{\text{less accurate}}}{4^n - 1} \tag{5-31}$$

where n is the order of the extrapolation and takes on values 1, 2, ...

A flow chart for Romberg integration is shown after the example which follows. The example should be studied prior to a study of the flow chart.

EXAMPLE 5-1

Using a pocket calculator, let us determine

$$\int_0^{\pi/2} \sin x \, dx$$

by Romberg integration. We will use up to third-order extrapolation, or until the value from the latest extrapolation completed differs from the most accurate value of the previous extrapolation by less than an epsilon value of 0.000001.

We begin by calculating trapezoid-rule results for a single strip, and for two strips, and form the following table:

Number of strips	*1*		*2*
Integral values from trapezoid rule	0.7853981634		0.948059449
First-order extrapolation value		1.002279878	

The extrapolated value shown is calculated from

$$\frac{4(0.948059449) - 0.7853981634}{4 - 1} = 1.002279878$$

Comparing 1.002279878 with 0.948059449, we find a difference whose absolute value exceeds the prescribed epsilon. Therefore, we find the integral value for four strips, and expand our table to the following:

Number of strips	*1*		*2*		*4*
Integral values	0.7853981634		0.948059449		0.987115801
First-order extrapolation values		1.002279878		1.000134585	
Second-order extrapolation value			0.9999915655		

The second-order extrapolation value is found from

$$\frac{4^2(1.000134585) - 1.002279878}{4^2 - 1} = 0.9999915655$$

The value 0.9999915655 differs from 1.000134585 by more than epsilon, so one more extrapolation is taken. This will be the last extrapolation regardless of the accuracy check. An integration value by the trapezoid rule with eight strips is calculated and the table expanded to

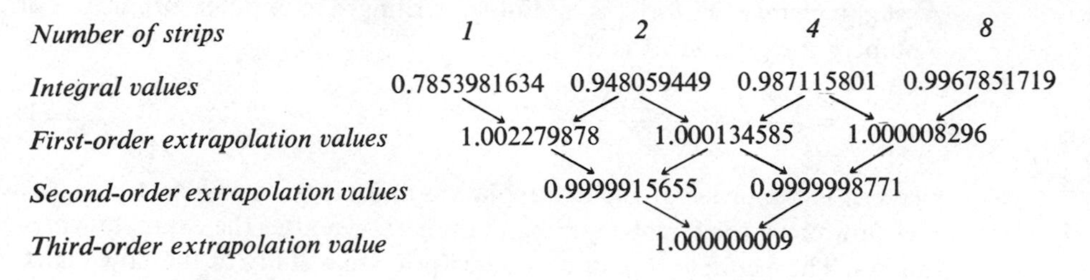

Number of strips	*1*		*2*		*4*		*8*
Integral values	0.7853981634		0.948059449		0.987115801		0.9967851719
First-order extrapolation values		1.002279878		1.000134585		1.000008296	
Second-order extrapolation values			0.9999915655		0.9999998771		
Third-order extrapolation value				1.000000009			

The third-order extrapolation value is obtained from

$$\frac{4^3(0.9999998771) - 0.9999915655}{4^3 - 1} = 1.000000009$$

This value is taken as our final answer. We note that it is an excellent result for using a maximum of only eight strips in our trapezoid integration.

In the computer, the table shown might be stored as a two-dimensional array in the form

$$\begin{matrix} R_{11} & R_{12} & R_{13} & R_{14} \\ R_{21} & R_{22} & R_{23} & \\ R_{31} & R_{32} & & \\ R_{41} & & & \end{matrix}$$

in which R_{41} is the final answer.

Flow Chart for Romberg Integration

The flow chart shown in Fig. 5-3 calls for the use of the subroutine subprogram TRAPZ from Sec. 5-2 for integration by the trapezoid rule. Integration begins with N strips, and this is doubled a maximum of six times. Because of a limitation in the subprogram TRAPZ, N should not be less than two. Thus the program may employ up to sixth-order extrapolation, unless the accuracy requirement is met sooner. The function to be integrated is defined in a function subprogram named F. Integration

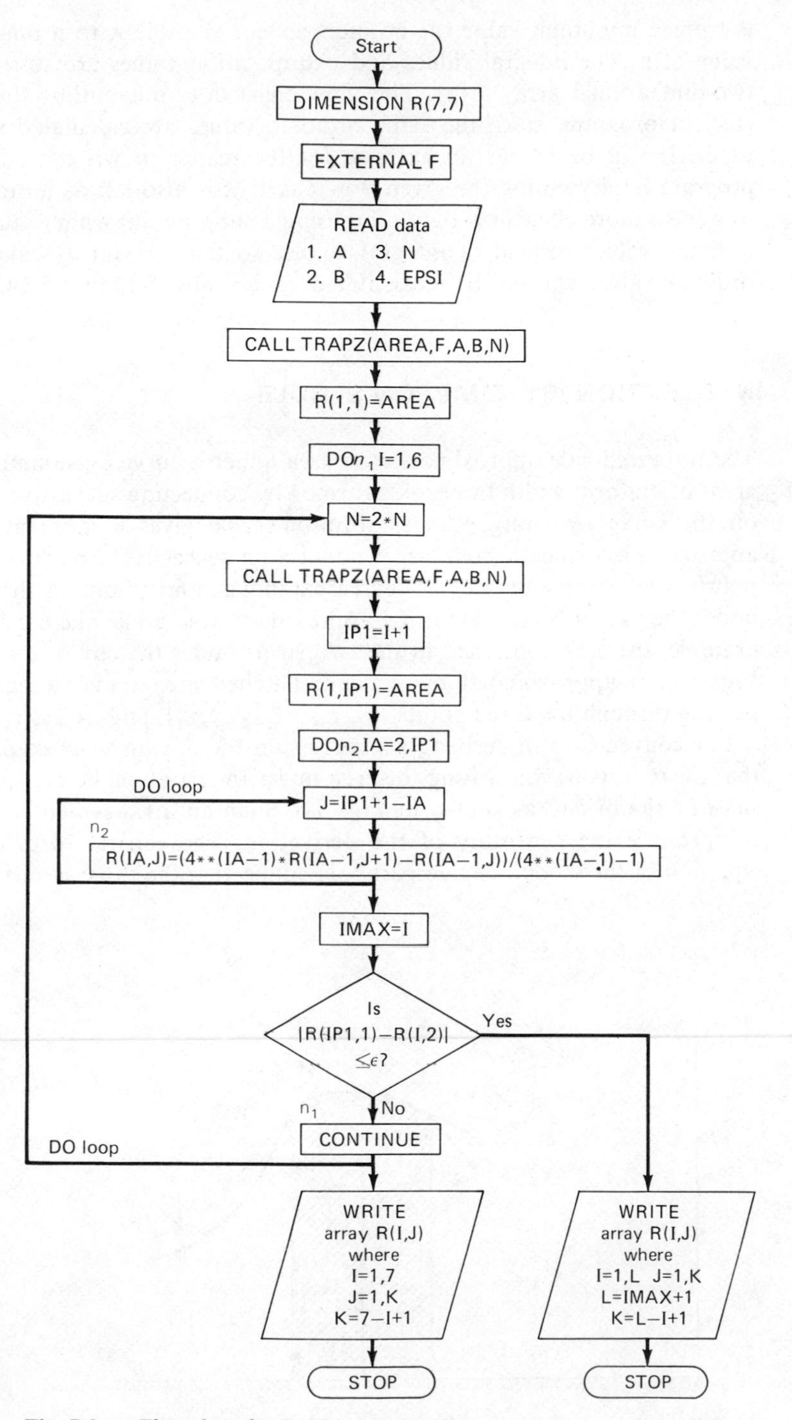

Fig. 5-3 / *Flow chart for Romberg integration.*

is from a minimum value of the independent variable A to a maximum value of B. The integral values and extrapolation values are stored in a two-dimensional array R(I,J). This flow chart does not outline the most efficient program, since the same ordinate values are calculated several times. It will be left as an exercise for the reader to write a FORTRAN program implementing the given flow chart. It is also left as an exercise to write a more efficient program by using a subprogram which calculates ordinate values instead of integral values, so that previously calculated ordinate values will not be recalculated. (See Probs. 5-13 and 5-14.)

5-4 / INTEGRATION BY SIMPSON'S RULE

The trapezoid rule approximates the area under a curve by summing the areas of uniform-width trapezoids formed by connecting successive points on the curve by straight lines. Simpson's rule gives a more accurate approximation since it consists of connecting successive groups of three points on the curve by second-degree parabolas, and summing the areas under the parabolas to obtain the approximate area under the curve. For example, the area contained in the two strips under the curve of $f(x)$, in Fig. 5-4, is approximated by the crosshatched area under a parabola passing through the three points (x_i, y_i), (x_{i+1}, y_{i+1}), and (x_{i+2}, y_{i+2}).

For convenience in deriving an expression for this area, let us assume that the two strips comprising the area under the parabola lie on opposite sides of the origin, as shown in Fig. 5-5. Such an arrangement will not compromise the generality of the derivation. The general form of the equation of the second-degree parabola connecting the three points is

$$y = ax^2 + bx + c \tag{5-32}$$

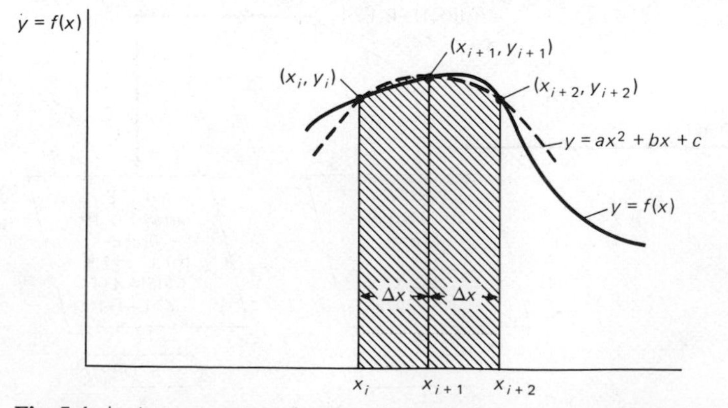

Fig. 5-4 / *Approximation of area under a curve by use of a second-degree parabola.*

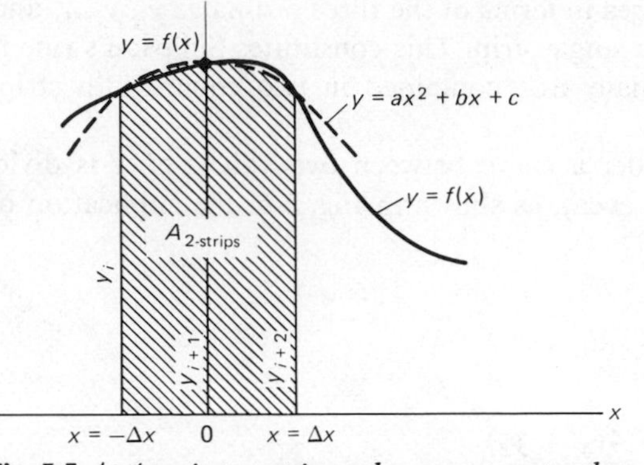

Fig. 5-5 / *Area in two strips under a curve centered on y axis.*

The integration of Eq. 5-32 from $-\Delta x$ to Δx gives the area contained in the two strips shown under the parabola. Hence

$$A_{2\text{ strips}} = \int_{-\Delta x}^{\Delta x} (ax^2 + bx + c)\, dx = \left[\frac{ax^3}{3} + \frac{bx^2}{2} + cx\right]_{-\Delta x}^{\Delta x} \tag{5-33}$$

Substituting the limits into Eq. 5-33 yields

$$A_{2\text{ strips}} = \tfrac{2}{3}a(\Delta x)^3 + 2c(\Delta x) \tag{5-34}$$

The constants a and c can be determined from the fact that points $(-\Delta x, y_i)$, $(0, y_{i+1})$, and $(\Delta x, y_{i+2})$ must all satisfy Eq. 5-32. The substitution of these three sets of coordinates into Eq. 5-32 yields

$$\begin{aligned} y_i &= a(-\Delta x)^2 + b(-\Delta x) + c \\ y_{i+1} &= c \\ y_{i+2} &= a(\Delta x)^2 + b(\Delta x) + c \end{aligned} \tag{5-35}$$

Solving these equations simultaneously for the constants a, b, and c, we find that

$$\begin{aligned} a &= \frac{y_i - 2y_{i+1} + y_{i+2}}{2(\Delta x)^2} \\ b &= \frac{y_{i+2} - y_i}{2(\Delta x)} \\ c &= y_{i+1} \end{aligned} \tag{5-36}$$

The substitution of the first and third parts of Eq. 5-36 into Eq. 5-34 yields

$$A_{2\text{ strips}} = \frac{\Delta x}{3}(y_i + 4y_{i+1} + y_{i+2}) \tag{5-37}$$

which gives the area in terms of the three ordinates y_i, y_{i+1}, and y_{i+2} and the width Δx of a single strip. This constitutes Simpson's rule for obtaining the approximate area contained in two equal-width strips under a curve.

If the area under a curve between two values of x is divided into n uniform strips (n even), as shown in Fig. 5-6, the application of Eq. 5-37 shows that

$$\begin{aligned} A_1 &= \frac{\Delta x}{3}(y_1 + 4y_2 + y_3) \\ A_2 &= \frac{\Delta x}{3}(y_3 + 4y_4 + y_5) \\ A_3 &= \frac{\Delta x}{3}(y_5 + 4y_6 + y_7) \\ \vdots \quad & \quad \vdots \qquad \qquad \vdots \\ A_{n/2} &= \frac{\Delta x}{3}(y_{n-1} + 4y_n + y_{n+1}) \end{aligned} \tag{5-38}$$

Summing these areas, we can write

$$\int_{x_1=0}^{x_{n+1}} f(x)\,dx = \sum_{i=1}^{i=n/2} A_i = \frac{\Delta x}{3}(y_1 + 4y_2 + 2y_3 + 4y_4 + 2y_5 + \cdots + 2y_{n-1} + 4y_n + y_{n+1})$$

or

$$\int_{x_1=0}^{x_{n+1}} f(x)\,dx = \frac{\Delta x}{3}\left(y_1 + 4\sum_{i=2,4,6}^{i=n} y_i + 2\sum_{i=3,5,7}^{i=n-1} y_i + y_{n+1}\right) \tag{5-39}$$

where n is even.

Equation 5-39 is called Simpson's one-third rule for obtaining the approximate area under a curve. It may be used when the area is divided into an *even* number of strips of width Δx.

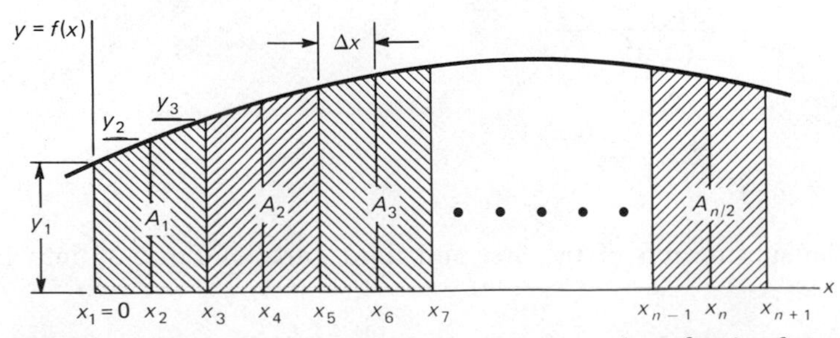

Fig. 5-6 / *Area under curve divided into an even number of strips for approximation by Simpson's rule.*

A subroutine subprogram utilizing Simpson's rule for integrating any FORTRAN-supplied function such as SIN(X), or any function defined by a function subprogram, is given below. The integration is from XMIN to XMAX, using N strips. The dummy function name DUMMYF is used, and it is replaced by the actual function name when the subprogram SIMPSN is called. The function name must appear in an EXTERNAL statement in the calling program.

```
      SUBROUTINE SIMPSN(AREA,DUMMYF,XMIN,XMAX,N)
C THIS SUBPROGRAM INTEGRATES A FUNCTION BY SIMPSON'S RULE
C DUMMYF IS A DUMMY NAME FOR THE FUNCTION INTEGRATED
      H=(XMAX-XMIN)/N
      SUM=0.0
      X=XMIN+H
      DO 4 I=2,N
      IF(MOD(I,2))2,2,3
    2 SUM=SUM+4.*DUMMYF(X)
      GO TO 4
    3 SUM=SUM+2.*DUMMYF(X)
    4 X=X+H
      AREA=H/3.*(DUMMYF(XMIN)+SUM+DUMMYF(XMAX))
      RETURN
      END
```

EXAMPLE 5-2[5]

Let us consider a U.S. Army M-14 rifle having a 0.6096 m barrel, a 7.62 mm bore, and 101.6 μm deep rifling grooves. The lead bullet used weighs 0.0956 *N*. The gas-pressure curve varies with the temperature of the barrel, but for practical purposes of illustration, we will use the gas-pressure data for a particular firing condition as shown in Fig. 5-7. The frictional resistance exerted on the bullet is to be considered small as compared with the propulsive force of the gas on the bullet, and the cross-sectional area of the rifling grooves is considered negligible. Our problem will be to determine the muzzle velocity of the bullet using Simpson's one-third rule.

The solution is begun by considering that the work done on the bullet while it is in the barrel is given by

$$\text{work} = \int_0^{x_{\text{final}}} F(x)\,dx \tag{5-40}$$

where $F(x)$ is the force of the gas acting on the bullet as a function of the displacement x. The work is thus represented by the area under the pressure-displacement curve of Fig. 5-7, multiplied by the cross-sectional

[5] The SI system of units is used in this example; see App. A.

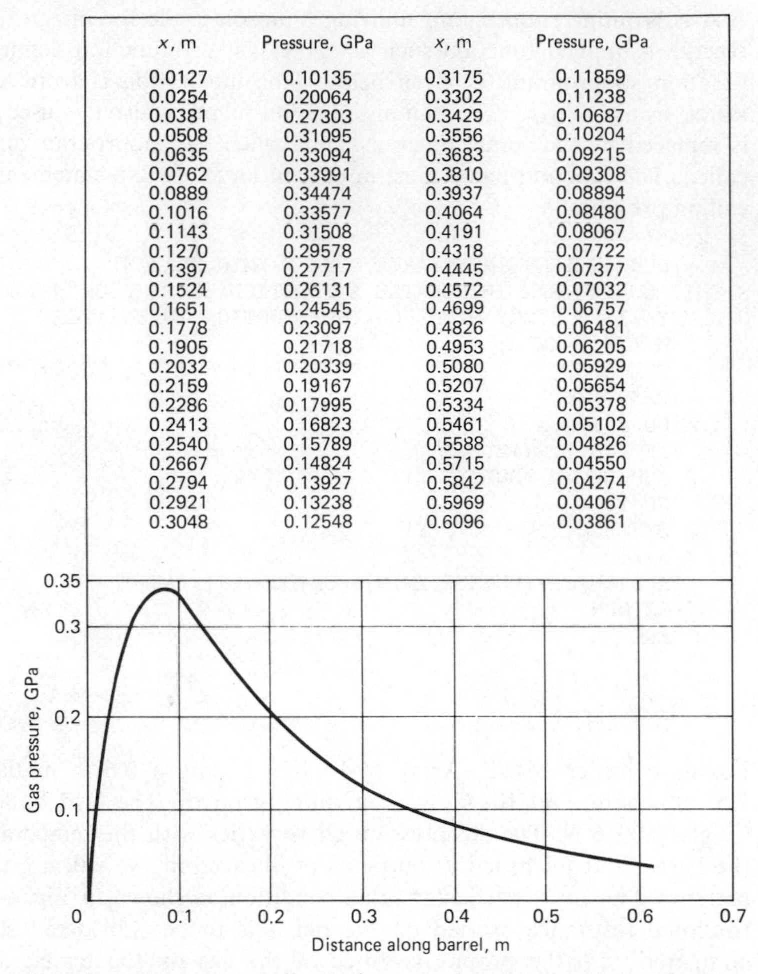

x, m	Pressure, GPa	x, m	Pressure, GPa
0.0127	0.10135	0.3175	0.11859
0.0254	0.20064	0.3302	0.11238
0.0381	0.27303	0.3429	0.10687
0.0508	0.31095	0.3556	0.10204
0.0635	0.33094	0.3683	0.09215
0.0762	0.33991	0.3810	0.09308
0.0889	0.34474	0.3937	0.08894
0.1016	0.33577	0.4064	0.08480
0.1143	0.31508	0.4191	0.08067
0.1270	0.29578	0.4318	0.07722
0.1397	0.27717	0.4445	0.07377
0.1524	0.26131	0.4572	0.07032
0.1651	0.24545	0.4699	0.06757
0.1778	0.23097	0.4826	0.06481
0.1905	0.21718	0.4953	0.06205
0.2032	0.20339	0.5080	0.05929
0.2159	0.19167	0.5207	0.05654
0.2286	0.17995	0.5334	0.05378
0.2413	0.16823	0.5461	0.05102
0.2540	0.15789	0.5588	0.04826
0.2667	0.14824	0.5715	0.04550
0.2794	0.13927	0.5842	0.04274
0.2921	0.13238	0.5969	0.04067
0.3048	0.12548	0.6096	0.03861

Fig. 5-7 / *Gas-pressure-bullet-displacement data for the M-14 rifle for a particular firing condition.*

area of the bore $[4.561(10)^{-5}\ \text{m}^2]$. The work done on the bullet by the gas must equal the change in kinetic energy (K.E.) of the bullet, since we are neglecting any energy loss due to frictional resistance. Therefore,

$$\int_0^{x_{\text{final}}} F(x)\,dx = \Delta\text{K.E.} = \tfrac{1}{2}m\dot{x}_{\text{final}}^2 - \tfrac{1}{2}m\dot{x}_{\text{initial}}^2 \tag{5-41}$$

Since the initial kinetic energy is zero, we may write from Eq. 5-41

$$\dot{x}_{\text{final}} = \sqrt{\frac{2}{m}\int_0^{x_{\text{final}}} F(x)\,dx} = \sqrt{\frac{2[4.56(10)^{-6}]}{0.0956/9.81}\int_0^{x_{\text{final}}} P(x)\,dx}$$

where $P(x)$ is the gas pressure in terms of the displacement, as given in Fig. 5-7, and

$$\int_0^{x_{\text{final}}} P(x)\,dx \tag{5-42}$$

is the area under the pressure-displacement curve, which we will obtain by dividing the area under the curve of Fig. 5-7 into 48 equal strips and summing the area contained in the 24 pairs of strips by the use of Eq. 5-39. This curve has been plotted from experimental data, and we do not have a mathematical equation for it. Therefore, the subroutine subprogram previously given for integration by Simpson's rule cannot be utilized. An equation could be obtained by the curve-fitting techniques discussed in Chap. 4, but since the data shown have little scatter, an integration program will be written which utilizes the raw data.

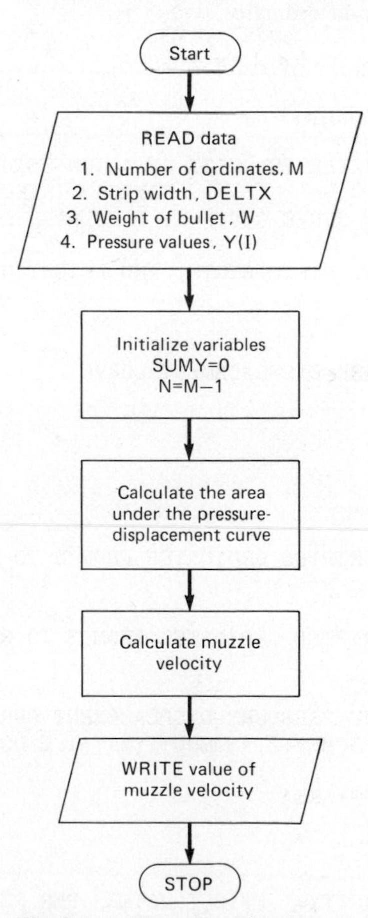

Fig. 5-8 / *Flow chart for determining muzzle velocity of a rifle bullet, Ex. 5-2.*

A flow chart for the computer solution is shown in Fig. 5-8. The FORTRAN names selected to represent the variable quantities are

FORTRAN Name	Quantity
N	Number of strips
M	Number of ordinates
DELTX	Width of each strip
W	Weight of bullet
Y(I)	Values of pressure readings
SUMY	Sum of ordinates
AREA	Area under pressure curve
XDSQ	Square of muzzle velocity
XD	Muzzle velocity
SUMEV	Sum of the even ordinates, $\sum_{i=2,4,6}^{i=n} y_i$
SUMOD	Sum of the odd ordinates, $\sum_{i=3,5,7}^{i=n-1} y_i$
K	Number of strips less 1, $(n-1)$

The FORTRAN IV program is as follows:

```
C PROGRAM TO DETERMINE THE MUZZLE VELOCITY OF A GUN USING
                                                SIMPSON INTEGRATION
C ORDINATE VALUES OF PRESSURE CURVE GIVEN BY TABULATED DATA
      DIMENSION Y(49)
C READ IN NUMBER OF ORDINATES, STRIP WIDTH, AND WEIGHT OF
                                                             BULLET
      READ(5,1) M,DELTX,W
    1 FORMAT(I2,2F10.0)
C READ IN ORDINATES OF PRESSURE-DISPLACEMENT CURVE
      READ(5,2) (Y(I),I=1,M)
    2 FORMAT(8F10.0)
      SUMEV=0.
      SUMOD=0.
      N=M-1
      K=N-1
C DETERMINE SUM OF EVEN SUBSCRIPTED ORDINATES FROM 2 TO N
      DO 4 I=2,N,2
    4 SUMEV=SUMEV+Y(I)
C DETERMINE SUM OF ODD SUBSCRIPTED ORDINATES FROM 3 TO N-1
      DO 5 I=3,K,2
    5 SUMOD=SUMOD+Y(I)
C CALCULATE THE AREA UNDER THE PRESSURE-DISPLACEMENT CURVE
      AREA=DELTX/3.*(Y(1)+4.*SUMEV+2.*SUMOD+Y(M))*1.E 09
C CALCULATE MUZZLE VELOCITY
      XDSQ=2.*0.0000456*9.81/W*AREA
      XD=SQRT(XDSQ)
C WRITE OUT THE MUZZLE VELOCITY
      WRITE(6,6)XD
    6 FORMAT('1','MUZZLE VELOCITY=',F7.2,' METERS PER SECOND')
      STOP
      END
```

The execution of the preceding program yielded this computer printout:

```
MUZZLE VELOCITY= 928.01 METERS PER SECOND
```

Truncation Error Estimation in Integration by Simpson's Rule

If the function $f(x)$ can be expressed as a continuous mathematical function having continuous derivatives, $f'(x)$ through $f^{IV}(x)$, the error resulting from approximating the true area in two strips under the graph of $f(x)$ between x_{i-1} and x_{i+1} by the area under a second-degree parabola can be shown to be

$$E_{T_i} = -\tfrac{1}{90} f^{IV}(\xi) h^5 \qquad x_{i-1} < \xi < x_{i+1} \tag{5-43}$$

where $h = \Delta x$. To show this, let us consider the graphs of $y' = f(x)$ and the integrated function y, plotted versus the independent variable x, as shown in Fig. 5-9. The crosshatched area in the two strips under the y' curve is equal to $y_{i+1} - y_{i-1}$.

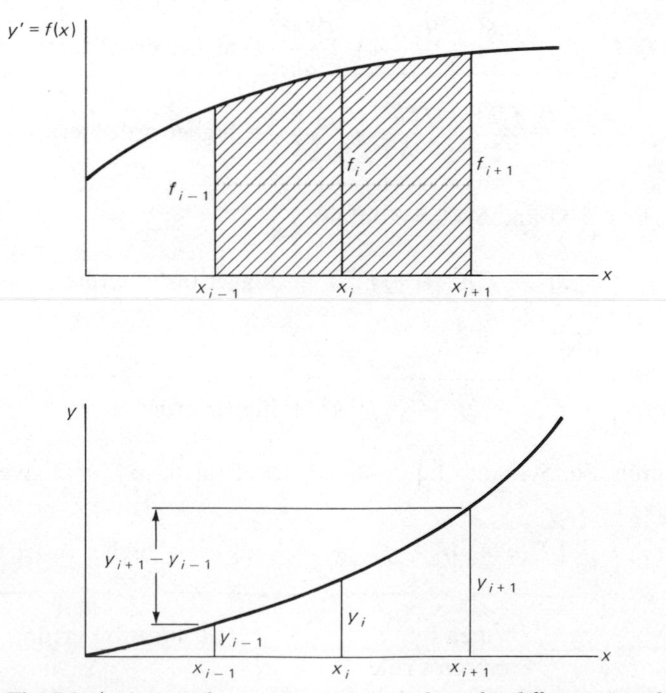

Fig. 5-9 / *Area under a curve represented as the difference in ordinate values of the integrated function.*

Expanding y_{i+1} in a Taylor series about x_i, we obtain

$$y_{i+1} = y_i + y_i'h + \frac{y_i''h^2}{2} + \frac{y_i'''h^3}{6} + \frac{y_i^{\text{IV}}h^4}{24} + \frac{y_i^{\text{V}}h^5}{120} + \text{higher order terms} \tag{5-44}$$

Expanding y_{i-1} about x_i, we obtain

$$y_{i-1} = y_i - y_i'h + \frac{y_i''h^2}{2} - \frac{y_i'''h^3}{6} + \frac{y_i^{\text{IV}}h^4}{24} - \frac{y_i^{\text{V}}h^5}{120} + \text{higher order terms} \tag{5-45}$$

Subtracting Eq. 5-45 from Eq. 5-44, we have

$$y_{i+1} - y_{i-1} = 2y_i'h + \frac{y_i'''h^3}{3} + \frac{y_i^{\text{V}}h^5}{60} + \text{higher order terms}$$

or

$$\text{exact area} = 2f_ih + \frac{f_i''h^3}{3} + \frac{f_i^{\text{IV}}h^5}{60} + \text{higher order terms} \tag{5-46}$$

Now, let us expand f_{i+1} and f_{i-1} in Taylor series about x_i.

$$f_{i+1} = f_i + f_i'h + \frac{f_i''h^2}{2} + \frac{f_i'''h^3}{6} + \frac{f_i^{\text{IV}}h^4}{24} + \text{higher order terms} \tag{5-47}$$

$$f_{i-1} = f_i - f_i'h + \frac{f_i''h^2}{2} - \frac{f_i'''h^3}{6} + \frac{f_i^{\text{IV}}h^4}{24} + \text{higher order terms} \tag{5-48}$$

Adding Eqs. 5-47 and 5-48, we obtain

$$f_{i+1} + f_{i-1} = 2f_i + f_i''h^2 + \tfrac{1}{12}f_i^{\text{IV}}h^4 + \text{higher order terms}$$

or

$$f_i''h^2 = f_{i+1} + f_{i-1} - 2f_i - \tfrac{1}{12}f_i^{\text{IV}}h^4 + \text{higher order terms} \tag{5-49}$$

Substituting Eq. 5-49 into Eq. 5-46 as part of the term $f_i''h^3/3$ gives

$$\text{exact area} = \underbrace{\frac{h}{3}[f_{i+1} + 4f_i + f_{i-1}]}_{\substack{\text{area by} \\ \text{Simpson's rule}}} \underbrace{- \frac{1}{90} f_i^{\text{IV}}h^5 + \text{higher order terms}}_{\text{truncation error}} \tag{5-50}$$

The truncation error of Simpson's rule (the error in each two strips) is

$$E_{T_i} = -\tfrac{1}{90} f_i^{IV} h^5 + \text{higher order terms} \tag{5-51}$$

which may be shown to be equivalent to

$$E_{T_i} = -\tfrac{1}{90} f^{IV}(\xi) h^5 \qquad x_{i-1} < \xi < x_{i+1} \tag{5-52}$$

This *truncation error* is the quantity which must be added to the approximate area in two strips, obtained by Simpson's one-third rule, to get the true area under the curve in this interval. The truncation-error term shown cannot usually be evaluated directly. However, a good estimate of its value for each two-strip interval can be obtained by assuming that f^{IV} is fairly constant over the interval (the higher derivatives are assumed to be negligible) and evaluating f^{IV} at $\xi = x_i$. The truncation-error estimate for the total integration is obtained by summing the two-strip estimates. If the total truncation-error estimate is larger than can be tolerated, smaller two-strip intervals should be used. Considering roundoff error, which is also present, there is an optimum strip width for obtaining a minimum total error in the integration. (See the error analysis of the trapezoid method, preceding.)

In evaluating the error in the total integration by summing up the error in each two strips,

$$E_T = -\tfrac{1}{90}[f^{IV}(\xi_1) + f^{IV}(\xi_2) + \cdots + f^{IV}(\xi_{n/2})]h^5 \tag{5-53}$$

we must be able to differentiate the function being integrated four times. In many cases this leads to an excessively long expression. To avoid having to differentiate the function being integrated, let us develop a method for estimating the error, or improving the result, based on integrating twice with different strip widths, as was done for the trapezoid rule.

Equation 5-53 contains an f^{IV} evaluation for each pair of strips, or $n/2$ of them for n strips. Defining f_{av}^{IV} as the arithmetic mean of these, we have

$$f_{av}^{IV} = \frac{f^{IV}(\xi_1) + f^{IV}(\xi_2) + \cdots + f^{IV}(\xi_{n/2})}{n/2} \tag{5-54}$$

Then

$$E_T = -\frac{1}{90}\frac{n}{2} f_{av}^{IV} h^5 \tag{5-55}$$

But $nh = (b - a)$, the interval of integration, so

$$E_T = -\tfrac{1}{180}(b - a) f_{av}^{IV} h^4 \tag{5-56}$$

Since the average fourth derivative will not vary greatly with different numbers of strips, if the number of strips is not too small, we can express the error as

$$E_T \cong Ch^4 \tag{5-57}$$

where C is a constant. For strip widths $\Delta x = h_1$ and $\Delta x = h_2$, we have

$$I \cong I_{h_1} + Ch_1^4$$

$$I \cong I_{h_2} + Ch_2^4$$

Subtracting gives

$$0 \cong I_{h_1} - I_{h_2} + C(h_1^4 - h_2^4)$$

or

$$C \cong \frac{I_{h_2} - I_{h_1}}{h_1^4 - h_2^4} \tag{5-58}$$

Then

$$\begin{aligned} I_{\text{improved}} &= I_{h_2} + \left[\frac{I_{h_2} - I_{h_1}}{h_1^4 - h_2^4}\right] h_2^4 \\ &= I_{h_2} + \frac{I_{h_2} - I_{h_1}}{(h_1/h_2)^4 - 1} \end{aligned} \tag{5-59}$$

If we run first with a strip width of h_1 and then halve the interval so that $h_1/h_2 = 2$, Eq. 5-59 becomes

$$I_{\text{improved}} = I_{\text{more accurate}} + \frac{I_{\text{more accurate}} - I_{\text{less accurate}}}{15} \tag{5-60}$$

where the second term on the right is the error estimate for the more accurate integration. If we only want an improved result without separating out the truncation-error estimate, Eq. 5-60 can be put in the form

$$I_{\text{improved}} = \frac{16I_{\text{more accurate}} - I_{\text{less accurate}}}{15} \tag{5-61}$$

Simpson's Three-Eighths Rule

In utilizing Simpson's rule for integrating a function, it might be desirable to use an odd number of strips to approximate the true area under the

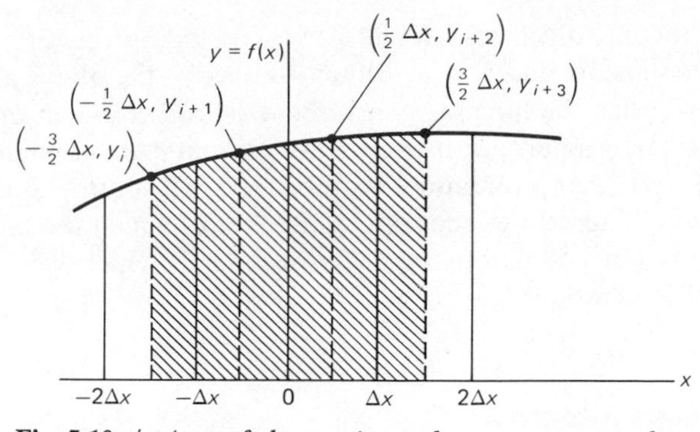

Fig. 5-10 / *Area of three strips under a curve centered on y axis.*

curve. If an odd number of strips is used, Simpson's three-eighths rule for obtaining the area contained in 3 strips under a curve can be used, in conjunction with Eq. 5-39, to find the total area. For example, if 97 strips were used, Simpson's three-eighths rule could be used to approximate the area under the curve occupied by the first 3 strips. The remaining 94 strips would then be summed, using Simpson's one-third rule. The derivation of the three-eighths rule is similar to that for the one-third rule, except that it determines the area under a *third*-degree parabola connecting four points on the given curve. The general form of the third-degree parabola is

$$y = ax^3 + bx^2 + cx + d \tag{5-62}$$

In the derivation, the constants are determined by requiring that the parabola pass through the four points indicated on the curve shown in Fig. 5-10. The range of integration is from $-3(\Delta x)/2$ to $3(\Delta x)/2$, resulting in

$$\begin{aligned} A_{3\text{ strips}} &= \int_{-3(\Delta x)/2}^{3(\Delta x)/2} (ax^3 + bx^2 + cx + d)\,dx \\ &= \frac{3(\Delta x)}{8}(y_i + 3y_{i+1} + 3y_{i+2} + y_{i+3}) \end{aligned} \tag{5-63}$$

which is Simpson's three-eighths rule.

5-5 / IMPROPER INTEGRALS

In our discussion of integration thus far, we have assumed that $f(x)$ in the integral

$$\int_a^b f(x)\,dx$$

is continuous for all values of x from a to b, including the end points. We have also assumed a and b to be finite. If either of the above assumptions is not true, then the integral given above is said to be an *improper integral.* Occasionally improper integrals are encountered in engineering computations, and their evaluation requires special consideration. We will discuss two different cases separately, the first in which the limits are finite but $f(x)$ is not continuous, and the second in which we have an infinite limit of integration.

Case I / Discontinuous Integrand

Problem 5-2 at the end of the chapter illustrates an example of a discontinuous integrand. One-quarter of the arc length of the lemniscate can be determined from

$$S = a \int_0^{\pi/4} \frac{d\theta}{\sqrt{\cos 2\theta}} \tag{5-64}$$

The integrand becomes infinite, however, at $\theta = \pi/4$. To solve the problem, we first break the integral into two parts

$$S = a \int_0^{\theta_1} \frac{d\theta}{\sqrt{\cos 2\theta}} + a \int_{\theta_1}^{\pi/4} \frac{d\theta}{\sqrt{\cos 2\theta}} \tag{5-65}$$

in which $(\pi/4 - \theta_1)$ is a small value. The first integral can be handled by a numerical method, and we would hope that it might be possible to obtain an estimate of the second integral. To accomplish this in this instance, let us expand the integrand in a power series about the θ value giving us difficulty, $\theta = \pi/4$. We will then attempt to integrate the series term by term and truncate the resulting series to obtain our estimate.

The Taylor series expansion for $\cos 2\theta$ about $\theta = \pi/4$ is

$$\cos 2\theta = 2\left(\frac{\pi}{4} - \theta\right) - \frac{4}{3}\left(\frac{\pi}{4} - \theta\right)^3 + \text{higher power terms}^6$$

or

$$\cos 2\theta = 2\left(\frac{\pi}{4} - \theta\right)\left[1 - \frac{2}{3}\left(\frac{\pi}{4} - \theta\right)^2 + \text{higher power terms}\right] \tag{5-66}$$

[6] Terms containing higher powers of $(\pi/4 - \theta)$.

Then

$$\sqrt{\cos 2\theta} = \sqrt{2}\left(\frac{\pi}{4} - \theta\right)^{1/2}\left[1 - \frac{2}{3}\left(\frac{\pi}{4} - \theta\right)^2 + \text{higher power terms}\right]^{1/2} \tag{5-67}$$

From the rules governing multiplication of infinite power series, we can easily verify that

$$\left[1 - \frac{2}{3}\left(\frac{\pi}{4} - \theta\right)^2 + \text{higher power terms}\right]^{1/2} = \left[1 - \frac{1}{3}\left(\frac{\pi}{4} - \theta\right)^2 + \text{higher power terms}\right]$$

Thus

$$\sqrt{\cos 2\theta} = \sqrt{2}\left(\frac{\pi}{4} - \theta\right)^{1/2}\left[1 - \frac{1}{3}\left(\frac{\pi}{4} - \theta\right)^2 + \text{higher power terms}\right] \tag{5-68}$$

Our integrand, $1/\sqrt{\cos 2\theta}$, becomes

$$\frac{1}{\sqrt{\cos 2\theta}} = \frac{1}{\sqrt{2}}\left(\frac{\pi}{4} - \theta\right)^{-1/2}\left[1 - \frac{1}{3}\left(\frac{\pi}{4} - \theta\right)^2 + \text{higher power terms}\right]^{-1}$$

From the rules for division of infinite power series, we can verify that

$$\left[1 - \frac{1}{3}\left(\frac{\pi}{4} - \theta\right)^2 + \text{higher power terms}\right]^{-1} = \left[1 + \frac{1}{3}\left(\frac{\pi}{4} - \theta\right)^2 + \text{higher power terms}\right]$$

Thus

$$\begin{aligned} a\int_{\theta_1}^{\pi/4} \frac{d\theta}{\sqrt{\cos 2\theta}} &= \frac{a}{\sqrt{2}}\int_{\theta_1}^{\pi/4}\left[\left(\frac{\pi}{4} - \theta\right)^{-1/2} + \frac{1}{3}\left(\frac{\pi}{4} - \theta\right)^{3/2} + \text{higher power terms}\right] d\theta \\ &= \frac{a}{\sqrt{2}}\left[-2\left(\frac{\pi}{4} - \theta\right)^{1/2} - \left(\frac{2}{5}\right)\left(\frac{1}{3}\right)\left(\frac{\pi}{4} - \theta\right)^{5/2} + \text{higher power terms}\right]_{\theta_1}^{\pi/4} \\ &= \sqrt{2}a\left(\frac{\pi}{4} - \theta_1\right)^{1/2} + \frac{\sqrt{2}}{5}a\left(\frac{1}{3}\right)\left(\frac{\pi}{4} - \theta_1\right)^{5/2} + \text{higher power terms} \end{aligned}$$

Since the quantity $(\pi/4 - \theta_1)$ is small, we can neglect all but the first term of the series for our estimate of

$$a \int_{\theta_1}^{\pi/4} \frac{d\theta}{\sqrt{\cos 2\theta}}$$

If we take θ_1 as 44.5 deg (0.77667 radian), the estimated value of this integral is $0.13211a$.

Another approach to this problem would be to approximate the length of lemniscate omitted in integrating from 0 to 44.5 deg instead of to 45 deg, with the chord ρ at 44.5 deg. This gives us

$$\rho = a\sqrt{\cos 2\theta} = 0.13211a$$

as before.

Still another approach that could be used on this problem would be to note that as θ_1 gets closer and closer to 45 deg, the omitted length becomes very small. Thus for a satisfactory solution to this problem, we could truncate the interval, integrating from 0 to, say, 44.9 deg by a numerical method. In general, truncating the interval is an unsafe practice, but in this instance we can see even without an estimate that we are omitting very little of the correct length.

If the singularity occurs in the interior of the integration interval, rather than at an end point, the interval must be broken into two parts and each handled as a problem where the discontinuity occurs at an end point of the interval.

Davis and Rabinowitz discuss other techniques for handling improper integrals of this type, such as change of variable, subtracting out the discontinuity, ignoring the discontinuity, and the use of integration by parts.[7]

Case II / Infinite Limit of Integration

This type of problem can often be handled by a change of variables to make the infinite limit become a zero limit for the new variable. For example, if we have the integral

$$\int_{1}^{\infty} \frac{dx}{1 + x^2 + x^3}$$

[7] P. J. Davis and P. Rabinowitz, *Numerical Integration* (Waltham, Mass.: Blaisdell Publishing Co., 1967), pp. 71–80.

we can make the upper limit zero by the substitution $x = t^n$ where $n \leq -1$. Taking $n = -1$, we have $dx = -t^{-2}\,dt$, and the integral becomes

$$-\int_1^0 \frac{dt}{t^2[1 + (1/t^2) + (1/t^3)]} = \int_0^1 \frac{t\,dt}{t^3 + t + 1}$$

which is a proper integral. If the transformation leads to case I, which was discussed earlier, in which an improper integral is obtained, the problem can then be approached by the methods discussed there.

5-6 / NUMERICAL DIFFERENTIATION

We will now direct our attention to ways of numerically differentiating functions which are defined only by tabulated data or by experimentally determined curves. Tabulated experimental data are generally not used directly in numerical differentiation because the scatter would cause serious accuracy problems. Whenever possible, an analytical expression for a smooth curve which fits the data should be determined (using techniques given in Chap. 4 on curve fitting). Analytical differentiation can then be used, or numerical differentiation with ordinate values obtained from the fitted curve. We will be interested here only in numerical differentiation. The method we shall discuss utilizes Taylor-series expansions. The Taylor series for a function $y = f(x)$ at $(x_i + h)$ expanded about x_i is

$$y(x_i + h) = y_i + y_i'h + \frac{y_i''h^2}{2!} + \frac{y_i'''h^3}{3!} + \cdots \tag{5-69}$$

where $h = \Delta x$ and y_i is the ordinate corresponding to x_i and $(x_i + h)$ is in the region of convergence. The function at $(x_i - h)$ is similarly given by

$$y(x_i - h) = y_i - y_i'h + \frac{y_i''h^2}{2!} - \frac{y_i'''h^3}{3!} + \cdots \tag{5-70}$$

Subtracting Eq. 5-70 from Eq. 5-69, we obtain

$$y_i' = \frac{y(x_i + h) - y(x_i - h)}{2h} - \left(\frac{1}{6}y_i'''h^2 + \cdots\right) \tag{5-71}$$

Looking at Fig. 5-11, we see that if we designate equally spaced points to the right of x_i as x_{i+1}, x_{i+2}, and so on, and those to the left of x_i as

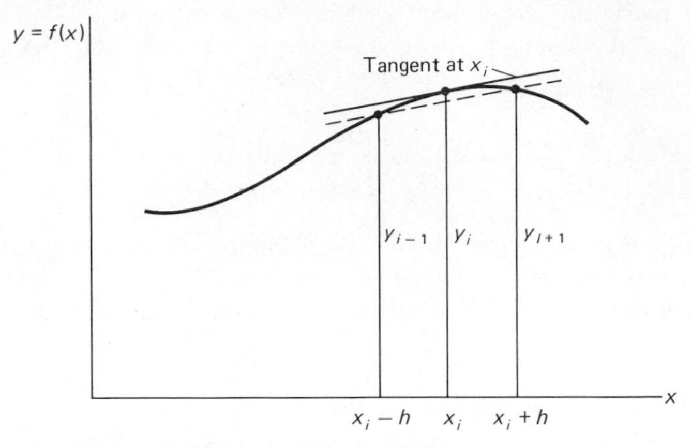

Fig. 5-11 / *Approximation of the derivative at x_i.*

x_{i-1}, x_{i-2}, and identify the corresponding ordinates as y_{i+1}, y_{i+2}, y_{i-1}, and y_{i-2}, respectively, Eq. 5-71 can be written in the form

$$y_i' = \frac{y_{i+1} - y_{i-1}}{2h} \tag{5-72}$$

with error of order h^2. Equation 5-72 is called the *central-difference* approximation of y' at x_i with errors, of order h^2. Graphically, the approximation represents the slope of the dashed line in Fig. 5-11. The actual derivative is represented by the solid line drawn tangent to the curve at x_i.

If we add Eqs. 5-69 and 5-70 and use the notation previously described, we may write the following expression for the second derivative:

$$y_i'' = \frac{y_{i+1} - 2y_i + y_{i-1}}{h^2} - \left(\frac{1}{12} y_i^{\text{IV}} h^2 + \cdots\right) \tag{5-73}$$

The approximate expression using only the first term to the right of the equal sign is the *central-difference* approximation of the second derivative of the function at x_i, with error of order h^2. As with the first derivative, approximations with error of higher order can also be derived. The approximate expression may be interpreted graphically as the slope of the line tangent to the curve at $x_{i+1/2}$ minus the slope of the line tangent to the curve at $x_{i-1/2}$ divided by h, where the slopes of the tangent lines are approximated by the expressions,

$$y_{i+1/2}' = \frac{y_{i+1} - y_i}{h}$$

$$y_{i-1/2}' = \frac{y_i - y_{i-1}}{h}$$

That is,

$$y_i'' = \frac{\dfrac{y_{i+1} - y_i}{h} - \dfrac{y_i - y_{i-1}}{h}}{h} = \frac{y_{i+1} - 2y_i + y_{i-1}}{h^2}$$

To obtain an expression for the third derivative, we use *four* terms on the right side of each of Eqs. 5-69 and 5-70. Subtracting Eq. 5-70 from Eq. 5-69 yields

$$y_{i+1} - y_{i-1} = 2y_i'h + \frac{2y_i'''h^3}{3!} \tag{5-74}$$

If we expand the Taylor series about x_i to obtain expressions for $y = f(x)$ at $(x_i + 2h)$ and $(x_i - 2h)$, respectively, we obtain

$$y(x_i + 2h) = y_i + y_i'(2h) + \frac{y_i''(2h)^2}{2!} + \frac{y_i'''(2h)^3}{3!} + \cdots \tag{5-75}$$

$$y(x_i - 2h) = y_i - y_i'(2h) + \frac{y_i''(2h)^2}{2!} - \frac{y_i'''(2h)^3}{3!} + \cdots \tag{5-76}$$

Subtracting Eq. 5-76 from Eq. 5-75, and using just the four terms of each expansion shown, gives

$$y_{i+2} - y_{i-2} = 4y_i'h + \tfrac{8}{3}y_i'''h^3 \tag{5-77}$$

The simultaneous solution of Eqs. 5-74 and 5-77 for the third derivative yields

$$y_i''' = \frac{y_{i+2} - 2y_{i+1} + 2y_{i-1} - y_{i-2}}{2h^3} \tag{5-78}$$

Equation 5-78 gives the first central-difference expression for the third derivative of y at x_i. By retaining two additional terms in the Taylor series expansions, we find that the error in this approximate expression is of order h^2.

Successively higher derivatives can be obtained by this method, but, since they require the solution of increasingly larger numbers of simultaneous equations, the process becomes quite tedious. The same technique may also be used to find more accurate expressions for the derivatives by using additional terms in the Taylor-series expansion. However, the derivations of these more accurate expressions, particularly for derivatives higher than the second, become very laborious because of the numbers of simultaneous equations which must be solved. Derivations of the more accurate expressions are not given here, but such expressions for several derivatives are included in the summary which follows this discussion. Derivations for the higher derivatives are accomplished with much greater facility and far less labor by using *difference*, *averaging*, and *derivative* operators. Such a method is outside the scope of this text, but it can be found in various books concerned with numerical analysis.[8]

It has been shown that the central-difference expressions for the various derivatives involve values of the function on both sides of the x value at which the derivative of the function is desired. By utilizing the appropriate Taylor-series expansions, one can easily obtain expressions for the derivatives which are entirely in terms of values of the function at x_i and points to the right of x_i. These are known as *forward-finite-difference* expressions. In a similar manner, derivative expressions which are entirely in terms of values of the function at x_i and points to the *left* of x_i can be found. These are known as *backward-finite-difference* expressions. In numerical differentiation, forward-difference expressions are used when data to the left of a point at which a derivative is desired are not available, and backward-difference expressions are used when data to the right of the desired point are not available. Central-difference expressions, however, are more accurate than either forward- or backward-difference expressions. This can be seen by noting the order of the error in the summary of the differentiation formulas which follows.

Central-Difference Expressions with Error of Order h^2

$$
\begin{aligned}
y_i' &= \frac{y_{i+1} - y_{i-1}}{2h} \\
y_i'' &= \frac{y_{i+1} - 2y_i + y_{i-1}}{h^2} \\
y_i''' &= \frac{y_{i+2} - 2y_{i+1} + 2y_{i-1} - y_{i-2}}{2h^3} \\
y_i'''' &= \frac{y_{i+2} - 4y_{i+1} + 6y_i - 4y_{i-1} + y_{i-2}}{h^4}
\end{aligned}
\tag{5-79}
$$

[8] M. G. Salvadori and M. L. Baron, *Numerical Methods in Engineering* (Englewood Cliffs, N.J.: Prentice-Hall, Inc., 1961).

Central-Difference Expressions with Error of Order h^4

$$y'_i = \frac{-y_{i+2} + 8y_{i+1} - 8y_{i-1} + y_{i-2}}{12h}$$

$$y''_i = \frac{-y_{i+2} + 16y_{i+1} - 30y_i + 16y_{i-1} - y_{i-2}}{12h^2}$$

$$y'''_i = \frac{-y_{i+3} + 8y_{i+2} - 13y_{i+1} + 13y_{i-1} - 8y_{i-2} + y_{i-3}}{8h^3} \tag{5-80}$$

$$y''''_i = \frac{-y_{i+3} + 12y_{i+2} - 39y_{i+1} + 56y_i - 39y_{i-1} + 12y_{i-2} - y_{i-3}}{6h^4}$$

Forward-Difference Expressions with Error of Order h

$$y'_i = \frac{y_{i+1} - y_i}{h}$$

$$y''_i = \frac{y_{i+2} - 2y_{i+1} + y_i}{h^2}$$

$$y'''_i = \frac{y_{i+3} - 3y_{i+2} + 3y_{i+1} - y_i}{h^3} \tag{5-81}$$

$$y''''_i = \frac{y_{i+4} - 4y_{i+3} + 6y_{i+2} - 4y_{i+1} + y_i}{h^4}$$

Forward-Difference Expressions with Error of Order h^2

$$y'_i = \frac{-y_{i+2} + 4y_{i+1} - 3y_i}{2h}$$

$$y''_i = \frac{-y_{i+3} + 4y_{i+2} - 5y_{i+1} + 2y_i}{h^2}$$

$$y'''_i = \frac{-3y_{i+4} + 14y_{i+3} - 24y_{i+2} + 18y_{i+1} - 5y_i}{2h^3} \tag{5-82}$$

$$y''''_i = \frac{-2y_{i+5} + 11y_{i+4} - 24y_{i+3} + 26y_{i+2} - 14y_{i+1} + 3y_i}{h^4}$$

Backward-Difference Expressions with Error of Order h

$$y_i' = \frac{y_i - y_{i-1}}{h}$$

$$y_i'' = \frac{y_i - 2y_{i-1} + y_{i-2}}{h^2}$$

$$y_i''' = \frac{y_i - 3y_{i-1} + 3y_{i-2} - y_{i-3}}{h^3} \tag{5-83}$$

$$y_i'''' = \frac{y_i - 4y_{i-1} + 6y_{i-2} - 4y_{i-3} + y_{i-4}}{h^4}$$

Backward-Difference Expressions with Error of Order h^2

$$y_i' = \frac{3y_i - 4y_{i-1} + y_{i-2}}{2h}$$

$$y_i'' = \frac{2y_i - 5y_{i-1} + 4y_{i-2} - y_{i-3}}{h^2}$$

$$y_i''' = \frac{5y_i - 18y_{i-1} + 24y_{i-2} - 14y_{i-3} + 3y_{i-4}}{2h^3} \tag{5-84}$$

$$y_i'''' = \frac{3y_i - 14y_{i-1} + 26y_{i-2} - 24y_{i-3} + 11y_{i-4} - 2y_{i-5}}{h^4}$$

EXAMPLE 5-3

In Ex. 2-4 the Newton-Raphson method was used to determine the output lever angles of a crank-and-lever four-bar linkage system for each 5 deg of rotation of the input crank. Now we will determine the angular velocity and the angular acceleration of the output lever of the same type of mechanism for each 5 deg of rotation of the input crank, with the latter rotating at a uniform angular velocity of 100 rad/s.

We can determine the output lever positions ϕ, corresponding to each 5 deg of crank rotation θ, by utilizing Freudenstein's equation and the Newton-Raphson method, as was done in Ex. 2-4. Such a set of values, in effect, gives us a series of points on the ϕ versus θ curve, and the ϕ values are stored in memory to provide data for the differentiation processes which follow. The slope of the ϕ-θ curve may be related to the angular velocity of the output lever $d\phi/dt$ if we realize that, with the crank rotating at a constant ω, its angular position is given by

$$\theta = \omega t$$

so that

$$\frac{d\phi}{d\theta} = \frac{1}{\omega}\frac{d\phi}{dt}$$

The angular acceleration of the output lever $d^2\phi/dt^2$ may be related to this curve by realizing that since

$$\frac{d\phi}{d\theta} = \frac{1}{\omega}\frac{d\phi}{dt}$$

it follows that

$$\frac{d^2\phi}{d\theta^2} = \frac{1}{\omega}\left(\frac{d^2\phi}{d\theta\, dt}\right)$$

But, since $d\theta = \omega\, dt$,

$$\frac{d^2\phi}{d\theta^2} = \frac{1}{\omega^2}\left(\frac{d^2\phi}{dt^2}\right)$$

Thus in performing the numerical differentiation indicated by Eqs. 5-72 and 5-73, using the ϕ values stored in memory (where θ is the independent variable), we multiply these expressions by ω and ω^2, respectively, to obtain the desired angular-velocity and acceleration values. A flow chart is shown in Fig. 5-12.

In the program which follows, the input crank angles and corresponding lever angles will be stored in memory in terms of degrees and will be printed out along with the angular-velocity and acceleration values. Subscripted variables are used for preserving these values in memory. The variable names used in the FORTRAN program are as follows:

FORTRAN Name	Quantity	Given Value
A	Length of crank a, in.	1
B	Length of crank b, in.	2
C	Length of crank c, in.	2
D	Length of crank d, in.	2
EPSI	Accuracy-check value ε, radians	0.00001
M	Number of positions of crank for which lever angle is determined	73
DELTH	Increment of crank angle $\Delta\theta$, deg	5
DELX	Increment of crank angle $\Delta\theta$, radians	
PHEST	Estimated value of ϕ corresponding to θ_{min}, deg	41
TH	Crank angle θ, deg	$\theta_0 = 0$
THMX	Maximum value of crank angle θ, deg	360
X	Crank angle θ, radians	
XMAX	Maximum value of crank angle θ, radians	
R1	d/c	

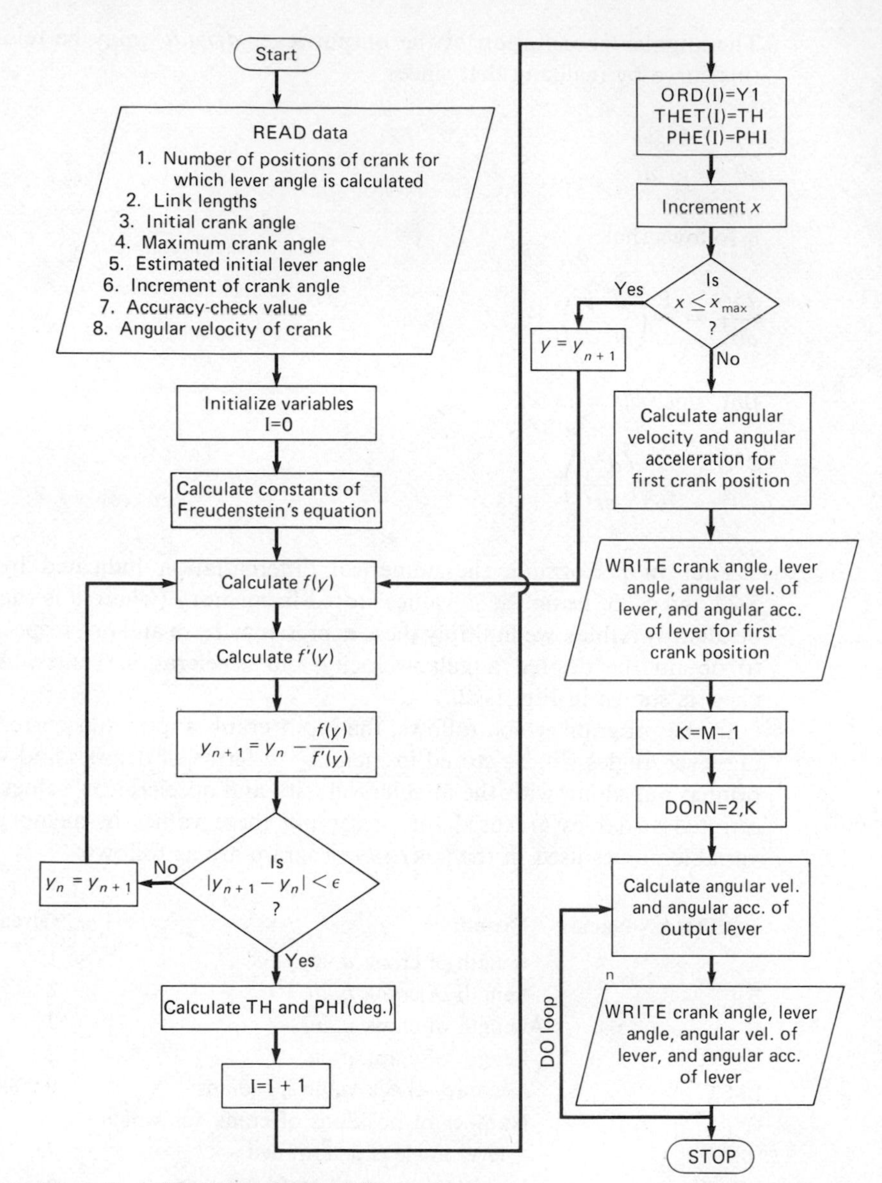

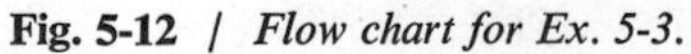
Fig. 5-12 / *Flow chart for Ex. 5-3.*

R2	d/a
R3	$(d^2 + a^2 - b^2 + c^2)/2ca$
FOFY	$f(y) = R_1 \cos x - R_2 \cos y + R_3 - \cos(x - y)$
DFOFY	$f'(y) = R_2 \sin y - \sin(x - y)$
PHI	Output lever angle ϕ, deg
Y	Output lever angle ϕ, radians

Y1	Improved value of output angle ϕ, for a given crank angle θ, radians	
OMEGA	Angular velocity of crank ω, rad/s	100
I	Subscript	
N	Subscript	
K	Number of crank positions less 1	
PHE(I)	Output-lever-angle values stored in memory, deg	
ORD(I)	Output-lever-angle values stored in memory, radians	
THET(I)	Crank-angle values stored in memory, deg	
ANG V	Angular velocity of output lever, rad/s	
ANG A	Angular acceleration of output lever, rad/s^2	

The FORTRAN program is as follows:

```
C THIS PROGRAM DETERMINES ANGULAR POSITION, VELOCITY, AND
                                                   ACCELERATION
C OF THE ROCKER LINK OF A CRANK AND ROCKER MECHANISM
C THE NEWTON-RAPHSON METHOD IS USED WITH FREUDENSTEINS
                                                       EQUATION
C TO DETERMINE THE ANGULAR POSITION OF THE ROCKER LINK
C ANGULAR VELOCITY AND ACCELERATION ARE DETERMINED BY
                                     NUMERICAL DIFFERENTIATION
      DIMENSION ORD(73),THET(73),PHE(73)
      WRITE(6,1)
    1 FORMAT('1','CRANK AND ROCKER KINEMATIC ANALYSIS'/)
      READ(5,2)M
    2 FORMAT(I3)
      READ(5,3)A,B,C,D,TH,THMAX,PHEST,DELTH,EPSI,OMEGA
    3 FORMAT(8F10.0)
      WRITE(6,4)A,B,C,D
    4 FORMAT(' ','A=',F6.3,1X,'B=',F6.3,1X,'C=',F6.3,1X,
                                                   'D=',F6.3/)
      WRITE(6,5)OMEGA
    5 FORMAT(' ','ANG VEL OF CRANK =',F5.0,' RAD PER SEC'//)
      WRITE(6,6)
      6 FORMAT(' ',3X,'THETA',4X,'PHI',4X,'ANG V',4X,'ANG
                                                         A'//)
      DELX=DELTH*.01745329
      XMAX=THMAX*.01745329
      I=0
      R1=D/C
      R2=D/A
      R3=(D*D+A*A-B*B+C*C)/(2.*C*A)
      X=TH*.01745329
      Y=PHEST*.01745329
```

```
    7 FOFY=R1*COS(X)-R2*COS(Y)+R3-COS(X-Y)
      DFOFY=R2*SIN(Y)-SIN(X-Y)
      Y1=Y-FOFY/DFOFY
      IF(ABS(Y1-Y)-EPSI)9,9,8
    8 Y=Y1
      GO TO 7
    9 TH=X/.01745329
      PHI=Y1/.01745329
      I=I+1
      ORD(I)=Y1
      THET(I)=TH
      PHE(I)=PHI
      X=X+DELX
      IF(X-XMAX)10,10,11
   10 Y=Y1
      GO TO 7
   11 ANG V=(ORD(2)-ORD(M-1))/(2.*DELX)*OMEGA
      ANG A=(ORD(2)-2.*ORD(1)+ORD(M-1))/(DELX*DELX)*OMEGA**2
      WRITE(6,12) THET(1),PHE(1),ANG V,ANG A
   12 FORMAT(' ',3X,F5.1,3X,F5.2,2X,F6.1,2X,F7.0)
      K=M-1
      DO 13 N=2,K
      ANG V=(ORD(N+1)-ORD(N-1))/(2.*DELX)*OMEGA
      ANG A=(ORD(N+1)-2.*ORD(N)+ORD(N-1))/(DELX*DELX)*OMEGA**2
   13 WRITE(6,12) THET(N),PHE(N),ANG V,ANG A
      STOP
      END
```

The computer printout of the results of the calculation is as follows:

CRANK AND ROCKER KINEMATIC ANALYSIS A= 1.000 B= 2.000 C= 2.000 D= 2.000
ANG VEL OF CRANK = 100. RAD PER SEC

THETA	PHI	ANG V	ANG A	THETA	PHI	ANG V	ANG A
0.0	41.41	33.3	2506.	100.0	87.09	42.8	−2882.
5.0	43.13	35.5	2442.	105.0	89.17	40.0	−3530.
10.0	44.96	37.6	2343.	110.0	91.09	36.6	−4280.
15.0	46.89	39.6	2220.	115.0	92.83	32.5	−5148.
20.0	48.92	41.4	2078.	120.0	94.34	27.6	−6156.
25.0	51.03	43.2	1913.	125.0	95.59	21.7	−7323.
30.0	52.23	44.8	1731.	130.0	96.51	14.7	−8662.
35.0	55.51	46.2	1530.	135.0	97.06	6.5	−10174.
40.0	57.85	47.4	1319.	140.0	97.16	−3.1	−11822.
45.0	60.25	48.5	1087.	145.0	96.75	−14.2	−13529.
50.0	62.70	49.3	847.	150.0	95.75	−26.7	−15126.
55.0	65.19	50.0	589.	155.0	94.08	−40.4	−16346.
60.0	67.70	50.4	319.	160.0	91.71	−54.9	−16803.
65.0	70.22	50.5	28.	165.0	88.60	−69.2	−16069.
70.0	72.75	50.4	−278.	170.0	84.79	−82.2	−13807.
75.0	75.26	50.0	−611.	175.0	80.37	−92.6	−10017.
80.0	77.75	49.3	−974.	180.0	75.52	−99.2	−5142.
85.0	80.19	48.3	−1373.	185.0	70.45	−101.5	−21.
90.0	82.58	46.9	−1817.	190.0	65.37	−99.6	4469.
95.0	84.88	45.1	−2314.	195.0	60.49	−94.2	7738.

THETA	PHI	ANG V	ANG A	THETA	PHI	ANG V	ANG A
200.0	55.95	-86.6	9638.	280.0	29.00	-2.1	2677.
205.0	51.83	-77.9	10353.	285.0	28.96	0.2	2576.
210.0	48.16	-68.9	10239.	290.0	29.02	2.4	2506.
215.0	44.93	-60.3	9618.	295.0	29.20	4.6	2465.
220.0	42.13	-52.3	8764.	300.0	29.48	6.7	2442.
225.0	39.71	-45.0	7835.	305.0	29.87	8.9	2440.
230.0	37.63	-38.6	6942.	310.0	30.37	11.0	2451.
235.0	35.85	-32.9	6128.	315.0	30.98	13.2	2476.
240.0	34.34	-27.8	5412.	320.0	31.69	15.3	2505.
245.0	33.07	-23.4	4802.	325.0	32.51	17.5	2538.
250.0	32.00	-19.4	4288.	330.0	33.44	19.8	2569.
255.0	31.13	-15.9	3857.	335.0	34.48	22.0	2596.
260.0	30.42	-12.6	3509.	340.0	35.64	24.3	2613.
265.0	29.86	-9.7	3220.	345.0	36.91	26.6	2616.
270.0	29.45	-7.0	2994.	350.0	38.30	28.8	2604.
275.0	29.16	-4.5	2813.	355.0	39.80	31.1	2572.

These results were compared with results obtained by a purely analytical method and found to compare very well. At the peak angular acceleration, the results obtained by numerical differentiation varied from the analytical result by approximately 0.6 percent.

Problems

5-1 The arc length BP of an ellipse is given by

$$BP = a \int_0^\phi \sqrt{1 - \left(\frac{a^2 - b^2}{a^2}\right) \sin^2 \phi} \, d\phi$$

where a and b are, respectively, the semimajor and semiminor axes of the ellipse, and ϕ is the angle shown in the accompanying figure.

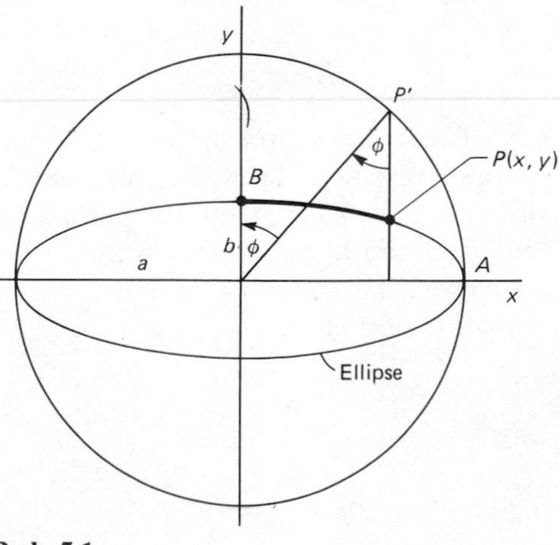

Prob. 5-1

It has been proved that the preceding integral cannot be evaluated in finite form in terms of elementary functions of ϕ.

Write a general FORTRAN program for determining, by the trapezoid rule, the arc length of an ellipse corresponding to any desired angle ϕ. Then use the program, with appropriate data input, to determine the circumference of an ellipse whose equation is given by $4x^2 + 9y^2 = 36$.

5-2 The equation of the lemniscate shown in the accompanying figure is

$$\rho^2 = a^2 \cos 2\theta$$

An element of arc length ds is equal to $\sqrt{(d\rho)^2 + (\rho\, d\theta)^2}$ from which

$$ds = \frac{a\, d\theta}{\sqrt{\cos 2\theta}}$$

and

$$s = \int_0^\theta \frac{a\, d\theta}{\sqrt{\cos 2\theta}}$$

Write a FORTRAN program for determining the total arc length of the lemniscate by use of the trapezoid rule. (See Sec. 5-5 for a discussion of this problem.)

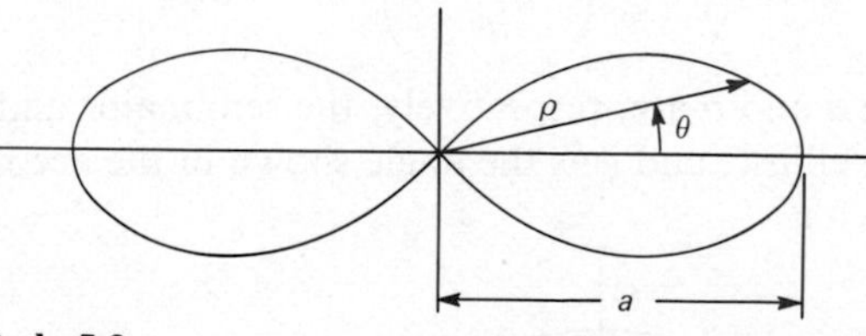

Prob. 5-2

5-3 A particle P of mass m is attracted toward a fixed point C by a force that varies inversely as the distance between them. It is desired to determine the time required for the particle to reach point C when starting from rest at a distance r from C.

Applying Newton's second law to the particle, we obtain the following equation of motion:

$$\frac{d^2x}{dt^2} = -\frac{k}{mx}$$

C P x

Prob. 5-3

The initial conditions are

$$t = 0 \left| \begin{array}{l} \dfrac{dx}{dt} = 0 \\ \\ x = r \end{array} \right.$$

Letting $v = dx/dt$, it can be seen that

$$\frac{dv}{dt} = \frac{d^2x}{dt^2} = \left(\frac{dv}{dx}\right)\left(\frac{dx}{dt}\right) = v\frac{dv}{dx}$$

Substituting the latter expression into the equation of motion yields

$$v\,dv = -\frac{k}{mx}\,dx$$

Integrating this equation, we obtain

$$\frac{v^2}{2} = -\frac{k}{m}\ln x + C$$

Application of the given initial conditions shows that

$$C = \frac{k}{m}\ln r$$

Substituting this expression for the constant of integration and the relation $v^2 = (dx/dt)^2$ into the equation resulting from the integration, we find that

$$t = -\frac{1}{\sqrt{2k/m}}\int_r^0 \frac{dx}{\sqrt{\ln(r/x)}}$$

Write a FORTRAN program for evaluating the integral by the trapezoid rule.

5-4 The system of the accompanying illustration is released from rest in the position shown. Each of the cylinders A and B weighs 32.2 lb, and C weighs 16.1 lb. The cord attached to C winds around cylinder B as cylinder B rotates about its pivot axis O. A light rod of negligible weight is attached to cylinders A and B.

a. Noting that the centroidal mass moment of inertia for each cylinder is $\bar{I}_A = \bar{I}_B = 0.125$ ft-lb sec^2, show that the equation of motion of the system is

$$\frac{d\dot{\theta}}{dt} = \frac{64.4 \cos\theta - 8.05}{4.375}$$

where θ is the angular displacement of the light rod and cylinders.

b. From the equation of motion, formulate the integral which may be used to calculate the time t for the system to rotate through the angle θ.

c. Using the integral of part b, write a FORTRAN program for obtaining the t versus θ values for one-half of a cycle of motion. Use either the trapezoid rule or Simpson's rule.

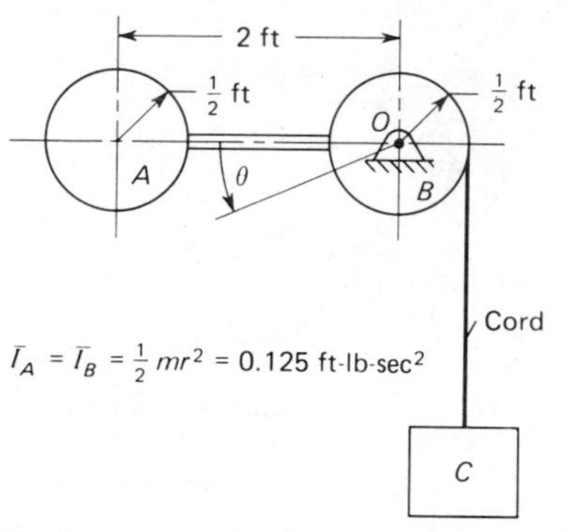

Prob. 5-4

5-5 An Alnico magnet A exerts a force on a steel block of weight W that is inversely proportional to the square of the distance between the mass centers of the magnet and the block. If the coefficient of friction between the block and the fixed surface is f, the equation of motion of the block is found to be

$$\frac{d\dot{x}}{dt} = -\frac{kg}{Wx^2} + fg$$

where k is a constant, and g is the acceleration of gravity.

a. Considering that the block is released from rest with an initial displacement of x_0, show that the expression for the velocity is

$$\dot{x} = \sqrt{\frac{2kg}{W}\left(\frac{1}{x} - \frac{1}{x_0}\right) + 2\,fg(x - x_0)}$$

b. Formulate the integral for calculating the time t for different values of x.

c. Write a FORTRAN program for obtaining the t versus x values. Use either the trapezoid rule or Simpson's rule. For a computer solution, take $f = 0.2$, $k = 1000$ lb-in.^{2}, $g = 386$ in./sec^2, $W = 10$ lb, and $x_0 = 12$ in.

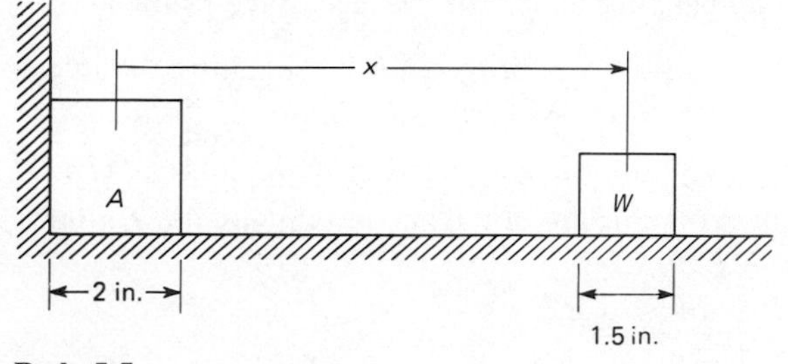

Prob. 5-5

5-6 Referring to Prob. 5-5, write a FORTRAN program for obtaining the t versus x values for coefficient-of-friction values of 0.1, 0.2, 0.3, 0.4, and 0.5. Use Simpson's rule.

5-7 A particle of mass m is released from rest at the edge of a hemispherical bowl such as the one shown in the accompanying figure. The particle then slides down the inside surface of the bowl which is assumed frictionless. It is desired to find the time required for the particle to reach the bottom of the bowl.

Applying Newton's second law in the tangential direction, write the differential equation of motion of the mass with θ as the dependent variable. Using the method of Prob. 5-3 as a reference, determine the integral which may be evaluated to find the required time for the mass to slide to the bottom of the bowl.

Write a FORTRAN program for evaluating the integral determined. Use the trapezoid rule.

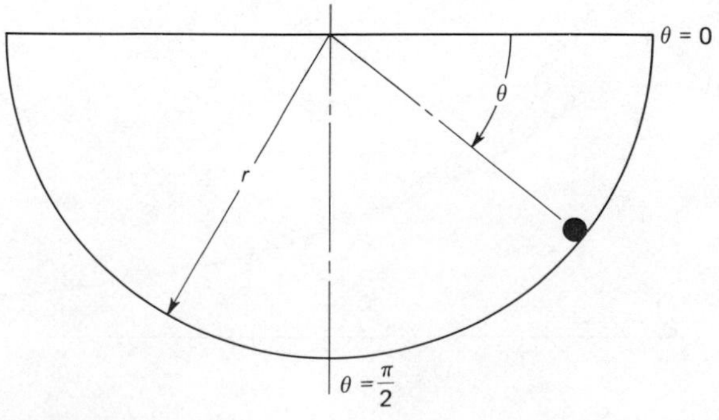

Prob. 5-7

5-8 Write a FORTRAN program for determining the arc length of an ellipse, corresponding to any angle ϕ, by the use of Simpson's rule. Refer to Prob. 5-1 and its accompanying figure for the necessary problem information.

5-9 Write a FORTRAN program for determining the total arc length of a lemniscate by the use of Simpson's rule. Refer to Prob. 5-2 and its accompanying figure for the necessary problem information.

5-10 Write a FORTRAN program for evaluating the integral of Prob. 5-3 by Simpson's rule.

5-11 The functions $f(x) = 1/(1 + x^2)$ and $g(x) = 1/(1 + e^{-x} + x^2)$ are shown graphically in the accompanying illustration. The evaluation

$$\int_0^\infty \frac{dx}{1 + x^2} = \frac{\pi}{2}$$

may be obtained by the direct integration of $f(x)$. Thus the integral

$$\int_0^\infty g(x)\, dx$$

may be evaluated by considering that

$$\int_0^\infty \frac{dx}{1 + e^{-x} + x^2} = \frac{\pi}{2} - A$$

where A is the area (see shaded area in accompanying illustration) between $f(x)$ and $g(x)$ from zero to infinity. Since $f(x)$ and $g(x)$

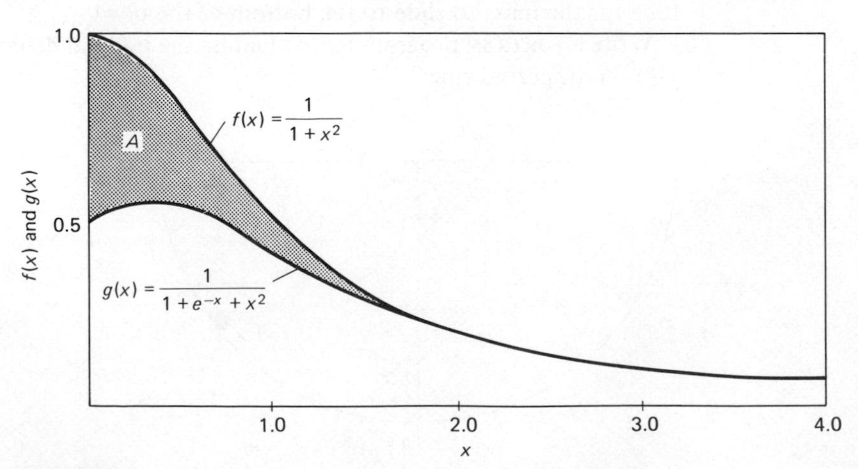

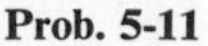
Prob. 5-11

approach each other rapidly for finite values of x, as shown in the figure, the area A may be evaluated quite accurately by using either the trapezoid rule or Simpson's rule, with integration over a finite range. Write a FORTRAN program for evaluating the area A and obtain the value of the integral

$$\int_0^\infty g(x)\,dx$$

by utilizing the relationship

$$\int_0^\infty g(x)\,dx = \frac{\pi}{2} - A$$

5-12 A satellite, having unequal principal moments of inertia, is subjected to a gravity-stabilizing torque which tends to align the axis of the minimum moment of inertia with the radius vector connecting the mass centers of the earth and the satellite. For the satellite shown in the accompanying figure, the minimum moment of inertia is about the x axis, and the gravity torque tends to align this axis with the radius vector r_0. However, since the inertia of the satellite

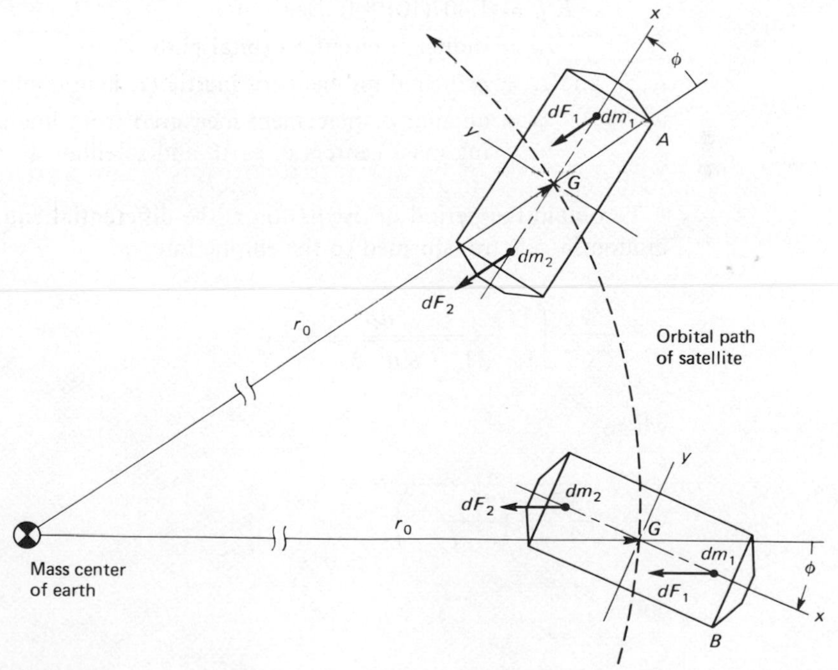

Prob. 5-12

causes it to overshoot the equilibrium position, thus introducing a torque in the opposite direction, the satellite oscillates very slowly as it moves on its orbital path.

The gravity torque can be visualized, in a qualitative sense, by considering the differential gravity forces dF_1 and dF_2, which are shown acting on the differential masses dm_1 and dm_2, respectively. If these differential masses are equidistant from the mass center G of the satellite, then $dF_2 > dF_1$, since dm_2 is closer to the mass center of the earth than is dm_1. Thus in position A the differential gravity torque due to dF_1 and dF_2 is clockwise. Similarly, in position B the differential gravity torque is counterclockwise. By integrating over the mass of the satellite, the differential equation of motion, describing the oscillation of a satellite moving in a circular orbit, is found to be

$$\frac{d\dot{\phi}}{dt} + \frac{3}{2}\frac{R^2 g}{r_0^3}\left(\frac{I_y - I_x}{I_z}\right)\sin 2\phi = 0$$

where

R = radius of earth, $20.9(10)^6$ ft
g = acceleration of gravity at earth's surface, 32.2 ft/sec^2
$R^2 g$ = $1.407(10)^{16}$ ft^3/sec^2
r_0 = radius of circular orbital path
I_x, I_y, and I_z = principal moments of inertia (I_x is minimum)
ϕ = angular displacement measured from line connecting mass centers of earth and satellite

To obtain the period of oscillation τ, the differential equation of motion in ϕ is transformed to the elliptic integral

$$\tau = \frac{4}{C\sqrt{2}}\int_0^{\pi/2}\frac{d\beta}{\sqrt{1 - \sin^2\phi_0 \sin^2\beta}}$$

where

$$C = \sqrt{\frac{3}{2}\frac{R^2 g}{r_0^3}\left(\frac{I_y - I_x}{I_z}\right)}$$

and

ϕ_0 = amplitude of oscillation

The elliptic integral is obtained from the differential equation of motion by utilizing the relations

$$\frac{d\phi}{dt} = \dot{\phi}$$

$$\cos 2\phi = 1 - \sin^2 \phi$$

$$\sin \phi = \sin \phi_0 \sin \beta$$

Write a FORTRAN program for obtaining the period of oscillation by integrating the given elliptic integral, by either the trapezoid rule or by Simpson's rule, with the following data:

$$r_0 = 1.1R$$

$$I_y = I_z$$

$$I_x = 0.5I_y$$

$$\phi_0 = 10, 20, 30, 40, 50, 60, 70, 80, \text{ and } 85 \text{ deg}$$

Using the computer results, plot the period τ as a function of ϕ_0. Compute the period τ for the above conditions, utilizing a table of values for elliptic integrals, and compare with the computer results.

5-13 Write a computer program for integrating any function defined by a function subprogram using Romberg integration. (Follow the flow chart given in Fig. 5-3 of the text.)

5-14 Improve the Romberg integration program of Prob. 5-13 by using a subprogram which calculates function ordinates rather than performing the complete trapezoidal integration. The trapezoidal integration will be a part of the Romberg program. Ordinate values which have been calculated once should not be recalculated when the number of strips is doubled. Their sum should be retained as the value of a variable.

5-15 In the slider-crank mechanism shown in part a of the accompanying figure, the slider B is driven by a Scotch-yoke mechanism shown in part b which gives it simple harmonic motion. Its displacement x is given by

$$x = R(1 - \cos \omega t)$$

where

ωt = crank angle of Scotch-yoke mechanism (see figure)

R = crank lengths of slider-crank and Scotch-yoke mechanisms

From geometry, x may be related to the crank angle θ by the formula

$$\begin{aligned} x &= R + L - R\cos\theta - L\cos\phi \\ &= R(1 - \cos\theta) + L(1 - \cos\phi) \\ &= R(1 - \cos\theta) + L(1 - \sqrt{1 - \sin^2\phi}) \\ &= R(1 - \cos\theta) + L[1 - \sqrt{1 - (R/L)^2\sin^2\theta}] \end{aligned}$$

where θ, ϕ, and L are shown in the figure.

Using the following data, determine the values of θ corresponding to values of ωt from 0 to 350 deg (use 10-deg increments), using the Newton-Raphson method. Then determine the values of $d\theta/d(\omega t)$ and $d^2\theta/d(\omega t)^2$ by numerical differentiation.

$R = 4$ in.

$L = 6$ in.

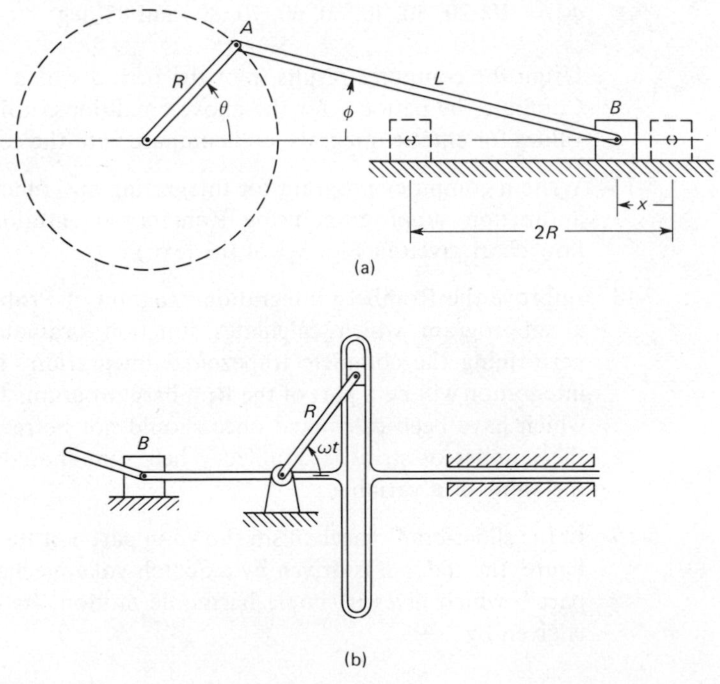

Prob. 5-15

5-16 The computer program shown in Ex. 5-3 was used for determining the angular velocity and acceleration of the output lever of a crank-and-lever mechanism using the first two numerical differentiation formulas of Eq. 5-79. Modify this program to utilize the first two differentiation formulas of Eq. 5-80, so that more accurate results may be obtained. Compare the solution obtained with the solution given in Ex. 5-3.

5-17 An automatic washer transmission uses the linkage shown in the accompanying figure to convert rotary motion from the drive motor to a large oscillating output of the agitator shaft *G*. The 79-tooth helical gear is driven by the 12-tooth helical pinion which rotates at a constant speed of 435 rpm. Links 2, 4, and 6 rotate or oscillate about fixed axes. Links 2, 3, and 4 (*DA*, *AB*, and *BC*) along with the fixed base constitute a four-bar linkage. Links 4, 5, and 6 (*CE*, *EF*, and *FG*) along with the fixed base constitute a second four-bar linkage in series with the first. Angle ϕ may be found for any angle θ using Freudenstein's equation and the Newton-Raphson root-finding technique as in Exs. 2-4 and 5-3. Angle α is equal to angle ϕ plus a constant. With angle α known, the corresponding angle β may be found, again using Freudenstein's equation and the Newton-Raphson method.

Write a FORTRAN program which determines and stores values of angle β for every degree of angle θ. Using these stored values of β and numerical differentiation, determine the angular velocity and angular acceleration of link 6 for every 5 deg of angle θ. Use Ex. 5-3 as a guide. For the numerical differentiation, use a $\Delta\theta$ value of 1 deg (or the corresponding value in radians). Print out values of θ and β in degrees and the angular velocity and angular acceleration of link 6 for every 5 deg of angle θ.

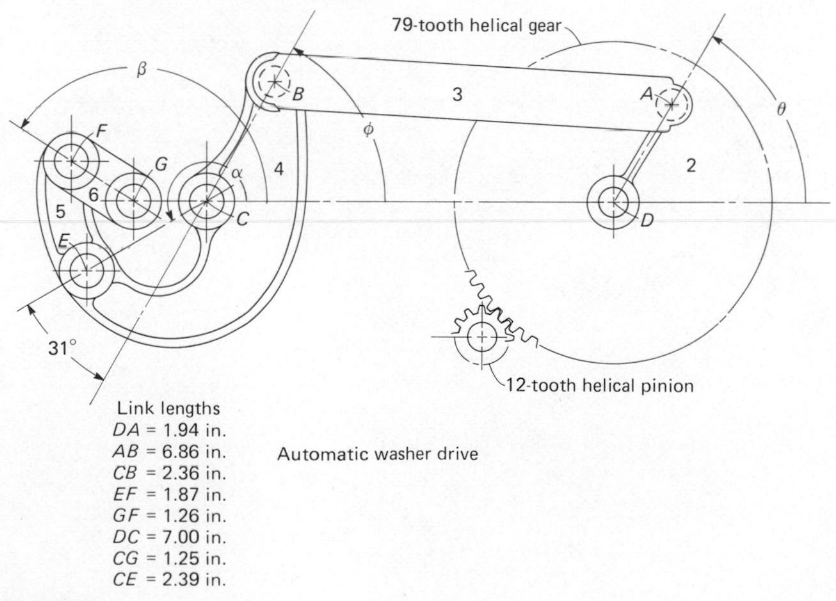

Automatic washer drive

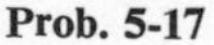

Prob. 5-17

The following problems involve SI units.

5-18 Work Prob. 5-4 where the 2-ft distance is to be taken as 0.5 m, and the diameter of cylinders A and B is 0.15 m. The weights and mass moments of inertia are

$$W_A = W_B = 100 \text{ N}$$

$$W_C = 50 \text{ N}$$

$$I_A = I_B = 0.15 \text{ kg-m}^2$$

5-19 Work Prob. 5-5 by converting the system data of the problem to the SI system of units.

5-20 Work Prob. 5-12 by converting the given data to the SI system of units.

6 Numerical Integration of Ordinary Differential Equations: Initial-Value Problems

6-1 / INTRODUCTION

Initial-value problems are problems in which the values of the dependent variable and the necessary derivatives are known at the point at which integration begins. Such a large number of integration methods are available to handle problems of this type that an engineer may have difficulty in deciding which to use. In this chapter the most widely used methods are presented in some detail, and a summary is included to aid the reader in selecting one of these methods for a particular application.

The methods discussed will vary in complexity, since, in general, the greater the accuracy of a method, the greater is its complexity. Included are Euler's, modified Euler, Runge-Kutta, Milne's, and Hamming's methods. The error analysis of each of these methods is explained in detail.

This chapter will be concerned primarily with the solution of first-order differential equations and of sets of simultaneous first-order differential equations, since, as will be seen later, an nth-order differential equation may be solved by transforming it to a set of n simultaneous first-order differential equations. Several examples will be given in which the methods developed for solving first-order differential equations are extended to solve equations of higher order.

6-2 / DIRECT NUMERICAL-INTEGRATION METHOD

Let us consider the cantilever beam shown in Fig. 6-1. The intensity of loading is some function of the displacement x along the beam, x being

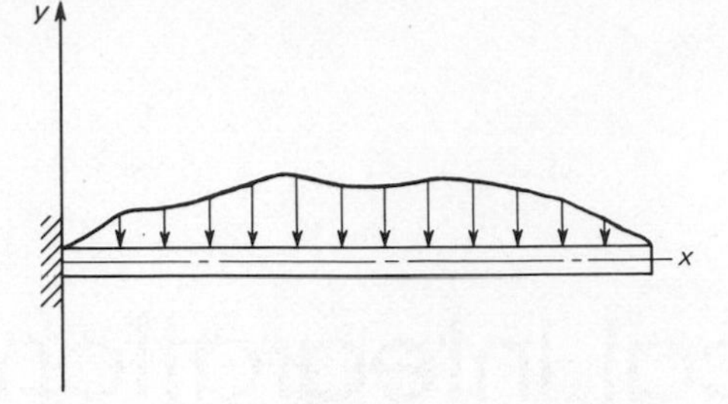

Fig. 6-1 / *Cantilever beam with intensity of loading indicated by length of arrows.*

measured positively from the left end of the beam. The bending moment M along the beam will also be some function of x, and the M/EI diagram will have the general form shown in Fig. 6-2.

The differential equation of the elastic curve of a beam with a uniform cross section is

$$\frac{d^2y}{dx^2} = \frac{M(x)}{EI} \tag{6-1}$$

where E and I are the modulus of elasticity and the moment of inertia of the cross section, respectively. The initial conditions for the cantilever beam are

$$x = 0 \left| \begin{array}{l} y = 0 \\ y' = 0 \end{array} \right.$$

The numerical solution of Eq. 6-1 presents no particular difficulty. Its solution involves the determination of the deflection y for a series of equally spaced x values and is accomplished by two successive integrations. The curve of y' (the slope of the elastic curve) versus x may easily be obtained, since the value of y' is known at $x = 0$, and the change in y' from x_i to x_{i+1} is represented by the area under the M/EI curve between the given x values. The general form of the y' versus x curve is shown in

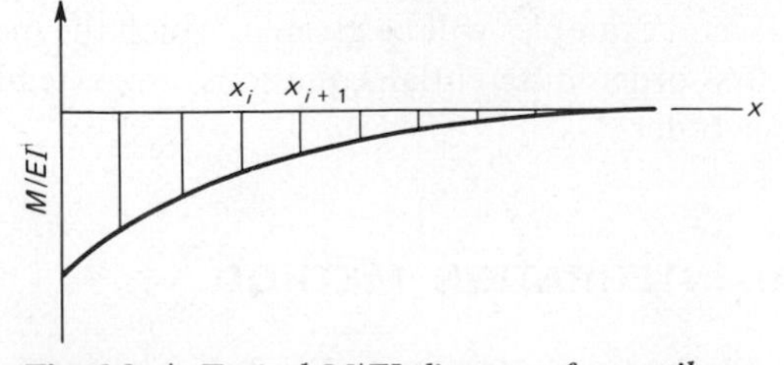

Fig. 6-2 / *Typical M/EI diagram of a cantilever beam.*

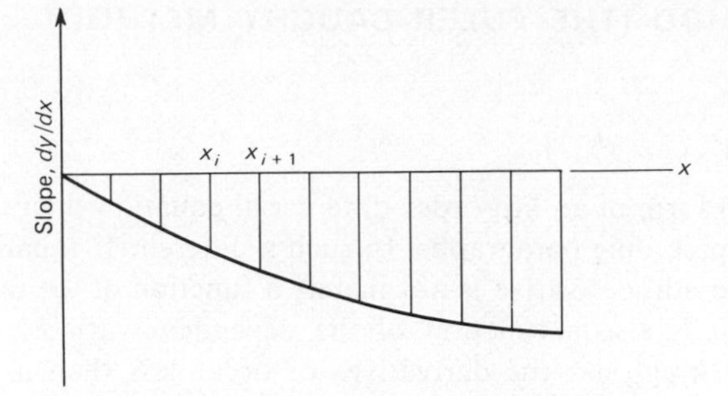

Fig. 6-3 / *Typical slope curve of a cantilever beam.*

Fig. 6-3. The y versus x curve, which is known as the elastic curve of the beam, is obtained in a similar manner, since the value of y is known at $x = 0$, and the change in y over an x interval is given by the corresponding area under the y' versus x curve. Either the trapezoid rule or Simpson's rule can be used for determining the areas mentioned. The general form of the elastic curve of a cantilever beam is shown in Fig. 6-4.

It now should be apparent that the numerical solution of a differential equation, in which the curve of the highest-order derivative is known, may be obtained by direct numerical integration, provided the necessary initial conditions are known. Denoting $y^n = d^n y/dx^n$, we may generalize that all nth-order differential equations of the form

$$y^n = f(x) \tag{6-2}$$

with the initial conditions

$$x = x_0 \left| \begin{array}{l} y = y_0 \\ y' = y_0' \\ \vdots \quad \vdots \\ y^{n-1} = y_0^{n-1} \end{array} \right.$$

can be solved in the manner discussed.

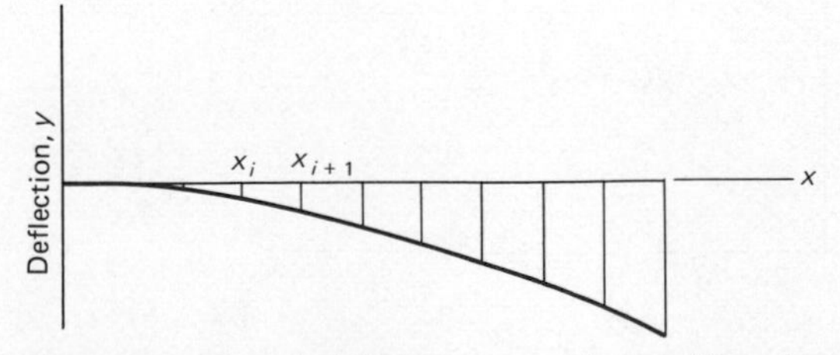

Fig. 6-4 / *Typical deflection curve of a cantilever beam.*

6-3 / EULER'S METHOD (THE EULER-CAUCHY METHOD)

The expression

$$y^n = f(x, y, y', y'', \ldots, y^{n-1}) \tag{6-3}$$

is a more general form of an nth-order differential equation than the one described in the preceding paragraphs. In such a differential equation the highest order (the nth) derivative is not merely a function of the independent variable but is also a function of the dependent variable, whose values we are seeking, and the derivatives of order less than n. *Direct* integration of such a differential equation, using the trapezoid rule or Simpson's rule, is not possible. To illustrate a simple method of obtaining an approximate solution of a differential equation of this type, let us consider the first-order equation

$$y' = f(x, y) \tag{6-4}$$

with the initial condition of $y = 0$ when $x = 0$. We wish to obtain the y versus x curve, which suggests using the area under the y' versus x curve. However, the latter curve is unknown, since y' is a function of y. To illustrate, let us assume that these curves have the general form of those shown in Fig. 6-5. Since the initial value of y is known, we can determine the

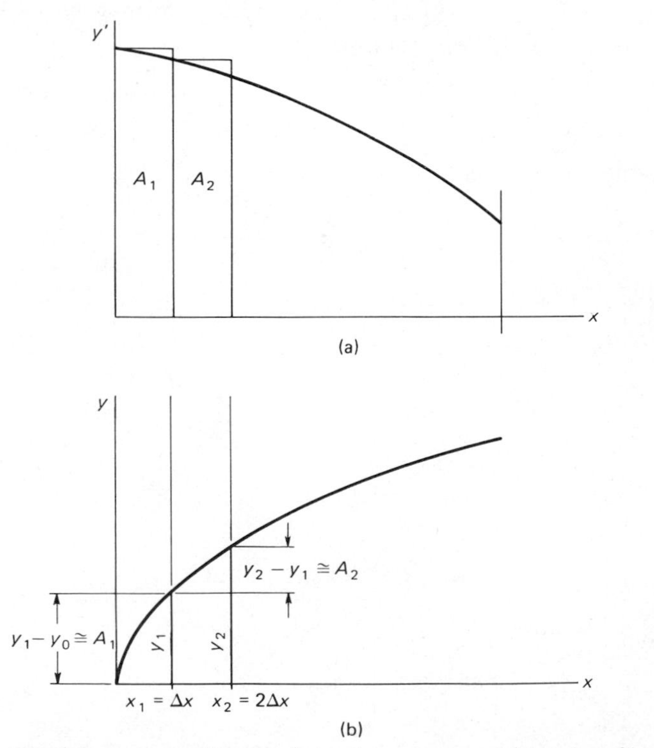

Fig. 6-5 / *Assumed curves of y' and y versus x satisfying $y' = f(x, y)$.*

initial value of y' from the given differential equation. The change in y from $x = 0$ to $x = \Delta x$ is represented by the area under the y' curve between the given values of x. An approximate value of this area can be obtained by assuming that it is equal to the rectangular area A_1, shown in Fig. 6-5(a). Letting $\Delta x = h$, this area may be expressed as

$$A_1 = y_1 - y_0 = y_0'h$$

from which

$$y_1 = y_0 + y_0'h \tag{6-5}$$

Having determined a close approximation of the value of y_1 by the use of Eq. 6-5 (by using small h values), we can get a good approximation of y_1' from the given differential equation, since

$$y_1' = f(x_1, y_1)$$

Then, since $y_2 - y_1$ is approximately equal to the area under the y' curve from x_1 to x_2, as shown by the rectangular area A_2 in Fig. 6-5,

$$y_2 = y_1 + y_1'h \tag{6-6}$$

Proceeding as before, y_2' may next be determined from the differential equation, by using the value of y_2 obtained from Eq. 6-6 and a value of $x_2 = 2h$. The subsequent y_i and y_i' values are found in a similar manner. Equations 6-5 and 6-6 may be expressed, in general form, as

$$y_{i+1} = y_i + y_i'h \tag{6-7}$$

Equation 6-7 is known as Euler's forward-integration equation. Euler's method is called a "self-starting" method, since it requires a value of the dependent variable at only one point to start the procedure.

A graphical representation of a portion of the function $y(x)$, as obtained by Euler's method, is shown in Fig. 6-6. Let us assume the initial condition of $y = 0$ at $x = 0$. With a known initial value of y, the initial slope of the curve can be determined from the differential equation. The first calculated y value, indicated in Fig. 6-6, can then be obtained from Euler's equation (Eq. 6-7). It can be seen from the figure that Euler's method, in effect, approximates the function $y(x)$ over the first-step interval by a straight line. The substitution of this calculated value of y into the differential equation then gives an approximation of the slope of the curve at $x = h$. A second application of Euler's equation yields a second calculated value of y, as shown in the figure, which is, in turn, substituted into the differential equation to obtain an approximation of the slope at $x = 2h$.

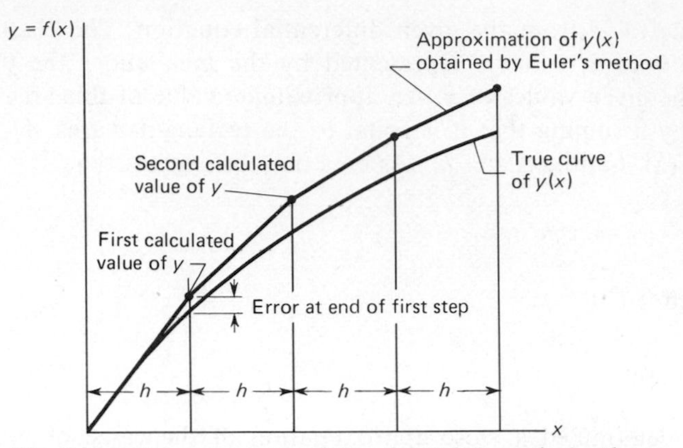

Fig. 6-6 / *Comparison of solution by Euler's method with true solution.*

This process is repeated over the desired range of integration, so that the true $y(x)$ curve is approximated by a series of straight-line segments.

Error Estimation for Euler's Method

Euler's formula is simply a truncated Taylor series in which terms containing h^2 and higher powers are neglected. The Taylor series for y_{i+1} expanded about x_i with the Lagrangian form of the remainder is

$$y_{i+1} = y_i + y_i'h + \frac{y''(\xi)h^2}{2}, \qquad x_i < \xi < x_{i+1} \tag{6-8}$$

The last term of Eq. 6-8 is the *truncation-error term* of Euler's method. Assuming that y'' is fairly constant over the ith-step interval, so that the higher derivative terms are negligible, an estimate of the truncation error E_T in the ith step can be obtained from

$$E_{T_i} \cong \frac{y_i''h^2}{2} \tag{6-9}$$

where y'' is evaluated at $\xi = x_i$. If a truncation-error analysis is made at each step, Eq. 6-9 may be used to estimate the truncation error at each step for comparison with some selected set of limiting values. If the estimate exceeds the bound of the maximum specified error, the step size can be reduced and the numerical process continued. On the other hand, if the estimate is smaller than the minimum limit, the step size can be increased and the numerical process continued. In either case the final value of y, calculated by using the old step size, is used as the initial value of y for the subsequent interval of integration using the new step size.

In an instance in which the differentiation of

$$y' = f(x, y) \tag{6-10}$$

to obtain y'' for use in estimating the per-step truncation error of Euler's method leads to a rather complicated function, a more practical approach can be made by using a constant step size h to obtain a solution, and then obtaining another solution using a smaller step size, say, $h/2$. If the two solutions yield results which are in good agreement throughout—to four figures, for example—it is fairly safe to assume that both are reasonably accurate to four figures, and that the one with the smaller step size has greater than four-figure accuracy. If the two solutions are not in good agreement throughout, one using a still smaller step size should be obtained for comparison with the solution obtained with the next higher step size, and so forth. If the initial step size chosen happened to be too small, so that roundoff error dominated, it would be necessary to follow the above procedure, but increase the step size each time. The programmer who does not wish to complicate his program by incorporating a variable step size over the range of integration will find the procedure just discussed a practical means of checking the accuracy of a solution obtained by any of the integration methods described in this chapter.

EXAMPLE 6-1

The rattrap shown in Fig. 6-7 consists of a movable jaw, trip arm, torsion spring, and trip pan. Upon release of the trip arm owing to a slight disturbance of the trip pan, the torque of the torsion spring closes the rotating jaw and kills the rodent.

The manufacturer has received complaints about the performance of the trap with respect to the closure time of the jaw and the impact force of the jaw on the rodent. To analyze the trap, it is desired to obtain data for drawing θ-t, ω-t, and ω-θ curves for the jaw of the trap.

The data for the trap are $A = 1.125$ in., $B = 0.5$ in., $R = 3.75$ in., and $I_O = 0.0006$ lb-ft-sec^2 (mass moment of inertia of the jaw about the pivot axis).

The torque-displacement characteristic of the linear torsion spring is shown in Fig. 6-8. When closed, the jaw exerts a force against the base which has a component perpendicular to the jaw of 2 lb, and a force of 9 lb is required to hold the jaw in the open position. From the given information, we find that

$\theta_k = 2.97$ radians or 170.4 deg (angular displacement of jaw upon contact with rat)

$\theta_c = 3.27$ radians or 187.7 deg (angular displacement of jaw in the closed position)

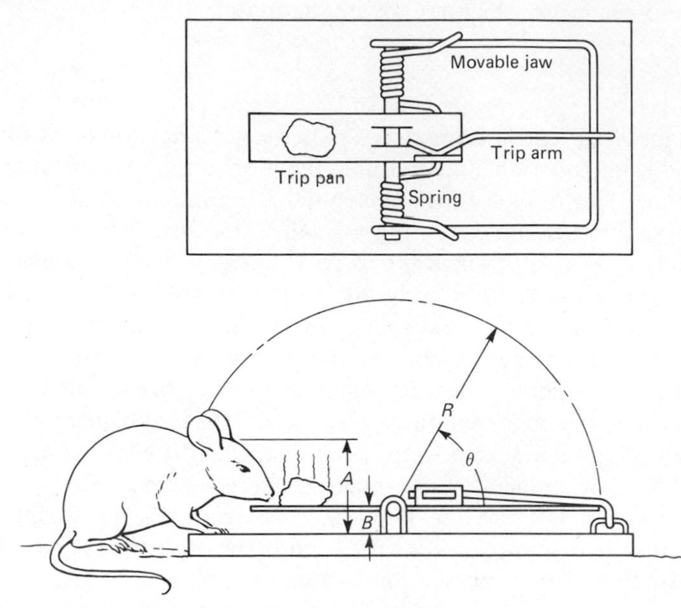

Fig. 6-7 / *Rattrap shown in open position.*

$T_O = 9(3.75)/12 = 2.81$ lb-ft (torque of spring in open position; $\theta = 0$)

$T_c = 2(3.75)/12 = 0.625$ lb-ft (torque of spring in closed position)

Upon release, the jaw is subjected to the torque T of the spring, as shown in Fig. 6-9. From elementary dynamics we know that the external torque T about point O equals $I_O\ddot{\theta}$. Taking θ and $\ddot{\theta}$ positive, as shown in the figure, the differential equation of motion is

$$\ddot{\theta} + \frac{k\theta}{I_O} = \frac{T_O}{I_O} \tag{6-11}$$

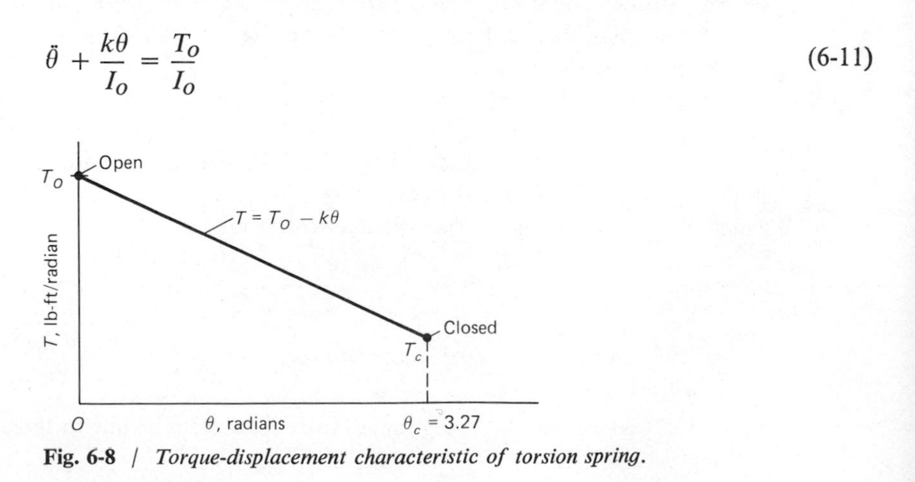

Fig. 6-8 / *Torque-displacement characteristic of torsion spring.*

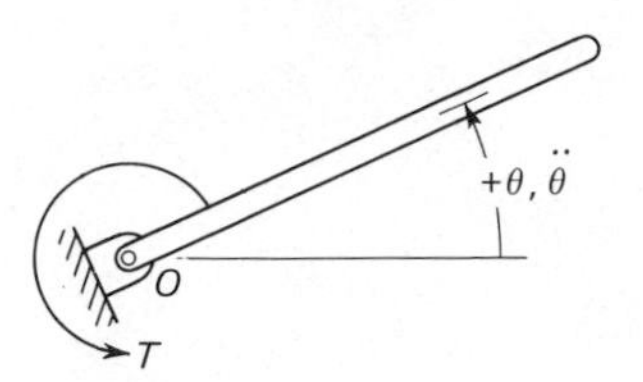

Fig. 6-9 / *Movable jaw after release.*

Referring to Fig. 6-8, it can be seen that the torsional spring constant may be expressed as

$$k = \frac{T_O - T_c}{\theta_c} = \frac{2.81 - 0.625}{3.27} = 0.669 \text{ lb-ft/radian} \tag{6-12}$$

Using the relationship given in Eq. 6-12, we may rewrite Eq. 6-11 as

$$\ddot{\theta} + \frac{k\theta}{I_O} = \frac{T_c + k\theta_c}{I_O} \tag{6-13}$$

In this example we will use Euler's forward-integration method to obtain the desired data for drawing θ-t, ω-t, and ω-θ curves for the trap jaw. We will also compare the data obtained for the θ-t curve by Euler's method with corresponding values determined from an analytical solution of Eq. 6-11.

Inspection of Eq. 6-13 reveals that we are considering a second-order differential equation of the form

$$\ddot{\theta} = f(\theta) \tag{6-14}$$

whereas the discussion thus far has dealt only with first-order differential equations. However, we can make the discussion pertinent to the solution of second-order equations, since the latter may be reduced to a set of two first-order equations which can be solved simultaneously. Realizing that

$$\dot{\theta} = \omega$$

and

$$\ddot{\theta} = \dot{\omega}$$

Eq. 6-13 can be reduced to the two first-order differential equations

$$\dot{\theta} = \omega \qquad \text{(a)}$$

$$\dot{\omega} + \frac{k\theta}{I_O} = \frac{T_c + k\theta_c}{I_O} \qquad \text{(b)} \tag{6-15}$$

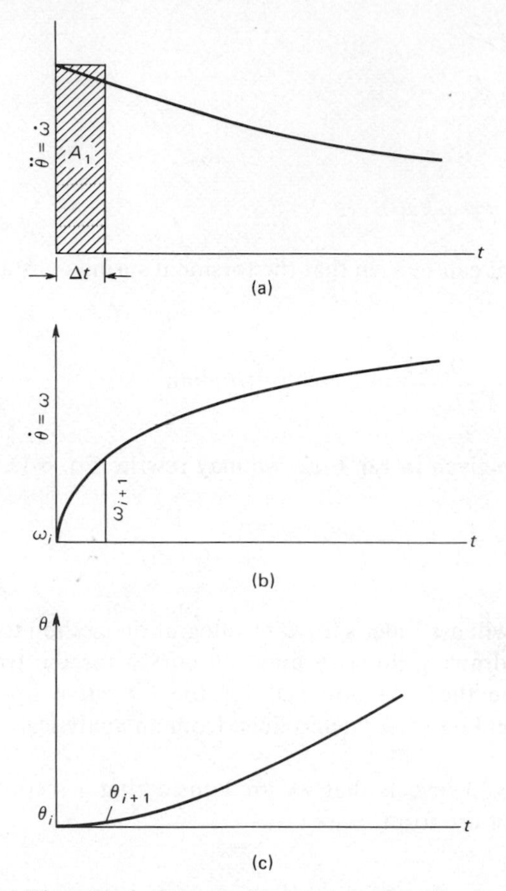

Fig. 6-10 / *Dynamic-characteristic curves for the rattrap jaw of Ex. 6-1.*

which can be solved simultaneously to obtain the desired data of the problem.

Using Euler's forward-integration equation (Eq. 6-7) with Eq. 6-15b, we can write

$$\omega_{i+1} = \omega_i + \dot{\omega}_i(\Delta t) \tag{6-16}$$

in which the term $\dot{\omega}_i(\Delta t)$ approximates the area in one strip under the $\dot{\omega}$-t curve. This approximate area is shown as the crosshatched area A_1 in Fig. 6-10(a). We next apply Euler's equation to Eq. 6-13a and obtain

$$\theta_{i+1} = \theta_i + \omega_i(\Delta t) \tag{6-17}$$

Let us now write Eqs. 6-15b, 6-16, and 6-17 as a set, so that we may examine their interdependence, writing Eq. 6-15b in the subscripted form in which it will be used in the numerical process.

$$\dot{\omega}_i = \left(\frac{T_c + k\theta_c}{I_O}\right) - \frac{k\theta_i}{I_O}$$

$$\omega_{i+1} = \omega_i + \dot{\omega}_i(\Delta t) \qquad (6\text{-}18)$$

$$\theta_{i+1} = \theta_i + \omega_i(\Delta t)$$

Selecting a value for Δt and remembering that the initial conditions are $\theta = \omega = 0$ when $t = 0$, we see that $\dot{\omega}_0$, ω_1, and θ_1 can be evaluated from Eq. 6-18. In doing this, the first-step results will contain the inherent errors of Euler's method. For example, we obtain the obviously incorrect value of $\theta_1 = 0$. If Δt is very small, this error will not be serious but it is apparent that a trapezoidal approximation to the area under the ω–t curve in the first step would give a much better value of θ_1. Use of a trapezoidal approximation will also improve the accuracy in all subsequent steps. Since ω_i and ω_{i+1} are both known before we are ready to calculate θ_{i+1}, we can use the trapezoid rule to obtain θ_{i+1}, as follows:

$$\theta_{i+1} = \theta_i + \left(\frac{\omega_i + \omega_{i+1}}{2}\right)\Delta t \qquad (6\text{-}19)$$

It can be seen that the following value will be obtained for θ_1

$$\theta_1 = \frac{\omega_1(\Delta t)}{2}$$

which is a closer approximation of the value of θ at the end of the first Δt interval.

Hence we will program the problem for a solution on the computer, using the following set of equations:

$$\dot{\omega}_i = \left(\frac{T_c + k\theta_c}{I_O}\right) - \frac{k\theta_i}{I_O}$$

$$\omega_{i+1} = \omega_i + \dot{\omega}_i(\Delta t) \qquad (6\text{-}20)$$

$$\theta_{i+1} = \theta_i + \left(\frac{\omega_i + \omega_{i+1}}{2}\right)\Delta t$$

where we are using Euler's forward-integration equation to solve Eq. 6-15b and the trapezoid rule to solve Eq. 6-15a.

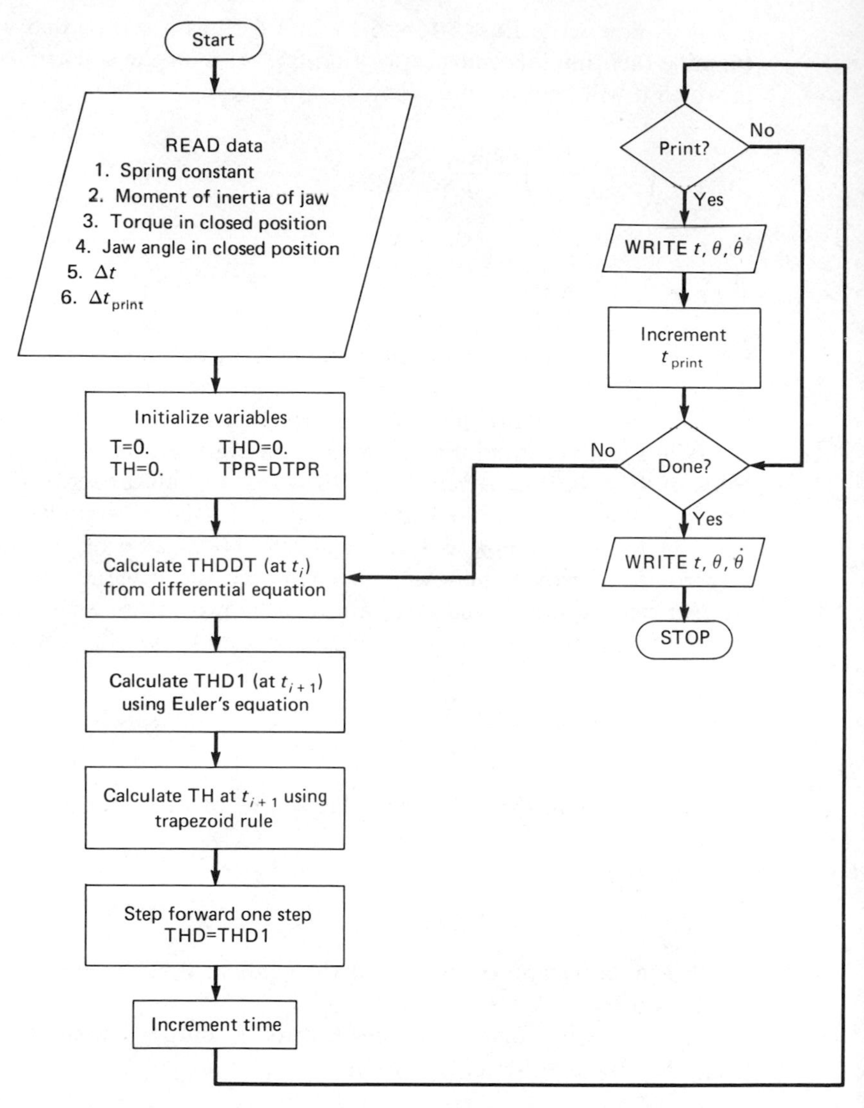

Fig. 6-11 / *Flow chart for numerical solution of rattrap problem of Ex. 6-1.*

This program is to be run on a pure binary computer, and we note that the time increment shown below, 0.0001 sec, cannot be expressed precisely in pure binary form, so that the 10 time increments will not add up exactly to the print increment of 0.001 sec. For a printout in this program, the time only need be within a small epsilon value (EPSI) of the print time. See the discussion following the listing of FORTRAN variable names in Ex. 6-4 regarding this problem on a pure binary computer.

The flow chart for the computer solution is shown in Fig. 6-11. The variable names selected for the program are as follows:

FORTRAN Name	Quantity	Numerical Value Used in Solution
SK	Spring constant k, lb-ft/radian	0.669
I	Moment of inertia I_0, lb-ft-sec^2	0.0006
TORCL	Torque in closed position T_c, lb-ft	0.625
THCL	Jaw angle θ in closed position θ_c, radians	3.270
DELT	Time increment Δt, sec	0.0001
DTPR	Print increment, Δt_{print}, sec	0.0010
T	Time t, sec	
TH	Jaw angle θ_i	
THD	Jaw angular velocity $\dot{\theta}_i$	
THDDT	Jaw angular acceleration $\ddot{\theta}_i$	
THD1	Jaw angular velocity one station ahead of THD, $\dot{\theta}_{i+1}$	
TPR	Time value at each printout	
EPSI	A small value used to assure a printout when T is supposed to exactly equal TPR but does not	0.00003

The FORTRAN IV program for the problem is as follows:

```
      REAL I
      WRITE(6,10)
   10 FORMAT('1',1X,'DYNAMIC CHARACTERISTICS OF A RATTRAP')
      WRITE(6,1)
    1 FORMAT('0',5X,' TIME',6X,'THETA',6X,'ANG V',//)
      READ(5,2) SK,I,TORCL,THCL,DELT,DTPR,EPSI
    2 FORMAT(7F10.0)
      T=0.
      TH=0.
      THD=0.
      TPR=DTPR
    3 THDDT=(TORCL+SK*THCL)/I-SK*TH/I
      THD1=THD+THDDT*DELT
      TH=TH+(THD+THD1)/2.*DELT
      THD=THD1
      T=T+DELT
      IF(ABS(T-TPR)-EPSI) 4,4,6
    4 WRITE(6,5) T,TH,THD
      TPR=TPR+DTPR
    5 FORMAT(' ',5X,F7.4,4X,F6.3,4X,F7.2)
    6 IF(TH-THCL) 3,7,7
    7 WRITE(6,5) T,TH,THD
      STOP
      END
```

The results obtained from the computer solution are given by the following printout:

```
DYNAMIC CHARACTERISTICS OF A RATTRAP
TIME            THETA           ANG V
0.0010          0.002            4.69
0.0020          0.009            9.37
0.0030          0.021           14.04
0.0040          0.037           18.70
0.0050          0.058           23.33
0.0060          0.084           27.94
0.0070          0.114           32.52
0.0080          0.149           37.07
0.0090          0.188           41.57
0.0100          0.232           46.02
0.0110          0.280           50.43
0.0120          0.333           54.78
0.0130          0.390           59.06
0.0140          0.451           63.29
0.0150          0.517           67.44
0.0160          0.586           71.52
0.0170          0.660           75.51
0.0180          0.737           79.43
0.0190          0.818           83.25
0.0200          0.904           86.99
0.0210          0.992           90.62
0.0220          1.085           94.16
0.0230          1.181           97.59
0.0240          1.280          100.91
0.0250          1.382          104.12
0.0260          1.488          107.21
0.0270          1.597          110.19
0.0280          1.708          113.04
0.0290          1.823          115.76
0.0300          1.940          118.36
0.0310          2.059          120.83
0.0320          2.181          123.16
0.0330          2.306          125.35
0.0340          2.432          127.40
0.0350          2.560          129.31
0.0360          2.691          131.08
0.0370          2.823          132.70
0.0380          2.956          134.18
0.0390          3.091          135.50
0.0400          3.227          136.67
0.0404          3.282          137.10
```

To check the accuracy of the numerical method used, the angular displacements were solved on the computer, using the equation

$$\theta = \frac{T_c + k\theta_c}{k}\left(1 - \cos\sqrt{\frac{k}{I_o}}\,t\right)$$

which is the analytical solution of Eq. 6-13. To save space, the complete printout for the latter solution is not shown, but some of the results are reproduced below, to afford a comparison with the numerical solution.

Time	Theta (Numerical)	Theta (Analytical)
0.0010	0.002	0.002
0.0050	0.058	0.058
0.0100	0.232	0.232
0.0200	0.904	0.903
0.0250	1.382	1.382
0.0300	1.940	1.939
0.0400	3.227	3.225
0.0404	3.282	3.279

It is apparent that the numerically obtained values compare quite closely with those determined analytically.

6-4 / MODIFIED EULER METHODS

The Self-Starting Modification

This is a self-starting method of the predictor-corrector type, having greater accuracy than Euler's method. It is known by a variety of names, the Euler-trapezoidal method, the Euler predictor-corrector method, the improved Euler method, and Heun's method. Some European authors refer to it as the Euler-Cauchy method, which the Euler method discussed in the preceding section is also sometimes called. We will call it the *self-starting modified Euler method* to differentiate it from another modification which is not self-starting.

We again consider a first-order differential equation of the form

$$y' = f(x, y) \tag{6-21}$$

where the value of y is known when $x = 0$. Let us assume that the curves of y' and y versus x are those shown in Fig. 6-12. If we substitute the known initial value of y into Eq. 6-21, we obtain the value of y' at $x = 0$. Next, a *predicted* value of y at $x = \Delta x$ is found by using Euler's equation

$$P(y_1) = y_0 + y_0'h \tag{6-22}$$

where $P(y_1)$ is the predicted value of y_1 and $h = \Delta x$. In using Eq. 6-22 to obtain the predicted value of y_1, the term $y_0'h$ is the rectangular area A_1 shown in Fig. 6-12(a). This area is obviously larger than the true area

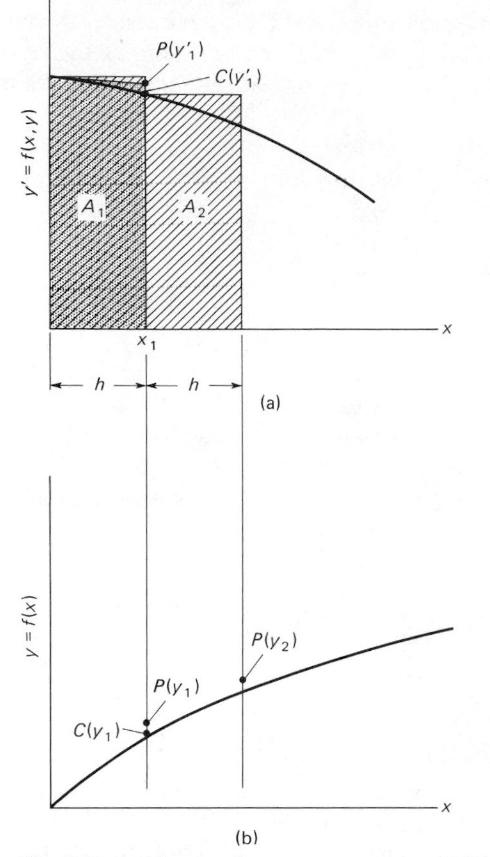

Fig. 6-12 / *The y′ and y versus x curves satisfying $y' = f(x, y)$.*

under the given curve, so the predicted value of y_1 obtained is too large by some small amount. However, if the predicted value of y_1 is substituted into the given differential equation (Eq. 6-21) along with $x = h$, an approximate value of y_1' may be obtained. Since this value of y_1' is based on the predicted value of y_1, $P(y_1)$, we will use the notation $P(y_1')$ to represent it. Then, using the shaded *trapezoidal* area shown in Fig. 6-12(a) as the approximation of the true area under the y' curve, a corrected value of y_1, which we will denote by $C(y_1)$, can be determined as

$$C(y_1) = y_0 + \left(\frac{y_0' + P(y_1')}{2}\right)h \tag{6-23}$$

Equation 6-23 is thus known as the *corrector* equation. The corrected value of y_1 is next substituted into Eq. 6-21 to obtain a corrected value of y_1'. The latter value is denoted by $C(y_1')$, as indicated in Fig. 6-12(a). The iteration process continues, by using $C(y_1')$ in place of $P(y_1')$ in Eq. 6-23, to obtain a still better value of y_1, and then using this improved value in Eq. 6-21 to get a further improved value of y_1'. The process is repeated until successive values of y_1 differ by less than some prescribed epsilon value selected to specify the accuracy desired.

With the desired value of y_1 obtained, we are ready to move ahead one step to determine the value of y_2. This begins by using the predictor equation

$$P(y_2) = y_1 + y_1' h \tag{6-24}$$

where $P(y_2)$ is the first predicted value of y_2, and y_1 and y_1' are the most accurate values obtained for these quantities in the preceding iteration. The iterative process described for determining an accurate value of y_1 is then repeated, and so on.

The general form of Eqs. 6-22 and 6-23, for application at any step, is

$$\text{(Predictor)} \qquad P(y_{i+1}) = y_i + y_i' h \qquad \text{(a)}$$

$$\text{(Corrector)} \qquad C(y_{i+1}) = y_i + \left[\frac{y_i' + P(y_{i+1}')}{2}\right] h \qquad \text{(b)} \qquad \text{(6-25)}$$

Error Analysis of the Self-Starting Modified Euler Method

Since the corrector equation of this method is based on using a trapezoidal area to represent the true area in one strip under the y' versus x curve, the per-step truncation error of this method is the same as the truncation error of the trapezoid rule, and is

$$E_{TC_i} = -\tfrac{1}{12} y'''(\xi) h^3 \qquad x_i < \xi < x_{i+1} \tag{6-26}$$

with order of h^3. Assuming that y''' is fairly constant over the ith step so that the higher derivatives are negligible, we can evaluate y''' at $\xi = x_i$ and obtain an estimate of the truncation error in computing y_{i+1} as

$$E_{TC_i} = -\tfrac{1}{12} y_i''' h^3 \tag{6-27}$$

As discussed under Euler's method, the error estimate given by Eq. 6-27 can be used for comparison with some selected set of limiting values to adjust the step size, when necessary, during the integration process.

If y''' is a rather complicated function, the determination of the truncation error at each step may become quite cumbersome if one uses the method just described. One alternative method of obtaining a solution having some desired degree of accuracy is to run separate solutions, each with a different step size, and compare the solutions obtained, decreasing (or possibly increasing) the step size at each run until consecutive solutions are in good agreement.

Another alternative is to base adjustment of the step size on a truncation-error estimate by Richardson's method,[1] in which two solutions, each with a different step size, are run simultaneously. A truncation-error estimate is obtained at each step in terms of the two y values obtained at that step. To develop this way of estimating the per-step truncation error for an rth-order method, consider that

$C_1(h)^{r+1}$ = the per-step truncation error in an rth-order method (a method having a per-step truncation error of order $r + 1$) using a step size of h

$C_2\left(\frac{h}{2}\right)^{r+1}$ = the per-step truncation error in the same rth-order method using a step size of $h/2$ [2]

$y^{(1)}$ = the y value obtained at x_{i+1} using a step size of h in stepping forward from x_i

$y^{(2)}$ = the y value obtained at x_{i+1} using two steps of size $h/2$ in stepping forward from x_i

Y = the true value of y at x_{i+1}

Isolating the truncation error of the step from the effects of previous error in the solution, in accordance with the definition given for per-step truncation error, and assuming that roundoff error in the step calculations is negligible, we can write

$$Y - y^{(1)} = C_1(h)^{r+1} \tag{6-28}$$

$$Y - y^{(2)} \cong 2C_2\left(\frac{h}{2}\right)^{r+1} \tag{6-29}$$

If the derivative y^{r+1} is fairly constant over the interval h, we may assume that $C_1 \cong C_2$. Subtracting Eq. 6-29 from Eq. 6-28 yields

$$\frac{y^{(2)} - y^{(1)}}{2^r - 1} \cong 2C_2\left(\frac{h}{2}\right)^{r+1} \tag{6-30}$$

[1] L. F. Richardson and J. A. Gaunt, "The Deferred Approach to the Limit," *Trans. Roy. Soc. London*, Vol. 226A (1927), p. 300.

[2] $$C_1 = cy^{r+1}(\xi_1) \qquad C_2 = cy^{r+1}(\xi_2) \qquad \begin{matrix} x_i < \xi_1 < x_{i+1} \\ x_i < \xi_2 < x_{i+1/2} \end{matrix}$$

Comparing Eqs. 6-30 and 6-29, it can be seen that the approximate truncation error in obtaining y at x_{i+1}, using two steps of size $h/2$ in stepping forward from x_i is given by

$$E_T \cong \frac{y^{(2)} - y^{(1)}}{2^r - 1} \tag{6-31}$$

For the modified Euler methods ($r = 2$), Eq. 6-31 becomes

$$E_T \cong \frac{y^{(2)} - y^{(1)}}{3} \tag{6-32}$$

An estimate of the per-step truncation error in the corrector equation of some predictor-corrector methods can also be determined, at each step, in terms of the predicted y value and the final corrected y value. When both the predictor equation and the corrector equation of a method have the same order of truncation error, it is relatively easy to obtain a truncation-error estimate at each step, in such terms, since the estimate can essentially be obtained as the by-product of calculations required in the predictor-corrector process.

The Non-Self-Starting Modified Euler Method

In the self-starting modified Euler method, the predictor equation has a truncation error of order h^2 (Eq. 6-9) whereas the corrector equation has a truncation error of order h^3 (Eq. 6-26). A truncation-error estimate at each step in terms of the predicted y value and final corrected y value is not possible. When an error estimate is to be made at each step in terms of these quantities, the non-self-starting modified Euler method should be used, since, as we will see, the predictor and corrector equations have truncation errors of the same order. The predictor equation of this method is

$$P(y_{i+1}) = y_{i-1} + 2y_i'h \tag{6-33}$$

The area in two strips under the y' versus x curve from x_{i-1} to x_{i+1} is approximated by the crosshatched area shown in Fig. 6-13. The corrector equation is the same as for the self-starting modified Euler method as given by Eq. 6-25b. Thus the complete set to be used in this method is

$$\left.\begin{array}{lll} \text{(Predictor)} & P(y_{i+1}) = y_{i-1} + 2y_i'h & \text{(a)} \\ \text{(Differential equation)} & P(y_{i+1}') = f[x_{i+1}, P(y_{i+1})] & \text{(b)} \\ \text{(Corrector)} & C(y_{i+1}) = y_i + \dfrac{h}{2}[y_i' + P(y_{i+1}')] & \text{(c)} \end{array}\right\} \quad (6\text{-}34)$$

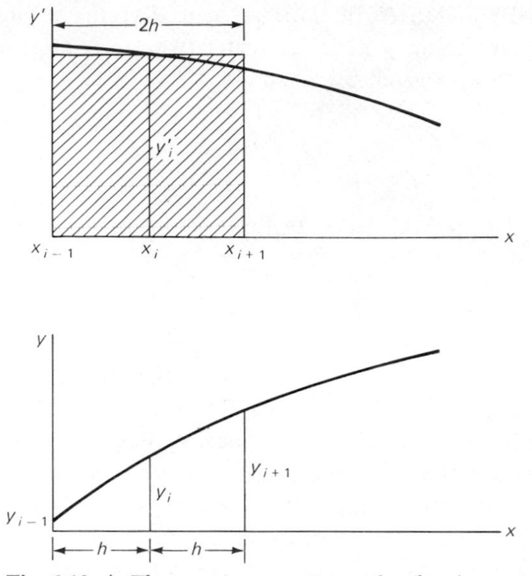

Fig. 6-13 / *The area in two strips under the y' versus x curve approximated by a rectangular area (the non-self-starting modified Euler method).*

Although we will be able to derive a way of easily estimating the per-step truncation error using this method, the convenient self-starting feature of the previously described modified Euler method is lost. In addition to the usual initial value of y required for solving a first-order differential equation, one additional value of y must be known to get the method started. Integration by this method is usually started by applying Eq. 6-25a and Eq. 6-25b for the first step, and then switching to the set given by Eq. 6-34. The additional required starting value can often be obtained quite easily with the use of a pocket calculator, but it can be calculated as part of the computer program if desired.

Error Analysis of the Non-Self-Starting Modified Euler Method

Since the corrector equation of this method is the same as that of the self-starting modification, the per-step truncation error is the same; it is

$$E_{TC_i} = -\tfrac{1}{12}y'''(\xi)h^3 \qquad x_i < \xi < x_{i+1} \tag{6-35}$$

However, as previously mentioned, y''' may turn out to be a complicated function, making the use of Eq. 6-35 quite cumbersome. Therefore, we will develop a technique for estimating the per-step truncation error in terms of the predicted and final corrected values of y at each step. This is possible because the predictor and corrector equations have errors of the same order.

First, examining the per-step truncation error of the predictor equation, we expand y_{i+1} in a Taylor series about x_i to obtain

$$y_{i+1} = y_i + y_i'h + \frac{y_i''h^2}{2!} + \frac{y'''(\xi_1)h^3}{3!}, \qquad x_i < \xi_1 < x_{i+1} \tag{6-36}$$

Similarly, expanding y_{i-1} about x_i yields

$$y_{i-1} = y_i - y_i'h + \frac{y_i''h^2}{2!} - \frac{y'''(\xi)h^3}{3!}, \qquad x_{i-1} < \xi_2 < x_i \tag{6-37}$$

Subtracting Eq. 6-37 from Eq. 6-36 gives

$$y_{i+1} - y_{i-1} = 2y_i'h + \frac{2y'''(\xi_3)h^3}{3!}, \qquad x_{i-1} < \xi_3 < x_{i+1} \tag{6-38}$$

Comparing Eq. 6-38 with Eq. 6-34a, we can write Eq. 6-38 as

$$y_{i+1} = \underbrace{y_{i-1} + 2y_i'h}_{\text{predictor equation}} + \underbrace{\frac{2y'''(\xi_3)h^3}{3!}}_{\text{truncation error}} \tag{6-39}$$

to see that the per-step truncation error of the predictor equation

$$E_{TP_i} = \frac{y'''(\xi_3)h^3}{3} \tag{6-40}$$

is of order h^3. This is of the same order as the per-step truncation error of the corrector equation given by Eq. 6-26.

Knowing the per-step truncation errors of both the predictor and the corrector equations, consider that

$P\binom{1}{y}$ = the predicted y value obtained at x_{i+1} in the integration process in the first and only application of the predictor equation at that step

$C\binom{n}{y}$ = the corrected y value obtained at x_{i+1} in the nth (final) application of the corrector equation at that step

Y = the true value of y at x_{i+1}

Isolating the truncation error of the ith step from the effects of previous errors, in accordance with the definition given of per-step truncation error, and assuming that roundoff error is negligible in the calculations of the step, we can write

$$\text{(Predictor)} \qquad Y = P\binom{1}{y} + \frac{1}{3}y'''(\xi_3)h^3 \qquad x_{i-1} < \xi_3 < x_{i+1} \tag{6-41}$$

$$\text{(Corrector)} \qquad Y = C\binom{n}{y} - \frac{1}{12}y'''(\eta)h^3 \qquad x_i < \eta < x_{i+1} \tag{6-42}$$

Subtracting Eq. 6-41 from Eq. 6-42, we obtain

$$0 = C\binom{n}{y} - P\binom{1}{y} - \frac{5}{12}y'''(\eta_1)h^3 \qquad x_{i-1} < \eta_1 < x_{i+1} \tag{6-43}$$

Comparing Eqs. 6-43 and 6-26, we can write an estimate of the per-step truncation error of the corrector equation, in terms of the predicted y value and the final corrected y value, as

$$E_{TC} \cong -\frac{1}{5}\left[C\binom{n}{y} - P\binom{1}{y}\right] \tag{6-44}$$

EXAMPLE 6-2

This example will illustrate the self-starting modified method applied to two simultaneous second-order differential equations.

When a paratrooper jumps from an aircraft which is in straight and level flight, he has an initial velocity equal to that of the aircraft, and he is a freely falling body during the time interval before opening his parachute. The velocity of the paratrooper during this free-fall interval may be described by a set of rectangular components, as shown in Fig. 6-14.

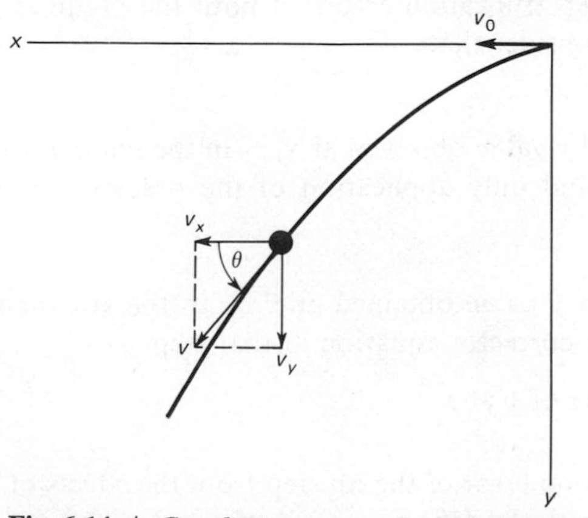

Fig. 6-14 / *Coordinate system of paratrooper problem.*

The decelerating force to which the paratrooper is subjected when the parachute opens is essentially proportional to the square of his free-fall velocity at the time the chute opens. With slow-flying aircraft, this decelerating force is not excessive. The paratrooper jumps, and a relatively short static line, attached to both the rip cord of the chute and the aircraft, opens the chute after a short free fall which is just sufficient to clear the aircraft. However, aircraft dropping paratroopers while flying at slow speeds present ideal targets for enemy ground fire. Therefore, it is desirable to drop the paratroopers from faster flying aircraft. Since the impact force on the paratrooper increases approximately as the *square* of the velocity at the time the chute opens, it is desirable to open the chute at the instant when the free-fall velocity is at a minimum.

It is a well-known fact that a body falling freely in a resisting medium, if allowed to fall far enough, will eventually reach a limiting or *terminal* velocity. This limiting velocity is reached when the resistive force, or drag due to the fluid medium, just balances the weight of the falling body. It has been claimed that, for any horizontal launching speed, the velocity of the paratrooper falling freely will pass through a *minimum* value *less* than either the *launching* or the terminal velocity.

In general, the drag force D on a body, owing to air resistance, is proportional to the projected area of the body normal to the direction of motion, the air density, and approximately the square of the velocity. Usually, the drag force D is expressed in the form

$$D = \frac{C\gamma A v^2}{2} \tag{6-45}$$

where

γ = mass density of the fluid medium, lb-sec^2/ft^4
A = projected area of the body normal to the direction of motion, ft^2
C = drag coefficient, dimensionless
v = velocity, ft/sec

If we assume that the projected area A of the paratrooper is essentially constant, that the drag coefficient C is constant over the range of velocities encountered, and that the mass density γ of the air is a constant, then the quantity $C\gamma A$ may be considered as a constant, so that

$$\frac{C\gamma A}{2} = \text{constant} = K$$

and, from Eq. 6-45,

$$D = Kv^2 \tag{6-46}$$

Referring to Fig. 6-14 and considering Newton's second law, the equations of motion in rectangular coordinates are readily found to be

$$\text{(x direction)} \qquad \frac{W}{g}\frac{dv_x}{dt} = -Kv^2 \cos\theta \tag{6-47}$$

and

$$\text{(y direction)} \qquad \frac{W}{g}\frac{dv_y}{dt} = W - Kv^2 \sin\theta \tag{6-48}$$

where

v_x = horizontal component of velocity, ft/sec
v_y = vertical component of velocity, ft/sec
$v = \sqrt{v_x^2 + v_y^2}$ = resultant velocity, ft/sec
W = weight of paratrooper and equipment, lb
g = acceleration of gravity = 32.2 ft/sec^2
K = coefficient of aerodynamic resistance, lb-sec^2/ft^2

Noting that $\cos\theta = v_x/v$ and that $\sin\theta = v_y/v$, Eqs. 6-47 and 6-48 may be written in the form

$$\frac{W}{g}\frac{dv_x}{dt} = -Kvv_x \tag{6-49}$$

$$\frac{W}{g}\frac{dv_y}{dt} = W - Kvv_y \tag{6-50}$$

Substituting

$$v = \sqrt{v_x^2 + v_y^2}$$

in Eqs. 6-49 and 6-50, expressing the time-dependent variables in Newtonian notation, and rearranging algebraically, we obtain the equations of motion in a form convenient for solution by numerical integration. They are

$$\text{(x direction)} \qquad \ddot{x} = -\frac{Kg}{W}\sqrt{\dot{x}^2 + \dot{y}^2}\,\dot{x} \tag{6-51}$$

and

$$\text{(y direction)} \qquad \ddot{y} = g - \frac{Kg}{W}\sqrt{\dot{x}^2 + \dot{y}^2}\,\dot{y} \tag{6-52}$$

where

$\dot{x}$ = horizontal component of velocity, ft/sec
$\dot{y}$ = vertical component of velocity, ft/sec
K = coefficient of aerodynamic resistance, lb-sec^2/ft^2
g = acceleration of gravity, ft/sec^2
W = weight of paratrooper and equipment, lb

The coefficient of aerodynamic resistance K can be determined from the fact that the average terminal velocity reached by paratroopers falling with unopened parachutes has been found to be approximately 160 ft/sec at low altitudes. Applying the equation of equilibrium in the vertical direction to a body falling at terminal velocity gives

$$W = K(V_t)^2$$

or

$$K = \frac{W}{V_t^2} \tag{6-53}$$

where V_t is the terminal velocity, in feet per second, and W is the weight of the body. Substituting this expression for K into Eqs. 6-51 and 6-52, we obtain

$$\ddot{x} = -\frac{g}{V_t^2}\sqrt{\dot{x}^2 + \dot{y}^2}\,\dot{x} \tag{6-54}$$

and

$$\ddot{y} = g - \frac{g}{V_t^2}\sqrt{\dot{x}^2 + \dot{y}^2}\,\dot{y} \tag{6-55}$$

With the origin of the x-y coordinate system located at the point of launch, as shown in Fig. 6-15, the initial conditions are

$$t = 0 \left| \begin{array}{l} x = 0 \\ \dot{x} = 440 \text{ ft/sec} \\ y = 0 \\ \dot{y} = 0 \end{array} \right.$$

The objective of our analysis will be to determine (1) the data necessary to plot the x-t, y-t, $\dot{x}$-t, $\dot{y}$-t, and V-t curves for a paratrooper launched horizontally with a velocity of 300 mph and (2) the optimum time for

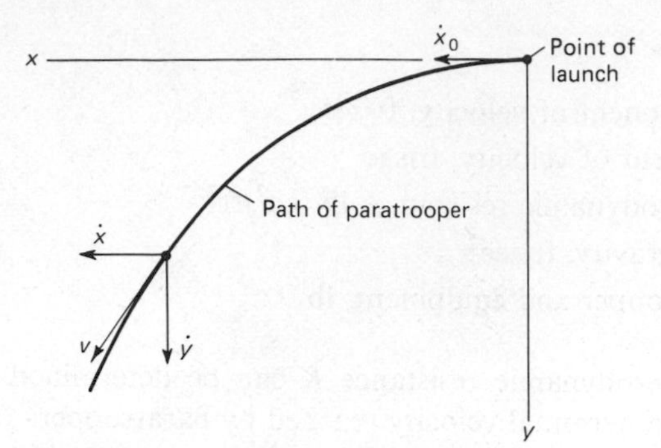

Fig. 6-15 / *Trajectory of paratrooper.*

opening the parachute after a period of free fall (the time when the velocity of the paratrooper is minimum).

In applying Euler's self-starting modified method to the simultaneous solution of Eqs. 6-54 and 6-55 for values of $\dot{x}$ and $\dot{y}$, the reader should realize that we are, in effect, solving the following two *first-order* differential equations

$$\dot{v}_x = -\frac{g}{V_t^2}\sqrt{v_x^2 + v_y^2}\, v_x \tag{6-56}$$

and

$$\dot{v}_y = g - \frac{g}{V_t^2}\sqrt{v_x^2 + v_y^2}\, v_y \tag{6-57}$$

for v_x and v_y, where the reduction was accomplished by the relations

$$v_x = \dot{x} \qquad v_y = \dot{y}$$

$$\dot{v}_x = \ddot{x} \qquad \dot{v}_y = \ddot{y}$$

However, in applying the predictor-corrector equations (Eq. 6-25), we will use the notation of Eqs. 6-54 and 6-55. After obtaining the desired values of $\dot{x}$ and $\dot{y}$ in each step, we will use these values with the trapezoid rule to solve the equations

$$\frac{dx}{dt} = \dot{x} \tag{6-58}$$

and

$$\frac{dy}{dt} = \dot{y} \tag{6-59}$$

to obtain values of the displacements x and y. Since x and y do not appear in explicit form in Eqs. 6-54 and 6-55, their solutions need not be obtained simultaneously with those of Eqs. 6-58 and 6-59. However, these equations will be solved simultaneously in the computer solution accompanying this example, to avoid the necessity of storing the calculated $\dot{x}$ and $\dot{y}$ values in memory for subsequent use in obtaining the desired x and y values.

The predictor-corrector equations, in a form applicable to the solution of Eqs. 6-54 and 6-55 for values of $\dot{x}$ and $\dot{y}$, are

$$\begin{aligned}
P(\dot{x}_{i+1}) &= \dot{x}_i + \ddot{x}_i(\Delta t) && \text{(a)}\\
C(\dot{x}_{i+1}) &= \dot{x}_i + \left[\frac{\ddot{x}_i + P(\ddot{x}_{i+1})}{2}\right]\Delta t && \text{(b)}\\
P(\dot{y}_{i+1}) &= \dot{y}_i + \ddot{y}_i(\Delta t) && \text{(c)}\\
C(\dot{y}_{i+1}) &= \dot{y}_i + \left[\frac{\ddot{y}_i + P(\ddot{y}_{i+1})}{2}\right]\Delta t && \text{(d)}
\end{aligned} \tag{6-60}$$

The equations for solving the x and y values by the trapezoid rule are

$$\begin{aligned}
x_{i+1} &= x_i + \left(\frac{\dot{x}_i + \dot{x}_{i+1}}{2}\right)\Delta t\\
y_{i+1} &= y_i + \left(\frac{\dot{y}_i + \dot{y}_{i+1}}{2}\right)\Delta t
\end{aligned} \tag{6-61}$$

The initial conditions supply us with values of x, y, $\dot{x}$, and $\dot{y}$ for $t = 0$, and we can determine values of $\ddot{x}$ and $\ddot{y}$ when $t = 0$ by substituting appropriate values in Eqs. 6-54 and 6-55. We next obtain predicted values of $\dot{x}_1$ and $\dot{y}_1$, using the predictor equations (Eq. 6-60a and c). In obtaining these predicted values, the true areas under the curves of $\ddot{x}$ and $\ddot{y}$ are approximated by the *rectangular crosshatched* areas shown in Fig. 6-16(a) and (d). The predicted values $P(\dot{x}_1)$ and $P(\dot{y}_1)$ are shown in Fig. 6-16(b) and (e). We next substitute $P(\dot{x}_1)$ and $P(\dot{y}_1)$ into Eqs. 6-54 and 6-55 to obtain approximate values of $\ddot{x}_1$ and $\ddot{y}_1$, which are shown as $P(\ddot{x}_1)$ and $P(\ddot{y}_1)$ in Fig. 6-16(a) and (d). Then, using the corrector equations (Eq. 6-60b and d), corrected values of $\dot{x}_1$ and $\dot{y}_1$ are obtained. These values, designated as $C(\dot{x}_1)$ and $C(\dot{y}_1)$, are shown in Fig. 6-16(b) and (e). The

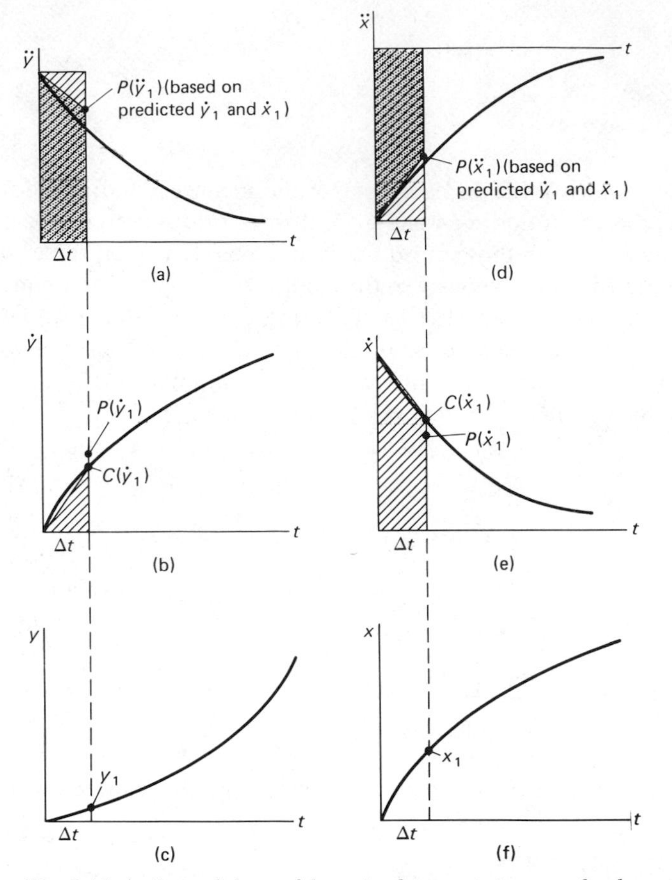

Fig. 6-16 / *General shape of dynamic-characteristic curves for the paratrooper of Ex. 6-2.*

corrector equation approximates the true areas under the $\ddot{x}$ and $\ddot{y}$ curves by the *shaded trapezoidal* areas shown in Fig. 6-16(a) and (d). The corrected values of $\dot{x}_1$ and $\dot{y}_1$ are next substituted into Eqs. 6-54 and 6-55 to obtain improved values of $\ddot{x}_1$ and $\ddot{y}_1$ which are, in turn, used in Eq. 6-60b and d to obtain still better values of $\dot{x}_1$ and $\dot{y}_1$. The process continues until successive values of $\dot{x}_1$ and $\dot{y}_1$, respectively, vary by less than some preassigned epsilon value specifying the accuracy desired. At this point we proceed with the determination of the values of x_1 and y_1, using Eq. 6-61, which approximates the true area under the $\dot{x}$ and $\dot{y}$ curves by the cross-hatched areas shown in Fig. 6-16(b) and (e).

In the computer program, shown later, the total velocity of the paratrooper is calculated at each step and compared with the velocity at the previous station to determine the lowest velocity calculated. As stated at the beginning of the problem, the time at which the lowest velocity is reached is considered the optimum parachute-opening time.

The flow chart is shown in Fig. 6-17. The variable names and the quantities which they represent in the FORTRAN program are as follows:

Variable Name	Quantity	Value Used in Problem
TERM	Terminal velocity, ft/sec	160
XD	Velocity in x direction $\dot{x}_i$, ft/sec	440 (initial)
DELT	Time increment Δt, sec	0.1
DTPR	Print increment Δt_{print}, sec	0.5
EPSI	Accuracy-check value, ft/sec	0.0001
TMAX	Maximum time value, sec	20
TPR	Time at which a print occurs, sec	
N	Integer variable used in program control	
T	Time, sec	
G	Acceleration of gravity, ft/sec^2	32.17
X	Displacement in x direction, ft	
Y	Displacement in y direction, ft	
YD	Vertical component of velocity, $\dot{y}_i$, ft/sec	
V	Total velocity, ft/sec	
XDD	Acceleration in x direction $\ddot{x}_i$, ft/sec^2	
YDD	Acceleration in y direction $\ddot{y}_i$, ft/sec^2	
XD1	Velocity in x direction one station ahead of t_i, $\dot{x}_{i+1}$, ft/sec	
YD1	Velocity in y direction one station ahead of t_i, $\dot{y}_{i+1}$, ft/sec	
V1	Total velocity one station ahead of t_i, ft/sec	
XDD1	Acceleration in x direction one station ahead of t_i, $\ddot{x}_{i+1}$, ft/sec^2	
YDD1	Acceleration in y direction one station ahead of t_i, $\ddot{y}_{i+1}$, ft/sec^2	
YD1C	Corrected value of y_{i+1}, ft/sec	
DIFXD	Difference in successive calculated values of $\dot{x}_{i+1}$ at a particular station, ft/sec	
DIFYD	Difference in successive calculated values of $\dot{y}_{i+1}$ at a particular station, ft/sec	
X1	Displacement in x direction one station ahead of t_i, x_{i+1}, ft	
Y1	Displacement in y direction one station ahead of t_i, y_{i+1}, ft	

The FORTRAN program follows.

```
C DYNAMIC CHARACTERISTICS OF FREE FALL OF A PARATROOPER
      READ(5,1) TERM,XD,DELT,DTPR,EPSI,TMAX
    1 FORMAT(6F10.0)
      WRITE(6,2) XD
```

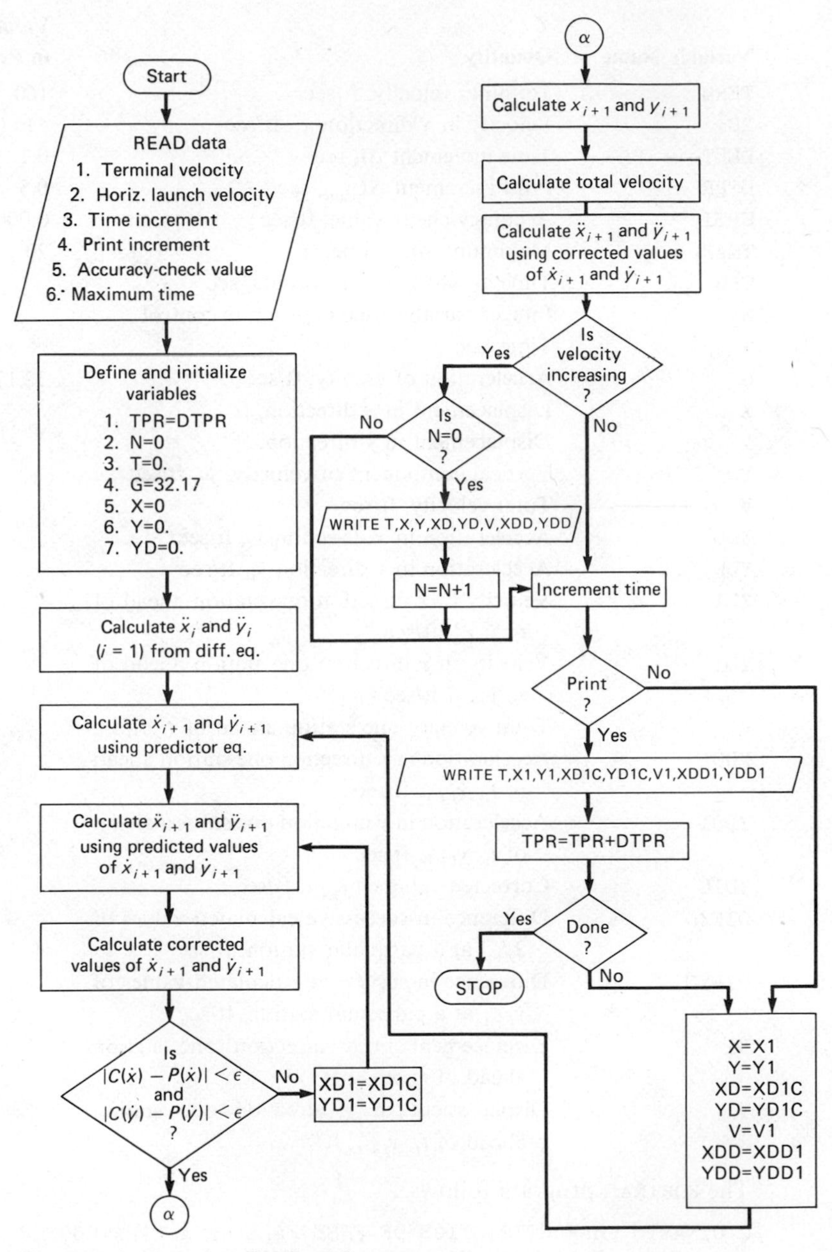

Fig. 6-17 / *Flow chart for Ex. 6-2.*

```
    2 FORMAT('1','HORIZONTAL LAUNCH VELOCITY=',
                                                    F7.2,' FT/SEC'//)
      WRITE(6,3)
    3 FORMAT(' ',2X,'T',6X,'X',7X,'Y',6X,'VX',6X,'VY',6X,'V'/)
      TPR=DTPR
      N=0
      T=0.
      G=32.17
      X=0.
      Y=0.
      YD=0.
      V=XD
      XDD=-G/TERM**2*V*XD
      YDD=G-G/TERM**2*V*YD
    4 XD1=XD+XDD*DELT
      YD1=YD+YDD*DELT
    5 V1=SQRT(XD1**2+YD1**2)
      XDD1=-G/TERM**2*V1*XD1
      YDD1=G-G/TERM**2*V1*YD1
      XD1C=XD+(XDD+XDD1)/2.*DELT
      YD1C=YD+(YDD+YDD1)/2.*DELT
      DIFXD=XD1C-XD1
      DIFYD=YD1C-YD1
      IF(ABS(DIFXD)-EPSI) 6,7,7
    6 IF(ABS(DIFYD)-EPSI) 8,7,7
    7 XD1=XD1C
      YD1=YD1C
      GO TO 5
    8 X1=X+(XD+XD1C)/2.*DELT
      Y1=Y+(YD+YD1C)/2.*DELT
      V1=SQRT(XD1C**2+YD1C**2)
      XDD1=-G/TERM**2*V1*XD1C
      YDD1=G-G/TERM**2*V1*YD1C
      IF(V1-V) 12,12,9
    9 IF(N) 12,10,12
   10 WRITE(6,11) T,X,Y,XD,YD,V
   11 FORMAT(' ',F4.1,4F8.1,F7.1)
      N=N+1
   12 T=T+DELT
      IF(ABS(T-TPR)-.03) 13,13,16
   13 WRITE(6,11) T,X1,Y1,XD1C,YD1C,V1
      TPR=TPR+DTPR
      IF(ABS(T-TMAX)-.03) 15,15,16
   15 STOP
   16 X=X1
      Y=Y1
      XD=XD1C
      YD=YD1C
      V=V1
      XDD=XDD1
      YDD=YDD1
      GO TO 4
      END
```

The results of the computer calculations are shown in the following printout, where the column headings VX and VY denote the velocity components XD and YD, respectively.

HORIZONTAL LAUNCH VELOCITY= 440.00 FT/SEC

T	X	Y	VX	VY	V
0.5	194.3	3.7	344.6	14.3	344.9
1.0	350.3	14.0	283.1	26.4	284.3
1.5	480.5	29.9	240.1	37.3	242.9
2.0	592.2	51.1	208.1	47.3	213.4
2.5	689.8	77.2	183.3	56.8	191.9
3.0	776.3	107.9	163.3	65.8	176.1
3.5	853.6	142.9	146.8	74.4	164.5
4.0	923.4	182.2	132.7	82.6	156.3
4.5	986.7	225.5	120.5	90.4	150.6
5.0	1044.2	272.5	109.8	97.7	146.9
5.5	1096.6	323.1	100.2	104.5	144.7
6.0	1144.5	376.9	91.5	110.8	143.7
6.4	1179.8	422.2	85.1	115.5	143.5
6.5	1188.2	433.8	83.6	116.6	143.5
7.0	1228.2	493.5	76.4	122.0	143.9
7.5	1264.7	555.7	69.8	126.8	144.7
8.0	1298.1	620.1	63.7	131.1	145.7
8.5	1328.5	686.7	58.1	134.9	146.9
9.0	1356.2	755.0	52.9	138.4	148.1
9.5	1381.5	825.0	48.2	141.4	149.4
10.0	1404.5	896.3	43.9	144.0	150.5
10.5	1425.4	968.9	39.9	146.3	151.6
11.0	1444.5	1042.6	36.3	148.3	152.7
11.5	1461.8	1117.2	32.9	150.0	153.6
12.0	1477.5	1192.6	29.9	151.5	154.4
12.5	1491.7	1268.6	27.1	152.8	155.2
13.0	1504.6	1345.3	24.6	153.9	155.9
13.5	1516.3	1422.5	22.3	154.8	156.4
14.0	1527.0	1500.1	20.2	155.6	156.9
14.5	1536.6	1578.1	18.3	156.3	157.4
15.0	1545.3	1656.4	16.6	156.9	157.8
15.5	1553.2	1735.0	15.0	157.4	158.1
16.0	1560.4	1813.8	13.6	157.8	158.4
16.5	1566.8	1892.8	12.3	158.2	158.6
17.0	1572.7	1972.0	11.1	158.4	158.8
17.5	1578.0	2051.2	10.1	158.7	159.0
18.0	1582.8	2130.6	9.1	158.9	159.2
18.5	1587.1	2210.1	8.3	159.1	159.3
19.0	1591.0	2289.7	7.5	159.2	159.4
19.5	1594.6	2369.4	6.8	159.4	159.5
20.0	1597.8	2449.1	6.1	159.5	159.6

The computer results show that a minimum velocity of 143.5 ft/sec is reached when $t = 6.4$ sec, which indicates the optimum time to open the parachute. The velocity data are plotted in Fig. 6-18.

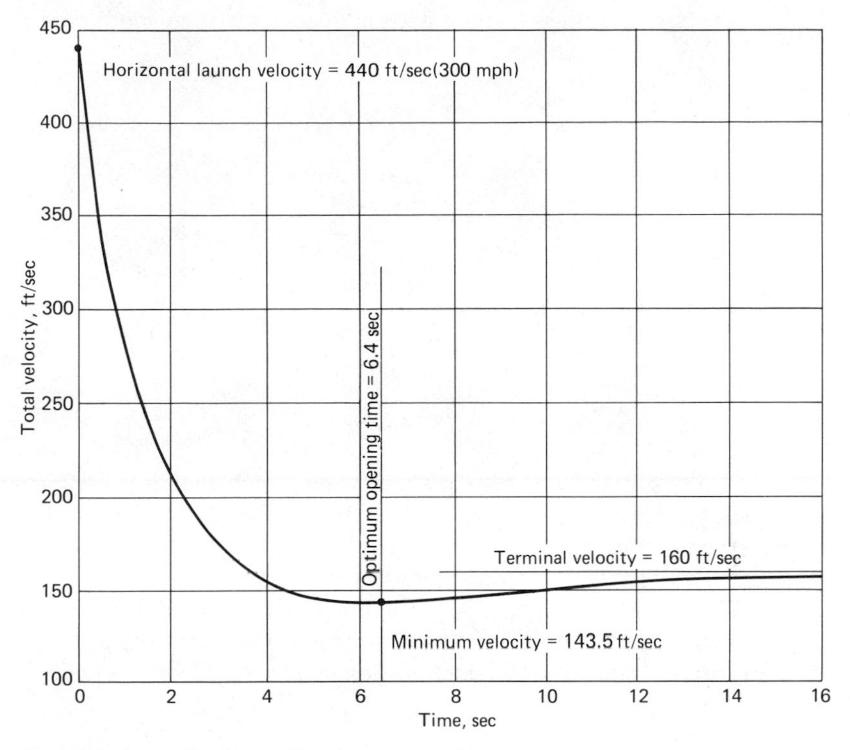

Fig. 6-18 / *Total velocity of paratrooper.*

6-5 / RUNGE-KUTTA METHODS

A Runge-Kutta method is one which employs a recurrence formula of the form

$$y_{i+1} = y_i + a_1k_1 + a_2k_2 + a_3k_3 + \cdots + a_nk_n \tag{6-62}$$

to calculate successive values of the dependent variable y of the differential equation

$$\frac{dy}{dx} = y' = f(x, y) \tag{6-63}$$

where

$$\begin{aligned}
k_1 &= (h)f(x_i, y_i) \\
k_2 &= (h)f(x_i + p_1h, y_i + q_{11}k_1) \\
k_3 &= (h)f(x_i + p_2h, y_i + q_{21}k_1 + q_{22}k_2) \\
&\cdots\cdots\cdots\cdots\cdots\cdots\cdots\cdots \\
k_n &= (h)f(x_i + p_{n-1}h, y_i + q_{n-1,1}k_1 + q_{n-1,2}k_2 + \cdots + q_{n-1,n-1}k_{n-1})
\end{aligned} \tag{6-64}$$

The above Runge-Kutta equations can be written more compactly as

$$y_{i+1} = y_i + \sum_{j=1}^{n} a_j k_j \tag{6-65}$$

where

$$k_j = hf(x_i + p_{j-1}h, y_i + \sum_{l=1}^{j-1} q_{j-1,l}k_l) \qquad (j = 1, 2, \ldots, n) \tag{6-66}$$

in which, by definition,

$$p_0 = 0 \qquad \text{and} \qquad \sum_{l=1}^{j-1} q_{j-1,l}k_l = 0 \qquad \text{for} \qquad j = 1$$

The a's, p's, and q's must assume values such that Eq. 6-62 accurately yields successive values of y. As will be shown later, these values are determined by making Eq. 6-62 equivalent to a certain specified number of terms of a Taylor-series expansion of y about x_i.

If we let m be the order of the Runge-Kutta method, with n the number of function evaluations per step (number of k values), there is a particular maximum value of m for each n. For n values up to 7, we have

$$1 \leq n \leq 4, m_{\max} = n$$
$$n = 5, m_{\max} = 4$$
$$n = 6, m_{\max} = 5$$
$$n = 7, m_{\max} = 6$$

Various Runge-Kutta methods are classified as (m, n) methods. A (5, 6) method, for example, would be a fifth-order method requiring six function evaluations per step.

Runge-Kutta methods are called *single-step* methods, since they require only knowledge of y_i to determine y_{i+1}. Thus these methods are *self-starting*. It is theoretically possible to develop methods of any order, but the developments get increasingly complex with increasing order. Runge-Kutta methods were among the earliest methods employed in the numerical solution of differential equations. and they are still widely used. As with any method, they possess certain advantages and disadvantages which must be weighed in considering their suitability for a particular application. The principal advantage of the Runge-Kutta methods is their self-starting feature and consequent ease of programming. One disadvantage is the requirement that the function $f(x, y)$ must be evaluated for several slightly different values of x and y in every step of the solution (in every incrementation of x by h). This repeated determination of

$f(x, y)$ may result in a less efficient method with respect to computing time than other methods of comparable accuracy in which previously determined values of the dependent variable are used in subsequent steps. Methods of the latter type mentioned are called multiple-step methods, and Hamming's and Milne's methods are examples of such. They are discussed later in this chapter. Another disadvantage of Runge-Kutta methods is that it is more difficult to estimate the per-step error for higher order solutions than for solutions obtained by some of the multiple-step methods.

Before illustrating the general procedure for developing a Runge-Kutta method, let us review (1) the definition of a *total differential* and (2) the expansion of a function of two variables in a *Taylor series*, since both will be pertinent to the development.

1 If $y' = f(x, y)$, the total differential dy' is

$$dy' = \frac{\partial f}{\partial x}\,dx + \frac{\partial f}{\partial y}\,dy$$

Upon dividing by dx, we obtain

$$\frac{dy'}{dx} = \frac{\partial f}{\partial x} + \frac{\partial f}{\partial y}\frac{dy}{dx} \tag{6-67}$$

Differentiating again with respect to x yields

$$\frac{d^2y'}{dx^2} = \frac{\partial[(\partial f/\partial x) + (\partial f/\partial y)(dy/dx)]}{\partial x} + \frac{\partial[(\partial f/\partial x) + (\partial f/\partial y)(dy/dx)]}{\partial y}\left(\frac{dy}{dx}\right) \tag{6-68}$$

Higher derivatives of y' may be determined in a similar manner.

2 Consider some general function of two variables such as $z = f(x, y)$. Such a function may be expanded about a point x_i, y_i in a Taylor series for functions of two variables, as follows:

$$z(x_i + l, y_i + j) = f(x_i, y_i) + l\left[\frac{\partial f}{\partial x}(x_i, y_i)\right] + j\left[\frac{\partial f}{\partial y}(x_i, y_i)\right]$$
$$+ \frac{1}{2!}\left\{l^2\left[\frac{\partial^2 f}{\partial x^2}(x_i, y_i)\right] + 2lj\left[\frac{\partial^2 f}{\partial x\,\partial y}(x_i, y_i)\right] + j^2\left[\frac{\partial^2 f}{\partial y^2}(x_i, y_i)\right]\right\} + \cdots \tag{6-69}$$

where l and j are increments of x and y, respectively.

We are now ready to proceed with the development of a *second-order* Runge-Kutta method ($n = 2$ in Eq. 6-62). Equation 6-62 becomes

$$y_{i+1} = y_i + a_1 k_1 + a_2 k_2 \tag{6-70}$$

in which

$$\left.\begin{aligned} k_1 &= (h)f(x_i, y_i) \\ k_2 &= (h)f(x_i + p_1 h, y_i + q_{11} k_1) \end{aligned}\right\} \tag{6-71}$$

Our problem is to determine values for a_1, a_2, p_1, and q_{11}, so that Eq. 6-70 yields an accurate value of y_{i+1}. A graphical interpretation of the k functions is illustrated in Fig. 6-19. The crosshatched area represents k_1, and k_2 is represented by the shaded area. (There would be similar rectangular areas for each k function of Eq. 6-62 for higher order Runge-Kutta equations.) For the second-order method being discussed, Eq. 6-62 could be written as

$$y_{i+1} = y_i + a_1 \text{ (crosshatched area)} + a_2 \text{ (shaded area)}$$

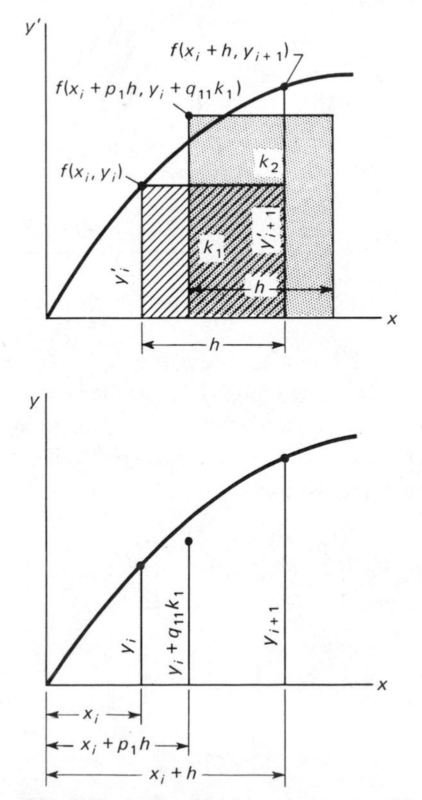

Fig. 6-19 / *Graphical interpretation of k functions.*

It should be evident that the size of the shaded area representing k_2 depends upon the values determined for p_1 and q_{11}.

We shall determine values for a_1, a_2, p_1, and q_{11} by making Eq. 6-70 equivalent to a truncated Taylor-series expansion of y about x_i. As the first step, let us expand y_{i+1} about x_i. We obtain

$$y_{i+1} = y_i + hy_i' + \frac{h^2}{2!}y_i'' + \cdots \tag{6-72}$$

Remembering that the given differential equation is $y' = f(x, y)$, we write

$$y_i' = f(x_i, y_i) \tag{6-73}$$

From Eq. 6-67 it can be seen that

$$y_i'' = \frac{\partial f}{\partial x}(x_i, y_i) + \left[\frac{\partial f}{\partial y}(x_i, y_i)\right]\left[f(x_i, y_i)\right] \tag{6-74}$$

Substituting Eqs. 6-73 and 6-74 into Eq. 6-72, we obtain the expansion in the following form:

$$\begin{aligned} y_{i+1} &= y_i + (h)f(x_i, y_i) \\ &\quad + \frac{h^2}{2!}\left\{\frac{\partial f}{\partial x}(x_i, y_i) + \left[\frac{\partial f}{\partial y}(x_i, y_i)\right]\left[f(x_i, y_i)\right]\right\} + \frac{h^3}{3!}\left\{\cdots\right\} + \cdots \end{aligned} \tag{6-75}$$

Looking at Eqs. 6-70, 6-71, and 6-75, we see that k_2 must be expressed in terms of $f(x_i, y_i)$, $\partial f/\partial x(x_i, y_i)$, and $\partial f/\partial y(x_i, y_i)$ if Eqs. 6-70 and 6-75 are to contain similar terms. This can be accomplished by expanding k_2 in a Taylor series for functions of two variables about x_i, y_i. Using the first three terms of Eq. 6-69 and realizing that $l = p_1 h$ and $j = q_{11}k_1$ (see Eq. 6-71), we may write

$$k_2 = h\left\{f(x_i, y_i) + p_1 h\left[\frac{\partial f}{\partial x}(x_i, y_i)\right] + q_{11}k_1\left[\frac{\partial f}{\partial y}(x_i, y_i)\right]\right\} \tag{6-76}$$

Substituting the first of Eq. 6-71 and Eq. 6-76 into Eq. 6-70, we obtain

$$\begin{aligned} y_{i+1} &= y_i + a_1(h)f(x_i, y_i) + a_2(h)f(x_i, y_i) \\ &\quad + a_2 h^2\left\{p_1\frac{\partial f}{\partial x}(x_i, y_i) + q_{11}\left[\frac{\partial f}{\partial y}(x_i, y_i)\right]\left[f(x_i, y_i)\right]\right\} \end{aligned} \tag{6-77}$$

Equating coefficients of similar terms in Eqs. 6-75 and 6-77, we obtain the following three independent equations:

$$\begin{aligned} a_1 + a_2 &= 1 \\ a_2 p_1 &= \tfrac{1}{2} \\ a_2 q_{11} &= \tfrac{1}{2} \end{aligned} \tag{6-78}$$

which contain four unknowns. By arbitrarily assigning a value to one unknown and then solving for the other three, we can obtain as many different sets of values as we desire and, in turn, as many different sets of Eqs. 6-70 and 6-71 as desired.

A solution obtained by the use of Eq. 6-70 in a step-by-step integration will have a per-step truncation error of order h^3, since terms containing h^3 and higher powers of h were neglected in the development. Thus this is known as a *second-order* Runge-Kutta method.

If a particular second-order method is defined by letting $a_1 = \frac{1}{2}$ in Eq. 6-78, then

$$\begin{aligned} a_2 &= \tfrac{1}{2} \\ p_1 &= 1 \\ q_{11} &= 1 \end{aligned}$$

Equations 6-70 and 6-71 then become

$$\left.\begin{aligned} &y_{i+1} = y_i + \tfrac{1}{2}(k_1 + k_2) && \text{(a)} \\ &\text{with} \\ &\left.\begin{aligned} k_1 &= (h)f(x_i, y_i) \\ k_2 &= (h)f(x_i + h, y_i + k_1) \end{aligned}\right\} && \text{(b)} \end{aligned}\right\} \tag{6-79}$$

This set of equations may be used to solve first-order differential equations with an accuracy comparable to that of Euler's modified methods. In fact, if Eq. 6-79b is substituted into Eq. 6-79a, the resulting equation

$$y_{i+1} = y_i + \tfrac{1}{2}\{(h)f(x_i, y_i) + (h)f[x_i + h, y_i + (h)f(x_i, y_i)]\} \tag{6-80}$$

is exactly equivalent to Euler's self-starting modified method with the iteration at each step omitted. This may be verified by examining the graphical representation of the equation in Fig. 6-20.

Because the second-order Runge Kutta method has only the order of accuracy of Euler's modified methods, let us consider Runge-Kutta

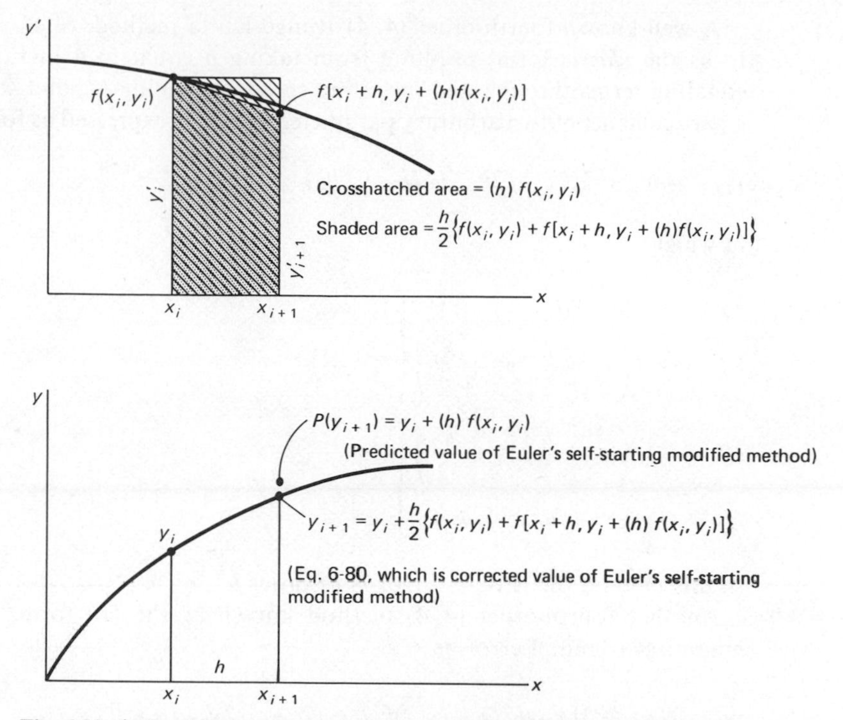

Fig. 6-20 / *Graphical representation of a second-order Runge-Kutta method.*

methods with higher orders of accuracy. Since the development of such methods rapidly increases in complexity with an increase in accuracy, and since the general procedure has just been demonstrated for a second-order method, the developments will not be given for higher orders. For example, in developing a third-order method, all Taylor-series terms containing h^3 must be retained; in developing a fourth-order method, all h^4 terms must be retained, and so forth.

If $n = 3$, in Eq. 6-62, and a procedure similar to the one previously outlined is followed, the equating of coefficients of terms containing h, h^2, and h^3 will yield a set of equations with six constraints on the parameters. With eight parameters involved, two of them can be assigned values arbitrarily. With a particular choice of values for two such parameters, the following *third-order* (3, 3) Runge-Kutta formula can be obtained:

$$y_{i+1} = y_i + \tfrac{1}{6}(k_1 + 4k_2 + k_3) \tag{6-81}$$

in which

$$\left.\begin{aligned} k_1 &= (h)f(x_i, y_i) \\ k_2 &= (h)f\left(x_i + \frac{h}{2}, y_i + \frac{k_1}{2}\right) \\ k_3 &= (h)f(x_i + h, y_i - k_1 + 2k_2) \end{aligned}\right\} \tag{6-82}$$

Since terms containing h^4 are neglected, the error is said to be of order h^4.

A well-known fourth-order (4, 4) Runge-Kutta method, often referred to as the *classic* form, resulting from taking n equal to 4 in Eq. 6-62, equating terms through and including those containing h^4, and selecting a particular set of two arbitrary parameter values, is expressed as follows:[3]

$$y_{i+1} = y_i + \tfrac{1}{6}(k_1 + 2k_2 + 2k_3 + k_4) \tag{6-83}$$

in which

$$\left.\begin{aligned} k_1 &= (h)f(x_i, y_i) \\ k_2 &= (h)f\left(x_i + \frac{h}{2}, y_i + \frac{k_1}{2}\right) \\ k_3 &= (h)f\left(x_i + \frac{h}{2}, y_i + \frac{k_2}{2}\right) \\ k_4 &= (h)f(x_i + h, y_i + k_3) \end{aligned}\right\} \tag{6-84}$$

In this method the per-step error is of order h^5.

Another fourth-order (4, 4) method known as the Gil form,[4] which minimizes roundoff error, is

$$y_{i+1} = y_i + \tfrac{1}{6}[k_1 + (2 - \sqrt{2})k_2 + (2 + \sqrt{2})k_3 + k_4] \tag{6-85}$$

where

$$\begin{aligned} k_1 &= (h)f(x_i, y_i) \\ k_2 &= (h)f\left(x_i + \frac{h}{2}, y_i + \frac{k_1}{2}\right) \\ k_3 &= (h)f\left[x_i + \frac{h}{2}, y_i + (\sqrt{2} - 1)\frac{k_1}{2} + (2 - \sqrt{2})\frac{k_2}{2}\right] \\ k_4 &= (h)f\left[x_i + h, y_i - \sqrt{2}\frac{k_2}{2} + \left(1 + \frac{\sqrt{2}}{2}\right)k_3\right] \end{aligned} \tag{6-86}$$

When the order of the method is 3 or 4, there are two arbitrary parameters to be chosen in the derivation of the various Runge-Kutta forms. When the order is 5 or more ($m = 5$), there are five or more arbitrary

[3] The interested reader will find a complete derivation in *Mathematical Methods for Digital Computers*, eds. Anthony Ralston and Herbert S. Wilf (New York: John Wiley & Sons, Inc., 1960).

[4] S. Gil, "A Process for the Step-by-Step Integration of Differential Equations in an Automatic Digital Computing Machine," *Proc. Cambridge Philos. Soc.*, Vol. 47 (1951), pp. 96–108.

parameters, leading to a very large number of different forms. Fifth-order methods have been found by numerical experiments to generally be highly desirable where quite high accuracy is required. A (5, 6) method by Butcher[5] has been found to be particularly desirable. The equations of this method are

$$y_{i+1} = y_i + \tfrac{1}{90}(7k_1 + 32k_3 + 12k_4 + 32k_5 + 7k_6) \tag{6-87}$$

where

$$\left.\begin{aligned}
k_1 &= (h)f(x_i, y_i)\\
k_2 &= (h)f\left(x_i + \frac{h}{4}, y_i + \frac{k_1}{4}\right)\\
k_3 &= (h)f\left(x_i + \frac{h}{4}, y_i + \frac{k_1}{8} + \frac{k_2}{8}\right)\\
k_4 &= (h)f\left(x_i + \frac{h}{2}, y_i - \frac{k_2}{2} + k_3\right)\\
k_5 &= (h)f(x_i + \tfrac{3}{4}h, y_i + \tfrac{3}{16}k_1 + \tfrac{9}{16}k_4)\\
k_6 &= (h)f(x_i + h, y_i - \tfrac{3}{7}k_1 + \tfrac{2}{7}k_2 + \tfrac{12}{7}k_3 - \tfrac{12}{7}k_4 + \tfrac{8}{7}k_5)
\end{aligned}\right\} \tag{6-88}$$

Numerical experiments by Waters[6] led to the conclusion that the above method was "best" of many methods tried on a system of 64 first-order differential equations in terms of accuracy and minimum computing time. The authors have found similar favorable results for this Butcher (5, 6) method in applying various fourth-, fifth-, and sixth-order methods to the orbit problem given at the end of this chapter (Prob. 6-22).

Error Analysis of Runge-Kutta Methods

One of the disadvantages of the higher order Runge-Kutta methods mentioned earlier is the difficulty encountered in estimating the per-step truncation error. An expression for the approximate per-step truncation error is given by Romonelli, but it is difficult to apply.[7] Richardson's method (p. 382) is, perhaps, the most practical one available. Richardson's

[5] J. C. Butcher, "On Runge-Kutta Processes of Higher Order," *J. Australian Math. Soc.*, Vol. 4 (1964), p. 179.

[6] J. Waters, "Methods of Numerical Integration Applied to a System Having Trivial Function Evaluations," *Commun. Ass. Comp. Mach.*, Vol. 9 (1966), p. 293.

[7] Anthony Ralston and Herbert S. Wilf, eds., *Mathematical Methods for Digital Computers* (New York: John Wiley & Sons, Inc., 1960), p. 116.

expression for the fourth-order Runge-Kutta method (Eq. 6-31 with $r = 4$) is

$$E_T \cong \frac{y^{(2)} - y^{(1)}}{15} \tag{6-89}$$

Collatz has suggested that, in using the fourth-order method, the step size used may be based upon the relationship

$$\left|\frac{k_2 - k_3}{k_1 - k_2}\right|$$

where the expressions for the k's are as given by Eq. 6-84.[8] When this quantity exceeds a few hundredths, the step size should be decreased accordingly.

The engineer who is unfamiliar with the intricacies of error analysis will find that a practical method of ensuring the use of a suitable step size in a Runge-Kutta method is to run several trial solutions, with different step sizes in each one, until consecutive solutions are obtained in which the results are in good agreement throughout. (This procedure was discussed in Sec. 6-3 in regard to the error analysis of Euler's method.)

Having discussed the general method of developing Runge-Kutta formulas and having shown several specific different-order Runge-Kutta equations, let us illustrate the use of the classic third-order method in solving a first-order differential equation.

EXAMPLE 6-3

The circuit shown in Fig. 6-21 contains a source of emf, an inductance, and a resistor, the magnitude of the latter varying with its temperature. Since the temperature of the resistor increases with increasing current in the circuit, the resistance is a function of the current. In the range of

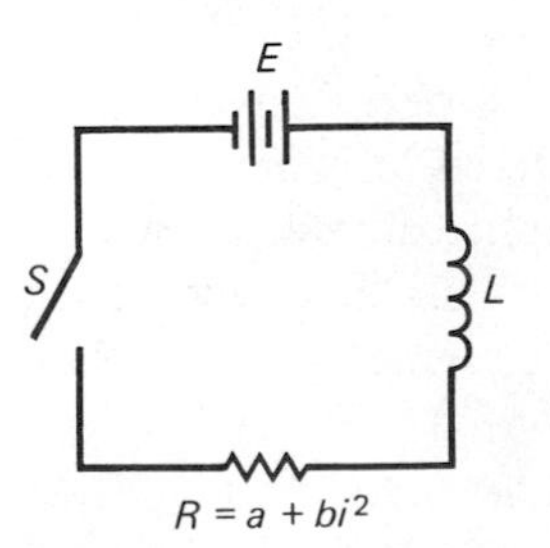

Fig. 6-21 / *Circuit diagram for Ex. 6-3.*

[8] L. Collatz, *Numerische Behandlung von Differentialgleichungen* (Berlin: Springer-Verlag, 1950), p. 34.

current flow i for this problem, the resistance can be expressed as $R = a + bi^2$. Switch S is closed at time $t = 0$, and the current flow is desired as a function of time for $t > 0$.

Applying Kirchhoff's voltage law (the algebraic sum of all voltage changes around a closed circuit is equal to zero) to the circuit loop, the following differential equation is obtained:

$$E - L\frac{di}{dt} - (a + bi^2)i = 0 \tag{6-90}$$

Equation 6-90 may be rearranged algebraically so that

$$\frac{di}{dt} = \frac{E}{L} - \frac{b}{L}i^3 - \frac{a}{L}i \qquad (\text{when } t = 0,\ i = 0) \tag{6-91}$$

Let us now write a FORTRAN program for determining $i(t)$, using Eqs. 6-81 and 6-82. The program will be kept general, so that various parameter values may be used for investigating different circuit combinations. We will assume that the parameter values for this particular example are

$E = 200$ volts
$L = 3$ henries
$a = 100$ ohms
$b = 50$ ohms/amp^2

The FORTRAN names used and the quantities which they represent are as follows:

FORTRAN Name	**Quantity**	**Numerical Values Used as Data**
AMP	Current, amps	Initially zero
A	Constant part of resistance, ohms	100
B	Coefficient of i^2 in resistance expression, ohms/amp^2	50
HENRY	Inductance, henries	3
E	Emf, volts	200
C	B/henry	
D	A/henry	
AK1, AK2, AK3	k_1, k_2, and k_3[9]	

[9] Since the independent variable t does not appear explicitly in Eq. 6-91, it is not necessary to provide incremented values of time to calculate k values within each step.

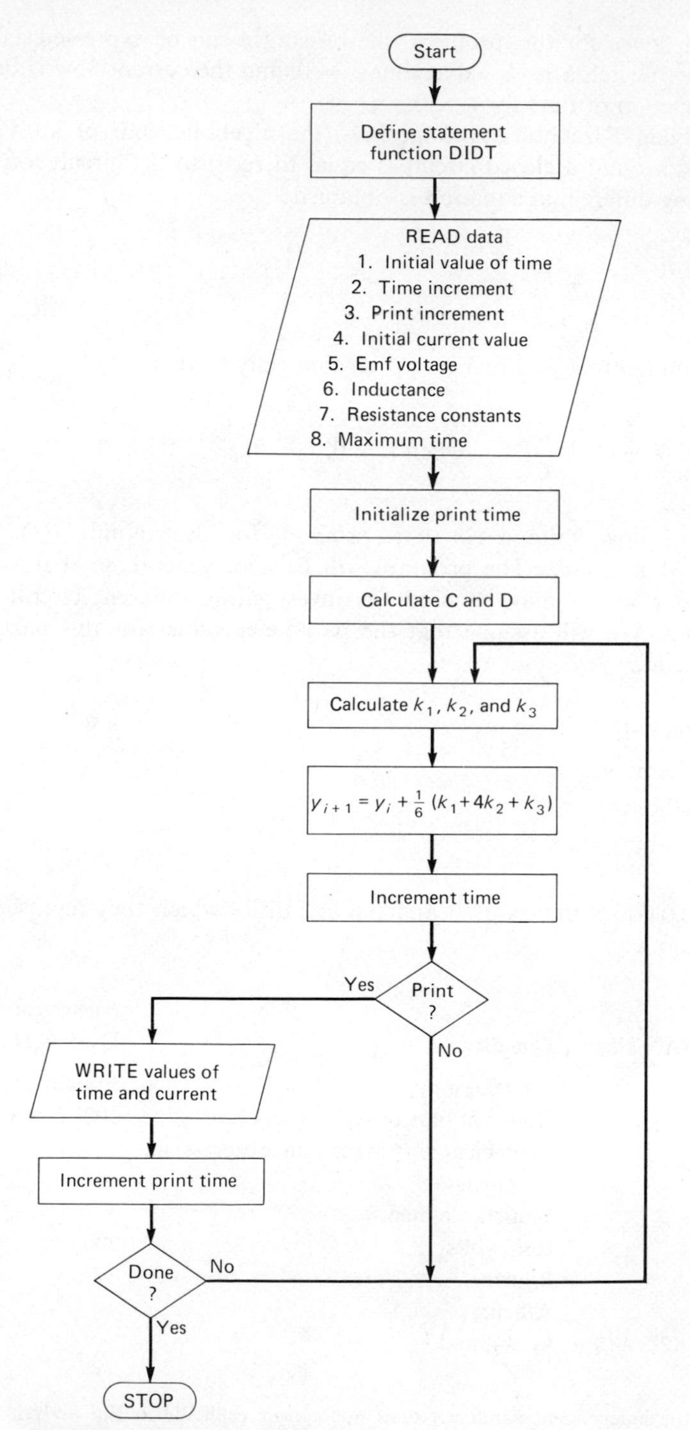

Fig. 6-22 / *Flow chart for Ex. 6-3.*

T	Time, sec	
TPR	Print time, sec	
DELT	Increment of time, sec	0.001
DTPR	Increment of print time, sec	0.002
TMAX	Maximum time, sec	0.05
DIDT	Statement function name for di/dt of Eq. 6-91, with AMP being the argument	

The flow chart for the problem is shown in Fig. 6-22. The FORTRAN IV program, written from the outline provided by the flow chart, is as follows:

```
  DIDT(AMP)=E/HENRY-C*AMP**3-D*AMP
  WRITE(6,1)
1 FORMAT(1H1,'CURRENT IN A RESISTANCE-INDUCTANCE CIRCUIT'//)
  WRITE(6,2)
2 FORMAT(11X,'TIME',11X,'CURRENT'//)
  READ(5,3)T,DELT,DTPR,AMP,E,HENRY,A,B,TMAX
3 FORMAT(9F8.0)
  TPR=DTPR
  C=B/HENRY
  D=A/HENRY
4 AK1=DELT*DIDT(AMP)
  AK2=DELT*DIDT(AMP+AK1/2.)
  AK3=DELT*DIDT(AMP-AK1+2.*AK2)
  AMP=AMP+(AK1+4.*AK2+AK3)/6.
  T=T+DELT
  IF(ABS(T-TPR)-.0001)5,5,4
5 WRITE(6,6)T,AMP
6 FORMAT(1H ,10X,F5.3,12X,F6.4)
  TPR=TPR+DTPR
  IF(ABS(T-TMAX)-.0001)7,7,4
7 STOP
  END
```

The data output obtained from the computer solution is

```
CURRENT IN A RESISTANCE-INDUCTANCE CIRCUIT

TIME              CURRENT

0.002             0.1290
0.004             0.2494
0.006             0.3613
0.008             0.4647
0.010             0.5593
0.012             0.6451
0.014             0.7221
0.016             0.7905
0.018             0.8506
0.020             0.9029
0.022             0.9480
0.024             0.9866
```

continued overleaf

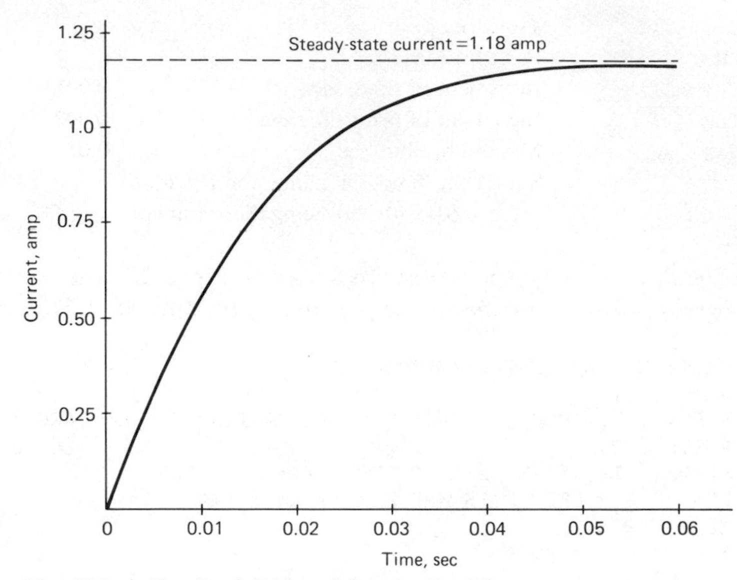

Fig. 6-23 / *Graphical display of data for Ex. 6-3.*

TIME	CURRENT
0.026	1.0193
0.028	1.0470
0.030	1.0701
0.032	1.0894
0.034	1.1055
0.036	1.1188
0.038	1.1298
0.040	1.1388
0.042	1.1462
0.044	1.1523
0.046	1.1573
0.048	1.1614
0.050	1.1647

The data are represented graphically by the curve of Fig. 6-23, illustrating the current buildup in the circuit with time.

6-6 / SOLUTION OF SIMULTANEOUS ORDINARY DIFFERENTIAL EQUATIONS BY RUNGE-KUTTA METHODS

It is frequently necessary to solve sets of simultaneous first-order differential equations in analyzing engineering systems. Such equations occur quite frequently in obtaining solutions of higher order differential equations which are transformed to sets of first-order differential equations as part of the solution process. Runge-Kutta methods are well suited for the solution of such equations, and their application will be discussed here.

Let us first consider the solution of two simultaneous first-order differential equations of the form

$$\begin{aligned} \frac{dy}{dx} &= f[x, y(x), u(x)] \\ \frac{du}{dx} &= F[x, y(x), u(x)] \end{aligned} \tag{6-92}$$

where the initial values ($y = y_0$, $u = u_0$, when $x = x_0$) are known. Runge-Kutta formulas can be used to solve Eq. 6-92. Using the classic fourth-order method already described (Eqs. 6-83 and 6-84), for example, the following *sets* of equations would be used:

$$y_{i+1} = y_i + \tfrac{1}{6}(k_1 + 2k_2 + 2k_3 + k_4) \tag{6-93}$$

where

$$\left.\begin{aligned} k_1 &= (h)f(x_i, y_i, u_i) \\ k_2 &= (h)f\left(x_i + \frac{h}{2}, y_i + \frac{k_1}{2}, u_i + \frac{q_1}{2}\right) \\ k_3 &= (h)f\left(x_i + \frac{h}{2}, y_i + \frac{k_2}{2}, u_i + \frac{q_2}{2}\right) \\ k_4 &= (h)f(x_i + h, y_i + k_3, u_i + q_3) \end{aligned}\right\} \tag{6-94}$$

and

$$u_{i+1} = u_i + \tfrac{1}{6}(q_1 + 2q_2 + 2q_3 + q_4) \tag{6-95}$$

where

$$\left.\begin{aligned} q_1 &= (h)F(x_i, y_i, u_i) \\ q_2 &= (h)F\left(x_i + \frac{h}{2}, y_i + \frac{k_1}{2}, u_i + \frac{q_1}{2}\right) \\ q_3 &= (h)F\left(x_i + \frac{h}{2}, y_i + \frac{k_2}{2}, u_i + \frac{q_2}{2}\right) \\ q_4 &= (h)F(x_i + h, y_i + k_3, u_i + q_3) \end{aligned}\right\} \tag{6-96}$$

Note that in the expressions for the k's and q's, y_i is incremented by a parameter times a k value, since k's are used in the recurrence formula for y values. Also, u_i is incremented by a parameter times a q value, since q's are used in the recurrence formula for u values. Also note that function $f(x, y, u)$ is used for calculating k values, since it is equal to dy/dx, and k's are used in the recurrence formula for y. The function $F(x, y, u)$ is used for calculating q values, since it is equal to du/dx, and q's are used in the recurrence formula for u values. This same general pattern holds for any number of simultaneous equations.

The solution of Eq. 6-92 begins by substituting the initial values of y and u, which must be known, into the given differential equations to obtain initial values of the functions f and F. Values of k_1 and q_1 are next obtained by multiplying the initial values of f and F, respectively, by h, as indicated in Eqs. 6-94 and 6-96 where $h = \Delta x$, the step increment. With values of k_1 and q_1 known, k_2 and q_2 are next evaluated, then k_3 and q_3, and finally k_4 and q_4. Then the recurrence formulas (Eqs. 6-93 and 6-95) are used to obtain values of y and u at $x = x_i + h$ (y_{i+1} and u_{i+1}). These new values of y and u are then used as beginning values in starting the procedure (just described) over again, to obtain values of y_{i+2} and u_{i+2} at $x = x_i + 2h$, and so on, until the desired range of integration has been covered.

Any number of simultaneous first-order differential equations may be solved by merely using a set of equations such as those shown for each dependent variable appearing in the set of simultaneous differential equations.

An nth-order differential equation can be solved by transforming the equation to a set of n simultaneous first-order differential equations and applying n Runge-Kutta formulas, as discussed in the preceding paragraphs. The procedure can be simplified by combining equations, as will be shown in the following paragraphs.

Consider the second-order differential equation

$$\frac{d^2x}{dt^2} = f\left(t, x, \frac{dx}{dt}\right) \tag{6-97}$$

Letting $v = dx/dt$, Eq. 6-97 can be transformed to the two first-order differential equations

$$\left.\begin{aligned} \frac{dv}{dt} &= f(t, x, v) \\ \frac{dx}{dt} &= v \end{aligned}\right\} \tag{6-98}$$

Referring to Eqs. 6-93 through 6-96, the following two fourth-order Runge-Kutta formulas could be used to solve Eq. 6-98:

$$v_{i+1} = v_i + \tfrac{1}{6}(k_1 + 2k_2 + 2k_3 + k_4) \tag{6-99}$$

where

$$\left.\begin{aligned} k_1 &= (h)f(t_i, x_i, v_i) \\ k_2 &= (h)f\left(t_i + \frac{h}{2}, x_i + \frac{q_1}{2}, v_i + \frac{k_1}{2}\right) \\ k_3 &= (h)f\left(t_i + \frac{h}{2}, x_i + \frac{q_2}{2}, v_i + \frac{k_2}{2}\right) \\ k_4 &= (h)f(t_i + h, x_i + q_3, v_i + k_3) \end{aligned}\right\} \tag{6-100}$$

and

$$x_{i+1} = x_i + \tfrac{1}{6}(q_1 + 2q_2 + 2q_3 + q_4) \tag{6-101}$$

where

$$\left.\begin{aligned} q_1 &= hF(v_i) = h(v_i) \\ q_2 &= hF\left(v_i + \frac{k_1}{2}\right) = h\left(v_i + \frac{k_1}{2}\right) \\ q_3 &= hF\left(v_i + \frac{k_2}{2}\right) = h\left(v_i + \frac{k_2}{2}\right) \\ q_4 &= hF(v_i + k_3) = h(v_i + k_3) \end{aligned}\right\} \tag{6-102}$$

However, we may obtain a more compact set of equations by substituting the expressions for the q's into both the expressions for the k's and the recurrence formula of Eq. 6-101. Performing these substitutions yields

$$\begin{aligned} x_{i+1} &= x_i + hv_i + \frac{h}{6}(k_1 + k_2 + k_3) \\ v_{i+1} &= v_i + \tfrac{1}{6}(k_1 + 2k_2 + 2k_3 + k_4) \end{aligned} \tag{6-103}$$

where

$$\left.\begin{aligned} k_1 &= (h)f(t_i, x_i, v_i) \\ k_2 &= (h)f\left[t_i + \frac{h}{2}, x_i + \frac{h}{2}v_i, v_i + \frac{k_1}{2}\right] \\ k_3 &= (h)f\left[t_i + \frac{h}{2}, x_i + \frac{h}{2}v_i + \frac{h}{4}k_1, v_i + \frac{k_2}{2}\right] \\ k_4 &= (h)f\left[t_i + h, x_i + hv_i + \frac{h}{2}k_2, v_i + k_3\right] \end{aligned}\right\} \tag{6-104}$$

EXAMPLE 6-4

To illustrate the application of Eqs. 6-103 and 6-104 in solving a second-order differential equation, consider the vibrating system shown in Fig. 6-24(a). A unit mass is attached to a spring having a spring constant of 1 lb/ft, and a damping device that exerts a force on the mass which depends upon both the displacement and the velocity of the mass. The damping force may be expressed mathematically as

$$F_d = \mu(x^2 - 1)\dot{x} \tag{6-105}$$

where μ is a damping coefficient. It is desired to obtain x-t and $\dot{x}$-t curves for studying the displacement and velocity characteristics of the system.

A free-body diagram of the mass is shown in Fig. 6-24(b). The gravity force acting on such a system is customarily not shown on the free body, since the amplitude of vibration x is measured from the static-equilibrium position of the mass, in which position the gravity force is equal to the spring force, and these forces cancel from the resulting differential equation.

Noting the positive direction assumed for x, $\dot{x}$, and $\ddot{x}$, as shown in the figure, and the positive displacement used in determining the free-body diagram, the application of Newton's second law yields

$$-kx - \mu(x^2 - 1)\dot{x} = m\ddot{x}$$

or, more simply,

$$\ddot{x} + \mu(x^2 - 1)\dot{x} + x = 0 \tag{6-106}$$

since $m = k = 1$.

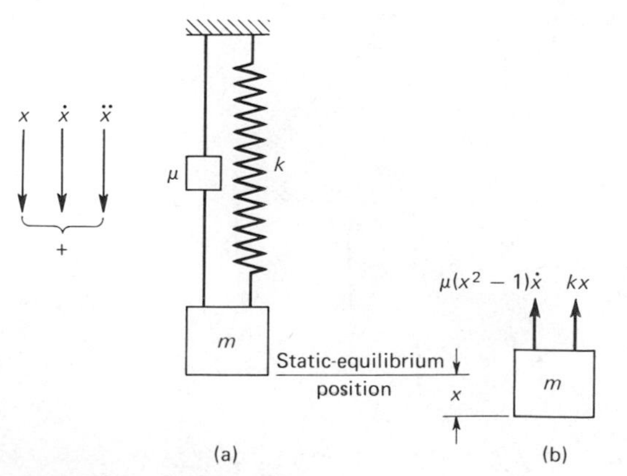

Fig. 6-24 / *System of Ex. 6-4.*

Equation 6-106 is a classical nonlinear differential equation known as *Van der Pol's equation.* Although the mechanical system from which Eq. 6-106 was derived is a hypothetical one, this equation does actually represent the behavior of a certain type of electronic oscillator.[10]

The reader who is familiar with vibration theory will recognize that if $|x| > 1$, the second term of Eq. 6-106 is a *positive-damping* term which represents a loss of energy from the system. Conversely, if $|x| < 1$ this term is a *negative-damping* term which represents an addition of energy to the system. Thus if we were initially to displace the mass less than 1 ft and then release it, the resulting vibrations would increase in amplitude until they reached some magnitude exceeding 1 ft, at which time energy would be removed from the system during the portion of each cycle in which $|x| > 1$. The buildup in amplitude would thus be limited, and it would seem reasonable to expect the amplitude of vibration to reach some stable state after some interval of time.

To program Eq. 6-106 for solution on the computer, we note that

$$\left.\begin{array}{l} \dot{x} = \dfrac{dx}{dt} \\ \\ \text{so that} \\ \\ \ddot{x} = \dfrac{d\dot{x}}{dt} \end{array}\right\} \qquad (6\text{-}107)$$

Utilizing these relationships, Eq. 6-106 may be transformed to the first-order differential equations

$$\left.\begin{array}{ll} \dfrac{d\dot{x}}{dt} = -\mu(x^2 - 1)\dot{x} - x & \text{(a)} \\ \dfrac{dx}{dt} = \dot{x} & \text{(b)} \end{array}\right\} \qquad (6\text{-}108)$$

Referring to Eqs. 6-103 and 6-104, and noting that the variable t does not appear explicitly in Eq. 6-108, we may write the fourth-order Runge-Kutta formulas

$$\begin{aligned} x_{i+1} &= x_i + (h)\dot{x}_i + \frac{h}{6}(k_1 + k_2 + k_3) \\ \dot{x}_{i+1} &= \dot{x}_i + \tfrac{1}{6}(k_1 + 2k_2 + 2k_3 + k_4) \end{aligned} \qquad (6\text{-}109)$$

[10] N. Minorsky, *Introduction to Nonlinear Mechanics* (Ann Arbor, Mich.: J. W. Edwards, Publisher, Inc., 1947), p. 293.

where

$$\left.\begin{aligned}
k_1 &= (h)f(x_i, \dot{x}_i) \\
k_2 &= (h)f\left[x_i + \frac{h}{2}\dot{x}_i, \dot{x}_i + \frac{k_1}{2}\right] \\
k_3 &= (h)f\left[x_i + \frac{h}{2}\dot{x}_i + \frac{h}{4}k_1, \dot{x}_i + \frac{k_2}{2}\right] \\
k_4 &= (h)f\left[x_i + (h)\dot{x}_i + \frac{h}{2}k_2, \dot{x}_i + k_3\right]
\end{aligned}\right\} \tag{6-110}$$

The numerical solution begins with the substitution of the initial values of x and $\dot{x}$ into Eq. 6-108a to obtain a value of this function for use in determining k_1. The successive k values are then determined for use in the recurrence formulas shown to obtain values of x_{i+1} and $\dot{x}_{i+1}$. The latter values obtained are then used in Eqs. 6-108a and 6-110 to obtain new k values for substitution into Eq. 6-109 to obtain values of x_{i+2} and $\dot{x}_{i+2}$, and so on. An outline of the procedure is shown in the flow chart of Fig. 6-25. The initial values used are

$$t = 0 \left| \begin{aligned} x &= 0.75 \text{ ft} \\ \dot{x} &= 0 \end{aligned} \right.$$

The FORTRAN variable names used in the source program and the quantities which they represent are as follows:

FORTRAN Name	Quantity	Values Used in Problem
COEFF	Damping coefficient, lb-sec/ft^3	4.0
X	Displacement of mass from static-equilibrium position, ft	$x_0 = 0.75$
XD	Velocity of mass, ft/sec	$x_0 = 0.0$
H	Time increment, sec	0.05
DTPR	Print increment, sec	0.10
TMAX	Time interval for which solution is desired, sec	20.0
T	Time, sec	
TPR	Print time, sec	
AK1, AK2, AK3, AK4	Quantities k_1, k_2, k_3, and k_4 of Runge-Kutta formulas	

The FORTRAN IV computer program used to implement the procedures outlined by the flow chart is shown below. The time increment has been

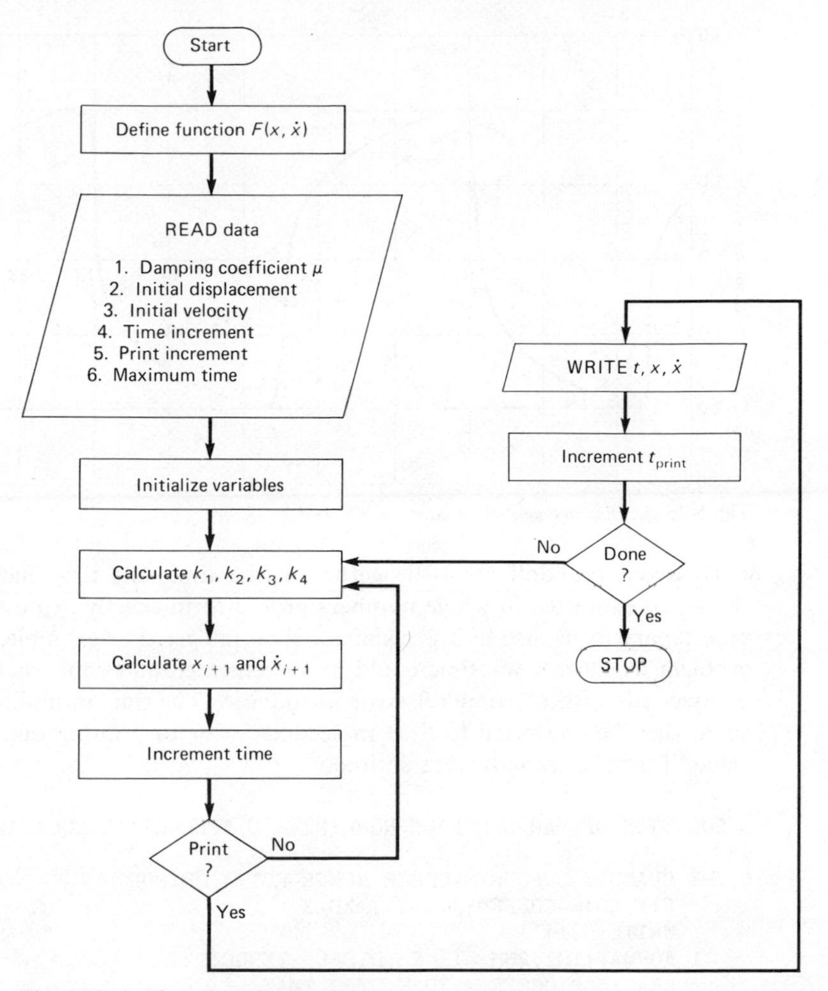

Fig. 6-25 / *Flow chart for Ex. 6-4.*

chosen as 0.05 sec. Since the program was run on a pure binary computer, and this quantity cannot be expressed exactly in pure binary form, some small roundoff error is to be expected in the time. For example, 20 time increments of 0.05 sec will not add up to exactly 1.0 sec. As a result, print times will not be exact integer multiples of the time increments as they would if no roundoff error were present. To obtain printout at the times specified in the program, it is required that the time be within one-tenth of a step of the print time. This is accomplished with the statement

```
IF(ABS(T-TPR)-.005)4,4,3
```

in the program.

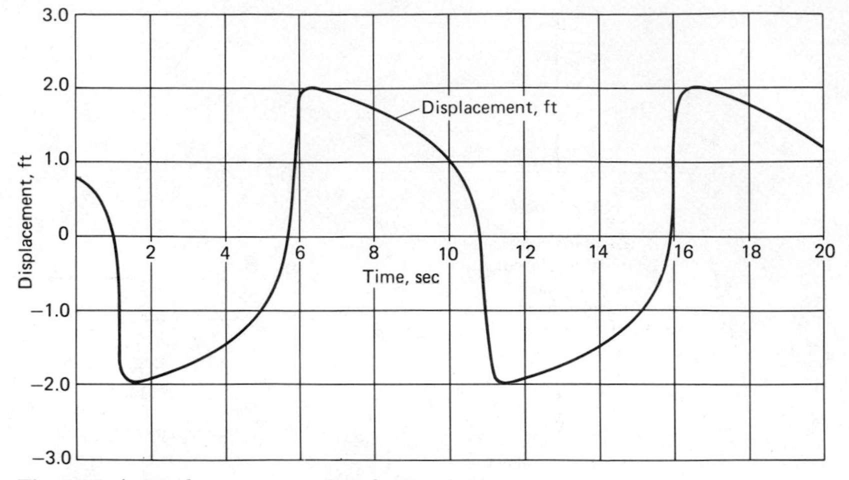

Fig. 6-26 / *Displacement-time data for Ex. 6-4.*

To avoid roundoff error altogether in the time, the time increment should be expressed in whole numbers or in a form exactly expressible in pure binary form such as $1/2^n$, where n is an integer. For example, in the problem which follows, time could be incremented and kept track of in milliseconds with no roundoff error introduced. The time in milliseconds could then be converted to time in seconds prior to printing out a time value, if time in seconds were desired.

```
C SOLUTION OF VAN DER POLS NONLINEAR DIFFERENTIAL EQUATION
                                                        USING
C THE CLASSICAL FOURTH-ORDER RUNGE-KUTTA METHOD
      F(X,XD)=-COEFF*(X*X-1.)*XD-X
      WRITE(6,1)
    1 FORMAT(1H1,26H  TIME  DISPL  VELOCITY/)
      READ(5,2)COEFF,X,XD,H,DTPR,TMAX
    2 FORMAT(6F10.0)
      T=0
      TPR=DTPR
    3 AK1=H*F(X,XD)
      AK2=H*F(X+H/2.*XD,XD+AK1/2.)
      AK3=H*F(X+H/2.*(XD+AK1/2.),XD+AK2/2.)
      AK4=H*F(X+H*(XD+AK2/2.),XD+AK3)
      X=X+H*(XD+(AK1+AK2+AK3)/6.)
      XD=XD+(AK1+2.*AK2+2.*AK3+AK4)/6.
      T=T+H
      IF(ABS(T-TPR)-.005)4,4,3
    4 WRITE(6,5)T,X,XD
    5 FORMAT(1H ,F6.2,3X,F6.3,4X,F7.3)
      TPR=TPR+DTPR
      IF(ABS(T-TMAX)-.005)6,6,3
    6 STOP
      END
```

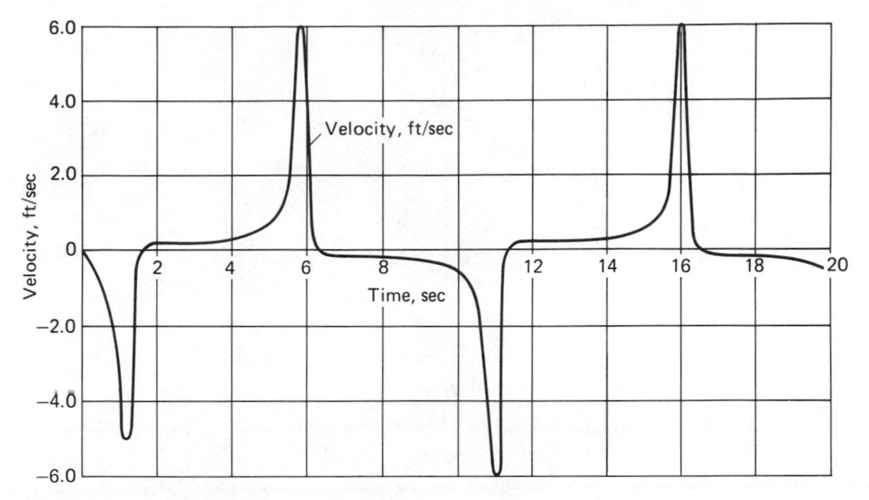

Fig. 6-27 / *Velocity-time data for Ex. 6-4.*

Since 200 data printouts were obtained over the integration interval of 20 sec, the computer printout is not shown, owing to space considerations. Instead, the data obtained are plotted in graphical form in Figs. 6-26 and 6-27. Note that the vibration becomes essentially periodic and reaches a stable state in the 20-sec interval considered.

6-7 / MILNE'S METHOD

As the step-by-step numerical integration of a differential equation progresses, the solution being obtained may tend to depart more and more from the true answer, owing to cumulative *per-step* error. If the solution must be obtained by using a relatively large number of steps, it is very important that the per-step error be kept small. The per-step error, using Euler's method, is of order h^2. The modified Euler methods have a per-step error of order h^3. It can be shown that Milne's method has a per-step error of order h^5. Thus Milne's method is considerably more accurate than either Euler's method or the modified Euler methods, when a relatively large number of steps is involved. However, Milne's method does have the disadvantage of not being self-starting, though, as we will see later, this is not a serious disadvantage.

The development of Milne's method begins by dividing the area under a given portion of a curve $y = f(x)$ into four h-width strips, as shown in Fig. 6-28. The true area under this portion of the curve is then approximated by considering the area of these four strips under a second-degree parabola having three coordinates in common with the actual curve, as indicated by the dashed line in Fig. 6-28. The crosshatched area is the approximate area obtained. To determine an expression for this cross-

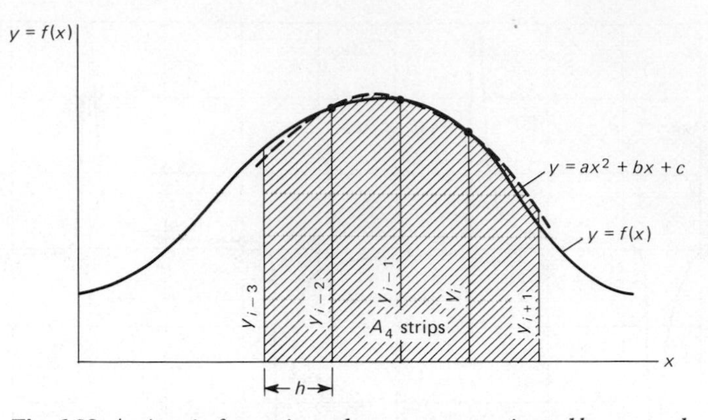

Fig. 6-28 | *Area in four strips under a curve approximated by area under a second-degree parabola.*

hatched area in terms of h and the appropriate y ordinates, it is convenient to consider the four strips as centered on the y axis, as shown in Fig. 6-29. This arrangement does not compromise the generality of the results obtained, and it has the advantage of simplifying the intermediate expressions involved in determining the desired form of the expression for the area of four such strips. The crosshatched area of Fig. 6-29 is given by

$$A_{4\,\text{strips}} = \int_{-2h}^{2h} (ax^2 + bx + c)\,dx \tag{6-111}$$

Integrating Eq. 6-111 and substituting the limits gives

$$A_{4\,\text{strips}} = \tfrac{16}{3}ah^3 + 4ch \tag{6-112}$$

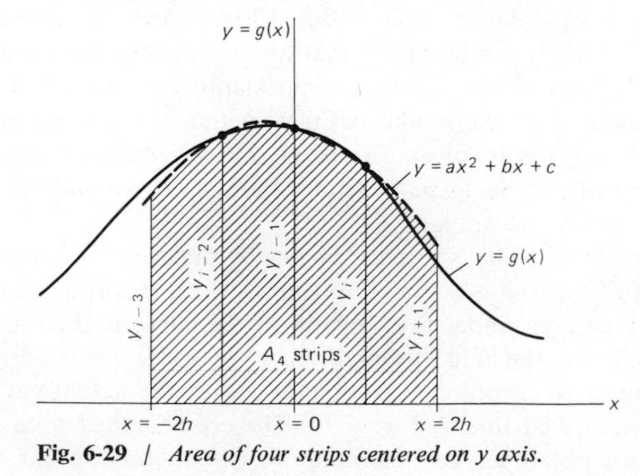

Fig. 6-29 | *Area of four strips centered on y axis.*

The constants a and c are determined in the manner explained on page 329. The appropriate expressions are

$$a = \frac{y_i - 2y_{i-1} + y_{i-2}}{2h^2} \tag{6-113}$$

$$c = y_{i-1}$$

Substituting Eq. 6-113 into Eq. 6-112 yields, for the area of the four strips in terms of h and the y ordinates shown,

$$A_{4\text{ strips}} = \tfrac{4}{3}h[2y_i - y_{i-1} + 2y_{i-2}] \tag{6-114}$$

This expression will be used later as part of the predictor equation.

Let us consider the application of Milne's method in integrating a first-order differential equation of the form

$$y' = f(x, y) \tag{6-115}$$

where the value of y is known for $x = 0$. This technique consists, basically, of obtaining approximate values of y by the use of a *predictor* equation and then correcting these values by the iterative use of a corrector equation. Milne's predictor equation

$$P(y_{i+1}) = y_{i-3} + \tfrac{4}{3}h[2y'_i - y'_{i-1} + 2y'_{i-2}] \tag{6-116}$$

utilizes the area of four strips under a parabolic approximation of a curve (see Eq. 6-114) to provide a predicted value for the successive y ordinates. Milne's *corrector* equation

$$C(y_{i+1}) = y_{i-1} + \frac{h}{3}[y'_{i-1} + 4y'_i + P(y'_{i+1})] \tag{6-117}$$

provides corrected y values by using Simpson's rule for determining the area of two strips under a curve (see p. 329).

Assuming that the resulting y and y' curves of Eq. 6-115 have the general form of the curves shown in Fig. 6-30, the first step is to obtain a predicted value of y_4. Utilizing Eq. 6-116 with $i = 3$,

$$P(y_4) = y_0 + \tfrac{4}{3}h[2y'_3 - y'_2 + 2y'_1] \tag{6-118}$$

The predicted value obtained is then substituted into the given differential equation (Eq. 6-115) along with the appropriate x value to obtain a predicted

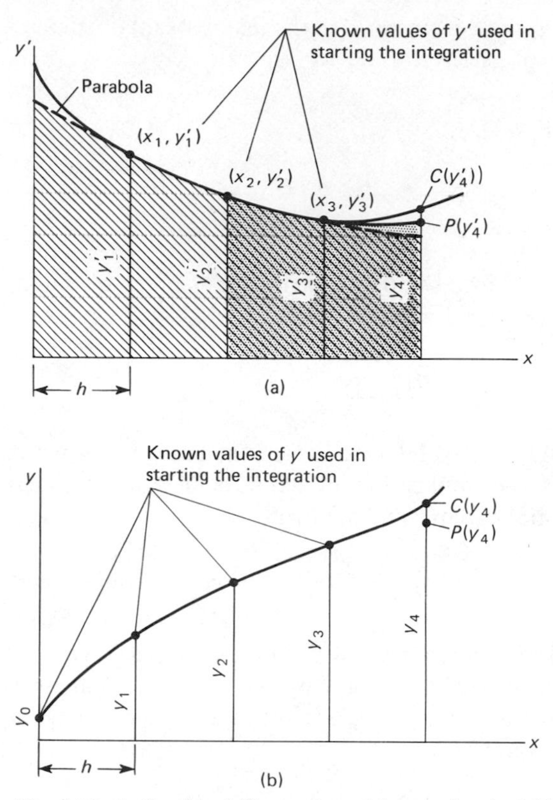

Fig. 6-30 / *Graphical illustration of integration by Milne's method.*

value of y'_4, which is designated as $P(y'_4)$. This value is then used in Eq. 6-117 to obtain a corrected value of y_4 as

$$C(y_4) = y_2 + \frac{h}{3}[y'_2 + 4y'_3 + P(y'_4)] \tag{6-119}$$

The three steps just outlined are the first steps in the application of Milne's method. However, inspection of Eqs. 6-118 and 6-119 reveals that, before $P(y_4)$ and $C(y_4)$ can be obtained, starting values must be determined, in some manner, for y'_1, y'_2, y'_3, and y_2. Furthermore, in stepping forward to obtain values of $P(y_5)$ and $C(y_5)$, starting values must also be known for y_1 and y_3.

The required starting values of y can be determined by the use of the following Taylor-series expansions:

$$\begin{aligned} y_1 &= y_0 + y'_0 h + \frac{y''_0 h^2}{2!} + \frac{y'''_0 h^3}{3!} + \cdots \\ y_2 &= y_0 + y'_0[2h] + \frac{y''_0[2h]^2}{2!} + \frac{y'''_0[2h]^3}{3!} + \cdots \\ y_3 &= y_0 + y'_0[3h] + \frac{y''_0[3h]^2}{2!} + \frac{y'''_0[3h]^3}{3!} + \cdots \end{aligned} \tag{6-120}$$

The numerical values of each of the derivatives, y_0', y_0'', y_0''', and so on, which are required to evaluate Eq. 6-120, are obtained by substituting $x = 0$ and $y = y_0$ into Eq. 6-115 and its derivatives y'', y''', and so forth. Having determined y_1, y_2, and y_3 from Eq. 6-120, the required values of y_1', y_2', and y_3' are next obtained by substituting the corresponding x and y values into Eq. 6-115. The number of terms used in each Taylor series should include terms up to and including h^4 or the last term which yields a numerical value compatible with the digit capacity allowed by the computer being used, whichever number of terms is the least.

Another way of starting Milne's method is to determine the required value of each of the ordinates y_1, y_2, and y_3 by the use of a Runge-Kutta formula. A value of each of the ordinates y_1', y_2', and y_3' may then be obtained, as before, by substituting the appropriate x and y values into Eq. 6-115. The fourth-order Runge-Kutta formula frequently used is

$$y_{i+1} = y_i + \tfrac{1}{6}(k_1 + 2k_2 + 2k_3 + k_4) \tag{6-121}$$

where the respective k values are determined by substituting appropriate values into Eq. 6-115, as follows:

$$k_1 = (h)f(x_i, y_i)$$

$$k_2 = (h)f\left(x_i + \frac{h}{2}, y_i + \frac{k_1}{2}\right)$$

$$k_3 = (h)f\left(x_i + \frac{h}{2}, y_i + \frac{k_2}{2}\right)$$

$$k_4 = (h)f(x_i + h, y_i + k_3)$$

The solution of Eq. 6-115 could be obtained in its entirety using only the Runge-Kutta method, and with the same degree of accuracy as Milne's method, since both have a per-step error of the order of h^5. However, Milne's method is considered to be more efficient (less machine time) than the Runge-Kutta method, and estimation of the per-step error is easier using Milne's method. It is often used for these reasons, even though the Runge-Kutta method has the advantage of being self-starting. Because the Runge-Kutta method is self-starting and has the same degree of accuracy as Milne's method, it is frequently used to start the latter method, as just described. In using either the Taylor series or the Runge-Kutta method for starting Milne's procedure, the required starting values may be obtained manually or by using a calculator. However, if it is desired to use the computer to obtain the starting values so that the program is self-contained, it is generally easier to use the Runge-Kutta method, since higher derivatives of the given differential equation are not required by this method. If the derivatives of the given differential equation become increasingly complex with each successive differentiation, the Runge-Kutta procedure should be utilized for obtaining starting values, whether done manually, by calculator, or by computer.

With the starting procedure established, the following equations provide an outline of the subsequent steps involved,

(Predictor)

$$P(y_{i+1}) = y_{i-3} + \tfrac{4}{3}h[2y'_i - y'_{i-1} + 2y'_{i-2}] \quad \text{(a)}$$

(Differential equation)

$$P(y'_{i+1}) = f[x_{i+1}, P(y_{i+1})] \quad \text{(b)}$$

(Corrector)

$$C(y_{i+1}) = y_{i-1} + \frac{h}{3}[y'_{i-1} + 4y'_i + P(y'_{i+1})] \quad \text{(c)} \quad \text{(6-122)}$$

(Differential equation)

$$C(y'_{i+1}) = f[x_{i+1}, C(y_{i+1})] \quad \text{(d)}$$

(Iterating corrector)

$$C(y_{i+1}) = y_{i-1} + \frac{h}{3}[y'_{i-1} + 4y'_i + C(y'_{i+1})] \quad \text{(e)}$$

where Eq. 6-122b and d indicate the forms in which the given differential equation is used. Assuming that the starting values (shown in Fig. 6-30) have been determined, the first step is to obtain a predicted value of y_4, using Eq. 6-122a. This equation predicts the value of y_4 by adding the value of y_0 to the value of the crosshatched area shown under the parabola in Fig. 6-30(a). Since this area is indicated as less than the true area, $P(y_4)$ will be smaller than the true value of y_4, as shown in Fig. 6-30(b). The next step is to substitute $x_4 = 4h$ and $P(y_4)$ into the given differential equation, as indicated by Eq. 6-122b to obtain a predicted value of y'_4, which is shown as $P(y'_4)$ in Fig. 6-30(a). This predicted value of y'_4 is then used in the corrector equation (Eq. 6-122c) to obtain a *corrected* value of y_4, which is shown as $C(y_4)$ in Fig. 6-30(b). This corrected value of y_4 is obtained as the sum of y_2 and the shaded area under the curve in Fig. 6-30(a), which area is found from Simpson's rule for the area of two strips. The corrected value of y_4 is next substituted into Eq. 6-122d to obtain an improved value of y'_4, which is shown as $C(y'_4)$ in Fig. 6-30(a). The latter value is then used in Eq. 6-122e to obtain a further improved value of y_4, which is, in turn, substituted back into Eq. 6-122d to obtain a further improved value of y'_4, and so on. The iteration process involving Eq. 6-122d and e then continues until successive values of y_4 differ by less than the value of some desired epsilon. With y_4 determined to the desired accuracy, the method steps forward one h increment, and the above process is repeated to obtain y_5, and so on. In programming the problem, it is not actually necessary to write Eq. 6-122e as a separate FORTRAN statement, since the program can be written so that Eq. 6-122c performs the function of Eq. 6-122e in the iteration process of each step.

Relieving the Instability of Milne's Method

As is pointed out in Sec. 6-8, Milne's method is *unstable*. This instability does not appear in the solution of many problems, but in those in which it does the effects can be disastrous. Since stability is discussed in Sec. 6-8, we will simply show in this section the method given by Milne and Reynolds[11] for damping out the extraneous oscillations which sometimes appear in Milne's method after a sufficient number of steps.

The corrector equation of Milne's method is based on Simpson's one-third rule (Eq. 6-117). It is this equation which under certain conditions can produce an error which fluctuates in sign and grows exponentially. The following procedure is shown by Milne and Reynolds as a procedure to damp out such oscillations.

After a finite number of steps, k, $y_{i+1} = y_k$ is recomputed using Simpson's three-eighths rule,

$$Y_k = y_{k-3} + \frac{3h}{8}(y'_k + 3y'_{k-1} + 3y'_{k-2} + y'_{k-3})$$

Then a new $C(y_k)$ is calculated to replace the original $C(y_k)$ as follows:

$$C(y_k)_{\text{new}} = \frac{C(y_k)_{\text{old}} + Y_k}{2}$$

The maximum value of k that will keep the solution stable depends on the step size h. In problems in which stabilizing is required, some experimenting may be required to determine a suitable k. Milne and Reynolds show an example in which stabilizing is carried out every 50 steps. They say that if instability is not removed for k as small as 10, it may be worth-

[11] See W. E. Milne and R. R. Reynolds, "Stability of a Numerical Solution of Differential Equations," *J. Assoc. Computing Machinery,* Vol. 6, No. 2 (April 1959), pp. 196–203, and "Stability of a Numerical Solution of Differential Equations—Part II," *J. Assoc. Computing Machinery*, Vol. 7, No. 1 (Jan. 1960), pp. 46–56.

while to reduce the step size h. Their rough rule of thumb states that cutting h in half permits the doubling of k.

Error Estimation in Milne's Method

Milne's method is a predictor-corrector type, and we will be interested in a means of estimating its per-step truncation error. To investigate this error, let us again assume that we are integrating the differential equation

$$y' = g[x, y(x)] \tag{6-123}$$

with the initial condition of $y = 0$ at $x = a$. Since y is a function of x, y' is actually some unknown function of x alone, which we will call $f(x)$. We will further assume that the graphs of $f(x)$ and y appear in the general form given in Fig. 6-31.

In Milne's method the approximate area under the $f(x)$ curve from x_{i-3} to x_{i+1} (the shaded area of Fig. 6-31) is added to y_{i-3} to obtain predicted value of y_{i+1}. To determine an estimate of the per-step truncation error in the predictor equation, we obtain an exact expression for the area in the four strips (the crosshatched area in Fig. 6-31) and compare

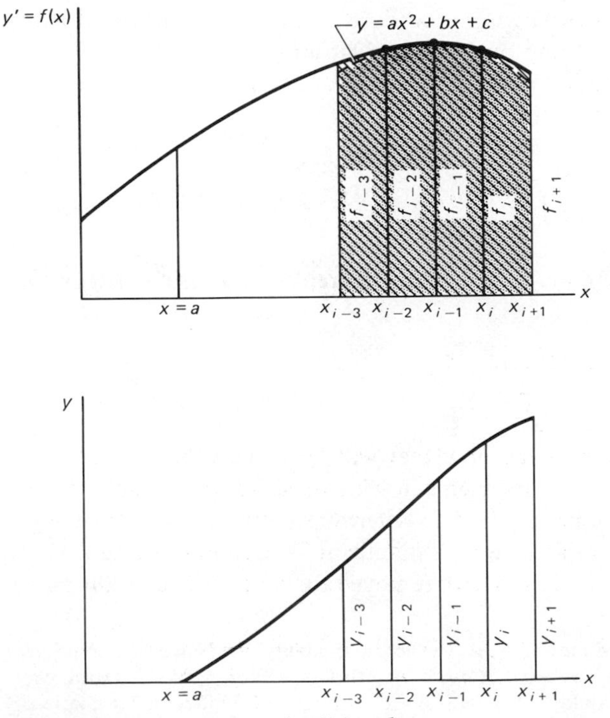

Fig. 6-31 / *Assumed curves for $f(x)$ and y.*

this expression with the expression for the approximate area used in the predictor equation.

The area under the graph of $f(x)$ from $x = a$ to any x value in Fig. 6-31 is a function of x and is equal to the value of y at the x value. That is,

$$y(x) = \int_a^x f(x)\,dx$$

from which

$$\begin{aligned} y_i' &= f_i \\ y_i'' &= f_i' \\ \vdots \quad &\quad \vdots \end{aligned} \tag{6-124}$$

To determine an exact expression for the area of the four strips shown in Fig. 6-31, we first expand y_{i+1} in a Taylor series about x_{i-1}. This expansion is

$$y_{i+1} = y_{i-1} + y_{i-1}'(2h) + \frac{y_{i-1}''(2h)^2}{2!} + \frac{y_{i-1}'''(2h)^3}{3!} + \frac{y_{i-1}^{\mathrm{IV}}(2h)^4}{4!} + \frac{y^{\mathrm{V}}(\xi_1)(2h)^5}{5!}$$

or

$$y_{i+1} - y_{i-1} = 2y_{i-1}'h + 2y_{i-1}''h^2 + \tfrac{4}{3}y_{i-1}'''h^3 + \tfrac{2}{3}y_{i-1}^{\mathrm{IV}}h^4 + \tfrac{4}{15}y^{\mathrm{V}}(\xi_1)h^5, \quad x_{i-1} < \xi_1 < x_{i+1} \tag{6-125}$$

Similarly, expanding y_{i-3} about x_{i-1} yields

$$y_{i-3} - y_{i-1} = -2y_{i-1}'h + 2y_{i-1}''h^2 - \tfrac{4}{3}y_{i-1}'''h^3 + \tfrac{2}{3}y_{i-1}^{\mathrm{IV}}h^4 - \tfrac{4}{15}y^{\mathrm{V}}(\xi_2)h^5, \quad x_{i-3} < \xi_2 < x_{i+1} \tag{6-126}$$

Subtracting Eq. 6-126 from Eq. 6-125 yields

$$y_{i+1} - y_{i-3} = 4y_{i-1}'h + \tfrac{8}{3}y_{i-1}'''h^3 + \tfrac{8}{15}y^{\mathrm{V}}(\xi_3)h^5, \qquad x_{i-3} < \xi_3 < x_{i+1} \tag{6-127}$$

which is an exact expression for the area of the four strips composing the crosshatched area of Fig. 6-31.

Substituting the relationships of Eq. 6-124 into Eq. 6-127, we can express the exact area as

$$y_{i+1} - y_{i-3} = 4f_{i-1}h + \tfrac{8}{3}f_{i-1}''h^3 + \tfrac{8}{15}f^{\mathrm{IV}}(\xi_3)h^5, \qquad x_{i-3} < \xi_3 < x_{i+1} \tag{6-128}$$

To put Eq. 6-128 in a form in which we can compare it with Milne's predictor equation, we must express $f''_{i-1}h^2$ in terms of f_i, f_{i-1}, and f_{i-2}. Using a Taylor series, we expand f_i and f_{i-2} about x_{i-1} as

$$f_i = f_{i-1} + f'_{i-1}h + \frac{f''_{i-1}h^2}{2!} + \frac{f'''_{i-1}h^3}{3!} + \frac{f^{\mathrm{IV}}(\xi_4)h^4}{4!}, \qquad x_{i-1} < \xi_4 < x_i \tag{6-129}$$

and

$$f_{i-2} = f_{i-1} - f'_{i-1}h + \frac{f''_{i-1}h^2}{2!} - \frac{f'''_{i-1}h^3}{3!} + \frac{f^{\mathrm{IV}}(\xi_5)h^4}{4!}, \qquad x_{i-2} < \xi_5 < x_{i-1} \tag{6-130}$$

Adding Eqs. 6-129 and 6-130 gives

$$f_i + f_{i-2} = 2f_{i-1} + f''_{i-1}h^2 + \tfrac{1}{12}f^{\mathrm{IV}}(\xi_6)h^4$$

from which

$$f''_{i-1}h^2 = (f_i - 2f_{i-1} + f_{i-2}) - \tfrac{1}{12}f^{\mathrm{IV}}(\xi_6)h^4, \qquad x_{i-2} < \xi_6 < x_i \tag{6-131}$$

Then, substituting Eq. 6-131 into Eq. 6-128, we obtain

$$\underbrace{y_{i+1} - y_{i-3} = \frac{4h}{3}(2f_i - f_{i-1} + 2f_{i-2})}_{\text{Milne's predictor equation}} + \underbrace{\frac{14}{45}f^{\mathrm{IV}}(\xi_7)h^5}_{\text{truncation error}}, \qquad x_{i-3} < \xi_7 < x_{i+1} \tag{6-132}$$

Equation 6-132 reveals that the per-step truncation error of Milne's predictor equation is given by

$$E_{TP} = \tfrac{14}{45}f^{\mathrm{IV}}(\eta)h^5 = \tfrac{14}{45}y^{\mathrm{V}}(\eta)h^5, \qquad x_{i-3} < \eta < x_{i+1} \tag{6-133}$$

The per-step truncation error of Milne's corrector equation is the truncation error of Simpson's one-third rule, which has been shown to be

$$E_{TC} = -\tfrac{1}{90}f^{\mathrm{IV}}(\xi)h^5 = -\tfrac{1}{90}y^{\mathrm{V}}(\xi)h^4, \qquad x_{i-1} < \xi < x_{i+1}$$

An estimate of the per-step truncation error of Milne's corrector equation can be obtained in terms of the predicted and corrected y values at a step from

$$E_{TC} \cong -\frac{1}{29}\left[C\binom{n}{y} - P\binom{1}{y}\right] \tag{6-134}$$

where

$P\binom{1}{y}$ = the predicted value of y at x_{i+1}

$C\binom{n}{y}$ = the corrected value of y obtained in the nth (final) iteration at x_{i+1}

Equation 6-134 is developed in a manner similar to that of Eq. 6-44, as discussed under the error analysis of the non-self-starting modified Euler method.

EXAMPLE 6-5

A small rocket having an initial weight of 3000 lb, including 2400 lb of fuel, is fired vertically upward. The rocket burns fuel at a constant rate of 40 lb/sec, which provides a thrust of 7000 lb. Instruments are carried by the rocket to record data from which acceleration-time, velocity-time, and displacement-time curves can be obtained.

We wish to provide a theoretical analysis of the rocket flight which will yield data enabling us to plot the same curves obtained from the data recorded by the rocket instruments. A comparison of these sets of data then can check the validity of the mathematical model used in the theoretical analysis, since the model is conditioned by certain necessary simplifying assumptions, as follows: (1) the drag force is proportional to the square of the velocity ($D = Kv^2$), and (2) the coefficient of aerodynamic resistance K has an average value of 0.008 lb-sec^2/ft^2.

A free-body diagram of the rocket appears in Fig. 6-32. The forces acting on the rocket are

T = thrust force, 7000 lb
W = total weight of rocket, fuel, and equipment at time t, $(3000 - 40t)$
D = drag, $0.008\dot{y}^2$ lb

Assuming y to be positive upward, as shown in the figure, we use Newton's second law to obtain the second-order nonlinear differential equation

$$\ddot{y} = \frac{gT}{W} - g - \frac{K\dot{y}^2 g}{W} \tag{6-135}$$

which has variable coefficients. The initial conditions of the problem are

$$t = 0 \left| \begin{array}{l} y = 0 \\ \dot{y} = 0 \end{array} \right.$$

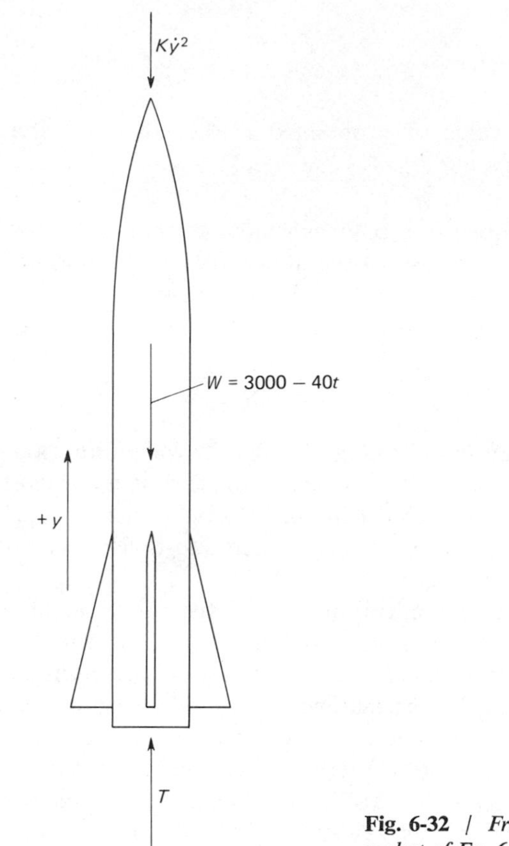

Fig. 6-32 / *Free-body diagram of rocket of Ex. 6-5.*

The solution of Eq. 6-135 will illustrate the application of Milne's method for obtaining the solution of a second-order differential equation. The required starting values of y, $\dot{y}$, and $\ddot{y}$, shown below, were obtained from the initial conditions, Taylor-series expansions, and the use of Eq. 6-135 by the procedure outlined in the preceding discussion.

$$y_0 = 0.0 \text{ ft} \qquad \dot{y}_0 = 0.0 \text{ ft/sec}$$

$$y_1 = 0.2146 \text{ ft} \qquad \dot{y}_1 = 4.294 \text{ ft/sec} \qquad \ddot{y}_1 = 42.99 \text{ ft/sec}^2$$

$$y_2 = 0.8591 \text{ ft} \qquad \dot{y}_2 = 8.598 \text{ ft/sec} \qquad \ddot{y}_2 = 43.08 \text{ ft/sec}^2$$

$$y_3 = 1.935 \text{ ft} \qquad \dot{y}_3 = 12.911 \text{ ft/sec} \qquad \ddot{y}_3 = 43.18 \text{ ft/sec}^2$$

In obtaining these values, a value of $\Delta t = 0.1$ was used, and the highest derivative used in the Taylor-series expansions was y^{IV}.

Remember that if the computer program is to be run a number of times, using different data in each run, the calculation of the starting values should be included in the program, so that their calculation becomes part of the computing process. As mentioned previously, the Runge-Kutta equation is more easily programmed for the calculation of starting values than are Taylor-series expansions. The Runge-Kutta formula (Eq. 6-121) can be used to determine starting values for the solution of a first-order differential equation. However, in applying the Runge-Kutta method to the solution of a second-order differential equation of the form $\ddot{y} = f(\dot{y}, y, t)$, as in this example, two formulas are required as discussed in Sec. 6-6. Although many variations of Runge-Kutta formulas exist, Eqs. 6-103 and 6-104 with y as a dependent variable instead of x will be found suitable for the determination of the required starting values. As in the formula given for solving first-order differential equations, the error in these formulas is of order h^5.

We will make use of Eq. 6-122 in the following form:

(Predictor)

$$P(\dot{y}_{i+1}) = \dot{y}_{i-3} + \tfrac{4}{3}h[2\ddot{y}_i - \ddot{y}_{i-1} + 2\ddot{y}_{i-2}] \qquad \text{(a)}$$

(Differential equation)

$$P(\ddot{y}_{i+1}) = f[t_{i+1}, P(\dot{y}_{i+1})] \qquad \text{(b)}$$

(Corrector)

$$C(\dot{y}_{i+1}) = \dot{y}_{i-1} + \frac{h}{3}[\ddot{y}_{i-1} + 4\ddot{y}_i + P(\ddot{y}_{i+1})] \qquad \text{(c)} \quad (6\text{-}136)$$

(Differential equation)

$$C(\ddot{y}_{i+1}) = f[t_{i+1}, C(\dot{y}_{i+1})] \qquad \text{(d)}$$

(Iterating corrector)

$$C(\dot{y}_{i+1}) = \dot{y}_{i-1} + \frac{h}{3}[\ddot{y}_{i-1} + 4\ddot{y}_i + C(\ddot{y}_{i+1})] \qquad \text{(e)}$$

The dynamic-characteristics curves of the rocket for t near zero are sketched in Fig. 6-33. A predicted value of $\dot{y}_4$ is the first value obtained. It is determined from Eq. 6-136a, with $i = 3$, as

$$P(\dot{y}_4) = \dot{y}_0 + \tfrac{4}{3}h[2\ddot{y}_3 - \ddot{y}_2 + 2\ddot{y}_1]$$

where the second term on the right side is the crosshatched area under the parabola in Fig. 6-33(a). A predicted value of $\ddot{y}_4$ is next obtained by substituting $t_4 = 4h$ and the predicted value of $\dot{y}_4$ into Eq. 6-135, as indicated by Eq. 6-136b. This predicted value of $\ddot{y}_4$ is then substituted

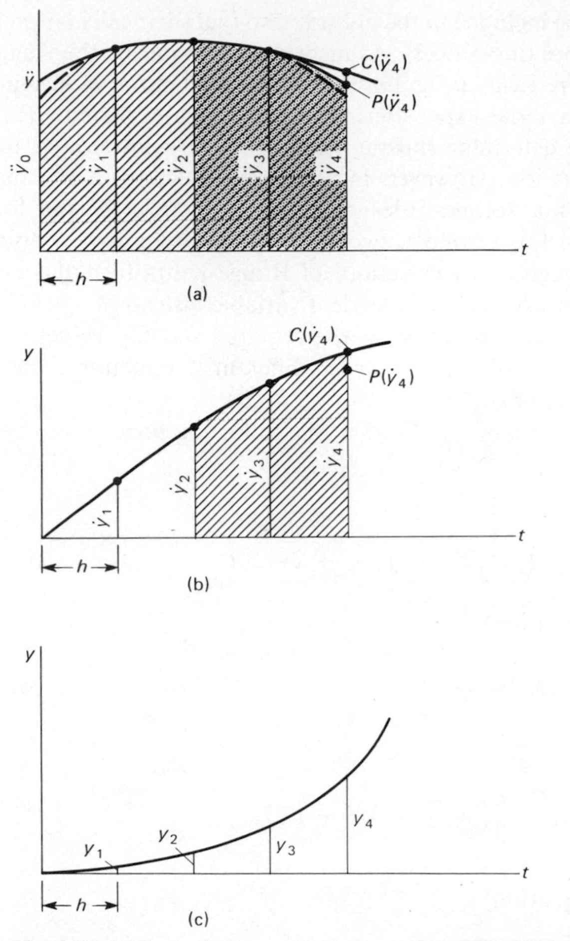

Fig. 6-33 / *General characteristic curves for rocket of Ex. 6-5.*

into Eq. 6-136c to obtain a corrected value of $\dot{y}_4$. The second term on the right side of Eq. 6-136c is the shaded area shown in Fig. 6-33(a). The corrected value $C(\dot{y}_4)$ is next substituted, along with $t_4 = 4h$, into Eq. 6-135 to obtain a corrected value of $\ddot{y}_4$, as indicated by Eq. 6-136d. This corrected value of $\ddot{y}_4$ is then substituted into Eq. 6-136e to obtain a further improved value of $\dot{y}_4$. This value is then substituted into Eq. 6-135 to obtain a further improved value of $\ddot{y}_4$, which is, in turn, substituted back into Eq. 6-136e to obtain an even-more-accurate value of $\dot{y}_4$. The iterative process involving Eq. 6-136d and e continues until successive values of $\dot{y}_4$ vary by less than some value epsilon chosen to specify the accuracy desired. In programming the problem, it is not actually necessary to write Eq. 6-136e as a separate FORTRAN statement, since the program can be written so that Eq. 6-136c performs the function of Eq. 6-136e in the iteration process of each step (see the FORTRAN program of this example).

Since the given differential equation (Eq. 6-135) does not contain y in explicit form, it is not necessary to calculate a value of y_4 until $\dot{y}_4$ has been determined with the accuracy desired.[12] The value of y_4 is then calculated, using Milne's corrector equation (Simpson's rule), as

$$y_4 = y_2 + \frac{h}{3}[\dot{y}_2 + 4\dot{y}_3 + \dot{y}_4]$$

where the second term on the right is the crosshatched area shown in Fig. 6-33b. After obtaining y_4, we move forward one step and repeat the process to determine values of $\dot{y}_5$ and y_5, and so on. This is continued until a time is reached corresponding to burnout (60 sec).

The flow chart for programming this problem is given in Fig. 6-34. The variable names, quantities, and values are as follows:

Variable Name	Quantity	Value
W1	Total initial weight of rocket, lb	3000
WF	Initial weight of fuel, lb	2400
TH	Thrust, lb	7000
FR	Fuel-consumption rate, lb/sec	40
DK	Coefficient of aerodynamic resistance, lb-sec^2/ft^2	0.008
G	Acceleration of gravity, ft/sec^2	32.17
DELT	Time increment, sec	0.1
DTPR	Print increment, sec	2.0
EPSI	Accuracy-check value, ft/sec	0.0001
Y	Displacement value at station $i - 3$, ft	(Initial) 0.0000
Y1	Displacement at station $i - 2$, ft	(Initial) 0.2146
Y2	Displacement at station $i - 1$, ft	(Initial) 0.8591
Y3	Displacement at station i, ft	(Initial) 1.9350
YD	Velocity value at station $i - 3$, ft/sec	(Initial) 0.0000
YD1	Velocity at station $i - 2$, ft/sec	(Initial) 4.294
YD2	Velocity at station $i - 1$, ft/sec	(Initial) 8.598
YD3	Velocity at station i, ft/sec	(Initial) 12.911
YDD1	Acceleration at station $i - 2$, ft/sec^2	(Initial) 42.99
YDD2	Acceleration at station $i - 1$, ft/sec^2	(Initial) 43.08
YDD3	Acceleration at station i, ft/sec^2	(Initial) 43.18
T	Time, sec	
TMAX	Time at burnout, sec	60
TPR	Time at which printout occurs, sec	
W	Weight of rocket, fuel, and equipment at time t, lb	
YD4P	Predicted value of velocity, ft/sec	
YDD4	Acceleration of rocket, ft/sec^2	
YD4C	Corrected value of velocity, ft/sec	
Y4	Displacement at station $i + 1$, ft	

[12] See brief discussion of the procedure for using Milne's method when y appears in explicit form in the differential equation following data shown at the end of this example.

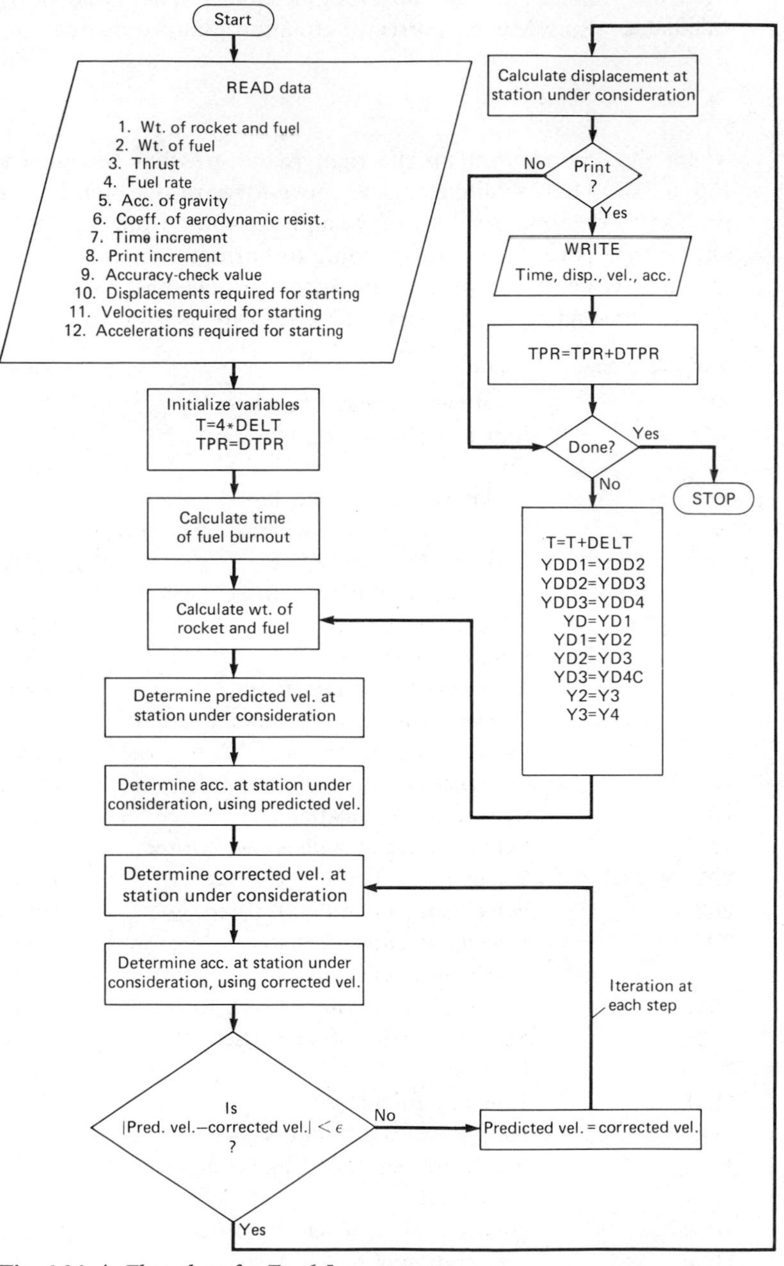

Fig. 6-34 / *Flow chart for Ex. 6-5.*

The FORTRAN IV program follows. Note that all parts of Eq. 6-136 do not appear explicitly in the program. Since Eq. 6-136c is used only once in a particular sequence and is then replaced by Eq. 6-136e, the program utilizes statement 6 to fulfill the function of both Eqs. 6-136c and e.

```
C     DYNAMIC CHARACTERISTICS OF A ROCKET
C     NUMERICAL INTEGRATION BY MILNE METHOD
      READ(5,1)W1,WF,TH,FR,G,DK,H,DTPR,EPSI
    1 FORMAT(11F5.0)
      READ(5,1)Y,Y1,Y2,Y3,YD,YD1,YD2,YD3,YDD1,YDD2,YDD3
      WRITE(6,2)
    2 FORMAT(1H1,5X,1HW,8X,1HF,7X,2HTH,7X,1HR,7X,1HG,7X,
     *1HK,5X,2H H,4X,3HDTP,5X,3HEPS/)
      WRITE(6,3)W1,WF,TH,FR,G,DK,H,DTPR,EPSI
    3 FORMAT(2F9.1,F8.0,2F8.2,F8.4,2F6.2,F9.5///)
      WRITE(6,4)
    4 FORMAT(7H   TIME,6X,6HDISPL.,6X,4HVEL.,7X,4HACC./)
      T=4.*H
      TMAX=WF/FR
      TPR=DTPR
    5 W=W1-FR*T
      YD4P=YD+4./3.*(2.*YDD1-YDD2+2.*YDD3)*H
      YDD4=G*TH/W-G-DK*G/W*(YD4P**2)
    6 YD4C=YD2+H/3.*(YDD2+4.*YDD3+YDD4)
      YDD4=G*TH/W-G-DK*G/W*(YD4C**2)
      IF(ABS(YD4C-YD4P)-EPSI)8,8,7
    7 YD4P=YD4C
      GO TO 6
    8 Y4=Y2+H/3.*(YD2+4.*YD3+YD4C)
      IF(ABS(T-TPR)-.01)9,9,11
    9 WRITE(6,10)T,Y4,YD4C,YDD4
   10 FORMAT(1H ,F7.2,F12.1,F11.2,F10.2)
      TPR=TPR+DTPR
   11 IF(ABS(T-TMAX)-.01) 12,12,13
   12 STOP
   13 T=T+H
      YDD1=YDD2
      YDD2=YDD3
      YDD3=YDD4
      YD=YD1
      YD1=YD2
      YD2=YD3
      YD3=YD4C
      Y2=Y3
      Y3=Y4
      GO TO 5
      END
```

The computer printout is shown in tabular form. The desired dynamic-characteristic curves, as plotted from the data printout, are given in

Fig. 6-35. These curves are available for comparison with the corresponding curves obtained from the instruments carried by the rocket.

W	F	TH	R	G	K	H	DTP	EPS
3000.0	2400.0	7000.	40.00	32.17	0.0080	0.10	2.00	0.00100

TIME	DISPL.	VEL.	ACC.
2.00	86.9	87.38	44.28
4.00	350.5	176.20	44.31
6.00	790.8	263.68	42.94
8.00	1402.5	347.09	40.29
10.00	2174.9	424.15	36.63
12.00	3093.7	493.20	32.35
14.00	4141.8	553.38	27.82
16.00	5301.2	604.55	23.39
18.00	6554.3	647.19	19.32
20.00	7884.8	682.17	15.75
22.00	9278.5	710.57	12.76
24.00	10723.5	733.57	10.33
26.00	12209.9	752.24	8.42
28.00	13730.1	767.55	6.96
30.00	15278.3	780.33	5.87
32.00	16850.2	791.24	5.08
34.00	18442.4	800.79	4.51
36.00	20052.6	809.38	4.11
38.00	21679.4	817.30	3.83
40.00	23321.5	824.75	3.64
42.00	24978.1	831.88	3.50
44.00	26648.8	838.78	3.41
46.00	28333.1	845.53	3.34
48.00	30030.8	852.16	3.29
50.00	31741.6	858.70	3.25
52.00	33465.5	865.16	3.21
54.00	35202.1	871.55	3.18
56.00	36951.6	877.89	3.15
58.00	38713.6	884.17	3.11
60.00	40488.1	890.39	3.08

If y had appeared explicitly in Eq. 6-135, a slightly modified procedure would have had to be used. The following set of equations would have been used:

(Predictor for $\dot{y}$)

$$P(\dot{y}_{i+1}) = \dot{y}_{i-3} + \tfrac{4}{3}h[2\ddot{y}_i - \ddot{y}_{i-1} + 2\ddot{y}_{i-2}] \qquad \text{(a)}$$

(Predictor for y)

$$P(y_{i+1}) = y_{i-1} + \frac{h}{3}(\dot{y}_{i-1} + 4\dot{y}_i + P(\dot{y}_{i+1})] \qquad \text{(b)}$$

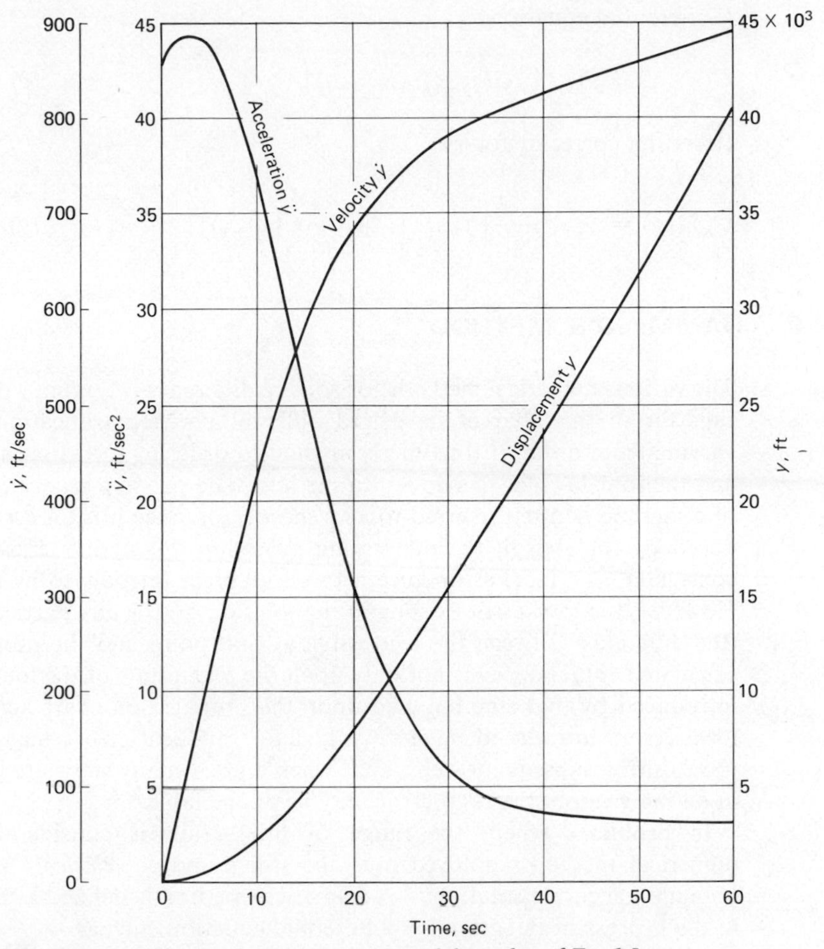

Fig. 6-35 / *Dynamic-characteristic curves of the rocket of Ex. 6-5.*

(Differential equation)

$$P(\ddot{y}_{i+1}) = f[t_{i+1}, P(y_{i+1}), P(\dot{y}_{i+1})] \qquad \text{(c)}$$

(Corrector for $\dot{y}$)

$$C(\dot{y}_{i+1}) = \dot{y}_{i-1} + \frac{h}{3}[\ddot{y}_{i-1} + 4\ddot{y}_i + P(\ddot{y}_{i+1})] \qquad \text{(d)} \quad \text{(6-137)}$$

(Iterating corrector for y)

$$C(y_{i+1}) = y_{i-1} + \frac{h}{3}[\dot{y}_{i-1} + 4\dot{y}_i + C(\dot{y}_{i+1})] \qquad \text{(e)}$$

(Differential equation)

$$C(\ddot{y}_{i+1}) = f[t_{i+1}, C(y_{i+1}), C(\dot{y}_{i+1})] \qquad \text{(f)}$$

(Iterating corrector for $\dot{y}$)

$$C(\dot{y}_{i+1}) = \dot{y}_{i-1} + \frac{h}{3}[\ddot{y}_{i-1} + 4\ddot{y}_i + C(\ddot{y}_{i+1})] \qquad \text{(g)}$$

6-8 / HAMMING'S METHOD

The various numerical methods for solving differential equations discussed thus far in this chapter have had different per-step truncation errors, varying from order h^2 for Euler's method to order h^5 for Milne's method and the fourth-order Runge-Kutta method. The per-step truncation error of a method is often referred to as a *local* error since it is an error introduced by the step in the integrating procedure. Roundoff error is also considered as a local error since it too is an error introduced by the step. However, the *total* error existing in the solution during any particular step (the difference between the true value at that point and the numerically calculated value) depends not only upon the magnitude of the local errors introduced by that step but also upon the propagation characteristics of local errors introduced in preceding steps, for local errors may tend to grow during subsequent steps, and, when a great many steps are involved in an integration, the total error may be appreciable.

In problems where the range of integration is considerable, the numerical method employed must be *stable* and/or *relatively stable* to obtain an accurate solution.[13] A numerical method is defined as stable if, in the process of integrating a differential equation such as

$$y' = f(x, y) \qquad \text{(6-138)}$$

where $\partial f/\partial y < 0$, the difference between the true solution and the numerical solution (the total error) tends to *decrease* in magnitude as the integration progresses. The stability of a method is not defined when considering the integration of a differential equation such as Eq. 6-138 where $\partial f/\partial y > 0$. The solution of such an equation will increase in an exponential manner, and the total error will generally increase in the same way. In solving such a differential equation over an extended range of integration, a method should be used which is *relatively stable*. A method

[13] Richard W. Hamming, *Numerical Methods for Scientists and Engineers* (New York: McGraw-Hill Book Company, 1962), p. 191. Anthony Ralston and Herbert Wilf, eds., *Mathematical Methods for Digital Computers* (New York: John Wiley & Sons, 1960), p. 103.

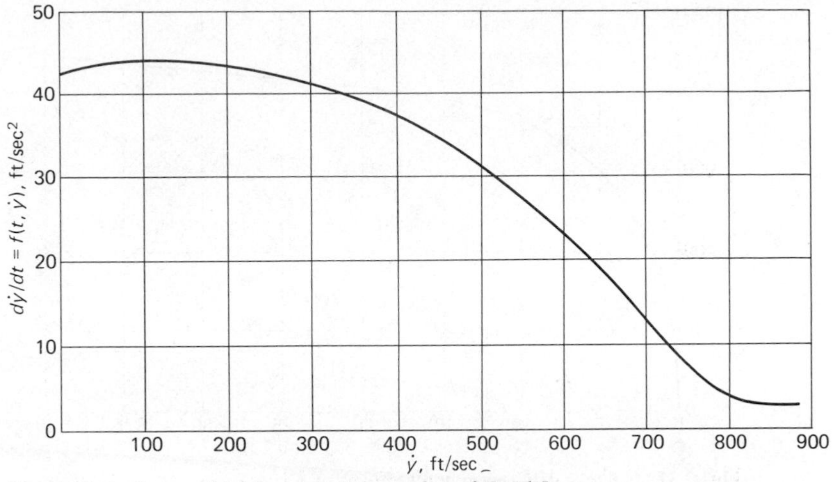

Fig. 6-36 / *Curve of $d\dot{y}/dt$ versus $\dot{y}$ for rocket of Ex. 6-5.*

is defined as relatively stable if the *rate of growth* of the total error during the process of integration is *less* than that of the solution. This may be expressed mathematically by stating that if the solution has the general form of

$$y = Ae^{x}$$

then the total error caused by the numerical method must be of the general form of

$$\varepsilon = Be^{mx}$$

with $m < 1$ for the method to be considered relatively stable.

The relative stability of a method is also important when used to solve a differential equation such as Eq. 6-138 for which $\partial f/\partial y < 0$ if the solution approaches zero asymptotically, for, if an accurate solution is desired involving a value of y very close to zero, the *rate of diminution* of the total error must be *greater* than that of the solution. Expressed mathematically, if the solution has the form of

$$y = Ae^{-x}$$

then the total error caused by the method must be of the general form of

$$\varepsilon = Be^{-mx}$$

with $m > 1$ for the method to be considered relatively stable.

Figure 6-35, accompanying the rocket problem of Ex. 6-5, provides the data necessary to plot a curve of $d\dot{y}/dt$ versus $\dot{y}$, as shown in Fig. 6-36.

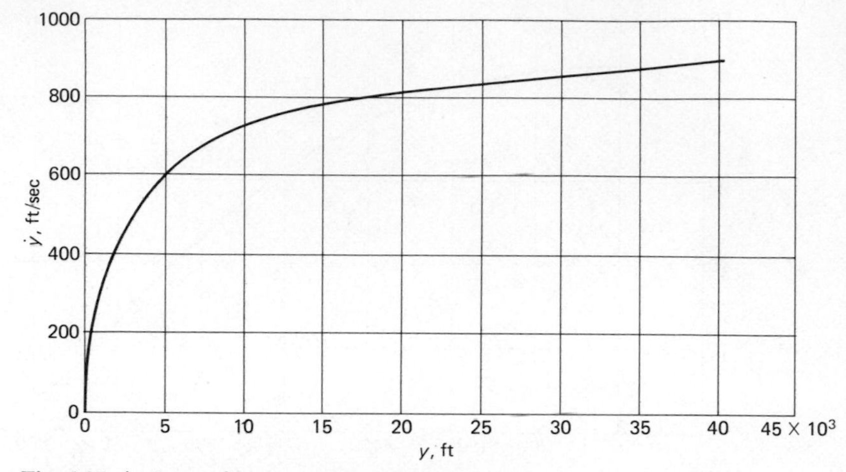

Fig. 6-37 | *Curve of ẏ versus y for rocket of Ex. 6-5.*

It can be seen, from this latter figure, that $\partial f/\partial \dot{y} < 0$ for most of the range of $\dot{y}$ under consideration. Therefore, if a stable numerical method is used to integrate

$$\frac{d\dot{y}}{dt} = f(\dot{y}, t)$$

there should be no danger of the total error growing to serious proportions, even though the integration might involve a large number of steps over an extended range of integration.

The curve of $\dot{y}$ versus y for the same problem is shown in Fig. 6-37. In this instance it can be seen that $\partial f/\partial y > 0$. Therefore, in the integration of

$$\dot{y} = f(y, t)$$

the stability of the method employed is not defined, as discussed previously. The solution y increases with time in a general exponential manner, as shown in Fig. 6-38. Here the use of a relatively stable method will ensure that the rate of growth of the total error is no greater than the rate of growth of the solution, even though both grow in an exponential manner.

The propagation of errors during an integration is not a serious obstacle in the solution of many practical engineering problems. Since we are often not interested in solutions requiring extended ranges of integration, the use of relatively small h values will provide sufficiently accurate answers, regardless of the stability or relative-stability characteristics of the numerical method employed. For example, the solutions provided by Milne's method (which is neither stable nor relatively stable) in Ex. 6-5

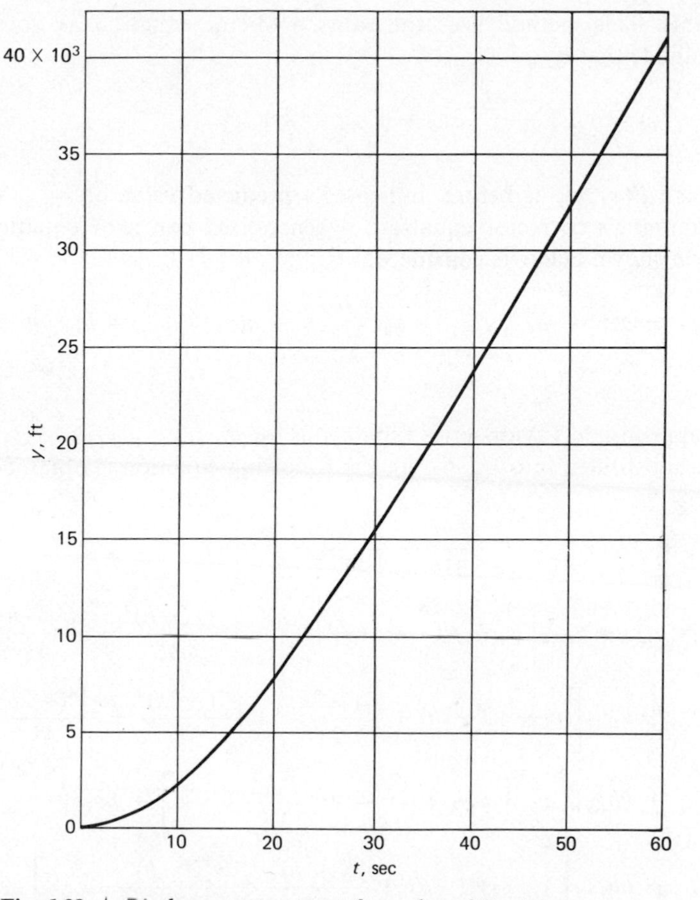

Fig. 6-38 / *Displacement y versus t for rocket of Ex. 6-5.*

agree very closely with the corresponding solutions provided by Hamming's method (which is both stable and relatively stable) in Ex. 6-6 which appears later in this section. However, if solutions were desired over a much longer range of integration than is required in these examples, Milne's method would not provide the same degree of accuracy as would Hamming's method, because Milne's method is unstable, and the total error would grow to proportions capable of introducing serious error during the latter stages of the integration; the discrepancy between the solutions would become more and more pronounced as the range of integration was arbitrarily extended. In certain types of equations, such as Van der Pol's equation, for example, instability leads to serious errors in the solution, even over a relatively short range of integration (see Prob. 6-17).

Hamming's method, as mentioned above, is both stable and relatively stable. This method uses the same predictor equation as does Milne's method; that is,

$$P(y_{i+1}) = y_{i-3} + \frac{4h}{3}[2y'_i - y'_{i-1} + 2y'_{i-2}] \tag{6-139}$$

where $P(y_{i+1})$, as before, indicates a predicted value of y_{i+1}. To derive Hamming's corrector equation, a generalized corrector equation of the form shown below is considered:

$$y_{i+1} = a_i y_i + a_{i-1} y_{i-1} + a_{i-2} y_{i-2} + h(b_{i+1} y'_{i+1} + b_i y'_i + b_{i-1} y'_{i-1}) \tag{6-140}$$

If appropriate Taylor-series expansions for $y_{i-2}, y_{i-1}, y_{i+1}, y'_{i-1}$, and y'_{i+1} are substituted into Eq. 6-140, the following equation is obtained:

$$\begin{aligned}
&\left[y_i + y'_i h + \frac{y''_i h^2}{2!} + \frac{y'''_i h^3}{3!} + \frac{y''''_i h^4}{4!} + \cdots\right] \\
&\quad = a_i y_i + a_{i-1}\left[y_i + y'_i(-h) + \frac{y''_i(-h)^2}{2!} + \frac{y'''_i(-h)^3}{3!} + \frac{y^{\text{IV}}_i(-h)^4}{4!} + \cdots\right] \\
&\quad + a_{i-2}\left[y_i + y'_i(-2h) + \frac{y''_i(-2h)^2}{2!} + \frac{y'''_i(-2h)^3}{3!} + \frac{y^{\text{IV}}_i(-2h)^4}{4!} + \cdots\right] \\
&\quad + hb_{i+1}\left[y'_i + y''_i h + \frac{y'''_i h^2}{2!} + \frac{y^{\text{IV}}_i h^3}{3!} + \cdots\right] + hb_i y'_i \\
&\quad + hb_{i-1}\left[y'_i + y''_i(-h) + \frac{y'''_i(-h)^2}{2!} + \frac{y^{\text{IV}}_i(-h)^3}{3!} + \cdots\right]
\end{aligned} \tag{6-141}$$

where the bracketed terms are the series expansions. Equating coefficients of y_i, y'_i, y''_i, y'''_i, and y^{IV}_i, respectively, on each side of Eq. 6-141 yields the following relationships:

$$\begin{aligned}
&a_i + a_{i-1} + a_{i-2} = 1 && \text{(a)} \\
&- a_{i-1} - 2a_{i-2} + b_{i+1} + b_i + b_{i-1} = 1 && \text{(b)} \\
&+ \tfrac{1}{2}a_{i-1} + 2a_{i-2} + b_{i+1} - b_{i-1} = \tfrac{1}{2} && \text{(c)} \\
&- \tfrac{1}{6}a_{i-1} - \tfrac{4}{3}a_{i-2} + \tfrac{1}{2}b_{i+1} + \tfrac{1}{2}b_{i-1} = \tfrac{1}{6} && \text{(d)} \\
&+ \tfrac{1}{24}a_{i-1} + \tfrac{2}{3}a_{i-2} + \tfrac{1}{6}b_{i+1} - \tfrac{1}{6}b_{i-1} = \tfrac{1}{24} && \text{(e)}
\end{aligned} \tag{6-142}$$

The five equations shown above contain six unknowns. A sixth equation necessary for determining the six unknowns could be obtained by adding one additional term to each of the Taylor-series expansions used in obtaining Eq. 6-142 and equating the coefficients of y_i^{V}. However, Hamming[14] found that the use of $a_{i-1} = 0$ led to a corrector equation that was both stable and relatively stable when certain conditions were imposed upon the magnitude of h. Using a value of $a_{i-1} = 0$, the parts of Eq. 6-142 can be solved simultaneously to yield the following values:

$$a_i = \tfrac{9}{8} \qquad b_{i+1} = \tfrac{3}{8}$$

$$a_{i-2} = -\tfrac{1}{8} \qquad b_i = \tfrac{3}{4}$$

$$b_{i-1} = -\tfrac{3}{8}$$

Substituting these values into Eq. 6-140 yields Hamming's corrector equation

$$C(y_{i+1}) = \tfrac{1}{8}\{9y_i - y_{i-2} + 3h[P(y'_{i+1}) + 2y'_i - y'_{i-1}]\} \tag{6-143}$$

where $C(y_{i+1})$ denotes a corrected value of y_{i+1}. Hamming's corrector equation is both stable and relatively stable for solving differential equations with $\partial f/\partial y < 0$ when

$$h < \frac{0.75}{|\partial f/\partial y|}$$

and the corrector equation is used to iterate to convergence. The corrector equation is relatively stable for solving differential equations in which $\partial f/\partial y > 0$ when

$$h < \frac{0.4}{\partial f/\partial y}$$

and the corrector equation is used to iterate to convergence. To keep the per-step error small, h would normally be less than either of the h values specified above. The per-step error of Hamming's method is of order h^5, which is of the same order as Milne's method.

One form of Hamming's method consists of using Eq. 6-139 as a predictor equation and then using the given differential equation in conjunction with Hamming's corrector equation (Eq. 6-143) to iterate to convergence, as was done in Milne's method. An alternate procedure, suggested by Hamming, saves iteration time on the computer. A study of the truncation errors of Eqs. 6-139 and 6-143 indicates that most of the

[14] Richard W. Hamming, "Stable Predictor-Corrector Methods for Ordinary Differential Equations," *J. Assoc. Computing Machinery*, Vol. 6, No. 1 (1959), pp. 37–47.

error in the predicted value of y_{i+1} can be eliminated by the use of the following *modifier* equation:

$$M(y_{i+1}) = P(y_{i+1}) - \tfrac{112}{121}[P(y_i) - C(y_i)] \tag{6-144}$$

where $M(y_{i+1})$ indicates a modified value of y_{i+1} and the other symbols indicate predicted and corrected values, as before. This modified value of y_{i+1} is next substituted into the given differential equation to obtain a modified value of y'_{i+1} which is designated as $M(y'_{i+1})$. This modified value is then used in the corrector equation

$$C(y_{i+1}) = \tfrac{1}{8}\{9y_i - y_{i-2} + 3h[M(y'_{i+1}) + 2y'_i - y'_{i-1}]\} \tag{6-145}$$

to obtain a corrected value of y_{i+1}, as shown. The study made of the truncation errors of Eqs. 6-139 and 6-143 also reveals that most of the error in the corrected value of y_{i+1} can be eliminated by using the equation

$$F(y_{i+1}) = C(y_{i+1}) + \tfrac{9}{121}[P(y_{i+1}) - C(y_{i+1})] \tag{6-146}$$

where $F(y_{i+1})$ indicates the final value of y_{i+1}. Thus in place of iterating the corrector equation to convergence at each step, Eqs. 6-139, 6-144, 6-145, and 6-146 are used just once (in that sequence) for each step. These equations are grouped here for the sake of convenience and are shown in the order in which they are used:

(Predictor)

$$P(y_{i+1}) = y_{i-3} + \frac{4h}{3}[2y'_i - y'_{i-1} + 2y'_{i-2}]$$

(Modifier)

$$M(y_{i+1}) = P(y_{i+1}) - \tfrac{112}{121}[P(y_i) - C(y_i)] \tag{6-147}$$

(Corrector)

$$C(y_{i+1}) = \tfrac{1}{8}\{9y_i - y_{i-2} + 3h[M(y'_{i+1}) + 2y'_i - y'_{i-1}]\}$$

(Final value)

$$F(y_{i+1}) = C(y_{i+1}) + \tfrac{9}{121}[P(y_{i+1}) - C(y_{i+1})]$$

When the procedure involving Eq. 6-147 is employed instead of the iteration procedure, a different condition must be imposed upon the magnitude of h to ensure stability and relative stability. In this case (with $\partial f/\partial y < 0$) the condition is

$$h < \frac{0.65}{|\partial f/\partial y|}$$

Hamming's method, as well as other predictor-corrector methods, can be further refined to save machine time by varying the magnitude of h during the progress of the integration. A change in the magnitude of h is dictated by the relationship existing between the magnitudes of the predicted and corrected values of y_{i+1} calculated. If $[P(y_{i+1}) - C(y_{i+1})]$ is less than some prescribed value, h is doubled; if the same expression is larger than some other prescribed value, h is halved. For a further discussion of adjusting step size, based on a truncation-error analysis at each step, see Sec. 6-9.

Error Estimation for Hamming's Method

The predictor equation used in Hamming's method is the same as the predictor equation of Milne's method, so the per-step truncation error in a predicted value is given by

$$E_{TP} = \tfrac{14}{45} f^{\mathrm{IV}}(\eta)h^5 = \tfrac{14}{45} y^{\mathrm{V}}(\eta)h^5, \qquad x_{i-3} < \eta < x_{i+1} \tag{6-148}$$

as developed earlier (Eq. 6-133).

To obtain the per-step truncation error of Hamming's corrector equation, let us return for a moment to the development of Hamming's method and read particularly the text from Eqs. 6-140 to 6-143. From that discussion it can be seen that the left and right sides of Eq. 6-141 have been made equal by the a and b values determined from Eq. 6-142 (with $a_{i-1} = 0$). It can also be seen that the left side of Eq. 6-141 is only an approximate expression for y_{i+1}, since y_{i+1} is expressed as a truncated Taylor series. If we add the remaining Taylor-series terms of the expansion of y_{i+1} (in a Lagrangian-remainder form) to both sides of Eq. 6-141 and insert Hamming's a and b values, we can obtain an expression for the *true* value of y_{i+1} as

$$\begin{aligned}
(y_{i+1})_{\text{true}} &= \left[y_i + y_i'h + \frac{y_i''h^2}{2!} + \frac{y_i'''h^3}{3!} + \frac{y_i^{\mathrm{IV}}h^4}{4!} + \frac{y^{\mathrm{V}}(\xi_1)h^5}{5!} \right] \\
&= \frac{9}{8}y_i - \frac{1}{8}\left[y_i + y_i'(-2h) + \frac{y_i''(-2h)^2}{2!} \right. \\
&\qquad\qquad \left. + \frac{y_i'''(-2h)^3}{3!} + \frac{y_i^{\mathrm{IV}}(-2h)^4}{4!} \right] \\
&\quad + \frac{3h}{8}\left[y_i' + y_i''h + \frac{y_i'''h^2}{2!} + \frac{y_i^{\mathrm{IV}}h^3}{3!} \right] + \frac{3h}{4}y_i' \\
&\quad - \frac{3h}{8}\left[y_i' + y_i''(-h) + \frac{y_i'''(-h)^2}{2!} + \frac{y_i^{\mathrm{IV}}(-h)^3}{3!} \right] + \frac{y^{\mathrm{V}}(\xi_1)h^5}{5!}
\end{aligned} \tag{6-149}$$

Now let us examine the expression for y_{i+1} which is obtained by using Hamming's corrector equation (Eq. 6-143). Considering only the truncation error of the step by assuming the solution to be correct up to the step considered, all the y and y' values shown on the right side of the equation will be *true* values. Therefore, substituting a complete Taylor series (in Lagrangian-remainder form) for y_{i-2}, y'_{i-1}, and y'_{i+1} in Eq. 6-143, we obtain

$$(y_{i+1})_{\text{Hamming}} = \frac{9}{8}y_i - \frac{1}{8}\left[y_i + y'_i(-2h) + \frac{y''_i(-2h)^2}{2!} + \frac{y'''_i(-2h)^3}{3!} + \frac{y^{\text{IV}}_i(-2h)^4}{4!} + \frac{y^{\text{V}}(\xi_2)(-2h)^5}{5!}\right]$$
$$+ \frac{3h}{8}\left[y'_i + y''_i h + \frac{y'''_i h^2}{2!} + \frac{y^{\text{IV}}_i h^3}{3!} + \frac{y^{\text{V}}(\xi_3)h^4}{4!}\right] + \frac{3h}{4}y'_i$$
$$- \frac{3h}{8}\left[y'_i + y''_i(-h) + \frac{y'''_i(-h)^2}{2!} + \frac{y^{\text{IV}}_i(-h)^3}{3!} + \frac{y^{\text{V}}(\xi_4)(-h)^4}{4!}\right] \tag{6-150}$$

Comparing the expression for the true value of y_{i+1} (given by Eq. 6-149) with the expression for y_{i+1} from Hamming's corrector equation (given by Eq. 6-150), it is apparent that Hamming's corrector equation does not yield a true value of y_{i+1}. The difference between these expressions is the per-step truncation error of Hamming's corrector equation. Subtracting Eq. 6-150 from Eq. 6-149, we obtain

$$E_{TC} = (y_{i+1})_{\text{true}} - (y_{i+1})_{\text{Hamming}}$$
$$= \frac{1}{8}\frac{y^{\text{V}}(\xi_2)(-2h)^5}{5!} - \frac{3}{8}\frac{y^{\text{V}}(\xi_3)h^5}{4!} + \frac{3}{8}\frac{y^{\text{V}}(\xi_4)h^5}{4!} + \frac{y^{\text{V}}(\xi_1)h^5}{5!} \tag{6-151}$$

Assuming that y^{V} is nearly constant in the interval from x_{i-2} to x_{i+1}, Eq. 6-151 reduces to

$$E_{TC} = -\tfrac{1}{40}y^{\text{V}}(\eta)h^5, \qquad x_{i-2} < \eta < x_{i+1} \tag{6-152}$$

Thus Hamming's corrector equation for a true value of y_{i+1} may be written as

$$y_{i+1} = \tfrac{1}{8}[9y_i - y_{i-2} + 3h(y'_{i+1} + 2y'_i - y'_{i-1})] - \tfrac{1}{40}y^{\text{V}}(\eta)h^5 \tag{6-153}$$

As discussed before, an estimate of the per-step truncation error can be obtained from Eq. 6-152 by assuming that y^{V} is fairly constant over the interval and evaluating the truncation-error term at $\eta = x_i$, although the estimate is generally not made in this manner, owing to the difficulty of obtaining y^{V}.

A more practical approach can be made by obtaining an estimate of the per-step truncation error in terms of the predicted and corrected values of y at each step. Consider that

$P\binom{1}{y}$ = the predicted value of y obtained at x_{i+1} in the integration process

$C\binom{n}{y}$ = the corrected value of y obtained at x_{i+1} in the nth (final) iteration at that step

Y = the true value of y at x_{i+1}

Isolating the truncation error of the ith step from the effects of previous errors, in accordance with the definition given for per-step truncation error, and assuming that roundoff error in the calculations of the step is negligible, we can write

$$Y = P\binom{1}{y} + \frac{14}{45}y^{V}(\eta_1)h^5, \qquad x_{i-3} < \eta_1 < x_{i+1} \tag{6-154}$$

$$Y = C\binom{n}{y} - \frac{1}{40}y^{V}(\eta_2)h^5, \qquad x_{i-2} < \eta_2 < x_{i+1} \tag{6-155}$$

Subtracting Eq. 6-154 from Eq. 6-155 yields

$$0 = C\binom{n}{y} - P\binom{1}{y} - \frac{121}{360}y^{V}(\eta_3)h^5, \qquad x_{i-3} < \eta_3 < x_{i+1}$$

or

$$C\binom{n}{y} - P\binom{1}{y} = \left(\frac{121}{9}\right)\left(\frac{1}{40}\right)y^{V}(\eta_3)h^5 \tag{6-156}$$

Comparing Eqs. 6-152 and 6-156, it can be seen that an estimate of the per-step truncation error, in the final application of Hamming's corrector equation, is

$$E_{TC} \cong -\frac{9}{121}\left[C\binom{n}{y} - P\binom{1}{y}\right] \tag{6-157}$$

EXAMPLE 6-6

Let us use Hamming's method to solve the rocket problem of Ex. 6-5. The reader should reread the first three paragraphs of Ex. 6-5 for a discussion of the problem and the assumptions involved in deriving the following differential equation of motion of the rocket:

$$\ddot{y} = \frac{gT}{W} - g - \frac{K\dot{y}^2 g}{W} \tag{6-158}$$

where

y = displacement of rocket, ft
$\dot{y}$ = velocity of rocket, ft/sec
$\ddot{y}$ = acceleration of rocket, ft/sec^2
T = thrust force, 7000 lb
W = total weight of rocket, fuel, and equipment at time t, $(3000 - 40t)$ lb
K = coefficient of aerodynamic resistance, 0.008 lb-sec^2/ft^2
g = acceleration of gravity, 32.17 ft/sec^2

The initial conditions of the problem, as before, are

$$t = 0 \left| \begin{array}{l} y = 0 \\ \dot{y} = 0 \end{array} \right.$$

The same starting values used in solving the problem by Milne's method are required here. They are repeated for convenience:

$y_0 = 0.0$ ft $\dot{y}_0 = 0.0$ ft/sec

$y_1 = 0.2146$ ft $\dot{y}_1 = 4.294$ ft/sec $\ddot{y}_1 = 42.99$ ft/sec^2

$y_2 = 0.8591$ ft $\dot{y}_2 = 8.598$ ft/sec $\ddot{y}_2 = 43.08$ ft/sec^2

$y_3 = 1.935$ ft $\dot{y}_3 = 12.911$ ft/sec $\ddot{y}_3 = 43.18$ ft/sec^2

In using Hamming's noniterative method to solve a second-order differential equation such as Eq. 6-158 in which y is not present in explicit form, we use Eq. 6-147 in the following form:

(Predictor)

$$P(\dot{y}_{i+1}) = \dot{y}_{i-3} + \frac{4h}{3}[2\ddot{y}_i - \ddot{y}_{i-1} + 2\ddot{y}_{i-2}] \tag{a}$$

(Modifier)

$$M(\dot{y}_{i+1}) = P(\dot{y}_{i+1}) - \tfrac{112}{121}[P(\dot{y}_i) - C(\dot{y}_i)] \qquad \text{(b)}$$

(Differential equation)

$$M(\ddot{y}_{i+1}) = f[t_{i+1}, M(\dot{y}_{i+1})] \qquad \text{(c)} \quad (6\text{-}159)$$

(Corrector)

$$C(\dot{y}_{i+1}) = \tfrac{1}{8}\{9\dot{y}_i - \dot{y}_{i-2} + 3h[M(\ddot{y}_{i+1}) + 2\ddot{y}_i - \ddot{y}_{i-1}]\} \qquad \text{(d)}$$

(Final value of $\dot{y}$)

$$F(\dot{y}_{i+1}) = C(\dot{y}_{i+1}) + \tfrac{9}{121}[P(\dot{y}_{i+1}) - C(\dot{y}_{i+1})] \qquad \text{(e)}$$

(Final value of $\ddot{y}$)

$$F(\ddot{y}_{i+1}) = f[t_{i+1}, F(\dot{y}_{i+1})] \qquad \text{(f)}$$

where Eq. 6-159c indicates the substitution of the modified value of $\dot{y}_{i+1}$ and the corresponding time value into the given differential equation to obtain the modified value of $\ddot{y}_{i+1}$ required in Eq. 6-159d. Since y does not appear explicitly in the differential equation being solved, y_{i+1} values are not required in determining $M(\ddot{y}_{i+1})$ values. Thus, final $\dot{y}_{i+1}$ values can be obtained without determining the corresponding y_{i+1} values. If displacement values are desired, they may be determined from Hamming's corrector equation as the final step in each cycle of calculation. This equation is

$$y_{i+1} = \tfrac{1}{8}\{9y_i - y_{i-2} + 3h[F(\dot{y}_{i+1}) + 2\dot{y}_i - \dot{y}_{i-1}]\} \qquad (6\text{-}160)$$

where the final value of $\dot{y}_{i+1}$ is used in place of the modified value to obtain the most accurate value of y_{i+1} possible. It may be noted, in the program which follows, that in the *first cycle* of Eq. 6-159 the starting value of $\dot{y}_i$ is used for both $P(\dot{y}_i)$ and $C(\dot{y}_i)$ in Eq. 6-159b.

If the differential equation to be solved contains y in explicit form, Eqs. 6-159 and 6-160 are not sufficient for obtaining a solution. A brief discussion of this case, with the necessary equations, is given at the conclusion of this example.

A computer solution of Eq. 6-158 designed to yield the data necessary to plot curves of the displacement, velocity, and acceleration of the rocket is outlined in the flow chart shown in Fig. 6-39. Note that the previously determined starting values are read in as data and that Hamming's method

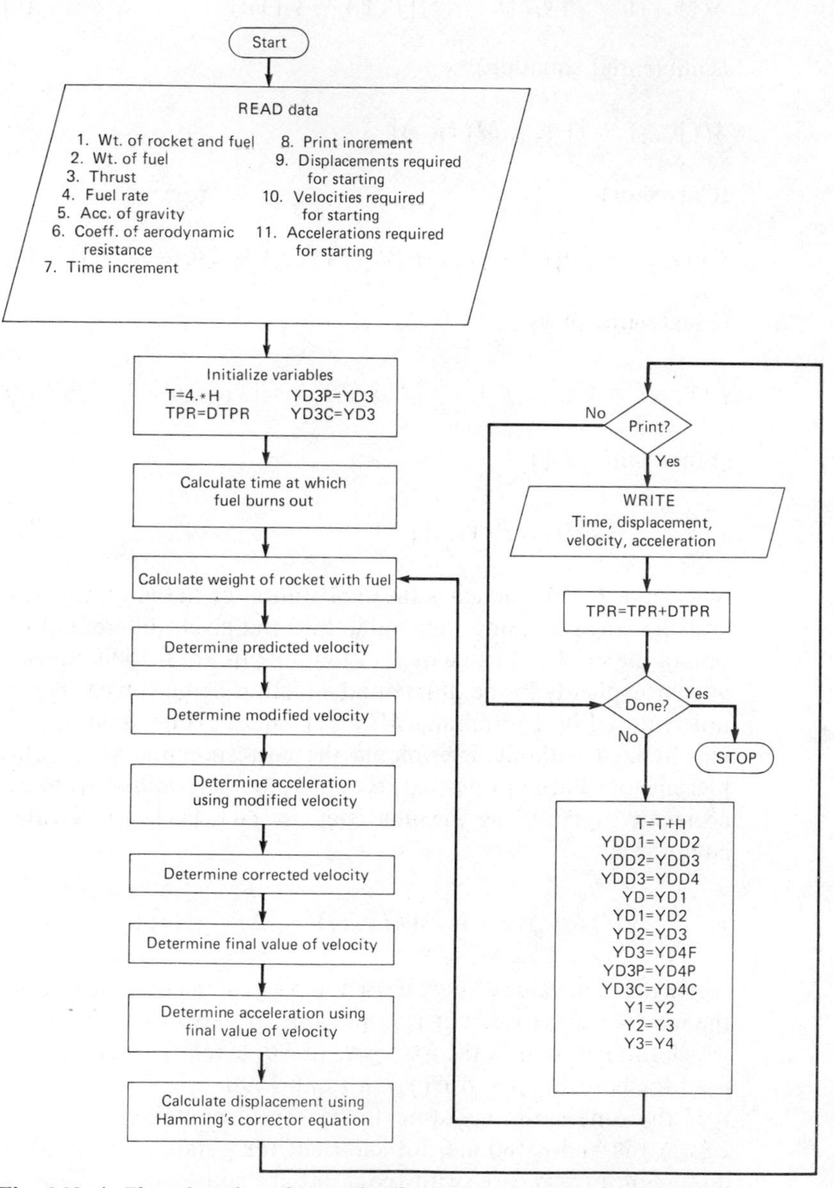

Fig. 6-39 / *Flow chart for solution of rocket problem by Hamming's method.*

utilizing Eqs. 6-159 and 6-160 is the method programmed. The variable names used in the program are as follows:

FORTRAN Name	Quantity	Value Used
W1	Initial weight of rocket, fuel, and equipment	3000 lb
WF	Initial weight of fuel	2400 lb
TH	Thrust	7000 lb
FR	Rate of fuel consumption	40 lb/sec
G	Acceleration of gravity	32.17 ft/sec^2
DK	Coefficient of aerodynamic resistance	0.008 lb-sec^2/ft^2
DELT	Time increment	0.1 sec
DTPR	Print increment	2.0 sec
Y,Y1,Y2,Y3	Successive values of displacement	(See starting values)
YD,YD1,YD2,YD3	Successive values of velocity	(See starting values)
YDD1,YDD2,YDD3	Successive values of acceleration	(See starting values)
T	Time, sec	
TMAX	Time at which fuel burnout occurs	
TPR	Time at which printout occurs, sec	
YD4P	Predicted value of $\dot{y}$ at time considered, ft/sec	
YD4M	Modified value of $\dot{y}$ at time considered, ft/sec	
W	Weight of rocket, fuel, and equipment at time t, lb	
YD4C	Corrected value of $\dot{y}$ at time considered, ft/sec	
YD4F	Final value of $\dot{y}$ at time considered, ft/sec	
YDD4	Acceleration at time considered, ft/sec^2	
Y4	Displacement at time considered, ft	
YD3P	Predicted value of $\dot{y}$ one time increment preceding time considered, ft/sec	
YD3C	Corrected value of $\dot{y}$ one time increment preceding time considered, ft/sec	

The computer program used is as follows:

```
C     DYNAMIC CHARACTERISTICS OF A ROCKET
C     NUMERICAL INTEGRATION BY HAMMING METHOD
      READ(5,1)W1,WF,TH,FR,G,DK,H,DTPR
    1 FORMAT(11F5.0)
      READ(5,1)Y,Y1,Y2,Y3,YD,YD1,YD2,YD3,YDD1,YDD2,YDD3
      WRITE(6,2)
```

```
  2 FORMAT(1H1,5X,1HW,8X,1HF,7X,2HTH,7X,1HR,7X,1HG,7X,
   *1HK,5X,2H H,4X,3HDTP/)
    WRITE(6,3)W1,WF,TH,FR,G,DK,H,DTPR
  3 FORMAT(2F9.1,F8.0,2F8.2,F8.4,2F6.2///)
    WRITE(6,4)
  4 FORMAT(7H   TIME,6X,6HDISPL.,6X,4HVEL.,7X,4HACC./)
    T=4.*H
    TMAX=WF/FR
    TPR=DTPR
    YD3P=YD3
    YD3C=YD3
  5 W=W1-FR*T
    YD4P=YD+4./3.*(2.*YDD1-YDD2+2.*YDD3)*H
    YD4M=YD4P-112./121.*(YD3P-YD3C)
    YDD4=G*TH/W-G-DK*G/W*(YD4M**2)
    YD4C=.125*(9.*YD3-YD1+3.*H*(YDD4+2.*YDD3-YDD2))
    YD4F=YD4C+9./121.*(YD4P-YD4C)
    YDD4=G*TH/W-G-DK*G/W*(YD4F**2)
    Y4=.125*(9.*Y3-Y1+3.*H*(YD4F+2.*YD3-YD2))
    IF(ABS(T-TPR)-.01)6,6,8
  6 WRITE(6,7)T,Y4,YD4F,YDD4
  7 FORMAT(1H ,F7.2,F12.1,F11.2,F10.2)
    TPR=TPR+DTPR
  8 IF(ABS(T-TMAX)-.01)9,9,10
  9 STOP
 10 T=T+H
    YDD1=YDD2
    YDD2=YDD3
    YDD3=YDD4
    YD=YD1
    YD1=YD2
    YD2=YD3
    YD3=YD4F
    YD3P=YD4P
    YD3C=YD4C
    Y1=Y2
    Y2=Y3
    Y3=Y4
    GO TO 5
    END
```

The computer results are shown in the following tabular form:

```
   W        F      TH       R       G       K        H      DTP
3000.0   2400.0  7000.   40.00   32.17   0.0080   0.10   2.00
TIME       DISPL.          VEL.         ACC.

  2.00        86.9        87.38        44.28
  4.00       350.5       176.20        44.31
  6.00       790.8       263.67        42.94
  8.00      1402.4       347.09        40.29
 10.00      2174.9       424.15        36.63
 12.00      3093.6       493.20        32.35
 14.00      4141.7       553.36        27.82
```

16.00	5301.0	604.53	23.40
18.00	6554.0	647.15	19.32
20.00	7884.3	682.13	15.76
22.00	9277.8	710.54	12.76
24.00	10722.4	733.53	10.34
26.00	12208.4	752.20	8.43
28.00	13728.3	767.51	6.97
30.00	15276.2	780.29	5.88
32.00	16847.7	791.20	5.09
34.00	18439.6	800.76	4.52
36.00	20049.5	809.35	4.12
38.00	21675.9	817.27	3.84
40.00	23317.7	824.72	3.64
42.00	24974.0	831.85	3.51
44.00	26644.3	838.76	3.42
46.00	28328.3	845.51	3.35
48.00	30025.7	852.14	3.30
50.00	31736.3	858.68	3.26
52.00	33459.8	865.14	3.22
54.00	35196.2	871.53	3.19
56.00	36945.4	877.87	3.16
58.00	38707.0	884.14	3.13
60.00	40481.2	890.36	3.10

A comparison of the results shown with those obtained by Milne's method in Ex. 6-5 reveals that they are nearly identical. As explained previously, the results obtained by these methods are very similar when the range of integration required is not extended and the stability or relative stability of the numerical method employed is not a critical factor in obtaining an accurate solution.

As stated before, Eq. 6-159 must be modified to obtain the solution of a second-order differential equation if y appears in explicit form in the equation. The following equations, shown in correct sequence, would then be applicable:

(Predictor for $\dot{y}$)

$$P(\dot{y}_{i+1}) = \dot{y}_{i-3} + \frac{4h}{3}[2\ddot{y}_i - \ddot{y}_{i-1} + 2\ddot{y}_{i-2}] \qquad \text{(a)}$$

(Modifier for $\dot{y}$)

$$M(\dot{y}_{i+1}) = P(\dot{y}_{i+1}) - \tfrac{112}{121}[P(\dot{y}_i) - C(\dot{y}_i)] \qquad \text{(b)}$$

(Predictor for y)

$$P(y_{i+1}) = \tfrac{1}{8}\{9y_i - y_{i-2} + 3h[M(\dot{y}_{i+1}) + 2\dot{y}_i - \dot{y}_{i-1}]\} \qquad \text{(c)}$$

(Differential equation)

$$M(\ddot{y}_{i+1}) = f[t_{i+1}, P(y_{i+1}), M(\dot{y}_{i+1})] \qquad \text{(d)} \quad \text{(6-161)}$$

(Corrector for $\dot{y}$)

$$C(\dot{y}_{i+1}) = \tfrac{1}{8}\{9\dot{y}_i - \dot{y}_{i-2} + 3h[M(\ddot{y}_{i+1}) + 2\ddot{y}_i - \ddot{y}_{i-1}]\} \qquad \text{(e)}$$

(Final value of $\dot{y}$)

$$F(\dot{y}_{i+1}) = C(\dot{y}_{i+1}) + \tfrac{9}{121}[P(\dot{y}_{i+1}) - C(\dot{y}_{i+1})] \qquad \text{(f)}$$

(Final value of y)

$$F(y_{i+1}) = \tfrac{1}{8}\{9y_i - y_{i-2} + 3h[F(\dot{y}_{i+1}) + 2\dot{y}_i - \dot{y}_{i-1}]\} \qquad \text{(g)}$$

(Final value of $\ddot{y}$)

$$F(\ddot{y}_{i+1}) = f[t_{i+1}, F(y_{i+1}), F(\dot{y}_{i+1})] \qquad \text{(h)}$$

Note that Eq. 6-161c, which is used as a predictor equation for y_{i+1}, is Hamming's corrector equation. An equation of the form of Eq. 6-161a could be used to obtain this predicted value, but, since $M(\dot{y}_{i+1})$ is known at this point, Eq. 6-161c can be used to obtain a better predicted value. Equation 6-161c provides a y_{i+1} value to substitute into the given differential equation in obtaining a modified value of $\ddot{y}_{i+1}$, as indicated by Eq. 6-161d. The final value of y_{i+1} is again determined by Hamming's corrector equation, as discussed in explaining Eq. 6-160.

6-9 / ERROR IN THE NUMERICAL SOLUTIONS OF DIFFERENTIAL EQUATIONS

The use of numerical methods for solving differential equations generally yields "solutions" which differ from the true solutions. The difference between the numerical solution and the true solution, at any given step is known as the *total* error at that step. If the numerical solution is to be of practical value, the total error obviously must be kept within reasonable limits over the desired solution interval. The primary purpose of error analysis is to provide a means of controlling this error. The total error at any step results from the following conditions:

1. A roundoff error is introduced in the integration process at a given step by performing the arithmetic operations of that step with numerical values having a limited number of significant digits. This is known as *per-step roundoff error*.
2. A truncation error is introduced in the integration process at a given step by the use of approximate formulas in the calculations of that step. For example, the corrector formula of the modified Euler methods has

a truncation error of order h^3, which means that series terms containing h^3 and higher powers are neglected in using a trapezoidal area to approximate the true area of one strip under the curve of the function being integrated. This is known as *per-step truncation error*.

3 An error is present at a given step because of errors introduced in preceding steps.

With perhaps eight significant digits used in the calculations, the roundoff error introduced at each step for any of the integration methods we have discussed is very small. However, if an *unstable* method is used and if the integration involves a large number of steps, the cumulative effect of the per-step roundoff errors and their magnification in calculating subsequent steps can lead to serious total error. The use of double-precision arithmetic, which is possible on most computers, is an effective means of controlling total error due to roundoff.

Obviously, the truncation error at each step is minimum in methods which employ formulas having truncation errors of high order. This error can be reduced, in any method, by reducing the step size. However, in reducing the per-step truncation error by decreasing the step size, a limit is reached at which further reduction in step size increases the total number of steps to a point where roundoff error becomes dominant, and the total error will increase with further reduction in step size. As with roundoff error, the cumulative effect of small per-step truncation errors and their magnification in calculating subsequent steps can lead to serious total error. The use of a *stable* method such as Hamming's, under certain conditions, ensures that an error introduced at a particular step will not be magnified in subsequent ones. Although unstable methods give no such guarantee, we have seen that they can be used quite satisfactorily if an excessive number of steps is not required.

Since the total error in a numerical-integration process depends on the step size used, the analyst is faced with the problem of selecting a step size to maintain the total error less than some specified bound. On the other hand, there is no point in making the solution more accurate than the data justify by choosing too small a step size, since this unnecessarily uses expensive machine time. Furthermore, as previously stated, there is a limit to the increase in accuracy which can be obtained by decreasing the step size, owing to the presence of roundoff error. On what basis, then, can the step size be selected? To answer this question let us further examine, with respect to step size, the effect of each of the per-step errors discussed on the total error.

Roundoff error is generally present to some degree in each step of a numerical-integration process. Since its magnitude in each step depends primarily upon the digital capacity of the computer being used, the per-step roundoff error is essentially independent of the step size used. The total error due to roundoff increases with the number of steps.

Truncation error also appears in each step of the numerical process. Unlike roundoff error, the per-step truncation error is a function of the step size, since it varies as the order of error, h^n, of the method being used. Thus the only means of controlling the overall per-step error of a method at a given step is by controlling the truncation error at that step. The total error is decreased as the per-step truncation error is decreased up to the point where the increasing number of steps and the presence of roundoff error causes an increase in the total error. Thus the total error can be kept to a minimum by maintaining the per-step truncation error within some appropriate limiting values or bounds. However, bounds are not always selected on the basis of optimum accuracy; they may be chosen to minimize the machine time required for a solution. This would be true in a case where the optimum accuracy would exceed the justifiable accuracy, as previously mentioned.

In initially selecting a step size for a method, the per-step truncation error, which, it is felt, can be tolerated with respect to the number of steps anticipated in the process, is defined by a set of limiting values as discussed in the preceding paragraph. The best possible estimate is then made of the general magnitude of the derivative function appearing in the truncation-error term of the method, and a step size is calculated to make the per-step truncation error fall within the defined limits. After some experience with different methods for various types of problems, the initial step size can usually be selected with little difficulty.

If the derivative in the truncation-error term of a method is of high order, an estimate of its value may be difficult to determine, and the initial step size chosen may need to be adjusted on the basis of results obtained in the numerical process. A truncation-error analysis can often be made at each step of a numerical process, and the results can be used to adjust the step size at that point, if necessary, for subsequent calculations. Such per-step truncation-error analyses have been discussed for each of the integration methods considered in this chapter. The use of such an analysis in adjusting the step size is discussed in the subsection on the error estimation of Euler's method (Sec. 6-3).

6-10 / SELECTING A NUMERICAL-INTEGRATION METHOD

Having discussed several different numerical methods of integrating differential equations and the error analysis of each of these methods, we are now in a position to compare the methods from the standpoint of selecting the best one for a particular application.

1 When it is obvious that the range of integration of a problem is relatively short, relatively small step sizes can be used without excessive computing time, and there is little reason to analyze the truncation

error at each step, or to complicate the solution by using one of the more accurate methods. Stability is also unlikely to be a problem over a short range of integration, so a simple self-starting method such as Euler's should be satisfactory.

2 When it is obvious that the range of integration of a problem is long enough to involve a large number of steps, as in a vibration problem where the motion is to be studied over a substantial number of oscillations, a method having a small per-step truncation error (such as Milne's or Hamming's) should be used in order to minimize the cumulative error. A per-step truncation-error analysis could be made at each step to minimize the computing time required, while at the same time attaining the desired accuracy, by controlling the step size. If stability is a problem, either Hamming's method should be used in preference to Milne's, or the stabilizing technique of Milne and Reynolds should be used with Milne's method. Starting values, in either case, could be determined by using a fourth-order Runge-Kutta method.

3 When a small per-step truncation error is desired, and computing time is not important enough to require a per-step truncation-error analysis, the classic fourth-order Runge-Kutta method is a convenient one to use. It is self-starting, stability is usually not a problem, and the equations are easy to program. If double-precision arithmetic is not available, or if for some other reason roundoff error must be minimized, Gill's fourth-order Runge-Kutta method may be used. For problems in which very high accuracy must be maintained, as, for example, in the orbit problems at the end of this chapter (Probs. 6-22 and 6-23), Butcher's fifth-order Runge-Kutta method has been found highly desirable.

4 When a problem involves some intermediate range of integration wherein the accumulation of error and computing time must be considered, but neither is a critical factor, the self-starting modified Euler method or a third-order Runge-Kutta method should prove satisfactory. If a per-step error estimate is desired for purposes of controlling step size, the non-self-starting modification of Euler's method should be used.

There are many numerical methods, other than those covered in this chapter, for solving differential equations, including techniques especially fitted for certain types or orders of differential equations. However, one of the methods described here should be suitable for most engineering problems of the initial-value type.

Problems

6-1 A spherical water tank of radius R is drained through a circular orifice of radius r at the bottom of the tank. An air hole at the top

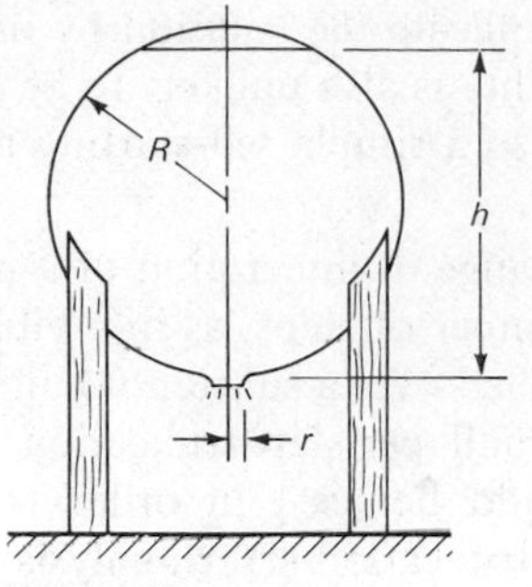

Prob. 6-1

of the tank allows atmospheric pressure to exist in the empty portion of the tank. It is desired to determine the curve of the water height h versus time, so that the time to drain the tank to any level from any other level may be determined.

The volume of the water in the tank for any height h is

$$V = \tfrac{1}{3}\pi h^2(3R - h) \tag{a}$$

The rate of change of volume of water in the tank is

$$\frac{dV}{dt} = -Av \tag{b}$$

where A is the area of the orifice and v is the velocity of the water flowing through the orifice. Equation b may be rewritten as

$$\frac{dV}{dt} = -\pi r^2\sqrt{2gh} \tag{c}$$

where g is the acceleration of gravity.

Differentiation of Eq. a will also yield an expression for dV/dt, which is

$$\frac{dV}{dt} = (2\pi hR - \pi h^2)\frac{dh}{dt} \tag{d}$$

Combining Eqs. c and d and solving for dh/dt yields the differential equation

$$\frac{dh}{dt} = \frac{-r^2\sqrt{2gh}}{2hR - h^2} \tag{e}$$

The data values for the problem are

$$R = 10 \text{ ft}$$
$$r = 0.125 \text{ ft}$$
$$g = 115{,}812 \text{ ft/min}^2$$
$$h_0 = 19.875 \text{ ft (note that if tank is full initially, the denominator of Eq. e is zero)}$$
$$\Delta t = 0.125 \text{ min}$$

Solve the first-order differential equation by the method assigned. Stop execution of the program when the tank is nearly empty. You may want to decrease the time increment when the water height gets below a certain level. If you do not program the problem to stop when some small h value is reached, execution will automatically stop when the value of h goes negative.

6-2 An arrow (assumed to be a long slender cylinder) is fired vertically upward so that it is subjected to a drag force D as shown on the figure and the gravity force shown. The drag force for a long slender cylinder moving in a fluid medium with a velocity v may be taken as

$$D = \frac{C\gamma A v^2}{2}$$

where

γ = mass per unit volume of fluid medium
A = cross-sectional area of cylinder
C = drag coefficient (approximately equal to 1 for a cylinder in coaxial flow for a Reynolds number in the region $10^3 < R < 10^5$)

Assuming a constant value of $C = 1$, the differential equation of motion for the arrow during vertical flight upward is found to be

$$\frac{d\dot{y}}{dt} + \frac{K}{m}\dot{y}^2 = -g$$

or

$$\frac{dv}{dt} + \frac{K}{m}v^2 = -g$$

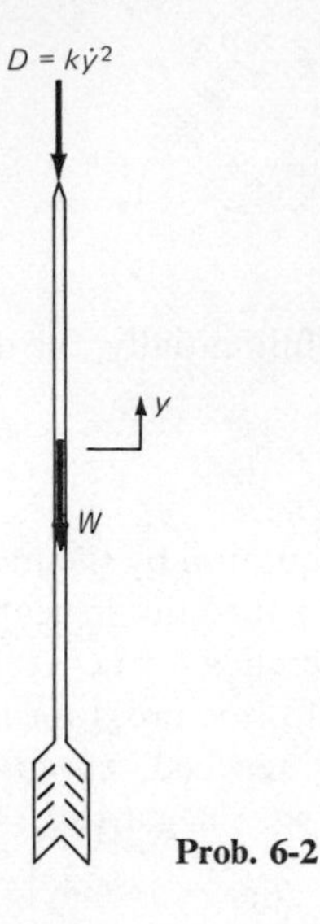

Prob. 6-2

where

$\dot{y} = v$
$K = \gamma A/2$ (coefficient of aerodynamic resistance)
$g =$ acceleration of gravity (32.2 ft/sec^2)
$m =$ mass of arrow (W/g)

The data values for the problem are

$\gamma = 0.00237$ lb-sec^2/ft^4
$A = 7(10)^{-4}$ ft^2
$W = 0.0573$ lb

Write a program to solve the first-order differential equation by the method assigned, and determine the time t_f to reach the maximum height h for an initial velocity of 200 ft/sec.

6-3* The Northwestern Natural Gas Co. has just discovered a new gas well, and they want to study the economics of connecting this well into their existing distribution system. Their geologists estimate that the well contains 8500 tons of natural gas at a pressure of 900 psia. The point at which they want to connect the well with the distributions system is at 400 psia, so with a direct connection the latter pressure is the discharge pressure P_d which the new well will initially sense.

* Based on an IBM short-course problem.

The discharge Q in tons/day through the outlet pipe from the well may be approximated by

$$Q = C(P^2 - P_d^2)^{0.8}$$

where

$C = 1.11151(10)^{-4}$
$P =$ internal pressure of gas in well, psia
$P_d =$ discharge pressure, psia

The rate of change of pressure in the gas well is assumed to be directly proportional to the discharge Q. That is,

$$\frac{dP}{dt} = -kQ \qquad \left(\text{where } k = \frac{900 \text{ psia}}{8500 \text{ tons}}\right)$$

Substituting the expression for Q into the latter equation, we obtain the first-order differential equation

$$\frac{dP}{dt} = -kC(P^2 - P_d^2)^{0.8}$$

with the initial condition that when $t = 0$, $P = 900$ psia.

As the gas pool is depleted, the pressure in the well will drop, and the discharge Q will diminish. When production drops below 2 tons/day, a constant pressure differential compressor is to be added to the system as shown in the figure. The compressor will be used to boost the pressure by 100 psia so that the discharge pressure sensed by the pipe from the well will change from 400 to 300 psia.

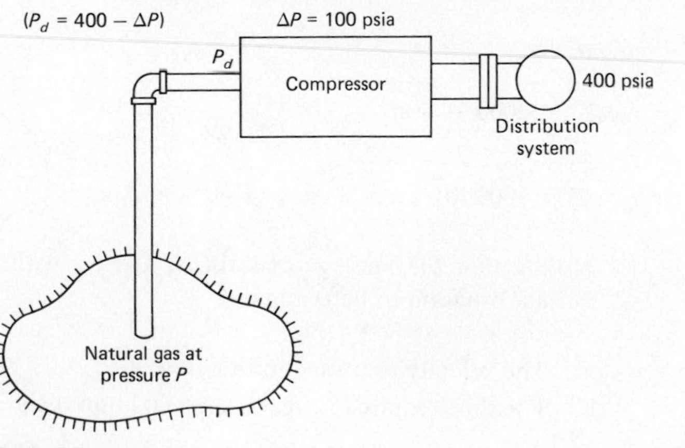

Prob. 6-3

It is planned that production will continue with this compressor until the discharge again falls below 2 tons/day. At this time a second compressor will be added which can boost the pressure an additional 100 psia, so that the pipe from the well will sense a pressure of 200 psia instead of 300 psia. Production is to continue with the two compressors in the system until the discharge falls below 1 ton/day, below which production is assumed unfeasible.

Determine the complete pressure-time curve for the well using the integration method assigned. Use $\Delta t = 0.5$ day. Also determine the total cumulative recovery of tons of natural gas by using trapezoidal integration of the $Q - t$ relationship.

6-4 Using the data of Prob. 6-3, evaluate an alternative plan using two 75 psia pressure differential compressors instead of the 100 psia compressors. Such a plan would result in an initially smaller investment by the company. Compare the total cumulative recovery with that obtained in Prob. 6-3.

6-5 A small rocket with an initial weight of 3000 lb, including 2400 lb of fuel, is fired vertically upward. The rocket burns fuel at the rate of 40 lb/sec, which develops a thrust T of 7000 lb. Considering that the drag force D is proportional to the square of the velocity $\dot{y}$, the differential equation of motion from Newton's second law is found to be

$$\frac{d\dot{y}}{dt} + \frac{K}{m}\dot{y}^2 = \frac{T}{m} - g$$

or

$$\frac{dv}{dt} + \frac{K}{m}v^2 = \frac{T}{m} - g$$

where

$$m = \frac{3000 - 40t}{g} \qquad 0 < t \le 60$$

$$T = 7000 \text{ lb} \qquad 0 < t \le 60$$

Noting that the mass m and thrust T vary with time as shown, write a program to determine

a. The velocity as a function of time.
b. The time required to reach the maximum height h when $v = 0$.

Assume that $g = 32.2$ ft/sec^2 and is constant, and that the coefficient of aerodynamic resistance is $K = 8(10)^{-3}$ lb-sec^2/ft^2.

Note that the velocity values obtained from the solution of the first-order differential equation can be compared with the values obtained in Ex. 6-5 in which the second-order differential equation for the rocket was used.

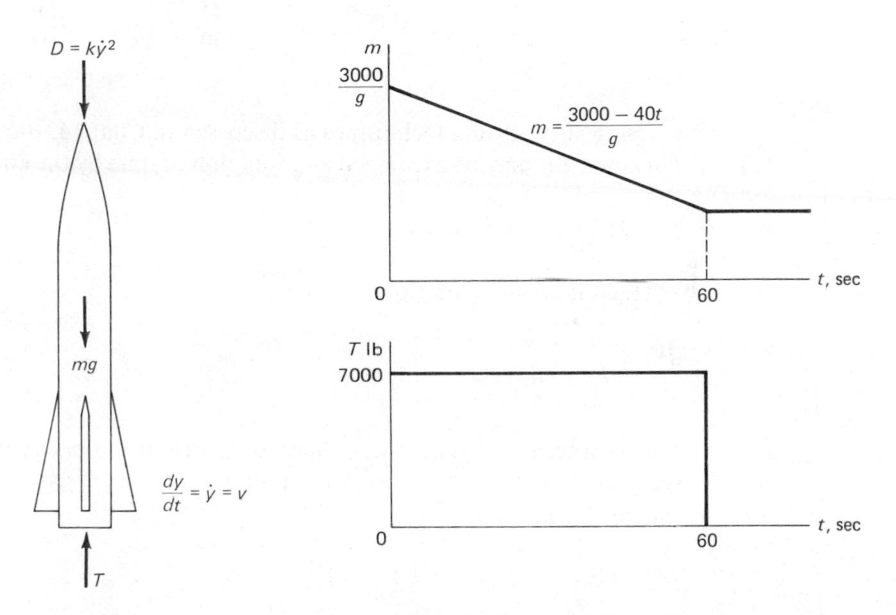

Prob. 6-5

6-6* A water conservation district has dammed a stream to create a small reservoir for irrigation and flood control. The rated capacity of the reservoir is 4000 acre-ft. When the total storage exceeds that amount, discharge is automatically initiated through the spillway of the dam. Since past records show that very heavy rainstorms occur on the average of about once in every 10 years, it is desired to determine the effectiveness of the reservoir if heavy flooding should occur. Weather data further reveal that a typical heavy rainstorm could be expected to cause an inflow into the reservoir in excess of normal flow for approximately 30 hours as shown in the following table:

* Based on an IBM short-course problem.

Time (hr)	Inflow (acre-ft/hr)	Time (hr)	Inflow (acre-ft/hr)
0	10	10	35
1	35	11	31
2	61	12	28
3	69	13	25
4	70	14	21
5	66	15	20
6	60	20	12
7	54	25	11
8	48	30	10
9	42		

Using curve-fitting techniques as discussed in Chap. 4, the inflow I in acre-ft/hr may be expressed as a function of time by the equation

$$I = 41.91te^{-0.273t} + 10$$

The differential equation

$$\frac{dV}{dt} = I - D$$

can be solved to obtain the volume of water in the reservoir as a function of time, and involves the inflow I and the discharge D as shown. In this equation

$$D = CK(V - V_d)^{1.5} \quad \text{if } V > V_d$$

$$D = 0 \qquad \text{if } V \leq V_d$$

Where V_d is the rated capacity of the reservoir (4000 acre-ft in this problem). The constant $K = 0.02$, and the discharge coefficient C is a function of the discharge D. Calibration tests of the gated spillway show that this function may be approximated as follows:

$$0 < D < 25 \qquad C = 0.8$$

$$25 \leq D < 50 \qquad C = 0.85$$

$$50 \leq D < 75 \qquad C = 0.95$$

$$75 \leq D \qquad C = 1.00$$

Using the self-starting modified Euler method, obtain the data from which plots of reservoir volume versus time and discharge versus time may be drawn. Use $\Delta t = 0.1$ hr, and study the flow for a period of 30 hr, starting at the beginning of the storm. Assume that the reservoir holds 3900 acre-ft of water when the storm begins.

6-7 Utilizing the relationship from elementary beam theory that

$$EI\frac{d^2y}{dx^2} = M$$

write the moment equation for the beam shown in the accompanying figure and write a program for determining the elastic curve of the beam.

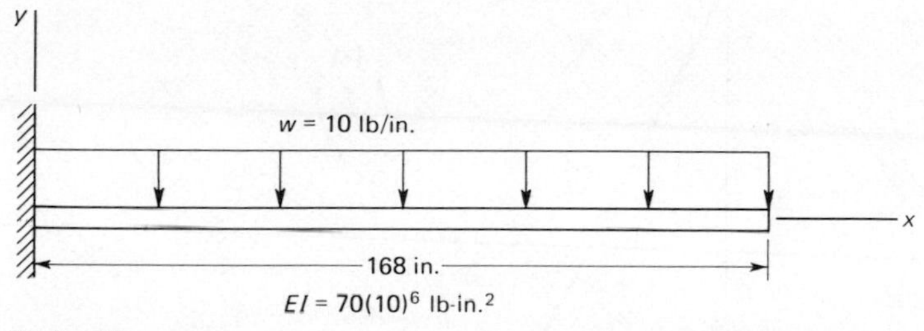

Prob. 6-7

6-8 Utilizing the relationship from elementary beam theory that

$$EI\frac{d^2y}{dx^2} = M$$

write the necessary moment equations and a program for determining the elastic curve of the beam shown in the accompanying figure.

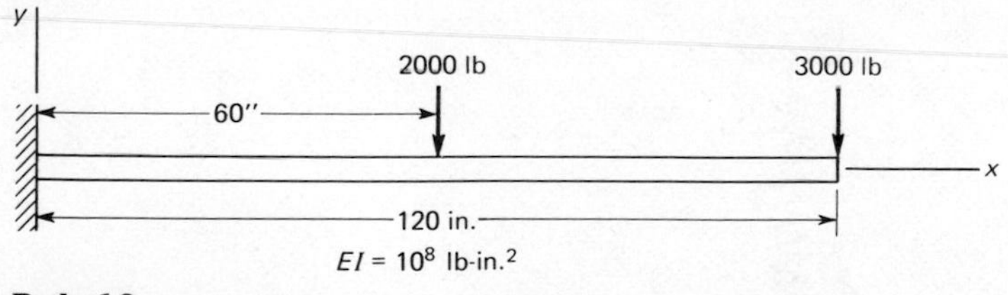

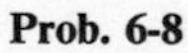

Prob. 6-8

6-9 Neglecting the mass of the slender rod, the differential equation of motion of the pendulum shown in the accompanying figure is

$$\ddot{\theta} + \frac{g}{l}\sin\theta = 0$$

Using the initial values

$$t = 0 \left| \begin{aligned} \theta &= \pi/4 \\ \dot{\theta} &= 0 \end{aligned} \right.$$

write a program for determining the θ-t curve for one period of oscillation for each of the following rod lengths: 40 in., 30 in., and 10 in. (frequency $\cong \sqrt{g/4l\pi^2}$ Hz).

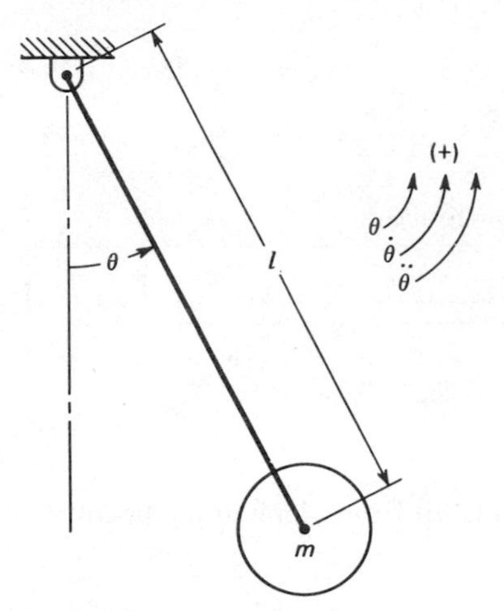

Prob. 6-9

6-10 The differential equation of motion of the spring-and-mass system shown is

$$\ddot{x} + p^2x = 0$$

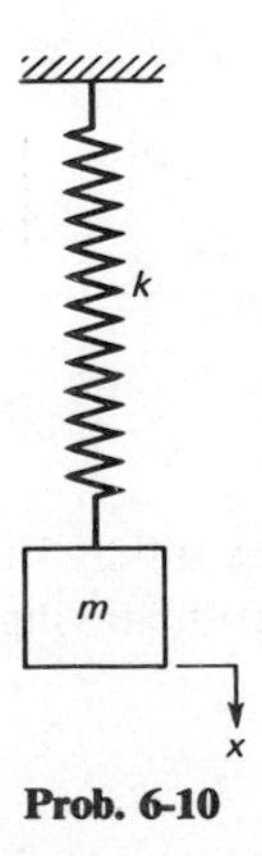

Prob. 6-10

where $p^2 = k/m$. If the weight of the mass is 10 lb, the spring constant $k = 1.5$ lb/in., and the initial values are

$$t = 0 \left| \begin{aligned} x &= 5 \text{ in.} \\ \dot{x} &= 0 \end{aligned} \right.$$

write a program for determining the x-t curve for one cycle of motion (frequency $= p/2\pi$ Hz).

6-11 The differential equation of motion of the viscously damped spring-and-mass system shown in the accompanying figure is

$$\ddot{x} + 2n\dot{x} + p^2x = 0$$

where $p^2 = k/m$, and $2n = c/m$. If the weight of the block is 15 lb, the spring constant $k = 2$ lb/in., and the initial values are

$$t = 0 \left| \begin{aligned} x &= 5 \text{ in.} \\ \dot{x} &= 0 \end{aligned} \right.$$

a. Write a program for determining the x-t and $\dot{x}$-t curves for three cycles of motion for each of the following damping ratios: $n/p = 0.05$, 0.10, 0.25, and 1.0 (frequency $= \sqrt{p^2 - n^2}/2\pi$ Hz).
b. Modify the program written for part a so that only the peak values of motion (the amplitude of each successive half-cycle) are printed out as data.

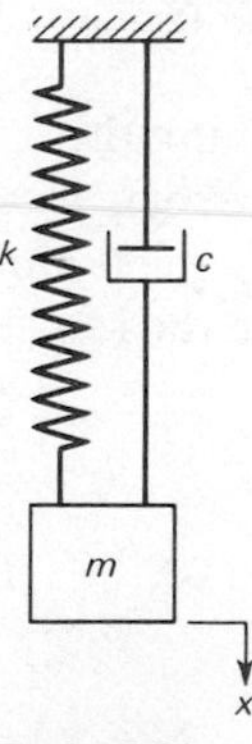

Prob. 6-11

6-12 The permanent magnet A exerts a force on the weight W that is inversely proportional to the square of the distance x between the mass center of W and the mass center of the magnet. If the coefficient of friction between W and the flat surface is f, show that the differential equation of motion of W is

$$\frac{d^2x}{dt^2} + \frac{k}{mx^2} - fg = 0$$

Write a FORTRAN program for obtaining the acceleration $\ddot{x}$, the velocity $\dot{x}$, and the displacement x in terms of time t. Consider that the weight W is released from rest with an initial value of $x_0 =$ 12 in. As additional data, it is given that $k/m = 7200 \text{ sec}^{-2}$, $f = 0.1$, and $g = 386 \text{ in./sec}^2$. It is suggested that the solution be obtained by using Euler's method.

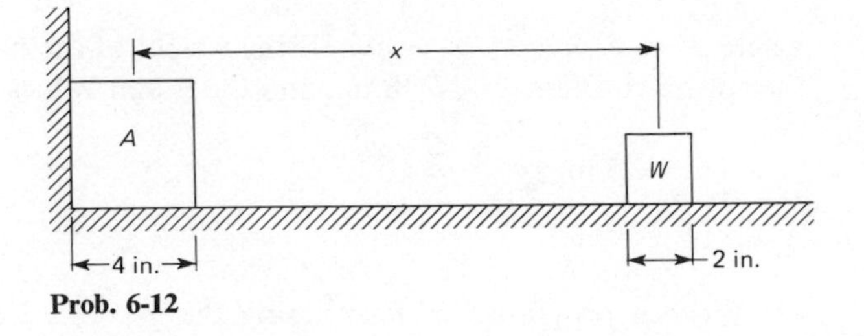

Prob. 6-12

6-13 A particle of mass m is released from rest at the edge of a hemispherical bowl, as shown in the accompanying figure, and slides down the inside surface of the bowl. If the coefficient of friction between the particle and the surface of the bowl is f, determine the differential *equations* of motion of the particle in the tangential direction. Note that since friction always opposes motion, different equations define the motion when $\dot{\theta}$ is positive and when $\dot{\theta}$ is negative.

Write a FORTRAN program for determining the θ-t and $\dot{\theta}$-t curves for the period of time necessary for the particle to return to rest on the side from which it was released. Draw the θ-t and $\dot{\theta}$-t curves for a bowl with $r = 1.0$ ft and $f = 0.20$. Use the self-starting modified Euler method.

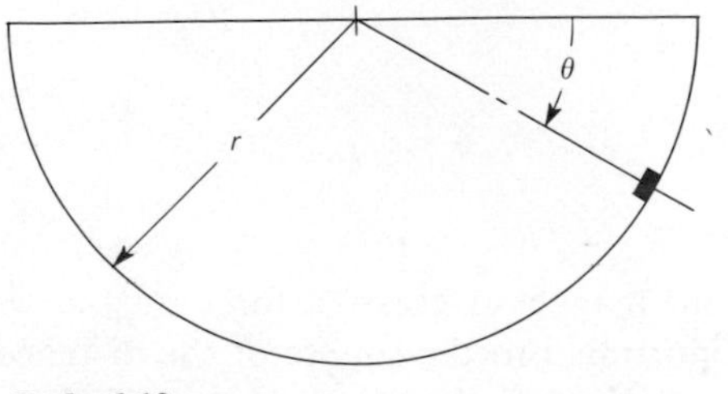

Prob. 6-13

6-14 The accompanying figure shows a projectile of weight W in motion in a coordinate system whose origin is at the point from which the projectile was fired. Assuming that the acceleration of gravity is constant and that the air resistance is proportional to the square of the *total* velocity, derive the differential equations of motion for the projectile (in the x and y directions) and write a FORTRAN program for determining the x-t, y-t, $\dot{x}$-t, $\dot{y}$-t, $\ddot{x}$-t, and $\ddot{y}$-t curves from the time of firing until the projectile strikes the ground. Use the self-starting modified Euler method.

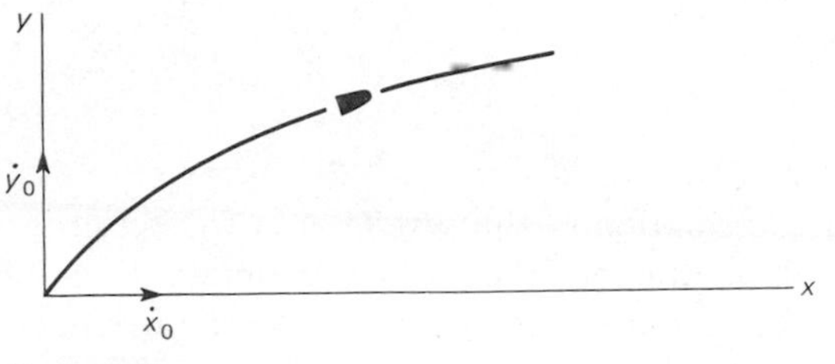

Prob. 6-14

6-15 Section 6-7 includes an explanation of Milne's method of numerically integrating ordinary differential equations. Also included is a discussion concerned with using the fourth-order Runge-Kutta method to obtain starting values for the numerical process, by means of the computer. In Ex. 6-5 a program is shown in which the starting values were precalculated manually and entered into the program as data input.

Revise the program shown by adding the program steps necessary to have the computer calculate the necessary starting values for the problem, in Ex. 6-5, by using the appropriate Runge-Kutta formulas.

6-16 Work the problem of Ex. 6-3 using Euler's method and compare results with the more accurate values obtained by the third-order Runge-Kutta method used in that example.

6-17 Rewrite the FORTRAN source program of Ex. 6-4, using fourth-order Runge-Kutta formulas for obtaining starting values before switching over to Milne's method. Does instability appear in the solution? If so, use Hamming's corrector equation in place of Milne's, and rerun the problem.

6-18 A system is released from rest in the position shown in part a of the accompanying figure. Each of the two cylinders shown weighs 32.2 lb and has a centroidal mass moment of inertia of 0.125 ft-lb-sec^2. Body C weighs 16.1 lb. The free-body diagram for dynamic equilibrium is shown in part b. Determine the differential

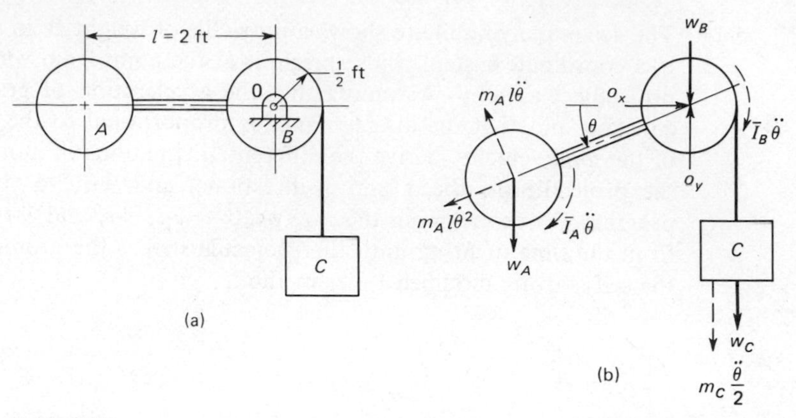

Prob. 6-18

equation of motion of the system, and write a program for solving the resulting differential equation to obtain $\dot{\theta}$-t and θ-t curves over a 2-sec interval after the system is released from rest.

6-19 The circuit shown in the accompanying figure consists of a coil wound around an iron core, a resistance R, a switch SW, and a voltage source E. The magnetization curve may be obtained from the equation

$$Ni = 0.5\phi + 0.003\phi^3$$

where

N = number of turns of coil
ϕ = flux in the core, kilolines
i = current, amp

From Kirchhoff's law, the impressed voltage E is

$$E = Ri + L\frac{di}{dt} = Ri + N\frac{d\phi}{dt}(10)^{-5}$$

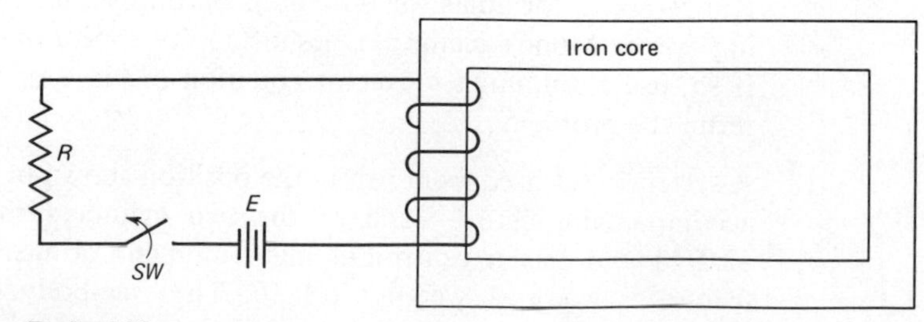

Prob. 6-19

where

L = self-inductance, henries
R = resistance, ohms
t = time, sec

If $N = 100$ turns and $R = 500$ ohms, show that the differential equation for the flux is

$$\frac{d\phi}{dT} = E - 2.5\phi - 0.015\phi^3$$

where T is in *milliseconds*. Determine solutions of this differential equation for (a) $E = 20$ volts and (b) $E = 40$ volts.

6-20 In the accompanying figure a rocket is anchored to a very rigid abutment by an elastic supporting structure of stiffness k lb/ft. The pertinent data for a static firing test are as follows:

m_0 = initial mass = 100 slugs
t_0 = burnout time = 100 sec
m_{bo} = mass at burnout = $0.1\ m_0$ = 10 slugs
m' = constant rate at which fuel is burned = 0.9 slug/sec
u = velocity of jet stream relative to rocket = 8000 ft/sec
T = constant thrust = $m'u$ = 7200 lb
m = mass of rocket for $0 \leq t \leq 100$ sec

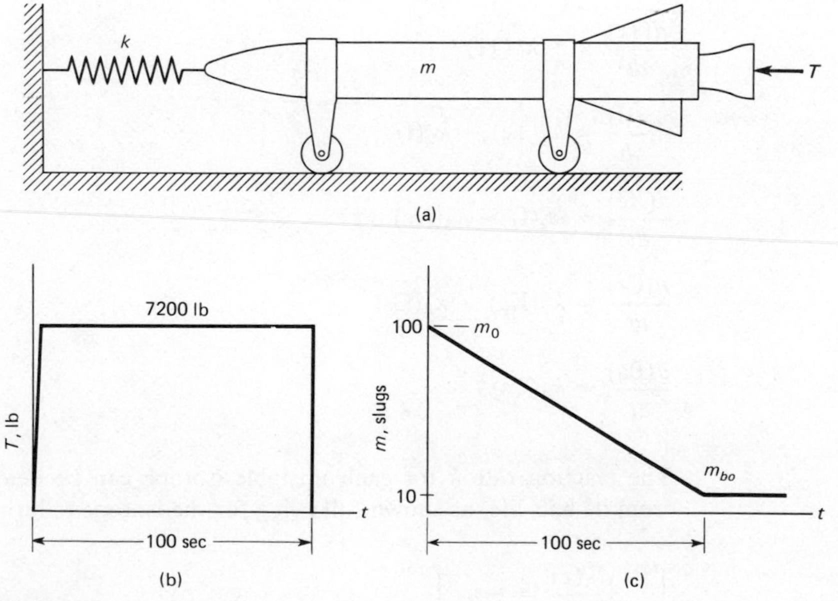

Prob. 6-20

Considering that the thrust T reaches its maximum value in a few milliseconds, the differential equation of motion of the rocket is $m\ddot{x} + kx = T$ where $m = m_0 - m't$. Thus the mathematical model of the system is a differential equation with a variable coefficient.

Program the equation to the computer and determine (a) the maximum acceleration $\ddot{x}$ and (b) the maximum displacement x for k/m_0 equal to 2000, 4000, 6000, 8000, and 10,000.

6-21 The decay of the unstable isotope ^{135}Te, with atomic weight of 135, results in the chemical-kinetics process

$$(^{135}\text{Te})^{k_1} \rightarrow (^{135}\text{I})^{k_2} \rightarrow (^{135}\text{Xe})^{k_3} \rightarrow (^{135}\text{Cs})^{k_4} \rightarrow \text{Ba}$$

where

Element	**Half-life**
Te = tellurium	2 min
I = iodine	6.7 hr
Xe = xenon	9.2 hr
Cs = cesium	$2(10)^4$ yr
Ba = barium	Stable

The set of simultaneous equations representing the transformations of the unstable isotopes to an end result of stable barium are

$$\frac{d(\text{Te})}{dt} = -k_1(\text{Te})$$

$$\frac{d(\text{I})}{dt} = k_1(\text{Te}) - k_2(\text{I})$$

$$\frac{d(\text{Xe})}{dt} = k_2(\text{I}) - k_3(\text{Xe})$$

$$\frac{d(\text{Cs})}{dt} = k_3(\text{Xe}) - k_4(\text{Cs})$$

$$\frac{d(\text{Ba})}{dt} = k_4(\text{Cs})$$

The reaction rate k for each unstable isotope can be determined from its half-life, as shown following for the isotope tellurium:

$$\int_1^{1/2} \frac{d(\text{Te})}{\text{Te}} = -k_1 \int_0^{120\ \text{sec}} dt$$

Write a computer program for determining the amount of each element present (a) at 5-sec intervals for the first 2 min, (b) at 1000-sec intervals for the next 9 hr, and (c) at $(10)^{10}$-sec intervals for the next 100 intervals. (*Hint:* Use the Runge-Kutta method and change the time increments for the different time intervals given above.)

6-22 The accompanying figure shows the earth with mass m_e, the moon with mass m_m, and a satellite with mass m. The satellite is small compared to the earth and the moon ($m \ll m_e$, $m \ll m_m$). The three masses are assumed to attract each other according to the inverse square law, but the influence of the satellite on the motion of the earth and the moon is considered to be negligible. The analysis of such a system is referred to as the restricted three-body problem.

The earth and the moon revolve about their common center of mass, which is located inside the earth as shown in the figure. The distance D between the mass center of the earth and that of the moon is assumed to be constant. The motion of the satellite is such that it lies in the plane of revolution of the earth and the moon.

The x_0-y_0 coordinate system shown is nonrotating, and its acceleration is assumed as negligible (Newton's second law is valid in this reference frame). Transforming the equation of motion of the satellite in the x_0 direction in the nonrotating set of axes to the rotating x-y coordinate system results in two equations of motion in the latter coordinate system. If the distance D between the mass centers of the earth and moon is chosen as unity, and the unit of time is chosen such that the angular velocity ω of the rotating x-y reference frame is unity ($\tau = 2\pi$), the equations of motion of the satellite in terms of the x-y coordinate system are

$$\ddot{x} = x + 2\dot{y} - \mu' \frac{x + \mu}{[(x + \mu)^2 + y^2]^{3/2}} - \mu \frac{x - \mu'}{[(x - \mu')^2 + y^2]^{3/2}}$$

$$\ddot{y} = y - 2\dot{x} - \mu' \frac{y}{[(x + \mu)^2 + y^2]^{3/2}} - \mu \frac{y}{[(x - \mu')^2 + y^2]^{3/2}}$$

where

$$\mu = \frac{m_m}{m_e + m_m}$$

$$\mu' = 1 - \mu$$

These equations may be solved numerically to obtain the x and y coordinates of the satellite for various values of the independent variable time. It is known that closed periodic orbits exist in the rotating x-y coordinate system.

Write a FORTRAN program for solving these equations simultaneously, using the fourth-order Runge-Kutta method. Using the following initial conditions and parameters, determine the path of the satellite in the x-y system:*

$$t = 0 \left| \begin{array}{l} x = 0.9940000000000000 \\ y = 0.0000000000000000 \\ \dot{x} = 0.0000000000000000 \\ \dot{y} = -2.1138987966945027 \end{array} \right.$$

$$\mu = 0.0122774710000000$$

$$\tau = 5.4367954392601900 \text{ time units}$$

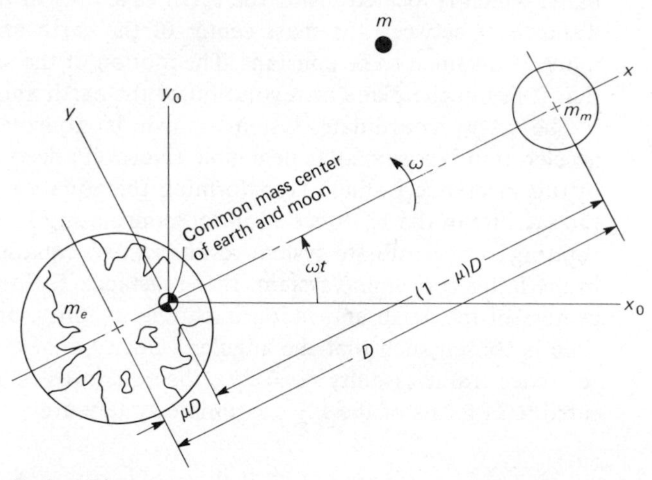

Prob. 6-22

Use the radius of curvature of the satellite's path in the x-y system to control the time increment in the computer program. If $\rho < 0.1$, use $\Delta t = 1/2^{13}$; if $0.1 \leq \rho < 0.7$, use $\Delta t = 1/2^{10}$; if $0.7 \leq \rho < 2.0$, use $\Delta t = 1/2^{7}$; and if $\rho \geq 2.0$, use $\Delta t = 1/2^{6}$. The given time increments are expressible exactly in binary form so that cumulative roundoff error in summing the time increments is avoided using a binary computer.

6-23 Using the computer program written for Prob. 6-22, and the

* These initial conditions result in a closed periodic orbit. See NASA Contractor Report CR-61139, "Study of the Methods for the Numerical Solution of Ordinary Differential Equations," prepared by O. B. Francis, Jr. et al. for the NASA-George C. Marshall Space Flight Center, June 7, 1966.

following initial conditions and parameters, determine the path of the satellite in the rotating x-y coordinate system:

$$t = 0 \left| \begin{aligned} x &= 0.9940000000000000 \\ y &= 0.0000000000000000 \\ \dot{x} &= 0.0000000000000000 \\ \dot{y} &= -2.0317326295573368 \end{aligned} \right.$$

$$\mu = 0.0122774710000000$$

$$\tau = 11.1243403372660851 \text{ time units}$$

Use the same time increments as suggested in Prob. 6-22.

The following problems involve SI units.

6-24 A narrow ring of radius $R = 0.1$ m has a positive charge of $q = 8.00(10)^{-18}$ coulomb uniformly distributed over its circumference. The resultant electric-field strength at a distance x from the ring along the axis shown in the accompanying figure is given by

$$E = \frac{q \cos \alpha}{4\pi\varepsilon_0 s^2}$$

where s and α are as defined in the figure. If $\varepsilon_0 = 8.85(10)^{-12}$ coulomb^2/newton-m^2, determine the differential equation of motion of an electron released from rest at a position x_0 along the axis of the ring.

Write a FORTRAN program for determining the x-t and $\dot{x}$-t curves for one cycle of motion of the electron, using Euler's method. The mass of an electron is approximately $9.11(10)^{-31}$ kg, and the charge on an electron is $1.6(10)^{-19}$ coulomb. The electron is released from rest at a distance $x_0 = 0.2$ m from the plane of the ring.

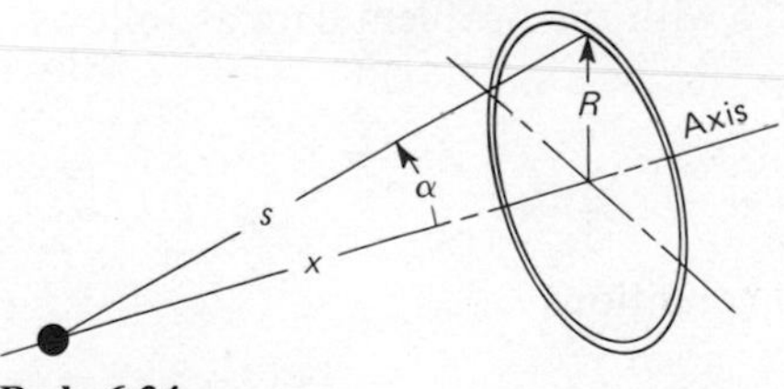

Prob. 6-24

6-25 An "infinitely long straight wire" has a uniform positive charge of $\lambda = 2.56(10)^{-17}$ coulomb/m. The resultant electric-field strength at a point a distance r from the wire is given by

$$E = \frac{\lambda}{2\pi\varepsilon_0 r}$$

where $\varepsilon_0 = 8.85(10)^{-12}$ coulomb²/newton-m². The charge on an electron is $1.6(10)^{-19}$ coulomb. Determine the differential equation of motion of an electron released from rest near the wire.

Write a FORTRAN program for determining the r-t and $\dot{r}$-t curves for one cycle of motion of the electron, using Euler's method. The electron is released from rest at a distance $r_0 = 0.1$ m from the wire. Assume that there is a very small hole in the wire through which the electron can pass. The mass of an electron is approximately $9.11(10)^{-31}$ kg.

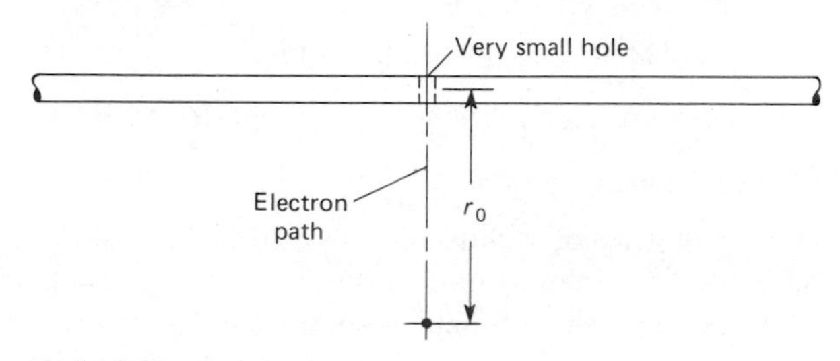

Prob. 6-25

6-26 Solve Prob. 6-1 in which the data values are

$$R = 3.5 \text{ m}$$
$$r = 40 \text{ mm}$$
$$g = 9.81 \text{ m/s}^2$$
$$h_0 = 6.9 \text{ m}$$
$$\Delta t = 7.5 \text{ s}$$

6-27 Refer to Prob. 6-2. Convert the data given in that problem to the SI system of units, and use this as input data to obtain a solution of the problem.

6-28 Solve Prob. 6-11 with the problem data as follows:

$$k = 250 \text{ N/m}$$

$$m = 5.2 \text{ kg}$$

with the initial conditions

$$t = 0 \left| \begin{array}{l} x = 0 \\ x = 0.2 \text{ m/s} \end{array} \right.$$

6-29 Solve Prob. 6-12 by converting the problem data given in that problem to the SI system of units for input data to your program.

6-30 Solve Prob. 6-18 by converting the problem data of that problem to the SI system of units.

7 Ordinary Differential Equations: Boundary-Value Problems

7-1 / INTRODUCTION

Problems in which the conditions to be satisfied by the solution or its derivatives are given at the point where the integration is to begin are usually referred to as *initial-value* problems, as explained in Chap. 6. We will now consider problems in which conditions are specified at both ends of the interval over which the integration is to occur. Ones of this type are known as *boundary-value* problems. Since these involve a minimum of two boundary conditions, we will obviously be considering differential equations of the second order or higher.

The numerical solution of boundary-value problems is somewhat more difficult than that of initial-value problems. Two elementary methods are used to solve engineering problems of this type: one consists of a trial-and-error procedure; the other requires the simultaneous solution of a set of algebraic equations.

7-2 / TRIAL-AND-ERROR METHOD

A trial-and-error method can be used to solve both linear and nonlinear differential equations of the second order or higher. The technique involves the assumption or approximate calculation of the initial values of the problem which are unknown. This, in effect, reduces the problem to an initial-value problem for which trial solutions can be obtained by using one of the integration methods discussed in Chap. 6. After a trial solution is obtained, the known boundary values at the end of the integration interval are compared with the corresponding values provided by the trial solution. If the solution values at this point do not agree with the known boundary values, new initial-condition values must be approximated and another trial solution made. This procedure continues until a trial solution is obtained which satisfies the known final boundary values.

It should be apparent that if it becomes necessary to approximate several different initial values, the number of trial solutions required could become very large. For this reason the trial-and-error approach is usually applied only to the solution of problems in which just one initial value is unknown. For example, if we consider the equation

$$y'' = f(x, y, y') \tag{7-1}$$

with the boundary conditions

$$x = a \left| \begin{array}{l} y = y_a \\ y' = ? \end{array} \right. \qquad x = b \left| \begin{array}{l} y = y_b \\ \end{array} \right.$$

we can assume a value for y' at $x = a$ and integrate as we normally would to obtain the solution of a second-order initial-value problem, using a method applicable to that type of problem. If the value of y at $x = b$, provided by the trial solution, is larger than the known boundary value at this point, a new initial value of y' is chosen which is smaller than the original approximate value used, and another trial solution is made. If the resulting value of y at $x = b$ is smaller than y_b, a larger initial value is chosen for y', and another trial solution is made. This procedure continues until a solution is obtained in which the calculated value of y at $x = b$ is approximately equal to the boundary value at that point, within some prescribed limit of accuracy.

In general, the relationship between the initial-trial values of y' and the resulting final values of y obtained from the trial y' values is not a linear one. However, linear interpolation is sometimes used to determine approximate initial values. In this procedure successive approximate values of y'_a and the corresponding values of y obtained at $x = b$ are used along with the known boundary value of y at $x = b$ to obtain a new value of y'_a for a subsequent trial solution.

A bisection method, similar to the bisection technique employed for finding roots of functions (Sec. 2-3), is frequently used and is somewhat easier to apply than linear interpolation. In this procedure, two trial values of y' at $x = a$ are selected which are quite sure to bracket the correct value of y'_a. Let these selected trial values be y'_{a_1} and y'_{a_2}. We then use

$$y'_a = 0.5(y'_{a_1} + y'_{a_2})$$

in an integration procedure to determine y at $x = b$ for this y'_a value. From the y_b value obtained, it can usually be determined whether the correct y'_a value is between y'_{a_1} and y'_a or between y'_a and y'_{a_2} by checking whether the value of y_b obtained is larger or smaller than the correct value known for y_b from the boundary condition. Suppose it is between y'_{a_1} and y'_a. In this case y'_a is renamed y'_{a_2}, and a new y'_a is calculated from

$$y'_a = 0.5(y'_{a_1} + y'_{a_2})$$

Integration is then performed using the new y'_a to find a new y_b. The latter value is then again compared to the known correct y_b value to determine if the correct y'_a value is between y'_{a_1} and y'_a or between y'_a and y'_{a_2}. Suppose that this time the correct value is between y'_a and y'_{a_2}. In this case y'_a is renamed y'_{a_1}, and a new y'_a value is calculated as before. The procedure continues until the y value at $x = b$ is sufficiently close to the known boundary value. The bisection technique is illustrated and further explained in the example which follows. In this example the second known boundary value is that of the first derivative at the second boundary.

EXAMPLE 7-1

Let us consider the problem of dissipating heat in space. In many engineering applications, heat must be dissipated from bodies. We are all aware of common examples of heat dissipation—from combustion engines, air-conditioning units, bearing surfaces, and so on. Often, heat is dissipated from a body by the simultaneous action of conductive and convective heat transfer to a fluid medium surrounding the body, and by thermal radiation. The dissipation of heat from a body deep in outer space is limited to thermal radiation, because of the negligible amount of atmosphere encountered there. Therefore, the thermal-radiation characteristics of materials and the geometries of radiating surfaces become very important in considering heat dissipation in such an environment.

In the power plant of a space station, for example, heat must be dissipated to maintain the plant at a normal operating temperature. If a working fluid is utilized to transfer heat from the power plant through coolant tubes which are exposed to surrounding space, and is then recycled through the power plant, the fluid must lose some heat during the cycle. Assuming that the material composing the coolant tubes is an excellent conductor, the amount of heat dissipated from the space station will depend primarily on the thermal-radiation characteristics of the tubing. Since the rate of heat transfer by thermal radiation is a function, among others, of the area of the radiating surface, the addition of area to the coolant tubing, in the form of numerous small fins, facilitates heat dissipation and reduces the amount of tubing required. Figure 7-1 is a simplified sketch of a section of coolant tubing with several such fins attached.

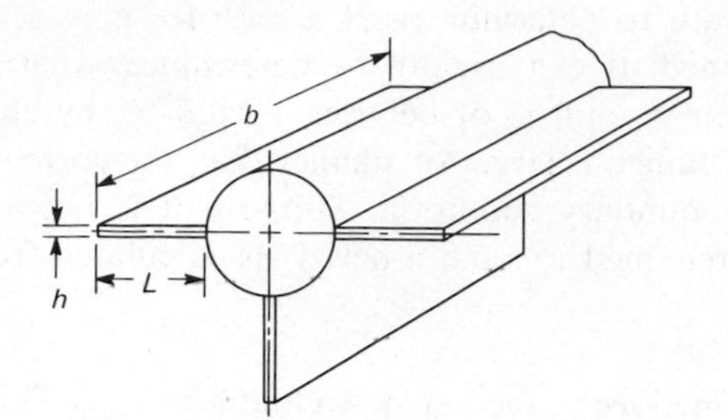

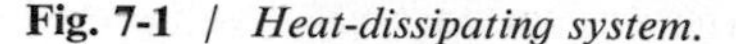

Fig. 7-1 / *Heat-dissipating system.*

We will be concerned with analyzing temperature distribution along fins with various geometries. With a constant operating temperature for the power plant, the problem is one of *steady* heat flow with the system in thermal equilibrium. Since the dissipation of heat by the fin being considered involves both conduction along the fin and radiation from the fin surface, a brief review of the basic principles of each type of heat transfer is in order, before proceeding to the development of the differential equation defining the temperature distribution along such a fin.

Conduction

Fourier's law of conduction states that the instantaneous rate of heat flow is proportional to the area A through which the heat flows at right angles, and to the temperature gradient along the flow axis. It is formulated as

$$\frac{\partial Q}{\partial \phi} = -kA\frac{\partial T}{\partial x} \tag{7-2}$$

where ϕ is time, A is area, $\partial T/\partial x$ is the temperature gradient and is negative when heat flow is in the positive x direction, and k is a proportionality factor which is a function of the physical properties of the conductor and is referred to as the *thermal conductivity* of the conductor. Since the temperature at any given location in a conductor does not vary with time for steady-state conduction, it follows that the temperature gradient and, consequently, the rate of heat flow are likewise independent of time. Hence the instantaneous rate of heat flow $\partial Q/\partial \phi$ becomes a constant rate of heat flow and is usually designated as q_c. Equation 7-2 then becomes, for steady heat flow,

$$q_c = -kA\frac{dT}{dx} \tag{7-3}$$

Equation 7-3, with the appropriate parameter values, will define the rate of heat transfer by conduction along the fin to be considered in our analysis.

Thermal Radiation

When a body is heated, radiant energy is emitted by the body at a rate and of a quality dependent on the temperature of the body. The *emissive power* W of a surface actually varies not only with the temperature but also with the roughness of the surface and, if the surface is metal, with the degree of surface oxidation. The quality of radiation is measured by its distribution in the spectrum, and is a function of the wavelength of emission and the temperature at which emission occurs. A surface which emits radiant energy will also absorb incident radiation. The fraction of the total incident radiation present which a surface is capable of absorbing is called the *absorptivity* of the surface; it is designated by α.

Consider two small surfaces, of area A_1 and A_2, in a large enclosure which is perfectly insulated from any external heat transfer. When the surfaces reach a state of thermal equilibrium, they will emit radiation at the respective rates A_1W_1 and A_2W_2 throughout the portion of the hemisphere above each element of surface. If the incident radiation impinging on the surfaces, owing to the radiation of the enclosure, is designated as I, and the surfaces have respective absorptivities of α_1 and α_2, they will absorb the incident radiation in that proportion, and an energy balance will show that

$$\alpha_1 I A_1 = A_1 W_1 \quad \text{and} \quad \alpha_2 I A_2 = A_2 W_2 \tag{7-4}$$

from which

$$\frac{W_1}{\alpha_1} = \frac{W_2}{\alpha_2} = \frac{W_3}{\alpha_3} = \cdots \frac{W_n}{\alpha_n} \tag{7-5}$$

The above generalization—that, at thermal equilibrium, the ratio of the emissive power of a surface to its absorptivity is the same for all bodies—is known as Kirchhoff's law. Since α is the fraction of incident radiation absorbed and cannot exceed unity, Eq. 7-4 shows that an upper limit exists on the emissive power of a body. This upper limit is usually denoted as W_B, and any surface having this upper-limiting emissive power is called a *perfect radiator*. Since such a surface must also have perfect absorptivity (α = unity), it will have zero reflectivity, and is thus commonly referred to as a *black body*.

The ratio of the emissive power of an actual surface (commonly called a *gray body*) to that of a black body is called the *emissivity* of the surface and is denoted by ε. Since the emissivity of an actual body is, in general, $W_n = W_B\varepsilon_n$, Eq. 7-5 may be rewritten as

$$\frac{W_B\varepsilon_1}{\alpha_1} = \frac{W_B\varepsilon_2}{\alpha_2} = \frac{W_B\varepsilon_3}{\alpha_3} = \cdots = \frac{W_B}{\alpha_B} \tag{7-6}$$

where W_B/α_B is the ratio for a black body. Since α_B is unity for a black body, Eq. 7-6 may be reduced to

$$\frac{\varepsilon_1}{\alpha_1} = \frac{\varepsilon_2}{\alpha_2} = \frac{\varepsilon_3}{\alpha_3} = \cdots = \text{unity} \tag{7-7}$$

Thus Kirchhoff's law may be restated to say that, at thermal equilibrium, the emissivity and absorptivity of a body are the same. Actually, this law is valid only when the body is receiving radiation from surroundings at its own temperature, since the emissivity of a body at temperature T_1 is not the same as its absorptivity when it is absorbing radiation from some other body at temperature T_2. However, owing to the high values of emissivity of most surfaces which are of engineering importance, and because of the small change of their emissivities with temperature, the emissivity and absorptivity of a body may be considered equal without introducing serious error. The emissivity value normally used is the value associated with the higher temperature.

Other factors involved in the transfer of heat by thermal radiation include the shape and relative positions of the heat-interchanging surfaces. Although the *intensity of radiation* of black bodies is independent of these factors, this is not true for actual bodies. For the latter, the intensity varies with the angle between the normals to the two surfaces. Two values of emissivity are therefore usually considered—*hemispherical* emissivity and *normal* emissivity. For nonmetals, the hemispherical emissivity is substantially the same as the normal emissivity. Fortunately, for most surfaces used in engineering, the values of normal emissivity may be used for hemispherical-emission conditions without introducing appreciable error.

The preceding discussion is a good example of the rational idealization which occurs in correlating physical phenomena with representative mathematical models.

Since the total emissive power of a black body depends only on its temperature, the second law of thermodynamics may be utilized to prove a proportionality between emissive power and the fourth power of the absolute temperature. This relation, the Stefan-Boltzmann law, is

$$W_B = \sigma T^4 \tag{7-8}$$

where σ is the Stefan-Boltzmann constant.

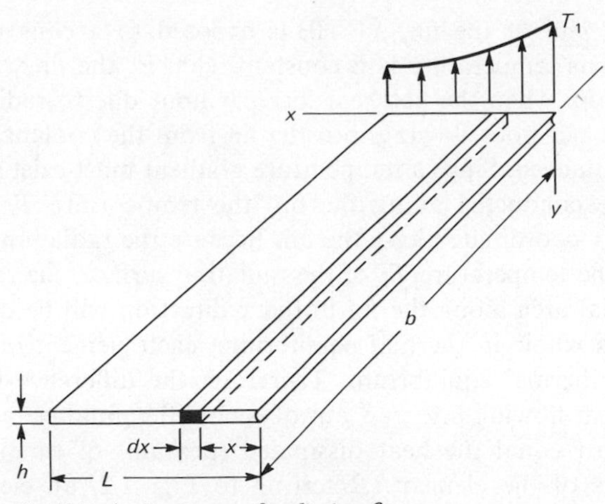

Fig. 7-2 / *Geometry of radiating fin.*

The net rate of loss of energy by radiation from a body at temperature T_1 in black surroundings at temperature T_2 is given by

$$q_r = \sigma A_r(\varepsilon_1 T_1^4 - \alpha_{1,2} T_2^4) \tag{7-9}$$

where A_r is the area of the radiating surface, ε_1 is the emissivity of the radiating body at temperature T_1, and T_1 and T_2 are the temperatures of the radiating and receiving bodies, respectively, measured in degrees Rankine (°R). The subscripts of the absorptivity factor $\alpha_{1,2}$ refer to the value of the absorptivity with different emitter and receiver temperatures T_1 and T_2. Referring to the previous discussion of the relationship between ε_1 and $\alpha_{1,2}$ for engineering applications, we may rewrite Eq. 7-9 as

$$q_r = \sigma A_r \varepsilon_1 (T_1^4 - T_2^4) \tag{7-10}$$

since $\varepsilon_1 \cong \alpha_{1,2}$.

With the preceding information as a background, let us consider the development of the differential equation defining the thermal-equilibrium state of one of the fins of Fig. 7-1. Such a fin is shown in Fig. 7-2, with the geometry defined. The parameters with which we will be concerned are as follows:

T_1 = absolute Fahrenheit temperature at any position along the fin, °R
T_2 = absolute Fahrenheit temperature of surrounding space, 0°R
σ = Stefan-Boltzmann constant, $0.173(10)^{-8}$ Btu/(ft)^2(hr)(°R)^4
ε_1 = emissivity of fin, 0.8
b = width of fin in z direction, 0.5 ft
h = thickness of fin, $0.005 \leq h \leq 0.01$ ft
k = thermal conductivity of fin, 25 Btu/(hr)(ft)(°R)
$T_1|_0$ = constant temperature of fin at root end ($x = 0$), 2000°R
L = length of fin, 0.25 ft

Since the root end of the fin, $x = 0$, is exposed to a constant heat source such that its temperature is a constant 2000°R, the fin will be in thermal equilibrium when the net heat loss per hour due to radiation is equal to the heat per hour flowing into the fin from the coolant tube to which the fin is attached. Since a temperature gradient must exist in order that heat can be conducted along the fin, the temperature T_1 will be different at each x coordinate along the fin. Because the radiation of heat is a function of the temperature T_1 of the radiating surface, the radiation of each differential area along the fin in the x direction will be different. With the fin as a whole in thermal equilibrium, each element of the fin must also be in thermal equilibrium. Therefore, the difference between the amount of heat flowing into and out of each differential element (per unit of time) must equal the heat dissipated (per unit of time) by the radiating surfaces of the element. Referring to Fig. 7-2, an element is selected which has a cross-sectional area bh, an effective radiating surface of $2b\,dx$, and a length of dx. Neglecting any radiation from the very small perimetrical surfaces of the fin and the curvature of the root area of the fin, we may write, from Eqs. 7-3 and 7-10,

$$\left[-kbh\left(\frac{dT_1}{dx}\right)_x\right] - \left[-kbh\left(\frac{dT_1}{dx}\right)_{x+dx}\right] = \sigma\varepsilon_1(2b\,dx)(T_1^4 - T_2^4)$$

from which

$$kbh\left[\frac{(dT_1/dx)_{x+dx} - (dT_1/dx)_x}{dx}\right] = 2\sigma\varepsilon_1 b(T_1^4 - T_2^4)$$

or

$$\frac{d^2T_1}{dx^2} = \frac{2\sigma\varepsilon_1}{kh}(T_1^4 - T_2^4) \tag{7-11}$$

Equation 7-11 is the differential equation defining the thermal-equilibrium state of the fin, the solution of which will yield the temperature distribution along the fin.

The pertinent boundary values are

$$x = 0 \left| \begin{array}{l} T_1 = 2000°\text{R} \\ \dfrac{dT_1}{dx} = \,? \end{array} \right. \qquad x = L \left| \; \frac{dT_1}{dx} = 0 \right. \tag{7-12}$$

To illustrate the trial-and-error solution of Eq. 7-11, we will use the self-starting modified Euler method which was discussed in Sec. 6-4. The equations outlining this method are as follows:

(Predictor for T')

$$P(T'_{i+1}) = T'_i + T''_i(\Delta x) \tag{a}$$

(Predictor for T)

$$P(T_{i+1}) = T_i + \left[\frac{T_i' + P(T_{i+1}')}{2}\right](\Delta x) \qquad \text{(b)}$$

(Differential equation)

$$P(T_{i+1}'') = f[P(T_{i+1})] \qquad \text{(c)}$$

(Corrector for T') (7-13)

$$C(T_{i+1}') = T_i' + \left[\frac{T_i'' + P(T_{i+1}'')}{2}\right](\Delta x) \qquad \text{(d)}$$

(Corrector for T)

$$C(T_{i+1}) = T_i + \left[\frac{T_i' + C(T_{i+1}')}{2}\right](\Delta x) \qquad \text{(e)}$$

(Differential equation)

$$C(T_{i+1}'') = f[C(T_{i+1})] \qquad \text{(f)}$$

where $P(T_{i+1}')$, $C(T_{i+1}')$, and $P(T_{i+1})$, $C(T_{i+1})$ are predicted and corrected values of the temperature gradients and temperatures, respectively. Note that T_1 is designated merely as T in Eq. 7-13 to facilitate the subscripting used. The symbols $P(T_{i+1}'')$ and $C(T_{i+1}'')$ indicate predicted and corrected values of (T_{i+1}'') which are obtained by substituting the appropriate T values into the given differential equation as indicated by Eq. 7-13c and f.

Inspection of Eq. 7-13a reveals that we cannot start the numerical integrating process until we have obtained an initial value of T_i'. The steps of a trial-and-error procedure employing bisection for arriving at a correct T_i' are listed below. It is suggested that the steps shown be studied in conjunction with the flow chart shown in Fig. 7-4.

1 Two initial values of T' are chosen, one of them being a very small negative value or zero and the other a very large negative value (T' is known to be negative at the root of the fin, $x = 0$) such that the correct initial value of T' is quite sure to be between the two values selected. In the program shown later, the name LOW is used for the small value and HIGH for the large value.

2 The T' value used for the first trial is the average of the two values discussed above (the T' interval is bisected). That is,

$$T' = 0.5(\text{HIGH} + \text{LOW}) \qquad \text{(7-14)}$$

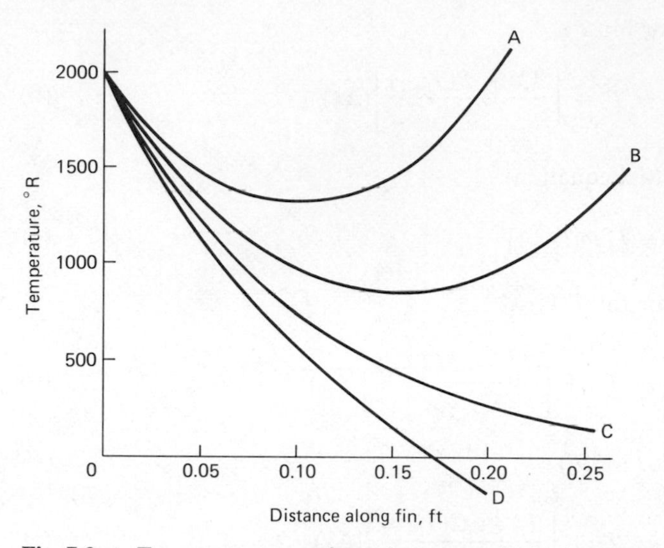

Fig. 7-3 / *Temperature curves for various assumed initial temperature gradients.*

3 Equation 7-13 is then used to integrate over the desired interval from $x = 0$ to $x = L$, using the T' value in step 2 as a starting value. Equation 7-11 is very sensitive with respect to the initial value of T'. For example, if T' is chosen as just a very little smaller negative value than the correct value, the T values can become very large before the integration interval is completed, even resulting in overflow in the computer. This is shown in curve A of Fig. 7-3 where the T curve heads for a large value before the end of the fin is reached. Conversely, if T' is chosen as just a very little larger negative value than the correct one, the T curve will go negative before the end of the fin is reached, as shown by curve D of Fig. 7-3. Physically it is known that T can never exceed the initial value of 2000°R, nor can it ever be less than 0°R. Therefore, if T exceeds 2000°R at any time during the integration, the integration is terminated. Since this indicates that T is becoming too large, the initial T' was too small. The correct initial T' value is then known to have a value between the T' value tried and HIGH. Therefore, the T' previously tried is assigned to the name LOW, and a new trial T' value equal to the average of LOW and HIGH is calculated, and the T values checked with this trial value to see if the values follow a curve like A or D. If they again result in a curve like A, the process just described is repeated. If the new trial value results in T values which go negative as in curve D, the integration is again terminated. This means that the initial T' value is too high, and that the correct T' value thus lies between the T' value used and LOW. Therefore, the T' previously tried is assigned to the name HIGH, and a new trial value of T' is calculated which is equal to the average of LOW and HIGH.

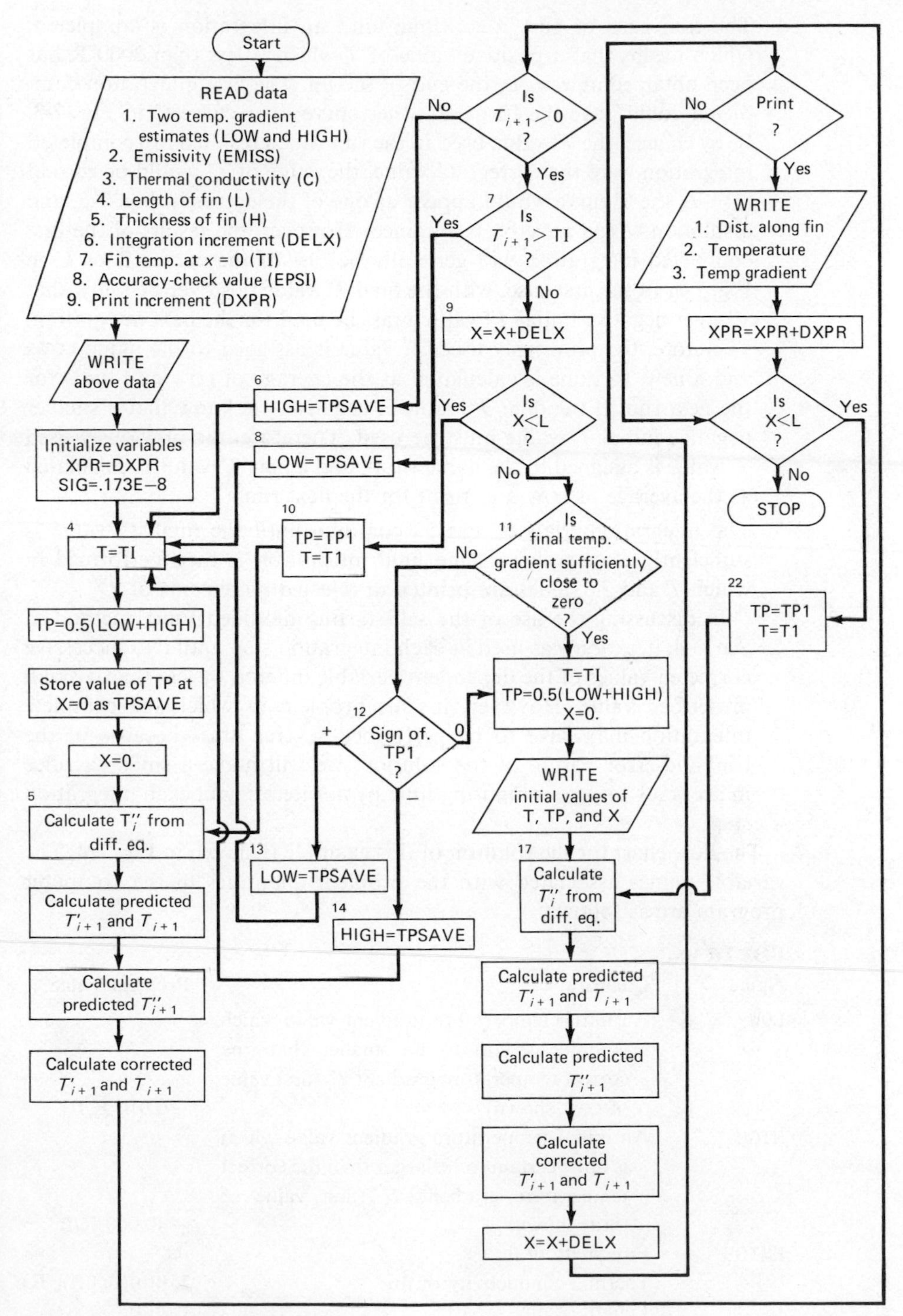

Fig. 7-4 / *Flow chart for Ex. 7-1.*

4 The trial runs of part 3 continue until an integration is completed, which means that a positive value of T which is less than 2000°R has been obtained at $x = L$, the end of the fin. The first integration completed could have the form of either curve B or curve C in Fig. 7-3. If, by chance, the T' value used in the run which resulted in a completed integration were the correct T' value, the value of T' would be zero at $x = L$, the T curve would appear as one of the curves in Fig. 7-5, and a solution would have been obtained. However, the results of the first completed integration will generally be one of the curves B or C in Fig. 7-3. In the first case, with the final T' value positive, we know that a larger negative initial T' value must be used for the next integration. Therefore, the previously used T' value is assigned to the name LOW, and a new T' value is calculated as the average of LOW and HIGH for the next run. If the final T' value is negative, we know that a smaller negative initial T' value must be used. Therefore, the previously used T' value is assigned to the name HIGH, and a new T' value is calculated as the average of LOW and HIGH for the next run.

5 The integration trials of part 4 continue until the final T' value is sufficiently close to zero. One final integration is then performed in which T and T' values are printed at selected increments of x.

In discussing the use of the self-starting modified Euler method in Sec. 6-4, iteration was used at each integration step until two successive corrected values of the dependent variable differed by less than a small prescribed value. However, in this problem in which the complete integration may have to be performed several times because of the trial-and-error nature of the solution, we will make a small sacrifice in accuracy to save computing time by not iterating at each integration step.

The flow chart for the solution of this example is shown in Fig. 7-4. The variable names associated with the problem quantities in the computer program are as follows:

FORTRAN Name	**Quantity**	**Problem Value**
LOW	An initial temperature gradient value which is quite certain to be smaller than the correct temperature gradient T' (first value of LOW shown)	−0.010°R/ft
HIGH	An initial temperature gradient value which is quite certain to be larger than the correct temperature gradient T' (first value of HIGH shown)	−40,000°R/ft
EMISS	Emissivity of fin	0.8
C	Thermal conductivity of fin	25Btu/(hr)(ft)(°R)
L	Length of fin	0.25 ft
H	Thickness of fin	0.005 ft
DELX	Increment of distance along fin	0.005 ft

TI	Known temperature of fin at $x = 0$	2000°R
EPSI	Accuracy-check value for final T'	1°R/ft
TPSAVE	Each initial value of T' used is stored under the name TPSAVE, and this value is later assigned to either HIGH or LOW	
XPR	x value at which a printout occurs	ft
DXPR	Increments of x used for printout	0.01 ft
X	Variable distance along fin	ft
SIG	Stefan-Boltzmann constant	$0.173(10)^{-8}$ Btu/(ft)^2(hr)(°R)^4
T	Temperature at points along fin length	°R
T1	Temperature one station in advance of that at which temperature is T	°R
TP	Temperature gradient at points along fin length	°R/ft
TP1	Temperature gradient one station in advance of that at which gradient is TP	°R/ft
TDP	Second derivative of temperature with respect to x, (T'')	°R/ft^2
TDP1	Second derivative of temperature, with respect to x, one station in advance of that at which TDP applies	°R/ft^2

The FORTRAN IV source program used to implement the flow chart is as follows:

```
C TEMPERATURE DISTRIBUTION ALONG A RADIATING FIN
      REAL LOW,L
      READ(5,1)LOW,HIGH,EMISS,C,L,H,DELX,TI,EPSI,DXPR
    1 FORMAT(10F7.0)
      WRITE(6,2)
    2 FORMAT(1H1,' LOW ',3X,'HIGH  EMISS',4X,'C',4X,'L',5X,
                                                     'H',4X,
     *'DELX',4X,'TI',4X,'EPSI',3X,'DXPR'/)
      WRITE(6,3)LOW,HIGH,EMISS,C,L,H,DELX,TI,EPSI,DXPR
    3 FORMAT(1H ,F5.3,F9.0,F4.1,F6.1,F5.2,F6.3,F7.3,F8.0,
                                                 F6.1,F7.3)
      XPR=DXPR
      SIG=.173E-8
    4 T=TI
      TP=0.5*(LOW+HIGH)
      TPSAVE=TP
      X=0.
    5 TDP=2.*SIG*EMISS*T**4/(C*H)
      TP1=TP+TDP*DELX
      T1=T+(TP+TP1)/2.*DELX
      TDP1=2.*SIG*EMISS*T1**4/(C*H)
      TP1=TP+(TDP+TDP1)/2.*DELX
      T1=T+(TP+TP1)/2.*DELX
      IF(T1)6,6,7
    6 HIGH=TPSAVE
      GO TO 4
```

```
  7 IF(T1-TI)9,9,8
  8 LOW=TPSAVE
    GO TO 4
  9 X=X+DELX
    IF(ABS(X-L)-.001)11,11,10
 10 TP=TP1
    T=T1
    GO TO 5
 11 IF(ABS(TP1)-EPSI)15,15,12
 12 IF(TP1)14,15,13
 13 LOW=TPSAVE
    GO TO 4
 14 HIGH=TPSAVE
    GO TO 4
 15 WRITE(6,16)
 16 FORMAT(1H0,15HDISPL ALONG FIN,6X,11HTEMPERATURE,7X,
   *13HTEMP GRADIENT/)
    T=TI
    TP=0.5*(LOW+HIGH)
    X=0.
    WRITE(6,19)X,T,TP
 17 TDP=2.*SIG*EMISS*T**4/(C*H)
    TP1=TP+TDP*DELX
    T1=T+(TP+TP1)/2.*DELX
    TDP1=2.*SIG*EMISS*T1**4/(C*H)
    TP1=TP+(TDP+TDP1)/2.*DELX
    T1=T+(TP+TP1)/2.*DELX
    X=X+DELX
    IF(ABS(X-XPR)-.001)18,18,20
 18 WRITE(6,19)X,T1,TP1
 19 FORMAT(1H ,4X,F6.3,13X,F7.0,13X,F9.0)
    XPR=XPR+DXPR
 20 IF(ABS(X-L)-.001)21,21,22
 21 STOP
 22 TP=T
    T=T1
    GO TO 17
    END
```

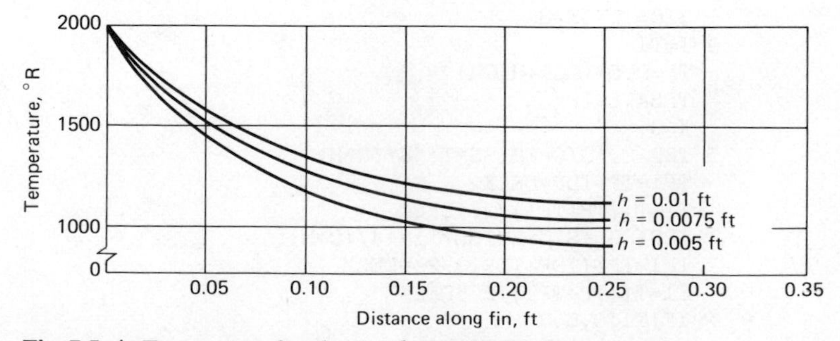

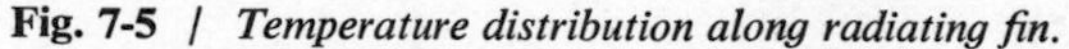

Fig. 7-5 / *Temperature distribution along radiating fin.*

A graphical plot of the results given by the computer printout (to follow) is shown in Fig. 7-5, along with similar curves for two other fin thicknesses for which the tabular data are not shown.

```
LOW    HIGH     EMISS C     L     H      DELX  TI     EPSI DXPR
-.010 -40000. 0.8   25.0 0.25 0.005 0.005 2000. 1.0  0.010
DISPL ALONG FIN          TEMPERATURE         TEMP GRADIENT
0.0                      2000.               -16639.
0.010                    1849.               -13598.
0.020                    1725.               -11336.
0.030                    1621.                -9601.
0.040                    1532.                -8234.
0.050                    1455.                -7132.
0.060                    1388.                -6227.
0.070                    1330.                -5472.
0.080                    1278.                -4831.
0.090                    1233.                -4280.
0.100                    1193.                -3801.
0.110                    1157.                -3380.
0.120                    1125.                -3005.
0.130                    1097.                -2668.
0.140                    1071.                -2362.
0.150                    1049.                -2082.
0.160                    1030.                -1824.
0.170                    1013.                -1583.
0.180                     998.                -1357.
0.190                     985.                -1143.
0.200                     975.                 -939.
0.210                     967.                 -742.
0.220                     960.                 -552.
0.230                     956.                 -365.
0.240                     953.                 -182.
0.250                     952.                    0.
```

7-3 / SIMULTANEOUS-EQUATION METHOD

We will now examine a method of solving ordinary differential equations of the boundary-value type which involves the solution of simultaneous algebraic equations. In this process, the differential equation to be solved is first put into finite-difference form. If the interval over which the integration is desired is then divided into equal increments, the finite-difference equation used to approximate the given differential equation must be satisfied at *each* of the stations dividing the interval. The finite-difference equation has the form of an algebraic equation at each station, and if n stations are used we obtain a set of $n - 2$ algebraic equations which can be solved simultaneously to obtain a desired solution. Such a set of equations can be solved by any of the applicable methods discussed in Chap. 3.

This approach to the solution of boundary-value problems is usually reserved for linear differential equations, since the resulting algebraic equations are also linear and are more easily solved than those resulting from nonlinear differential equations. The procedure involved is illustrated by the following example.

EXAMPLE 7-2

Let us determine the elastic curve of the nonuniform shaft described in Fig. 7-6(a).[1] From mechanics of materials, we know that the equation defining the elastic curve (the deflection of the beam at any point) can be obtained by two successive integrations beginning with the integration of the differential equation

$$\frac{d^2y}{dx^2} = \frac{M}{EI} \tag{7-15}$$

where

y = beam deflection at any point, in.
x = variable distance along the beam, in.
M = internal bending moment at any point along the beam, lb-in.
E = modulus of elasticity, lb/in.^{2}
I = area moment of inertia of beam cross section, in.^{4}

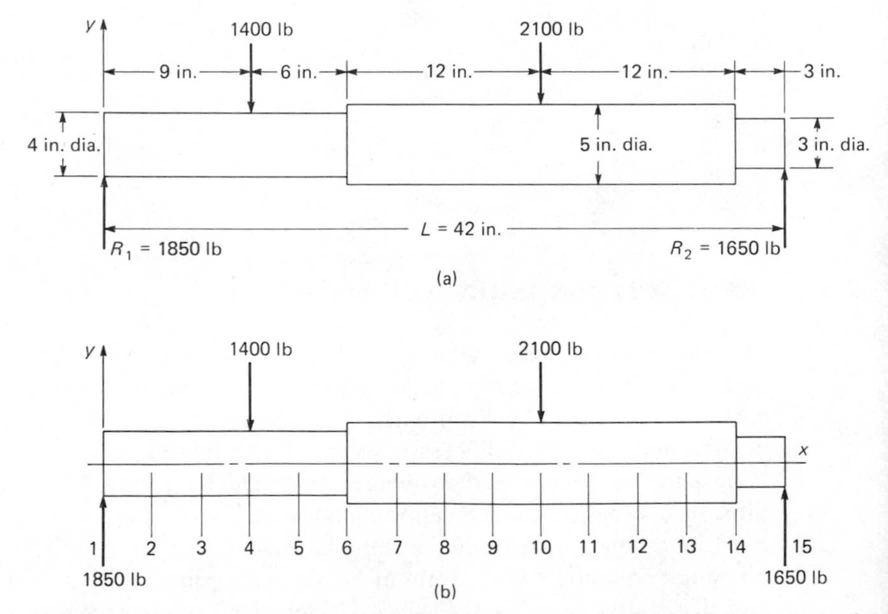

Fig. 7-6 / *Simply supported beam of Ex. 7-2. (a) Beam loading and dimensions of shaft. (b) Stations along beam.*

[1] Merl D. Creech, Robert L. Ammons, and James I. McPherson, "Deflections of Complex Beams," *Prod. Eng.*, Aug. 14, 1961, pp. 47–49.

In general, the expression M/EI is a function of the variable distance x along the beam. When this relationship can be easily expressed by a single continuous expression over the length of the beam (the desired range of integration), the *double-integration* method, mentioned above, provides a convenient means of analytically determining the elastic curve of the beam. However, when M/EI cannot be represented by a single elementary function over the desired range of x, an analytical solution becomes somewhat involved. For example, the determination of an analytical solution yielding the elastic curve of the shaft in Fig. 7-6(a), by the double-integration method, would require the solution of five differential equations, since five different expressions would be required to define M/EI over the length of the shaft, because of the abrupt changes in cross section and the loading shown. The double integration required would thus involve ten equations and ten attendant constants of integration. Although a solution could be obtained, the process would be rather tedious. The *moment-area* method or *Castigliano's* method could also be used but would be equally tedious for solving a beam or shaft problem of this type.

The reduction of Eq. 7-15 to finite-difference form, with the resulting set of simultaneous linear algebraic equations, provides a convenient method of programming the problem to the digital computer. Using the central-difference expression with error of order h^2 for d^2y/dx^2 (see Eq. 5-79), Eq. 7-15 becomes

$$\frac{y_{i+1} - 2y_i + y_{i-1}}{(\Delta x)^2} = \frac{M_i}{EI_i}$$

or

$$y_{i+1} - 2y_i + y_{i-1} = \frac{M_i(\Delta x)^2}{EI_i} \tag{7-16}$$

where the subscripts on M and I remind us that each is a function of x. Applying Eq. 7-16 at stations 2 through 14, as indicated by Fig. 7-6(b), we obtain the following set of algebraic equations:

$$\begin{aligned}
y_1 - 2y_2 + y_3 + (0)y_4 + \cdots \quad \cdots + (0)y_{15} &= \frac{M_2}{EI_2}(\Delta x)^2 \\
(0)y_1 + y_2 - 2y_3 + y_4 + (0)y_5 + \cdots \quad \cdots + (0)y_{15} &= \frac{M_3}{EI_3}(\Delta x)^2 \\
(0)y_1 + (0)y_2 + y_3 - 2y_4 + y_5 + (0)y_6 + \cdots + (0)y_{15} &= \frac{M_4}{EI_4}(\Delta x)^2 \\
\vdots \qquad\qquad & \qquad \vdots \\
(0)y_1 + (0)y_2 + \cdots \quad \cdots + (0)y_{12} + y_{13} - 2y_{14} + y_{15} &= \frac{M_{14}}{EI_{14}}(\Delta x)^2
\end{aligned} \tag{7-17}$$

This set of 13 linear algebraic equations contains the 15 deflection terms y_1 through y_{15}. However, since $y_1 = y_{15} = 0$ as known boundary values, Eq. 7-17 constitutes a set of 13 independent equations containing 13 unknowns, the simultaneous solution of which will yield unique solutions for the desired deflections.

The value of M and I at each station must be available as data for solving Eq. 7-17. The computation of these values can either be incorporated in the computer program or be determined manually beforehand and introduced into the computer program as data input, as is done in this example. The following values were determined, using an E value of $30(10)^6$ lb/in.^{2} for the modulus of elasticity of the steel. It should be noted that the I values, at the stations where the cross section changes, are obtained by averaging the I values of the abutting sections.

Station	M_i *(lb-in.)*	I_i *(in.^{4})*	$\frac{M_i}{EI_i}(\Delta x)^2$ *(in.)*
1	0	12.57	0.0
2	5,550	12.57	0.0001325
3	11,100	12.57	0.0002649
4	16,650	12.57	0.0003974
5	18,000	12.57	0.0004296
6	19,350	21.63	0.0002684
7	20,700	30.68	0.0002024
8	22,050	30.68	0.0002156
9	23,400	30.68	0.0002288
10	24,750	30.68	0.0002420
11	19,800	30.68	0.0001936
12	14,850	30.68	0.0001452
13	9,900	30.68	0.0000968
14	4,950	17.33	0.0000857
15	0	3.98	0.0

The computer program used to obtain the data shown is the Gauss-Jordan method flow-charted in Fig. 3-3.

The computer printout of the solution is as follows:

```
SOLUTION OF SIMULTANEOUS EQUATIONS BY GAUSS-JORDAN
         METHOD WITH PARTIAL PIVOTING
         X( 1)=        -0.1656066E-02
         X( 2)=        -0.3179636E-02
         X( 3)=        -0.4438303E-02
         X( 4)=        -0.5299572E-02
         X( 5)=        -0.5731229E-02
         X( 6)=        -0.5894497E-02
         X( 7)=        -0.5855359E-02
         X( 8)=        -0.5600613E-02
         X( 9)=        -0.5117066E-02
         X(10)=        -0.4391517E-02
         X(11)=        -0.3472364E-02
         X(12)=        -0.2408010E-02
         X(13)=        -0.1246855E-02
```

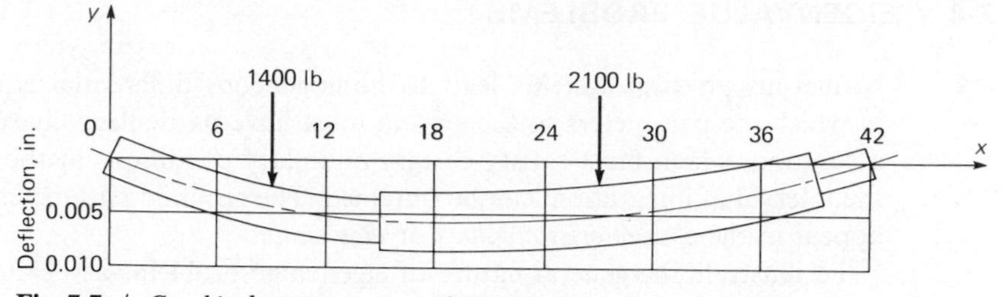

Fig. 7-7 / *Graphical representation of the deflection of the steel shaft of Ex. 7-2.*

A graphical representation of the same results is shown in Fig. 7-7.

The displacement shown at station 10, $y_{10} = 0.005117$ in., compares with a value of 0.005196 in. for the same station obtained by the use of Castigliano's method.[2] It can be seen that the accuracy obtained with the finite-difference method, using just 15 stations, is sufficient for most practical problems. Increasing the number of stations will increase the accuracy of the solution, but the use of an excessive number of stations with the elimination method can lead to a loss in accuracy owing to the large number of calculations introduced in solving the many more simultaneous equations. The authors have solved problems of this type, using a large number of stations, by the Gauss-Seidel iteration method discussed in Sec. 3-7. However, convergence is generally relatively slow, and considerable computing time is thus involved. For many practical boundary-value problems the number of stations can be kept small and an elimination method employed, as was done in this example, to obtain solutions quickly with sufficient accuracy.

The procedure used to solve Eq. 7-15 in this example is said to be of the *third order*. This means that such a method is capable of yielding exact results if $y(x)$ is a polynomial of the third degree or less. For the shaft used in this example, an analytical solution will show that the displacements can be expressed by five different third-degree polynomials, each applicable to a particular segment of the shaft. However, exact results are not obtained by a third-order numerical method unless $y(x)$ can be expressed over the entire interval of integration with a single polynomial of the third degree or less.

For problems in which $y(x)$ is known to be a polynomial of degree higher than 3, more accurate results can be expected if a numerical procedure of higher order than 3 is used. Methods, known as *fifth-* and *seventh-order* methods, have been developed for use with problems of this type.[3] However, in most engineering problems the degree of accuracy required in the solutions does not warrant the use of higher order procedures, and the third-order method, illustrated in this example, can be used to obtain the desired solutions.

[2] *Ibid.*

[3] F. B. Hildebrand, *Introduction to Numerical Analysis* (New York: McGraw-Hill Book Company, 1974).

7-4 / EIGENVALUE PROBLEMS

Numerous physical systems lead to homogeneous differential equations in which the parameters of the system must have particular values before a solution, which must satisfy certain boundary conditions at the end of the interval of integration, can be obtained. The required parameter values appear in the *characteristic values* or *eigenvalues.*

To illustrate the general nature of eigenvalue problems associated with differential equations, let us consider a slender column of length l subjected to an axial load P, as shown in Fig. 7-8. From beam theory the basic relationship between the curvature d^2y/dx^2 and the internal moment M for the axes shown is

$$\frac{d^2y}{dx^2} = \frac{M}{EI} \tag{7-18}$$

where

y = deflection at any point, in.
x = variable distance along column, in.
E = modulus of elasticity, lb/in.^{2}
I = area moment of inertia of column cross section, in.^{4}

Since y is negative for the configuration shown, and, by convention, the bending moment at any section is positive for this configuration, the bending moment M must be equal to $-Py$. With $M = -Py$, Eq. 7-18 becomes

$$\frac{d^2y}{dx^2} + \frac{P}{EI}y = 0 \tag{7-19}$$

or

$$\frac{d^2y}{dx^2} + \lambda y = 0 \tag{7-19a}$$

where $\lambda = P/EI$.

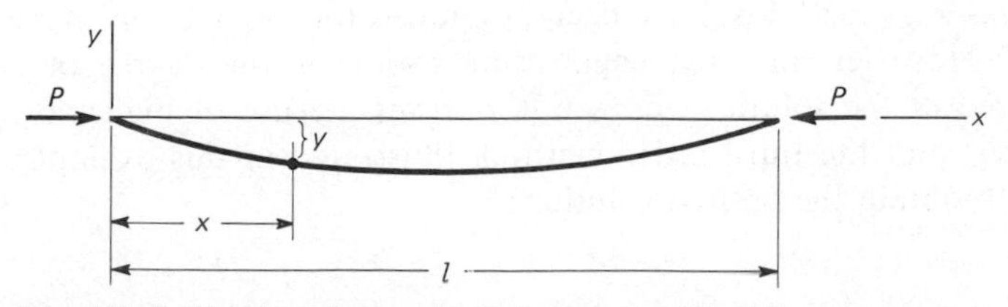

Fig. 7-8 / *Slender column subjected to an axial load P.*

The solution of Eq. 7-19a is

$$y = A \cos \sqrt{\lambda} x + B \sin \sqrt{\lambda} x \tag{7-20}$$

where A and B are constants. Equation 7-20 is easily verified as being a solution by substituting it and its second derivative into Eq. 7-19a.

The *known* boundary conditions for the column are

$$y \,|_{x=0} = y \,|_{x=l} = 0 \tag{7-21}$$

Since the slope dy/dx is not known at $x = 0$, only one *initial* condition is known, and therefore one constant, A or B, will be indeterminate. The parameter $\lambda = P/EI$ must have certain values to satisfy the condition that $y = 0$ at $x = l$. The required values of λ necessary to satisfy this condition are eigenvalues. To determine these values of λ, let us now consider the boundary conditions given in Eq. 7-21. Using $y = 0$ at $x = 0$, we find, from Eq. 7-20, that

$$A = 0 \tag{7-22}$$

Similarly, using $y = 0$ at $x = l$, we find that

$$B \sin \sqrt{\lambda} l = 0 \tag{7-23}$$

Since B cannot equal zero for a nontrivial solution, Eq. 7-23 can be satisfied only if

$$l\sqrt{\lambda} = n\pi \qquad n = 1, 2, 3, \ldots$$

Thus the eigenvalues are

$$\lambda = \frac{n^2\pi^2}{l^2} = \frac{P}{EI} \qquad n = 1, 2, 3, \ldots \tag{7-24}$$

The configurations of the column for the first three eigenvalues, determined by substituting the eigenvalues into Eq. 7-20 with $A = 0$, are shown in Fig. 7-9.

The values of P which satisfy Eq. 7-24, for a given value of EI, are known as the *buckling* or *critical* loads. From a practical point of view, the smallest value of P is of primary importance, since the configurations associated with the larger values of P cannot generally be obtained without failure first occurring under the action of the lowest value of P. The

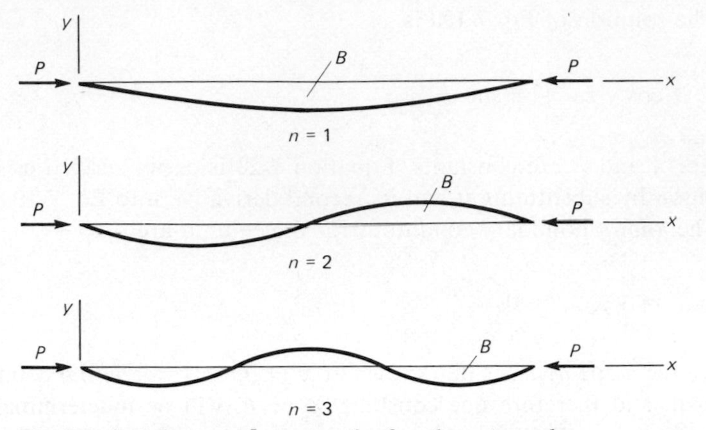

Fig. 7-9 / *Column configurations for first three eigenvalues.*

eigenvalue associated with the smallest buckling load is obtained with $n = 1$, so the smallest buckling load is

$$P_{\text{cr}} = \frac{\pi^2 EI}{l^2} \tag{7-25}$$

If EI is not a constant but is some function of x, then the solution of Eq. 7-19, given by Eq. 7-20, is not applicable. In most cases in which such a differential equation contains variable coefficients, analytical solutions cannot be obtained, and it becomes necessary to resort to some numerical procedure for finding the eigenvalues.

The remainder of this chapter will be devoted to the discussion of a finite-difference method for obtaining the solution of homogeneous differential equations such as that given by Eq. 7-19. Such a differential equation can be readily expressed in general finite-difference form as a homogeneous algebraic equation. The application of this general equation for a set of values of the independent variable (at uniformly spaced stations along the column) yields a *set* of homogeneous algebraic equations. The eigenvalues of the set are approximations of the eigenvalues of the differential equation.

Two methods (previously discussed in Chap. 3 for determining eigenvalues of sets of homogeneous algebraic equations), the polynomial method and the iteration method, can be applied in the solution of the algebraic equations resulting from the finite-difference form of the differential equations.

General Numerical Procedure

To develop the general numerical procedure, let us first consider Eq. 7-19a with $\lambda = P/EI$ a constant. Using a *central-difference* expression (see

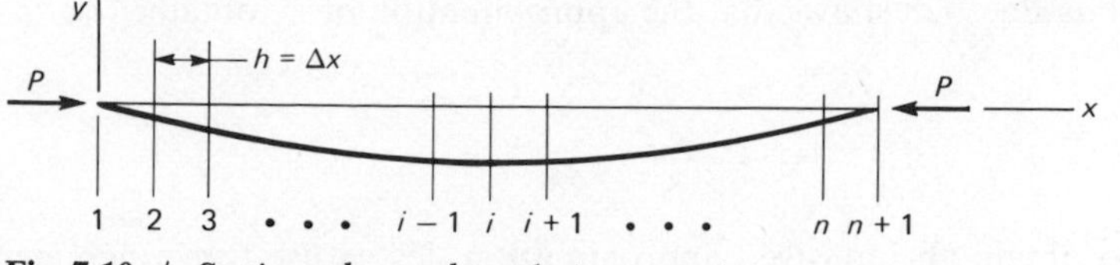

Fig. 7-10 / *Stations along column (n increments).*

Eq. 5-73) for the second derivative, we may express Eq. 7-19a, in finite difference form, as

$$\frac{y_{i+1} - 2y_i + y_{i-1}}{h^2} + \lambda y_i = 0$$

or as

$$y_{i-1} - (2 - h^2\lambda)y_i + y_{i+1} = 0 \tag{7-26}$$

where $h = \Delta x$ is the distance between successive stations along the column (see Fig. 7-10). Applying Eq. 7-26 at stations 2 through n along the column, we can write the system of homogeneous algebraic equations

$$\begin{aligned}
\overset{0}{\cancel{y_1}} - (2 - h^2\lambda)y_2 + y_3 + (0)y_4 + \cdots\cdots + (0)y_n &= 0 \\
y_2 - (2 - h^2\lambda)y_3 + y_4 + (0)y_5 + \cdots + (0)y_n &= 0 \\
y_3 - (2 - h^2\lambda)y_4 + y_5 + (0)y_6 + \cdots + (0)y_n &= 0 \\
\cdots\cdots\cdots\cdots\cdots\cdots\cdots\cdots\cdots\cdots \\
y_{n-1} - (2 - h^2\lambda)y_n + \overset{0}{\cancel{y_{n+1}}} &= 0
\end{aligned} \tag{7-27}$$

With n increments along the column, Eq. 7-27 yields $n - 1$ equations, the solution of which by the polynomial method involves the expansion of an $n - 1$ by $n - 1$ determinant. From Sec. 3-8 we know that the expansion of the determinant yields an $n - 1$ degree polynomial in λ, the roots of which are eigenvalues. Since we used only a finite number of stations along the column, the $n - 1$ eigenvalues obtained will be only approximations of the true eigenvalues of the column. Fortunately, the lowest root or eigenvalue, which, in this problem, is the most important, is the most accurately approximated root value obtained. Improved approximations of all the eigenvalues determined can be obtained by increasing the number of increments. To illustrate, let us first take $h = l/2$ and obtain a first approximation of λ_1. The exact value of λ_1 was found to be $9.87/l^2$ (see Eq. 7-24). With $h = l/2$ (only three stations), and using the boundary conditions $y_1 = y_3 = 0$, we can write, from Eq. 7-27,

$$0 - \left[2 - \left(\frac{l}{2}\right)^2 \lambda\right] y_2 + 0 = 0 \tag{7-28}$$

Thus Eq. 7-28 shows that the approximation of λ_1 obtained for $h = l/2$ is

$$\lambda_1 = \frac{8}{l^2}$$

To obtain an improved approximation, let us use three increments (four stations) so that $h = l/3$. From Eq. 7-27

$$\left.\begin{aligned} \left[2 - \left(\frac{l}{3}\right)^2 \lambda\right] y_2 - y_3 &= 0 \\ -y_2 + \left[2 - \left(\frac{l}{3}\right)^2 \lambda\right] y_3 &= 0 \end{aligned}\right\} \tag{7-29}$$

Setting the determinant $|\mathbf{D}|$ of the coefficients of the y's equal to zero gives

$$|\mathbf{D}| = \begin{vmatrix} \left[2 - \left(\frac{l}{3}\right)^2 \lambda\right] & -1 \\ -1 & \left[2 - \left(\frac{l}{3}\right)^2 \lambda\right] \end{vmatrix} = 0$$

Expanding the determinant and simplifying yields the quadratic

$$\lambda^2 - \frac{36}{l^2}\lambda + \frac{243}{l^4} = 0$$

which has the roots

$$\lambda_1 = \frac{9}{l^2} \qquad \text{and} \qquad \lambda_2 = \frac{27}{l^2}$$

Thus the approximation $\lambda_1 = 9/l^2$ is considerably closer to the exact value of $9.87/l^2$ than the approximation $8/l^2$ which was obtained by using only two increments. The value $\lambda_2 = 27/l^2$ is an approximation of the second eigenvalue which has the exact value $4(9.87)/l^2$ (see Eq. 7-24). It should now be apparent that the use of four increments ($h = l/4$) would yield a still better approximation of λ_1 and λ_2 as well as a rough approximation of λ_3.

In general, a very good approximation of the smallest eigenvalue can be obtained by using a reasonable number of increments—say, ten, for example. However, fairly accurate approximations of the higher eigenvalues are obtained only by using a large number of increments. Therefore, unless higher eigenvalues are desired, the required number of increments depends upon the desired accuracy of the approximation of λ_1 (the smallest eigenvalue).

One procedure for qualitatively evaluating the convergence of the approximation of λ_1 to the true value is to compute λ_1 by using successively larger numbers of increments. As the value of λ_1 converges to the true value, an increase in the number of increments will cause very little change in successively computed values of λ_1.

Polynomial Method

Let us consider a column having a variable cross section. If the cross section of the column is not uniform, then the coefficient P/EI in the differential equation (see Eq. 7-19) is a variable, since the moment of inertia of the cross section varies along the column.

Denoting the value of I at station i as I_i, Eq. 7-26 becomes

$$y_{i-1} - \left(2 - h^2 \frac{P}{EI_i}\right) y_i + y_{i+1} = 0 \tag{7-30}$$

Since the eigenvalues must appear as constants in each of the simultaneous equations resulting from Eq. 7-30, they must be defined in terms of the moment of inertia at some particular station. For example, if we relate the eigenvalues to the moment of inertia at station 1, we write

$$\frac{P}{EI_i} = \left(\frac{I_1}{I_i}\right) \frac{P}{EI_1} = \alpha_i \lambda \tag{7-31}$$

where the variable coefficient $\alpha_i = I_1/I_i$ and $\lambda = P/EI_1$.

Using Eq. 7-30 and the notation of Eq. 7-31, we can write the system of the simultaneous equations as

$$\begin{aligned}
\cancelto{0}{y_1} - (2 - h^2\alpha_2\lambda)y_2 + y_3 + (0)y_4 + \cdots + (0)y_n &= 0 \\
y_2 - (2 - h^2\alpha_3\lambda)y_3 + y_4 + (0)y_5 + \cdots + (0)y_n &= 0 \\
y_3 - (2 - h^2\alpha_4\lambda)y_4 + y_5 + (0)y_6 + \cdots + (0)y_n &= 0 \\
\cdots\cdots\cdots\cdots\cdots\cdots\cdots\cdots\cdots\cdots\cdots\cdots & \\
y_{n-1} - (2 - h^2\alpha_n\lambda)y_n + \cancelto{0}{y_{n+1}} &= 0
\end{aligned} \tag{7-32}$$

Dividing these equations through by the negatives of $h^2\alpha_2, h^2\alpha_3, \ldots, h^2\alpha_n$, respectively, we can express the determinant of the coefficient matrix in the form $|\mathbf{A} - \lambda\mathbf{I}|$ which is suitable for obtaining the characteristic polynomial by the Faddeev-Leverrier method discussed in Chap. 3. The resulting determinant is

$$|\mathbf{D}| = \begin{vmatrix} \left(\frac{2}{h^2\alpha_2} - \lambda\right) & -\frac{1}{h^2\alpha_2} & 0 & 0 & \cdots & 0 \\ -\frac{1}{h^2\alpha_3} & \left(\frac{2}{h^2\alpha_3} - \lambda\right) & -\frac{1}{h^2\alpha_3} & 0 & \cdots & 0 \\ 0 & -\frac{1}{h^2\alpha_4} & \left(\frac{2}{h^2\alpha_4} - \lambda\right) & -\frac{1}{h^2\alpha_4} & \cdots & 0 \\ \cdots & \cdots & \cdots & \cdots & \cdots & \cdots \\ 0 & 0 & 0 & \cdots & -\frac{1}{h^2\alpha_n} & \left(\frac{2}{h^2\alpha_n} - \lambda\right) \end{vmatrix} = 0 \tag{7-33}$$

With n increments along the column, the expansion of this $n - 1$ by $n - 1$ determinant yields an $n - 1$ degree polynomial in λ. The *smallest* root λ_1 of the polynomial will be a close approximation of the smallest value of P/EI_1. The buckling load may then be determined from

$$\lambda_1 = \frac{P}{EI_1} \tag{7-34}$$

In considering columns of variable cross section with uniform thickness (see Fig. 7-11), the equation

$$I = I_1\left(\frac{x'}{a}\right)^m \qquad x' \leq a + \frac{l}{2} \tag{7-35}$$

might be used to define the moment of inertia of the column about a centroidal axis perpendicular to the xy plane at any cross section of the column, where

x = distance along column (independent variable)

x' = $a + x$ (as shown in Fig. 7-11)

I = moment of inertia about centroidal axis perpendicular to xy plane (plane of bending) anywhere along column

I_1 = moment of inertia about centroidal axis perpendicular to plane of bending at $x' = a$

By using different values for the parameter a and the exponent m, Eq. 7-35 may be used for expressing the moment of inertia at any section of many

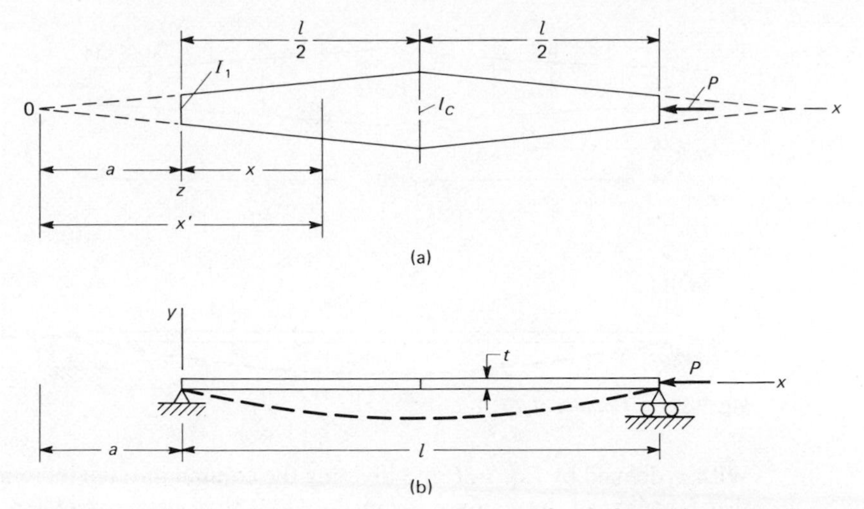

Fig. 7-11 / *Coordinate system for simply supported column of variable cross section.*

differently shaped columns of uniform thickness. (For the column shown in Fig. 7-11, $m = 1$.)[4]

EXAMPLE 7-3

Let us consider the column shown in Fig. 7-12. This column has a uniform thickness and a cross section which varies as shown. The shape of this parabolic column is defined by choosing

$a = 206.19$ in.

$m = 2$

$I_c = 1.0$ in.^{4}

Since the column of Fig. 7-12 is symmetrical about $x = l/2$, the values of I, computed from Eq. 7-35 for the stations in the region of $0 < x < l/2$, may also be used for corresponding stations in the region of $x > l/2$.

Referring to Eqs. 7-31 and 7-35, we see that

$$\alpha_i = \frac{I_1}{I_i} = \left(\frac{a}{x'_i}\right)^m$$

or

$$\alpha_i = \left(\frac{a}{a + x_i}\right)^m \tag{7-36}$$

[4] S. P. Timoshenko and J. M. Gere, *Theory of Elastic Stability* (New York: McGraw-Hill Book Company, 1961).

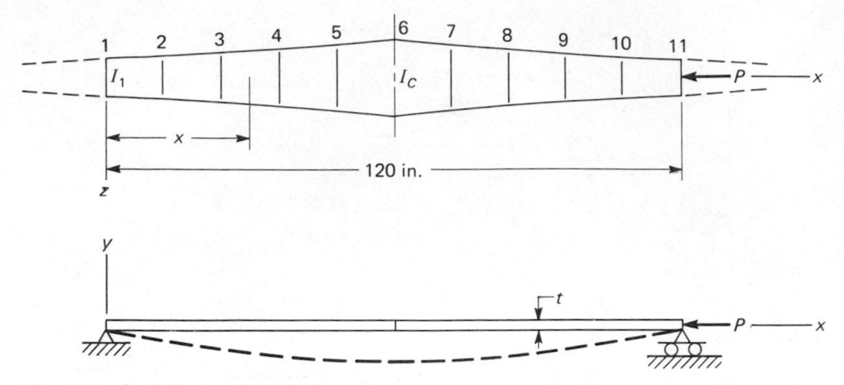

Fig. 7-12 / *Column of Ex. 7-3.*

With α_i defined by Eq. 7-36 and dividing the column into ten increments so that h is defined, we can see (from Eq. 7-33) that we are ready to write a program for calculating the elements of the matrix **A** in the determinant of Eq. 7-33, $|\mathbf{A} - \lambda\mathbf{I}| = 0$. We will need to do no further programming, since these elements will then be used as data for the program developed in Chap. 3 for generating the characteristic polynomial coefficients by the Faddeev-Leverrier method. The roots of the characteristic polynomial will then be determined by Bairstow's method, using the computer program developed in Chap. 2.

The procedure for generating the elements of the matrix for which we wish to determine the eigenvalues is outlined in the flow chart of Fig. 7-13. Note that an even number of column increments should always be used, since there is a whole number of increments in each half of the column. The elements of the matrix are calculated and then printed out by columns for visual inspection, and they are also used as data for the program concerned with generating the characteristic polynomial coefficients.

The FORTRAN names used in the program, the quantities they represent, and the data values used in this example are as follows:

FORTRAN Name	Quantity	Data Values Used in Problem
N	Number of increments into which column is divided for finite-difference representation of the differential equation $$\frac{d^2y}{dx^2} = \frac{M}{EI}$$	10
M	Exponent m of Eq. 7-36 (the exponent describes the column shape)	2
CL	Length of column, in.	120

FORTRAN Name	Quantity	Data Values Used in Problem
A1	Distance from point of zero column width (if column were extended to $x' = 0$) to actual end of column, in.	206.19
IC	I_c, the moment of inertia of the cross section of midpoint of column, in.^{4}	1.0
I1	I_1, the moment of inertia of the cross section at end of column, in.^{4}	
RATIO	I_1/I_c	
H	h ($h = \Delta x$)	
X	Variable distance x along the axis of column measured from column end	
NO2P1	$n/2 + 1$	
ALPHA(I)	$\alpha_i = [a/(a + x_i)]^m$	
NO2	$n/2$	
K	Quantity $(n - i + 2)$, where $i = 2, 3, \ldots, n/2$	
L	Order of the **A** matrix, equal to $(n - 1)$	
A(I,J)	Elements of the matrix for which we want the eigenvalues λ	

The FORTRAN source program, written to implement the flow chart shown in Fig. 7-13, is as follows:

```
C TAPERED COLUMN PROBLEM - GENERATION OF MATRIX A
      DIMENSION ALPHA(16),A(15,15)
      REAL IC,I1
      READ(5,2)N,M
    2 FORMAT(2I2)
      READ(5,3)CL,A1,IC
    3 FORMAT(3F10.0)
      I1=IC*(A1/(CL/2.+A1))**M
      RATIO=I1/IC
      H=CL/N
      X=0.
      NO2P1=N/2+1
      NO2=N/2
      L=N-1
      DO4I=2,NO2P1
      X=X+H
    4 ALPHA(I)=(A1/(A1+X))**M
      DO5I=2,NO2
      K=N-I+2
    5 ALPHA(K)=ALPHA(I)
      DO12J=1,L
      DO12I=1,L
      IF(I-J)6,9,6
```

Continued on page 507

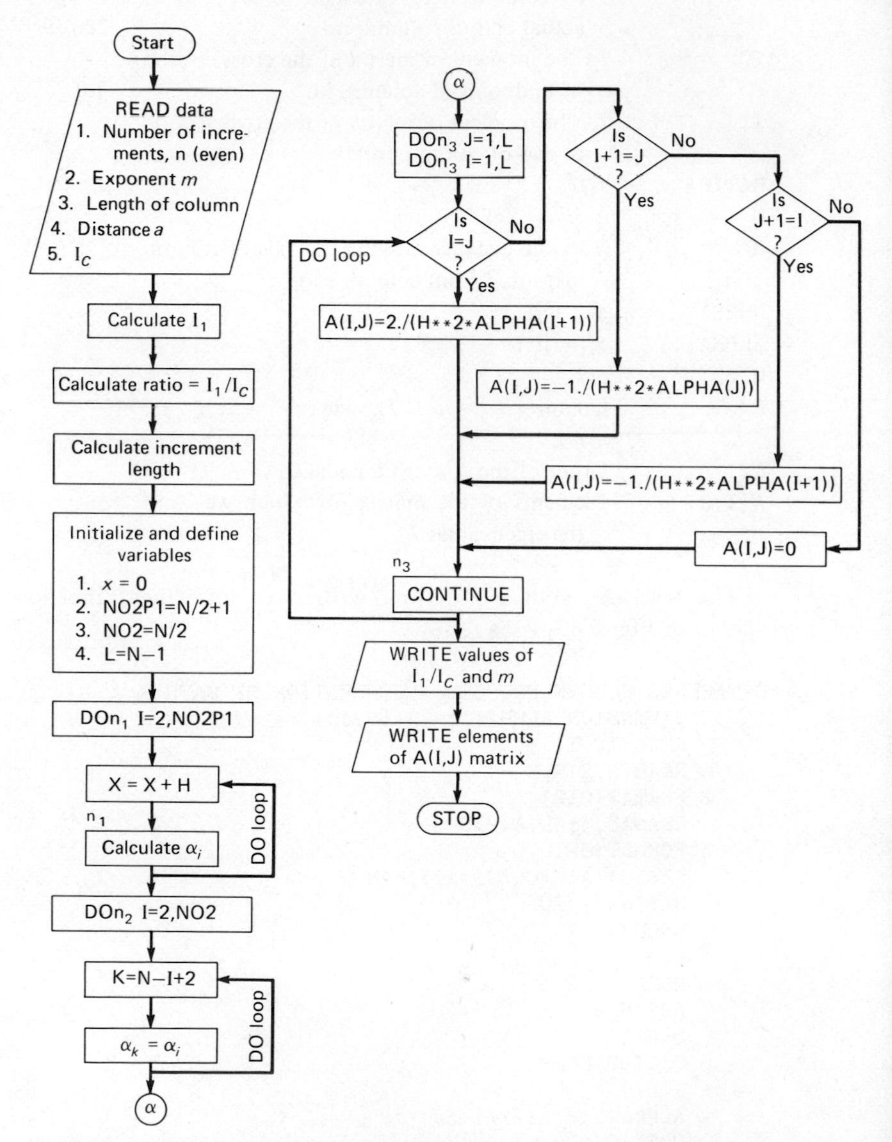

Fig. 7-13 / *Flow chart for the program to generate the matrix* **A** *of the characteristic polynomial* $|\mathbf{A} - \lambda\mathbf{I}| = 0$ *for the tapered-column problem.*

```
  6 IF(I+1-J)7,10,7
  7 IF(J+1-I)8,11,8
  8 A(I,J)=0.
    GO TO 12
  9 A(I,J)=2./(H**2*ALPHA(I+1))
    GO TO 12
 10 A(I,J)=-1./(H**2*ALPHA(J))
    GO TO 12
 11 A(I,J)=-1./(H**2*ALPHA(I+1))
 12 CONTINUE
    WRITE(6,13)RATIO,M
 13 FORMAT(1H1,'I1/IC=',F6.4,6X,'M= ',I1//)
    WRITE(6,14)
 14 FORMAT(1H ,'FIRST FIVE COLUMNS OF GENERATED MATRIX A'/)
    WRITE(6,15)((A(I,J),J=1,5),I=1,9)
 15 FORMAT(1H ,5E15.7)
    WRITE(6,16)
 16 FORMAT(1H0,/,' LAST FOUR COLUMNS OF GENERATED MATRIX A'/)
    WRITE(6,18)((A(I,J),J=6,9),I=1,9)
 18 FORMAT(1H ,4E15.7)
    WRITE(7,17)((A(I,J),J=1,9),I=1,9)
 17 FORMAT(5E14.7)
    STOP
    END
```

The following computer printout gives the elements of the matrix **A** for which we must determine the eigenvalues:

```
I1/IC=0.6000          M= 2
FIRST FIVE COLUMNS OF GENERATED MATRIX A
 0.1555257E-01 -0.7776283E-02  0.0            0.0
-0.8655164E-02  0.1731033E-01 -0.8655164E-02  0.0
 0.0           -0.9581085E-02  0.1916217E-01 -0.9581085E-02
 0.0            0.0           -0.1055405E-01  0.2110811E-01
 0.0            0.0            0.0           -0.1157406E-01
 0.0            0.0            0.0            0.0
 0.0            0.0            0.0            0.0
 0.0            0.0            0.0            0.0
 0.0            0.0            0.0            0.0

 0.0
 0.0
 0.0
-0.1055405E-01
 0.2314812E-01
-0.1055405E-01
 0.0
 0.0
 0.0
```

```
LAST FOUR COLUMNS OF GENERATED MATRIX A
 0.0              0.0              0.0              0.0
 0.0              0.0              0.0              0.0
 0.0              0.0              0.0              0.0
 0.0              0.0              0.0              0.0
-0.1157406E-01    0.0              0.0              0.0
 0.2110811E-01   -0.1055405E-01    0.0              0.0
-0.9581085E-02    0.1916217E-01   -0.9581085E-02    0.0
 0.0             -0.8655164E-02    0.1731033E-01   -0.8655164E-02
 0.0              0.0             -0.7776283E-02    0.1555257E-01
```

Using the coefficient matrix (which is punched on cards during execution of the program on pp. 505 and 507) as data for the program to determine the characteristic polynomial coefficients by the Faddeev-Leverrier method (see Sec. 3-10), we obtain the following computer output:

```
THE CHARACTERISTIC POLYNOMIAL COEFFICIENTS
P(1) THROUGH P(N)
 0.1694143E 00
-0.1198142E-01
 0.4595143E-03
-0.1038747E-04
 0.1407972E-06
-0.1112883E-08
 0.4738090E-11
-0.9169037E-14
 0.5344283E-17
```

These coefficients are substituted into Eq. 3-121 to obtain the characteristic polynomial where Eq. 3-121 is of the form

$$\lambda^9 - p_1\lambda^8 - p_2\lambda^7 \cdots p_8\lambda - p_9 = 0 \tag{7-37}$$

Using the characteristic polynomial coefficients as input data to the program developed in Chap. 2 for obtaining the roots of polynomials by Bairstow's method, the following is obtained as computer printout:

```
                 REAL PART          IMAGINARY PART         ITERATIONS
CONVERGENCE IS SLOW
U=-0.802700E-02  V= 0.6873272E-05  DELU= 0.1227109E-02
                                              DELV=-0.1186479E-05
X( 9) =      0.003447            0.0                        163
X( 8) =      0.000980            0.0                        163
X( 7) =      0.008185            0.000577                     8
X( 6) =      0.008185           -0.000577                     8
X( 5) =      0.010393            0.007246                     6
X( 4) =      0.010393           -0.007246                     6
X( 3) =      0.009921            0.023195                     4
X( 2) =      0.009921           -0.023195                     4
X( 1) =     -0.230838            0.0                          1
```

From this computer printout we can see that the smallest root of the polynomial is 0.000980. Thus we write for the *smallest* eigenvalue

$$\lambda_1 = \frac{P_{cr}}{EI_1} = 0.000980 \text{ in.}^{-2}$$

For our particular problem $I_c = 1.0$ and $I_1/I_c = 0.6$, as given by the computer printout of the coefficient matrix. Hence $I_1 = 0.6$. For a steel column in which $E = 30 \times 10^6$ psi, the critical or buckling load is

$$P_{cr} = (0.000980)(30 \times 10^6)(0.6) \\ = 17{,}640 \text{ lb}$$

The smallest eigenvalues for tapered columns of the form described by Eq. 7-35 have been determined analytically by Timoshenko.[5] He tabulates values of the dimensionless factor f in the formula

$$P_{cr} = \frac{fEI_c}{l^2}$$

for different values of I_1/I_c and the exponent m (see Eq. 7-35). For $I_1/I_c = 0.6$ and $m = 2$ (the values used in this example), Timoshenko gives $f = 8.51$. The numerical solution by the polynomial method yields

$$f = \frac{Pl^2}{EI_c} = \frac{P}{EI_1}\left(\frac{I_1}{I_c}\right)l^2 = (0.000980)(0.6)(120)^2 \\ = 8.47$$

Thus the numerical solution is well within the limits of engineering accuracy. Furthermore, it has the advantage of being applicable to any kind of variation of cross section. In addition, approximations to the higher eigenvalues—and thus higher buckling loads—may also be obtained from the polynomial method. For example, the second largest eigenvalue (next to the smallest) shown in the computer printout is $\lambda_2 = 0.003447$. Thus

$$\lambda_2 = \frac{P_{cr}}{EI_1} = 0.003447 \text{ in.}^{-2}$$

For a steel column ($E = 30 \times 10^6$), the critical load corresponding to λ_2 is

$$P_{cr} = (0.003447)(30 \times 10^6)(0.6) \\ = 62{,}046 \text{ lb}$$

[5] *Ibid.*

The general configuration of the column corresponding to λ_2 is shown in Fig. 7-14.

As previously discussed, the polynomial method results in the best approximation of the smallest eigenvalue. Therefore, we know that the approximation of $\lambda_2 = 0.003447$ is slightly less accurate than the approximation $\lambda_1 = 0.000980$.

Using the program of Sec. 3-10 for finding the linearly independent eigenvectors associated with an eigenvalue of a matrix **A**, we find that there is just one eigenvector associated with each eigenvalue in this problem.

The components of the eigenvector associated with $\lambda_1 = 0.000980$ are determined as

```
COMPONENTS OF LINEARLY INDEPENDENT EIGENVECTORS

-0.3403448E 00
-0.6377981E 00
-0.8630368E 00
-0.1000000E 01
-0.1044306E 01
-0.1000188E 01
-0.8631992E 00
-0.6379184E 00
-0.3404090E 00
```

and the components of the eigenvector associated with $\lambda_2 = 0.003447$ are found to be

```
COMPONENTS OF LINEARLY INDEPENDENT EIGENVECTORS

-0.1119923E 01
-0.1743418E 01
-0.1672582E 01
-0.1000000E 01
 0.0
 0.9986140E 00
 0.1671890E 01
 0.1743670E 01
 0.1120086E 01
```

In this problem the components of the eigenvectors are beam deflections, and it is apparent that these eigenvectors conform to the first two modes of bending, as shown in Fig. 7-9.

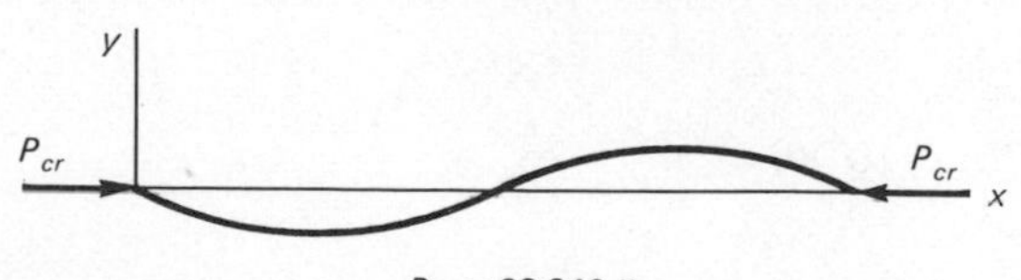

Fig. 7-14 / *General configuration of tapered column corresponding to $\lambda_2 = 0.003447$.*

Iteration Method

The iteration method discussed in Sec. 3-11 and Sec. 3-12 may also be used for solving eigenvalue problems associated with differential equations of the type discussed in the introductory paragraph of this section. As pointed out in Chap. 3, an advantage of the iteration procedure is that the eigenvectors are obtained simultaneously with the associated eigenvalues, rather than requiring separate operations as in the polynomial method. The following example will illustrate the iteration procedure discussed in Sec. 3-12 in which the sweeping technique is used to obtain the lowest eigenvalues and associated eigenvectors in ascending order.

EXAMPLE 7-4

Let us again consider the column of Fig. 7-12, which was analyzed in Ex. 7-3 by the polynomial method. In this example we will use the program developed in Sec. 3-12 to obtain the two lowest eigenvalues (lowest buckling loads) and associated eigenvectors. This column has a uniform thickness and a variable cross section for which the moment of inertia varies according to Eq. 7-35. A set of n simultaneous equations describing such a column with a variable cross section is given by Eq. 7-32. Dividing these equations through by $h^2\alpha_2, h^2\alpha_3, \ldots, h^2\alpha_n$, respectively, we obtain the equations in a matrix form, suitable for the iteration method, as

$$\begin{bmatrix} \left(\frac{2}{h^2\alpha_2}-\lambda\right) & -\frac{1}{h^2\alpha_2} & 0 & 0 & \cdots & 0 \\ -\frac{1}{h^2\alpha_3} & \left(\frac{2}{h^2\alpha_3}-\lambda\right) & -\frac{1}{h^2\alpha_3} & 0 & \cdots & 0 \\ 0 & -\frac{1}{h^2\alpha_4} & \left(\frac{2}{h^2\alpha_4}-\lambda\right) & -\frac{1}{h^2\alpha_4} & \cdots & 0 \\ \cdots & \cdots & \cdots & \cdots & \cdots & \cdots \\ 0 & 0 & 0 & \cdots & -\frac{1}{h^2\alpha_n} & \left(\frac{2}{h^2\alpha_n}-\lambda\right) \end{bmatrix} \begin{Bmatrix} y_2 \\ y_3 \\ y_4 \\ \cdot \\ y_n \end{Bmatrix} = \mathbf{0}$$

The above may be written more compactly as

$$[\mathbf{A} - \lambda\mathbf{I}]\mathbf{Y} = \mathbf{0} \tag{7-38}$$

and we see that the equations are in the form for which the iteration procedure will converge to the largest eigenvalue of the matrix **A**. However, for convergence to the lowest eigenvalue λ_1, we recall from Sec. 3-12 that the iteration procedure must be applied to equations of the form

$$\mathbf{A}^{-1}\mathbf{X} = \frac{1}{\lambda}\mathbf{X}$$

Using these, convergence is to the lowest eigenvalue λ_1, and then to higher eigenvalues $\lambda_2, \lambda_3, \ldots$, in ascending order, as lower eigenvalues are swept out of the equations.

Thus the matrix **A** of Eq. 7-38 must be inverted for input into the program developed in Sec. 3-12 if the smallest eigenvalues are desired.

The computer output resulting from the application of this program to the column problem of Ex. 7-3, with $\mathbf{A}^{-1}$ as input, is

```
LAMBDA( 1)= 0.9796550E-03
THE ASSOCIATED EIGENVECTOR COMPONENTS ARE
0.1000000E 01
0.1874022E 01
0.2535933E 01
0.2938549E 01
0.3068406E 01
0.2938548E 01
0.2535933E 01
0.1874023E 01
0.1000001E 01

LAMBDA( 2)= 0.3447913E-02

THE ASSOCIATED EIGENVECTOR COMPONENTS ARE
 0.1000000E 01
 0.1556616E 01
 0.1493131E 01
 0.8923209E 00
-0.2788382E-05
-0.8923259E 00
-0.1493137E 01
-0.1556622E 01
-0.1000006E 01
```

Thus the two smallest eigenvalues found from the iteration procedure are $\lambda_1 = 0.00097965$ and $\lambda_2 = 0.0034479$. These compare with values of $\lambda_1 = 0.000980$ and $\lambda_2 = 0.003447$, respectively, obtained by the polynomial method. Therefore, the smallest buckling load is found to be 17,634 lb as compared with 17,640 lb from the polynomial method. The small difference is obviously insignificant from a design standpoint.

Problems

7-1 In Ex. 7-1 the temperatures and temperature gradients along a given radiating fin were found. Having found the temperature gradient at $x = 0$, it is possible to determine the amount of heat per hour flowing into the fin by using Fourier's law of heat conduction, $q = -kA\, dT/dx$. Using this we obtain

$$q = -25(0.0025)(-16{,}639) = 1040 \text{ Btu/hr}$$

In this problem we wish to design a fin of length $L = 0.25$ ft and width $b = 0.5$ ft which will dissipate 1500 Btu/hr. The design variable will be the fin thickness h, with an upper constraint on the thickness of 0.015 ft. The other parameters are the same as in Ex. 7-1. Use a trial-and-error approach employing bisection as explained in Ex. 7-1, modifying the program of that example as required. Choose two values of h which are quite sure to bracket the correct fin thickness you are trying to determine. Suggested values would be 0.0 and 0.015, the latter being the upper constraint. Assign them to LOW and HIGH, respectively. For an initial trial run use h equal to the average of LOW and HIGH. With the h determined, calculate an initial temperature gradient from Fourier's law of heat conduction, and then integrate as in Ex. 7-1. After an integration, the value of h used will be assigned to the name LOW or HIGH, depending on the results of the integration, and a new h determined as the average of LOW and HIGH. For example, if the final temperature gradient comes out positive, a larger negative initial temperature gradient is needed. This will require a smaller fin thickness for a given q value, so h will be assigned to the name HIGH and a new h calculated as the average of LOW and HIGH. Trials should continue until the final temperature gradient is within 1°R of zero. Print out the final fin thickness to the nearest 0.0001 ft, and the final temperatures and temperature gradients along the fin.

7-2 Work the problem of Ex. 7-2 by the trial-and-error method discussed in Sec. 7-2, and compare the results obtained with the deflections found in that example. Use the self-starting modified Euler method for integration. Two initial shaft slopes must be chosen which will be certain to bracket the correct negative slope. Suggested values are 0.0 and -0.002 in./in. Assign the first to the variable name LOW and the second to the variable name HIGH. For an initial dy/dx, use the average of LOW and HIGH. After an integration, the dy/dx value used is assigned to either LOW or HIGH, depending on the results of the integration. For example, if the final y value comes out positive, a larger initial negative slope is required. Therefore, the initial dy/dx value used is assigned to the variable name LOW, and a new initial dy/dx value is determined as the average of LOW and HIGH. Use a Δx value of 0.1 in. and require that the final y value be within 0.0000005 in. of zero. After an essentially correct initial slope has been determined, print out the displacements at 3-in. intervals of beam length. Note that one of three different expressions for bending moment M as a function of x must be used, depending on the value of x, and that I varies with the section of the shaft in which the x value lies.

7-3 Example 7-1 is concerned with the digital-computer solution of the radiating-fin problem. The following differential equation, derived

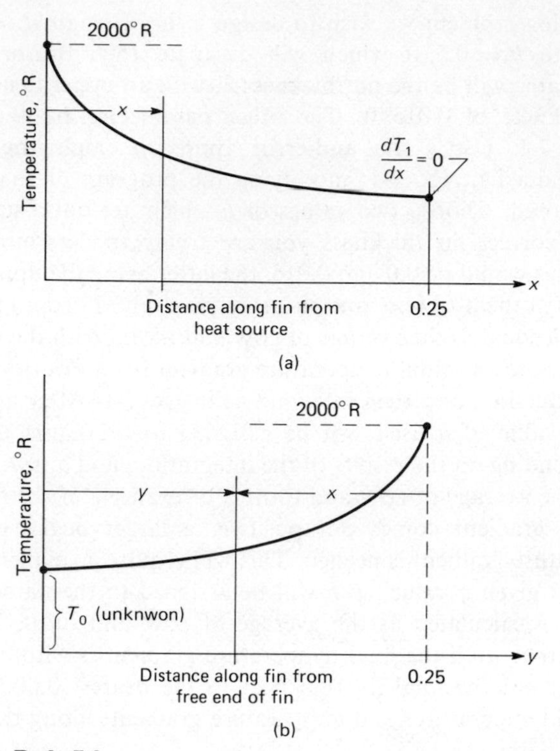

Prob. 7-3

in Ex. 7-1, refers to the coordinate system shown in part a of the accompanying figure.

$$\frac{d^2T_1}{dx^2} = \frac{2\sigma\varepsilon_1}{kh} T_1^4$$

where the pertinent initial and final boundary conditions are

$$x = 0 \left| \begin{array}{l} T_1 = 2000°\text{R} \\ \dfrac{dT_1}{dx} = \,? \end{array} \right. \qquad x = L \left| \; \frac{dT_1}{dx} = 0 \right.$$

The use of a different set of coordinate axes, such as shown in part b of the figure, with the new independent variable y, results in a new differential equation

$$\frac{d^2T_1}{dy^2} = \frac{2\sigma\varepsilon_1}{kh} T_1^4$$

which describes the system.

Considering T_1 as a function of y, determine the pertinent initial and final boundary conditions required to obtain a numerical solution by the trial-and-error approach discussed in Sec. 7-2.

Write a FORTRAN program for solving the differential equation associated with the new set of axes so that the temperature distribution along the fin may be determined. Using this program and the data given in Ex. 7-1, obtain a computer solution of the temperature distribution along the fin.

7-4 The simply supported beam of the accompanying figure is subjected to a uniformly distributed lateral load of $q = 20$ lb/in. and an axial load P. The differential equation of the beam is

$$EI\frac{d^2y}{dx^2} = -Py + \frac{ql}{2}x - \frac{q}{2}x^2$$

where

E = modulus of elasticity of beam
I = moment of inertia of beam cross section

Since the beam is simply supported, the y deflection of the beam is zero at $x = 0$ and $x = l$, and the slope (dy/dx) of the beam is unknown at both $x = 0$ and $x = l$.

Write a FORTRAN program for determining the deflections which define the elastic curve of the beam for various axial loads P. Include in the program the steps necessary to obtain the information required to plot the maximum deflection of the beam for each lateral load P as a function of P.

Using the program written, obtain computer solutions for axial loads of 10,000 lb, 20,000 lb, 30,000 lb, and 40,000 lb applied along with the lateral loading specified above. Is the maximum deflection of the beam a linear function of the axial loads P applied?

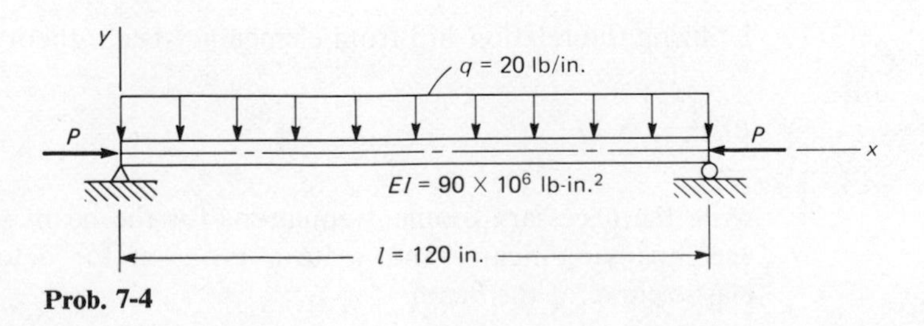

Prob. 7-4

7-5 The analytical solution of the differential equation of the beam shown in the figure accompanying Prob. 7-4 is

$$y = \frac{qEI}{P^2}\left[(1 - \cos kx) + \left(\frac{\cos kl - 1}{\sin kl}\right)\right] \sin kx + \frac{ql}{2P}x - \frac{q}{2P}x^2$$

where

$$k^2 = \frac{P}{EI}$$

Write a FORTRAN program for evaluating the analytical solution shown. Using the program written and the values of E, I, l, q, and P given in Prob. 7-4, determine the deflections defining the elastic curve of the beam for each value of P.

If Prob. 7-4 has been worked prior to working this problem, select a Δx increment corresponding to that used in Prob. 7-4 and compare the solutions obtained in the two problems at various points along the beam.

7-6 Utilizing the relationship from elementary beam theory

$$EI\frac{d^2y}{dx^2} = M$$

write the necessary moment equations for the beam shown in the accompanying figure, and write a program for determining the elastic curve of the beam.

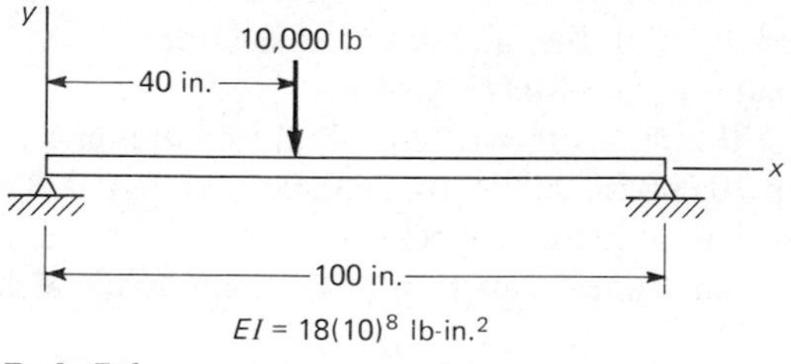

Prob. 7-6

7-7 Utilizing the relationship from elementary beam theory

$$EI\frac{d^2y}{dx^2} = M$$

write the necessary moment equations for the beam shown in the accompanying figure, and write a program for determining the elastic curve of the beam.

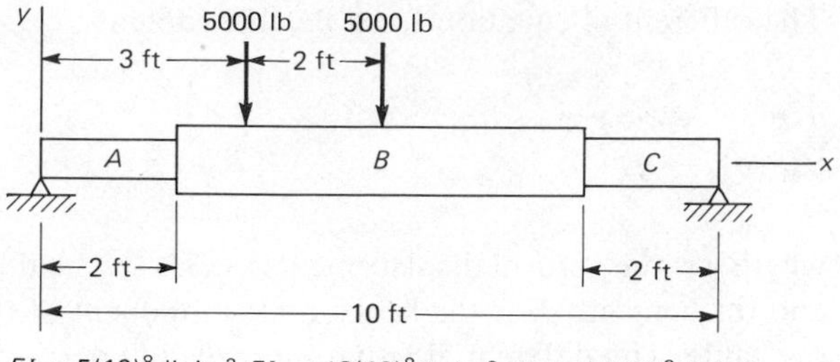

$EI_A = 5(10)^8$ lb-in.^{2} $EI_B = 18(10)^8$ lb-in.^{2} $EI_C = 11(10)^8$ lb-in.^{2}

Prob. 7-7

7-8 The beam shown in part a of the accompanying figure is fixed at A and pinned at B and is thus statically indeterminate. However, looking at the free-body diagram of the beam in part b, we can take moments about an axis through B and obtain the following relationship between the moment at the fixed end and the reaction there:

$$\frac{M_A}{100} + 5000 = R_A$$

Utilizing the relationship from elementary beam theory

$$EI\,\frac{d^2y}{dx^2} = M$$

write the moment equation of the beam, in terms of M_A, and write a program for determining the elastic curve of the beam, the moment at the fixed end, and the vertical reactions at each end.

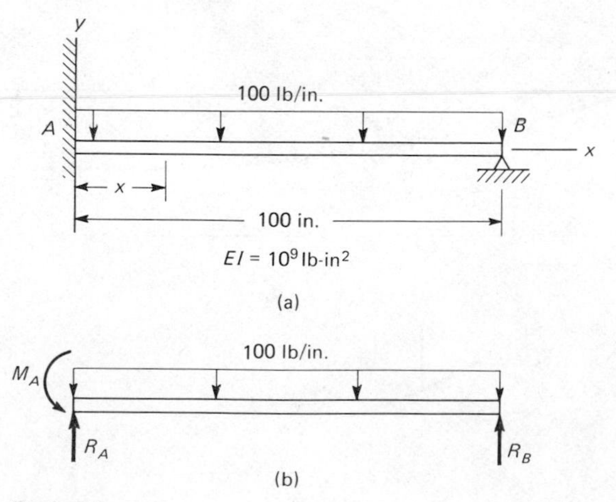

Prob. 7-8

7-9 The differential equation for a flexible cable is

$$\frac{d^2y}{dx^2} = -\frac{w(x)}{H} \tag{1}$$

where y is the vertical displacement, $w(x)$ is the load per unit length, and the constant H is the horizontal component of the tension T in the cable. The differential equation is readily derived by setting the summation of forces in the x and y directions equal to zero for a differential segment, as shown in the accompanying illustration. That is,

$$T_1 \sin \theta_1 - T_2 \sin \theta_2 = w(x)\, dx \tag{2}$$

$$H = T_1 \cos \theta_1 = T_2 \cos \theta_2 \qquad \text{(a constant)} \tag{3}$$

Eliminating T_1 and T_2 from the first equation gives

$$H(\tan \theta_1 - \tan \theta_2) = w(x)\, dx \tag{4}$$

where $\tan \theta_1 = dy/dx\,|_x$ and $\tan \theta_2 = dy/dx\,|_{x+dx}$. Since

$$\frac{dy/dx\,|_x - dy/dx\,|_{x+dx}}{dx} = -\frac{d^2y}{dx^2}$$

Equation 4 yields the differential equation, Eq. 1.

Write a FORTRAN program for obtaining, by the trial-and-error method, the y displacements of the cable shown in the accompanying

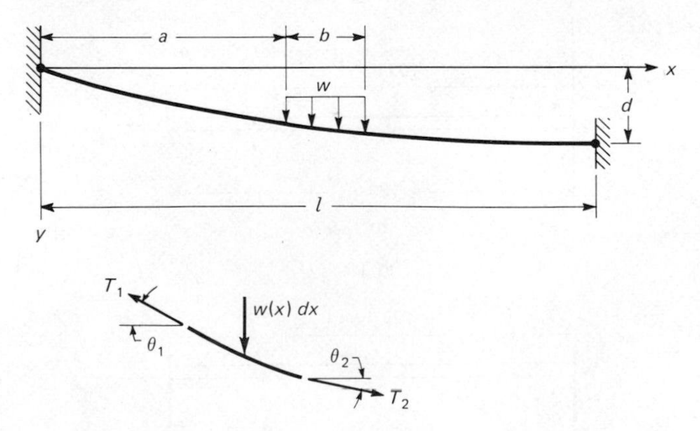

Prob. 7-9

illustration. The right end is anchored a distance d below the left end. The pertinent data are as follows:

$l = 200$ ft $\quad d = 10$ ft $\quad a = 100$ ft

$b = 10$ ft $\quad H = 10{,}000$ lb $\quad w = 20$ lb/ft

7-10 Referring to Prob. 7-9, write a FORTRAN program for obtaining, by the finite-difference–simultaneous-equation method, the y displacements of the cable.

7-11 Referring to Prob. 7-9, write a FORTRAN program for obtaining the y displacements of the cable and the tension T in the cable at various values of x along the cable. Do not make the assumptions that $\sin\theta = \tan\theta = \theta$ for obtaining the values for the tension T.

7-12 The differential equation for the radial displacement u of a point in a thick-walled cylinder is

$$\frac{d^2u}{d\rho^2} + \frac{1}{\rho}\frac{du}{d\rho} - \frac{u}{\rho^2} = 0$$

The radial displacement of point A to A', as shown in the accompanying illustration, is a result of the tangential and radial strains which may be induced by subjecting the inside and outside boundaries of the cylinder to changes in temperature and/or pressures. The relation between the tangential strain ε_t, of a fiber at a distance ρ from the center, and the radial displacement u is $\varepsilon_t = u/\rho$. Assuming that the tangential strains ε_t have been measured on the inside and outside boundaries of a cylinder subjected to an internal pressure and known temperature changes, write a FORTRAN program for obtaining the radial displacements u throughout the cylinder.

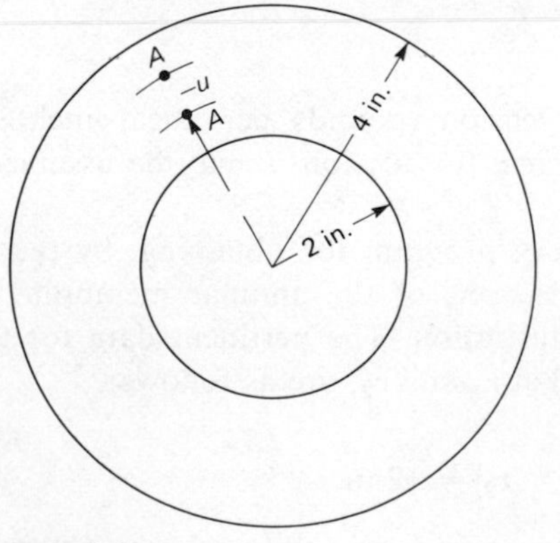

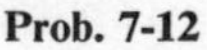
Prob. 7-12

The tangential strains on the boundaries, as measured with SR-4 strain gauges, are

$$\varepsilon_t \mid_{\rho=2\text{ in.}} = 1000 \times 10^{-6}\text{ in./in.}$$

$$\varepsilon_t \mid_{\rho=4\text{ in.}} = 200 \times 10^{-6}\text{ in./in.}$$

7-13 In Prob. 7-12 the differential equation for the radial displacement u in a thick-walled cylinder was given as

$$\frac{d^2u}{d\rho^2} + \frac{1}{\rho}\frac{du}{d\rho} - \frac{u}{\rho^2} = 0$$

Assume a solution to this differential equation of the form

$$u = \sum \alpha_n \rho^n$$

and show that the solution is

$$u = \alpha_1 \rho + \frac{\alpha_2}{\rho}$$

where α_1 and α_2 are constants of integration. If a numerical solution has been obtained for Prob. 7-12, compare these results with the analytical solution.

7-14 The deflection z of a circular membrane which is loaded uniformly with a pressure p is described by the differential equation

$$\frac{d^2z}{dr^2} + \frac{1}{r}\frac{dz}{dr} = -\frac{p}{T}$$

where T is the tension (pounds per linear inch). For a highly stretched membrane, the tension T may be assumed constant for small deflections.

Write a FORTRAN program for obtaining, by the trial-and-error method, the deflections of the annular membrane shown in the accompanying illustration. The pertinent data for the membrane, which is fastened at r_1 and r_2, are as follows:

$r_1 = 6$ in. $\quad r_2 = 12$ in.

$T = 100$ lb/in. $\quad p = 5$ psi (uniformly distributed)

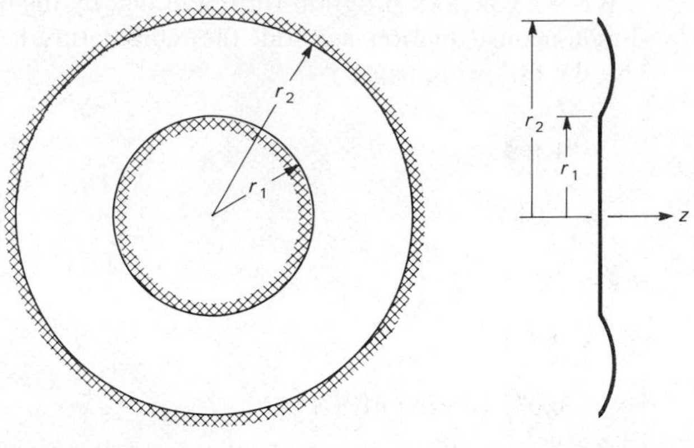

Prob. 7-14

7-15 Fluid flowing in a conduit flows by a thin wire of length l and radius r, as shown in the accompanying illustration (part a). The temperature of the fluid flowing by the wire is U, and the temperatures at the ends of the wire in the walls of the conduit are $u(0) = u_0$ and $u(l) = u_e$, as shown in part b of the figure. Thus heat flows along the wire by conduction, and from the surface of the wire by convection and conduction. Equating the heat flow q_1 into the element to the heat flow $q_2 + q_3$ from the element, as shown in part c of the figure, the differential equation for the temperature u in the wire is found to be

$$\frac{d^2u}{dx^2} = \frac{2K}{kr}u - \frac{2K}{kr}U$$

where

k = thermal conductivity of wire material
r = radius of wire
K = effective conductance of wire surface for heat transfer by convection and conduction

The conductance K depends to some extent upon the Reynolds number, which is proportional to the velocity of the fluid. Since the wire cuts across the fluid stream at a bend in the conduit, the velocity of the fluid varies along the length of the wire. Considering that the velocity along the wire increases with x, a reasonable assumption for the variation of K along the wire is

$$K = K_0 + cx$$

where c is a constant, and K_0 is the value of K at $x = 0$.

Write a FORTRAN program for obtaining, by the finite-difference–simultaneous-equation method, the temperature u along the wire. Use the following data:

$$l = 1 \text{ ft}$$

$$r = 0.01 \text{ ft}$$

$$u_0 = 100°\text{F}$$

$$u_e = 250°\text{F}$$

$$k = 0.035 \text{ Btu/(sec)(ft)(°F)}$$

$$K_0 = 3.8 \text{ Btu/(sec)(ft}^2\text{)(°F)}$$

$$c = 2 \text{ Btu/(sec)(ft)}^3\text{(°F)}$$

$$U = 300°\text{F}$$

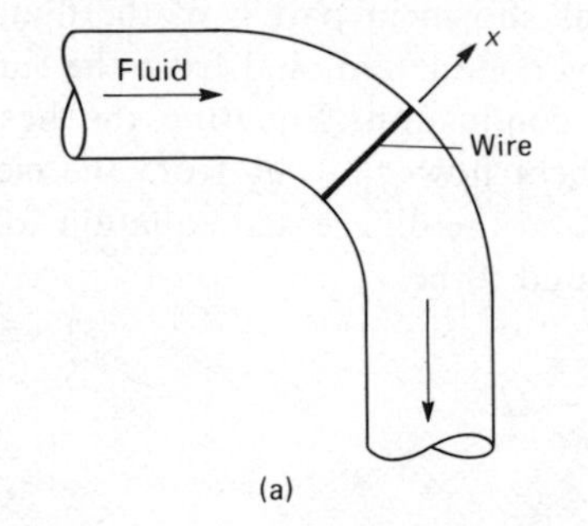

(a)

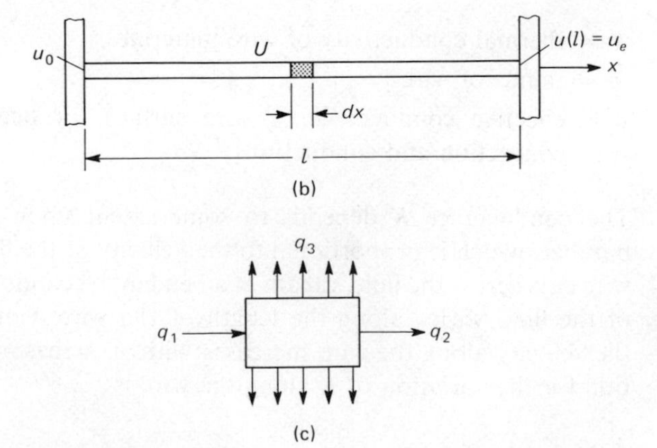

(b)

(c)

Prob. 7-15

The following problems involve SI units.

7-16 Utilizing the relationship from elementary beam theory

$$EI \frac{d^2y}{dx^2} = M$$

write the necessary moment equations for the beam shown in the accompanying figure. Then write a program for determining the elastic curve of the beam. Express the deflection y in units of millimeters.

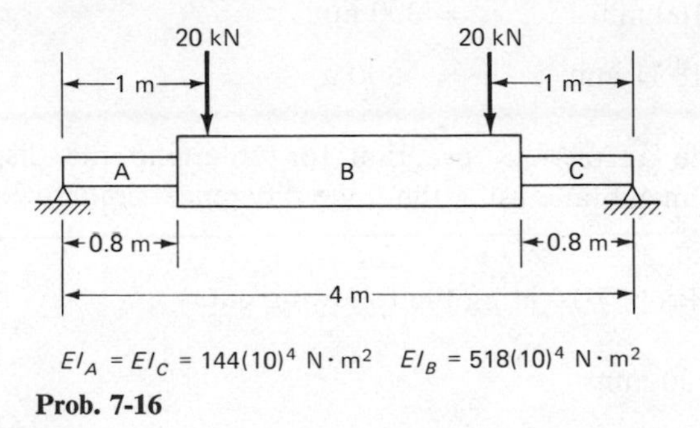

Prob. 7-16

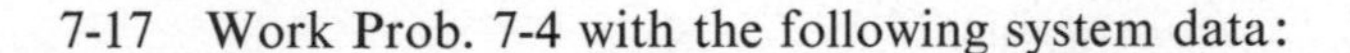

7-17 Work Prob. 7-4 with the following system data:

$$q = 4000 \text{ N/m}$$

$$EI = 26(10)^4 \text{ N} \cdot \text{m}^2$$

$$l = 3.0 \text{ m}$$

Computer solutions are to be obtained for axial loads, P, of 45 kN, 90 kN, 135 kN, and 180 kN. Express the results in millimeters.

7-18 Refer to Prob. 7-9 and determine the y displacements of the cable using the following system data as input to the computer program:

$l = 65$ m $\quad d = 4$ m $\quad a = 25$ m

$b = 7$ m $\quad H = 45$ kN $\quad w = 300$ N/m

Express the results in millimeters.

7-19 Refer to the annular membrane shown in the figure accompanying Prob. 7-14. The deflection z of the membrane is described by the equation

$$\frac{d^2z}{dr^2} + \frac{1}{r}\frac{dz}{dr} = -\frac{p}{T}$$

in which p is a uniform pressure (pascals), and T is the tension (newtons per linear meter). For a highly stretched membrane the tension may be assumed constant for small deflections. The data for the membrane which is fastened at r_1 and r_2 are as follows:

$r_1 = 160$ mm $\qquad r_2 = 300$ mm

$T = 18$ N/mm $\qquad p = 35$ kPa

Write a FORTRAN program for obtaining the displacement z of the membrane using the finite-difference–simultaneous-equation method.

7-20 Solve Prob. 7-15 using the following data:

$l = 30$ mm

$r = 300$ mm

$u_0 = 40$°C (Celsius)

$u_e = 40$°C (Celsius)

$k = 190$ J/(S)(m)(°C)

$K_0 = 60(10)^3$ J/(S)(m^2)(°C)

$c = 90(10)^3$ J/(S)(m^3)(°C)

$U = 150$°C

8 Partial Differential Equations

8-1 / INTRODUCTION

We will now consider the numerical solution of partial differential equations of the general form

$$a \frac{\partial^2 u}{\partial x^2} + b \frac{\partial^2 u}{\partial x \, \partial y} + c \frac{\partial^2 u}{\partial y^2} = f \tag{8-1}$$

where, in general, the coefficients a, b, and c are functions of x and y, and f is a function of x, y, u, $\partial u/\partial x$, and $\partial u/\partial y$. Such equations arise in engineering work involving heat transfer, boundary-layer flow, vibrations, elasticity, and so on. It will not be possible, within the scope of this text, to present a comprehensive coverage of the various numerical methods for solving partial differential equations. However, the material presented will provide an introduction to the subject by illustrating some of the most easily understandable and useful methods. The various forms which Eq. 8-1 may take are classified as *elliptic*, *parabolic*, or *hyperbolic*. The solution of each type will be illustrated by an engineering example.

8-2 / ELLIPTIC PARTIAL DIFFERENTIAL EQUATIONS

Elliptic partial differential equations are found in *equilibrium*-type boundary-value problems. In this kind of problem the differential equation will have coefficients such that

$$b^2 - 4ac < 0$$

where a, b, and c are the coefficients associated with Eq. 8-1.[1] Typical examples of this type of differential equation include Laplace's equation

$$\frac{\partial^2 u}{\partial x^2} + \frac{\partial^2 u}{\partial y^2} = 0 \tag{8-2}$$

and Poisson's equation

$$\frac{\partial^2 u}{\partial x^2} + \frac{\partial^2 u}{\partial y^2} = f(x, y) \tag{8-3}$$

where the function $u(x, y)$ must satisfy both the differential equation over a *closed domain* and the boundary conditions on the closed boundary of the domain. A closed solution domain (as opposed to an open-ended domain) is characteristic of elliptic partial differential equations. This is illustrated graphically in Fig. 8-1. The solution domains of parabolic and hyperbolic differential equations are fundamentally different in nature, as they are open-ended domains.

To illustrate a physical situation in which an elliptic partial differential equation arises, let us consider the problem of determining the temperature distribution in a thin homogeneous rectangular plate, insulated perfectly on both faces, with prescribed boundary conditions maintained along the edges of the plate. It will be assumed that the specific heat and thermal conductivity of the plate material do not vary throughout the plate and that the plate is thin enough to consider the heat flow in a direction normal to the insulated faces as negligible.

Establishing an x-y coordinate system on the plate, as shown in Fig. 8-2, we first consider a small element of the plate with dimensions of Δx and Δy. Such an element is shown greatly exaggerated in size in the figure.

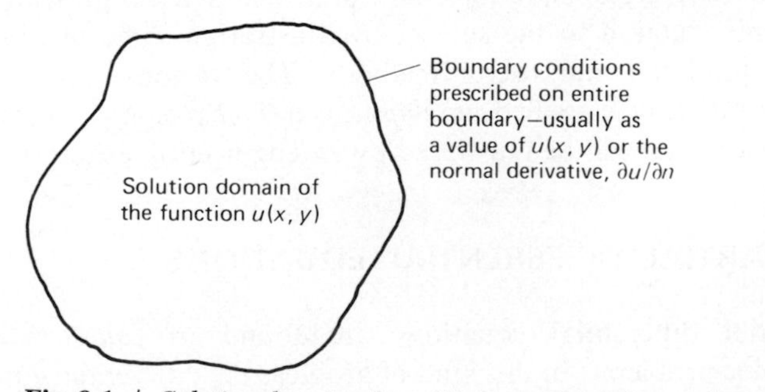

Fig. 8-1 / *Solution domain of an elliptic partial differential equation.*

[1] Stephen H. Crandall, *Engineering Analysis* (New York: McGraw-Hill Book Company, 1956), pp. 353ff.

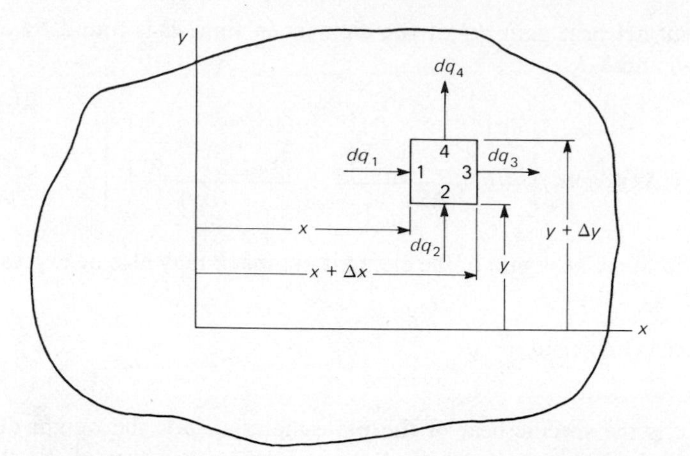

Fig. 8-2 / *Heat entering and leaving an element of a plate in time dt.*

The quantity of heat entering side 1 of the element in time dt is given by Fourier's law of heat conduction as

$$dq_1 = -k(\Delta y)(d)\frac{\partial u}{\partial x}\bigg|_x dt \tag{8-4}$$

where k is the coefficient of thermal conductivity, d is the thickness of the plate, and $\partial u/\partial x|_x$ is the average temperature gradient over side 1 of the element in the x direction. The heat leaving side 3 in time dt is

$$dq_3 = -k(\Delta y)(d)\frac{\partial u}{\partial x}\bigg|_{x+\Delta x} dt \tag{8-5}$$

The *net* heat gain of the element in time dt due to heat flow in the x direction is thus

$$dq_1 - dq_3 = \left[-k(\Delta y)(d)\frac{\partial u}{\partial x}\bigg|_x + k(\Delta y)(d)\frac{\partial u}{\partial x}\bigg|_{x+\Delta x}\right]dt$$

or

$$dq_1 - dq_3 = k(\Delta y)(d)(\Delta x)\left[\frac{\dfrac{\partial u}{\partial x}\bigg|_{x+\Delta x} - \dfrac{\partial u}{\partial x}\bigg|_x}{(\Delta x)}\right]dt \tag{8-6}$$

Similarly, the net heat gain in time dt due to heat flow in the y direction is

$$dq_2 - dq_4 = k(\Delta x)(d)(\Delta y)\left[\frac{\dfrac{\partial u}{\partial y}\bigg|_{y+\Delta y} - \dfrac{\partial u}{\partial y}\bigg|_y}{(\Delta y)}\right]dt \tag{8-7}$$

The *total* net heat gain dQ of the element in time dt is found by adding Eqs. 8-6 and 8-7,

$$dQ = k(\Delta x)(\Delta y)(d)\left[\frac{\left.\frac{\partial u}{\partial x}\right|_{x+\Delta x} - \left.\frac{\partial u}{\partial x}\right|_{x}}{(\Delta x)} + \frac{\left.\frac{\partial u}{\partial y}\right|_{y+\Delta y} - \left.\frac{\partial u}{\partial y}\right|_{y}}{(\Delta y)}\right] dt \tag{8-8}$$

The total net heat gain of the element in time dt may also be expressed as

$$dQ = c(\Delta x)(\Delta y)(d)\rho \frac{\partial u}{\partial t}\, dt \tag{8-9}$$

where c is the specific heat of the plate material, ρ is the weight density, and $\partial u/\partial t$ is the average rate of change of the temperature of the element with respect to time.

Equating Eqs. 8-8 and 8-9 and dividing both sides by $(\Delta x)(\Delta y)(dt)(d)$ gives

$$k\left[\frac{\left.\frac{\partial u}{\partial x}\right|_{x+\Delta x} - \left.\frac{\partial u}{\partial x}\right|_{x}}{(\Delta x)} + \frac{\left.\frac{\partial u}{\partial y}\right|_{y+\Delta y} - \left.\frac{\partial u}{\partial y}\right|_{y}}{(\Delta y)}\right] = c\rho \frac{\partial u}{\partial t} \tag{8-10}$$

Letting both (Δx) and (Δy) approach zero and rearranging terms,

$$\frac{\partial u}{\partial t} = \frac{k}{c\rho}\left[\frac{\partial^2 u}{\partial x^2} + \frac{\partial^2 u}{\partial y^2}\right] \tag{8-11}$$

Equation 8-11 is the *two-dimensional heat equation.*

If we are interested in the *equilibrium* or *steady-state* condition, we realize that $\partial u/\partial t = 0$, and Eq. 8-11 reduces to

$$\frac{\partial^2 u}{\partial x^2} + \frac{\partial^2 u}{\partial y^2} = 0 \tag{8-12}$$

The latter is Laplace's partial differential equation which was mentioned earlier in this section as an example of an elliptic partial differential equation. The temperature $u(x, y)$ throughout the plate must satisfy this equation, as well as the boundary conditions along the entire boundary of the plate when the plate is in thermal equilibrium. Usually, the boundary is either insulated in part, in which case the *normal derivative* of the temperature at the boundary is equal to zero, and/or the temperature is maintained at a specified value over part or all of the boundary.

We will next develop a numerical method for obtaining the solution of Eq. 8-12. Let us superimpose a lattice or grid with a mesh size of

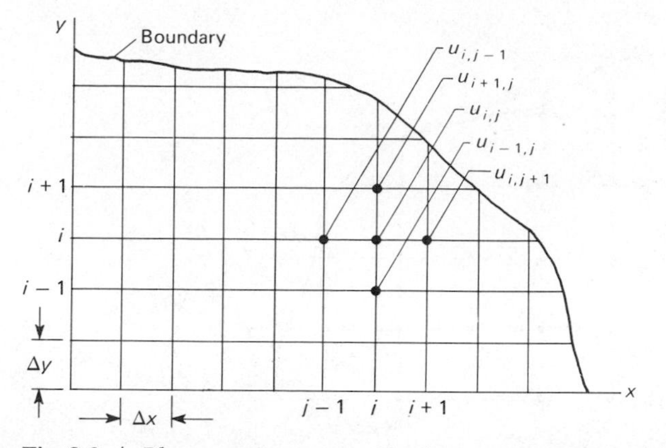

Fig. 8-3 / *Plate temperatures at grid points near intersection of the ith row and jth column of the points.*

Δx by Δy upon the plate, as shown in Fig. 8-3. The numerical solution will consist of determining the temperatures at the finitely spaced grid points shown. Figure 8-3 shows the temperatures at the grid points in the neighborhood of the intersection of the *i*th row and *j*th column. From Eq. 5-79 we can write *central-difference* approximations for the second partial derivatives of the temperature with respect to distance, in the x and y directions, as follows:

$$\frac{\partial^2 u}{\partial x^2} = \frac{u_{i,j+1} - 2u_{i,j} + u_{i,j-1}}{(\Delta x)^2} \tag{8-13}$$

$$\frac{\partial^2 u}{\partial y^2} = \frac{u_{i+1,j} - 2u_{i,j} + u_{i-1,j}}{(\Delta y)^2}$$

Substituting these finite-difference expressions into Eq. 8-12,

$$\frac{u_{i,j+1} - 2u_{i,j} + u_{i,j-1}}{(\Delta x)^2} + \frac{u_{i+1,j} - 2u_{i,j} + u_{i-1,j}}{(\Delta y)^2} = 0 \tag{8-14}$$

If we use a square mesh so that $\Delta x = \Delta y = h$, Eq. 8-14 reduces to

$$u_{i,j+1} + u_{i,j-1} + u_{i+1,j} + u_{i-1,j} - 4u_{i,j} = 0 \tag{8-15}$$

and is referred to as the *Laplacian* difference equation. This equation must hold at every *interior* grid point on the plate.

As mentioned previously, usually the temperatures at the boundary points are known, or the boundary is considered to be perfectly insulated. Insulated boundaries are handled by developing boundary equations. Figure 8-4 shows a *half-element* lying on an *insulated* left boundary of

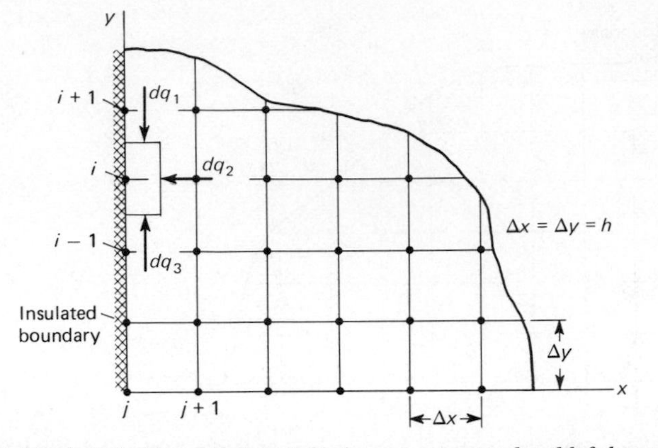

Fig. 8-4 / *Heat flow into half-element on an insulated left boundary.*

the plate. The *net* heat flowing into this element must equal zero when we are considering a steady-state condition for the plate. The quantity of heat flowing into the three faces of the half-element in time dt are given by the following equations:

$$\begin{aligned} dq_1 &= \left(\frac{khd}{2}\right)\frac{\partial u}{\partial y}\bigg|_{i+1/2} dt = \left(\frac{khd}{2}\right)\left(\frac{u_{i+1,j} - u_{i,j}}{h}\right)dt \\ dq_2 &= (khd)\,\frac{\partial u}{\partial x}\bigg|_{j+1/2} dt = (khd)\left(\frac{u_{i,j+1} - u_{i,j}}{h}\right)dt \\ dq_3 &= -\left(\frac{khd}{2}\right)\frac{\partial u}{\partial y}\bigg|_{i-1/2} dt = -\left(\frac{khd}{2}\right)\left(\frac{u_{i,j} - u_{i-1,j}}{h}\right)dt \end{aligned} \tag{8-16}$$

Substituting Eq. 8-16 into the heat-balance equation for the half-element

$$dq_1 + dq_2 + dq_3 = 0$$

yields

$$\text{(Left)} \qquad u_{i+1,j} + 2u_{i,j+1} + u_{i-1,j} - 4u_{i,j} = 0 \tag{8-17}$$

Similarly, the difference equations which apply on the right, upper, and lower *insulated* boundaries of a rectangular plate are

$$\text{(Right)} \qquad u_{i+1,j} + 2u_{i,j-1} + u_{i-1,j} - 4u_{i,j} = 0 \tag{8-18}$$

$$\text{(Upper)} \qquad u_{i,j-1} + 2u_{i-1,j} + u_{i,j+1} - 4u_{i,j} = 0 \tag{8-19}$$

$$\text{(Lower)} \qquad u_{i,j-1} + 2u_{i+1,j} + u_{i,j+1} - 4u_{i,j} = 0 \tag{8-20}$$

Thus in obtaining a solution for the steady-state heat flow in a rectangular plate, Eq. 8-15 must hold at every *interior* grid point, and Eqs. 8-17 through 8-20 must be satisfied at any appropriate insulated boundaries. (A method for handling irregular or curved boundaries will be discussed on p. 537.) Reviewing the pertinent equations, it can be seen that the problem has been reduced to obtaining the simultaneous solution of a set of linear algebraic equations in the unknown grid-point temperatures u, the total number of equations depending upon the extent of the insulated boundaries and the number of grid points used. If the number of grid points required is not too large, the resulting equations may be solved by the *elimination* methods, discussed in Chap. 3. Usually, however, a rather large number of grid points is desirable, and the number of equations which must be solved simultaneously becomes too large to solve accurately by the elimination methods. In this case the Gauss-Seidel iteration method (see Sec. 3-7) can be employed to obtain a more accurate solution. (In connection with the solution of elliptic partial differential equations, the Gauss-Seidel method is often referred to as Liebmann's method.) The equations which must be solved are of the type which have a dominant coefficient for a different unknown in each equation. In addition, there are many zero coefficients in each equation (only four or five unknowns in each equation have nonzero coefficients). The existence of a dominant coefficient in each equation, as mentioned, is a necessary feature for convergence with the equations which are to be solved by the Gauss-Seidel method, and the presence of zero coefficients is a very convenient feature.

Following the procedure outlined for the Gauss-Seidel method in Sec. 3-7, we first arrange each equation in a form convenient for solving for the unknown with the *largest* coefficient in that equation. In the case we are considering, this unknown will always be $u_{i,j}$. For example, from Eq. 8-15 we obtain

$$u_{i,j} = \frac{u_{i,j+1} + u_{i,j-1} + u_{i+1,j} + u_{i-1,j}}{4} \tag{8-21}$$

All unknown temperatures at the grid points are then assigned initial values on the basis of the best estimate available (the better the estimate, the more rapid the convergence to a solution). In calculating $u_{i,j}$ the values of the temperatures on the right side of Eq. 8-21 consist initially of estimated values or known boundary values. However, as soon as an approximate temperature at a grid point is calculated, this calculated value supersedes the estimated value and is used as the temperature at that point until it is, in turn, superseded by a new calculated value. Thus the latest calculated values of the temperatures we are seeking are always used in calculating newer and better values. The applicable equations are used at points on the plate which are selected in some systematic manner, usually either by rows or by columns. The following example illustrates the procedure. (A discussion of a technique for speeding up convergence follows the example.)

EXAMPLE 8-1

Utilizing the digital computer to perform the calculations required by the Gauss-Seidel method, let us determine the temperature distribution in a square plate which has both faces insulated and also has several portions of the edges insulated, as shown in Fig. 8-5. The upper edge and the upper portions of the left and right edges are held at a constant temperature U_T. The bottom edge is maintained at a constant temperature U_B. A grid having ten rows and ten columns of points is superimposed upon the plate, as shown. The left edge is insulated from the bottom up to row $IA = 5$, and the right edge is insulated from the bottom up to row $IB = 8$.

Referring to Eqs. 8-15 through 8-20 and the explanation accompanying Eq. 8-21, we can write the applicable interior and boundary difference equations for this problem as follows:

(Interior)

$$u_{i,j} = \frac{u_{i,j+1} + u_{i,j-1} + u_{i+1,j} + u_{i-1,j}}{4} \qquad \begin{matrix} 1 < i < 10 \\ 1 < j < 10 \end{matrix} \qquad \text{(a)}$$

(Insulated portion of left boundary)

$$u_{i,j} = \frac{u_{i+1,j} + 2u_{i,j+1} + u_{i-1,j}}{4} \qquad 1 < i < IA, \quad j = 1 \qquad \text{(b)}$$

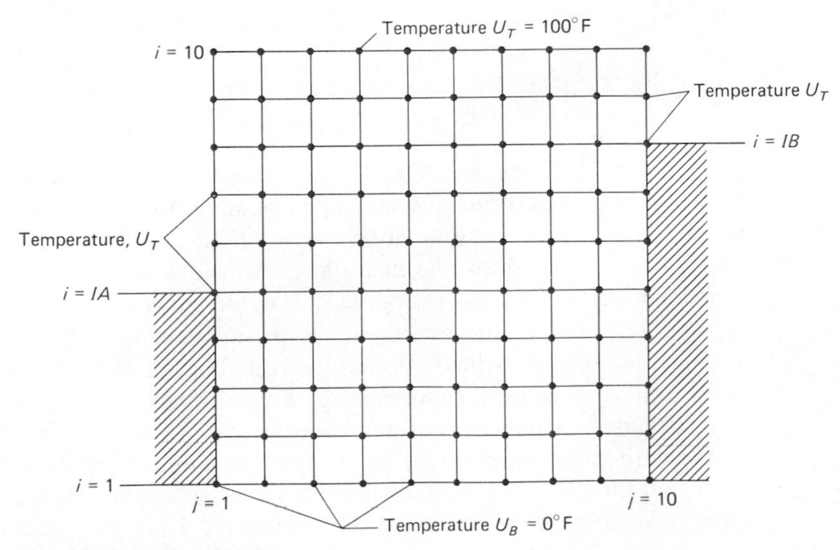

Fig. 8-5 / *Plate of Ex. 8-1.*

(Insulated portion of right boundary)

$$u_{i,j} = \frac{u_{i+1,j} + 2u_{i,j-1} + u_{i-1,j}}{4} \qquad 1 < i < IB, \ \ j = 10 \qquad \text{(c)} \quad \text{(8-22)}$$

(Boundaries with temperature U_T)

$$u_{i,j} = U_T \qquad 1 \le j \le 10, \quad i = 10 \qquad \text{(d)}$$

$$IA \le i < 10, \ \ j = 1$$

$$IB \le i < 10, \ \ j = 10$$

(Boundary with temperature U_B)

$$u_{i,j} = U_B \qquad 1 \le j \le 10, \ \ i = 1 \qquad \text{(e)}$$

Equation 8-22, for all i and j values, must be solved simultaneously to obtain the equilibrium temperature over the plate. The initial temperatures at the grid points where the temperatures are unknown are estimated by assuming a linear temperature gradient between U_B at the bottom of the plate and U_T at the top of the plate. The reader should review the steps outlined for employing the Gauss-Seidel method (see p. 214) before studying the flow chart for this problem (Fig. 8-6). The FORTRAN names used for this problem and the quantities they represent are as follows:

FORTRAN Name	**Quantity**	**Value Used in Problem**
U(I,J)	Temperatures at grid points, °F	
IA	Row at which insulation stops on left boundary	5
IB	Row at which insulation stops on right boundary	8
NMAX	Maximum number of iterations desired	100
UT	Temperature at top of plate and portions of left and right boundaries, °F	100
UB	Temperature at bottom of plate, °F	0
EPSI	Accuracy-check value, °F	0.05
I	Grid row number	
J	Grid column number	
N	Number of iterations completed	
K	Number of grid points at which accuracy check fails	
UTEMP	Temporary name for value of U(I,J), °F	
DIFF	Difference in calculated temperatures at a grid point for two successive iterations, °F	

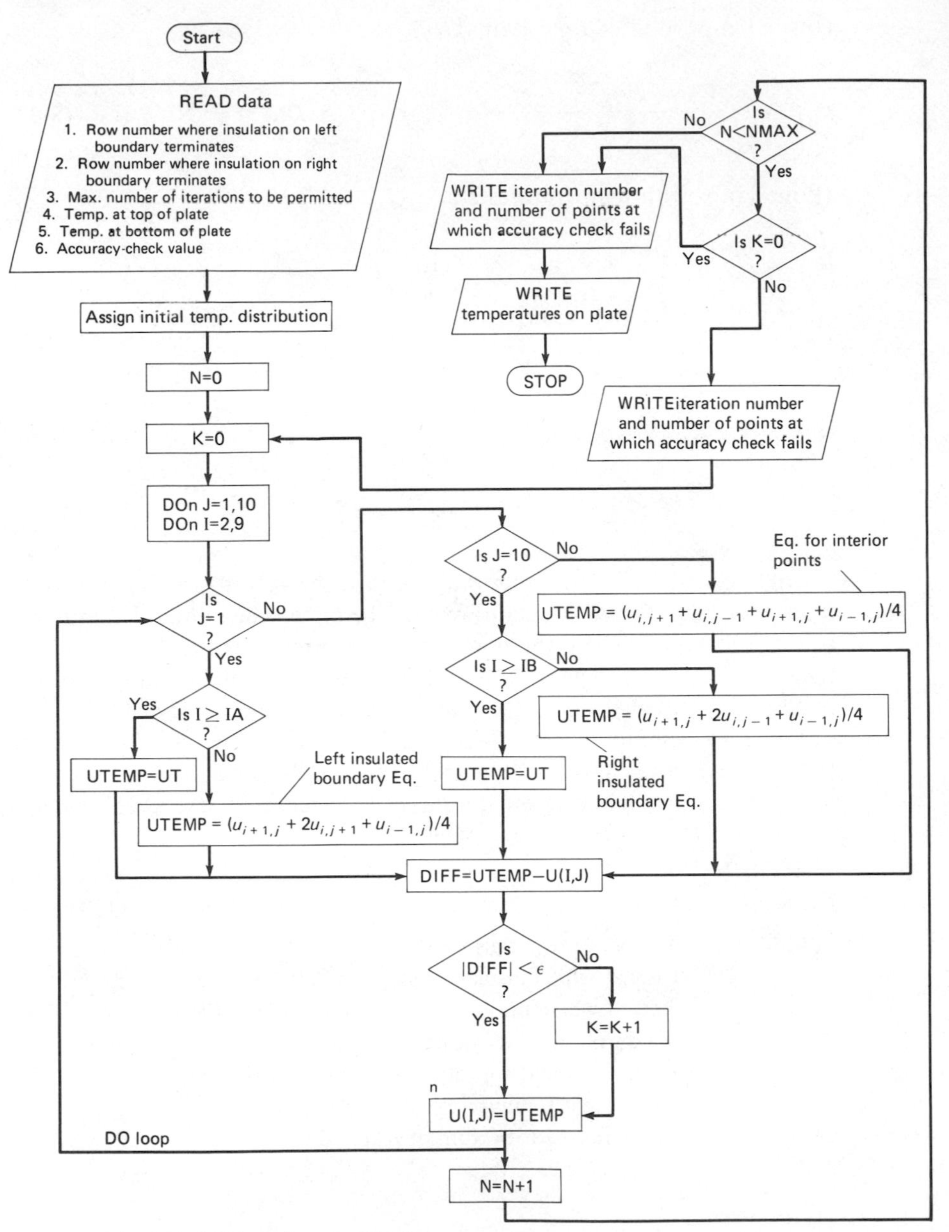

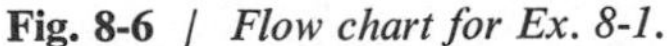
Fig. 8-6 / *Flow chart for Ex. 8-1.*

Referring now to the flow chart, it can be seen that one iteration is complete when an approximate value has been computed for each grid point whose temperature is sought. In all iterations the *latest* calculated temperature of each of these grid points is used in each equation in which it appears. After each step in the iteration, the computer is instructed to check the difference between successive calculated values for that grid point. At the end of each iteration, the computer is then instructed to print out the number of grid points at which the value of the temperature has changed from the preceding iteration by more than some predetermined value, epsilon, which defines the accuracy desired. At this time the computer is also instructed to print out the number of iterations completed. When the temperature change at *all* grid points between successive iterations is less than or equal to the prescribed epsilon value, the computer is programmed to print out these temperatures in a tabular array corresponding to the location of the grid points on the plate. The program also has a provision to instruct the computer to print out the temperatures in the form stated if a certain maximum number of iterations is completed before all the grid-point temperatures vary by less than the prescribed epsilon value. This procedure is incorporated to avoid excessive computing time in case the convergence to a solution is slow. Either printout mentioned terminates the program.

The FORTRAN IV computer program used to implement the flow chart shown is as follows:

```
C TEMPERATURE DISTRIBUTION IN A SQUARE PLATE
C SOLUTION BY GAUSS-SEIDEL ITERATION METHOD
      DIMENSION U(10,10)
      READ(5,2) IA,IB,NMAX,UT,UB,EPSI
    2 FORMAT(2I3,I4,2F6.0,F5.2)
      WRITE(6,3)
    3 FORMAT('1','TEMP DIST IN A PLATE BY GAUSS-SEIDEL
                                                     ITERATION'/)
      WRITE(6,30)
   30 FORMAT(' ',8X,'ITERATION NUMBER',6X,'K VALUE'/)
      DO 4 I=1,10
      DO 4 J=1,10
    4 U(I,J)=UB+(UT-UB)*(I-1.)/9.
      N=0
    5 K=0
      DO 16 J=1,10
      DO 16 I=2,9
      IF(J-1) 9,6,9
    6 IF(IA-I) 7,7,8
    7 UTEMP=UT
      GO TO 14
    8 UTEMP=(U(I+1,J)+2.*U(I,J+1)+U(I-1,J))/4.
      GO TO 14
    9 IF(J-10) 13,10,13
   10 IF(I-IB) 12,11,11
   11 UTEMP=UT
      GO TO 14
   12 UTEMP=(U(I+1,J)+2.*U(I,J-1)+U(I-1,J))/4.
      GO TO 14
   13 UTEMP=(U(I,J+1)+U(I,J-1)+U(I+1,J)+U(I-1,J))/4.
   14 DIFF=UTEMP-U(I,J)
      IF(ABS(DIFF)-EPSI) 16,15,15
```

```
   15 K=K+1
   16 U(I,J)=UTEMP
      N=N+1
      IF(N-NMAX) 17,20,20
   17 IF(K) 18,20,18
   18 WRITE(6,19) N,K
   19 FORMAT(' ',13X,I4,14X,I4)
      GO TO 5
   20 WRITE(6,19) N,K
   21 FORMAT('0',10F7.2)
      DO 22 M=1,10
      I=11-M
   22 WRITE(6,21) (U(I,J),J=1,10)
      STOP
      END
```

The computer results are shown in the tabular printout below:

TEMP DIST IN A PLATE BY GAUSS-SEIDEL ITERATION

ITERATION NUMBER	K VALUE	ITERATION NUMBER	K VALUE
1	32	20	68
2	39	21	67
3	52	22	65
4	62	23	62
5	65	24	61
6	68	25	58
7	70	26	56
8	70	27	51
9	73	28	46
10	73	29	43
11	73	30	38
12	73	31	37
13	73	32	34
14	71	33	31
15	71	34	24
16	71	35	17
17	71	36	13
18	71	37	3
19	69	38	0

100.00	100.00	100.00	100.00	100.00	100.00	100.00	100.00	100.00	100.00
100.00	97.66	95.64	94.07	93.03	92.51	92.62	93.56	95.84	100.00
100.00	95.03	90.81	87.65	85.53	84.43	84.40	85.80	89.78	100.00
100.00	91.65	84.98	80.20	77.06	75.30	74.78	75.49	77.49	80.08
100.00	86.62	77.29	71.16	67.27	65.00	63.98	63.96	64.63	65.33
100.00	77.59	66.45	59.96	55.95	53.53	52.25	51.79	51.84	51.99
63.44	57.31	51.05	46.35	43.11	41.02	39.80	39.20	39.01	39.00
39.19	37.23	34.15	31.35	29.21	27.73	26.81	26.30	26.09	26.04
18.93	18.29	17.04	15.77	14.72	13.97	13.48	13.20	13.07	13.04
0.0	0.0	0.0	0.0	0.0	0.0	0.0	0.0	0.0	0.0

In recent years much study has been given to ways for speeding the convergence of iterative methods in solving computational problems. The

Gauss-Seidel method is iterative, and it has been studied extensively in this regard. A technique called *overrelaxation* has been found to be extremely profitable in speeding the convergence of the Gauss-Seidel method, when applied to the solution of elliptic partial differential equations (the Liebmann process). In place of Eq. 8-21, the equation

$$(u_{i,j})_{\text{new}} = \omega\left(\frac{u_{i,j+1} + u_{i,j-1} + u_{i+1,j} + u_{i-1,j}}{4}\right) + (1 - \omega)(u_{i,j})_{\text{old}} \tag{8-21a}$$

is used.[2] The parameter ω is known as the *relaxation parameter*. For *overrelaxation* the value of ω lies between 1 and 2; for *underrelaxation*, between 0 and 1. For the Gauss-Seidel method, overrelaxation must be used to speed up convergence. It should be noted that when $\omega = 1$, Eq. 8-21a reduces to Eq. 8-21.

Forsythe and Wasow show that for a square solution domain with a 45-by-45 mesh (1936 interior points), the optimum value of ω is approximately 1.870.[3] They further point out that, in using optimal relaxation in this case, convergence is approximately 30 times faster than for the usual Leibmann process ($\omega = 1$).

Estimating ω_{opt} is an advanced problem beyond the scope of this text, but, for any given situation, it can be determined experimentally (see Prob. 8-2). The value of $\omega = 1.870$ can be used as a guide in choosing a value of ω for a square solution domain where the number of grid points is "large."

Grid Points Near Irregular Boundaries

Since it is obviously not possible to have a rectangular grid fit the boundaries of an irregular domain, we must be concerned, in such instances, with writing equations which are applicable to interior grid points that lie near irregular or curved boundaries.

Consider such a point P in a square mesh, as shown in Fig. 8-7(a). Equation 8-15, which was developed earlier for the interior grid points of a square mesh, does not apply at a grid point such as P, since the grid points C and D are not the defined mesh interval h away from point P. The simplest way to obtain an equation defining the temperature at P, when the boundary temperatures are known, is to make use of linear interpolation. Referring to Fig. 8-7(b), we see that the interpolation may be made either between points A and C or between points B and D. An

[2] George E. Forsythe and Wolfgang R. Wasow, *Finite Difference Methods for Partial Differential Equations* (New York: John Wiley & Sons, 1960), p. 247, eq. 22.4.
[3] *Ibid.*, p. 256.

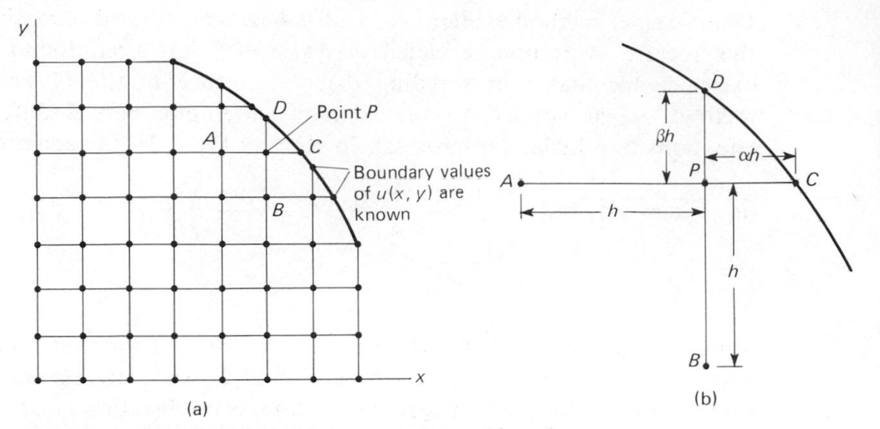

Fig. 8-7 / *Domain with grid points near a curved boundary.*

average of these two interpolations may also be used. For example, if we linearly interpolate between grid points A and C, we find that

$$u_P = u_A + (u_C - u_A)\left(\frac{h}{h + \alpha h}\right)$$

or

$$u_P = \left(\frac{\alpha}{1 + \alpha}\right)u_A + \left(\frac{1}{1 + \alpha}\right)u_C \tag{8-23}$$

Similarly, interpolating between grid points B and D yields

$$u_P = \left(\frac{\beta}{1 + \beta}\right)u_B + \left(\frac{1}{1 + \beta}\right)u_D \tag{8-24}$$

An equation such as Eq. 8-23 or 8-24 must be used for each grid point lying close to an irregular boundary.

If the normal derivative of the function $u(x, y)$ is known at the boundary rather than the temperature $u(x, y)$ itself, the problem of handling irregular boundaries is, in general, more difficult. For example, consider point P near an irregular boundary for which $\partial u/\partial N$ is specified, as shown in Fig. 8-8. To solve the problem it is necessary to write equations defining the temperature at all grid points such as P near the boundary. The value of the normal derivative at G, $\partial u/\partial N|_G$, is known and can be expressed approximately in the following finite-difference form:

$$\left.\frac{\partial u}{\partial N}\right|_G = \frac{u_P - u_F}{\overline{FP}} \tag{8-25}$$

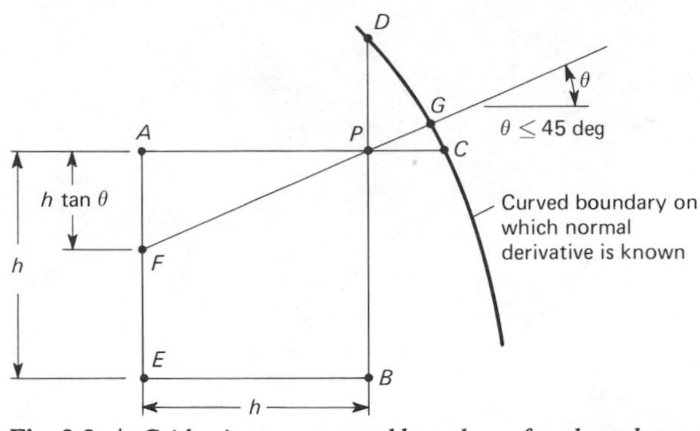

Fig. 8-8 / *Grid points near curved boundary of a plate along which the normal derivative is known.*

When the angle θ is less than 45 deg, a value of u_F can be obtained by linearly interpolating between the vertically spaced grid points A and E such that

$$u_F = u_A + (u_E - u_A)\frac{h \tan \theta}{h} \tag{8-26}$$

Combining Eqs. 8-25 and 8-26 and expressing the distance $\overline{FP}$ as $h/\cos \theta$, we obtain

$$u_P = \left.\frac{\partial u}{\partial N}\right|_G \left(\frac{h}{\cos \theta}\right) + u_A(1 - \tan \theta) + u_E \tan \theta \tag{8-27}$$

An equation of this type must be used for each grid point close to the irregular boundary. The value of θ for each such grid point will be determined from the geometry of the irregular boundary.

When the angle θ is greater than 45 deg, linear interpolation can be used along a horizontal grid line, such as line EB in Fig. 8-9, to determine an expression for the temperature at a grid point such as F. Combining the expression derived from this interpolation with Eq. 8-25 yields the following equation, defining the temperature at grid point P:

$$u_P = \left.\frac{\partial u}{\partial N}\right|_G \left(\frac{h}{\sin \theta}\right) + u_B(1 - \cot \theta) + u_E \cot \theta \tag{8-28}$$

In considering problems with irregular boundaries where the boundary conditions are known normal derivatives, the boundary equations developed previously for rectangular domains (Eqs. 8-17 through 8-20) are obviously not applicable. The equations developed for grid points near irregular boundaries in the preceding paragraphs (Eqs. 8-27 and 8-28), in effect, replace the boundary equations used for rectangular

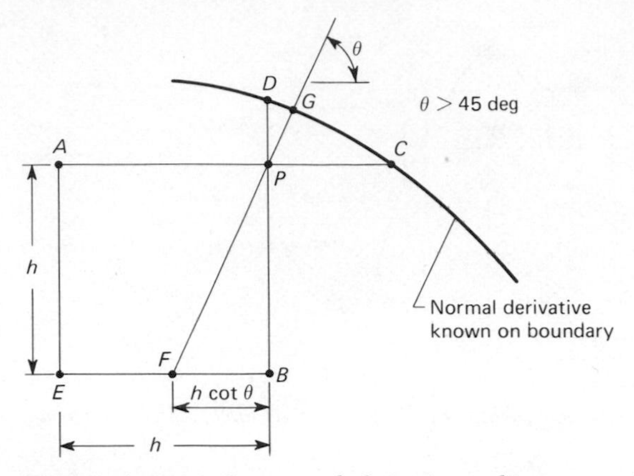

Fig. 8-9 / *Case of a normal derivative making an angle greater than 45 deg with horizontal.*

domains. If it is desired to obtain the temperatures at the boundary grid points, they may be calculated by using linear interpolation between the boundary point desired and adjacent grid points whose temperatures were calculated in the solution obtained.

More elaborate (and more accurate) methods have been developed for handling grid points near curved boundaries by the use of two-dimensional Taylor series.[4] Such methods greatly increase the complexity of computer programming, if there are many such grid points to be considered. The resulting increase in accuracy is usually insignificant in most engineering problems.

For some nonrectangular domains it is possible to use non-Cartesian grids to simplify the solution of the problem. For example, polar coordinates may be used with circular domains and triangular and skew coordinates with appropriately shaped domains.[5] However, a discussion of these methods is beyond the scope of this text.

8-3 / PARABOLIC PARTIAL DIFFERENTIAL EQUATIONS

Parabolic partial differential equations arise in what are called *propagation* problems. In this type of problem the solution advances outward indefinitely from known initial values, always satisfying the known boundary conditions as the solution progresses. This *open-ended* type of solution domain is illustrated in Fig. 8-10, where the dependent variable is u and the independent variables are x and t. The solution u must satisfy

[4] Crandall, *op. cit.*, pp. 262–264.

[5] M. G. Salvadori and M. L. Baron, *Numerical Methods in Engineering* (Englewood Cliffs, N.J.: Prentice-Hall, Inc., 1961), pp. 237ff.

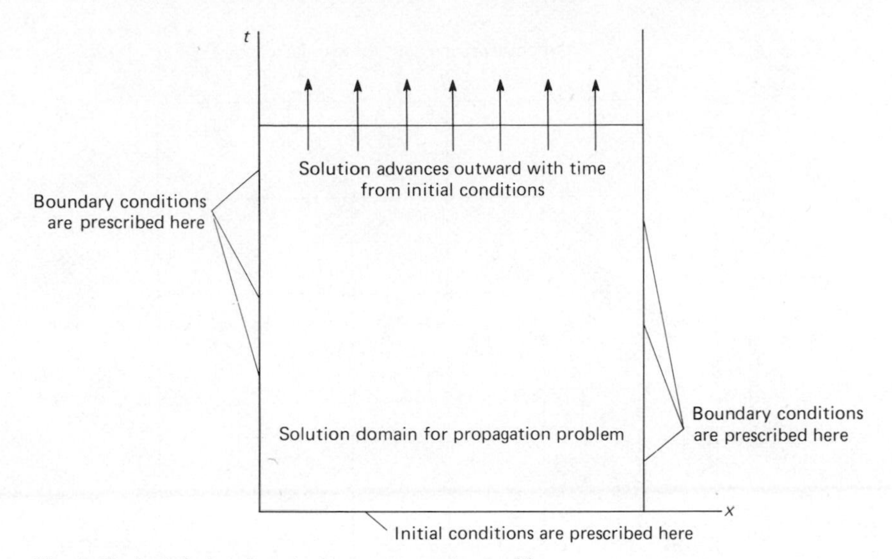

Fig. 8-10 / *Solution domain for a propagation problem.*

the partial differential equation throughout the open domain, as well as the initial and boundary conditions.

The general partial differential equation, as given by Eq. 8-1, is *parabolic* when the coefficients a, b, and c have values such that

$$b^2 - 4ac = 0$$

An example of a parabolic partial differential equation is found in considering the one-dimensional *transient*-heat-flow problem which is defined by the differential equation

$$\frac{\partial u}{\partial t} = \frac{k}{c\rho}\frac{\partial^2 u}{\partial x^2} \tag{8-29}$$

which may be obtained from the two-dimensional heat equation (Eq. 8-11).

As mentioned previously, the solution $u(x, t)$ propagates with time in a space-time plane, as shown in Fig. 8-10. If we consider a space-time grid, such as the one shown in Fig. 8-11, a solution of Eq. 8-29 will consist of determining the temperature u at each grid point used. To utilize the grid shown, we must put Eq. 8-29 in finite-difference form. Referring to Eqs. 5-79 and 5-81, we find it convenient to substitute a *forward-difference* expression for the first partial such that

$$\frac{\partial u}{\partial t} = \frac{u_{i+1,j} - u_{i,j}}{\Delta t} \tag{8-30}$$

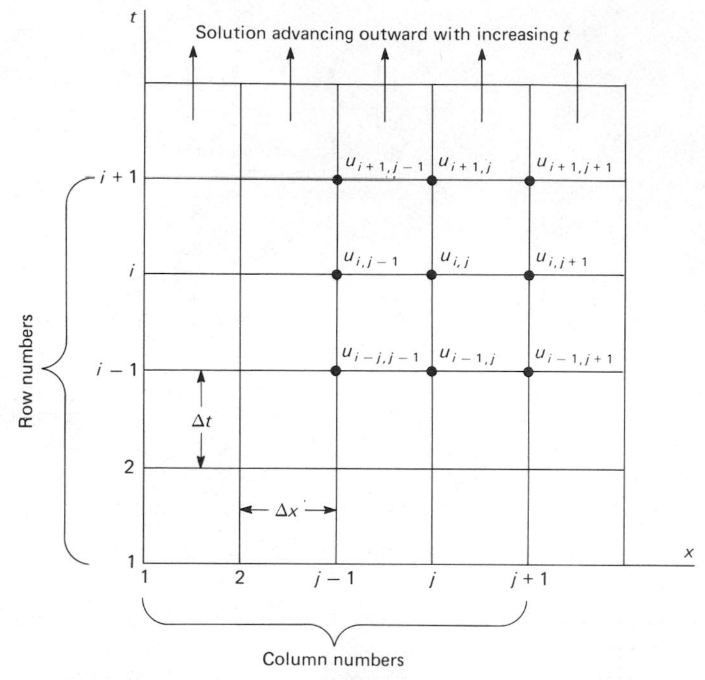

Fig. 8-11 / *Temperature at grid points in the solution domain near ith row and jth column of points.*

and a *central-difference* expression for the second partial such that

$$\frac{\partial^2 u}{\partial x^2} = \frac{u_{i,j+1} - 2u_{i,j} + u_{i,j-1}}{(\Delta x)^2} \tag{8-31}$$

Substituting these difference equations into Eq. 8-29,

$$u_{i+1,j} = u_{i,j} + \frac{k}{c\rho}\frac{(\Delta t)}{(\Delta x)^2}(u_{i,j+1} - 2u_{i,j} + u_{i,j-1}) \tag{8-32}$$

It can be seen, from Eq. 8-32, that we can obtain the temperature at a particular grid point with coordinates x and $(t + \Delta t)$ in terms of the temperatures at adjacent grid points $(x - \Delta x)$, x, and $(x + \Delta x)$ at a time t. Known initial temperatures and boundary temperatures provide the values necessary to start the calculations which then proceed *row* by *row*, the end points of each row satisfying the given boundary conditions, until some final temperature state (with time) is approximately satisfied as the solution approaches a steady state. This row-by-row progression with time, which continues indefinitely, illustrates the open-ended nature of the solution domain of a parabolic-type partial differential equation.

In solving Eq. 8-32, the stability of the solution obtained is of great importance. It can be shown that the solution will be stable and non-oscillatory for boundary conditions that remain constant with time if[6]

$$\frac{k}{c\rho}\frac{(\Delta t)}{(\Delta x)^2} \leq 0.25$$

The solution obtained will be stable if

$$\frac{k}{c\rho}\frac{(\Delta t)}{(\Delta x)^2} \leq 0.50$$

The following example shows in detail the application of Eq. 8-32.

EXAMPLE 8-2

The problem is to determine the transient temperature distribution along a slender rod. The rod, which is shown in Fig. 8-12, is made of aluminum and is 1 ft long. It is assumed to be perfectly insulated at all boundaries except the left end. The pertinent material properties are

$$k = 0.0370 \text{ Btu/(sec)(ft)(°F)}$$

$$c = 0.212 \text{ Btu/(lb)(°F)}$$

$$\rho = 168 \text{ lb/ft}^3$$

and the rod is divided into 12 equal increments by the 13 stations shown.

Initially, the rod is in a state of thermal equilibrium at a temperature of 100°F. The temperature at the uninsulated end is then suddenly reduced to 0°F, at which time the temperature distribution in the rod assumes a transient state. This state exists until a time at which the temperature everywhere in the rod approaches the final equilibrium state of 0°F. The boundary conditions stated mathematically are

$$u(0, t) = 0°\text{F}$$

$$\frac{\partial u}{\partial x}(l, t) = 0°\text{F/ft}$$

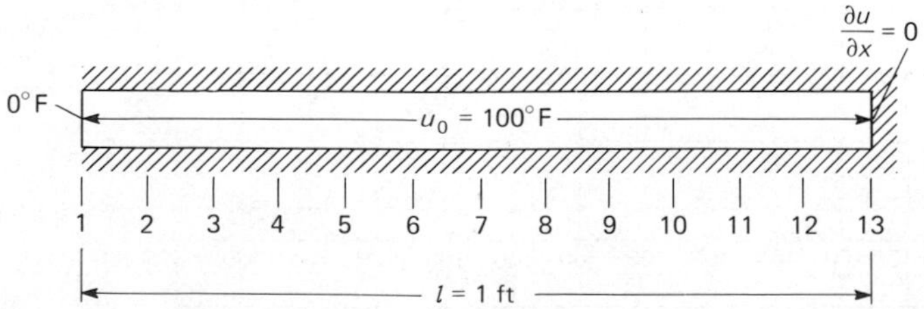

Fig. 8-12 / *Insulated aluminum bar.*

[6] *Ibid.*, p. 261.

and the initial condition is

$$u(x, 0) = 100°\text{F}$$

In the preceding discussion it was shown that the one-dimensional transient-heat-flow equation (Eq. 8-29) can be approximated by the following finite-difference equation:

$$u_{i+1,j} = u_{i,j} + \frac{k}{c\rho}\frac{(\Delta t)}{(\Delta x)^2}\left[u_{i,j+1} - 2u_{i,j} + u_{i,j-1}\right] \tag{8-33}$$

The given values of k, c, and ρ establish a thermal diffusivity $(k/c\rho)$ of 0.00104 ft^2/sec. With the number of stations shown in Fig. 8-12, the value of Δx is 1/12 ft. Using these values and referring to the stability criterion discussed earlier, which established that

$$\frac{k}{c\rho}\frac{(\Delta t)}{(\Delta x)^2} \leq 0.25$$

for a stable and nonoscillatory solution, we see that the use of a convenient value of $\Delta t = 1.0$ sec will satisfy the requirement for such a solution.

Figure 8-13 shows a space-time grid superimposed upon the solution domain. A careful study of the information contained in this figure will give an excellent insight into the numerical procedures required to obtain a solution of the problem. The temperature is known for each station at $t = 0$, as indicated. An *average* value of 50°F is assumed at station 1 for

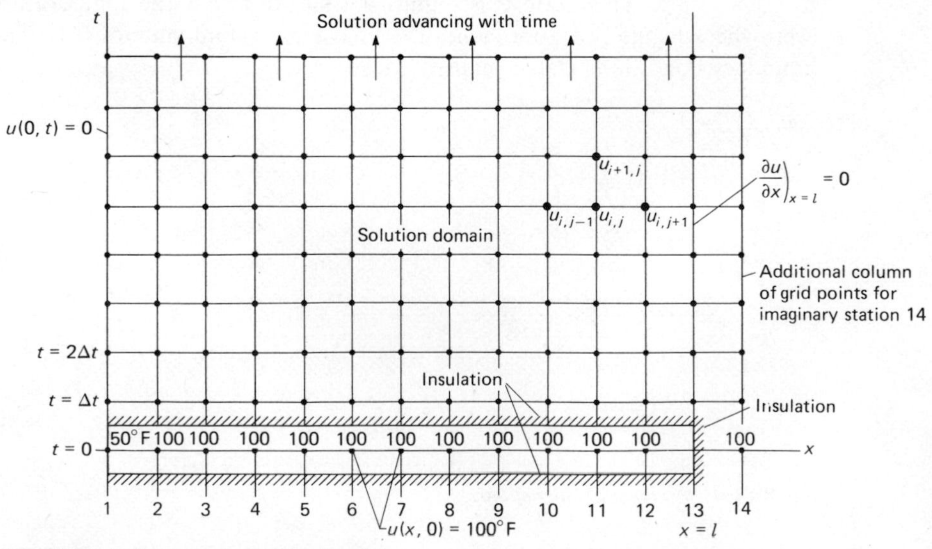

Fig. 8-13 / *Space-time grid for Ex. 8-2*

$t = 0$, as shown, since the temperature at this station is discontinuous at $t = 0$, changing very suddenly from 100°F to 0°F when the transient state is introduced. This average value of 50°F is used only in calculating the temperature at the grid point above station 2 in the row corresponding to $t = \Delta t$. In calculating the temperature at all subsequent grid points above station 2, the *boundary* value of 0°F is used for the temperature at the left-boundary grid points. An additional column of grid points corresponding to an *imaginary* station 14 is included in the space-time grid to provide a means of satisfying the boundary condition imposed at the insulated end of the rod. This condition may be expressed, in finite-difference form, as

$$\frac{u_{i,14} - u_{i,12}}{2(\Delta x)} = 0$$

from which

$$u_{i,14} = u_{i,12}$$

Applying Eq. 8-33, the temperature at each grid point from 2 through 13 is determined in the row corresponding to $t = \Delta t$, using the known initial temperatures. The grid point in that row associated with station 14 is then assigned the value calculated for the grid point associated with station 12, thus approximately establishing a zero temperature gradient at station 13. Equation 8-33 is then used again to determine the temperature at grid points 2 through 13 in the next row corresponding to $t = 2\Delta t$, using the grid-temperature values calculated for the *preceding* row. The grid point in this row corresponding to station 14 is then assigned the value calculated for station 12, and the solution advances to the next row, where $t = 3\Delta t$. This procedure is repeated for each subsequent row of grid points until the temperatures at *all* the grid points in a row differ by less than a small predetermined value epsilon from the known final equilibrium temperature of 0°F. Thus the solution advances outward with time, the number of rows ultimately calculated in the solution depending upon the time rate at which the temperature of the stations approaches the final equilibrium temperature and the temperature limit desired for the solution, as specified by the value of epsilon chosen. Since the temperature at each station along the rod approaches the final equilibrium temperature of 0°F asymptotically in this problem, too small a value of epsilon can result in excessive computing time.

The flow chart for programming this problem for a computer solution is shown in Fig. 8-14. The FORTRAN names and the problem quantities which they represent immediately follow Fig. 8-14.

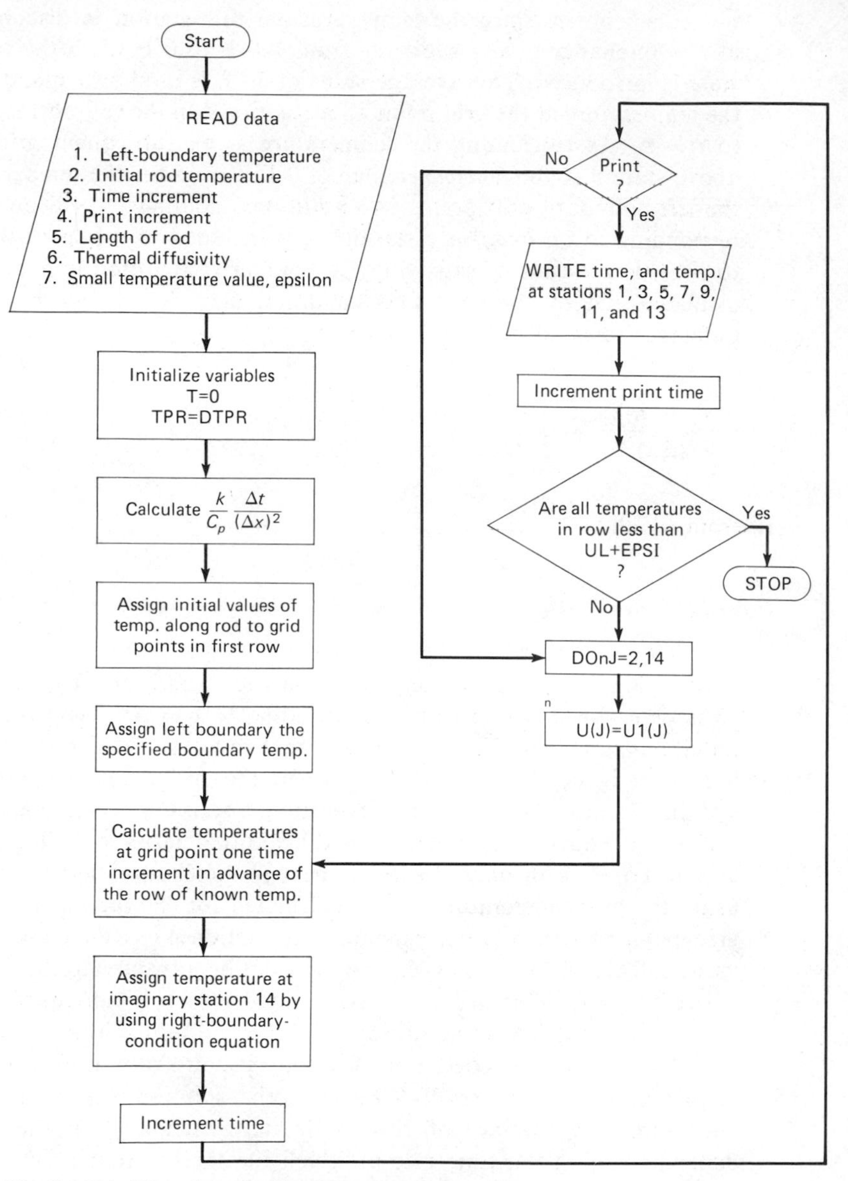

Fig. 8-14 / *Flow chart for Ex. 8-2.*

FORTRAN Name	**Quantity**	**Value Used in Problem**
`U(J)`	Temperature along a row of grid points corresponding to time t, °F	
`U1(J)`	Temperature along a row of grid points corresponding to time $t + \Delta t$, °F	

UL	Temperature at which left end of rod is held	0°F
UX	Initial temperature along rod	100°F
DELT	Time increment	1 sec
EL	Length of rod	1 ft
TPR	Time at which printout occurs, sec	
DTPR	Print increment	100 sec
THDIF	Thermal diffusivity, k/cp	0.00104 ft^2/sec
B	$k(\Delta t)/c\rho(\Delta x)^2$	
EPSI	Small value of temperature selected to terminate program. Program terminates when temperature at each station in row is less than UL + EPSI	0.05°F

The FORTRAN IV computer program used to solve the problem is as follows:

```
      DIMENSION U(14),U1(14)
      READ(5,2) UL,UX,DELT,DTPR,EL,THDIF,EPSI
    2 FORMAT(5F4.0,F7.0,F4.0)
      WRITE(6,3)
    3 FORMAT(' ',14X,'TRANSIENT HEAT-FLOW PROBLEM')
      WRITE(6,4)
    4 FORMAT(' ',13X,'(PARABOLIC PARTIAL DIFF EQN.)'//)
      WRITE(6,5)
    5 FORMAT('    TIME',6X,'1',5X,'3',6X,'5',6X,'7',6X,'9',6X,
                                                            '11',5X,
     *'13'/)
      T=0.
      TPR=DTPR
      B=THDIF*DELT/(EL/12.)**2
      U(1)=(UL+UX)/2.
      DO 6 J=2,14
    6 U(J)=UX
      U1(1)=UL
    7 DO 8 J=2,13
    8 U1(J)=U(J)+B*(U(J+1)-2.*U(J)+U(J-1))
      U1(14)=U1(12)
      T=T+DELT
      IF(T-TPR) 12,9,9
    9 WRITE(6,10) T,(U1(J),J=1,13,2)
   10 FORMAT(' ',F6.0,7F7.2)
      TPR=TPR+DTPR
      DO 11 J=2,13
      DIFF=U1(J)-(UL+EPSI)
      IF(DIFF) 11,12,12
   11 CONTINUE
      STOP
   12 DO 13 J=2,14
   13 U(J)=U1(J)
      U(1)=UL
      GO TO 7
      END
```

The computer results are shown in the following tabular printout:

TRANSIENT HEAT-FLOW PROBLEM
(PARABOLIC PARTIAL DIFF EQN.)

TIME	1	3	5	7	9	11	13
100.	0.0	28.52	53.49	72.60	85.26	92.16	94.32
200.	0.0	20.02	38.52	54.18	66.00	73.31	75.78
300.	0.0	15.29	29.52	41.72	51.06	56.92	58.92
400.	0.0	11.81	22.81	32.26	39.50	44.06	45.61
500.	0.0	9.13	17.65	24.96	30.56	34.09	35.29
600.	0.0	7.07	13.65	19.31	23.65	26.37	27.30
700.	0.0	5.47	10.56	14.94	18.30	20.41	21.13
800.	0.0	4.23	8.17	11.56	14.15	15.79	16.34
900.	0.0	3.27	6.32	8.94	10.95	12.21	12.65
1000.	0.0	2.53	4.89	6.92	8.47	9.45	9.78
1100.	0.0	1.96	3.78	5.35	6.56	7.31	7.57
1200.	0.0	1.52	2.93	4.14	5.07	5.66	5.86
1300.	0.0	1.17	2.27	3.20	3.92	4.38	4.53
1400.	0.0	0.91	1.75	2.48	3.04	3.39	3.51
1500.	0.0	0.70	1.36	1.92	2.35	2.62	2.71
1600.	0.0	0.54	1.05	1.48	1.82	2.03	2.10
1700.	0.0	0.42	0.81	1.15	1.41	1.57	1.62
1800.	0.0	0.33	0.63	0.89	1.09	1.21	1.26
1900.	0.0	0.25	0.49	0.69	0.84	0.94	0.97
2000.	0.0	0.19	0.38	0.53	0.65	0.73	0.75
2100.	0.0	0.15	0.29	0.41	0.50	0.56	0.58
2200.	0.0	0.12	0.23	0.32	0.39	0.43	0.45
2300.	0.0	0.09	0.17	0.25	0.30	0.34	0.35
2400.	0.0	0.07	0.13	0.19	0.23	0.26	0.27
2500.	0.0	0.05	0.10	0.15	0.18	0.20	0.21
2600.	0.0	0.04	0.08	0.11	0.14	0.16	0.16
2700.	0.0	0.03	0.06	0.09	0.11	0.12	0.12
2800.	0.0	0.02	0.05	0.07	0.08	0.09	0.10
2900.	0.0	0.02	0.04	0.05	0.06	0.07	0.07
3000.	0.0	0.01	0.03	0.04	0.05	0.06	0.06
3100.	0.0	0.01	0.02	0.03	0.04	0.04	0.04

The computer printout provides the temperatures at stations 1, 3, 5, 7, 9, 11, and 13 at intervals of 100 sec. The computer printout for stations 3, 5, and 13 is shown in graphical form in Fig. 8-15.

8-4 / HYPERBOLIC PARTIAL DIFFERENTIAL EQUATIONS

The general partial differential equation (Eq. 8-1) is hyperbolic when the coefficients a, b, and c have values such that

$$b^2 - 4ac > 0$$

This type of partial differential equation, like the parabolic type, arises in the solution of so-called *propagation* problems. The solution domain of

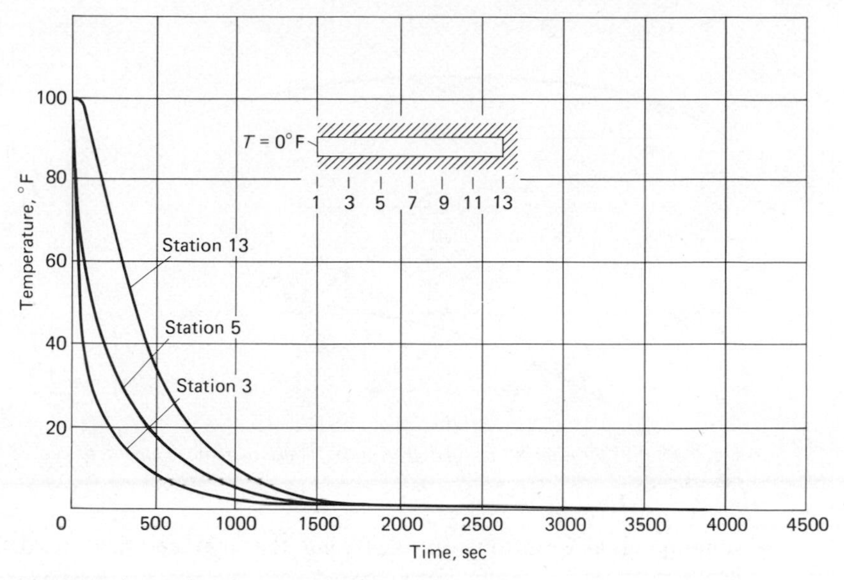

Fig. 8-15 / *Results of numerical solution of Ex. 8-2.*

the hyperbolic differential equation has the same open-ended characteristic encountered in the solution of parabolic differential equations, consisting of a solution which advances outward indefinitely from known initial conditions while always satisfying specified boundary conditions as the solution progresses.

The *one-dimensional wave equation*

$$\frac{\partial^2 u}{\partial t^2} = a^2 \frac{\partial^2 u}{\partial x^2} \tag{8-34}$$

which is frequently encountered in the areas of physics and engineering, is an example of a hyperbolic partial differential equation. This equation, with appropriate variables and associated physical constants, describes the motion of various types of systems, including the torsional vibrations of cylindrical rods, the longitudinal vibrations of slender rods, the transmission of sound in a column of air, and the transverse vibrations of flexible members in a stressed state.

In the last category mentioned, let us consider the free vibrational characteristics of a length of string tightly stretched between two fixed points such as O and E in Fig. 8-16(a). The general string shape shown (with greatly exaggerated y displacements) represents one vibrational configuration of the string at some instant in time. Figure 8-16(b) shows an enlarged view of the small Δs length element PQ. In deriving the wave equation for the vibrating string, we will make use of the following

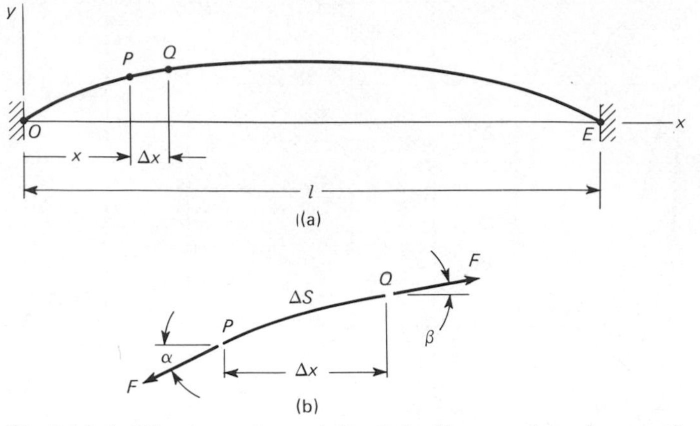

Fig. 8-16 / *Vibrating string and free-body diagram of an element of the string.*

simplifying assumptions in idealizing the mathematical model which results:

1. For small displacements, $\Delta x = \Delta s$.
2. For small displacements, the additional tension introduced in the string due to its displacement is negligible compared to the initial tension introduced in stretching the string between the supporting points O and E; that is, the tension in the string is considered as constant, with a magnitude equal to the initially applied tension F.
3. The gravity force acting upon the string is negligible compared to the tension in the string.
4. The string is flexible enough to consider the bending and shear stresses developed as negligible compared to the tension in the string.

With these assumptions in mind, the forces acting upon the element are shown in Fig. 8-16(b). Applying Newton's second law of motion in the y direction

$$F \sin \beta - F \sin \alpha = \frac{w}{g}(\Delta x)\frac{\partial^2 y}{\partial t^2} \tag{8-35}$$

where w is the weight per unit length of the string. Since α and β are very small angles, we may use the approximations $\sin \alpha = \tan \alpha$ and $\sin \beta = \tan \beta$ in rewriting Eq. 8-35 as

$$F \tan \beta - F \tan \alpha = \frac{w}{g}(\Delta x)\frac{\partial^2 y}{\partial t^2}$$

Noting that tan α and tan β represent the respective slopes at the ends of the element, we may write

$$F\left[\frac{\left.\dfrac{\partial y}{\partial x}\right|_{x+\Delta x} - \left.\dfrac{\partial y}{\partial x}\right|_{x}}{(\Delta x)}\right] = \frac{w}{g}\frac{\partial^2 y}{\partial t^2}$$

Letting Δx approach zero and rearranging terms,

$$\frac{\partial^2 y}{\partial t^2} = \frac{Fg}{w}\frac{\partial^2 y}{\partial x^2} \tag{8-36}$$

which has the form of Eq. 8-34, where $Fg/w = a^2$.

The solution of Eq. 8-36 exists in a general space-time plane such as the one shown in Fig. 8-17. A numerical solution begins by superimposing a rectangular grid over the solution domain, a portion of which is shown in the figure. The value of the dependent variable y is then determined at all grid points over the desired time interval. The numerical procedure itself begins with the use of the known initial-condition values and proceeds from there in a *row-by-row* progression with time, always satisfying the specified *boundary* conditions as the solution progresses. (This procedure will be discussed more specifically in the next example.)

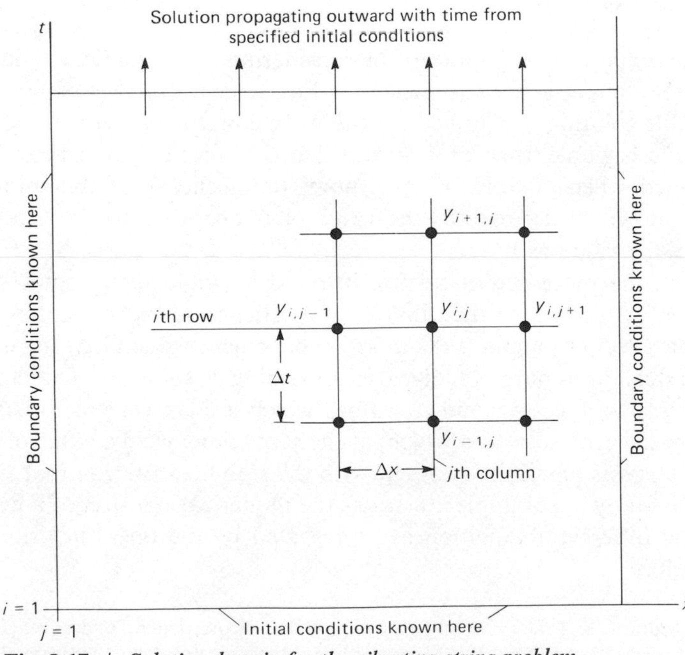

Fig. 8-17 / *Solution domain for the vibrating-string problem.*

Equation 8-36 may be expressed, in finite-difference form, by using the following *central-difference* approximations (see Eq. 5-79) for the two second partial derivatives which appear in the equation:

$$\frac{\partial^2 y}{\partial t^2} = \frac{y_{i+1,j} - 2y_{i,j} + y_{i-1,j}}{(\Delta t)^2}$$

and

$$\frac{\partial^2 y}{\partial x^2} = \frac{y_{i,j+1} - 2y_{i,j} + y_{i,j-1}}{(\Delta x)^2}$$

Substituting these two finite-difference expressions into Eq. 8-36 yields

$$\frac{y_{i+1,j} - 2y_{i,j} + y_{i-1,j}}{(\Delta t)^2} = \frac{Fg}{w}\left(\frac{y_{i,j+1} - 2y_{i,j} + y_{i,j-1}}{(\Delta x)^2}\right)$$

from which

$$y_{i+1,j} = 2y_{i,j} - y_{i-1,j} + C[y_{i,j+1} - 2y_{i,j} + y_{i,j-1}] \tag{8-37}$$

where

$$C = \frac{Fg}{w}\frac{(\Delta t)^2}{(\Delta x)^2}$$

The value of C is important in considering the numerical solution of a partial differential equation, such as Eq. 8-36. It has been shown that an unstable solution is obtained for the wave equation when $C > 1$, that the solution is stable when $C \leq 1$, and that a theoretically correct solution is obtained when $C = 1$.[7] Furthermore, the accuracy of the solution has been shown to decrease as the value of C decreases further and further below the value of 1.

With the preceding discussion in mind it would seem logical to select values for Δx and Δt such that $C = 1$. However, experience reveals that such a selection often results in an inconvenient value of Δt for use in the numerical procedures involved in obtaining a solution. Therefore, it is desirable to select Δx and Δt values which are convenient for use in the numerical procedure and which, at the same time, yield a value of C which is as large as possible consistent with the stability criterion that $C \leq 1$.

The following example illustrates the numerical solution of a hyperbolic partial differential equation as represented by the one-dimensional wave equation.

[7] R. Courant, K. Friedrichs, and H. Lewy, "Über die partiellen Differenzengleichungen der Mathematischen Physik," *Mathematische Annalen*, Vol. 100 (1928), pp. 32–74.

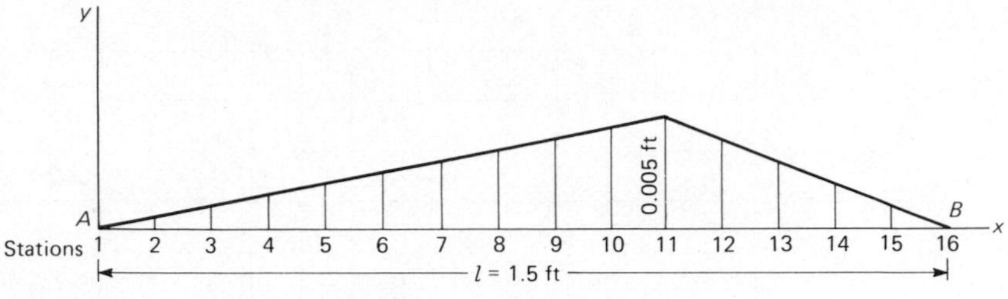

Fig. 8-18 / *Configuration of string at $t = 0$ for Ex. 8-3.*

EXAMPLE 8-3

A string 1.5 ft in length, weighing 0.02 lb/ft, is stretched between supports *A* and *B*, as shown in Fig. 8-18, so that the string has an initial tension of 6 lb. The string is then plucked at the two-thirds point, yielding the initial configuration shown in the figure. It is desired to determine various intermediate configurations of the string during the first 0.2 sec after the string is released from rest in the configuration shown. To determine the desired configurations we must calculate the y displacements at various points along the string at various time intervals.

Selecting a value of $\Delta x = 0.1$ ft as a convenient interval for determining stations along the length of the string, we find that for

$$C = \frac{Fg}{w} \frac{(\Delta t)^2}{(\Delta x)^2} = 1$$

a value of $\Delta t = 0.00102$ sec must be used. A more convenient value is $\Delta t = 0.001$ sec, the selection of which will yield a value of C only slightly less than 1, which will still assure us of a stable and accurate solution. We next determine that $Fg/w = 9651\ \text{ft}^2/\text{sec}^2$ from the values given for the problem parameters. This latter value, along with the Δx and Δt values selected, is used as input data in the computer program which is shown near the end of this example.

We next sketch a space-time grid such as the one in Fig. 8-19. Since the string is released from rest, the initial conditions are

$$y(x, 0) = \begin{Bmatrix} 0.005x \\ -0.01x + 0.015 \end{Bmatrix} \qquad \begin{matrix} (0 \le x \le 1.0) \\ (1.0 < x \le 1.5) \end{matrix}$$

$$\frac{\partial y}{\partial t}(x, 0) = 0$$

The boundary conditions which must be satisfied are

$$y(0, t) = 0$$

$$y(l, t) = 0$$

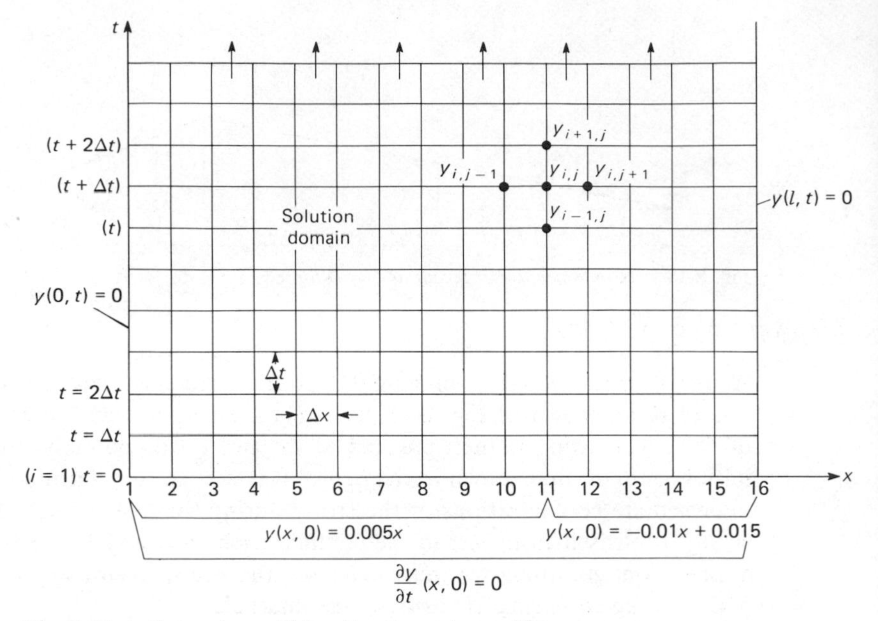

Fig. 8-19 / *Space-time grid for vibrating-string problem.*

The second initial condition shown can be expressed in finite-difference form as

$$\frac{y_{i+1,j} - y_{i-1,j}}{2(\Delta t)} = 0$$

from which

$$y_{i+1,j} = y_{i-1,j} \tag{8-38}$$

This relationship implies the use of an *imaginary* row of grid points preceding the row identified by $t = 0$ in Fig. 8-19. If we substitute Eq. 8-38 into Eq. 8-37, we obtain the following equation:

$$y_{i+1,j} = y_{i,j} + \frac{C}{2}[y_{i,j+1} - 2y_{i,j} + y_{i,j-1}] \tag{8-39}$$

Equation 8-39 is used to calculate the displacements at the grid points in the row corresponding to $t = \Delta t$, since its use yields values which satisfy the initial condition that the string is released from rest.

The first step in the numerical procedure, then, is to calculate the displacements at grid points 2 through 15 in the row corresponding to $t = \Delta t$ by the use of Eq. 8-39, utilizing the known initial displacements to start the calculating process. Equation 8-37 is then employed to calculate the displacements at grid points 2 through 15 for all subsequent rows in a *row-by-row* progression. The grid points in the vicinity of the ith row and the jth column, shown in Fig. 8-19, illustrate the calculating process, the displacement at each grid point under consideration ($y_{i+1,j}$) being calculated from the previously calculated values at the adjacent grid points shown. The number of rows of grid points calculated obviously depends upon the time interval over which a solution is desired. Thus the computer solution is usually terminated by adding the Δt increments used and specifying a termination when their sum corresponds to the time interval desired.

In writing the computer program it will not be necessary to use a double-subscripted variable name for identifying the displacements at the various grid points, because all the values calculated need not be stored in memory if we specify a printout of the desired displacements as they are calculated. The numerical procedure discussed shows that it is necessary to retain in memory only the displacements of the last two rows calculated to furnish sufficient data for the ensuing calculation (see Fig. 8-19). Therefore, we will use single-subscripted variable names for identifying the grid points in the pertinent rows, as shown in the following list.

FORTRAN Name	**Quantity**	**Value Used in Problem**
Y(J), Y1(J), Y2(J)	Displacements corresponding to rows of grid points for times t, $t + \Delta t$, and $t + 2\Delta t$, respectively, ft.	
ASQR	$a^2 = Fg/w$	9651 ft^2/sec^2
DELT	Time increment, Δt	0.001 sec
DELX	Increment of string length, Δx	0.1 ft
TPR	Time at which a printout occurs, sec	
DTPR	Print time increment	0.005 sec
T	Time, sec	
C	$C = \dfrac{Fg}{w}\dfrac{(\Delta t)^2}{(\Delta x)^2}$	
TFIN	Length of time desired for solution interval	0.2 sec

The flow chart for the problem is shown in Fig. 8-20. The FORTRAN IV source program, written from the outline provided by the flow chart, is as follows:

```
C THIS PROGRAM DETERMINES DISPLACEMENTS IN A TIGHTLY
                                        STRETCHED PLUCKED STRING
      DIMENSION Y(16),Y1(16),Y2(16)
      READ(5,3) (Y1(J),J=1,16)
    3 FORMAT(8F10.0)
      READ(5,3) ASQR,DELT,DELX,DTPR,TFIN,EPSI
```

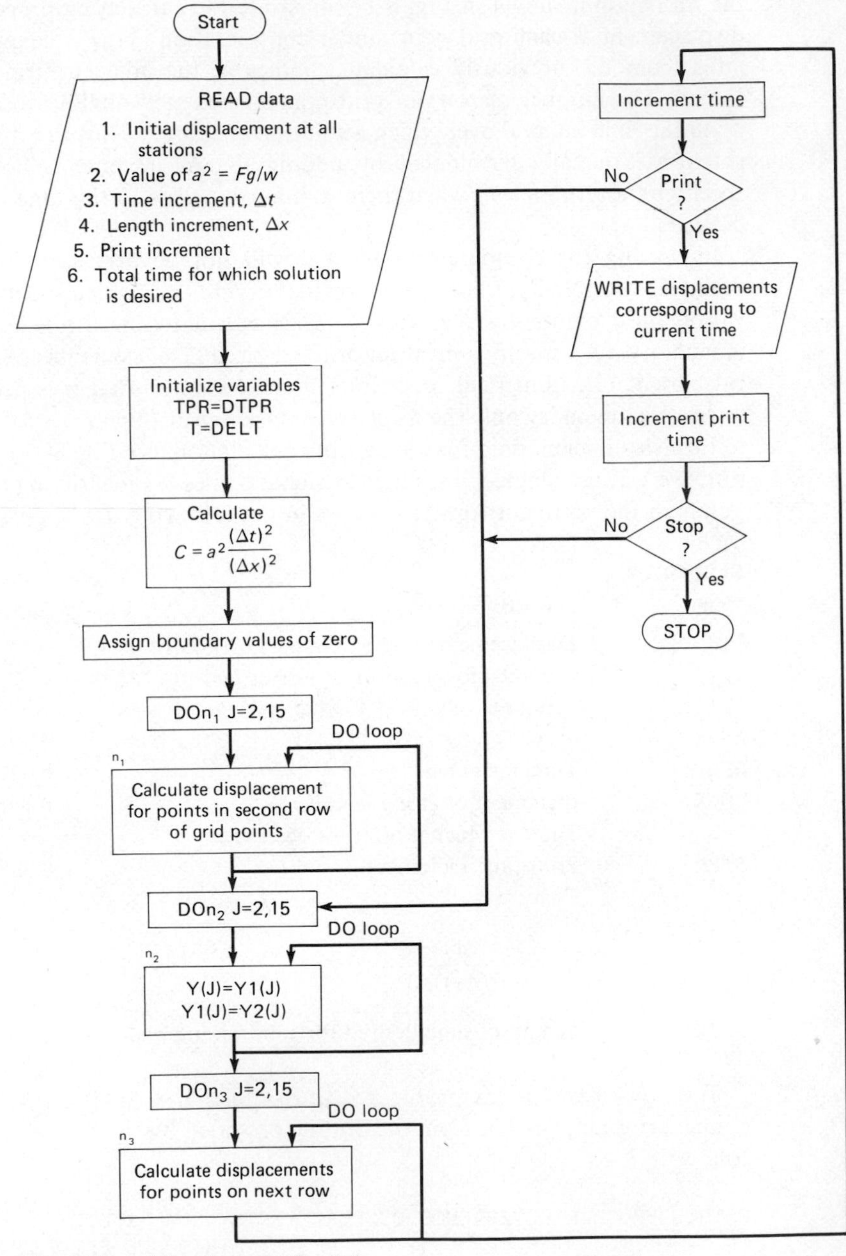

Fig. 8-20 / *Flow chart for Ex. 8-3.*

```
C INITIALIZE VARIABLES
      TPR=DTPR
      T=DELT
      C=ASQR*DELT**2/DELX**2
      WRITE(6,4)
    4 FORMAT('1',' TIME',11X,'DISPLACEMENTS AT STATIONS
                                                  1,3,5,7,9,11,13,
     *15,16'/)
C ASSIGN BOUNDARY VALUES OF ZERO
      Y2(1)=0.
      Y2(16)=0.
C CALCULATE DISPLACEMENTS FOR POINTS IN SECOND ROW OF
                                                       GRID POINTS
      DO 5 J=2,15
    5 Y2(J)=Y1(J)+.5*C*(Y1(J+1)-2.*Y1(J)+Y1(J-1))
    6 DO 7 J=2,15
      Y(J)=Y1(J)
    7 Y1(J)=Y2(J)
C CALCULATE DISPLACEMENTS FOR POINTS IN SUBSEQUENT ROWS
      DO 8 J=2,15
    8 Y2(J)=2.*Y1(J)-Y(J)+C*(Y1(J+1)-2.*Y1(J)+Y1(J-1))
      T=T+DELT
      IF(ABS(T-TPR)-EPSI) 9,9,6
    9 WRITE(6,10) T,(Y2(J),J=1,15,2),Y2(16)
   10 FORMAT(' ',F6.4,9F8.4)
      TPR=TPR+DTPR
      IF(ABS(T-TFIN)-EPSI) 11,11,6
   11 STOP
      END
```

It should be noted in the program that the displacements at the grid points of the row corresponding to $t = \Delta t$ in Fig. 8-19 are calculated in the DO5J=2,15 loop using Eq. 8-39, while the displacements at the grid points of subsequent rows are calculated in the DO8J=2,15 loop using Eq. 8-37, but that all grid points in the process of *being calculated* are identified by the Y2(J) designation.

The computer printout of the displacement-time data for the stations selected is shown on page 558.

Figure 8-21 shows a sketch of the *approximate* string configurations for the various times indicated with the y displacements greatly exaggerated. Although the graphical plot shown was done manually, plotters are available which will display digital readout in graphical form, plotting from either the computer output or magnetic-tape storage.

The solution shown reveals that the period of vibration of the string is approximately 0.03 sec, since the displacements at each station repeat at approximately this interval. If it were desired to plot the displacement-time curve for a particular station over one cycle, a printout increment of 0.001 sec should be used to obtain enough values to plot a reasonably accurate curve.

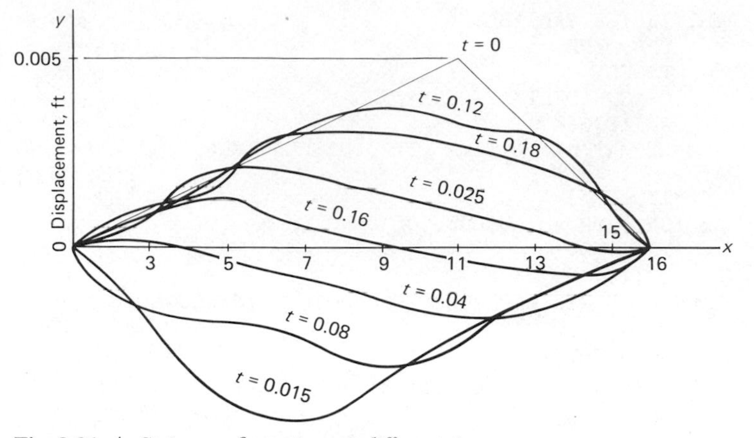

Fig. 8-21 / *String configurations at different times.*

TIME	DISPLACEMENTS AT STATIONS 1,3,5,7,9,11,13,15,16							
0.0050	0.0	0.0010	0.0020	0.0024	0.0019	0.0014	0.0009	0.0004	0.0
0.0100	0.0	-0.0004	-0.0009	-0.0014	-0.0019	-0.0024	-0.0014	-0.0005	0.0
0.0150	0.0	-0.0020	-0.0041	-0.0046	-0.0035	-0.0025	-0.0015	-0.0005	0.0
0.0200	0.0	-0.0008	-0.0013	-0.0018	-0.0023	-0.0025	-0.0015	-0.0005	0.0
0.0250	0.0	0.0010	0.0021	0.0019	0.0014	0.0010	0.0005	-0.0002	0.0
0.0300	0.0	0.0010	0.0020	0.0031	0.0041	0.0046	0.0030	0.0010	0.0
0.0350	0.0	0.0010	0.0020	0.0028	0.0022	0.0017	0.0012	0.0008	0.0
0.0400	0.0	0.0001	-0.0004	-0.0010	-0.0015	-0.0020	-0.0016	-0.0006	0.0
0.0450	0.0	-0.0019	-0.0040	-0.0045	-0.0035	-0.0026	-0.0016	-0.0005	0.0
0.0500	0.0	-0.0013	-0.0016	-0.0021	-0.0028	-0.0026	-0.0016	-0.0005	0.0
0.0550	0.0	0.0010	0.0020	0.0016	0.0010	0.0004	0.0000	-0.0004	0.0
0.0600	0.0	0.0011	0.0020	0.0030	0.0042	0.0042	0.0031	0.0010	0.0
0.0650	0.0	0.0010	0.0021	0.0030	0.0026	0.0022	0.0016	0.0009	0.0
0.0700	0.0	0.0006	-0.0001	-0.0006	-0.0010	-0.0016	-0.0016	-0.0004	0.0
0.0750	0.0	-0.0021	-0.0037	-0.0042	-0.0036	-0.0026	-0.0015	-0.0005	0.0
0.0800	0.0	-0.0017	-0.0021	-0.0026	-0.0032	-0.0025	-0.0015	-0.0005	0.0
0.0850	0.0	0.0011	0.0015	0.0011	0.0007	0.0002	-0.0005	-0.0005	0.0
0.0900	0.0	0.0010	0.0020	0.0029	0.0041	0.0039	0.0031	0.0009	0.0
0.0950	0.0	0.0010	0.0020	0.0030	0.0032	0.0025	0.0022	0.0010	0.0
0.1000	0.0	0.0008	0.0003	-0.0001	-0.0007	-0.0012	-0.0014	-0.0004	0.0
0.1050	0.0	-0.0022	-0.0034	-0.0039	-0.0037	-0.0024	-0.0015	-0.0005	0.0
0.1100	0.0	-0.0018	-0.0026	-0.0031	-0.0033	-0.0025	-0.0015	-0.0005	0.0
0.1150	0.0	0.0011	0.0012	0.0007	0.0002	-0.0002	-0.0009	-0.0005	0.0
0.1200	0.0	0.0009	0.0020	0.0031	0.0038	0.0034	0.0028	0.0011	0.0
0.1250	0.0	0.0010	0.0020	0.0030	0.0036	0.0027	0.0026	0.0010	0.0
0.1300	0.0	0.0010	0.0009	0.0002	-0.0003	-0.0007	-0.0013	-0.0007	0.0
0.1350	0.0	-0.0020	-0.0030	-0.0035	-0.0036	-0.0025	-0.0016	-0.0005	0.0
0.1400	0.0	-0.0020	-0.0029	-0.0034	-0.0035	-0.0026	-0.0015	-0.0005	0.0
0.1450	0.0	0.0010	0.0009	0.0004	-0.0003	-0.0007	-0.0013	-0.0005	0.0
0.1500	0.0	0.0011	0.0019	0.0032	0.0035	0.0030	0.0024	0.0012	0.0
0.1550	0.0	0.0011	0.0021	0.0030	0.0038	0.0034	0.0028	0.0009	0.0
0.1600	0.0	0.0010	0.0013	0.0006	0.0002	-0.0004	-0.0008	-0.0008	0.0
0.1650	0.0	-0.0019	-0.0026	-0.0030	-0.0034	-0.0026	-0.0015	-0.0004	0.0
0.1700	0.0	-0.0021	-0.0033	-0.0039	-0.0036	-0.0025	-0.0015	-0.0005	0.0
0.1750	0.0	0.0009	0.0003	-0.0002	-0.0005	-0.0012	-0.0014	-0.0005	0.0
0.1800	0.0	0.0010	0.0019	0.0031	0.0031	0.0027	0.0021	0.0012	0.0
0.1850	0.0	0.0009	0.0020	0.0031	0.0038	0.0040	0.0029	0.0011	0.0
0.1900	0.0	0.0010	0.0016	0.0010	0.0006	-0.0000	-0.0003	-0.0005	0.0
0.1950	0.0	-0.0017	-0.0021	-0.0027	-0.0031	-0.0026	-0.0014	-0.0006	0.0
0.2000	0.0	-0.0020	-0.0038	-0.0043	-0.0035	-0.0024	-0.0015	-0.0005	0.0

Problems

8-1 Write a FORTRAN program for determining the temperature distribution over the triangular plate shown in the accompanying figure. The left boundary is insulated up to and including row $I = IA$, and the right diagonal boundary is completely insulated. A temperature of $U = U_B$ is maintained along the bottom edge of the plate, and a temperature of $U = U_L$ is maintained along the uninsulated portion of the left edge.

Note that the applicable equation on the insulated diagonal boundary is

$$u_{i,j} = \frac{2u_{i,j-1} + 2u_{i-1,j}}{4}$$

This equation can be obtained by considering an imaginary row of grid points having temperatures which are the reflections of temperatures at points situated perpendicularly across the boundary, as shown in the figure. Such a consideration reveals that

$$u_{i-1,j} = u_{i,j+1}$$

and

$$u_{i,j-1} = u_{i+1,j}$$

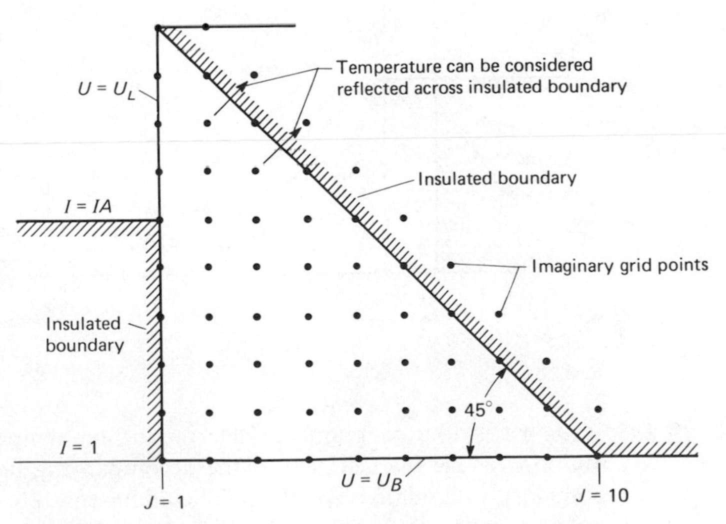

Prob. 8-1

These relationships are substituted into the general finite-difference form of the differential equation

$$u_{i,j} = \frac{u_{i,j+1} + u_{i,j-1} + u_{i+1,j} + u_{i-1,j}}{4}$$

to obtain the diagonal-boundary equation given.

8-2 Rewrite the program of Ex. 8-1 to utilize overrelaxation (Eq. 8-21a in place of Eq. 8-21). Run the problem for a range of values of ω between 1 and 2, and determine an approximate value of ω_{opt} for this example.

8-3 Write a FORTRAN program for determining the temperature distribution over the plate shown in the accompanying figure. The left boundary is insulated up to and including row $I = IA$, and the right boundary is completely insulated. A temperature of $U = U_B$ is maintained along the bottom edge of the plate, and a temperature of $U = U_T$ is maintained along the upper edge of the plate and along the upper portion of the left edge, as shown in the figure. Use a grid having a space of $h = l/9$.

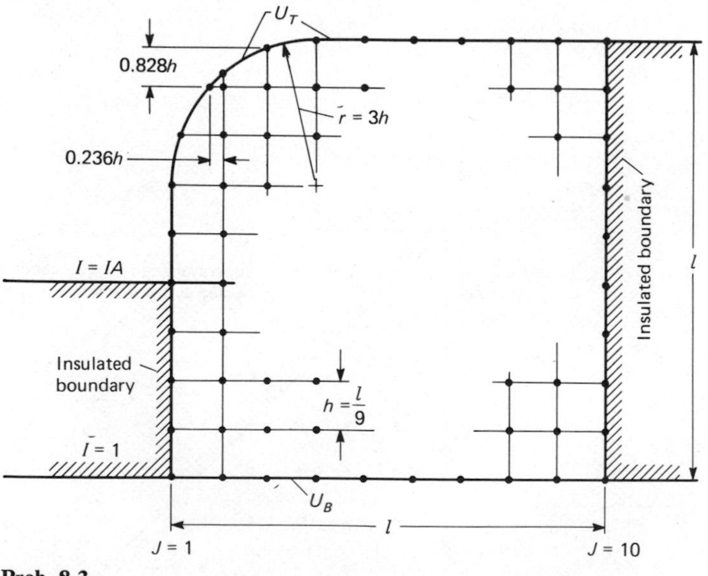

Prob. 8-3

8-4 Write a FORTRAN program for determining the temperature distribution over the plate shown in the accompanying figure. The left boundary is insulated up to and including row $I = IA$, and the right boundary is completely insulated, as shown in the figure. A temperature of $U = U_B$ is maintained along the bottom edge of the plate, and a temperature of $U = U_T$ is maintained along the portions of the left and upper edges, as shown in the figure. Use a grid spacing of $h = l/9$.

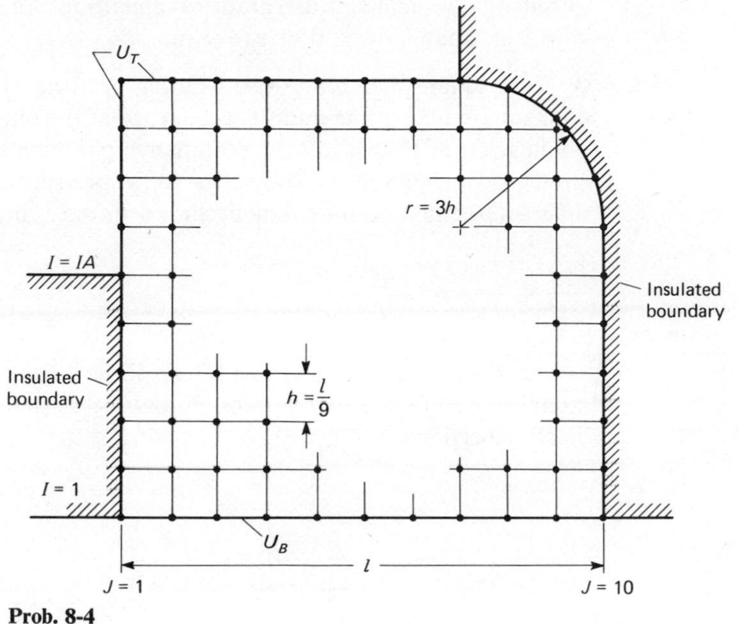

Prob. 8-4

8-5 As shown in the accompanying figure, a 6-in.-square tube, with a 2-in.-square hole in the center portion, carries a fluid having a temperature of 300°F. The outside boundary of the tube is 75°F.

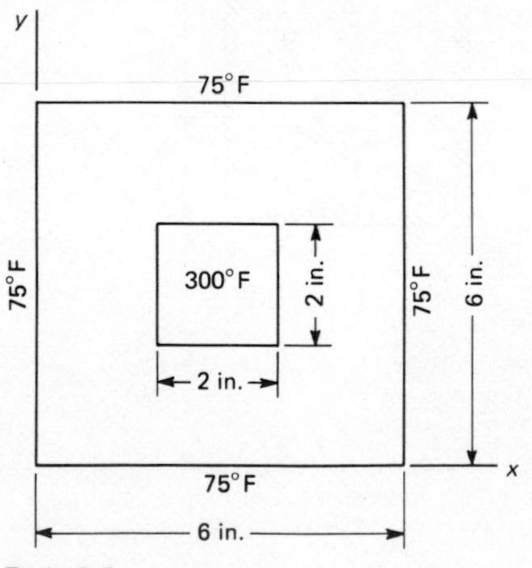

Prob. 8-5

Write a FORTRAN program for obtaining the temperature distribution across the tube. Owing to the symmetry of the temperature distribution, it is necessary to calculate the temperatures over only one-fourth of the tube cross section. In using only one-fourth of the tube for obtaining the temperature distribution, it is necessary to satisfy the temperature-gradient conditions of $\partial u/\partial x = 0$ at $x = 3$ in. and $\partial u/\partial y = 0$ at $y = 3$ in.

8-6 A 12-in.-square membrane (no bending or shear stresses), with a 4-in.-square hole in the middle, is fastened at the outside and inside boundaries, as shown in the accompanying illustration. If a highly stretched membrane is subjected to a pressure p, the partial differential equation for the deflection w in the z direction is

$$\frac{\partial^2 w}{\partial x^2} + \frac{\partial^2 w}{\partial y^2} = -\frac{p}{T}$$

where T is the tension (pounds per linear inch). For a highly stretched membrane, the tension T may be assumed constant for small deflections.

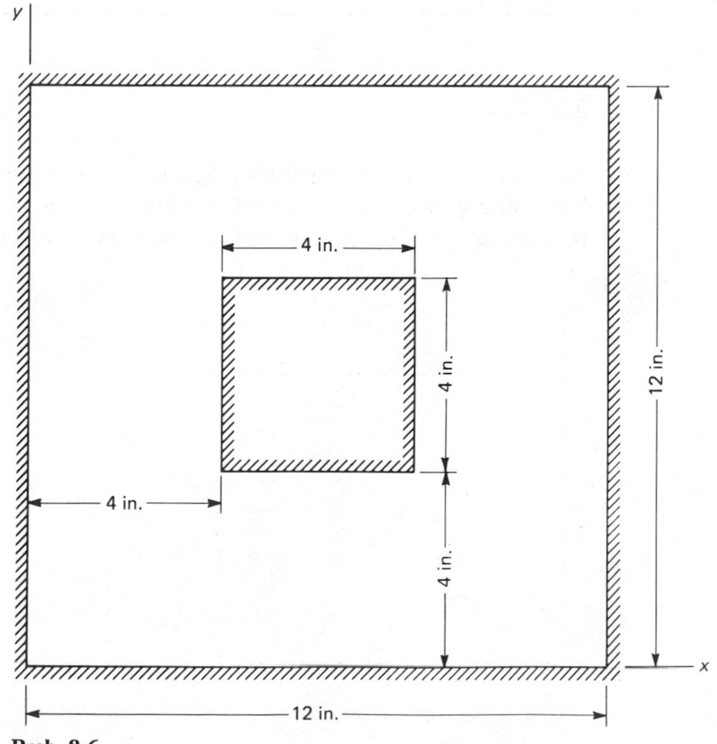

Prob. 8-6

Utilizing the finite-difference expressions of Eq. 8-13, write a FORTRAN program and obtain the deflections w for a membrane in which

$p = 5$ psi (uniformly distributed)

$T = 100$ lb/in.

It should be noted that, owing to symmetry, it is necessary to consider only one-fourth of the membrane. In calculating the deflection over one-fourth of the symmetrical membrane, it is necessary to satisfy the conditions of $\partial w/\partial y = 0$ at $y = 6$ in. and $\partial w/\partial x = 0$ at $x = 6$ in. (This is analogous to an insulated boundary in the temperature-distribution problem discussed in Sec. 8-2.)

8-7 Fluid is flowing through a conduit having a cross section as shown in the accompanying illustration. The partial differential equation for the steady-state velocity ϕ of the fluid is

$$\frac{\partial^2 \phi}{\partial x^2} + \frac{\partial^2 \phi}{\partial y^2} = -\frac{c}{\mu}$$

where c is the absolute value of the pressure gradient in the direction of flow, and μ is the viscosity of the fluid. Noting that the velocity of the fluid is zero on the boundaries, write a FORTRAN program for obtaining the velocity distribution of the fluid. In addition to obtaining the velocities of the fluid flow, the program is to include the calculation of the flow rate Q (volume/sec) where

$$Q = \int_A \phi \, dA$$

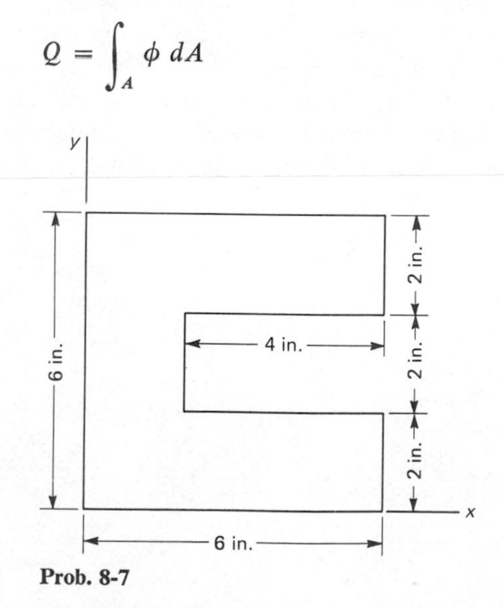

Prob. 8-7

The values for c and μ are as follows:

$$c = 0.02 \text{ lb/in.}^2\text{/in. (constant over the cross section)}$$

$$\mu = 0.25 \times 10^{-6} \text{ lb-sec/in.}^2$$

8-8 Example 8-3 (Sec. 8-4) discusses the numerical solution, on the digital computer, of a vibrating-string problem. The analytical solution of this problem is

$$y(x, t) = \frac{2bl^2}{c(l - c)\pi^2} \sum_{n=1}^{\infty} \frac{1}{n^2} \sin \frac{n\pi c}{l} \cos \frac{n\pi at}{l} \sin \frac{n\pi x}{l}$$

where b is the initial displacement of the string at the point where the string is plucked, and c is the x value at which the string is plucked. The analytical solution is evaluated by summing a number of terms of the series shown, the number of terms summed depending upon the accuracy desired and on the rate of convergence of the series.

Write a FORTRAN program for evaluating the analytical solution for the displacements of the string during one cycle at stations 3, 5, 7, 11, 13, and 15, as indicated by Fig. 8-18 in Ex. 8-3, using the data given in that example. Use all terms in each summation which have a value greater than 0.000001 ft. Use Δt increments of 0.0001 sec.

Compare the results obtained with the results given for corresponding stations in the computer printout included in Ex. 8-3.

8-9 A steel bar has one end fixed, as shown in the accompanying figure. The unrestrained end of the bar is suddenly subjected to a constant force F which is applied as shown in the figure. Sections at various locations along the bar will be displaced longitudinally by the application of such a force, and the displacements $u(x, t)$ at these sections will be functions of both time and location along the bar.

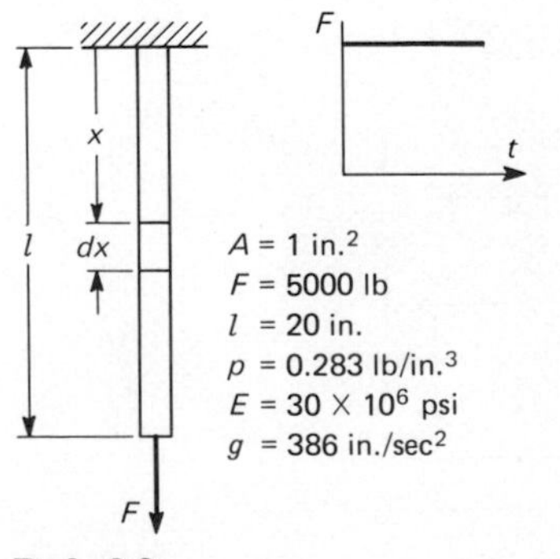

Prob. 8-9

Applying Newton's second law to an element of the bar of length dx, determine the partial differential equation describing the displacement of any section along the bar at any time t where

A = cross-sectional area of bar
E = modulus of elasticity of bar
ρ = weight density of bar
l = length of bar

Recall, from strength of materials or theory of elasticity, that the strain in the x direction is given by

$$\varepsilon_x = \frac{\partial u}{\partial x}$$

It is left to the reader to determine the initial and boundary conditions.

Write a FORTRAN program for determining, by a numerical method, the displacements of the bar at equally spaced stations along the bar at equally spaced time intervals over a total time interval equal to at least one period of the fundamental mode of vibration of the bar. It can be shown that the fundamental period is given by

$$\tau = \frac{4l}{\sqrt{Eg/\rho}}$$

where g is the acceleration of gravity.

Obtain a solution using the values shown in the accompanying figure.

8-10 The analytical solution of Prob. 8-9 is

$$u(x, t) = \frac{8glF}{a^2\pi^2\rho A} \sum_{n=1,3,5,\ldots}^{\infty} \frac{(-1)^{(n-1)/2}}{n^2} \sin \frac{n\pi x}{2l} \left(1 - \cos \frac{n\pi at}{2l}\right)$$

where

$$a = \sqrt{\frac{Eg}{\rho}}$$

and the other parameters are as defined in Prob. 8-9 and its accompanying figure.

Write a FORTRAN program for evaluating the series expression for $u(x, t)$. Obtain a solution using the values shown in the figure accompanying Prob. 8-9.

If you have obtained a solution of Prob. 8-9, check the correspondence of the two solutions.

8-11 A circular shaft of length l has a rigid disk attached to each end, as shown in the accompanying figure, and is supported by "frictionless" bearings (not shown). When free torsional vibrations are initiated in the shaft, the twisting moments on the ends of the elements of length dx (shown in the figure) are

$$GJ\frac{\partial\theta}{\partial x} \qquad \text{and} \qquad GJ\left(\frac{\partial\theta}{\partial x} + \frac{\partial^2\theta}{\partial x^2}\,dx\right)$$

where

G = modulus of rigidity of shaft, lb/in.^{2}
J = polar moment of inertia of shaft cross section, in.^{4}
ρ = weight density of shaft material, lb/in.^{3}
θ = angle of twist at any cross section, radians

By applying the equation $T = I(\partial^2\theta/\partial t^2)$ to the element considered as a free body, show that the partial differential equation describing the angular motion of the shaft for any position x at any time t is

$$\frac{\partial^2\theta}{\partial t^2} = \frac{Gg}{\rho}\frac{\partial^2\theta}{\partial x^2}$$

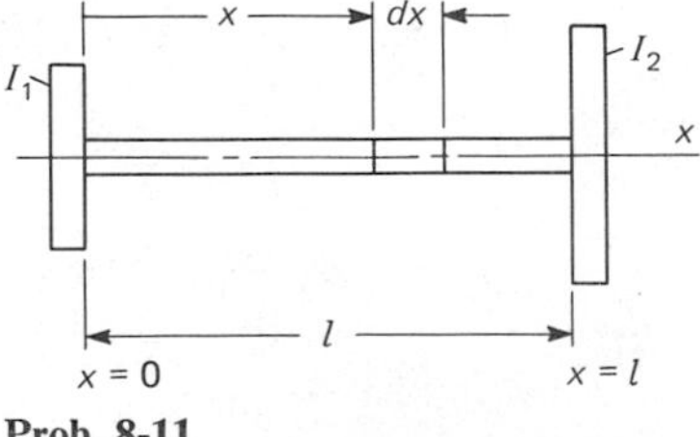

Prob. 8-11

When the shaft is released from rest with a specified initial configuration, the initial conditions are given by the equations

$$\theta(x, 0) = f(x)$$

$$\frac{\partial\theta}{\partial t}(x, 0) = 0$$

By considering the disks at the ends of the shaft as free bodies and applying the equation $T = I(\partial^2\theta/\partial t^2)$, show that the boundary conditions at the ends of the shaft are given by

$$I_1\left(\frac{\partial^2\theta}{\partial t^2}\right)_{x=0} = GJ\left(\frac{\partial\theta}{\partial x}\right)_{x=0}$$

and

$$I_2\left(\frac{\partial^2\theta}{\partial t^2}\right)_{x=l} = -GJ\left(\frac{\partial\theta}{\partial x}\right)_{x=l}$$

where I_1 and I_2 are the mass moments of inertia of the disks about the x axis, as indicated on the accompanying figure.

Write a FORTRAN program for determining θ for equally spaced increments of x and t over a time interval equal to at least one period of the fundamental mode. The period of the fundamental mode is given by

$$\tau = \frac{2\pi l}{\beta a}$$

In the latter equation $a = \sqrt{Gg/\rho}$ and β is the first positive root of the transcendental equation

$$\tan\beta = \frac{(m+n)\beta}{mn\beta^2 - 1}$$

where

$$n = I_2/I_O$$

$$m = I_1/I_O$$

$I_O = \rho lJ/g$ (mass moment of inertia of shaft about longitudinal axis)

Include in your overall program the steps necessary to determine the fundamental period of the system, and use the value determined as the criterion for stopping the overall program.

As a hint for applying the boundary-condition equations in the solution of the problem, put the boundary equations in difference form by using a *central-difference* expression for each second derivative and *forward-* and *backward-difference* expressions for the respective first derivatives. Use the resulting equations to solve for boundary values of θ as the solution progresses with time.

8-12 Using the program written for Prob. 8-11, determine values of θ for equally spaced x values over a period of time approximately equal to the fundamental period of the system discussed in Prob. 8-11. The shaft is 24 in. long and 2 in. in diameter. The disk on the left end has a mass moment of inertia of 0.25 lb-in.-sec^2, and the disk on the right end has a mass moment of inertia of 0.5 lb-in.-sec^2. The shaft is made of steel such that

$$G = 12(10)^6 \text{ lb/in.}^2$$

$$\rho = 0.286 \text{ lb/in.}^3$$

An initial configuration is given to the shaft by holding the left end fixed and twisting the right end through an angle of 0.04 radian.

The following problems involve SI units.

8-13 Refer to Ex. 8-2 and consider that the temperature of the rod at the left end is 0°C and that the initial temperature along the rod is 100°C. Determine whether there are any necessary changes in the program shown if the values for the rod length (l = 1 ft) and thermal diffusivity ($k/c\rho$ = 0.00104 ft^2/sec) are expressed in terms of the SI units.

Also consider your answer for $l \leq 1$ ft.

8-14 Refer to Prob. 8-6 and consider a 0.3-m square membrane with a 0.1-m square hole in the middle in which

$$p = 40 \times 10^3 \text{ N/m}^2 \text{ (uniformly distributed)}$$

$$T = 20 \times 10^3 \text{ N/m}$$

8-15 In Prob. 8-7, the 6-in. dimension is to be taken as 0.3 m, the 4-in. dimension as 0.2 m, and the 2-in. dimension as 0.1 m. Determine the steady-state flow velocity ϕ in m/s.

8-16 Refer to Probs. 8-11 and 8-12 in which the system data are as follows:

$$l = 0.5 \text{ m (length of shaft)}$$

$$d = 0.05 \text{ m (shaft diameter)}$$

$$I_1 = 0.03 \text{ kg} \cdot \text{m}^2 \text{ (mass moment of inertia)}$$

$$I_2 = 0.06 \text{ kg} \cdot \text{m}^2 \text{ (mass moment of inertia)}$$

$$G = 82.7 \text{ GPa (modulus of rigidity)}$$

$$\rho = 77.63 \times 10^3 \text{ N/m}^3$$

9 Introduction to Digital Computer Simulation Using CSMP (Continuous System Modeling Program)

9-1 / INTRODUCTION

Before beginning the study of the CSMP language, let us note the level at which this language fits into the hierarchy of programming languages presently available. The most elementary language, and the language into which programs written in higher level languages must ultimately be translated, is the *machine language* of the computer used. Since a machine-language program written for one particular make and model of computer would not run on another computer, and because of the tedium involved in machine-language coding, engineers seldom write programs in machine language.

The next level of programming language involves the *assembler* or *symbolic* languages. In these, operation codes are symbolic, such as M for "multiply" and A for "add." Addresses may be either numeric or symbolic. Assembler language gives the programmer the advantages of machine language plus some additional advantages such as providing mechanisms for setting up branches to and from assembler-language subroutines. The assembler-language program is translated into machine language by a program called an *assembler*. This language is used primarily by professional programmers and is seldom used by engineers.

At the next level are the so-called *procedural*, or general-purpose, programming languages such as FORTRAN, ALGOL, and PL–1. A program written in one of the procedural languages is translated into machine language by a *compiler* for that particular language. FORTRAN is the most popular of the procedural languages with engineers at present.

Recently, still higher level languages called special-purpose languages, or problem-oriented languages, have been developed. These languages use words in their vocabulary that are part of the language of the particular discipline for which the program is written. COGO is one example of such a language. It is designed specifically for the solution of surveying problems in civil engineering (COGO is an acronym for coordinate geometry). CSMP is a special-purpose language written specifically for simulating continuous systems on IBM 1130 and 360 computers. System 360 CSMP[1] will be discussed in this chapter. It is only one of several digital simulation languages currently being used by engineers and is chosen for description here because of the wide availability of IBM 360 computers and the excellent features of the language. The study of one simulation language such as CSMP provides a good background for the study of many of the other simulation languages.

CSMP statements describing the physical system are interpreted and converted by a translator into a FORTRAN IV subroutine called UPDATE, which in turn is compiled and executed with an integration routine (fourth-order Runge-Kutta or some other selected routine) to effect the simulation. No knowledge of integration routines is required to use CSMP. However, an understanding of FORTRAN is very helpful and is assumed in the discussion of CSMP which follows. A knowledge of FORTRAN permits a programmer to supplement CSMP programs extensively by including FORTRAN statements as part of a CSMP program, and/or by the use of FORTRAN subprograms in conjunction with a CSMP program.

9-2 / GENERAL NATURE OF A CSMP PROGRAM

Before studying the details of CSMP statements and model structure, let us look at an example problem solved by CSMP to obtain a general idea of the nature of a CSMP program. Figure 9-1 shows a spring-and-mass system, viscously damped by the dashpot shown, and containing the dead spaces represented by the gaps a and b between the mass and the respective springs. When $t = 0$, the mass has zero displacement in the position shown in the figure and a velocity of 8 in./sec, as shown. Considering the mass m as a free body, and applying Newton's second law, the differential equation of motion is obtained as

$$m\ddot{x} + c\dot{x} + f(x) = 0 \tag{9-1}$$

where $f(x)$ represents the spring forces acting on the mass. Examining the system, it can be seen that $f(x)$ is given by

$$f(x) = k_2(x - b) \qquad x > b$$

$$f(x) = 0 \qquad a \le x \le b$$

$$f(x) = k_1(x - a) \qquad x < a$$

[1] *System/360 Continuous System Modeling Program, Users' Manual, CH20-0367-4*, 5th ed. (White Plains, N.Y.: IBM Corporation, Data Processing Division, 1972).

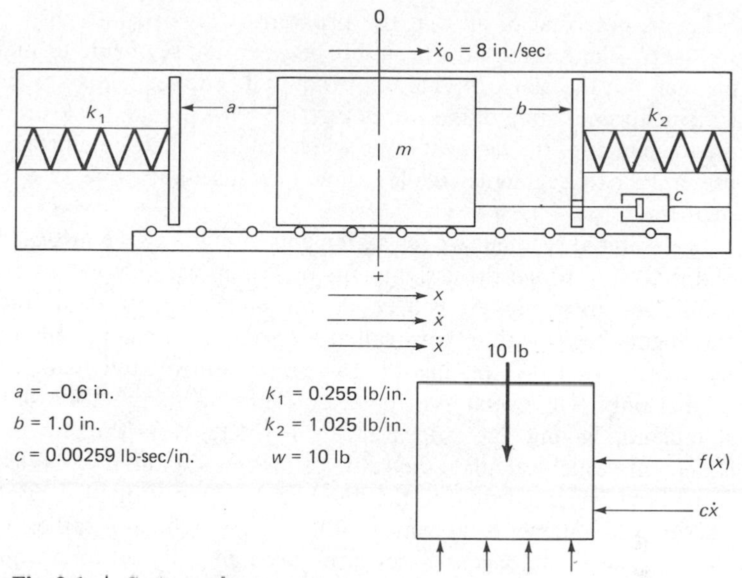

Fig. 9-1 / *Spring-and-mass system.*

A complete CSMP program for simulating the system shown in Fig. 9-1 follows:

```
INITIAL
 CONSTANT W=10.0,C=.00259,A=-0.6,B=1.0,K1=.255,K2=1.025
 INCON X0=0.0,XDOT0=8.0
      M=W/386.
      COEF=C/M
DYNAMIC
 NOSORT
      IF(X.GE.0.) GO TO 2
      FOFX=K1*DEADSP(A,0.,X)
      GO TO 3
  2   FOFX=K2*DEADSP(0.,B,X)
  3   CONTINUE
 SORT
      XDDOT=-COEF*XDOT-FOFX/M
      XDOT=INTGRL(XDOT0,XDDOT)
      X=INTGRL(X0,XDOT)
TERMINAL
      KE=0.5*M*XDOT**2
      WRITE(6,4)KE
  4   FORMAT(' ',E16.7)
 TIMER DELT=.05,OUTDEL=0.2,FINTIM=60.
 METHOD RKSFX
 LABEL SPRING MASS VISCOUSLY DAMPED WITH DEADSPACE
 PRTPLOT XDOT,X,FOFX
 END
 STOP
ENDJOB
```

Let us note first of all that the program is divided into three parts or *segments*. Blank lines have been left between the segments to make them distinct, but the blank lines are not required. The segments are headed by statements consisting of the words INITIAL, DYNAMIC, and TERMINAL. These are the names of the segments. We will learn later that some programs may use only two segments, while some may use only one, the DYNAMIC segment.

In the initial segment we see statements containing the *labels* CONSTANT and INCON, in which the assignments of system variable values and initial conditions are made. As will be shown later, all CSMP data and control statements begin with a word called a *label*, which enables the translator to identify what the statement is for. Some control statements consist of a label only. The CONSTANT and INCON statements are followed by two statements having the appearance of FORTRAN arithmetic-assignment statements which calculate the value of the mass M and a variable with the name COEF which is equal to the damping coefficient divided by the mass. These two statements appear in the INITIAL segment rather than the DYNAMIC segment because they are executed only at time equal zero. Statements appearing in the DYNAMIC segment are executed at each interval of time from the start of the simulation to the finish time. Note that M is a *real* variable name and not *integer* as it would be by the predefined type specification of FORTRAN.

In the DYNAMIC segment we see first an "unsorted" section consisting of statements placed between NOSORT and SORT statements. Statements in an unsorted section will not be reordered by the translator. It should be noted that this section contains a FORTRAN logical IF statement and a FORTRAN unconditional GO TO, as well as arithmetic-assignment statements and a FORTRAN CONTINUE statement. Full use of FORTRAN IV is permissible in an unsorted section.

The DYNAMIC segment in general is sorted, and if any part of it is not to be sorted it must be placed between NOSORT and SORT statements as illustrated. The statements in a sorted section or segment are reordered if necessary by the translator into proper computational sequence, and the user is freed of the responsibility of putting statements in the proper sequence. However, full use of FORTRAN is not permissible in a sorted section of a program.

We may also note in the DYNAMIC segment that the function names DEADSP and INTGRL are not standard supplied FORTRAN functions. They are supplied CSMP functions, and there are 34 such functions available. XDDOT, XDOT, and X are the CSMP variable names for $\ddot{x}$, $\dot{x}$, and x, respectively, and are names selected by the programmer.

In the TERMINAL segment we see a calculation made for the kinetic energy of the system (CSMP variable name is KE). This calculation is in the TERMINAL segment because it is made at the end of the simulation using the final value of XDOT. This statement is followed by ordinary FORTRAN

```
SPRING MASS VISCOUSLY DAMPED WITH DEADSPACE                          PAGE    1

                           MINIMUM                 XDOT    VERSUS TIME              MAXIMUM
                        -7.6828E 00                                               8.0000E 00
 TIME             XDOT          I                                                  I
0.0            8.0000E 00     -------------------------------------------------+
2.0000E-01     7.0260E 00     ----------------------------------------------+
4.0000E-01    -1.2600E 00     --------------------+
6.0000E-01    -7.6546E 00     +
8.0000E-01    -7.5684E 00     +
1.0000E 00    -6.4429E 00     ---+
1.2000E 00    -2.9556E 00     ---------------+
1.4000E 00     1.5774E 00     -----------------------------+
1.6000E 00     5.4257E 00     ------------------------------------------+
1.8000E 00     7.1513E 00     -----------------------------------------------+
2.0000E 00     7.0579E 00     ----------------------------------------------+
2.2000E 00     4.4652E 00     --------------------------------------+
2.4000E 00    -3.6666E 00     ------------+
2.6000E 00    -6.8152E 00     --+
2.8000E 00    -6.6794E 00     ---+
3.0000E 00    -5.2724E 00     -------+
3.2000E 00    -1.9046E 00     -------------------+
3.4000E 00     2.1149E 00     -------------------------------+
3.6000E 00     5.2571E 00     ------------------------------------------+
3.8000E 00     6.3542E 00     --------------------------------------------+
4.0000E 00     6.2284E 00     --------------------------------------------+
4.2000E 00     3.7195E 00     ------------------------------------+
4.4000E 00    -3.4711E 00     -------------+
4.6000E 00    -6.0133E 00     -----+
4.8000E 00    -5.8943E 00     -----+
5.0000E 00    -4.9065E 00     --------+
5.2000E 00    -2.0984E 00     -----------------+
5.4000E 00     1.4455E 00     -----------------------------+
5.6000E 00     4.3741E 00     --------------------------------------+
5.8000E 00     5.5949E 00     ------------------------------------------+
6.0000E 00     5.4951E 00     ------------------------------------------+
6.2000E 00     4.6239E 00     ---------------------------------------+
6.4000E 00    -1.2296E 00     ---------------------+
6.6000E 00    -5.2816E 00     -------+
6.8000E 00    -5.2013E 00     -------+
7.0000E 00    -4.9431E 00     --------+
7.2000E 00    -3.1837E 00     --------------+
7.4000E 00    -2.5824E-01     -----------------------+
7.6000E 00     2.7067E 00     ---------------------------------+
7.8000E 00     4.5920E 00     ---------------------------------------+
8.0000E 00     4.8491E 00     ---------------------------------------+
8.2000E 00     4.7531E 00     ---------------------------------------+
8.4000E 00     2.8427E 00     ---------------------------------+
8.6000E 00    -2.6444E 00     ----------------+
8.8000E 00    -4.5890E 00     ---------+
9.0000E 00    -4.4981E 00     ----------+
9.2000E 00    -4.2242E 00     -----------+
9.4000E 00    -2.6129E 00     ----------------+
9.6000E 00    -4.7999E-02     ------------------------+
9.8000E 00     2.4842E 00     --------------------------------+
1.0000E 01     4.0293E 00     -------------------------------------+
```

Fig. 9-2 / *A portion of the print-plot output of* XDOT *versus time.*

WRITE and FORMAT statements. Since the TERMINAL segment is unsorted, full FORTRAN is permissible in this segment. After these FORTRAN statements comes the statement with the label TIMER. This card is called a TIMER card and is used in this program to assign values to the integration increment of the independent variable, the output increment, and the finish time for the simulation. The next card has the label METHOD and specifies in this program that a fourth-order Runge-Kutta method with fixed step size is to be used in the integration. The next statement with the label LABEL specifies what the output page headings will be. The next statement with the label PRTPLOT specifies that the variables XDOT, X, and FOFX are to be print-plotted. A portion of the output from this program is shown in Fig. 9-2. Unlike a FORTRAN program, the END statement is not the last statement in a CSMP program. In general, there may be more than one END statement, and the last one is followed by STOP and then ENDJOB unless there are FORTRAN subprograms used with the CSMP program, in which case the ENDJOB card follows the END card of the last subprogram. In the sections which follow, the statements and structure of CSMP programs will be discussed in detail.

9-3 / CSMP STATEMENTS

There are three kinds of statements in CSMP. They are

1 Structure statements
2 Data statements
3 Control statements

Structure statements describe the functional relationships between the variables of the program and define the system being simulated. Data statements are used to assign values to variables which are to remain fixed during a particular run, as, for example, an initial condition for the integration of a differential equation. Control statements are used to control certain operations and quantities associated with the translation, execution, and output of the program. An example would be the step size in the integration routine. Control statements are classified as (a) translation control statements, (b) execution control statements, and (c) output control statements.

We will first discuss the rules for writing each of the three kinds of CSMP statements and look at examples of each, and later we will see how complete programs are formed using all three kinds of statements.

9-4 / STRUCTURE STATEMENTS

CSMP structure statements are like the arithmetic assignment statements of FORTRAN. They consist of a variable on the left of an equal sign and an expression on the right. The expression may be like an ordinary FORTRAN arithmetic expression. That is, it may consist of a single variable, a single constant, a function, or a combination of these separated by the FORTRAN arithmetic operators. However, in addition to the supplied functions available in FORTRAN such as SIN, COS, SQRT, and so on, there are 34 special CSMP functions available. One of these, the integrator function, might be used in a structure statement as follows:

```
Y=INTGRL(IC,X)
```
(9-2)

Mathematically this is equivalent to

$$Y = \int_0^t X\,dt + IC \tag{9-3}$$

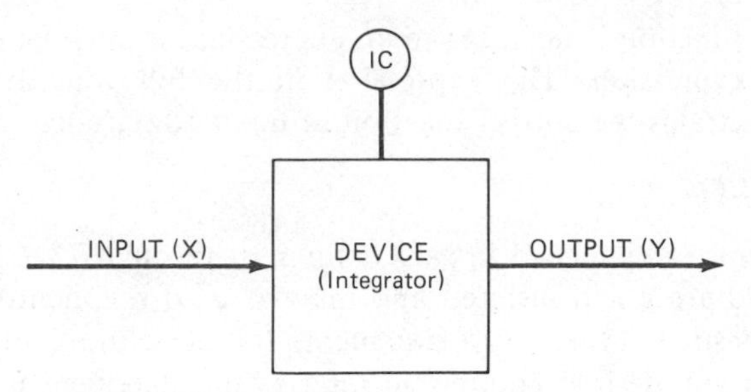

Fig. 9-3 / *Block diagram.*

The X in Eq. 9-2 is called an *input* to the function. In general, inputs can be expressions (single variables, constants, functions, or combinations of these). IC is the initial condition on Y and is a single variable name or constant. In the functions given in Figs. 9-8 through 9-13, X's are always used for inputs, IC for initial conditions, and P for parameters. Like initial conditions, parameters cannot be general expressions but are single variables or constants. If the system to be simulated is shown in block diagram form, INTGRL is thought of as defining the particular *device*, with Y the output from the function block, X the input, and IC the initial condition as shown in Fig. 9-3.

CSMP programs may be written either from a block diagram representation of the system or from a set of ordinary differential equations. CSMP includes a basic set of functional blocks (functions) similar to the integrator shown in Fig. 9-3 which may be used to represent the components of a system. These functions are given in the section which follows. Examples of programming from a block diagram representation of a system and from ordinary differential equations are given in Sec. 9-9.

Additional examples of structure statements are

```
H=(A+B)/2.
XDOT=INTGRL(XDOTO,-(K*X+C*XDOT)/M)
```

In general, FORTRAN rules apply to the writing of the structure statements and the elements of which they are formed. There are a few differences, however, which are brought out in the following rules governing numerical constants, variables, expressions, and structure statements.

1 In general, numerical constants are considered to be real, regardless of whether or not a decimal point is used, and variables are considered to be real regardless of the initial letter of a variable name. It takes a special control statement (the FIXED statement) to make a variable integer.

2 Real constants are restricted to 12 characters total. For example, -126.592E-02 contains the maximum number of characters allowed.

3 If an INTGRL function is included in an expression, it must be the last part of the expression. The expression in the following structure statement illustrates the correct location of the INTGRL function.

```
Y=A*B+INTGRL(IC,X)
```

4 CSMP statements are punched in card columns 1 through 72. Columns 73 through 80 are not translated and may be used for identification purposes if desired. Most CSMP statements (all structure statements) may begin in any desired column so long as the statement is within columns 1 through 72. Column 6 is not reserved for continuation as in FORTRAN statements.

5 Most structure statements can be continued (an exception will be noted later under "MACRO functions"), and this is done by concluding the card to be continued with three consecutive decimal points. None of the decimal points can be in columns 73 through 80. Cards should not be continued in the middle of variables or numerical constants. Up to eight continuation cards can be used for a maximum of nine cards for a statement.

6 Comment cards may be used in a CSMP program for the same purpose for which they are used in a FORTRAN program, to insert explanations in the program to make it more readily understood. In CSMP an asterisk in card column 1 denotes a comment card.

7 Cards with a slash in card column 1 are not translated but are transferred (with the slash removed) directly to the FORTRAN IV subroutine (called UPDATE) which the translator forms. Up to ten such cards can be used. They are inserted at the beginning of the subroutine and so can be used to insert FORTRAN specification statements (a DIMENSION statement, for example) into the subroutine. These statements must be written according to FORTRAN rules rather than CSMP rules. However, CSMP variables cannot be EQUIVALENCE'd.

8 Blank cards may be used to space statements in the listing.

9-5 / CSMP FUNCTIONS

Since CSMP functions are an essential part of structure statements, several of the CSMP functions given in Figs. 9-8 through 9-13 will be explained in more detail than is given by the brief definitions of the functions accompanying the figures.

The IMPLICIT Function

This function, which has the general form

```
Y=IMPL(IC,P,FOFY)
```

is used to solve iteratively an equation of the form

$$y = f(y)$$

for y such that

$$|y - f(y)| \leq p|y|$$

where IC is the first estimate of the value of Y which must be supplied by the programmer, P is the error bound and also supplied by the programmer, and FOFY is the name appearing to the left of the equal sign in the last statement of the algebraic loop definition of $f(y)$. As an example, suppose we have an equation of the form

$$x = Ce^{-2t}B \sin x$$

which must be solved for x. The following CSMP statements could be used to accomplish this:

```
X=IMPL(X0,ERROR,FUNC)
PART1=C*EXP(-2.*TIME)
PART2=B*SIN(X)
FUNC=PART1*PART2
```

The Arbitrary Function Generator

The arbitrary function generator with linear interpolation can be illustrated by a structure statement having the general form

```
Y=AFGEN(FUNCT,X)
```

FUNCT is the name of a function which has been entered into the computer by means of a data statement called a FUNCTION statement. This data statement will be discussed later. At this time it will suffice to say that the function is entered by giving pairs of values of the independent and dependent variables. Thus there are values of the dependent variable Y for discrete values of X, but not for a continuous range of X values as would be the case if Y were given as a mathematical function of X. Use of the AFGEN function will give the value of Y for *any value* of X in the range of X values entered as data, using linear interpolation. The arguments of AFGEN are first the name of the function for which a value of the dependent variable is to be obtained, and second the value of the independent variable in the function when the value of the dependent variable is obtained.

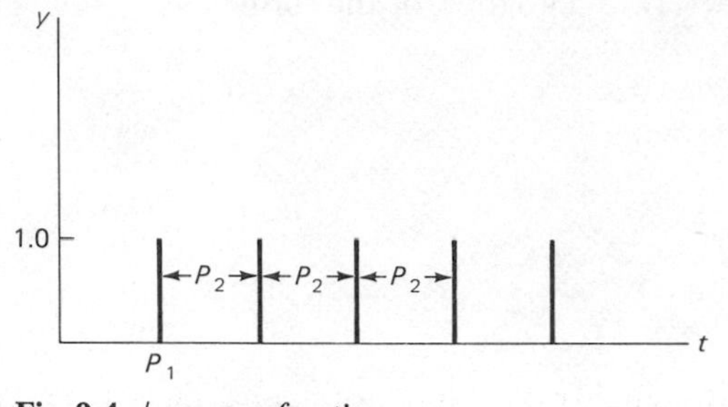

Fig. 9-4 / IMPULS *function.*

Nonlinear Function Generator

If quadratic interpolation is desired instead of linear interpolation, the nonlinear function generator NLFGEN is used instead of AFGEN. A structure statement using this function would have the form

```
Y=NLFGEN(FUNCT,X)
```

where the arguments are the same as given in the preceding paragraph.

The Impulse Generator

The impulse generator can be used to generate a series of impulses having a value of 1.0, starting at time equal to P_1 and continuing at times equal to $P_1 + kP_2$, where $k = 1, 2, 3, \ldots$. The general form of the function is illustrated by the statement

```
Y=IMPULS(P1,P2)
```

The function is represented graphically in Fig. 9-4. If the spike is to have a value of 2.0 instead of 1.0, the IMPULS function can be multiplied by 2.0. For example,

```
Y=2.*IMPULS(0.,P)
```

gives Y as shown in Fig. 9-5.

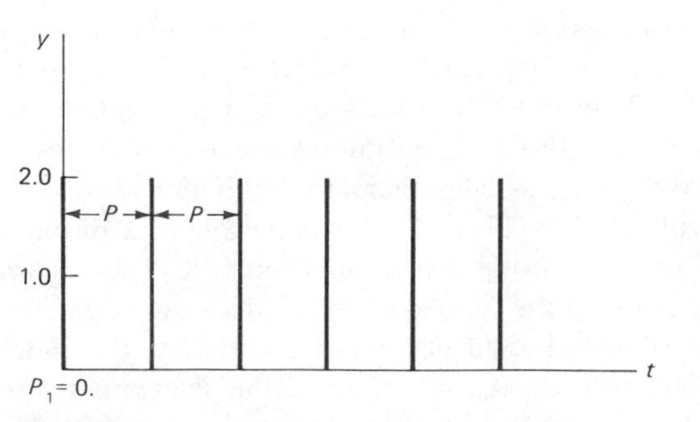

Fig. 9-5 / IMPULS *function.*

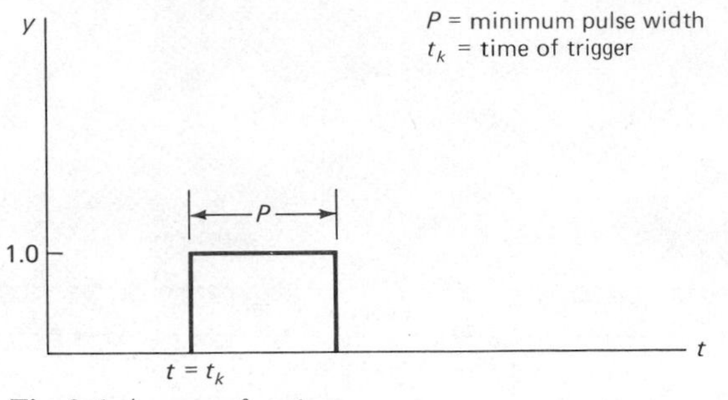

Fig. 9-6 / PULSE *function.*

The Pulse Generator

The PULSE function can be used to generate a pulse starting at any time a "trigger" is greater than zero and continuing for a minimum width of P. If after the pulse has continued for time P, the trigger value is zero or negative, the pulse will end. The function has the general form

```
Y=PULSE(P,X)
```

The input X, which is usually some selected function of time, triggers the pulse whenever its value is greater than zero, providing a pulse is not already being generated. Graphically the function is as shown in Fig. 9-6.

The IMPULS function can be used to trigger the PULSE function. For example, the statement

```
Y=PULSE(1.,IMPULS(0.,2.))
```

will give a square-wave function as shown in Fig. 9-7.

Other important CSMP functions include the first- and second-order lag functions, the limiter function, the dead-space function, the step function, and the ramp function. These, and others, are shown in Figs. 9-8 through 9-13, which follow.

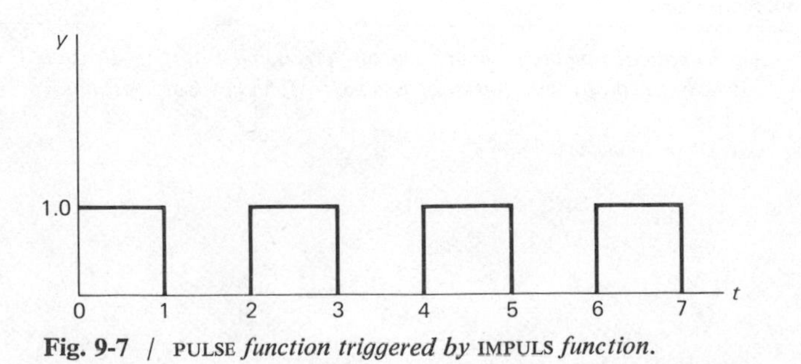

Fig. 9-7 / PULSE *function triggered by* IMPULS *function.*

General Form	Function
Y=INTGRL(IC,X) Y(0)=IC Integrator	$Y = \int_0^t X\,dt + IC$ Equivalent Laplace transform: $\frac{1}{S}$
Y=DERIV(IC,X) $\dot{X}$ (t=0) = IC Derivative	$Y = \frac{dX}{dt}$ Equivalent Laplace transform: S
Y=DELAY(N,P,X) P = delay time N = number of points sampled in interval P (integer constant) Dead time (delay)	$Y(t) = X(t - P) \quad t > P$ $Y = 0 \quad t < P$ Equivalent Laplace transform: e^{-PS}
Y=ZHOLD ($X_1 X_2$) Zero-order hold	$Y = X_2 \quad X_1 > 0$ Y = last output $\quad X_1 \leq 0$ Y(0)=0 Equivalent Laplace transform: $\frac{1}{S}(1 - e^{St})$
Y=IMPL(IC,P,FOFY) IC = first guess P = error bound FOFY = output name of last statement in algebraic loop definition Implicit function	Y = funct (Y) $\lvert Y - \text{funct}(Y)\rvert \leq P\lvert Y\rvert$

Fig. 9-8 / *Mathematical functions. Reprinted by permission from Application Program H20-0240-2. © 1967, 1968 by International Business Machines Corporation.*

General Form	Function	
Y=MODINT(IC,X_1,X_2,X_3) Mode-controlled integrator	$Y = \int_0^t X_3\,dt + IC$ $Y = IC$ Y = last output	$X_1 > 0$, any X_2 $X_1 \leq 0, X_2 > 0$ $X_1 \leq 0, X_2 \leq 0$
Y=REALPL(IC,P,X) Y(0)=IC 1st order lag (real pole)	$P\dot{Y} + Y = X$ Equivalent Laplace transform: $\frac{1}{PS+1}$	
Y=LEDLAG(P_1, P_2,X) Lead-lag	$P_2\dot{Y} + Y = P_1\dot{X} + X$ Equivalent Laplace transform: $\frac{P_1 S + 1}{P_2 S + 1}$	
Y=CMPXPL(IC_1,IC_2,P_1,P_2,X) Y(0)=IC_1 $\dot{Y}$(0)=IC_2 2nd order lag (complex pole)	$\ddot{Y} + 2P_1P_2\dot{Y} + P_2^2Y = X$ Equivalent Laplace transform: $\frac{1}{S^2 + 2P_1P_2S + P_2^2}$	

Fig. 9-9 / *System Macros. Reprinted by permission from Application Program H20-0240-2. © 1967, 1968 by International Business Machines Corporation.*

General Form	Function	
Y=FCNSW(X_1,X_2,X_3,X_4) Function switch	$Y = X_2$ $Y = X_3$ $Y = X_4$	$X_1 < 0$ $X_1 = 0$ $X_1 > 0$
Y=INSW(X_1,X_2,X_3) Input switch (relay)	$Y = X_2$ $Y = X_3$	$X_1 < 0$ $X_1 \geq 0$
Y_1,Y_2=OUTSW(X_1,X_2) Output switch	$Y_1 = X_2, Y_2 = 0$ $Y_1 = 0, Y_2 = X_2$	$X_1 < 0$ $X_1 \geq 0$
Y=COMPAR(X_1,X_2) Comparator	$Y = 0$ $Y = 1$	$X_1 < X_2$ $X_1 \geq X_2$
Y=RST(X_1,X_2,X_3) Resettable flip-flop	$Y = 0$ $Y = 1$ $Y = 0$ $Y = 1$ $Y = 0$ $Y = 1$	$X_1 > 0$ $X_2 > 0, X_1 \leq 0$ $X_1 \leq 0, X_2 \leq 0$, $\begin{cases} X_3 > 0, Y_{n-1} = 1 \\ X_3 > 0, Y_{n-1} = 0 \\ X_3 \leq 0, Y_{n-1} = 0 \\ X_3 \leq 0, Y_{n-1} = 1 \end{cases}$

Fig. 9-10 / *Switching functions. Reprinted by permission from Application Program H20-0240-2. © 1967, 1968 by International Business Machines Corporation.*

General Form	Function		
Y=AFGEN(FUNCT,X) Arbitrary function generator (linear interpolation)	Y = FUNCT (X)		
Y=NLFGEN(FUNCT,X) Arbitrary function generator (quadratic interpolation)	Y = FUNCT (X)	$X_0 \leq X \leq X_n$	
Y=LIMIT(P_1,P_2,X) Limiter	$Y = P_1$ $Y = P_2$ $Y = X$	$X < P_1$ $X > P_2$ $P_1 \leq X \leq P_2$	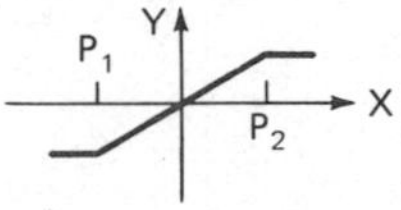
Y=QNTZR(P,X) Quantizer	Y = kP	$(k - 1/2)P < X \leq (k + 1/2)P$ $k = 0, \pm 1, \pm 2, \pm 3 \ldots$	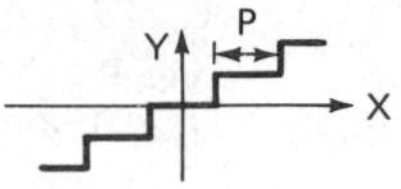
Y=DEADSP(P_1,P_2,X) Dead space	$Y = 0$ $Y = X - P_2$ $Y = X - P_1$	$P_1 \leq X \leq P_2$ $X > P_2$ $X < P_1$	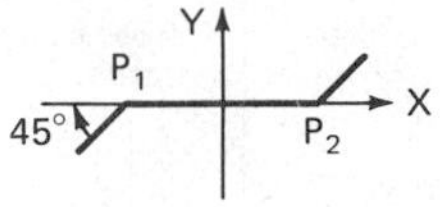
Y=HSTRSS(IC,P_1,P_2,X) Y(0)=IC Hysteresis loop	$Y = X - P_2$ $Y = X - P_1$ otherwise	$(X - X_{n-1}) > 0$ and $Y_{n-1} \leq (X - P_2)$ $(X - X_{n-1}) < 0$ and $Y_{n-1} \geq (X - P_1)$ Y = last output	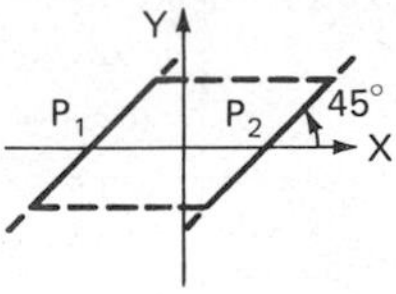

Fig. 9-11 / *Function generators. Reprinted by permission from Application Program H20-0240-2. © 1967, 1968 by International Business Machines Corporation.*

General Form	Function	
Y=STEP(P) Step function	$Y = 0$ $t < P$ $Y = 1$ $t \geq P$ 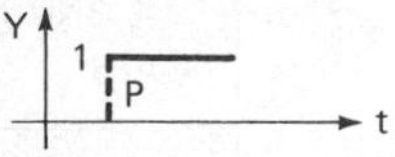	Y, 1, P, t
Y=RAMP(P) Ramp function	$Y = 0$ $t < P$ $Y = t - P$ $t \geq P$ 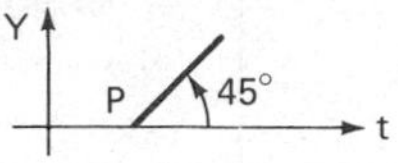	Y, P, 45°, t
$Y=IMPULS(P_1,P_2)$ Impulse generator	$Y = 0$ $t < P_1$ $Y = 1$ $(t - P_1) = kP_2$ $Y = 0$ $(t - P_1) \neq kP_2$ $k = 0, 1, 2, 3. \ldots$ 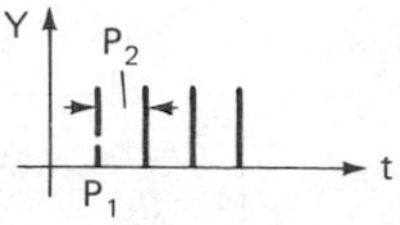	Y, P_2, P_1, t
Y=PULSE(P,X) P = minimum pulse width Pulse generator with $X > 0$ as trigger)	$Y = 1$ $T_k \leq t < (T_k + P)$ or $X > 0$ $Y = 0$ otherwise T_k = time of trigger 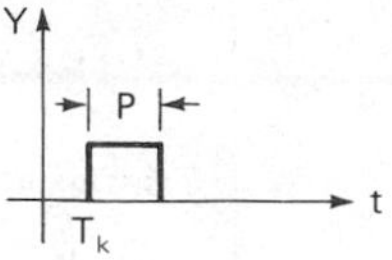	Y, P, t, T_k
$Y=SINE(P_1,P_2,P_3)$ P_1 = delay P_2 = frequency (radians per unit time) P_3 = phase shift in radians Trigonometric sine wave with delay, frequency and phase parameters	$Y = 0$ $t < P_1$ $Y = \sin(P_2(t - P_1) + P_3)$ $t \geq P_1$ 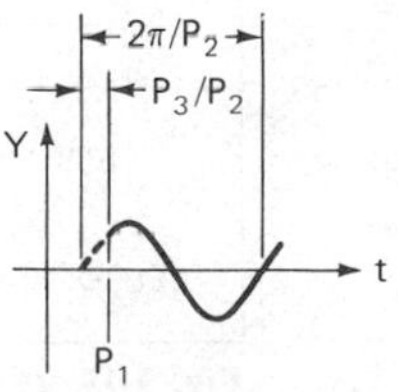	$2\pi/P_2$, P_3/P_2, Y, t, P_1
$Y=GAUSS(P_1,P_2,P_3)$ P_1 = any odd integer P_2 = mean P_3 = standard deviation Noise (random number) generator with normal distribution	Normal distribution of variable Y p(Y) = probability density function 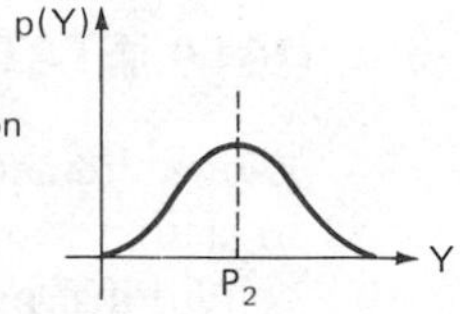	p(Y), Y, P_2
Y=RNDGEN(P) P = any odd integer Noise (random number) generator with uniform distribution	Uniform idstribution of variable Y p(Y) = probability density function 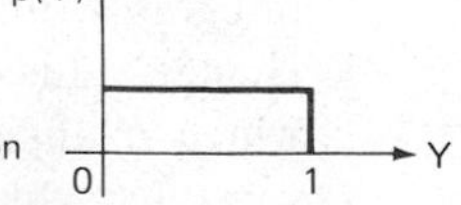	p(Y), Y, 0, 1

Fig. 9-12 / *Signal sources. Reprinted by permission from Application Program H20-0240-2. © 1967, 1968 by International Business Machines Corporation.*

General Form	Function	
Y=AND(X_1, X_2) And	Y = 1 Y = 0	$X_1 > 0, X_2 > 0$ otherwise
Y=NAND (X_1, X_2) Not and	Y = 0 Y = 1	$X_1 > 0, X_2 > 0$ otherwise
Y=IOR (X_1,X_2) Inclusive or	Y = 0 Y = 1	$X_1 \leq 0, X_2 \leq 0$ otherwise
Y=NOR(X_1,X_2) Not or	Y = 1 Y = 0	$X_1 \leq 0, X_2 \leq 0$ otherwise
Y=EOR(X_1,X_2) Exclusive or	Y = 1 Y = 1 Y = 0	$X_1 \leq 0, X_2 > 0$ $X_1 > 0, X_2 \leq 0$ otherwise
Y=NOT(X) Not	Y = 1 Y = 0	$X \leq 0$ $X > 0$
Y=EQUIV (X_1,X_2) Equivalent	Y = 1 Y = 1 Y = 0	$X_1 \leq 0, X_2 \leq 0$ $X_1 > 0, X_2 > 0$ otherwise

Fig. 9-13 / *Logic functions. Reprinted by permission from Application Program H20-0240-2. © 1967, 1968 by International Business Machines Corporation.*

9-6 / DATA STATEMENTS

Data statements are punched in card columns 1 through 72. They are used to assign values to the variable names constituting the various initial conditions, parameters, and constants of the simulation. In most simulation languages initial conditions, parameters, and constants[2] are all considered as variables. Initial conditions, which are used with the integrator function and several others, are variables which may or may not change values for different runs of a simulation. Parameters are variables which change values for different runs of a simulation. Constants are variables which do not change values for different runs of a simulation. These three types of variables can be used interchangeably in CSMP.

[2] Note that the constants discussed here are not the numerical constants of CSMP which are similar to FORTRAN constants, and which were discussed earlier in this chapter. Thus the word "constant" is used in two different ways in CSMP—first, to describe numerical constants consisting of a string of digits, and second, to describe a variable whose value remains constant during a simulation study.

INCON, PARAMETER, and CONSTANT cards would usually be used to assign values to variables used as initial conditions, parameters, and constants, respectively. These labels are used merely to make the source program easily understood. However, the programmer may, for example, use a PARAMETER card to assign values to variables used as initial conditions if he so desires.

As an example of entering initial conditions with an INCON card, the data statement

```
INCON XO=2.5, XDOTO=0.0
```

assigns the value 2.5 to the name XO and 0.0 to the name XDOTO. At least one blank must follow any card label such as INCON. The label can begin in any card column. Only the first four characters of data and control card labels are actually examined (except for COMMON and ENDJOB), so the labels given may be shortened if desired.

As mentioned earlier, parameter and constant cards use the labels PARAMETER and CONSTANT, respectively. Examples are

```
PARAMETER VT=160., VL=440., CD=0.7
CONSTANT PI=3.14159, SIGMA=0.173E-8
```

As many variables as space permits can be assigned values on a data card. If it is necessary to continue to another card, it is done with three successive decimal points in the same manner that structure statements are continued. That is, any card concluding with three decimal points consecutively is considered to be followed by a continuation card. As many continuation cards as required may be used. Variable names and constants should not be split in continuing from one card to another.

Several runs of a program using different values of a variable can be specified by enclosing in parentheses the several desired values of the variable, separated by commas. For example, if the statement

```
PARAMETER I=.0006, SK=(.669,.671,.680,.692)
```

were used in a program, the program would run four times, with SK having the value .699 in the first run, .671 in the second run, and so forth. If the successive values of the variable have equal incremental changes for each run, such as L = 3.5, 3.6, 3.7, . . . , 4.5, a statement such as

```
PARAMETER I=.0006, L=(3.5,10*0.1)
```

could be used. The value of L in the first run would be 3.5, and there would be ten additional runs with L taking on values of 3.6 through 4.5 with equal increments of 0.1. The repeat constant (10 in the above statement) should be written without a decimal point as shown. A sequence of simulation runs may also be specified using several repeat constants and increments, along with given values of the parameter. For example,

```
PARAMETER TERM=(150.,155.,5*1.0,163.,2*2.0)
```

is equivalent to

```
PARAMETER TERM=(150.,155.,156.,157.,158.,159.,160.,
                                       163.,165.,167.)
```

The maximum number of runs possible in any sequence is 50, and only one multiple-value parameter should be used for each sequence. A sequence of runs due to a multiple-value parameter is terminated by an END card. After an END card, more runs can be specified by entering new variable values, as will be explained later in the chapter under "Control Statements" and "CSMP Examples—Run Control," Secs. 9-8 and 9-9, respectively. This is the advantage of using data statements to assign numerical values to variable names instead of using structure statements for that purpose. Assigned in this way, variable values can be changed automatically between successive runs or sequences of runs of a program.

Another important data statement is the FUNCTION statement. It is used for entering into a simulation pairs of values of the independent and dependent variables of a function. Thus if $y = f(x)$ is available in tabular or graphic form rather than in equation form, the function is specified at discrete intervals by entering pairs of values of the coordinates x and y into the simulation using the FUNCTION statement. These coordinates are for use by the arbitrary function generator (AFGEN) and the nonlinear function generator (NLFGEN) as explained earlier under "CSMP Functions." As an example, a function named YOFX might be entered as follows:

```
FUNCTION YOFX=(0.0,4.0),(1.0,4.3),(2.0,5.0),(3.0,6.1),...
(4.0,7.3),(5.0,8.7),(6.0,10.1),(7.0,11.0)
```

In this example values of x and y are paired off with parentheses. However, the parentheses are optional and may be omitted. The first number of each pair shown is the value of the independent variable, the second that of the corresponding dependent variable. The increments of the independent variable need not be of equal size as they are in the above example, but they must be listed in order of *algebraically* increasing size. FUNCTION is the card label and must be followed by at least one blank. Any number of continuation cards may be used, and any number of points per function may be specified. There is no restriction on the number of functions defined per simulation.

Let us discuss how the function YOFX, as defined above, could be used in the arbitrary function generator. Suppose that the structure statement

```
Y=AFGEN(YOFX,X)
```

appears in a CSMP program in which the above FUNCTION statement has been used. Y is assigned the value of the function YOFX for whatever value X has in the program. If the value of X is between values given in the FUNCTION statement, the corresponding value of Y is obtained by linear interpolation when AFGEN is used. For example, if X had the value 1.5, then Y would be assigned the value 4.65 with YOFX defined as in the FUNCTION statement above.

Data statements can be in any order and may be mixed with structure statements if desired. However, many CSMP programmers prefer to put data statements near the beginning of the source program, except for those that follow an END statement. Data statements which follow an END statement cause the values of variables to be changed in preparation for another run, or another sequence of runs if a multiple-valued parameter is used. The details of changing parameters for new runs by placing new data statements after an END card will be discussed under "Control Statements" and "CSMP Examples—Run Control."

9-7 / MODEL STRUCTURE

Before discussing control statements, let us consider the structure of a CSMP simulation. The "heart" of a continuous system simulator is the built-in process for integration of the differential equations that describe the system. One of the integration methods of numerical analysis must be used. Thus the independent variable (TIME in a CSMP program unless the name is changed by one of the control statements to be discussed later) varies by increments from zero to the finish time. Calculations of the integration routine, and perhaps others as well, must be performed for each increment of time. However, there may be some calculations which need to be made just once for each run, perhaps prior to the run or after the completion of each run. For example, in the differential equation

$$\frac{d^2\theta}{dt^2} = \frac{T_c + k\theta_c}{I_0} - \frac{k}{I_0}\theta$$

the first term on the right of the equal sign is made up of four basic parameters which will not change during a run. There is no need to evaluate this term for each increment of time. It need be done only once prior to the run using a structure statement such as

```
TERM1=(TCL+SK*THCL)/EYE
```

Similarly, there are problems in which calculations must be made at the conclusion of a run using final values of some of the system variables. To satisfy these requirements, CSMP provides the possibility of writing a program in three *segments*, Initial, Dynamic, and Terminal. The Initial segment is intended for those structure statements that are to be executed only once prior to the run. Many programmers also include data statements in the Initial segment if one is used. The Dynamic segment contains the statements specifying integration of the differential equations, along with any other structure statements which must be executed at each increment of time during the run. The Terminal segment is used for calculations made at the conclusion of a run. In some simulations only a Dynamic segment need be used.

The segments described above consist of one or more *sections*. For many simulations, the Dynamic segment will consist of a single section which has what is called *parallel structure*. In other simulations, one or more *procedural* sections may be required in the Dynamic segment, along with parallel sections.

In an actual physical system, most phenomena occur in parallel. For example, in a spring-mass-damper system the displacement x, velocity $\dot{x}$, and acceleration $\ddot{x}$ are all changing simultaneously. When such a system is simulated on the analog computer, these quantities can all be obtained as output simultaneously. However, with a digital computer these quantities are obtained sequentially, and in a FORTRAN program which is solving such a system, care must be exercised to write the statements in the proper order. Thus the FORTRAN program has what is called procedural structure.

Digital computer simulation, however, is desired to achieve the same effect as the simulation on an analog computer. This is accomplished in a parallel section by means of a sorting algorithm which arranges computations into proper sequence, allowing the programmer to write his statements in any order he desires. This type of section is also called a *sorted* section. The Initial segment of a CSMP simulation is sorted. The Dynamic segment is also sorted, but in some simulations it is necessary to have one or more unsorted sections in this segment. In an unsorted section full use of FORTRAN IV, including the use of branching, input, and output statements is permitted. For such statements FORTRAN rules apply, in general. However, the statements may begin in any column, and column 6 need not be reserved for continuation purposes. If continuation of FORTRAN statements in an unsorted section is required, it is done by CSMP rules; by the use of three successive periods. How an unsorted section may be specified will be explained under "Control Statements." The Terminal segment has procedural structure (is unsorted), and all structure statements are considered to be in the proper order. This allows the full use of FORTRAN IV in the Terminal segment, including the use of FORTRAN branching statements, and input and output statements if desired. This standard structure may be overridden, as will be explained later under "Control Statements."

9-8 / CONTROL STATEMENTS

Control statements make specifications having to do with the translation, execution, and output of the simulation, such as specifying the increment of the independent variable of the integration routine, the finish time for the simulation run, and the variables whose values are to be printed out. In general, these statements may appear in any order and may be mixed with structure and data statements. They may begin in any card column (except for the ENDJOB, COMMON, and ENDDATA statements) as long as card columns 73 through 80 are not used. In this section of the text we will discuss the most commonly used Control statements of the following three types: translation control, execution control, and output control.

Translation Control Statements

The INITIAL *Statement.* As previously discussed under "Model Structure," a CSMP program may be divided into three segments: Initial, Dynamic, and Terminal. A statement with just the label INITIAL on it marks the beginning of the Initial segment.

The DYNAMIC *Statement.* This statement consists of just the label DYNAMIC, and it marks the beginning of the Dynamic segment and the end of the Initial segment. The Dynamic segment is terminated by the TERMINAL card unless no Terminal segment is used. In this case the END or CONTINUE statement terminates the Dynamic segment.

The TERMINAL *Statement.* This statement consists of just the label TERMINAL, and it marks the beginning of the Terminal segment and the end of the Dynamic segment. The Terminal segment is terminated by the first END or CONTINUE statement.

The END *Statement.* The END card permits the simulation to accept new data and control cards, and if such cards follow the END card, a new run is initiated using these new statements. The independent variable (TIME unless the name has been changed) is reset to zero, and initial conditions are reset for the new run. The last END card is followed by a STOP card, which means that a new run will not be initiated.

The CONTINUE *Statement.* When an END statement is followed by data and/or control cards rather than a STOP card, the run (or sequence of runs if a multiple-value parameter is used) terminated by the END statement is followed by another simulation using the new variable values. The independent variable TIME is reset to zero, the initial conditions are reset, and a new run is initiated using the new values. However, if the programmer wishes to change a control statement during a simulation, and then wants to continue on from that point in time and not go back to TIME equal zero, he must use a CONTINUE statement instead of an END statement.

For example, the programmer might want to change the increment of the independent variable at some point in a simulation. He does not want the solution to start again from TIME equal to zero, so he would use a CONTINUE statement preceding his new control card rather than an END statement. The use of END and CONTINUE statements for run control will be illustrated in the section "Run Control," The FORTRAN statement CONTINUE, which might also be used in a CSMP program, should always be numbered so that no confusion exists between the CSMP CONTINUE and the FORTRAN CONTINUE. Multivalue parameters should not be used in conjunction with the CONTINUE statement.

The STOP *Statement.* This statement consists of the label STOP only. The STOP card follows the last END card and indicates that no new run is to be initiated.

The ENDJOB *Statement.* The ENDJOB card follows the STOP card unless user-supplied subroutines are used, in which case it follows the last subroutine. The statement consists of the label ENDJOB only, and it must begin in card column 1.

The RENAME *Statement.* In CSMP there are certain reserved names such as the name of the independent variable which has the reserved name TIME. The increment of the independent variable has the reserved name DELT. If the integration routine used adjusts the step size based on an error analysis, a minimum step size can be specified by assigning a value to DELMIN. The maximum value of the independent variable is specified by assigning a value to FINTIM. The independent variable increment for printing is called PRDEL, and for print-plot output OUTDEL is used. If any of these quantities are used in structure statements, or on a TIMER card (an execution control card on which all but TIME may be assigned values) in the program, these reserved names are used. However, any or all of these six reserved names can be changed by using a RENAME card. For example, if the independent variable of the problem is a linear distance x rather than time, the programmer might prefer to use the names X, XMAX, and DELTX instead of TIME, FINTIM, and DELT, respectively. A RENAME card to accomplish these changes would be

```
RENAME TIME=X, FINTIM=XMAX, DELT=DELTX
```

RENAME is the card label, and it must be followed by at least one blank. Commas separate the rename specifications. The substitute names X, XMAX, and DELTX replace the reserved names. For example, if the programmer uses the independent variable name in a statement in his program, he would use X instead of TIME. Also, in the printed output the name X will appear as a column heading instead of TIME. A RENAME card cannot be continued, but more than one RENAME card can be used if needed. When used, a RENAME card is usually the first card of a CSMP source program, and it *must* appear before the TIMER card.

The FIXED *Statement.* All variable names in CSMP are considered real unless they appear in a FIXED statement. This rule also applies to FORTRAN statements appearing in a CSMP program. If it should be necessary to use integer variables in the program with names such as N and JOB, the FIXED card would be punched as follows:

```
FIXED N,JOB
```

The label FIXED must be followed by at least one blank.

The COMMON *Statement.* When the FORTRAN subroutine UPDATE is formed by the CSMP translator, a COMMON block is established which includes TIME, DELT, DELMIN, FINTIM, PRDEL, and OUTDEL; the variables listed on PARAMETER, CONSTANT, and INCON cards; and the variables defined in CSMP structure statements. If the programmer includes user-supplied FORTRAN subprograms along with his CSMP program (as explained later), he can make these variables available to any of his subprograms by placing CSMP COMMON cards at the beginning of the subprograms. The translator replaces these cards with the FORTRAN COMMON statements needed. The CSMP COMMON card consists of just the label COMMON punched in card columns 1 through 6.

MACRO *Functions.* Cards with the labels MACRO and ENDMAC are used to precede and follow a group of cards, respectively, identifying them as defining a MACRO function. MACRO functions are one of three types of user-defined functions which may be used in a CSMP program. The other two are PROCEDURE functions and FORTRAN subprograms, both of which are discussed later.

Let us first consider a simple MACRO function definition to help clarify the rules concerning these functions which follow. Suppose that in a CSMP program it is necessary to determine the orthogonal components of several vector quantities. Instead of writing several groups of similar statements, a MACRO function could be used which might consist of the following statements:

```
MACRO VECTRX,VECTRY=COMPO(VECTOR,THETA)
              VEXTRX=VECTOR*COS(THETA)
              VECTRY=VECTOR*SIN(THETA)
ENDMAC
```

In the above statements

```
VECTRX,VECTRY=COMPO(VECTOR,THETA)
```

is called the *canonical* statement in the MACRO definition. Once having defined the MACRO function, it can be used (referenced) as often as desired.

Before illustrating the referencing of the MACRO function within the simulation structure statements, let us list the rules which must be observed in writing the MACRO definition.

1 VECTRX and VECTRY are dummy names for the outputs of the function, the x and y components of the vector quantity. There may be any number of outputs. When there are several outputs, as in this example, the dummy names for the outputs are to the left of the equal sign in the MACRO statement and separated by commas. If there is only one output, the single dummy name for the variable is also to the left of the equal sign.
2 The name assigned by the user to the above function is COMPO, and it appears just to the right of the equal sign in the MACRO statement. When the function is referenced in structure statements, the name COMPO must be used.
3 VECTOR and THETA are dummy names for the arguments of the function. They represent, respectively, the magnitude of the vector quantity and the angle in radians which locates the vector with respect to the x axis. In general, they may represent initial conditions, parameters, or other input variables used in the definition of the function outputs.
4 The statements between the MACRO and ENDMAC translation control cards define the output or outputs of the function (define the MACRO). They are self-explanatory in the above example. In general, they may contain any CSMP functions—INTGRL, for example—but FORTRAN logical, branching, or input/output statements are not to be used unless they are within a PROCEDURE function which is embedded within the MACRO function. There should be no data and control cards within the MACRO, and structure statements within a MACRO definition may not be continued. The canonical statement in the MACRO definition may be continued for a total of only four cards. The statements of a MACRO may be written in any order since the translator will automatically sort them to give the correct computational sequence, if this is required. If the MACRO is within a NOSORT section, however, the automatic sorting will not be done. If a PROCEDURE function is embedded within a MACRO function, the statements in the PROCEDURE will be moved as a single entity, but there is no internal sorting of statements within a PROCEDURE.
5 The card with the label ENDMAC marks the end of the MACRO definition.
6 The group of cards defining the MACRO are placed at the beginning of the CSMP source program ahead of any structure statements in the INITIAL or DYNAMIC segments.

Having seen a simple MACRO definition, and having discussed the rules to be observed for writing such a definition, let us next discuss the referencing or use of the MACRO function within the structure statements of a program.

The following structure statement uses the MACRO function COMPO.

```
VX,VY=COMPO(VELOC,THRAD)
```

Note that this statement has the same format as the canonical statement in the MACRO definition previously discussed. VELOC and THRAD are inputs to, or the real arguments of, the function COMPO. VX and VY are the outputs of the function. These four names replace the dummy names VECTOR, THETA, VECTRX, and VECTRY, respectively, in the statements of the MACRO definition which calculates the output values.

If the above statement appeared in the Dynamic segment of a CSMP program, values of VX and VY, the x and y components of VELOC, would be calculated at each time interval of the simulation, and they would be available for use in other statements of the simulation. In order for the MACRO to be a time-saving feature, it would have to be referenced more than once in the simulation.

PROCEDURE *Functions*. The PROCEDURE function is a type of user-defined function in which FORTRAN IV may be employed along with CSMP structure statements. There is no sorting done within a PROCEDURE, and all executable FORTRAN statements, including branching and input/output statements, may be freely used. We will first look at an example of a PROCEDURE function used in a CSMP program and then point out the rules that must be observed in defining such a function.

The CSMP program containing the PROCEDURE which we will examine is used to determine the dynamic characteristics of a small rocket fired vertically upward. The differential equation of motion of the rocket will include a constant thrust force until burnout, at which time the thrust drops to zero. The vertical acceleration $\ddot{y}$ must be expressed differently during the burn and after burnout. Therefore, $\ddot{y}$ (CSMP variable name YDDOT) is used as the output of a PROCEDURE function, and FORTRAN logic is employed to transfer control to the correct arithmetic assignment statement defining $\ddot{y}$. The complete CSMP program is as follows:

```
INITIAL
  CONSTANT THRUST=7000.,DK=.008,G=32.17,BURNTM=60.
  INCON Y0=0.0,YDOT0=0.0
  A=G*THRUST
  B=DK*G


DYNAMIC
PROCEDURE YDDOT=FUNCT(A,G,BURNTM,TIME,B,YDOT)
      IF(TIME.GE.BURNTM) GO TO 1
      W=3000.-40.*TIME
      YDDOT=A/W-G-B/W*YDOT**2
      GO TO 2
    1 YDDOT=-G-B/(3000.-40.*BURNTM)*SIGN(YDOT**2,YDOT)
    2 CONTINUE
```

```
ENDPRO
  YDOT=INTGRL(YDOTO,YDDOT)
  Y=INTGRL(YO,YDOT)

  TIMER DELT=0.1,OUTDEL=2.0,FINTIM=200.
  METHOD RKSFX
  LABEL DYNAMIC CHARACTERISTICS OF A ROCKET
  PRTPLOT YDDOT,YDOT,Y
  END
  STOP
ENDJOB
```

The PROCEDURE consists of the group of statements beginning with the translation control card having the label PROCEDURE, and it ends with the translation control card having the label ENDPRO. This group of statements may be inserted anywhere in the Dynamic segment since it will be sorted as a group, but there is no internal sorting of statements within the PROCEDURE.

The following rules govern the writing of PROCEDURE functions:

1 The first card of the PROCEDURE is a translation control card with the label PROCEDURE. The label is followed by at least one required blank separating the label from the outputs of the PROCEDURE. In this example there is only one output, YDDOT, but in general there may be any number of outputs. If there are several outputs, they are separated by commas like the outputs in the canonical statement of a MACRO definition. A PROCEDURE card on which two outputs are given might appear as

   ```
   PROCEDURE SDDOT,THDOT=FOFV(D,M,V,THETA)
   ```

   The output, or outputs, of the PROCEDURE are to the left of the equal sign, while the function name and the input variables of the function are to the right of the equal sign. The input variables are enclosed in parentheses and are separated by commas. The function name in the example being discussed is FUNCT, and the input variables are A, G, BURNTM, TIME, B, and YDOT. A CSMP PROCEDURE function name is just a dummy name because it is not used to reference the function and appears only on the card with the label PROCEDURE. It appears to serve no purpose, but it is required. Any arbitrary name will serve the purpose.

2 In the example program the variable YDDOT is defined within the PROCEDURE. Note that YDDOT appears on the PROCEDURE card as an output. If the function had more than one output, all the variable names for the output variables would be defined in statements following the PROCEDURE card. Variables other than outputs may also be defined in a PROCEDURE, but they are not available for output by use of the

CSMP PRINT or PRTPLOT (print-plot) output controls statements. Variables which are defined in a PROCEDURE must be listed as outputs of the PROCEDURE function if they are to appear on output control cards.
3 The PROCEDURE is terminated by a translation control card with the label ENDPRO.

Note that a PROCEDURE function is used differently from a MACRO function. The statements defining a MACRO are placed ahead of the INITIAL segment, and then the function is referenced as often as desired in the structure statements that follow by using a statement having the same format as the canonical statement of the MACRO definition. The statements defining a PROCEDURE are generally placed right where they are used since a PROCEDURE function is not referenced. It is generally used just once (once at each time interval of the simulation if it appears in the Dynamic segment). The only way it can be used more than once is to embed it in a MACRO definition which may then be referenced a number of times. By embedding a PROCEDURE within a MACRO, it becomes possible to use executable FORTRAN statements, including branching and input/output statements, in the MACRO. However, if a FORTRAN output or input statement is used in a PROCEDURE embedded in a MACRO, the accompanying FORMAT statement should be placed in the Terminal segment, since a FORMAT statement cannot be used in a MACRO. Although a PROCEDURE can be embedded in a MACRO, it should be noted that one PROCEDURE cannot be embedded within another PROCEDURE.

The NOSORT *Statement.* This statement defines the beginning of a section in which statements are not reordered by the translator. Full use of FORTRAN, including the use of branching and input and output, is permissible in an unsorted section. Continuation of the FORTRAN statements, if necessary, is done by CSMP rules, however. An unsorted section would be used in the Initial or Dynamic segments since they are normally sorted. An unsorted section is terminated by a SORT statement or by the statement indicating the beginning of the next segment (Dynamic or Terminal).

If an unsorted section appears in the middle of the Dynamic segment, the Dynamic segment will be separated into three distinct parts, a first part prior to the NOSORT statement, the part following this statement up to a SORT statement, and the part following the SORT statement. The first and third parts will be sorted, but the sorting is confined to within the parts; the translator will not reorder the statements so that they move from the first part to the third part or vice versa.

The SORT *Statement.* This statement defines the beginning of a section in which statements are sorted by the translator. The section is terminated by a NOSORT statement or by the end of the segment.

DATA *and* ENDDATA *Statements.* If the programmer uses a FORTRAN READ statement (usually placed in a NOSORT section of the Initial segment), he must of course have data cards available for reading. These data cards are placed between CSMP cards having the labels DATA and ENDDATA, and this group of cards follows the END statement. The DATA statement informs the translator that the cards which follow are not CSMP cards and are to be skipped over by the translator. The label ENDDATA must be punched in card columns 1 through 7.

Execution Control Statements

These statements are concerned with the actual execution of the simulation, such as specifying the step size for the integration routine, the finish time for the run, and so forth. They consist of a card label followed by at least one blank, and then by certain assignments that are made on that particular card. An execution control card called a TIMER card might appear as follows:

```
TIMER DELT=0.1, PRDEL=0.2, FINTIM=20.0
```

TIMER is the label, while DELT=0.1, PRDEL=0.2, and FINTIM=20.0 are assignments assigning values to the step size for integration, the print increment for output, and the total run time, respectively. As can be seen, successive assignments are separated by commas. Execution control statement labels are TIMER, FINISH, RELERR, ABSERR, and METHOD. Each is explained in the following discussion.

The TIMER *Statement.* This statement allows the programmer to specify the values of certain system variables as illustrated in the preceding paragraph. These variables all have *reserved* names and will be discussed in turn below.

DELT This is the name reserved for the increment of the independent variable. The programmer should assign a value to DELT which is a submultiple of the output increment PRDEL and OUTDEL. This would normally be done by choosing a value for DELT first and then making PRDEL and OUTDEL multiples of the chosen value of DELT. If the programmer does not make DELT a submultiple of PRDEL and OUTDEL, the translator adjusts DELT to be a submultiple of the smaller of the two output increments. If neither of the latter has been specified, DELT is made a submultiple of FINTIM/100. If DELT is not specified by the programmer, the translator will assign it a value equal to one-sixteenth of the smaller of PRDEL or OUTDEL, if either or both of the latter have been specified. In general, the programmer should always assign a value of DELT rather than depending upon the translator to assign the value. Roundoff error in the time increment can be avoided by assigning DELT a value equal to $1/2^n$, where n is an integer. Such values (as, for example, .015625 or .0078125) are expressed exactly in binary form.

DELMIN Two of the integration routines which may be used (the fifth-order predictor-corrector Milne method and the fourth-order Runge-Kutta with variable interval) have an adjustable step size, the adjustment being made based on the results of an error analysis at each step. DELMIN is used to specify a minimum value of DELT. That is, DELT will not be adjustable to less than the value of DELMIN.

FINTIM This is the maximum value of the independent variable specified for the simulation. In general, FINTIM should be assigned a value which is a multiple of the output increments PRDEL and/or OUTDEL. If this rule is not observed, the translator will adjust its value to the largest multiple of the smallest output increment which is less than the specified value of FINTIM. A value must always be specified for FINTIM, even though the simulation may be terminated before that value is reached. For if a value is not specified, FINTIM is set equal to zero by the translator, and the simulation cannot get started. The independent variable may not always attain the value of FINTIM if the simulation is halted by a FINISH condition (see discussion of the FINISH statement later in this section).

PRDEL In a simulation study the values of certain variables are normally required as output at specified increments of the independent variable. If these variable values are to be just printed out (not plotted), the output increment has the reserved name PRDEL, and it is assigned a value on the TIMER card. The smaller of PRDEL and OUTDEL should be a submultiple of the larger, or the smaller will be adjusted so that this is true by the translator. A value of PRDEL should normally always be specified if it is to be used, but if the programmer neglects to do so, its value will be set equal to the value specified for OUTDEL. If a value has not been specified for the latter, the value of PRDEL will be set to FINTIM/100.

OUTDEL If any of the variables are to have their output values print-plotted, the output increment is called OUTDEL. The programmer should assign a value to OUTDEL if there is a PRTPLOT card specifying that certain variables are to have their values print-plotted. If an OUTDEL value is required, but the programmer fails to assign a value, OUTDEL is assigned the value of PRDEL if a value has been specified for the latter. If not, OUTDEL is set equal to FINTIM/100 by the translator. If both PRDEL and OUTDEL are used, the smaller of the two should be a submultiple of the larger, or the smaller will be adjusted by the translator to make it so.

The FINISH *Statement.* Utilizing a FINISH card, the user can specify conditions for terminating the run other than having the independent value reach a value equal to that of FINTIM. For example, in the rocket problem on page 593 which was used to illustrate a procedure function, a value of 200 sec was used for FINTIM. If the programmer preferred to stop the run when the altitude reached 43,000 ft if this occurred before 200 sec, the FINISH card could be punched as

```
FINISH Y=43000.
```

Several finish conditions can be specified if desired. Using the rocket problem again as an example, the user might decide to stop the run when the altitude reached 43,000 ft or when the value of $\dot{y}$ became negative. The FINISH card might then appear as

```
FINISH Y=43000.,YDOT=-.0001
```

Note that YDOT is set equal to a small negative value rather than zero. This is done because the initial value of YDOT is zero, and the FINISH condition would stop the simulation immediately if YDOT were specified as zero in the FINISH statement. As soon as one of the variables Y or YDOT reaches or crosses the specified bound, the run will terminate.

A variable can be used on the right of the equal sign instead of a constant. The specification

```
FINISH X1=X2
```

would cause the run to terminate when X1 and X2 were equal or when their difference changed sign. Up to ten finish conditions can be specified. Finish specifications for one run can be nullified for a subsequent run by means of a RESET card placed after the END or CONTINUE card. The RESET card will be discussed subsequently.

The RELERR *and* ABSERR *Statements.* In the fourth-order Runge-Kutta integration routine with variable step size, the step size is reduced until the following criterion is satisfied:

$$\frac{Y_{t+\Delta t} - Y^s}{A + R|Y_{t+\Delta t}|} \leq 1$$

where Y^s is $Y_{t+\Delta t}$ calculated by Simpson's rule and where A is the *absolute error* and R is the *relative error*. In Milne's fifth-order predictor-corrector method, the step size is reduced until the following criteria are satisfied:

$$\frac{.04|Y^c - Y^p|}{R|Y^c|} \leq 1 \qquad \text{for} \qquad |Y^c| > 1$$

or

$$\frac{.04|Y^c - Y^p|}{R} \leq 1 \qquad \text{for} \qquad |Y^c| \leq 1$$

where Y^p is the predicted value of $Y_{t+\Delta t}$, Y^c is the corrected value, and R is the relative error. The absolute error A and the relative error R may be specified by the programmer. If RELERR is not specified, a value of .0001 will be used. ABSERR is used only with the fourth-order Runge-Kutta method. If it is not specified, .001 is used. The following statements illustrate the specification of relative error and absolute error values.

```
ABSERR XDOT=.0005,X=.0003
RELERR XDOT=.0002,X=.0002
```

The METHOD *Statement.* A method card consists of the label METHOD followed by a required blank and the name of the integration routine which the programmer wants used for the simulation. One of the following names may be used.

- `MILNE` This specifies the use of Milne's fifth-order predictor-corrector method with variable step size.
- `RKSFX` This name specifies that the fourth-order Runge-Kutta method with fixed step size will be used.
- `RKS` This specifies the fourth-order Runge-Kutta method with variable integration interval. It is the default method which is used if the METHOD card is omitted.
- `SIMP` A method based on Simpson's integration rule is specified by this name.
- `TRAPZ` This name specifies the use of a second-order method variously known as a modified Euler method, an improved Euler method, or the Euler-Gauss method in which a predicted value is obtained by Euler's method (rectangular integration) and a corrected value by trapezoidal integration.
- `ADAMS` This specifies the use of the second-order Adams integration routine.
- `RECT` This specifies the use of Euler's method, which employs rectangular integration.

Output Control Statements

Output control statements specify such things as the variables whose values are to be printed and/or print-plotted at specified increments of the independent variable, and page headings for printed or print-plotted output. Their detailed description follows.

The PRINT *Statement.* If the programmer wants values of certain variables printed at each PRDEL interval, he lists those variables, separated by commas, on a PRINT card as follows:

```
PRINT X,XD,Y,YD
```

The label PRINT is followed by at least one required blank. TIME should not be listed in the PRINT statement since it is always printed out automatically when other variables are listed for printing. Listing TIME in the PRINT statement will result in values of TIME being printed *twice* at each interval.

If eight or fewer variables are listed in the PRINT statement, the output will be in column form with the variable names printed as column headings. The independent variable (usually TIME) will be printed in a column at the left side of the page. If 9 to 49 variables are listed in the PRINT statement, the output will be in equation form, such as

```
X=2.2000E 00
```

Output values in both column form and equation form have the most significant digit to the left of the decimal point and four digits following the decimal point. Following the number is an E and then a power of 10 by which the number is multiplied.

Only one PRINT statement should be used for each run. If more than one is included, only the last will be used. The PRINT statement can be continued to additional cards by the use of three periods, as previously explained for other kinds of statements. Only real variables should have their values printed out with a PRINT statement. If integer variables are used in a program and their values are to be printed out, their values should be assigned to real variables and the real variable names should appear in the PRINT statement.

The TITLE *Statement.* This statement is used to specify page headings for each page of printed output. It consists of the label TITLE, at least one required blank, and then the heading the user wants to have printed at the top of each page. A typical TITLE statement might be

```
TITLE DYNAMIC CHARACTERISTICS OF A MECHANICAL SYSTEM
                                                  WITH BACKLASH
```

No card columns beyond number 72 can be used and a TITLE statement cannot be continued. However, several TITLE cards can be used, and this will result in an equal number of lines in the page headings.

The PRTPLOT *Statement.* Using this statement, the programmer specifies the variables whose values are to be plotted versus time, as well as printed.

Fig. 9-2 shows a portion of a PRTPLOT output. The PRTPLOT statement used was

```
PRTPLOT XDOT,X,FOFX
```

XDOT versus time is first plotted, with numerical values of TIME and XDOT printed in columns at the left of the page. It is a portion of this first print-plot which is shown in Fig. 9-2. This print-plot is followed by print-plots of X versus time and FOFX versus time.

A statement

```
PRTPLOT XDOT(X,FOFX),XDDOT
```

would cause XDOT to be print-plotted versus time and values of X and FOFX to be printed in columns to the right of the plot. A print-plot of XDDOT versus time would follow. Up to a maximum of three variables can be enclosed in parentheses and printed alongside a print-plot. If constants appear inside parentheses, they specify lower and upper bounds for the print-plot. For example, if the statement

```
PRTPLOT XDOT(-6.0,6.0,X)
```

appeared in the program on page 571, the XDOT output would appear as shown in Fig. 9-14. Note that the plot of XDOT is "clipped" at 6.0 and −6.0. Comparing this plot with the print-plot of XDOT in Fig. 9-2, it can be seen that the scale for XDOT has been altered so that the clipped plot uses the same page width as the unclipped plot. The printed values of XDOT are the same in the two figures; that is, the printed values are not subject to the bound specified for the plot. If no lower bound is desired, but an upper bound is to be specified, the statement would appear as

```
PRTPLOT XDOT(,6.0,X)
```

If neither bound were specified, the statement would be

```
PRTPLOT XDOT(X)
```

PRTPLOT cards cannot be continued, but several of them may be used, if required, up to a maximum of ten cards for each simulation.

The LABEL *Statement.* The LABEL card or cards specify page headings for print-plot output. LABEL cards cannot be continued, and multiple LABEL cards cannot be used to obtain multiple lines of page headings. If more than one LABEL card is used, the first will specify the page headings for the print-plot specified by the first PRTPLOT card, the second will be associated with the second PRTPLOT card, and so forth. If there are more PRTPLOT cards than LABEL cards, the print-plots specified by the excess

```
SPRING MASS VISCOUSLY DAMPED WITH DEADSPACE                                    PAGE   1

                                MINIMUM                XDOT    VERSUS TIME              MAXIMUM
                              -6.0000E 00                                              6.0000E 00
   TIME            XDOT          I                                                        I       X
0.0              8.0000E 00      -------------------------------------------------*        0.0
2.0000E-01       7.0260E 00      -------------------------------------------------*        1.5652E 00
4.0000E-01      -1.2600E 00      ------------------+                                       2.2287E 00
6.0000E-01      -7.6546E 00     *                                                          1.1950E 00
8.0000E-01      -7.5684E 00     *                                                         -3.3390E-01
1.0000E 00      -6.4429E 00     *                                                         -1.7783E 00
1.2000E 00      -2.9556E 00      -----------+                                             -2.7490E 00
1.4000E 00       1.5774E 00      ------------------------------+                          -2.8899E 00
1.6000E 00       5.4257E 00      ----------------------------------------------+          -2.1644E 00
1.8000E 00       7.1513E 00      -------------------------------------------------*       -8.6312E-01
2.0000E 00       7.0579E 00      -------------------------------------------------*        5.6198E-01
2.2000E 00       4.4652E 00      ------------------------------------------+               1.8429E 00
2.4000E 00      -3.6666E 00      ---------+                                                1.9319E 00
2.6000E 00      -6.8152E 00     *                                                          7.3890E-01
2.8000E 00      -6.6794E 00     *                                                         -6.1059E-01
3.0000E 00      -5.2724E 00      ---+                                                     -1.8461E 00
3.2000E 00      -1.9046E 00      -----------------+                                       -2.5871E 00
3.4000E 00       2.1149E 00      --------------------------------+                        -2.5640E 00
3.6000E 00       5.2571E 00      ---------------------------------------------+           -1.8005E 00
3.8000E 00       6.3542E 00      -------------------------------------------------*       -5.9935E-01
4.0000E 00       6.2284E 00      -------------------------------------------------*        6.5887E-01
4.2000E 00       3.7195E 00      ---------------------------------------+                  1.7724E 00
4.4000E 00      -3.4711E 00      ----------+                                               1.7982E 00
4.6000E 00      -6.0133E 00     *                                                          7.2534E-01
4.8000E 00      -5.8943E 00      +                                                        -4.6538E-01
5.0000E 00      -4.9065E 00      ----+                                                    -1.5809E 00
5.2000E 00      -2.0984E 00      ----------------+                                        -2.3043E 00
5.4000E 00       1.4455E 00      ------------------------------+                          -2.3706E 00
5.6000E 00       4.3741E 00      -------------------------------------------+             -1.7677E 00
5.8000E 00       5.5949E 00      -----------------------------------------------+         -7.3634E-01
6.0000E 00       5.4951E 00      ----------------------------------------------+           3.7373E-01
6.2000E 00       4.6239E 00      -------------------------------------------+              1.4395E 00
6.4000E 00      -1.2296E 00      -------------------+                                      1.8297E 00
6.6000E 00      -5.2816E 00      --+                                                       1.0749E 00
6.8000E 00      -5.2013E 00      ---+                                                      2.4212E-02
7.0000E 00      -4.9431E 00      ----+                                                    -1.0017E 00
7.2000E 00      -3.1837E 00      -----------+                                             -1.8415E 00
7.4000E 00      -2.5824E-01      -----------------------+                                 -2.1964E 00
7.6000E 00       2.7067E 00      ------------------------------------+                    -1.9422E 00
7.8000E 00       4.5920E 00      -------------------------------------------+             -1.1867E 00
8.0000E 00       4.8491E 00      --------------------------------------------+            -2.2138E-01
8.2000E 00       4.7531E 00      -------------------------------------------+              7.3881E-01
8.4000E 00       2.8427E 00      ------------------------------------+                     1.5889E 00
8.6000E 00      -2.6444E 00      -------------+                                            1.6096E 00
8.8000E 00      -4.5890E 00      -----+                                                    7.9124E-01
9.0000E 00      -4.4981E 00      ------+                                                  -1.1744E-01
9.2000E 00      -4.2242E 00      -------+                                                 -1.0023E 00
9.4000E 00      -2.6129E 00      --------------+                                          -1.7088E 00
9.6000E 00      -4.7999E-02      ------------------------+                                -1.9831E 00
9.8000E 00       2.4842E 00      -----------------------------------+                     -1.7303E 00
1.0000E 01       4.0293E 00      -----------------------------------------+               -1.0561E 00
```

Fig. 9-14 / *Print-plot with upper and lower bounds.*

PRTPLOT cards will not have page headings. To nullify a LABEL card of a previous run prior to a subsequent run, a RESET statement is used, as will be explained later.

The RANGE *Statement*. If the user does not need a complete printout or print-plot of one or more variables, but is interested only in the maximum and/or minimum values reached by the variables and the times at which these values are reached, these can be obtained with the use of a RANGE card. For example, in the paratrooper problem of Ex. 6-2, the most pertinent information is the time at which the minimum velocity of the paratrooper is reached. A RANGE statement can be used to obtain this. A PRINT or PRTPLOT statement need not appear in the program. A CSMP program for solving the paratrooper problem is given here, and it is suggested that the reader study the example problem referred to and correlate it with the program, which is

```
INCON XDO=440.,YDO=0.0
CONSTANT TERM=160.,G=32.17
XDD=-G/TERM**2*SQRT(XD**2+YD**2)*XD
YDD=G-G/TERM**2*SQRT(XD**2+YD**2)*YD
XD=INTGRL(XDO,XDD)
YD=INTGRL(YDO,YDD)
V=SQRT(XD**2+YD**2)
TIMER DELT=0.1,FINTIM=20.0
METHOD TRAPZ
RANGE V
END
STOP
ENDJOB
```

This program is an example of one which has only one segment, the Dynamic segment. In such a case, a card with the label DYNAMIC is not required. The printout from this program is

```
              PROBLEM DURATION 0.0           TO  2.0000E 01
VARIABLE      MINIMUM            TIME          MAXIMUM             TIME
 V           1.4351E 02     6.4000E 00     4.4000E 02     0.0
```

RANGE cards can be continued by the use of three decimal points, and up to 100 variables can appear in a RANGE statement. In the case of a sequence of runs due to a multiple-value parameter, the values provided by the RANGE statement are for the entire sequence, and the accompanying time value applies just to one run of the sequence. There is nothing in the printout indicating in which run the maximum or minimum occurred. If a CONTINUE statement is used with a RANGE card following the CONTINUE card, the values provided are the minimum and maximum values reached during the period following the first finish time. If the RANGE card precedes the CONTINUE card, maximum and minimum values will be provided for the periods preceding and following the first finish time. If no range printout is desired for the period following the first finish time, a RESET RANGE statement can be used following the CONTINUE statement to nullify the RANGE specification. In addition, a RANGE specification for one run can be nullified for a subsequent run by use of a RESET card. Ranges (minimum and maximum values) are automatically printed for all PRTPLOT variables.

The RESET *Statement*. If a program is to run several times because of the use of several END cards, the user can nullify certain specifications used for one or more of the runs by the use of a RESET statement. Then subsequent runs will have that particular specification canceled and the user can enter new specifications if this is desired. RESET can be used to nullify PRINT, PRTPLOT, RANGE, LABEL, RELERR, ABSERR, and FINISH specifications. The statement

```
RESET PRTPLOT
```

would nullify the print-plot specifications of previous runs *and* also all previous LABEL specifications. However, the statement

```
RESET LABEL
```

nullifies only the previous LABEL specifications.

If a card with just the label RESET on it is used, all previous PRINT, PRTPLOT, RANGE, and LABEL specifications are canceled.

TITLE specifications cannot be nullified by RESET but use of a new TITLE card after an END or CONTINUE card will cancel all previous TITLE specifications.

The RESET card should follow immediately after the END or CONTINUE card. Use of RESET will be illustrated under "CSMP Examples—Run Control."

9-9 / CSMP EXAMPLES—RUN CONTROL

We have previously mentioned several methods for automatically producing a sequence of runs with different data or control statements. We will next consider some examples illustrating these methods, as well as other CSMP features.

The simplest method for obtaining a sequence of runs where only one parameter, initial condition, or constant is to be changed in value is to specify the multiple values of the quantity on a data card. The following example illustrates this procedure.

EXAMPLE 9-1

Control systems are often presented in block-diagram form in which the transfer functions of the various system components are shown in blocks.

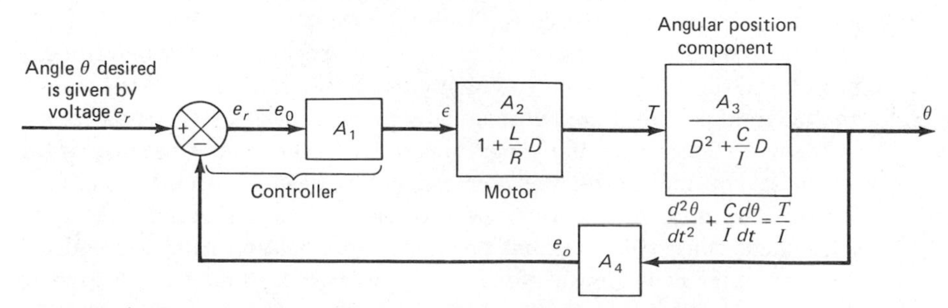

Fig. 9-15 / *Block diagram of an angular-position controller servomechanism.*

This is illustrated by the diagram of an angular-position controller servomechanism[3] as shown in Fig. 9-15. Following is the data for the system:

$L/R = 0.01$ sec

$c = 0.09$ lb-ft/(radians/sec)

$I = 0.1$ lb-sec^2-ft

$A_1 = 5.0$

$A_2 = 0.01$ lb–ft/volt

$A_3 = 1/I = 10.0$ lb^{-1}-sec^{-2}-ft^{-1}

$A_4 = 20.0, 10.0, 5.0$ volts/radian

It is desired to find the response of the system to a sudden change of the set point voltage, e_r, from 0 to 2 volts and a value of A_4 for suitable system stability. Therefore A_4 is given multiple values on the PARAMETER card. There is a System Macro, the first order lag function (REALPL), which gives the output of a block such as the one representing the motor for a given input and initial condition. There is no System Macro corresponding to the block for the angular position component, so the second-order differential equation for this component is obtained from the transfer function and the INTGRL function used twice to obtain the output θ. In the derivation of the transfer function for the angular position component, it is shown that A_3 is equal to $1/I$. This is the coefficient of torque T in the differential equation for the system component shown in Fig. 9-15. The name $A3$ is not used in the CSMP program below but appears as $1/I$ in the first statement of the Dynamic Segment.

The variable names and the quantities which they represent in the CSMP program are as follows:

Variable Name	**Quantity**
ESUBR	Set point voltage, e_r, volts
EOUT	Feedback voltage, e_0, which is proportional to the output angle θ, volts
TAU2	$\tau_2 = L/R$, sec
THETA	Output angle, θ, radians
A1,A2,A4	Component gains
RPOLE	Variable name to which the value of the function REALPL is assigned

[3] For derivation of the transfer functions for these components, see M. L. James, G. M. Smith, and J. C. Wolford, *Analog Computer Simulation of Engineering Systems* (New York: International Textbook Company, 1971), pp. 229–230.

COEFF	Variable name to which the value of c/I is assigned
C	Damping constant, lb-ft/(radians/sec)
I	Mass moment of inertia of the angular-position component, lb-sec^2-ft
TORQUE	Torque output of motor, lb-ft
EPSI	A voltage proportion to $e_r - e_0$
TH2DOT	Angular acceleration of angular-position component, radians/sec^2
THDOT	Angular velocity of angular position component, radians/sec
ST	Name to which the value of the step function is assigned
RS	Name to which value of TORQUE/I is assigned

CSMP RKSFX

TIME	TORQUE	ERROR	EPSI	EOUT	ESUBR
0.0	0.0	0.0	0.0	0.0	0.0
5.0000E-02	0.0	0.0	0.0	0.0	0.0
1.0000E-01	0.0	0.0	0.0	0.0	0.0
1.5000E-01	0.0	0.0	0.0	0.0	0.0
2.0000E-01	0.0	0.0	0.0	0.0	0.0
2.5000E-01	0.0	0.0	0.0	0.0	0.0
3.0000E-01	0.0	0.0	0.0	0.0	0.0
3.5000E-01	0.0	0.0	0.0	0.0	0.0
4.0000E-01	0.0	0.0	0.0	0.0	0.0
4.5000E-01	0.0	0.0	0.0	0.0	0.0
5.0000E-01	0.0	2.0000E 00	1.0000E 01	0.0	2.0000E 00
5.5000E-01	9.8720E-02	1.9832E 00	9.9162E 00	1.6752E-02	2.0000E 00
6.0000E-01	9.6789E-02	1.9206E 00	9.6030E 00	7.9400E-02	2.0000E 00
6.5000E-01	9.1872E-02	1.8135E 00	9.0675E 00	1.8650E-01	2.0000E 00
7.0000E-01	8.4898E-02	1.6663E 00	8.3314E 00	3.3372E-01	2.0000E 00
7.5000E-01	7.6123E-02	1.4841E 00	7.4206E 00	5.1587E-01	2.0000E 00
8.0000E-01	6.5832E-02	1.2729E 00	6.3643E 00	7.2715E-01	2.0000E 00
8.5000E-01	5.4336E-02	1.0388E 00	5.1938E 00	9.6125E-01	2.0000E 00
9.0000E-01	4.1962E-02	7.8845E-01	3.9422E 00	1.2116E 00	2.0000E 00
9.5000E-01	2.9049E-02	5.2869E-01	2.6434E 00	1.4713E 00	2.0000E 00
1.0000E 00	1.5931E-02	2.6619E-01	1.3310E 00	1.7338E 00	2.0000E 00
1.0500E 00	2.9386E-03	7.4730E-03	3.7365E-02	1.9925E 00	2.0000E 00
1.1000E 00	-9.6169E-03	-2.4129E-01	-1.2065E 00	2.2413E 00	2.0000E 00
1.1500E 00	-2.1446E-02	-4.7441E-01	-2.3721E 00	2.4744E 00	2.0000E 00
1.2000E 00	-3.2287E-02	-6.8682E-01	-3.4341E 00	2.6868E 00	2.0000E 00
1.2500E 00	-4.1915E-02	-8.7411E-01	-4.3706E 00	2.8741E 00	2.0000E 00
1.3000E 00	-5.0141E-02	-1.0327E 00	-5.1636E 00	3.0327E 00	2.0000E 00
1.3500E 00	-5.6820E-02	-1.1599E 00	-5.7994E 00	3.1599E 00	2.0000E 00
1.4000E 00	-6.1850E-02	-1.2537E 00	-6.2686E 00	3.2537E 00	2.0000E 00
1.4500E 00	-6.5171E-02	-1.3133E 00	-6.5663E 00	3.3133E 00	2.0000E 00
1.5000E 00	-6.6770E-02	-1.3384E 00	-6.6919E 00	3.3384E 00	2.0000E 00
1.5500E 00	-6.6675E-02	-1.3298E 00	-6.6490E 00	3.3298E 00	2.0000E 00
1.6000E 00	-6.4954E-02	-1.2891E 00	-6.4454E 00	3.2891E 00	2.0000E 00
1.6500E 00	-6.1715E-02	-1.2185E 00	-6.0923E 00	3.2185E 00	2.0000E 00
1.7000E 00	-5.7094E-02	-1.1208E 00	-5.6042E 00	3.1208E 00	2.0000E 00
1.7500E 00	-5.1259E-02	-9.9967E-01	-4.9983E 00	2.9997E 00	2.0000E 00
1.8000E 00	-4.4401E-02	-8.5881E-01	-4.2941E 00	2.8588E 00	2.0000E 00
1.8500E 00	-3.6727E-02	-7.0249E-01	-3.5125E 00	2.7025E 00	2.0000E 00
1.9000E 00	-2.8456E-02	-5.3514E-01	-2.6757E 00	2.5351E 00	2.0000E 00
1.9500E 00	-1.9813E-02	-3.6126E-01	-1.8063E 00	2.3613E 00	2.0000E 00
2.0000E 00	-1.1026E-02	-1.8537E-01	-9.2685E-01	2.1854E 00	2.0000E 00
2.0500E 00	-2.3129E-03	-1.1843E-02	-5.9214E-02	2.0118E 00	2.0000E 00
2.1000E 00	6.1150E-03	1.5518E-01	7.7589E-01	1.8448E 00	2.0000E 00
2.1500E 00	1.4063E-02	3.1186E-01	1.5593E 00	1.6881E 00	2.0000E 00
2.2000E 00	2.1357E-02	4.5479E-01	2.2740E 00	1.5452E 00	2.0000E 00
2.2500E 00	2.7843E-02	5.8101E-01	2.9050E 00	1.4190E 00	2.0000E 00
2.3000E 00	3.3394E-02	6.8808E-01	3.4404E 00	1.3119E 00	2.0000E 00
2.3500E 00	3.7913E-02	7.7415E-01	3.8708E 00	1.2258E 00	2.0000E 00
2.4000E 00	4.1327E-02	8.3794E-01	4.1897E 00	1.1621E 00	2.0000E 00
2.4500E 00	4.3599E-02	8.7875E-01	4.3937E 00	1.1213E 00	2.0000E 00
2.5000E 00	4.4716E-02	8.9649E-01	4.4825E 00	1.1035E 00	2.0000E 00
2.5500E 00	4.4697E-02	8.9164E-01	4.4582E 00	1.1084E 00	2.0000E 00
2.6000E 00	4.3586E-02	8.6519E-01	4.3260E 00	1.1348E 00	2.0000E 00
2.6500E 00	4.1455E-02	8.1864E-01	4.0932E 00	1.1814E 00	2.0000E 00
2.7000E 00	3.8394E-02	7.5392E-01	3.7696E 00	1.2461E 00	2.0000E 00

Fig. 9-16 / *Printed output resulting from the* PRINT *statement.*

The CSMP program from the block diagram is as follows:

```
INITIAL
 COEFF=C/I
 PARAMETER TAU2=.01,A1=5.,A2=.01,A4=(20.,10.,5.),C=.09,I=.1
DYNAMIC
      RS=TORQUE/I
      TH2DOT=RS-COEFF*THDOT
      THDOT=INTGRL(0.0,TH2DOT)
      TH=INTGRL(0.0,THDOT)
      EOUT=TH*A4
      ERROR=ESUBR-EOUT
      ST=STEP(.5)
      ESUBR=2.*ST
      EPSI=A1*ERROR
      RPOLE=REALPL(0.0,TAU2,EPSI)
      TORQUE=A2*RPOLE
```

```
REMOTE POSITION CONTROL SYSTEM                                      PAGE   1

                          MINIMUM                 TH     VERSUS TIME          MAXIMUM
                          0.0                     A3    = 2.0000E 01         5.6188E-01
 TIME          TH          I                                                      I
0.0           0.0          +
5.0000E-02    0.0          +
1.0000E-01    0.0          +
1.5000E-01    0.0          +
2.0000E-01    0.0          +
2.5000E-01    0.0          +
3.0000E-01    0.0          +
3.5000E-01    0.0          +
4.0000E-01    0.0          +
4.5000E-01    0.0          +
5.0000E-01    0.0          +
5.5000E-01    8.3759E-04   +
6.0000E-01    3.9700E-03   +
6.5000E-01    9.3251E-03   +
7.0000E-01    1.6686E-02   -+
7.5000E-01    2.5794E-02   --+
8.0000E-01    3.6357E-02   ---+
8.5000E-01    4.8062E-02   ----+
9.0000E-01    6.0578E-02   -----+
9.5000E-01    7.3566E-02   ------+
1.0000E 00    8.6690E-02   -------+
1.0500E 00    9.9626E-02   --------+
1.1000E 00    1.1206E-01   ---------+
1.1500E 00    1.2372E-01   -----------+
1.2000E 00    1.3434E-01   -----------+
1.2500E 00    1.4371E-01   ------------+
1.3000E 00    1.5164E-01   -------------+
1.3500E 00    1.5799E-01   --------------+
1.4000E 00    1.6269E-01   --------------+
1.4500E 00    1.6566E-01   --------------+
1.5000E 00    1.6692E-01   --------------+
1.5500E 00    1.6649E-01   --------------+
1.6000E 00    1.6445E-01   --------------+
1.6500E 00    1.6092E-01   --------------+
1.7000E 00    1.5604E-01   -------------+
1.7500E 00    1.4998E-01   -------------+
1.8000E 00    1.4294E-01   ------------+
1.8500E 00    1.3512E-01   ------------+
1.9000E 00    1.2676E-01   -----------+
1.9500E 00    1.1806E-01   ----------+
2.0000E 00    1.0927E-01   ---------+
2.0500E 00    1.0059E-01   --------+
2.1000E 00    9.2241E-02   --------+
2.1500E 00    8.4407E-02   -------+
2.2000E 00    7.7260E-02   ------+
2.2500E 00    7.0950E-02   ------+
2.3000E 00    6.5596E-02   -----+
2.3500E 00    6.1292E-02   -----+
2.4000E 00    5.8103E-02   -----+
2.4500E 00    5.6063E-02   ----+
2.5000E 00    5.5175E-02   ----+
```

Fig. 9-17 | *Print-plot output resulting from the* PRTPLOT *statement.*

```
METHOD RKSFX
TIMER DELT=.01,FINTIM=15.,OUTDEL=.05,PRDEL=.05
PRINT TORQUE,ERROR,EPSI,EOUT,ESUBR
LABEL REMOTE POSITION CONTROL SYSTEM
PRTPLT TH
END
STOP
ENDJOB
```

Portions of the printed output and the print-plot output are shown in Figs. 9-16 and 9-17. Although CSMP rules do not prohibit multiplying the System Macro REALPL by a variable, experience reveals that when this is done, the FORTRAN compiler may produce a message indicating that it is an unresolved external reference. Thus the value of REALPL is first assigned to the variable RPOLE, and the RPOLE is multiplied by A2.

EXAMPLE 9-2

A second method of obtaining sequential runs is by using several END cards followed by new data and/or control specifications. Each run obtained by this method could be a sequence of runs obtained by a multiple-value parameter. To illustrate this, let us consider the following problem. Figure 9-18 represents a simplified model of an automobile suspension system where the automobile is traveling over a bumpy road approximated by a sine curve. The motion of the car body, x_1, is desired for various values of damping between the car and the wheel for various automobile speeds.

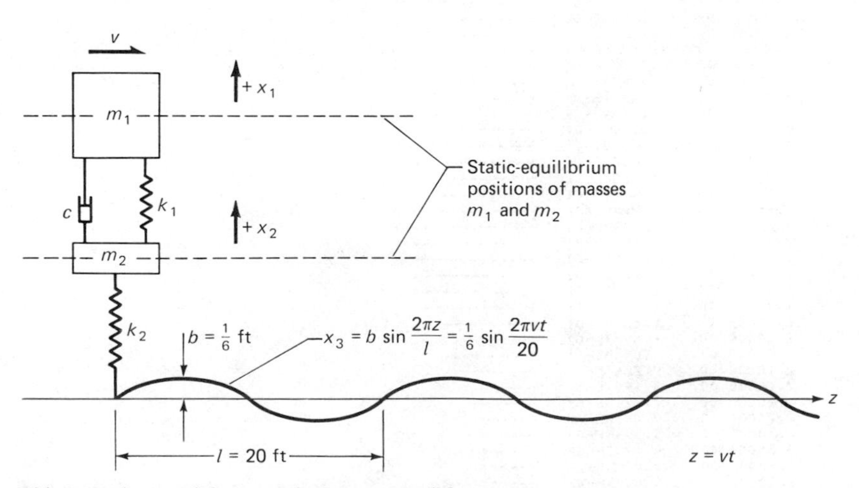

Fig. 9-18 / *Simplified model of an automobile suspension system.*

Application of Newton's second law to masses m_1 and m_2 results in the following simultaneous second-order differential equations describing the system

$$\ddot{x}_1 + \frac{c}{m_1}\dot{x}_1 + \frac{k_1}{m_1}x_1 = \frac{c}{m_1}\dot{x}_2 + \frac{k_1}{m_1}x_2 \tag{9-4}$$

$$\ddot{x}_2 + \frac{c}{m_2}\dot{x}_2 + x_2\left(\frac{k_1 + k_2}{m_2}\right) = \frac{c}{m_2}\dot{x}_1 + \frac{k_1}{m_1}x_1 + \frac{k_2}{6m_2}\sin\frac{2\pi vt}{20} \tag{9-5}$$

Displacements x_1 and x_2 are measured from the static equilibrium positions of masses m_1 and m_2. The initial conditions for the problem are

$$t = 0.0 \left| \begin{aligned} x_1 &= 0.0 \\ \dot{x}_1 &= 0.0 \\ x_2 &= 0.0 \\ \dot{x}_2 &= 0.0 \end{aligned} \right.$$

The problem data are given as

m_1 = $\frac{1}{4}$ the mass of the automobile = 25 slugs

m_2 = mass of one wheel = 2 slugs

k_1 = spring constant = 1000 lb/ft

k_2 = linear spring constant of tire = 4500 lb/ft

c = damping constant of dashpot = 0, 25, 50, and 75 lb-sec/ft

x_1 = displacement of automobile body from equilibrium position, ft

x_2 = displacement of wheel from equilibrium position, ft

x_3 = variable describing the profile of the roadway, ft

v = velocity of car = 30, 60, 90, and 120 ft/sec

The automobile velocity will be used as a multiple-value parameter. Each sequence of runs (due to the multiple-value parameter) except the last will be terminated by an END card followed by a new specification for the damping constant c. A RESET LABEL card will be used to nullify the LABEL specification of the previous run and a new LABEL will be specified for the subsequent sequence of runs. The last sequence with the damping coefficient equal to 75 lb-sec/ft will be terminated by an END card followed by a STOP card which indicates that no more runs are to follow. The CSMP program to solve this problem is:

```
INITIAL
    CONSTANT PI=3.14159
    PARAMETER C=0.,M1=25.,M2=2.,K1=1000.,K2=4500.,...
    V=(30.,60.,90.,120.)
       COEFF1=C/M1
       QUOT1=K1/M1
       COEFF2=C/M2
       QUOT2=(K1+K2)/M2
       A=K2/(6.*M2)
DYNAMIC
       XDDOT1=COEFF1*XDOT2+QUOT1*X2-COEFF1*XDOT1-QUOT1*X1
       XDDOT2=COEFF2*XDOT1+QUOT1*X1+A*SIN(2.*PI*V*TIME/20.)...
       -COEFF2*XDOT2-QUOT2*X2
       XDOT1=INTGRL(0.0,XDDOT1)
       XDOT2=INTGRL(0.0,XDDOT2)
       X1=INTGRL(0.0,XDOT1)
       X2=INTGRL(0.0,XDOT2)

    METHOD RKSFX
    TIMER DELT=.01,FINTIM=15.0,OUTDEL=.05
    LABEL MOTION OF AUTOMOBILE BODY WITH DAMPING=0.0 LB SEC/FT
    PRTPLOT X1
    END
    RESET LABEL
    PARAMETER C=25.
    LABEL MOTION OF AUTOMOBILE BODY WITH DAMPING
                                              =25.0 LB SEC/FT
    END
    RESET LABEL
    PARAMETER C=50.0
    LABEL MOTION OF AUTOMOBILE BODY WITH DAMPING
                                              =50.0 LB SEC/FT
    END
    RESET LABEL
    PARAMETER C=75.0
    LABEL MOTION OF AUTOMOBILE BODY WITH DAMPING
                                              =75.0 LB SEC/FT
    END
    STOP
ENDJOB
```

Shown at the right is the program output for the first 2.5 sec, with zero damping and the car traveling at 30 ft/sec.

MOTION OF AUTOMOBILE BODY WITH DAMPING=0.0 LB SEC/FT

MINIMUM X1 VERSUS TIME MAXIMUM

-2.7211E-01 V = 3.0000E 01 2.6799E-01

TIME	X1
0.0	0.0
5.0000E-02	3.1110E-04
1.0000E-01	6.1430E-03
1.5000E-01	2.3641E-02
2.0000E-01	5.1020E-02
2.5000E-01	8.7638E-02
3.0000E-01	1.2470E-01
3.5000E-01	1.5383E-01
4.0000E-01	1.6644E-01
4.5000E-01	1.5357E-01
5.0000E-01	1.1482E-01
5.5000E-01	4.9954E-02
6.0000E-01	-3.1307E-02
6.5000E-01	-1.1762E-01
7.0000E-01	-1.9521E-01
7.5000E-01	-2.4835E-01
8.0000E-01	-2.6887E-01
8.5000E-01	-2.4887E-01
9.0000E-01	-1.9174E-01
9.5000E-01	-1.0471E-01
1.0000E 00	-6.1260E-04
1.0500E 00	1.0242E-01
1.1000E 00	1.9056E-01
1.1500E 00	2.4779E-01
1.2000E 00	2.6799E-01
1.2500E 00	2.4888E-01
1.3000E 00	1.9516E-01
1.3500E 00	1.1923E-01
1.4000E 00	3.2475E-02
1.4500E 00	-4.8193E-02
1.5000E 00	-1.1269E-01
1.5500E 00	-1.5235E-01
1.6000E 00	-1.6433E-01
1.6500E 00	-1.5329E-01
1.7000E 00	-1.2379E-01
1.7500E 00	-8.7648E-02
1.8000E 00	-5.1863E-02
1.8500E 00	-2.4188E-02
1.9000E 00	-8.2353E-03
1.9500E 00	-1.5050E-03
2.0000E 00	-2.1689E-03
2.0500E 00	-1.4102E-03
2.1000E 00	4.9249E-03
2.1500E 00	2.2030E-02
2.2000E 00	5.1045E-02
2.2500E 00	8.7065E-02
2.3000E 00	1.2558E-01
2.3500E 00	1.5485E-01
2.4000E 00	1.6765E-01
2.4500E 00	1.5582E-01
2.5000E 00	1.1605E-01

EXAMPLE 9-3

To illustrate the use of the CONTINUE statement in run control, let us use the paratrooper problem of Ex. 6-2. Reviewing the problem briefly, we recall that when a paratrooper jumps from an aircraft which is in straight and level flight, his initial velocity is that of the aircraft. However, after a few seconds it drops below terminal velocity and then slowly increases up to terminal velocity. The optimum time for opening the parachute would generally be at the time of minimum velocity.

In a solution by a numerical method, most accurate results would be obtained by using a rather small integration step size for the first few seconds while the velocity is changing rapidly. After the minimum velocity is reached, and the velocity is changing much more slowly, a larger integration step size could be used if it were desired to extend the solution over a period of 20 or 30 sec.

Changing the integration interval, or any other control specifications or data, in the middle of a run is easily accomplished in CSMP. Suppose we wished to use 0.05 sec as an initial value of DELT, and after 10 sec change to a DELT value of 0.5 sec. Also suppose that that an initial value of OUTDEL equal to 0.2 sec were desired, with a change to 1.0 sec made after 10 sec. FINTIM is initially given the value 10 sec, DELT a value of 0.05 sec, and OUTDEL a value of 0.2 sec. Then, instead of an END card, a CONTINUE card would be used. The simulation would stop after 10 sec to accept new data and/or control specifications, but TIME would not be reset to zero as it would be with an END card. New DELT, OUTDEL, and FINTIM values would be assigned on a parameter card following the CONTINUE card. This would be followed by an END card, and then a STOP card if no further runs were desired. The print-plots obtained in this manner would be in two parts, one covering the period from 0 to 10 secs, and the other covering the period from 10 to 30 sec. In general, the dependent variable axis would be to a different scale for the two parts.

The simultaneous second-order differential equations which must be solved, and which are derived on pages 387, 388, and 389, are

$$\ddot{x} = -\frac{g}{V_t^2}\sqrt{\dot{x}^2 + \dot{y}^2}\,\dot{x} \tag{9-6}$$

$$\ddot{y} = g - \frac{g}{V_t^2}\sqrt{\dot{x}^2 + \dot{y}^2}\,\dot{y} \tag{9-7}$$

with initial conditions

$$t = 0 \left| \begin{array}{l} x = 0.0 \\ \dot{x} = 440 \text{ ft/sec} \\ y = 0.0 \\ \dot{y} = 0.0 \end{array} \right.$$

The variable names used and the quantities which they represent in the CSMP program are

Variable Name	Quantity
TERM	Terminal velocity, ft/sec
G	Acceleration of gravity, ft/sec^2
XDD	Acceleration in x direction, ft/sec^2
YDD	Acceleration in y direction, ft/sec^2
XD	Velocity in x direction, ft/sec
YD	Velocity in y direction, ft/sec
X	Displacement in x direction, ft
Y	Displacement in y direction, ft
XDO,YDO,XO,YO	Initial values of XD,YD,X, and Y, respectively

The CSMP program is as follows:

```
INCON XDO=440.,XO=0.0,YDO=0.0,YO=0.0
CONSTANT TERM=160.,G=32.17
XDD=-G/TERM**2*SQRT(XD**2+YD**2)*XD
YDD=G-G/TERM**2*SQRT(XD**2+YD**2)*YD
XD=INTGRL(XDO,XDD)
YD=INTGRL(YDO,YDD)
V=SQRT(XD**2+YD**2)
X=INTGRL(XO,XD)
Y=INTGRL(YO,YD)
TIMER DELT=0.05,FINTIM=10.0,OUTDEL=0.2
PRTPLOT V,XD,YD,X,Y
METHOD TRAPZ
CONTINUE
TIMER DELT=0.5,FINTIM=30.0,OUTDEL=1.0
END
STOP
ENDJOB
```

The print-plot of velocity versus time for the period from 10 to 30 sec is shown in Fig. 9-19.

```
                                                                        PAGE    1

                               MINIMUM           V        VERSUS TIME               MAXIMUM
                              1.5054E 02                                           1.5999E 02
 TIME            V             I                                                       I
1.0000E 01    1.5054E 02       +
1.1000E 01    1.5266E 02       ------------+
1.2000E 01    1.5442E 02       ----------------------+
1.3000E 01    1.5583E 02       ------------------------------+
1.4000E 01    1.5692E 02       -------------------------------------+
1.5000E 01    1.5775E 02       ------------------------------------------+
1.6000E 01    1.5837E 02       ----------------------------------------------+
1.7000E 01    1.5882E 02       ------------------------------------------------+
1.8000E 01    1.5916E 02       --------------------------------------------------+
1.9000E 01    1.5940E 02       ---------------------------------------------------+
2.0000E 01    1.5957E 02       ----------------------------------------------------+
2.1000E 01    1.5970E 02       -----------------------------------------------------+
2.2000E 01    1.5979E 02       -----------------------------------------------------+
2.3000E 01    1.5985E 02       ------------------------------------------------------+
2.4000E 01    1.5989E 02       ------------------------------------------------------+
2.5000E 01    1.5993E 02       ------------------------------------------------------+
2.6000E 01    1.5995E 02       ------------------------------------------------------+
2.7000E 01    1.5996E 02       ------------------------------------------------------+
2.8000E 01    1.5997E 02       ------------------------------------------------------+
2.9000E 01    1.5998E 02       ------------------------------------------------------+
3.0000E 01    1.5999E 02       ------------------------------------------------------+
```

Fig. 9-19 / *Velocity-time print-plot for Ex. 9-3.*

EXAMPLE 9-4

This example will illustrate another method of run control in which the TERMINAL segment is used to change parameter values, and the statement

```
CALL RERUN
```

is used to run the simulation again with the new parameter values. A two-point boundary-value problem which is solved in Ex. 7-2 on page 492 by the finite-difference method will be solved here by the trial-and-error method. The program will illustrate, in addition to the use of the CALL RERUN statement for run control, the use of the FUNCTION data statement and arbitrary function generator, the use of the RENAME statement, and the use of FORTRAN in a NOSORT section of the DYNAMIC segment and in the TERMINAL segment.

For the sake of convenience, the problem is restated here. It is desired to determine the elastic curve of the loaded nonuniform circular shaft shown in Fig. 9-20. The differential equation of the elastic curve of the shaft is

$$\frac{d^2y}{dx^2} = \frac{M}{EI}$$

with the following boundary conditions:

$$x = 0.0, \; y = 0.0$$

$$x = 42.0, \; y = 0.0$$

The initial value of dy/dx is not known. A value of $(dy/dx)|_{x=0}$ must be determined such that the second boundary condition above is satisfied, or very nearly so. The technique used will be to guess at minimum and maximum values of the slope at $x = 0.0$ which we are quite sure will bound the actual slope. We are sure the slope cannot be smaller in magnitude than 0.0, and we are reasonably sure the slope will not be greater (in magnitude) than -0.002. These values will be assigned to the variables MIN and MAX, respectively. The slope used at $x = 0.0$ is calculated

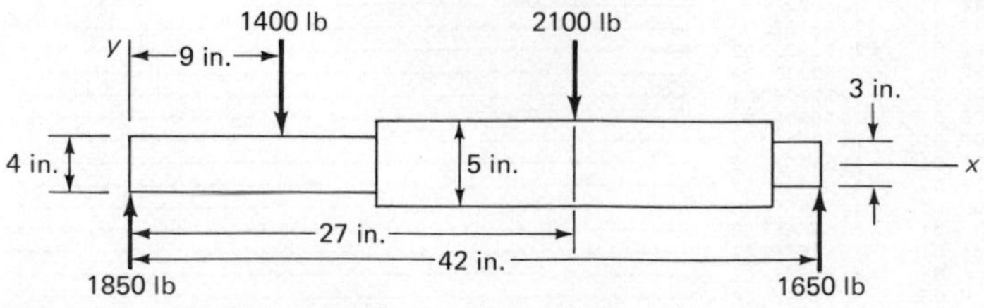

Fig. 9-20 / *Loaded nonuniform circular shaft.*

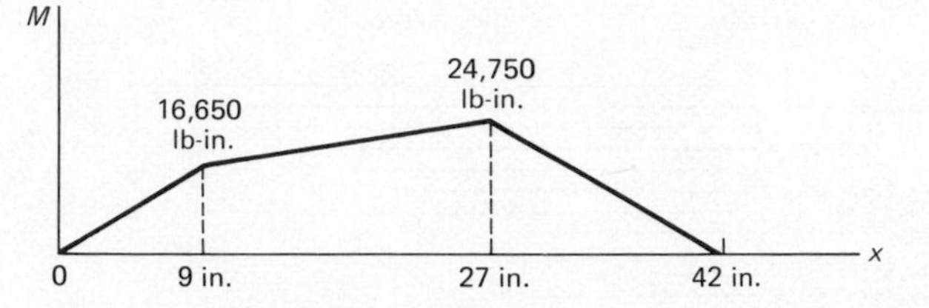

Fig. 9-21 / *Bending-moment diagram for shaft shown in Fig. 9-20.*

in the INITIAL segment as the average of MIN and MAX. Then if y at $x = 42.0$ turns out to be greater than 0.0 as the result of a run, the initial slope must be made larger negatively. This is done by making the initial slope used in the run just completed the MIN value for the next run. If y at $x = 42.0$ is less than 0.0, the initial slope must be made smaller negatively. This is done by making the initial slope used in the run just completed the MAX value for the next run. The FORTRAN statements for making these adjustments will be in the TERMINAL segment. Then a CALL RERUN statement will initiate a new run using the new value of MAX or MIN.

When a run is completed where $|y|$ is less than 0.00002 at $x = 42.0$, the adjustment algorithm described above and CALL RERUN will be by-passed and a new run initiated by an END statement. For this final run, a print-plot will be specified.

The bending-moment diagram for the shaft is shown in Fig. 9-21. The moment values at $x = 0, 9, 27$, and 42 in. can be entered using a FUNCTION statement, and the arbitrary function generator will give the correct moment value at any x value using linear interpolation.

The moment of inertia of the cross-sectional area of the shaft will have different values corresponding to the three different diameters. A NOSORT section in the Dynamic segment is used to assign the correct I value for any particular x value. Where there is a sudden change in the cross section of the shaft, the average I value is used.

The CSMP program listing for solving this problem follows, and the print-plot for shaft displacement versus x is shown in Fig. 9-22. Note in the program listing, CALL RERUN in the Terminal segment just precedes the FORTRAN CONTINUE statement to which transfer is made when CALL RERUN is to be by-passed. Experience indicates that for best results, CALL RERUN should be placed in this position. (See the flow chart accompanying Prob. 9-14.)

```
   RENAME TIME=X,DELT=DELX,FINTIM=XMAX
INITIAL
   DYDX0=0.5*(MIN+MAX)
   CONSTANT E=30000000.,MIN=0.0,MAX=-0.002,EPSI=.05
   FUNCTION MOMENT=(0.0,0.0),(9.0,16650.),(27.,24750.),
                                                (42.0,0.0)
DYNAMIC
```

```
                                                                                         PAGE    1

                                      MINIMUM                    Y        VERSUS X                   MAXIMUM
                                    -6.0251E-03                                                    5.3859E-06
    X                Y                 I                                                                I
 0.0               0.0                 --------------------------------------------------+
 1.0000E 00     -5.6560E-04            ---------------------------------------------+
 2.0000E 00     -1.1263E-03            -----------------------------------------+
 3.0000E 00     -1.6772E-03            ------------------------------------+
 4.0000E 00     -2.2133E-03            --------------------------------+
 5.0000E 00     -2.7298E-03            ---------------------------+
 6.0000E 00     -3.2218E-03            -----------------------+
 7.0000E 00     -3.6844E-03            -------------------+
 8.0000E 00     -4.1126E-03            ----------------+
 9.0000E 00     -4.5016E-03            -------------+
 1.0000E 01     -4.8470E-03            ----------+
 1.1000E 01     -5.1471E-03            -------+
 1.2000E 01     -5.4006E-03            -----+
 1.3000E 01     -5.6064E-03            ---+
 1.4000E 01     -5.7633E-03            --+
 1.5000E 01     -5.8700E-03            -+
 1.6000E 01     -5.9414E-03            +
 1.7000E 01     -5.9912E-03            +
 1.8000E 01     -6.0190E-03            +
 1.9000E 01     -6.0243E-03            +
 2.0000E 01     -6.0067E-03            +
 2.1000E 01     -5.9656E-03            +
 2.2000E 01     -5.9005E-03            -+
 2.3000E 01     -5.8110E-03            -+
 2.4000E 01     -5.6965E-03            --+
 2.5000E 01     -5.5567E-03            ---+
 2.6000E 01     -5.3909E-03            -----+
 2.7000E 01     -5.1987E-03            -------+
 2.8000E 01     -4.9800E-03            ---------+
 2.9000E 01     -4.7362E-03            -----------+
 3.0000E 01     -4.4691E-03            -------------+
 3.1000E 01     -4.1804E-03            ---------------+
 3.2000E 01     -3.8721E-03            ------------------+
 3.3000E 01     -3.5459E-03            ---------------------+
 3.4000E 01     -3.2035E-03            -----------------------+
 3.5000E 01     -2.8468E-03            --------------------------+
 3.6000E 01     -2.4775E-03            -----------------------------+
 3.7000E 01     -2.0975E-03            ---------------------------------+
 3.8000E 01     -1.7085E-03            ------------------------------------+
 3.9000E 01     -1.3123E-03            ---------------------------------------+
 4.0000E 01     -8.9611E-04            -------------------------------------------+
 4.1000E 01     -4.5227E-04            ----------------------------------------------+
 4.2000E 01      5.3859E-06            --------------------------------------------------+
```

Fig. 9-22 / *Print-plot of Ex. 9-4.*

```
     NOSORT
        IF(X.LT.15.)I=12.57
        IF(ABS(X-15.).LT.EPSI)I=21.63
        IF(X.LT.39.0.AND.X.GT.15.+EPSI)I=30.68
        IF(ABS(X-39.0).LT.EPSI)I=17.33
        IF(X.GT.39.0+EPSI)I=3.98
     SORT
     M=AFGEN(MOMENT,X)
     D2YDX2=M/(E*I)
     DYDX=INTGRL(DYDX0,D2YDX2)
     Y=INTGRL(0.0,DYDX)
  TERMINAL
        IF(ABS(Y).LT.0.00002)GO TO 3
        IF(Y.GT.0.)GO TO 1
        MAX=DYDX0
        GO TO 2
      1 MIN=DYDX0
      2 CALL RERUN
      3 CONTINUE
     TIMER DELX=0.1,XMAX=42.0
     METHOD TRAPZ
     END
     TIMER OUTDEL=1.0
     PRTPLOT Y,DYDX
     END
     STOP
  ENDJOB
```

9-10 / FORTRAN SUBPROGRAMS USED WITH CSMP PROGRAMS

Ordinary FORTRAN subprograms can be referenced by a CSMP program. The subprogram cards are placed between the STOP and the ENDJOB cards. If a subprogram has only one output, a function subprogram is generally used. A function subprogram is referenced by using the function name in a CSMP structure statement. To illustrate this, let us again use the shaft-deflection problem considered previously in Ex. 9-4. In the program shown in Sec. 9-9, the bending moment in the shaft at any x value is obtained by using the arbitrary function generator with the CSMP function MOMENT as the first argument, and x as the second. In the program which follows, the magnitude of the bending moment for any x value is obtained from a FORTRAN function subprogram having the name MOMENT. The statement referencing this subprogram is

```
M=MOMENT(X)
```

The complete listings of the CSMP program and the FORTRAN subprogram used with it follow.

```
   RENAME TIME=X,DELT=DELX,FINTIM=XMAX
INITIAL
   DYDX0=0.5*(MIN+MAX)
   CONSTANT E=30000000.,MIN=0.0,MAX=-0.002,EPSI=.05
DYNAMIC
   NOSORT
      IF(X.LT.15.)I=12.57
      IF(ABS(X-15.).LT.EPSI)I=21.63
      IF(X.LT.39.0.AND.X.GT.15.+EPSI)I=30.68
      IF(ABS(X-39.0).LT.EPSI)I=17.33
      IF(X.GT.39.0+EPSI)I=3.98
   SORT
   M=MOMENT(X)
   D2YDX2=M/(E*I)
   DYDX=INTGRL(DYDX0,D2YDX2)
   Y=INTGRL(0.0,DYDX)
TERMINAL
      IF(ABS(Y).LT.0.00002)GO TO 3
      IF(Y.GT.0.)GO TO 1
      MAX=DYDX0
      GO TO 2
    1 MIN=DYDX0
    2 CALL RERUN
    3 CONTINUE
   TIMER DELX=0.1,XMAX=42.0
   METHOD TRAPZ
   END
   TIMER OUTDEL=1.0
   PRTPLOT Y,DYDX
   END
   STOP
```

```
      FUNCTION MOMENT(X)
      REAL MOMENT
      IF(X.GT.9.)GO TO 2
      MOMENT=1850.*X
      RETURN
    2 IF(X.GT.27.)GO TO 3
      MOMENT=1850.*X-1400.*(X-9.)
      RETURN
    3 MOMENT=1650.*(42.-X)
      RETURN
      END
ENDJOB
```

If a subprogram has more than one output, a subroutine subprogram is used. Instead of using a FORTRAN CALL statement to call the subprogram, a statement similar to the canonical statement of a MACRO definition is used. In the example which follows (again using the shaft problem of Ex. 9-4), the subroutine subprogram BMOMI determines the bending moment and moment of inertia of the shaft cross section for any x value. The subprogram is called with the CSMP statement

```
M,I=BMOMI(X)
```

in which the two outputs are on the left of the equal sign and separated by commas. Any number of outputs is permissible. The translator translates this CSMP statement into the FORTRAN statement

```
CALL BMOMI(X,M,I)
```

which is indicated by the arrow in the subroutine UPDATE (written by the translator) shown below.

```
      SUBROUTINE UPDATE
      COMMON  ZZ9901(5),IZ9901,ZZ9902,IZ9902,ZZ9903,IZ9903,ZZ9991(54)
      COMMON X
     1,DELX  ,DELMIN,XMAX  ,PRDEL ,OUTDEL,DYDX  ,Y     ,D2YDX2,ZZ0005
     1,DYDX0 ,ZZ0003,E     ,MIN   ,MAX   ,EPSI  ,M     ,I     ,ZZ0004
      COMMON ZZ9992(7981),NALARM,IZ9993,ZZ9994(417),KEEP,ZZ9995(489)
     $,IZ0000,ZZ9996(824),IZ9997,IZ9998,ZZ9999(  59)
      REAL        MIN
     1,MAX   ,M     ,I
      GO TO(39995,39996,39997,39998),IZ0000
C    SYSTEM SEGMENT OF MODEL
39995 CONTINUE
      IZ9993=  20
      IZ9997=   2
      IZ9998=  19
      READ(5,39990)(ZZ9999(IZ9999),IZ9999=1,  59)
39990 FORMAT(18A4)
      IZ9901=   160013
      IZ9902=   190017
      IZ9903=       59
      GO TO 39999
C   INITIAL SEGMENT OF MODEL
```

```
39996 CONTINUE
      DYDX0=0.5*(MIN+MAX)
      GO TO 39999
C   DYNAMIC SEGMENT OF MODEL
39997 CONTINUE
----> CALL  BMOMI (X,M,I)
      D2YDX2=M/(E*I)
C     DYDX         =INTGRL      (DYDX0        ,D2YDX2        )
C     Y            =INTGRL      (ZZ0003       ,DYDX          )
      ZZ0005=DYDX
      GO TO 39999
C  TERMINAL SEGMENT OF MODEL
39998 CONTINUE
      IF(ABS(Y).LT.0.00002)GO TO 3
      IF(Y.GT.0.)GO TO 1
      MAX=DYDX0
      GO TO 2
1     MIN=DYDX0
2     CALL RERUN
3     CONTINUE
39999 CONTINUE
      RETURN
      END
```

In forming the CALL statement, the translator places the two outputs in the rightmost portion of the argument list with no change in their order. The dummy arguments of the subroutine subprogram must agree in number, order, and mode with the real arguments in the CALL statement generated by the translator. The complete program follows for solving the shaft problem, with a subroutine subprogram to determine the bending moment and moment of inertia.

```
   RENAME TIME=X,DELT=DELX,FINTIM=XMAX
INITIAL
   DYDX0=0.5*(MIN+MAX)
   CONSTANT E=30000000.,MIN=0.0,MAX=-0.002,EPSI=.05
DYNAMIC
   M,I=BMOMI(X)
   D2YDX2=M/(E*I)
   DYDX=INTGRL(DYDX0,D2YDX2)
   Y=INTGRL(0.0,DYDX)
TERMINAL
      IF(ABS(Y).LT.0.00002)GO TO 3
      IF(Y.GT.0.)GO TO 1
      MAX=DYDX0
      GO TO 2
    1 MIN=DYDX0
    2 CALL RERUN
    3 CONTINUE
   TIMER DELX=0.1,XMAX=42.0
   METHOD TRAPZ
   END
   TIMER OUTDEL=1.0
   PRTPLOT Y,DYDX
   END
   STOP
```

```
      SUBROUTINE BMOMI(X,M,I)
      REAL M,I
      IF(X.LT.15.)I=12.57
      IF(ABS(X-15.).LT.EPSI)I=21.63
      IF(X.LT.39.0.AND.X.GT.15.+EPSI)I=30.68
      IF(ABS(X-39.0).LT.EPSI)I=17.33
      IF(X.GT.39.0+EPSI)I=3.98
      IF(X.GT.9.)GO TO 2
      M=1850.*X
      RETURN
    2 IF(X.GT.27.)GO TO 3
      M=1850.*X-1400.*(X-9.)
      RETURN
    3 M=1650.*(42.-X)
      RETURN
      END
ENDJOB
```

If a subroutine subprogram has only one output (say, the bending moment in the shaft considered for any x value), the form of the CSMP calling statement would have to be

```
M=BNDMOM(X)
```

with the single output on the left of the equal sign. However, the translator will not recognize this as a statement to be translated into a CALL statement because of the single output. The subprogram BNDMOM could be called with a FORTRAN CALL statement providing it is placed in an unsorted section, as shown below.

```
   RENAME TIME=X,DELT=DELX,FINTIM=XMAX
INITIAL
   DYDX0=0.5*(MIN+MAX)
   CONSTANT E=30000000.,MIN=0.0,MAX=-0.002,EPSI=.05
DYNAMIC
   NOSORT
      IF(X.LT.15.)I=12.57
      IF(ABS(X-15.).LT.EPSI)I=21.63
      IF(X.LT.39.0.AND.X.GT.15.+EPSI)I=30.68
      IF(ABS(X-39.0).LT.EPSI)I=17.33
      IF(X.GT.39.0+EPSI)I=3.98
----> CALL BNDMOM(X,M)
   SORT
   D2YDX2=M/(E*I)
   DYDX=INTGRL(DYDX0,D2YDX2)
   Y=INTGRL(0.0,DYDX)
TERMINAL
      IF(ABS(Y).LT.0.00002)GO TO 3
      IF(Y.GT.0.)GO TO 1
      MAX=DYDX0
      GO TO 2
```

```
  1 MIN=DYDXO
  2 CALL RERUN
  3 CONTINUE
TIMER DELX=0.1,XMAX=42.0
METHOD TRAPZ
END
TIMER OUTDEL=1.0
PRTPLOT Y,DYDX
END
STOP
```

An attempt to call the subprogram BNDMOM with a FORTRAN CALL statement placed in a sorted section will result in a diagnostic message as shown below:

```
                RENAME TIME=X,DELT=DELX,FINTIM=XMAX
            INITIAL
                DYDXO=0.5*(MIN+MAX)
                CONSTANT E=30000000.,MIN=0.0,MAX=-0.002,EPSI=.05
            DYNAMIC
                NOSORT
                    IF(X.LT.15.)I=12.57
                    IF(ABS(X-15.).LT.EPSI)I=21.63
                    IF(X.LT.39.0.AND.X.GT.15.+EPSI)I=30.68
                    IF(ABS(X-39.0).LT.EPSI)I=17.33
                    IF(X.GT.39.0+EPSI)I=3.98
                SORT
          ————————→CALL BNDMOM(X,M)
****CSMP STATEMENT INCORRECTLY WRITTEN****
```

Generally, when a subprogram is to have a single output, a function subprogram is used.

Problems

9-1 Write the necessary CSMP statements to generate the function shown in which FINTIM = 30.

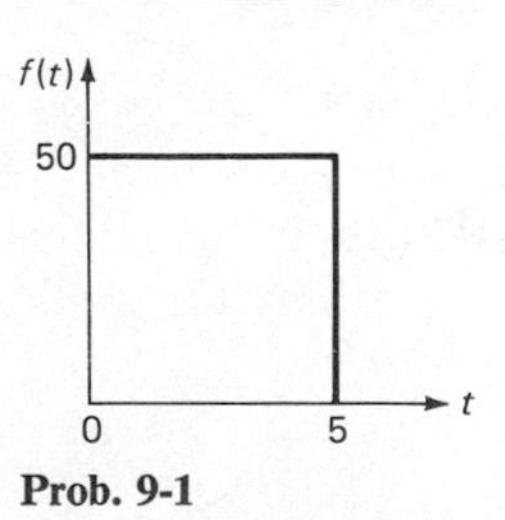

Prob. 9-1

9-2 In Ex. 9-4, the differential equation for the elastic curve was taken as

$$\frac{d^2y}{dx^2} = M/EI \qquad \text{(small deflection theory)}$$

This equation will normally give very accurate results for the deflection y for most practical problems. However, for beam problems involving large deflections, in which dy/dx is not small compared to 1, the following differential equation for the elastic curve should be used

$$\frac{d^2y}{dx^2} = \frac{M}{EI}\left[1 + \left(\frac{dy}{dx}\right)^2\right]^{3/2} \qquad \text{(large deflection theory)}$$

Modify the CSMP program of Ex. 9-4 so that the large deflection theory equation will be used for calculation of the beam deflections.

9-3 Write the necessary CSMP statements to generate the function shown in which $a = 0.5$ sec and FINTIM $= 4.5$ sec.

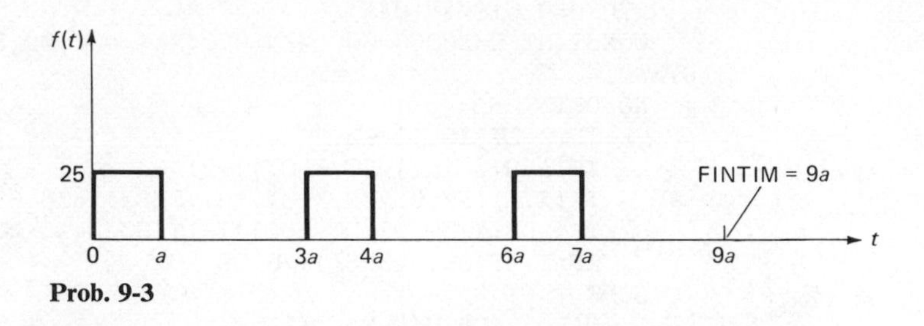

Prob. 9-3

9-4 Write the necessary CSMP statements to generate the ramp function shown.

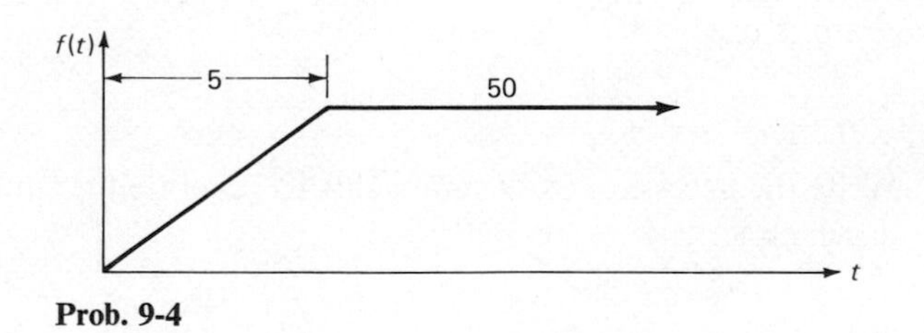

Prob. 9-4

9-5 The mass m slides on a surface with a coefficient of friction of μ. Since the friction force, $F = \mu W$, acts in the opposite direction to the velocity $\dot{x}$, the differential equations for motion to the right and left, respectively, are

$$\ddot{x} + \frac{k}{m}x = -\mu g \qquad (\dot{x} > 0) \text{ (right)}$$

$$\ddot{x} + \frac{k}{m}x = \mu g \qquad (\dot{x} < 0) \text{ (left)}$$

where g = acceleration of gravity, 386 in./sec^2. Use the appropriate switching function to obtain either μg or $-\mu g$ and write a CSMP program for a sequence of nine runs starting with $\mu = 0.1$, and followed by 0.1 increments for subsequent runs. The mass m is to be released from rest so that the initial value of x is sufficient to overcome the maximum friction force for $k/m = 100$. Obtain print-plots of x.

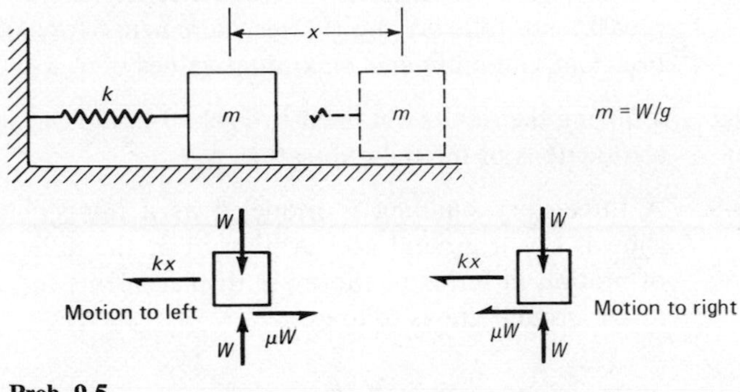

Prob. 9-5

9-6 The spring and mass system is subjected to a pulse of magnitude F_0 and time duration τ. From the free-body diagram shown and Newton's second law, the differential equation of motion is found to be

$$\ddot{x} + \frac{k_1}{m}x + \frac{[T + k_2(\sqrt{x^2 + l^2} - l)]}{m\sqrt{x^2 + l^2}}x = \frac{F(t)}{m}$$

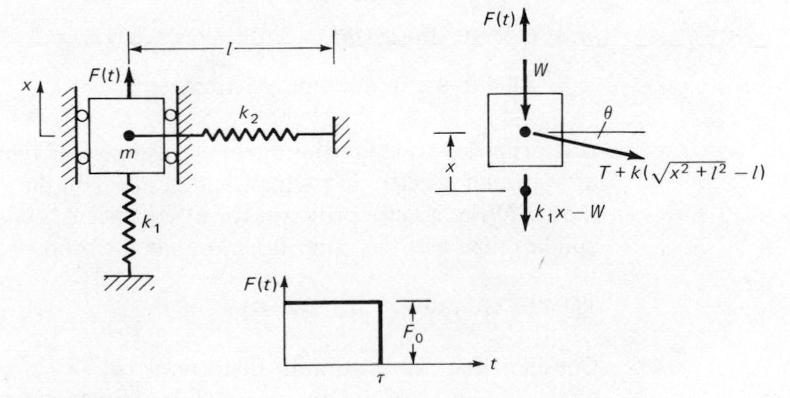

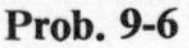
Prob. 9-6

where T is the tension in k_2 when $x = 0$. The system parameter values are as follows:

$k_1 = k_2 = 39.4$ lb/in. $\qquad l = 10$ in.

$m = 1.0$ lb-sec^2/in. $\qquad T = 394$ lb

$F_0 = 100$ lb $\qquad 0.1 \leq \tau \leq 1.0$ sec

Write a CSMP program for a sequence of ten runs starting with $\tau = 0.1$ sec followed by 0.1-sec increments for subsequent runs. Print out minimum and maximum values of x, $\dot{x}$, and $\ddot{x}$.

9-7 Utilizing the results obtained in Prob. 9-6, plot x_{max}, $\dot{x}_{max}$, and $\ddot{x}_{max}$ as functions of the pulse duration τ.

9-8 A three-story building is modeled as a lumped mass system as shown. For a ground acceleration of $\ddot{y}$, the differential equations of motion in terms of the mass displacements (q_1, q_2, q_3) relative to the ground are as follows:

$$\ddot{q}_1 + \frac{2c}{m_1}\dot{q}_1 - \frac{c}{m_1}\dot{q}_2 + \frac{(k_1 + k_2)}{m_1}q_1 - \frac{k_2}{m_1}q_2 = -\ddot{y}$$

$$\ddot{q}_2 + \frac{2c}{m_2}\dot{q}_2 - \frac{c}{m_2}(\dot{q}_1 + \dot{q}_3) - \frac{k_2}{m_2}q_1 + \frac{(k_2 + k_3)}{m_2}q_2 - \frac{k_3}{m_2}q_3 = -\ddot{y}$$

$$\ddot{q}_3 + \frac{c}{m_3}\dot{q}_3 - \frac{c}{m_3}\dot{q}_2 - \frac{k_3}{m_3}q_2 + \frac{k_3}{m_3}q_3 = -\ddot{y}$$

where

$k = 2 \times 10^6$ lb/in. (all columns)

$m = 6 \times 10^3$ lb-sec^2/in.

$c = 2400$ lb-sec/in. (damping parameter)

It is proposed to study the dynamic response of the structure due to a ground acceleration which is described by the ramp function shown. Write a CSMP program for a sequence of runs in which the buildup time τ of the ramp function has the following values:

```
PARAMETER TAU=(0.4, 9*0.4)
```

Consider that the maximum distortions ($u_1 = q_1$, $u_2 = q_2 - q_1$, $u_3 = q_3 - q_2$) are not to exceed 3 in. for no serious damage to occur. Is there any serious damage to the building?

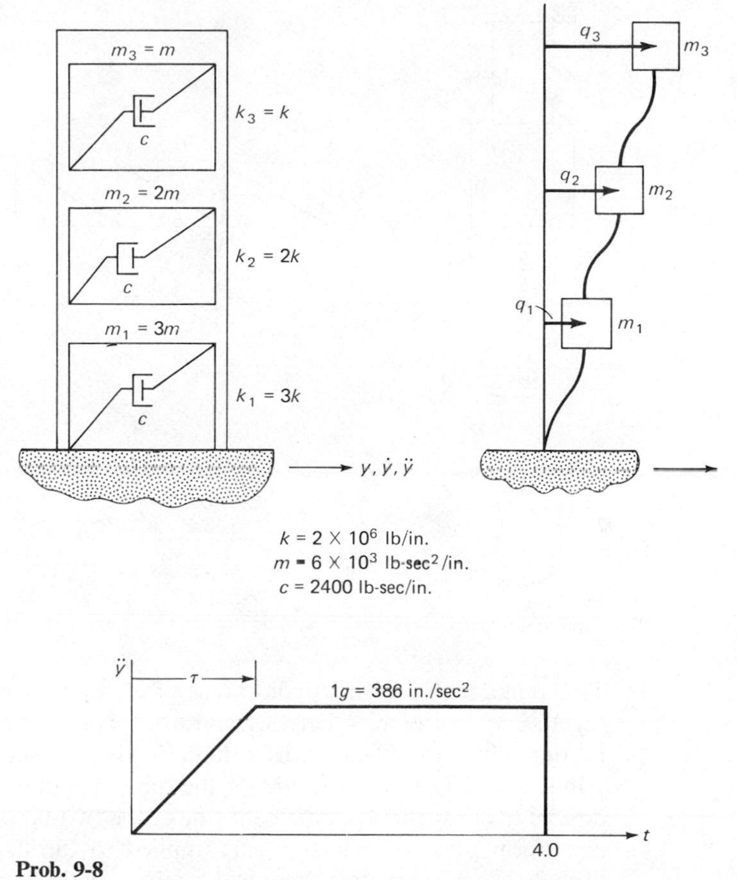

Prob. 9-8

9-9 Due to the shape and uniform rotation of the cam, the displacement y of the follower is essentially a sawtooth function with an amplitude of 1 in. and a period of 1 sec (fig., p. 626). From Newton's second law and the assumption that the follower remains in contact with the cam at all times, the differential equation of motion is found to be

$$\ddot{x} + \frac{c}{m}\dot{x} + \frac{2k}{m}x = \frac{k}{m}y$$

Write a CSMP program with a FINTIM of 10 sec and obtain the dynamic response of the mass m for the following system variable values:

$m = 0.1$ lb-sec^2/ft

$k = 20$ lb/ft

$c = 0.2$ lb-sec/ft

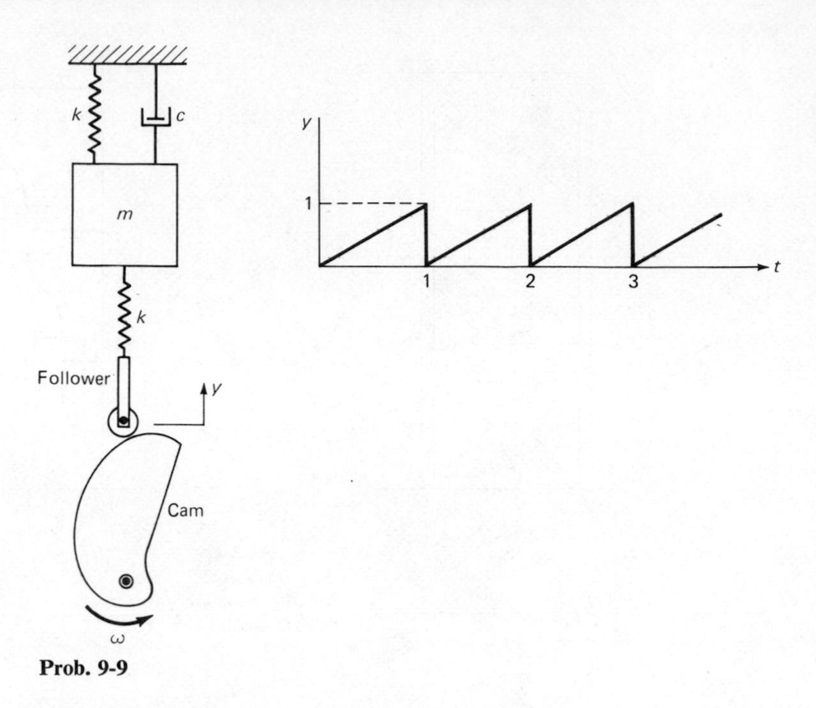

Prob. 9-9

9-10 Two tanks with cross-sectional areas of A_1 and A_2 are connected together with a pipe of length l and area A_0. The tanks and connecting pipe are filled with a fluid having a mass density of γ lb-sec^2/ft^4. The initial height of the fluid is h in each tank. It is desired to study the dynamic behavior of the two-tank system when a uniform pressure p is suddenly applied to the fluid surface of tank 2. The pressure p is described by the function

$$p = p_0 f(t)$$

where

$$f(t) = 1 \qquad 0 < t < t_0$$

$$f(t) = 0 \qquad t > t_0$$

Due to the blast pressure, the fluid in tank 1 will experience an upward displacement of x_1 and the fluid in tank 2 will experience a downward displacement of x_2. If x_1 and x_2 are not small in comparison to h, the nonlinear differential equation for the system, in terms of x_1 (neglecting friction) is

$$\ddot{x}_1 + \frac{[1 - (A_1/A_2)^2]}{2D}\dot{x}_1^2 + g\frac{[1 + A_1/A_2]}{D}x_1 = \frac{p_0 f(t)}{\gamma D}$$

where

$$D = h[1 + A_1/A_2] + [1 - (A_1/A_2)^2]x_1 + \frac{lA_1}{A_0}$$

$$g = 32.2 \text{ ft/sec}^2 \text{ (acceleration of gravity)}$$

Write a CSMP program for simulation runs for the following parameter values:

$$A_1/A_2 = 0.25, 2.0, 5.0$$

$$h = 10 \text{ ft}, \; l = 1 \text{ ft}, \; A_1/A_0 = 10$$

$$p_0/\gamma = 100 \text{ ft}^2/\text{sec}^2, \; t_0 = 2 \text{ sec}$$

Print-plots are to be obtained for x_1, $\dot{x}_1$, and $\ddot{x}_1$.

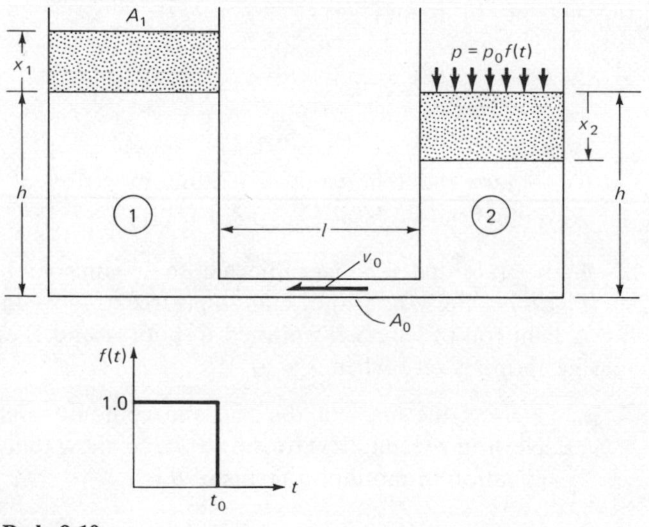

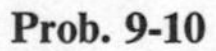
Prob. 9-10

9-11 Derive the differential equation given in Prob. 9-10. (*Hint:* Write the kinetic energy T and potential V of the system and use Lagrange's equation.) Lagrange's equation is

$$\frac{d}{dt}\left(\frac{\partial T}{\partial \dot{q}_i}\right) - \frac{\partial T}{\partial q_i} + \frac{\partial V}{\partial q_i} = Q_i \qquad i = 1, 2, 3, \ldots$$

where

q_i = generalized coordinates (x_1 for Prob. 9-10)
T = kinetic energy of the system
V = potential energy of the system
Q_i = generalized nonpotential force associated with the coordinate q_i

The generalized force Q_i includes all forces that add or remove energy from the system. The increment of work dW done by the force Q_i for a differential change dq_i of one of the q_i coordinates is $Q_i\, dq_i$. For the tank system of Prob. 9-10,

$$dW = p_0 f(t) A_2\, dx_2$$

Since $A_1 x_1 = A_2 x_2$, $dx_2 = (A_1/A_2)\, dx_1$ so that $Q_1 = p_0 f(t) A_1$. The kinetic energy T and potential energy V are

$$T = \frac{\gamma A_1}{2}(h + x_1)\dot{x}_1^2 + \frac{\gamma A_2}{2}(h - x_2)\dot{x}_2^2 + \frac{\gamma A_0}{2} l v_0^2$$

$$V = \gamma g A_2 x_2 \frac{(x_2 + x_1)}{2}$$

To obtain the differential equation in terms of x_1, note that $A_1 x_1 = A_2 x_2 = A_0 v_0$.

9-12 Body A of mass m is subjected to a sinusoidal exciting force $F_0 \sin \omega t$. Body B, also of mass m, is free to move in a vertical slot. A light rod of length l is pinned to both A and B and the force in the spring is zero when $\theta = 0$.

a. Neglect the mass of the rod and coulomb friction and apply Newton's second law to A and B and show that the differential equation of motion in terms of θ is

$$\ddot{\theta} + \frac{c}{m} \cos^2 \theta \dot{\theta} + \frac{g}{l} \sin \theta + \frac{k}{m} \sin \theta \cos \theta = \frac{F_0}{ml} \cos \theta \sin \omega t$$

For small values of θ, the linearized differential equation of motion is

$$\ddot{\theta} + \frac{c}{m}\dot{\theta} + \left(\frac{g}{l} + \frac{k}{m}\right)\theta = \frac{F_0}{ml} \sin \omega t$$

b. Write a CSMP program to obtain the maximum steady-state amplitude for the nonlinear equation for $\omega = \omega_n$, where

$$\omega_n = \sqrt{\frac{g}{l} + \frac{k}{m}}$$

The exciting frequency $\omega = \omega_n$ corresponds to resonance for the linear equation of motion. The values for k/m and g/l are 75 and 25, respectively. Thus $\omega_n = 10$ radians/sec. The response is to be obtained for the following damping values:

$$\frac{c}{m} = 2\xi\omega_n = 4, 8, 12, 16, 20$$

where ξ is the damping factor in which $\xi = 1$ corresponds to a critically damped system for small values of θ.

The value for F_0/ml is to be selected so that the steady-state amplitude for the linearized equation does not exceed $\pi/6$ (30 deg) for $\omega = \omega_n$. That is, from vibration theory

$$\theta_{max} = \frac{\pi}{6} = \frac{F_0/ml}{\omega_n^2 2\xi}$$

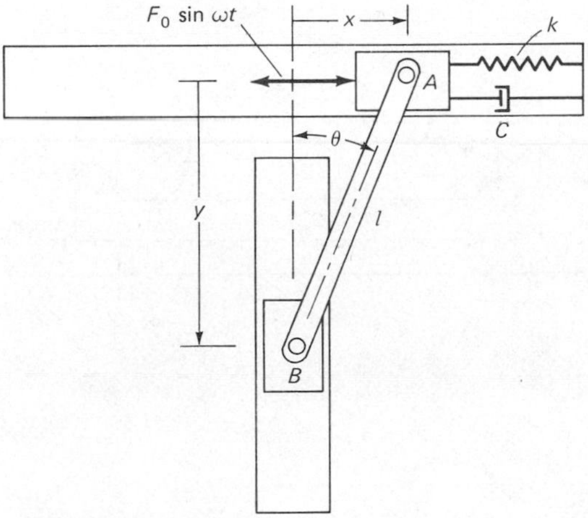

Prob. 9-12

9-13 The cylinder of radius r, centroidal moment of inertia $\bar{I}$, and mass m rolls without slipping on the moving support. The differential equation of motion of the cylinder, in terms of the displacement u relative to the moving support, is

$$\ddot{u} + \frac{2c}{3m}\dot{u} + \frac{2k}{3m}u = -\frac{2}{3}\ddot{y}$$

or

$$\ddot{u} + 2\xi\left(\frac{2\pi}{\tau}\right)\dot{u} + \left(\frac{2\pi}{\tau}\right)^2 u = -\frac{2}{3}\ddot{y}$$

where

ξ = damping factor as a fraction of critical damping ($\xi_c = 1$)
τ = natural undamped period of the cylinder and spring system, sec
$\ddot{y}$ = acceleration of moving support

Write a CSMP program in order to obtain the maximum response u_{max} in which the acceleration of the support is

$$\ddot{y} = -100 \sin 2\pi t \qquad 0 \le t \le 1$$
$$\ddot{y} = 0 \qquad 1 < t$$

The maximum response u_{max} is to be obtained for values of the natural period τ ranging from 0.1 to 2 sec for a damping factor of $\xi = 0.1$. A plot of u_{max} as a function of τ is called a *shock spectrum.*

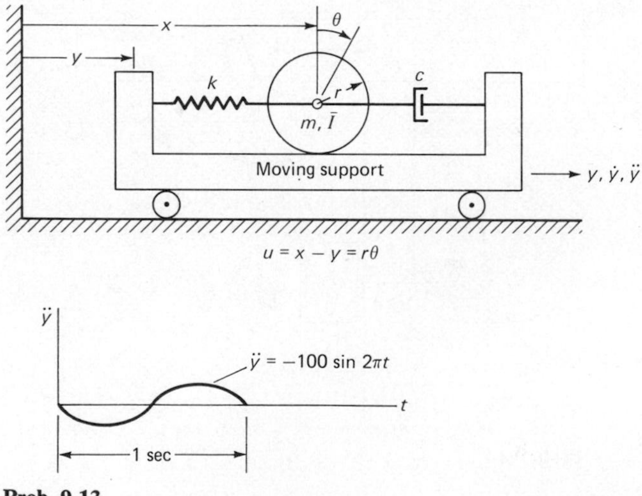

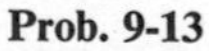
Prob. 9-13

9-14 A study is to be made of the trajectory of a pilot ejected from an airplane at various altitudes and various airplane velocities (based on a problem from an IBM CSMP short course). It is important that the ejected pilot and seat mechanism clear the vertical stabilizer by a safe margin.

Let the origin of a fixed x'-y' coordinate system be at the center of mass of the pilot and ejection mechanism at the instant ejection begins. In part a of the accompanying figure, the forces which act on the pilot and seat mechanism are shown at some time t after ejection has occurred.

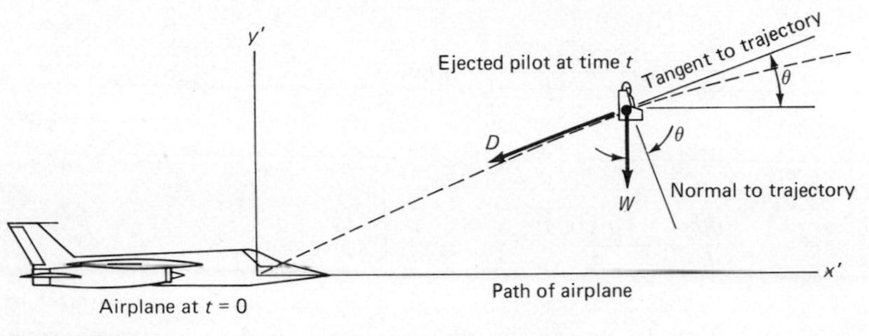

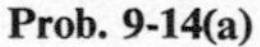

Prob. 9-14(a)

Applying Newton's second law along a tangent to the trajectory, with the direction of the velocity taken as the positive direction, we obtain

$$-D - W \sin \theta = \dot{V}M \tag{1}$$

in which D is the drag force, W is the weight of the pilot and seat, $\dot{V}$ is the tangential acceleration, and M is the mass of the pilot and seat. The drag force D is given by

$$\tfrac{1}{2}\gamma C_D S V^2$$

where γ is the mass density of the air, C_D is the drag coefficient, and S is the projected area of the pilot and seat normal to the direction of motion. The air density γ will vary with the altitude.

Applying Newton's second law in the normal direction and taking the direction toward the center of curvature as the positive direction, we obtain

$$W \cos \theta = \frac{MV^2}{\rho} \tag{2}$$

in which V is the velocity of the pilot and seat and ρ is the radius of curvature of the trajectory. The quantity V^2/ρ is the acceleration in the normal direction.

Replacing one of the V's in Eq. 2 by $-\rho\dot{\theta}$ ($V = -\rho\dot{\theta}$ since V is positive for a decreasing angle θ), we obtain

$$W \cos \theta = -M\dot{\theta}V \tag{3}$$

Dividing Eqs. 1 and 3 through by M, and replacing W/M with g, we obtain

$$\frac{dV}{dt} = -\frac{\frac{1}{2}\gamma C_D S V^2}{M} - g \sin \theta \tag{4}$$

and

$$\frac{d\theta}{dt} = \frac{-g \cos \theta}{V} \tag{5}$$

Now let us consider the trajectory of the pilot and seat relative to the airplane as shown in part b of the accompanying figure. Let the x-y coordinate system shown move with the airplane in horizontal flight. The y displacement of the pilot and seat may be obtained at any time t by integrating

$$\frac{dy}{dt} = V \sin \theta \tag{6}$$

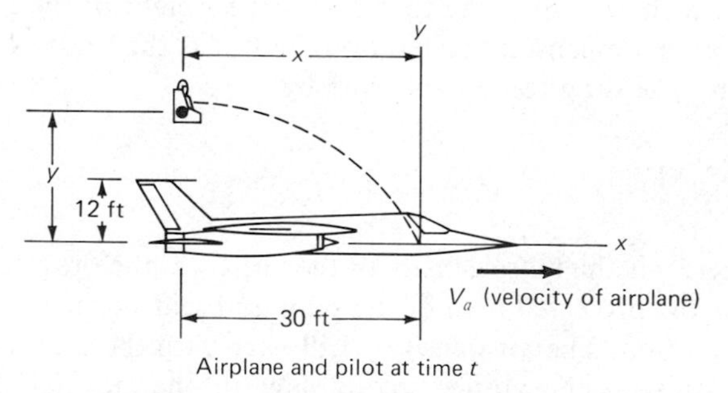

Prob. 9-14(b)

The x displacement may be obtained by integrating

$$\frac{dx}{dt} = V \cos \theta - V_a \tag{7}$$

in which V_a is the velocity of the airplane.

Equations 4, 5, 6, and 7 constitute a set of four first-order differential equations to be solved simultaneously. It should be noted, however, that Eqs. 4 and 5 are valid only after the seat has cleared the rails ($y > 4$ ft for this problem). Prior to that time

$$\frac{dV}{dt} = 0 \tag{8}$$

and

$$\frac{d\theta}{dt} = 0 \tag{9}$$

are assumed to hold. We are assuming that the pilot and seat reach the ejection velocity relative to the plane, V_e, in a very short time, and that this relative velocity remains constant until the seat clears the rails.

It is suggested that a PROCEDURE be used in the DYNAMIC segment to define dV/dt and $d\theta/dt$ by either Eqs. 4 and 5 or Eqs. 8 and 9, depending upon the value of y. The PROCEDURE is preferred to the use of NOSORT since a PROCEDURE can be moved around as an entity by the translator while a NOSORT section cannot.

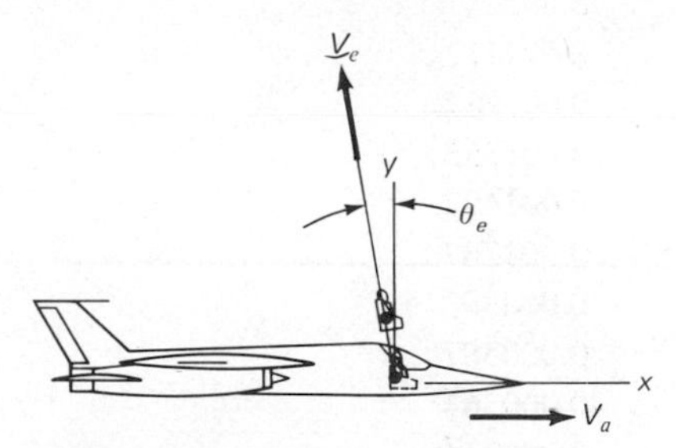

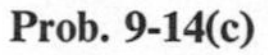
Prob. 9-14(c)

The initial conditions are

$$t = 0 \quad \left| \begin{aligned} x &= 0 \\ y &= 0 \\ V &= [(V_a - V_e \sin \theta_e)^2 + (V_e \cos \theta_e)^2]^{1/2} \\ \theta &= \tan^{-1} \frac{V_e \cos \theta_e}{V_a - V_e \sin \theta_e} \end{aligned} \right. \tag{10}$$

The angle θ_e is shown in part c of the accompanying figure.

The trajectory study should include velocities, V_a, ranging from 100 to 1000 ft/sec in increments of 100 ft/sec, and altitudes ranging from sea level to 60,000 ft in increments of 1000 ft, and determining the maximum velocity at each altitude for a safe ejection. A safe ejection will be considered one in which y is greater than 20 ft when x equals minus 30 ft. This will result in the center of mass of the pilot and seat passing more than 8 ft above the top of the vertical stabilizer (see part b of the accompanying figure).

The following data values are to be used:

M = mass of pilot and seat, 7.0 slugs
V_e = ejection velocity relative to plane, 40.0 ft/sec
C_D = drag coefficient, 1.0 (dimensionless)
θ_e = angle of ejection, 15.0 deg
S = projected area, pilot and seat, 10.0 ft^2
g = acceleration of gravity, 32.2 ft/sec^2

The mass density of the air at various altitudes is as follows:

Altitude (ft)	γ(lb-sec^2/ft^4)
0	0.002377
1,000	0.002308
2,000	0.002241
4,000	0.002117
6,000	0.001987
10,000	0.001755
15,000	0.001497
20,000	0.001267
30,000	0.000891
40,000	0.000587
50,000	0.000364
60,000	0.000234

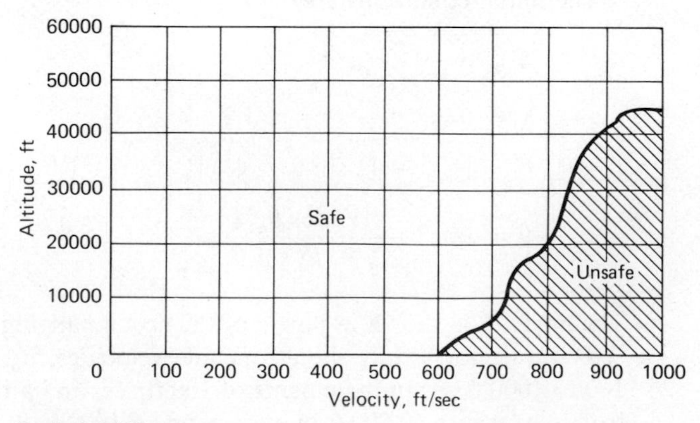

Prob. 9-14(d)

These values can be entered into the program using a CSMP FUNCTION statement, and the air density at any altitude obtained using the arbitrary function generator.

Begin the study at sea level and a velocity of 100 ft/sec. Use a FINTIM of 3.0 sec and a finish condition which will stop the integration when x becomes algebraically less than -30. Thus when $x \cong -30$ ft, integration will cease, and the TERMINAL segment will be entered. In the TERMINAL segment, the y value should be checked to see if it is greater than 20.0 ft, which indicates a safe ejection. If a safe ejection is indicated, write out the altitude, velocity of the airplane, and the y value. If an ejection is safe at some particular altitude and velocity, it will be safe at all higher altitudes up to that velocity because of the decreasing drag force with higher altitudes. Therefore, after a safe ejection at a given altitude and velocity is indicated, the velocity should be increased by 100 ft/sec and a CALL RERUN used to run again. Eventually an unsafe ejection condition will occur. When it does, the altitude of the airplane should

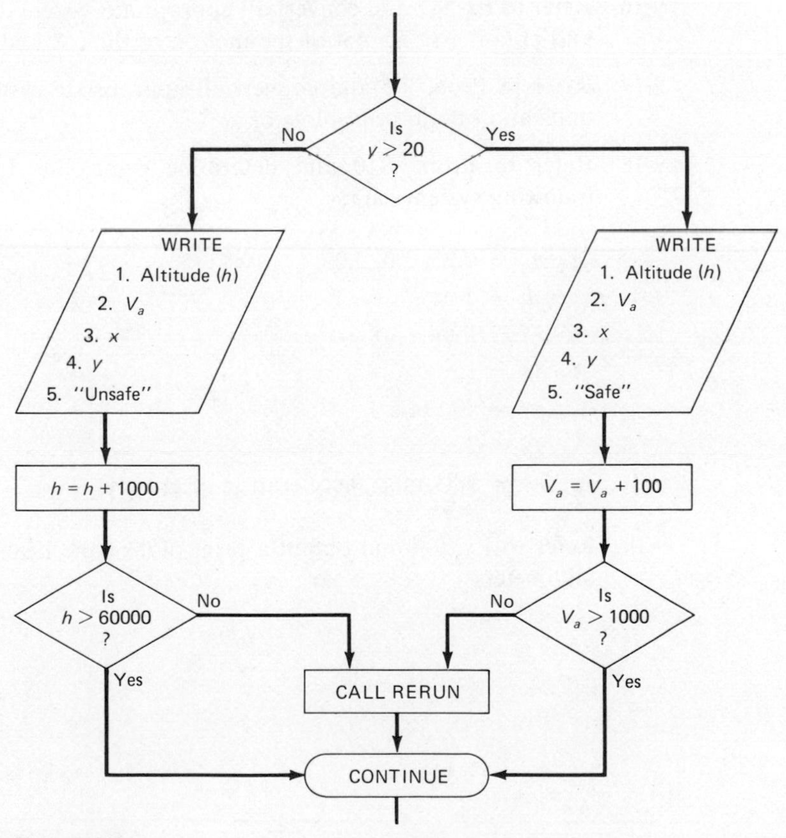

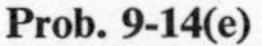
Prob. 9-14(e)

be incremented by 1000 ft with the velocity unchanged. A CALL RERUN statement then causes another run. This continues until a safe ejection again occurs, and the velocity is then incremented as explained earlier. Reruns are made until either the incremented altitude exceeds 60,000 ft or the incremented velocity exceeds 1000 ft/sec. With the data obtained, the altitude-velocity region shown in part d of the accompanying figure can be separated into safe and unsafe parts. With the technique outlined, it will not be necessary to run for all combinations of altitude and velocity.

Use a step size of 0.01 sec and trapezoidal integration. A suggested flow chart for the TERMINAL segment is shown in part e of the accompanying figure.

The following problems involve SI units.

9-15 Refer to Ex. 9-1 and convert all appropriate system data to SI units and obtain a print-plot of the output angle θ. Compare the print-plot results with those shown in Fig. 9-17.

9-16 Refer to Ex. 9-2 and convert all appropriate system data to SI units and obtain a print-plot of the motion of the car body x_1.

9-17 Refer to Prob. 9-5 and convert all appropriate system data to SI units and obtain print-plots of x.

9-18 Refer to Prob. 9-10 and determine print-plots for x_1 for the following system data:

$$\begin{aligned} A_1/A_2 &= 0.25,\ 2.0,\ 5.0 \\ h &= 4\ \text{m} \\ l &= 1\ \text{m} \\ A_1/A_0 &= 10 \\ A_1 p_0/\gamma &= 300\ \text{m/s}^2 \\ t_0 &= 2\ \text{s} \\ g &= 9.81\ \text{m/s}^2\ \text{(acceleration of gravity)} \end{aligned}$$

9-19 Refer to Ex. 9-4 and obtain a print-plot of the beam deflection in millimeters.

APPENDIXES

A International System of Units (SI Units)

A-1 / INTRODUCTION

The International System of Units (abbreviated SI in all languages) was adopted at the 1960 meeting of the International General Conference of Weights and Measures. As of 1976 all the industrial nations of the world, with the exception of the United States, had adopted the SI system of units. Therefore, scientists, engineers, educators, technicians, and the like, in the United States can be fairly certain that they will be involved in the use of the SI metric units in the not too distant future. Although the United States had not officially adopted the SI system of units at the time that this book went to press, some U.S. industries have for the past several years been in the process of moving toward the use of the SI system.

It should be emphasized that the SI system is not the CGS system (centimeter-gram-second) commonly used by physicists nor the MKS gravitational system (meters-kilogram-second) in which mass is a derived unit. Physical quantities in any system of units must be consistent with Newton's second law (force = mass × acceleration). The SI system is an absolute system in which the kilogram (kg) is selected as the basic unit of mass and the derived unit of force is the *newton*, such that

$$1 \text{ newton} = (1 \text{ kg})(1 \text{ m/s}^2) \tag{A-1}$$

In Eq. A-1, the unit of length is the meter m, and s is the approved SI designation of time in seconds. For a freely falling body in a vacuum, the acceleration of gravity g at the surface of the earth is 9.81 m/s^2 and

Newton's second law ($F = Ma$) gives

$$F = W = Mg \tag{A-2}$$

where W is the weight of the body. Thus a 1-kg mass weighs 9.81 newtons at the surface of the earth since

$$W = (1\ \text{kg})(9.81\ \text{m/s}^2) \tag{A-3}$$

A-2 / NOMENCLATURE OF SI UNITS AND QUANTITIES

In the following table the basic units are length, mass, time, temperature, and current. All other named units and unnamed quantities are derived from fundamental definitions or concepts.

Units or Quantities	Name	SI Notation
length	meter	m
mass	kilogram	kg
time	second	s
temperature	kelvin	K
current	ampere	A
force	newton	N ($\text{kg} \cdot \text{m/s}^2$)
stress, pressure	pascal	Pa (N/m^2)
energy, work	joule	J ($\text{N} \cdot \text{m}$)
power	watt	W (J/s)
frequency	hertz	Hz (1/s)
potential difference	volt	V (W/A)
resistance	ohm	Ω (V/A)
velocity (linear)	—	m/s
velocity (angular)	—	rad/s
acceleration (linear)	—	m/s^2
acceleration (angular)	—	rad/s^2
moment of force	—	$\text{N} \cdot \text{m}$ ($\text{kg} \cdot \text{m}^2/\text{s}^2$)
moment of inertia (area)	—	m^4
moment of inertia (mass)	—	$\text{kg} \cdot \text{m}^2$
impulse (linear)	—	$\text{N} \cdot \text{s}$ ($\text{kg} \cdot \text{m/s}$)
impulse (angular)	—	$\text{N} \cdot \text{m} \cdot \text{s}$ ($\text{kg} \cdot \text{m}^2/\text{s}$)
momentum (linear)	—	$\text{N} \cdot \text{s}$ ($\text{kg} \cdot \text{m/s}$)
momentum (angular)	—	$\text{N} \cdot \text{m} \cdot \text{s}$ ($\text{kg} \cdot \text{m}^2 \cdot \text{s}$)
weight (specific)	—	N/m^3 $\text{kg}/(\text{s}^2 \cdot \text{m}^2)$
density (mass)	—	kg/m^3
density (weight)	—	N/m^3
viscosity (absolute)	—	$\text{N} \cdot \text{s/m}^2$ ($\text{Pa} \cdot \text{s}$)

A-3 / CONVERSION OF U.S. CUSTOMARY UNITS TO SI UNITS

Multiply (U.S. Customary)	By	To Obtain (SI)
Acceleration (ft/sec^2)	3.048×10^{-1}	m/s^2 ($meter/second^2$)
Acceleration ($in./sec^2$)	2.54×10^{-2}	m/s^2 ($meter/second^2$)
Area (ft^2)	9.2903×10^{-2}	m^2 ($meter^2$)
Area ($in.^2$)	6.4516×10^{-4}	m^2 ($meter^2$)
Density (lb/ft^3)	1.5708×10^2	N/m^3 ($newton/meter^3$)
Density ($lb/in.^3$)	2.7145×10^5	N/m^3 ($newton/meter^3$)
Force (lb)	4.4482	N (newton)
Length (in.)	2.54×10^{-2}	m (meter)
Length (ft)	3.048×10^{-1}	m (meter)
Mass ($lb\text{-}sec^2/ft$)	1.4594×10	kg (kilogram)
Power (horsepower—550 ft-lb/sec)	7.4569×10^2	W (watt)
Stress, pressure (psi)	6.8947×10^3	N/m^2 or Pa (pascal)
Velocity (ft/sec)	3.048×10^{-1}	m/s (meter/second)
Volume (ft^3)	2.8317×10^{-2}	m^3 ($meter^3$)
Volume ($in.^3$)	1.6387×10^{-5}	m^3 ($meter^3$)
Work, energy (ft-lb)	1.3558	J (joule)
Work, energy (Btu)	1.0551×10^3	J (joule)
Work, energy (kw·h)	3.60×10^6	J (joule)

A-4 / PREFIXES

The following prefixes are commonly used when very small or very large numbers are involved:

Prefix	Numerical Value	SI Designation
nano	10^{-9}	n
micro	10^{-6}	μ
milli	10^{-3}	m
kilo	10^3	k
mega	10^6	M
giga	10^9	G

For example, the modulus of elasticity E of steel is 2.068×10^{11} Pa (pascals) (30×10^6 psi), which may also be written as 206.8 GPa (giga-pascals).

B Matrix Algebra

B-1 / MULTIPLICATION

The product of two matrices $\mathbf{A} = [a_{ij}]$ and $\mathbf{B} = [b_{ij}]$ may be written as

$$[a_{ij}][b_{ij}] = [c_{ij}]$$

or, more simply, as

$$\mathbf{AB} = \mathbf{C}$$

The elements c_{ij} of the $\mathbf{C}$ matrix are obtained from

$$c_{ij} = \sum_{k=1} a_{ik} b_{kj} \tag{B-1}$$

where k identifies the kth element in the ith row of $\mathbf{A}$ and the kth element in the jth column of $\mathbf{B}$. To illustrate Eq. B-1, consider the equation

$$\begin{bmatrix} a_{11} & a_{12} & a_{13} \\ a_{21} & a_{22} & a_{23} \\ a_{31} & a_{32} & a_{33} \end{bmatrix} \begin{bmatrix} b_{11} & b_{12} & b_{13} \\ b_{21} & b_{22} & b_{23} \\ b_{31} & b_{32} & b_{33} \end{bmatrix} = \begin{bmatrix} c_{11} & c_{12} & c_{13} \\ c_{21} & c_{22} & c_{23} \\ c_{31} & c_{32} & c_{33} \end{bmatrix}$$

From Eq. B-1, the product of **A** and **B** gives

$$c_{11} = a_{11}b_{11} + a_{12}b_{21} + a_{13}b_{31}$$

$$c_{21} = a_{21}b_{11} + a_{22}b_{21} + a_{23}b_{31}$$

....

....

....

$$c_{33} = a_{31}b_{13} + a_{32}b_{23} + a_{33}b_{33}$$

It should now be apparent that matrix multiplication consists of a *row-on-column* multiplication sequence.

Matrix multiplication is not commutative. That is, **AB** ≠ **BA**. In addition, the product of two matrices **AB** is defined only if the number of columns in **A** is equal to the number of rows in **B**.

As shown in the following example, the multiplication of a matrix **A** by a *column* matrix **B** results in a column matrix:

$$\begin{bmatrix} 1 & 2 & 3 \\ 3 & 4 & 1 \\ 1 & 2 & 1 \end{bmatrix} \begin{Bmatrix} 1 \\ 2 \\ 3 \end{Bmatrix} = \begin{Bmatrix} (1)(1) + (2)(2) + (3)(3) \\ (3)(1) + (4)(2) + (1)(3) \\ (1)(1) + (2)(2) + (1)(3) \end{Bmatrix} = \begin{Bmatrix} 14 \\ 14 \\ 8 \end{Bmatrix}$$

The *identity* matrix **I**, or *unit* matrix, is an *n*-by-*n* matrix with *ones* on the main diagonal and zeros everywhere else. The identity matrix is to matrix algebra what the identity number (one) is to ordinary algebra. That is,

$$\mathbf{IA} = \mathbf{AI} = \mathbf{A} \tag{B-2}$$

Matrix multiplication is associative. That is,

$$\mathbf{ABC} = (\mathbf{AB})\mathbf{C} = \mathbf{A}(\mathbf{BC}) \tag{B-3}$$

Premultiplying the product **AB** by **C** means

$$\mathbf{CAB} = \mathbf{D}$$

If **AB** is to be *postmultiplied* by **C**, then

$$\mathbf{ABC} = \mathbf{E}$$

B-2 / MATRIX INVERSION (Also see Section 3-6.)

The inverse of the matrix **A** is denoted as $\mathbf{A}^{-1}$. Also,

$$\mathbf{A}^{-1}\mathbf{A} = \mathbf{I}$$

To invert the matrix

$$\mathbf{A} = \begin{bmatrix} 1 & 1 & 0 \\ 1 & 0 & 1 \\ 1 & 2 & 2 \end{bmatrix}$$

first write the *augmented* matrix

$$\begin{bmatrix} 1 & 1 & 0 & 1 & 0 & 0 \\ 1 & 0 & 1 & 0 & 1 & 0 \\ 1 & 2 & 2 & 0 & 0 & 1 \end{bmatrix}$$

The elements b_{ij} of the inverse matrix $\mathbf{A}^{-1}$ may be obtained by *successive* applications of the following equations on the augmented matrix:

$$b_{i-1,j-1} = a_{ij} - \frac{a_{1j}a_{i1}}{a_{11}} \qquad \begin{cases} 1 < i \leq n \\ 1 < j \leq m \\ a_{11} \neq 0 \end{cases} \tag{B-4}$$

$$b_{n,j-1} = \frac{a_{1j}}{a_{11}} \qquad \begin{cases} 1 < j \leq m \\ a_{11} \neq 0 \end{cases} \tag{B-5}$$

where

m = maximum column number
n = maximum row number
i = row number of old matrix **A**
j = column number of old matrix **A**
a = an element of old matrix **A**
b = an element of new matrix **B**

Three successive applications of Eqs. B-4 and B-5 (starting with the augmented matrix) yield

$$\mathbf{A}^{-1} = \begin{bmatrix} \frac{2}{3} & \frac{2}{3} & -\frac{1}{3} \\ \frac{1}{3} & -\frac{2}{3} & \frac{1}{3} \\ -\frac{2}{3} & \frac{1}{3} & \frac{1}{3} \end{bmatrix}$$

B-3 / TRANSPOSE OF A MATRIX

The transpose $\mathbf{A}^{\mathrm{T}}$ of $\mathbf{A}$ is obtained by interchanging the rows and columns of $\mathbf{A}$. That is, if

$$\mathbf{A} = \begin{bmatrix} a_{11} & a_{12} & a_{13} \\ a_{21} & a_{22} & a_{23} \\ a_{31} & a_{32} & a_{33} \end{bmatrix}$$

then

$$\mathbf{A}^{\mathrm{T}} = \begin{bmatrix} a_{11} & a_{21} & a_{31} \\ a_{12} & a_{22} & a_{32} \\ a_{13} & a_{23} & a_{33} \end{bmatrix}$$

If $\mathbf{A}$ is a *symmetric* matrix ($a_{ij} = a_{ji}$), then

$$\mathbf{A} = \mathbf{A}^{\mathrm{T}}$$

Also,

$$[\mathbf{AB}]^{\mathrm{T}} = \mathbf{B}^{\mathrm{T}}\mathbf{A}^{\mathrm{T}} \tag{B-6}$$

B-4 / ORTHOGONALITY PRINCIPLE OF SYMMETRIC MATRICES

The matrix equation for a set of n homogeneous algebraic equations may be written as

$$[\mathbf{A} - \lambda\mathbf{I}]\{X\} = \mathbf{0}$$

or

$$\mathbf{AX} = \lambda\mathbf{X} \tag{B-7}$$

where $\mathbf{A}$ is a square symmetric matrix and λ is the eigenvalue corresponding to the eigenvector $\mathbf{X}$. For the ith and jth eigenvalues and corresponding eigenvectors, we may write from Eq. B-7 the following:

$$\mathbf{AX}_i = \lambda_i\mathbf{X}_i \tag{B-8}$$

and

$$\mathbf{AX}_j = \lambda_j\mathbf{X}_j \tag{B-9}$$

Postmultiplying the transpose of Eq. B-8 by X_j, gives

$$[\mathbf{AX}_i]^{\mathrm{T}}\mathbf{X}_j = \lambda_i \mathbf{X}_i^{\mathrm{T}}\mathbf{X}_j \tag{B-10}$$

From Eq. B-6, we note that $[\mathbf{AX}_i]^{\mathrm{T}} = \mathbf{X}_i^{\mathrm{T}}\mathbf{A}^{\mathrm{T}}$. Thus Eq. B-10 may be written as

$$\mathbf{X}_i^{\mathrm{T}}\mathbf{A}^{\mathrm{T}}\mathbf{X}_j = \lambda_i \mathbf{X}_i^{\mathrm{T}}\mathbf{X}_j \tag{B-11}$$

Premultiplying Eq. B-9 by $\mathbf{X}_i^{\mathrm{T}}$ gives

$$\mathbf{X}_i^{\mathrm{T}}\mathbf{AX}_j = \lambda_j \mathbf{X}_i^{\mathrm{T}}\mathbf{X}_j \tag{B-12}$$

Noting that $\mathbf{A}^{\mathrm{T}} = \mathbf{A}$, since $\mathbf{A}$ is a symmetric matrix, we obtain upon subtracting Eq. B-12 from Eq. B-11

$$(\lambda_i - \lambda_j)\mathbf{X}_i^{\mathrm{T}}\mathbf{X}_j = 0$$

Since $\lambda_i \neq \lambda_j$, it follows that

$$\mathbf{X}_i^{\mathrm{T}}\mathbf{X}_j = 0 \tag{B-13}$$

which states that the eigenvectors $\mathbf{X}_i$ and $\mathbf{X}_j$ are orthogonal. In summary,

$$\mathbf{X}_i^{\mathrm{T}}\mathbf{X}_j = \begin{Bmatrix} 0 & i \neq j \\ M_{ii} & i = j \end{Bmatrix} \tag{B-14}$$

With $i = j$, M_{ii} is simply equal to the sum of the squares of the components of $\mathbf{X}_i$.

B-5 / ORTHOGONALITY PRINCIPLE OF THE FORM AX = λBX

For many physical problems, the matrix form of a set of homogeneous equations does not occur in the standard form of Eq. B-7, but instead in the form

$$\mathbf{AX} = \lambda\mathbf{BX} \tag{B-15}$$

where **A** and **B** are square symmetric matrices. By following the procedure of B-4, it is easy to show that the orthogonality of the eigenvectors for matrices of the form of Eq. B-15 is

$$\mathbf{X}_i^{\mathrm{T}}\mathbf{BX}_j = \begin{Bmatrix} 0 & i \neq j \\ G_{ii} & i = j \end{Bmatrix} \tag{B-16}$$

C Interpolating Polynomials and Application to Numerical Integration and Differentiation

C-1 / INTRODUCTION TO INTERPOLATION

Suppose that we have some tabulated values of a function over a certain range of its independent variable such as shown in Fig. C-1. The problem of *interpolation* is to find the value of the function for some intermediate value of x not included in the table. This x value is assumed to lie within the range of tabulated abscissas. If it does not, the problem of finding the corresponding function value is called *extrapolation*.

Tabular values may be plotted to give us a graphical picture of points through which the function must pass. Connecting the points with a smooth curve gives us a graphical representation of the function. If only a rough approximation of the function value is required for some intermediate x value, it can be obtained by reading the function value directly from the graph. This procedure is called *graphical interpolation*. If the given values of the independent variable are close together, a sufficiently good graphical approximation to the function might be obtained by connecting the points with straight-line segments as shown in Fig. C-2. An intermediate function value could then be obtained by reading it from the graph. An intermediate function value could also be obtained analytically by a method based on this piecewise linear approximation of the function. From similar triangles in Fig. C-2,

$$\frac{f(x) - f(x_0)}{x - x_0} = \frac{f(x_1) - f(x_0)}{x_1 - x_0}$$

or

$$f(x) = f(x_0) + \frac{f(x_1) - f(x_0)}{x_1 - x_0}(x - x_0) \tag{C-1}$$

x	$f(x)$
0	0
0.2	0.199
0.4	0.389
0.6	0.565
0.8	0.717
1.0	0.841

Fig. C-1 / *Particular x values with f(x).*

for x values between x_0 and x_1. The use of Eq. C-1 is called linear interpolation, which is familiar to everyone who has used log tables.

A smooth curve passing through the given points can be obtained analytically by assuming the function to be approximated by a polynomial function. This was done in a piecewise manner in the derivation of Simpson's rule in Sec. 5-4 where it was assumed that the actual function could be approximated by second-degree parabolas connecting sets of three successive points. In general, a unique polynomial function of degree n can be determined which passes through $n + 1$ given distinct points. This can be done in a manner similar to that employed in Sec. 5-4 in determining a second-degree parabola which passes through three given points. That is, the nth-degree polynomial passing through the $n + 1$ points may be expressed as

$$f(x) = a_1x^n + a_2x^{n-1} + a_3x^{n-2} + \cdots + a_nx + a_{n+1} \tag{C-2}$$

where the a's are unknown coefficients. The substitution of the $n + 1$ x values and corresponding function values into this equation will give $n + 1$ simultaneous equations in the $n + 1$ unknown a values. This approach was the most simple and direct for deriving Simpson's rule, but

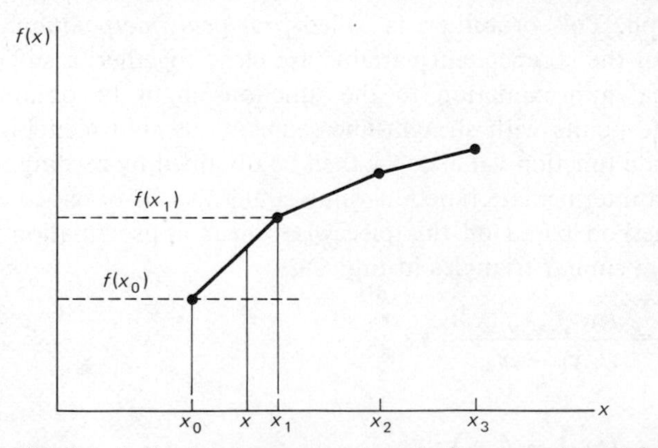

Fig. C-2 / *Graphical approximation of a function with straight-line segments.*

for determining higher degree interpolating polynomials, more efficient methods will be derived. However, before proceeding with these derivations, it will be necessary to discuss some basic definitions.

C-2 / DEFINITIONS

Forward Differences

Let $f(x)$ be some function of x. An increment Δx of the independent variable will be denoted by h. The corresponding increment of the function is

$$\Delta f(x) = f(x + h) - f(x)$$

This increment is called the *first forward difference* of $f(x)$ with respect to h. The operator Δ always implies this operation on any function of x on which it operates. Since $\Delta f(x)$ is a function of x, it can be operated on by the operator Δ giving

$$\Delta[\Delta f(x)] = \Delta^2 f(x) = \Delta[f(x + h) - f(x)]$$

Hence

$$\begin{aligned}\Delta^2 f(x) &= [f(x + 2h) - f(x + h)] - [f(x + h) - f(x)] \\ &= f(x + 2h) - 2f(x + h) + f(x)\end{aligned}$$

The function $\Delta^2 f(x)$ is called the *second forward difference* of $f(x)$. Third, fourth, and higher differences are similarly obtained. In summary, the forward difference expressions are

$$\begin{aligned}\Delta f(x_i) &= f(x_i + h) - f(x_i) \\ \Delta^2 f(x_i) &= f(x_i + 2h) - 2f(x_i + h) + f(x_i) \\ \Delta^3 f(x_i) &= f(x_i + 3h) - 3f(x_i + 2h) + 3f(x_i + h) - f(x_i) \\ \Delta^4 f(x_i) &= f(x_i + 4h) - 4f(x_i + 3h) + 6f(x_i + 2h) - 4f(x_i + h) + f(x_i) \\ &\vdots \\ \Delta^n f(x_i) &= f(x_i + nh) - nf[x_i + (n - 1)h] + \frac{n(n - 1)}{2!} f[x_i + (n - 2)h] \\ &\quad - \frac{n(n - 1)(n - 2)}{3!} f[x_i + (n - 3)h] + \cdots + (-1)^n f(x_i)\end{aligned}$$

in which x_i denotes any specific value for x such as x_0, x_1, and so forth.

The nth difference is often written as

$$\Delta^n f(x_i) = f(x_i + nh) - \binom{n}{1} f[x_i + (n-1)h] + \binom{n}{2} f[x_i + (n-2)h]$$

$$- \cdots + (-1)^k \binom{n}{k} f[x_i + (n-k)h] + \cdots + (-1)^n f(x_i)$$

where the symbol

$$\binom{n}{k} = \frac{n(n-1)(n-2)(n-3)\cdots(n-k+1)}{k!}$$

is the familiar symbol used for binomial coefficients.

If we have a table of function values such as shown in Fig. C-1, a difference table can be formed having entries such as those shown in Fig. C-3. In this table, $f(x_0 + sh)$ is denoted by f_s and $\Delta^k f(x_0 + sh)$ by $\Delta^k f_s$. Each difference is obtained by subtracting the table entry to its upper left from the table entry to its lower left.

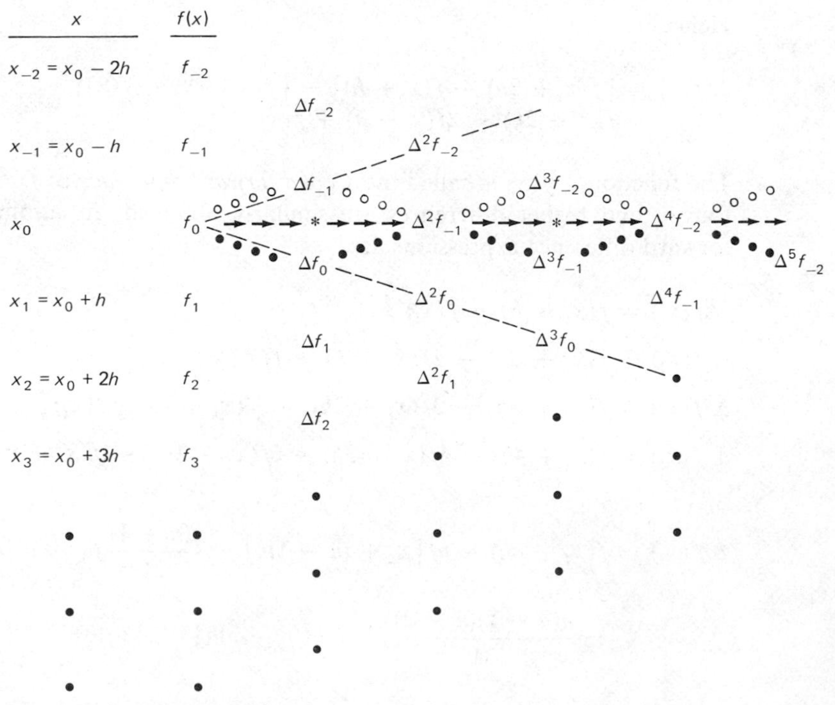

Fig. C-3 / *Difference table.*

Backward Differences

The first backward difference of $f(x)$ with respect to increment h is defined as

$$\nabla f(x) = f(x) - f(x - h)$$

The operator ∇ always implies this operation on the function of x on which it operates, so that

$$\begin{aligned}\nabla[\nabla f(x_i)] &= \nabla^2 f(x_i) = [f(x_i) - f(x_i - h)] - [f(x_i - h) - f(x_i - 2h)] \\ &= f(x_i) - 2f(x_i - h) + f(x_i - 2h)\end{aligned}$$

where x_i denotes any specific values for x such as x_0, x_1, and so forth. In general, $\nabla^n f(x_i) = \nabla[\nabla^{n-1} f(x_i)]$, and the backward differences of $f(x_i)$ are

$$\begin{aligned}\nabla f(x_i) &= f(x_i) - f(x_i - h) \\ \nabla^2 f(x_i) &= f(x_i) - 2f(x_i - h) + f(x_i - 2h) \\ \nabla^3 f(x_i) &= f(x_i) - 3f(x_i - h) + 3f(x_i - 2h) - f(x_i - 3h) \\ \nabla^4 f(x_i) &= f(x_i) - 4f(x_i - h) + 6f(x_i - 2h) - 4f(x_i - 3h) + f(x_i - 4h) \\ \vdots \quad & \quad \vdots \\ \nabla^n f(x_i) &= f(x_i) - \binom{n}{1} f(x_i - h) + \binom{n}{2} f(x_i - 2h) - \cdots \\ &\quad + (-1)^n f(x_i - nh)\end{aligned}$$

We may also note that

$$\begin{aligned}\nabla f(x_i) &= \Delta f(x_i - h) \\ \nabla^2 f(x_i) &= \Delta^2 f(x_i - 2h) \\ \vdots \quad & \quad \vdots \\ \nabla^n f(x_i) &= \Delta^n f(x_i - nh)\end{aligned}$$

If we let $x_i = x_0$, and use the simplified notation used in Fig. C-3, the above relationships between the forward- and backward-difference expressions become

$$\begin{aligned}\nabla f_0 &= \Delta f_{-1} \\ \nabla^2 f_0 &= \Delta^2 f_{-2} \\ \vdots \quad & \quad \vdots \\ \nabla^n f_0 &= \Delta^n f_{-n}\end{aligned}$$

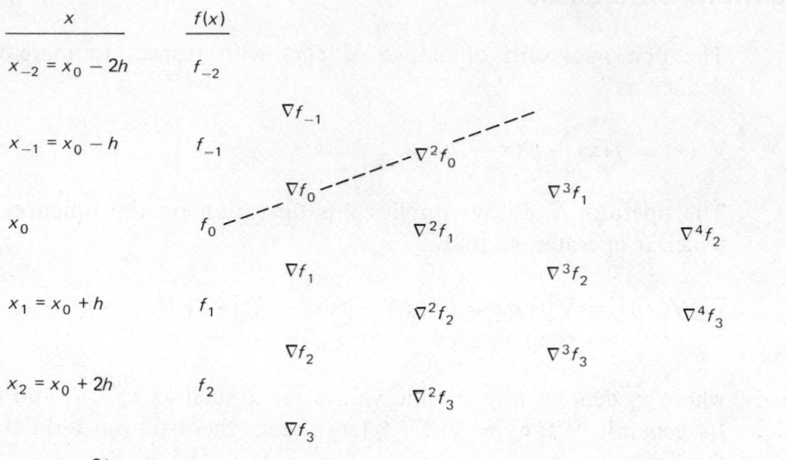

Fig. C-4 / *Table of Fig C-3 written in backward-difference notation.*

or, in general,

$$\nabla^k f_s = \Delta^k f_{s-k} \qquad (k = 1, 2, 3, \ldots)$$

Thus the table in Fig. C-3 can be written with backward-difference notation as shown in Fig. C-4. The numerical values of the table entries would be identical to those in the table of Fig. C-3 for the same function and arguments, and the differences are found as described for that table. The table entries in Fig. C-4 are merely expressed in different notation.

Central Differences

The tables in Figs. C-3 and C-4 can be written in central-difference notation by introduction of the central-difference operator δ defined by

$$\delta^{2n} f(x) = \Delta^{2n} f(x - nh)$$

$$\delta^{2n+1} f(x + h/2) = \Delta^{2n+1} f(x - nh)$$

Using these definitions, the table in Fig. C-3 can be converted to the notation shown in Fig. C-5. Numerical values in this table would be identical to those in the tables of Figs. C-3 and C-4 for the same function and arguments. The tables differ only in the notation used to represent the table entries. For example, the tables of Figs. C-3 through C-5 would all appear as shown in Fig. C-6 for the given function values.

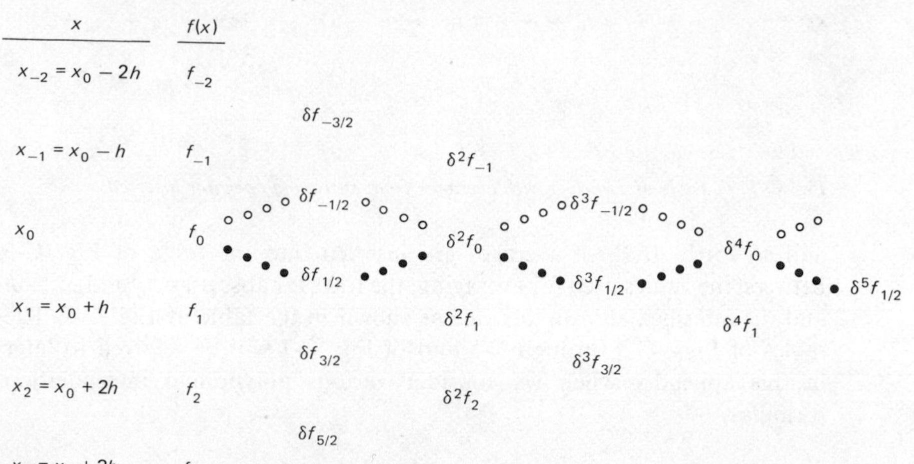

Fig. C-5 / *Table of Fig. C-3 written in central-difference notation.*

The Averaging Operator

The averaging operator is defined as

$$\mu f(x) = \tfrac{1}{2}[f(x + h/2) + f(x - h/2)]$$

Then,

$$\mu\delta f_0 = \tfrac{1}{2}[\delta f_{1/2} + \delta f_{-1/2}]$$

$$\mu\delta^2 f_{1/2} = \tfrac{1}{2}[\delta^2 f_1 + \delta^2 f_0]$$

$$\mu\delta^3 f_0 = \tfrac{1}{2}[\delta^3 f_{1/2} + \delta^3 f_{-1/2}]$$

x	$f(x)$					
0	0					
		0.03125				
0.5	0.03125		0.9375			
		0.96875		4.6875		
1.0	1.00000		5.6250		7.50	
		6.59375		12.1875		3.75
1.5	7.59375		17.8125		11.25	
		24.40625		23.4375		
2.0	32.00000		41.2500			
		65.65625				
2.5	97.65625					

Fig. C-6 / *Numerical values for Figs. C-3, C-4, and C-5.*

x_0	f_0 → → →		$\mu\delta f_0$ → → →		$\delta^2 f_0$ → → →		$\mu\delta^3 f_0$ → → →		$\delta^4 f_0$ → → →		
		$\mu f_{1/2}$		$\delta f_{1/2}$		$\mu\delta^2 f_{1/2}$		$\delta^3 f_{1/2}$		$\mu\delta^4 f_{1/2}$	$\delta^5 f_{1/2}$
$x_0 + h$	f_1		$\mu\delta f_1$		$\delta^2 f_1$		$\mu\delta^3 f_1$		$\delta^4 f_1$		

Fig. C-7 / *Table of Fig. C-5 with averages from averaging operator inserted.*

and so forth. If these averages are inserted into the table of Fig. C-5 between the values used for averaging, the rows for abscissas x_0 and $x_0 + h$ and the intermediate row become as shown in the table of Fig. C-7. The tables of Figs. C-3 through C-5 and of Fig. C-7 will be referred to later in this appendix when we consider various polynomial interpolation formulas.

Difference Quotients

The operator $\underset{h}{\Delta}$ is defined as follows:

$$\underset{h}{\Delta} f(x) = \frac{f(x + h) - f(x)}{h}$$

The function $\underset{h}{\Delta} f(x)$ is called the *first difference quotient* of $f(x)$. The second difference quotient is

$$\underset{h}{\Delta}{}^2 f(x) = \underset{h}{\Delta}\left[\underset{h}{\Delta} f(x)\right]$$

$$= \frac{\{[f(x + 2h) - f(x + h)]/h\} - \{[f(x + h) - f(x)]/h\}}{h}$$

$$= \frac{f(x + 2h) - 2f(x + h) + f(x)}{h^2}$$

The nth difference quotient is

$$\underset{h}{\Delta}{}^n f(x) = \underset{h}{\Delta}\left[\underset{h}{\Delta}{}^{n-1} f(x)\right]$$

It is apparent that

$$\Delta^n f(x) = h^n \underset{h}{\Delta}{}^n f(x)$$

Divided Differences

The divided difference of $f(x)$ for the arguments x_k and x_0 will be written $[x_k x_0]$ and is defined as

$$[x_k x_0] = \frac{f(x_k) - f(x_0)}{x_k - x_0}$$

Since

$$[x_0 x_k] = \frac{f(x_0) - f(x_k)}{x_0 - x_k}$$

it is clear that

$$[x_k x_0] = [x_0 x_k]$$

Divided differences of three arguments are defined as follows:

$$[x_k x_0 x_1] = \frac{[x_k x_0] - [x_0 x_1]}{x_k - x_1}$$

or since

$$[x_0 x_1] = [x_1 x_0]$$

$$[x_k x_0 x_1] = \frac{[x_k x_0] - [x_1 x_0]}{x_k - x_1}$$

Divided differences of any number of arguments are defined similarly. Thus for $n + 1$ arguments

$$[x_k x_0 x_1 \cdots x_{n-1}] = \frac{[x_k x_0 x_1 \cdots x_{n-2}] - [x_0 x_1 x_2 \cdots x_{n-1}]}{x_k - x_{n-1}} \tag{C-3}$$

C-3 / POLYNOMIAL APPROXIMATION AND INTERPOLATION

With the definitions previously discussed, we are now in a position to derive many of the well-known interpolating polynomials. We will derive three of these, and give others with a reference to which the interested reader may refer for the derivations. It should be emphasized that there is only one polynomial of degree n passing through $n + 1$ given points. Although the forms of the various polynomials presented will differ, they all yield the same polynomial if the same $n + 1$ points are used.

Newton's General Interpolating Polynomial

If we let $x_k = x$ in Eq. C-3, we have

$$[xx_0x_1 \cdots x_n] = \frac{-[x_0x_1x_2 \cdots x_n]}{x - x_n} + \frac{[xx_0x_1 \cdots x_{n-1}]}{x - x_n}$$

$$[xx_0x_1 \cdots x_{n-1}] = \frac{-[x_0x_1x_2 \cdots x_{n-1}]}{x - x_{n-1}} + \frac{[xx_0x_1 \cdots x_{n-2}]}{x - x_{n-1}}$$

$$[xx_0x_1 \cdots x_{n-2}] = \frac{-[x_0x_1x_2 \cdots x_{n-2}]}{x - x_{n-2}} + \frac{[xx_0x_1 \cdots x_{n-3}]}{x - x_{n-2}}$$

$$\vdots \qquad\qquad \vdots$$

$$[xx_0x_1] = \frac{-[x_0x_1]}{x - x_1} + \frac{[xx_0]}{x - x_1}$$

$$[xx_0] = \frac{-f(x_0)}{x - x_0} + \frac{f(x)}{x - x_0}$$

Note that the numerator of the second term of the first equation is given by the second equation, the numerator of the second term of the second equation is given by the third equation, and so forth. Making these substitutions we obtain

$$\begin{aligned}[xx_0x_1 \cdots x_n] = {} & \frac{-[x_0x_1x_2 \cdots x_n]}{x - x_n} - \frac{[x_0x_1x_2 \cdots x_{n-1}]}{(x - x_n)(x - x_{n-1})} \\ & - \frac{[x_0x_1x_2 \cdots x_{n-2}]}{(x - x_n)(x - x_{n-1})(x - x_{n-2})} - \cdots \\ & - \frac{[x_0x_1]}{(x - x_n)(x - x_{n-1}) \cdots (x - x_1)} \\ & - \frac{f(x_0)}{(x - x_n)(x - x_{n-1}) \cdots (x - x_0)} \\ & + \frac{f(x)}{(x - x_n)(x - x_{n-1}) \cdots (x - x_0)}\end{aligned}$$

Solving for $f(x)$ yields

$$\begin{aligned}f(x) = {} & f(x_0) + (x - x_0)[x_0x_1] + (x - x_0)(x - x_1)[x_0x_1x_2] + \cdots \\ & + (x - x_0)(x - x_1)(x - x_2) \cdots (x - x_{n-1})[x_0x_1x_2 \cdots x_n] \\ & + R(x)\end{aligned} \tag{C-4}$$

where the remainder term, $R(x)$, is given by

$$R(x) = (x - x_0)(x - x_1) \cdots (x - x_n)[xx_0x_1 \cdots x_n] \tag{C-5}$$

If $f(x)$ is a polynomial of degree n, the remainder term is zero for all x. This can be shown as follows. If $f(x)$ is a polynomial of degree n, then

$$[xx_0] = \frac{f(x) - f(x_0)}{x - x_0}$$

is a polynomial of degree $n - 1$. Taking the divided difference of a polynomial lowers the degree by one. Thus for a polynomial of degree n, $[xx_0x_1]$ is of degree $n - 2$, $[xx_0x_1x_2]$ is of degree $n - 3$, and so forth. Therefore, $[xx_0x_1 \cdots x_{n-1}]$ is of degree zero (is a constant). Hence $[xx_0x_1 \cdots x_n]$ in Eq. C-5 is equal to zero, and

$$R(x) \equiv 0$$

If $f(x)$ is not a polynomial, we note from Eq. C-5 that the remainder term is zero at $x = x_0$, $x = x_1, \ldots, x = x_n$, but is not in general elsewhere. If we omit the remainder term from Eq. C-4, we have a polynomial of degree n which coincides with $f(x)$ at $x = x_0, x_1, \ldots, x_n$. This polynomial is called an interpolating polynomial, and will be denoted by $p_n(x)$. The basis for using it is the assumption that $p_n(x)$ will be very nearly equal to $f(x)$ for intermediate x values.

Omitting the remainder term in Eq. C-4 gives

$$p_n(x) = f(x_0) + \sum_{S=0}^{n-1} (x - x_0)(x - x_1) \cdots (x - x_S)[x_0x_1 \cdots x_{S+1}] \tag{C-6}$$

which is known as Newton's general interpolating polynomial.

EXAMPLE C-1

Determine the equation of a third-degree polynomial passing through the points (0, 5), (1, 14), (2, 41), and (3, 98) using Newton's general interpolating polynomial. In this example, the abscissas are equally spaced, but they need not be. Also, x_0, x_1, x_2, and x_3 need not be abscissa values necessarily in ascending or descending order. The abscissas, ordinates, and divided differences will first be tabulated in the scheme shown overleaf.

$$
\begin{array}{llccc}
x_0 & f(x_0) & & & \\
 & & [x_0x_1] & & \\
x_1 & f(x_1) & & [x_0x_1x_2] & \\
 & & [x_1x_2] & & [x_0x_1x_2x_3] \\
x_2 & f(x_2) & & [x_1x_2x_3] & \\
 & & [x_2x_3] & & \\
x_3 & f(x_3) & & &
\end{array}
$$

Using the scheme above, we obtain

$$
\begin{array}{llccc}
x & f(x) & & & \\
0 & 5 & & & \\
 & & 9 & & \\
1 & 14 & & 9 & \\
 & & 27 & & 2 \\
2 & 41 & & 15 & \\
 & & 57 & & \\
3 & 98 & & &
\end{array}
$$

Equation C-6, for $n = 3$, is

$$
\begin{aligned}
p_3(x) = f(x_0) &+ (x - x_0)[x_0x_1] + (x - x_0)(x - x_1)[x_0x_1x_2] \\
&+ (x - x_0)(x - x_1)(x - x_2)[x_0x_1x_2x_3]
\end{aligned}
$$

Substituting values from the table above into Eq. C-6 yields

$$
\begin{aligned}
p_3(x) &= 5 + (x - 0)9 + (x - 0)(x - 1)9 + (x - 0)(x - 1)(x - 2)2 \\
&= 2x^3 + 3x^2 + 4x + 5
\end{aligned}
$$

The Newton-Gregory Forward-Interpolating Polynomial

Newton's general interpolating polynomial of degree n can be put in a convenient form involving 1st, 2nd, . . . , nth forward differences if the abscissa increment is uniform. By definitions previously given,

$$
[x_0x_1] = \frac{f(x_0) - f(x_1)}{x_0 - x_1} = \underset{h}{\Delta} f(x_0) = \frac{\Delta f(x_0)}{h}
$$

If the arguments are equally spaced so that $x_1 = x_0 + h$, $x_2 = x_0 + 2h$, and so forth, then

$$[x_0x_1x_2] = \frac{\{[f(x_0) - f(x_1)]/[x_0 - x_1]\} - \{[f(x_1) - f(x_2)]/[x_1 - x_2]\}}{x_0 - x_2}$$

$$= \frac{f(x_0) - 2f(x_1) + f(x_2)}{2h^2}$$

$$= \frac{1}{2!}\,\underset{h}{\Delta^2} f(x_0)$$

$$= \frac{1}{2!}\,\frac{\Delta^2 f(x_0)}{h^2}$$

and, in general, if we let $x_S = x_0 + Sh$, we have that

$$[x_0x_1x_2 \cdots x_S] = \frac{1}{S!}\,\frac{\Delta^S f(x_0)}{h^S} \qquad \text{for } S = 1, 2, \ldots, n \tag{C-7}$$

and

$$[x_0x_1x_2 \cdots x_n] = \frac{1}{n!}\,\frac{\Delta^n f(x_0)}{h^n} \tag{C-8}$$

Letting $x_S = x_0 + Sh$ in Eq. C-4, and using the expressions given by Eq. C-7 to substitute for the divided differences in Eq. C-4, we obtain

$$f(x) = f(x_0) + (x - x_0)\,\frac{\Delta f(x_0)}{h} + \frac{(x - x_0)(x - x_0 - h)}{2!}\,\frac{\Delta^2 f(x_0)}{h^2} + \cdots$$
$$+ \frac{(x - x_0)(x - x_0 - h) \cdots (x - x_0 - [n - 1]h)}{n!}\,\frac{\Delta^n f(x_0)}{h^n} + R(x) \tag{C-9}$$

where the remainder, $R(x)$, is given by Eq. C-5 and is

$$R(x) = (x - x_0)(x - x_1) \cdots (x - x_n)[xx_0x_1 \cdots x_n]$$

Equation C-9 can be simplified somewhat if we define

$$u = \frac{x - x_0}{h}$$

That is, u is the ratio of the difference between x and x_0 to the increment of the tabulated argument values. This ratio is called the *phase*. Rearranging the equation, we obtain

$$\begin{aligned} x - x_0 &= uh \\ x - x_0 - h &= uh - h = h(u-1) \\ x - x_0 - 2h &= uh - 2h = h(u-2) \\ &\vdots \\ x - x_0 - (n-1)h &= uh - (n-1)h = h(u-n+1) \end{aligned} \tag{C-10}$$

Making these substitutions into Eq. C-9 and denoting $f(x_0)$ as f_0, $\Delta f(x_0)$ as Δf_0, and so forth, gives

$$f(x) = f_0 + u\Delta f_0 + \frac{u(u-1)}{2!}\Delta^2 f_0 + \frac{u(u-1)(u-2)}{3!}\Delta^3 f_0 + \cdots + \frac{u(u-1)\cdots(u-n+1)}{n!}\Delta^n f_0 + R(x) \tag{C-11}$$

Omitting the remainder term of Eq. C-11 gives the Newton-Gregory forward-interpolating polynomial which may be expressed more compactly by using the binomial coefficient notation; it is

$$p_n(x) = f_0 + \binom{u}{1}\Delta f_0 + \binom{u}{2}\Delta^2 f_0 + \binom{u}{3}\Delta^3 f_0 + \cdots + \binom{u}{n}\Delta^n f_0 \tag{C-12}$$

Note that the function value f_0 and the differences $\Delta f_0, \Delta^2 f_0, \ldots, \Delta^n f_0$ used in Eq. C-12 appear on a line of long dashes running diagonally downward through the difference table in Fig. C-3.

It may be shown[1] that the remainder term $R(x)$ in Eq. C-11 is equal to

$$\frac{(x-x_0)(x-x_0-h)(x-x_0-2h)\cdots(x-x_0-nh)}{(n-1)!}f^{n+1}(\xi)$$

where $f^{n+1}(\xi)$ is the $(n+1)$st derivative of $f(x)$ evaluated at ξ where $x_0 < \xi < x_n$. Then using Eq. C-10,

$$\begin{aligned} R(x) &= \frac{(uh)(h[u-1])(h[u-2])\cdots(h[u-n])}{(n+1)!}f^{n+1}(\xi) \\ &= \binom{u}{n+1}h^{n+1}f^{n+1}(\xi) \end{aligned} \tag{C-13}$$

[1] L. M. Milne-Thompson, *The Calculus of Finite Differences* (London: Macmillan & Co. Ltd., 1960), pp. 5, 6.

The function $f(x)$ is usually not known, so its $(n + 1)$st derivative would also be unknown. Furthermore, the exact location at which the derivative is to be evaluated is unknown. Thus the expression for the remainder, which is the difference between the actual function and the interpolating polynomial, is of limited usefulness. One possible use is to compare the coefficient $\binom{u}{n+1}$ of the remainder term with the coefficient of the remainder term of other interpolating polynomials for which one or more different points have been used. The interpolating polynomial with the smallest coefficient in the remainder term usually gives the best approximation of $f(x)$ at the particular x value for which the comparison is made. If the function is known, an upper bound on the error can be established.

EXAMPLE C-2

Using the data given in the table of Fig. C-1, form a difference table. Then using the Newton-Gregory forward-interpolating polynomial (Eq. C-12), approximate $f(.30)$. The difference table for the data from Fig. C-1 is

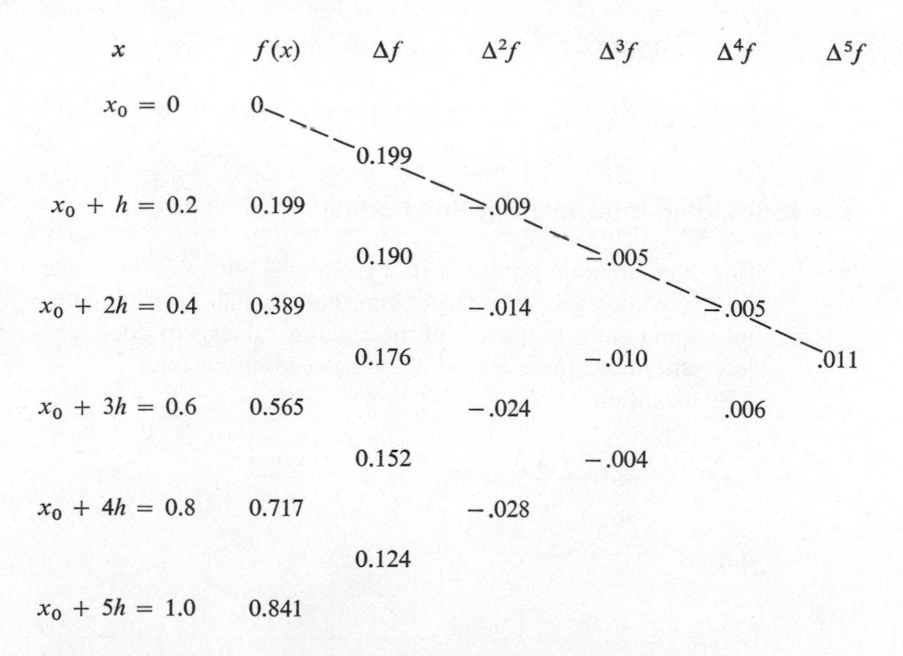

x	$f(x)$	Δf	$\Delta^2 f$	$\Delta^3 f$	$\Delta^4 f$	$\Delta^5 f$
$x_0 = 0$	0					
		0.199				
$x_0 + h = 0.2$	0.199		.009			
		0.190		.005		
$x_0 + 2h = 0.4$	0.389		−.014		.005	
		0.176		−.010		.011
$x_0 + 3h = 0.6$	0.565		−.024		.006	
		0.152		−.004		
$x_0 + 4h = 0.8$	0.717		−.028			
		0.124				
$x_0 + 5h = 1.0$	0.841					

Selecting the first tabulated value for x as x_0, the function f_0 and the differences Δf_0, $\Delta^2 f_0$, and so forth, appear on a diagonal running downward to the right as shown by the dashed line in the table.

Substituting values from the table into Eq. C-12, and noting that $x = 0.3$ so that

$$u = \frac{.3 - 0}{.2} = 1.5$$

we obtain

$$p_5(x) = 0 + \binom{1.5}{1}(.199) + \binom{1.5}{2}(-.009) + \binom{1.5}{3}(-.005)$$

$$+ \binom{1.5}{4}(-.005) + \binom{1.5}{5}(.011)$$

$$= 0 + (1.5)(.199) + \frac{(1.5)(.5)}{2!}(-.009) + \frac{(1.5)(.5)(-.5)}{3!}(-.005)$$

$$+ \frac{(1.5)(.5)(-.5)(-1.5)}{4!}(-.005)$$

$$+ \frac{(1.5)(.5)(-.5)(-1.5)(-2.5)}{5!}(.011)$$

$$= 0.2952$$

The function for this example is $f(x) = \sin x$, and $\sin(.30 \text{ radian}) = 0.2955$.

The Lagrangian Interpolating Polynomial

This polynomial is expressed in a form which is more convenient to use than Newton's general interpolating polynomial. Like the latter, it does not require uniform spacing of the abscissa values, nor does $x_0, x_1, x_2, \ldots$ necessarily need to be ascending or descending x values.

By definition,

$$[xx_0] = \frac{f(x) - f(x_0)}{x - x_0} = \frac{f(x)}{x - x_0} + \frac{f(x_0)}{x_0 - x}$$

and

$$[xx_0x_1] = \frac{[xx_0] - [x_0x_1]}{x - x_1} = \frac{f(x)}{(x - x_0)(x - x_1)} + \frac{f(x_0)}{(x_0 - x)(x - x_1)}$$

$$- \frac{f(x_0)}{(x_0 - x_1)(x - x_1)} - \frac{f(x_1)}{(x_1 - x_0)(x - x_1)}$$

Combining the second and third terms, and rewriting the fourth term, gives

$$[xx_0x_1] = \frac{f(x)}{(x - x_0)(x - x_1)} + \frac{f(x_0)}{(x_0 - x)(x_0 - x_1)} + \frac{f(x_1)}{(x_1 - x)(x_1 - x_0)}$$

Proceeding in this manner, we finally obtain

$$\begin{aligned}[xx_0x_1 \cdots x_n] = &\frac{f(x)}{(x - x_0)(x - x_1)(x - x_2) \cdots (x - x_n)} \\ &+ \frac{f(x_0)}{(x_0 - x)(x_0 - x_1)(x_0 - x_2) \cdots (x_0 - x_n)} \\ &+ \frac{f(x_1)}{(x_1 - x)(x_1 - x_0)(x_1 - x_2) \cdots (x - x_n)} + \cdots \\ &+ \frac{f(x_n)}{(x_n - x)(x_n - x_0)(x_n - x_1) \cdots (x_n - x_{n-1})}\end{aligned}$$

Solving for $f(x)$,

$$\begin{aligned}f(x) = &f(x_0)\frac{(x - x_1)(x - x_2) \cdots (x - x_n)}{(x_0 - x_1)(x_0 - x_2) \cdots (x_0 - x_n)} \\ &+ f(x_1)\frac{(x - x_0)(x - x_2)(x - x_3) \cdots (x - x_n)}{(x_1 - x_0)(x_1 - x_2)(x_1 - x_3) \cdots (x_1 - x_n)} \\ &+ f(x_2)\frac{(x - x_0)(x - x_1)(x - x_3) \cdots (x - x_n)}{(x_2 - x_0)(x_2 - x_1)(x_2 - x_3) \cdots (x_2 - x_n)} + \cdots \\ &+ f(x_n)\frac{(x - x_0)(x - x_1)(x - x_2) \cdots (x - x_{n-1})}{(x_n - x_0)(x_n - x_1)(x_n - x_2) \cdots (x_n - x_{n-1})} + R(x)\end{aligned} \tag{C-14}$$

where $R(x) = (x - x_0)(x - x_1)(x - x_2) \cdots (x - x_n)[xx_0x_1 \cdots x_n]$.

The remainder of Eq. C-14 is identical with that of Eq. C-4. With the remainder omitted from Eq. C-14, and $p_n(x)$ substituted for $f(x)$, we have the Lagrangian interpolating polynomial

$$\begin{aligned}p_n(x) = &f(x_0)\frac{(x - x_1)(x - x_2) \cdots (x - x_n)}{(x_0 - x_1)(x_0 - x_2) \cdots (x_0 - x_n)} \\ &+ f(x_1)\frac{(x - x_0)(x - x_2)(x - x_3) \cdots (x - x_n)}{(x_1 - x_0)(x_1 - x_2)(x_1 - x_3) \cdots (x_1 - x_n)} \\ &+ f(x_2)\frac{(x - x_0)(x - x_1)(x - x_3) \cdots (x - x_n)}{(x_2 - x_0)(x_2 - x_1)(x_2 - x_3) \cdots (x_2 - x_n)} + \cdots \\ &+ f(x_n)\frac{(x - x_0)(x - x_1)(x - x_2) \cdots (x - x_{n-1})}{(x_n - x_0)(x_n - x_1)(x_n - x_2) \cdots (x_n - x_{n-1})}\end{aligned} \tag{C-15}$$

EXAMPLE C-3

Given the data

x	$f(x)$
1.20	3.3201
1.70	5.4739
1.80	6.0496
2.00	7.3891

use Eq. C-15 to approximate $f(1.95)$.

Substituting into Eq. C-15 with $n = 3$, gives

$$\begin{aligned} p_3(1.95) &= 3.3201\frac{(1.95 - 1.70)(1.95 - 1.80)(1.95 - 2.00)}{(1.20 - 1.70)(1.20 - 1.80)(1.20 - 2.00)} \\ &\quad + 5.4739\frac{(1.95 - 1.20)(1.95 - 1.80)(1.95 - 2.00)}{(1.70 - 1.20)(1.70 - 1.80)(1.70 - 2.00)} \\ &\quad + 6.0496\frac{(1.95 - 1.20)(1.95 - 1.70)(1.95 - 2.00)}{(1.80 - 1.20)(1.80 - 1.70)(1.80 - 2.00)} \\ &\quad + 7.3891\frac{(1.95 - 1.20)(1.95 - 1.70)(1.95 - 1.80)}{(2.00 - 1.20)(2.00 - 1.70)(2.00 - 1.80)} \\ &= 7.0290 \end{aligned}$$

The data given is from the function $f(x) = e^x$, and the value of $e^{1.95}$ to four places is 7.0287, so the error made in using the interpolating polynomial is $-.0003$. It is important to emphasize that the error is the quantity which must be added to the approximation to obtain the exact answer.

C-4 / OTHER INTERPOLATING FORMULAS

Interpolating polynomials can be expressed in many different forms. Three forms have been derived. Other classical forms are given here for reference without derivation. For the interested reader, their derivations are given by Milne-Thomson[2] and others.

Newton-Gregory Backward Formula

When interpolating near the end of a table, in order to make sure that the differences required are available in the table, the last abscissa is often

[2] *Ibid.*

designated as x_0, with the previous abscissas designated x_{-1}, x_{-2}, x_{-3}, and so forth. Then the Newton-Gregory backward formula is used. Given in terms of backward-difference notation, it is

$$p_n(x) = f_0 + \binom{u}{1}\nabla f_0 + \binom{u+1}{2}\nabla^2 f_0 + \binom{u+2}{3}\nabla^3 f_0 + \cdots$$
$$+ \binom{u+k-1}{k}\nabla^k f_0 + \cdots + \binom{u+n-1}{n}\nabla^n f_0$$

As before, the phase u is defined as $(x - x_0)/h$, and will be negative for the situation described.

Expressed in forward-difference notation, without the use of the binomial coefficient symbols, the *same* formula is

$$p_n(x) = f_0 + u\Delta f_{-1} + \frac{u(u+1)}{2!}\Delta^2 f_{-2} + \frac{u(u+1)(u+2)}{3!}\Delta^3 f_{-3}$$
$$+ \frac{u(u+1)(u+2)(u+3)}{4!}\Delta^4 f_{-4} + \cdots$$
$$+ \frac{u(u+1)(u+2)\cdots(u+k-1)}{k!}\Delta^k f_{-k} + \cdots$$
$$+ \frac{u(u+1)(u+2)\cdots(u+n-1)}{n!}\Delta^n f_{-n}$$

Note that the differences used in the two different forms of this backward formula lie on a diagonal line running upward and to the right from f_0 in the tables of Figs. C-3 and C-4 (shown with a line of short dashes) whereas in the forward formulas (see Eq. C-12) the differences were found to lie on a diagonal line running downward and to the right as shown in Fig. C-3 with a line of long dashes.

Gauss's Forward Formula

In forward-difference notation, this formula is

$$p_n(x) = f_0 + \binom{u}{1}\Delta f_0 + \binom{u}{2}\Delta^2 f_{-1} + \binom{u+1}{3}\Delta^3 f_{-1} + \binom{u+1}{4}\Delta^4 f_{-2}$$
$$+ \binom{u+2}{5}\Delta^5 f_{-2} + \cdots + \binom{u+k-1}{2k}\Delta^{2k} f_{-k}$$
$$+ \binom{u+k}{2k+1}\Delta^{2k+1} f_{-k} + \cdots$$

In central-difference notation, it becomes

$$p_n(x) = f_0 + \binom{u}{1}\delta f_{1/2} + \binom{u}{2}\delta^2 f_0 + \binom{u+1}{3}\delta^3 f_{1/2} + \binom{u+1}{4}\delta^4 f_0$$
$$+ \binom{u+2}{5}\delta^5 f_{1/2} + \cdots + \binom{u+k-1}{2k}\delta^{2k} f_0$$
$$+ \binom{u+k}{2k+1}\delta^{2k+1} f_{1/2} + \cdots$$

The differences used in this formula lie on zigzag paths in the tables of Figs. C-3 and C-5, which are indicated there by black dots.

Gauss's Backward Formula

In forward-difference notation, this formula is expressed as

$$p_n(x) = f_0 + \binom{u}{1}\Delta f_{-1} + \binom{u+1}{2}\Delta^2 f_{-1} + \binom{u+1}{3}\Delta^3 f_{-2}$$
$$+ \binom{u+2}{4}\Delta^4 f_{-2} + \cdots + \binom{u+k}{2k}\Delta^{2k} f_{-k}$$
$$+ \binom{u+k}{2k+1}\Delta^{2k+1} f_{-k-1} + \cdots$$

In central-difference notation it is

$$p_n(x) = f_0 + \binom{u}{1}\delta f_{-1/2} + \binom{u+1}{2}\delta^2 f_0 + \binom{u+1}{3}\delta^3 f_{-1/2}$$
$$+ \binom{u+2}{4}\delta^4 f_0 + \cdots + \binom{u+k}{2k}\delta^{2k} f_0$$
$$+ \binom{u+k}{2k+1}\delta^{2k+1} f_{-1/2} + \cdots$$

The differences for this formula lie on zigzag paths in the tables in Figs. C-3 and C-5, which are indicated there by small circles.

Stirling's Formula

Stirling's formula is obtained by averaging Gauss's forward and backward formulas. Thus in forward-difference notation we have

$$p_n(x) = f_0 + \binom{u}{1}\frac{\Delta f_0 + \Delta f_{-1}}{2} + \frac{\binom{u}{2} + \binom{u+1}{2}}{2}\Delta^2 f_{-1}$$

$$+ \binom{u+1}{3}\frac{\Delta^3 f_{-1} + \Delta^3 f_{-2}}{2} + \frac{\binom{u+1}{4} + \binom{u+2}{4}}{2}\Delta^4 f_{-2} + \cdots$$

In central-difference notation with the averaging operator, it becomes

$$p_n(x) = f_0 + \binom{u}{1}\mu\delta f_0 + \frac{u}{2}\binom{u}{1}\delta^2 f_0 + \binom{u+1}{3}\mu\delta^3 f_0 + \cdots$$

$$+ \frac{u}{2k}\binom{u+k-1}{2k-1}\delta^{2k} f_0 + \binom{u+k}{2k+1}\mu\delta^{2k+1} f_0 + \cdots$$

For Stirling's formula the differences used lie on horizontal lines in the tables shown in Figs. C-3 and C-7, which are indicated there by lines of short arrows. At points indicated by an asterisk in Fig. C-3, the averages of the differences above and below are used to obtain the quantities $(\Delta f_0 + \Delta f_{-1})/2$, $(\Delta^3 f_{-1} + \Delta^3 f_{-2})/2$, and so forth.

The Lozenge Diagram

We have seen that the differences for various interpolation formulas can be obtained by taking various paths through a difference table. The functions of u used with the differences depends on the path. A lozenge diagram may be used to illustrate the various paths and resulting differences and functions of u for various interpolation formulas. Such a diagram is shown in Fig. C-8 in forward-difference notation.

Interpolation formulas are determined by starting at one of the entries in the first column, usually f_0, and following some path across to the right column. The path may follow a diagonal line, or it may cut horizontally across a lozenge. The rules to be followed for writing the interpolation formula are as follows:

1 The entry in the left column at which the path starts is the first term, and a term is added each time a new column is reached.

2 Steps may be taken from right to left if the terms are subtracted instead of added as in rule 1. However, this is seldom done.

3 The path consists of steps through the diagram. If the slope of a step is positive, the term to be added consists of the product of the difference reached by the step and the function of u immediately below the

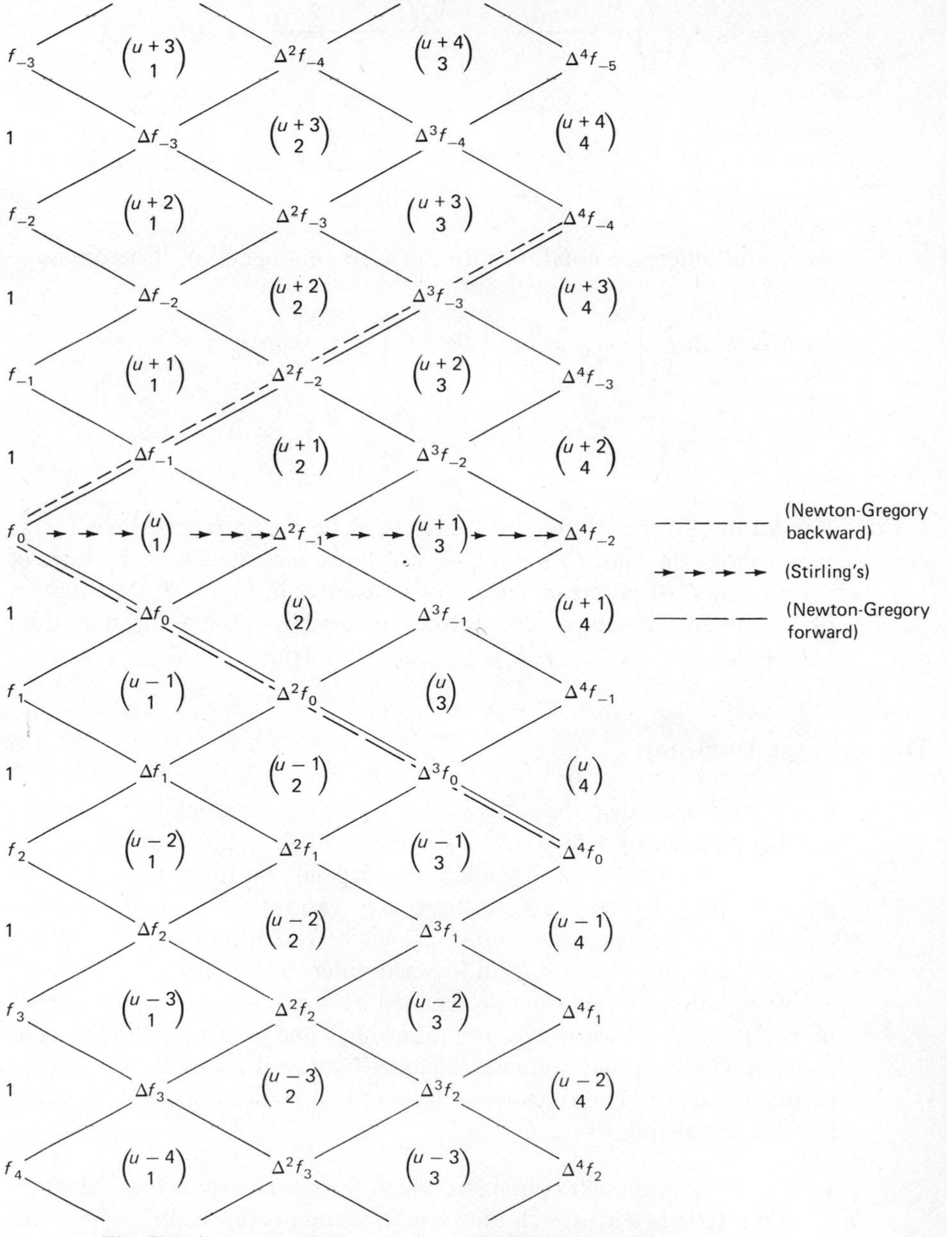

Fig. C-8 / *Lozenge diagram in forward-difference notation.*

difference. If the slope of the step is negative, the term to be added consists of the product of the difference reached by the step and the function of u immediately above the difference reached. If the step is horizontal and reaches a function of u, the term added consists of the function of u times the average of the differences above and below. If the step is horizontal and reaches a difference, the term added consists of the product of the difference and the average of the function of u above and below it.

Following the line of long dashes from f_0 in Fig. C-8, for example, gives the Newton-Gregory forward formula, while following the horizontal line of arrows gives Stirling's formula.

Hamming[3] gives a proof that any path chosen will result in a valid interpolation formula, and that if two paths end at the same point, the formulas give identical polynomials even though the entry points into the diagram may not be the same.

C-5 / INVERSE INTERPOLATION

Given a table of x values and corresponding $f(x)$ values, interpolation involves finding $f(x)$ for any nontabulated x values which fall within the range of tabulated values. Inverse interpolation, on the other hand, involves finding the value of x corresponding to some given nontabulated $f(x)$ value. For simplicity, let us designate our given function by y. That is,

$$y = f(x)$$

One approach to the problem of inverse interpolation is to consider our x-y relationship to be

$$x = f^{-1}(y)$$

Then the table in Fig. C-1, rewritten for inverse interpolation, is as shown in Fig. C-9. For some nontabulated y value, the corresponding $f^{-1}(y)$ value can be found by one of the interpolation methods previously discussed. Usually the Lagrangian form is used for this purpose since the y values are, in general, not uniformly spaced.

EXAMPLE C-4

Using the data given in Fig. C-9, find the value of x corresponding to a value of 0.25 for y. Since we have six data points given, a fifth-degree

[3] Richard W. Hamming, *Numerical Methods for Scientists and Engineers* (New York: McGraw-Hill Book Co., Inc., 1962), pp. 111–115.

	y	$f^{-1}(y)$
$y_0 =$	0.0	0.0
$y_1 =$	0.199	0.2
$y_2 =$	0.389	0.4
$y_3 =$	0.565	0.6
$y_4 =$	0.717	0.8
$y_5 =$	0.841	1.0

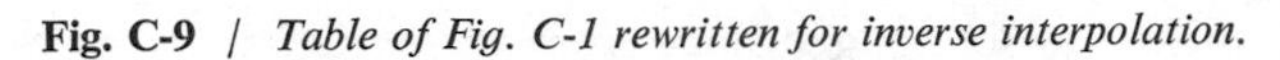

Fig. C-9 / *Table of Fig. C-1 rewritten for inverse interpolation.*

interpolating polynomial can be used. Equation C-15, the Lagrangian form, is used with x replaced by y, x_0 by y_0, x_1 by y_1, and so forth. Doing this, we obtain

$$
\begin{aligned}
p_5(y) = 0 &+ 0.2\frac{(.25-0.)(.25-.389)(.25-.565)(.25-.717)(.25-.841)}{(.199-0.)(.199-.389)(.199-.565)(.199-.717)(.199-.841)} \\
&+ 0.4\frac{(.25-0.)(.25-.199)(.25-.565)(.25-.717)(.25-.841)}{(.389-0.)(.389-.199)(.389-.565)(.389-.717)(.389-.841)} \\
&+ 0.6\frac{(.25-0.)(.25-.199)(.25-.389)(.25-.717)(.25-.841)}{(.565-0.)(.565-.199)(.565-.389)(.565-.717)(.565-.841)} \\
&+ 0.8\frac{(.25-0.)(.25-.199)(.25-.389)(.25-.565)(.25-.841)}{(.717-0.)(.717-.199)(.717-.389)(.717-.565)(.717-.841)} \\
&+ 1.0\frac{(.25-0.)(.25-.199)(.25-.389)(.25-.565)(.25-.717)}{(.841-0.)(.841-.199)(.841-.389)(.841-.565)(.841-.717)} \\
&= 0.253
\end{aligned}
$$

The table in Fig. C-1 is based on the sine function, so that

$$x = \sin^{-1}(.25) = 0.253$$

to three-digit accuracy, verifying the result obtained by interpolation.

In some cases, the inverse function, $x = f^{-1}(y)$, cannot be well represented by a polynomial even though $y = f(x)$ can be approximated accurately by a polynomial. In such cases, the method just discussed would lead to inaccurate results. A more accurate result could be obtained by determining the polynomial $p_n(x)$ which approximates $f(x)$, replacing $p_n(x)$ with the given value of $f(x)$, and, for the resulting equation, finding the root nearest the approximately known value of x for known $f(x)$ by one of the methods of Chap. 2.

EXAMPLE C-5

Given the data of Ex. C-3, repeated here for convenience, find the value

x	$f(x)$
1.20	3.3201
1.70	5.4739
1.80	6.0496
2.00	7.3891

of x corresponding to a value of 4.0000 for $f(x)$. Since the x values are not equally spaced, we use the Lagrangian form of the interpolating polynomial and determine the third-degree polynomial through the four given points. It is

$$
\begin{aligned}
p_3(x) = {} & 3.3201\frac{(x-1.70)(x-1.80)(x-2.00)}{(1.20-1.70)(1.20-1.80)(1.20-2.00)} \\
& + 5.4739\frac{(x-1.20)(x-1.80)(x-2.00)}{(1.70-1.20)(1.70-1.80)(1.70-2.00)} \\
& + 6.0496\frac{(x-1.20)(x-1.70)(x-2.00)}{(1.80-1.20)(1.80-1.70)(1.80-2.00)} \\
& + 7.3891\frac{(x-1.20)(x-1.70)(x-1.80)}{(2.00-1.20)(2.00-1.70)(2.00-1.80)} \\
= {} & 0.899x^3 - 1.810x^2 + 3.830x - 0.223
\end{aligned}
$$

Setting the value of this polynomial equal to 4.000 gives the equation

$$0.899x^3 - 1.810x^2 + 3.830x - 4.223 = 0$$

We know from visual inspection of the tabulated data that x will probably be somewhere in the neighborhood of 1.35. Using an incremental-search technique with a pocket calculator, starting at $x = 1.35$, we find that $x = 1.386$ to four significant digits. Since the data for the table are based on $f(x) = e^x$, this is the correct answer to four significant digits.

C-6 / APPLICATION OF POLYNOMIAL APPROXIMATION TO THE DERIVATION OF NUMERICAL INTEGRATION FORMULAS

The Trapezoid Rule

If we let $n = 1$ in Eq. C-11 with the remainder term as given by Eq. C-13, we have

$$f(x) = f_0 + u\,\Delta f_0 + \frac{u(u-1)}{2}h^2 f''(\xi) \qquad x_0 < \xi < x_1$$

Now suppose we integrate this function from x_0 to $x_1 = x_0 + h$. The result is equal to the crosshatched area in the strip under the graph of $f(x)$ shown in Fig. C-10. Since $u = (x - x_0)/h$, we have $dx = h\,du$, and for $x = x_0$, we see that $u = 0$, while for $x = x_1$, $u = 1$. Therefore,

$$\int_{x=x_0}^{x=x_1} f(x)\,dx = \int_{u=0}^{u=1} f(x)h\,du$$

Then, assuming that $f''(\xi)$ is essentially a constant for very small h as u varies from 0 to 1,

$$h\int_{u=0}^{u=1}\left[f_0 + u\,\Delta f_0 + \frac{u(u-1)}{2}\,h^2 f''(\xi)\right]du$$

$$= h\left[uf_0 + \frac{u^2}{2}\Delta f_0 + \left(\frac{u^3}{6} - \frac{u^2}{4}\right)h^2 f''(\xi)\right]_0^1$$

$$= h\left[f_0 + \frac{1}{2}\Delta f_0\right] - \frac{1}{12}h^3 f''(\xi)$$

$$= \underbrace{\frac{h}{2}[f_0 + f_1]}_{\text{first term}} - \underbrace{\frac{1}{12}h^3 f''(\xi)}_{\text{second term}} \qquad x_0 < (\xi) < x_1$$

Using only the first term of this equation, we have the trapezoid rule for the approximate area in one strip under the graph of a function. The second term then becomes the error. This is in agreement with the result obtained from Taylor-series expansions in Chap. 5. The error is said to be "of order h^3", usually written $O(h^3)$.

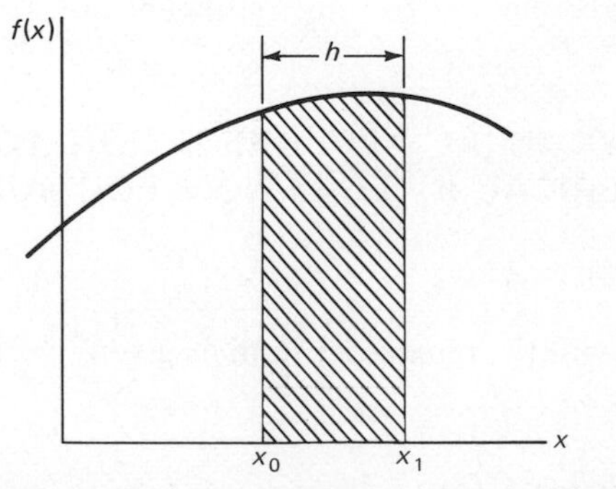

Fig. C-10 / *Graph of function based on Eqs. C-11 and C-13.*

More Accurate Formulas for the Area in One Strip Under a Curve

Formulas for approximating the area in one strip under a curve with error of any order desired may be derived by letting n take on higher values in Eq. C-11. For example, if we let $n = 2$ in Eq. C-11, and integrate from $x = x_0$ to $x = x_1$, we obtain

$$\int_{x=x_0}^{x=x_1} f(x)\,dx = \int_{u=0}^{u=1} f(x)h\,dx$$

$$= h\int_0^1 \left[f_0 + u\,\Delta f_0 + \frac{u(u-1)}{2!}\Delta^2 f_0 + \frac{u(u-1)(u-2)}{3!}h^3 f'''(\xi) \right] du$$

$$= \frac{h}{12}[5f_0 + 8f_1 - f_2] + O(h^4)$$

If we let $n = 3$ in Eq. C-11, and integrate over one strip, we obtain

$$\int_{x=x_0}^{x=x_1} f(x)\,dx = \int_{u=0}^{u=1} f(x)h\,du = \frac{h}{24}[9f_0 + 19f_1 - 5f_2 + f_3] + O(h^5)$$

Simpson's Rule

If we take $n = 3$ in Eq. C-11, and integrate the resulting function over two strips, from $x = x_0$ to $x = x_0 + 2h = x_2$, we get the expression for Simpson's rule for two strips, and the error made by using Simpson's rule for the area in those two strips.

$$\int_{x=x_0}^{x=x_2} f(x)\,dx = \int_{u=0}^{u=2} f(x)h\,du$$

$$= h\int_0^2 \left[f_0 + u\,\Delta f_0 + \frac{u^2 - u}{2}\Delta^2 f_0 + \frac{u^3 - 3u^2 + 2u}{6}\Delta^3 f_0 + \frac{u^4 - 6u^3 + 11u^2 - 6u}{24}h^4 f''''(\xi) \right] du$$

Assuming that $f''''(\xi)$ is essentially a constant for very small h as u varies from 0 to 2, we obtain

$$\int_{x_0}^{x_2} f(x)\,dx = h\left[uf_0 + \frac{u^2}{2}\Delta f_0 + \left(\frac{u^3}{6} - \frac{u^2}{4}\right)\Delta^2 f_0 + \left(\frac{u^4}{24} - \frac{3u^3}{18} + \frac{2u^2}{12}\right)\Delta^3 f_0 + \left(\frac{u^5}{120} - \frac{6u^4}{96} + \frac{11u^3}{72} - \frac{6u^2}{48}\right)h^4 f''''(\xi) \right]_0^2 \tag{C-16}$$

Substituting limits gives

$$\int_{x_0}^{x_2} f(x)\,dx = \underbrace{\frac{h}{3}[f_0 + 4f_1 + f_2]}_{\substack{\text{Simpson's rule}\\ \text{expression for}\\ \text{the approximate}\\ \text{area in two strips}\\ \text{under a curve}}} - \underbrace{\frac{1}{90}h^5 f''''(\xi)}_{\text{error term}} \qquad x_0 < (\xi) < x_2 \qquad \text{(C-17)}$$

Equation C-17 is in agreement with the results obtained in Chap. 5. Because the coefficient of $\Delta^3 f_0$ in Eq. C-16 turns out to be zero when the limits are substituted, we need only three ordinate values (we actually use a second-degree parabola) to obtain the accuracy of a third-degree polynomial. This is one of the features of Simpson's rule which makes it appealing.

Newton-Cotes Formulas

If the ordinates of a function $f(x)$ are known for $n + 1$ equally spaced abscissas, we can form an nth-degree polynomial which matches all of the tabulated values. We can then integrate this polynomial over different limits to obtain approximations to the area in different numbers of strips under the curve of $f(x)$. Newton-Cotes formulas result from integration over the limits of abscissa values for which the polynomial matches the tabulated values (x_0 to x_n). Thus the trapezoid rule and Simpson's rule are special cases of Newton-Cotes formulas for two and three points, respectively. As a further example, let us assume that we have seven points whose abscissas are equally spaced. To obtain a Newton-Cotes formula, we let $n = 7$ in Eq. C-11 and integrate over six strips. After integrating and substituting limits, it turns out that the coefficient of $\Delta^7 f_0$ is zero, so we actually use a sixth-degree polynomial having the accuracy of a seventh-degree polynomial, similar to what occurred in the derivation of Simpson's rule. We obtain the following very accurate seven-point Newton-Cotes formula with error of order h^9 (this compares with an error of order h^5 for Simpson's rule):

$$\int_{x_0}^{x_6} f(x)\,dx = \int_{u=0}^{u=6} f(x)h\,du$$

$$= \frac{h}{140}[41f_0 + 216f_1 + 27f_2 + 272f_3 + 27f_4 + 216f_5 + 41f_6] + O(h^9) \qquad \text{(C-18)}$$

The error of order h^9 comes from integrating the remainder term in the Newton-Gregory forward formula with $n = 7$.

A table of Newton-Cotes coefficients and orders of error for formulas up to and including the seven-point formula is shown below. This table is a modified version of one given by Milne-Thomson[4] which was taken from Pascal's *Repertorium*. The various coefficients are defined as follows:

$$\int_{x_0}^{x_n} f(x)\,dx = Ch[a_0f_0 + a_1f_1 + a_2f_2 + a_3f_3 + \cdots + a_nf_n] + O(h^k)$$

No. of Points n + 1	C	a_0	a_1	a_2	a_3	a_4	a_5	a_6	k
2	$\frac{1}{2}$	1	1		(trapezoid rule)				3
3	$\frac{1}{3}$	1	4	1	(Simpson's one-third rule)				5
4	$\frac{3}{8}$	1	3	3	1	(Simpson's three-eighths rule)			5
5	$\frac{2}{45}$	7	32	12	32	7			7
6	$\frac{5}{288}$	19	75	50	50	75	19		7
7	$\frac{1}{140}$	41	216	27	272	27	216	41	9

There is little to be gained in accuracy in using a four-point formula over a three-point formula (they have errors of the same order), or a six-point formula over a five-point formula. Odd-point formulas are preferred unless it is necessary to integrate over an odd number of strips. For example, Simpson's three-eighths rule ($n + 1 = 4$) is used to integrate over three strips, but with little more accuracy than Simpson's one-third rule over two strips.

Weddle's Formula

Weddle's formula is a seven-point formula, but it has simpler coefficients than the seven-point Newton-Cotes formula. It may be derived as follows. By definition,

$$\Delta^6 f_0 = f_0 - 6f_1 + 15f_2 - 20f_3 + 15f_4 - 6f_5 + f_6$$

It may also be shown[5] that

$$\Delta^6 f_0 = h^6 f^6(\xi) \qquad x_0 < (\xi) < x_6$$

[4] Milne-Thomson, p. 170.
[5] *Ibid.*, p. 56.

If we add the first expression given above for $\Delta^6 f_0$ to the bracketed expression in Eq. C-18, and then, to compensate, subtract the second expression given above for $\Delta^6 f_0$, multiplied by $h/140$, from the error term of Eq. C-18, we obtain Weddle's formula, which is

$$\int_{x_0}^{x_6} f(x)\,dx = \frac{3}{10}h[f_0 + 5f_1 + f_2 + 6f_3 + f_4 + 5f_5 + f_6] + O(h^7) \quad \text{(C-19)}$$

The penalty for the simpler coefficients is an error of order h^7 instead of order h^9 obtained with the Newton-Cotes seven-point formula.

Open Integration Formulas

The Newton-Cotes formulas and others which have been presented thus far are called "closed" formulas. In the case of closed formulas, ordinate values are known at the limits of the integration. To obtain "open" formulas, the range of integration extends beyond the abscissa values for which the ordinates are known. Formulas of the open type are useful in the numerical solution of differential equations.

As an example of the derivation of an open formula, let us derive Eq. 6-114 used in Milne's predictor equation. We let $n = 2$ in Eq. C-11 and integrate from $x = x_0 - h$ to $x = x_0 + 3h$:

$$\int_{x=x_{-1}}^{x=x_3} f(x)\,dx = h\int_{u=-1}^{u=3} f(x)\,du$$

$$= h\int_{-1}^{3}\left[f_0 + u\,\Delta f_0 + \frac{u(u-1)}{2!}\Delta^2 f_0 + \frac{u(u-1)(u-2)}{3!}h^3 f'''(\xi)\right]du$$

$$= h\left[uf_0 + \frac{u^2}{2}(f_1 - f_0) + \left(\frac{u^3}{6} - \frac{u^2}{4}\right)(f_2 - 2f_1 + f_0)\right]_{-1}^{3} + O(h^4)$$

$$= \frac{4}{3}h[2f_0 - f_1 + 2f_2] + O(h^4) \quad \text{(C-20)}$$

C-7 / APPLICATION OF POLYNOMIAL APPROXIMATION TO THE DERIVATION OF NUMERICAL DIFFERENTIATION FORMULAS

The Newton-Gregory forward-interpolation formula (Eq. C-11) with remainder (Eq. C-13) is

$$
\begin{aligned}
f(x) = f_0 + u\,\Delta f_0 &+ \frac{u(u-1)}{2!}\Delta^2 f_0 + \frac{u(u-1)(u-2)}{3!}\Delta^3 f_0 \\
&+ \frac{u(u-1)(u-2)(u-3)}{4!}\Delta^4 f_0 + \cdots \\
&+ \frac{u(u-1)(u-2)\cdots(u-n+1)}{n!}\Delta^n f_0 \\
&+ \frac{u(u-1)(u-2)\cdots(u-n)}{(n+1)!}h^{n+1}f^{n+1}(\xi) \qquad \text{(C-21)}
\end{aligned}
$$

Approximations for the derivative of a function will be determined by determining $f'(x)$. Let us note that

$$f'(x) = \frac{df(x)}{du}\frac{du}{dx}$$

and that

$$\frac{du}{dx} = \frac{1}{h}$$

Thus we can differentiate $f(x)$ with respect to u and multiply by $1/h$ to obtain

$$
\begin{aligned}
f'(x) = \frac{1}{h}\Bigg\{0 + \Delta f_0 &+ \frac{1}{2}(2u-1)\Delta^2 f_0 + \frac{1}{6}(3u^2 - 6u + 2)\Delta^3 f_0 \\
&+ \frac{1}{24}(4u^3 - 18u^2 + 22u - 6)\Delta^4 f_0 + \cdots \\
&+ \frac{d}{du}\left[\binom{u}{n+1}h^{n+1}f^{n+1}(\xi)\right]\Bigg\} \qquad \text{(C-22)}
\end{aligned}
$$

Since $f^{n+1}(\xi)$ is an unknown function of u, differentiation of the remainder term as a product cannot be carried out except for $u = 0$, which causes the term containing the derivative of $f^{n+1}(\xi)$ to vanish. Even then the derivative of the remainder term cannot be evaluated exactly because $f^{n+1}(\xi)$ cannot be evaluated. Thus we will let the derivative of the remainder term constitute our error term which will be $O(h^n)$.

If the derivative is to be evaluated at x_0, then $u = 0$, and

$$f'(x) = \frac{1}{h}\left[\Delta f_0 - \frac{1}{2}\Delta^2 f_0 + \frac{1}{3}\Delta^3 f_0 - \frac{1}{4}\Delta^4 f_0 + \cdots \pm \frac{1}{n}\Delta^n f_0\right] + O(h^n) \tag{C-23}$$

Differentiating Eq. C-21 with $n = 1$, and letting $u = 0$, only the first term in brackets in Eq. C-23 will remain so that

$$f'(x_0) = \frac{f_1 - f_0}{h} + O(h)$$

This equation agrees with the first of Eq. 5-81 which was derived using the Taylor series. Similarly, if we let $n = 2$ in Eq. C-21, the first two terms of Eq. C-23 will remain, giving

$$\begin{aligned} f'(x_0) &= \frac{1}{h}\left[\Delta f_0 - \frac{1}{2}\Delta^2 f_0\right] + O(h^2) \\ &= \frac{1}{h}\left[f_1 - f_0 - \frac{1}{2}(f_2 - 2f_1 + f_0)\right] + O(h^2) \\ &= \frac{-f_2 + 4f_1 - 3f_0}{2h} + O(h^2) \end{aligned}$$

This agrees with the first of Eq. 5-82, which was derived using the Taylor series.

If we differentiate Eq. C-21 with $n = 2$, and take $u = 1$ to obtain the derivative at $x = x_1$, we obtain

$$\begin{aligned} f'(x_1) &= \frac{1}{h}\left[\Delta f_0 + \frac{1}{2}(1)\Delta^2 f_0\right] + O(h^2) \\ &= \frac{1}{h}\left[f_1 - f_0 + \frac{1}{2}(f_2 - 2f_1 + f_0)\right] + O(h^2) \\ &= \frac{f_2 - f_0}{2h} + O(h^2) \end{aligned}$$

Then for the derivative at x_0, we have

$$f'(x_0) = \frac{f_1 - f_{-1}}{2h} + O(h^2)$$

which agrees with the first of Eq. 5-79.

If we differentiate Eq. C-21 with $n = 4$, and take $u = 2$, we obtain the derivative at $x = x_2$ as

$$f'(x_2) = \frac{1}{h}\left[\Delta f_0 + \frac{3}{2}\Delta^2 f_0 + \frac{1}{3}\Delta^3 f_0 - \frac{1}{12}\Delta^4 f_0\right] + O(h^4)$$

$$= \frac{1}{12h}[f_0 - 8f_1 + 8f_3 - f_4] + O(h^4)$$

It now follows that the derivative at x_0 is

$$f'(x_0) = \frac{1}{12h}[f_{-2} - 8f_{-1} + 8f_1 - f_2] + O(h^4)$$

This agrees with the first of Eq. 5-80.

To obtain expressions for the second derivative of $f(x)$, the Newton-Gregory forward equation is differentiated twice. Noting that

$$f''(x) = \frac{d^2 f(x)}{du^2}\left(\frac{du}{dx}\right)^2$$

we obtain

$$f''(x) = \frac{1}{h^2}\left[0 + \frac{1}{2}(2)\Delta^2 f_0 + \frac{1}{6}(6u - 6)\Delta^3 f_0 + \cdots\right] + O(h^{n-1}) \quad \text{(C-24)}$$

If we take $n = 2$ in Eq. C-21, differentiate twice, and let $u = 0$ to get the second derivative at $x = x_0$, we have only the second term in brackets of Eq. C-24 remaining. Thus

$$f''(x_0) = \frac{1}{h^2}[f_2 - 2f_1 + f_0] + O(h)$$

This agrees with the second of Eq. 5-81.

If we take $n = 3$ and let $u = 0$, we obtain

$$f''(x_0) = \frac{1}{h^2}[f_2 - 2f_1 + f_0 - (f_3 - 3f_2 + 3f_1 - f_0)] + O(h^2)$$

$$= \frac{1}{h^2}[-f_3 + 4f_2 - 5f_1 + 2f_0] + O(h^2) \qquad \text{(C-25)}$$

This agrees with the second of Eq. 5-82.

Proceeding in this manner, the finite-difference approximations for the derivatives given in Chap. 5 in summary form can all be derived.

Index

Printer and Binder: Halliday Lithograph Corporation

81 82 10